# Water and Wastewater Calculations Manual

## ABOUT THE AUTHOR

Shun Dar Lin is an Emeritus Faculty of University of Illinois and in Taiwan. A registered professional engineer in Illinois, he has published nearly 100 papers, articles, and reports related to water and wastewater engineering. Dr. Lin brings to the book a background in teaching, research, and practical field experience spanning nearly 50 years. Dr. Lin received his Ph.D. in Sanitary Engineering from Syracuse University, an M.S. in Sanitary Engineering from the University of Cincinnati, and a B.S. in Civil Engineering from National Taiwan University. He has taught and conducted research since 1960 at the Institute of Public Health of National Taiwan University. In 1986, Dr. Lin received the Water Quality Division Best Paper Award for "*Giardia lamblia* and Water Supply" from the American Water Works Association. He developed the enrichment-temperature acclimation method for recovery enhancement of stressed fecal coliform. The method has been adopted in the *Standard Methods for the Examination of Water and Wastewater* since the 18th edition (1990). Dr. Lin is a life member of the American Society of Civil Engineers, the American Water Works Association, and the Water Environment Federation. He is a consultant to the governments of Taiwan and the United States and for consultant firms.

*Note:* This book was written by Dr. Shun Dar Lin in his private capacity.

# Water and Wastewater Calculations Manual

**Shun Dar Lin**

**C. C. Lee**
Editor of *Handbook of Environmental Engineering Calculations*

**Second Edition**

New York Chicago San Francisco Lisbon London Madrid
Mexico City Milan New Delhi San Juan Seoul
Singapore Sydney Toronto

The McGraw·Hill Companies

**Library of Congress Cataloging-in-Publication Data**

Lin, Shun Dar.
Water and wastewater calculations manual / Shun Dar Lin, C. C. Lee.—2nd ed.
p. cm.
Includes index.
ISBN-13: 978-0-07-147624-9
ISBN-10: 0-07-147624-5 (alk. paper)
1. Water-supply—Mathematics—Handbooks, manuals, etc. 2. Water—Purification—Mathematics—Handbooks, manuals, etc. 3. Hydrology—Mathematics—Handbooks, manuals, etc. 4. Sewage disposal—Mathematics—Handbooks, manuals, etc. 5. Sewage—Purification—Mathematics—Handbooks, manuals, etc. I. Lee, C. C. II. Title.
TD351.L55 2007
628.101′51—dc22

2007020112

McGraw-Hill books are available at special quantity discounts to use as premiums and sales promotions, or for use in corporate training programs. For more information, please write to the Director of Special Sales, McGraw-Hill Professional, Two Penn Plaza, New York, NY 10121-2298. Or contact your local bookstore.



1 2 3 4 5 6 7 8 9 0 DOC/DOC 0 1 2 1 0 9 8 7

ISBN-13: 978-0-07-147624-9
ISBN-10: 0-07-147624-5

This book is printed on acid-free paper.

**Sponsoring Editor**
Larry S. Hager

**Editorial Supervisor**
David E. Fogarty

**Project Manager**
Gita Raman

**Copy Editor**
Yumnam Ojen

**Proofreader**
Surendra Nath Shivam

**Indexer**
Steve Ingle

**Production Supervisor**
Richard C. Ruzycka

**Composition**
International Typesetting and Composition

**Art Director, Cover**
Jeff Weeks

# Contents

# Preface

This manual presents the basic principles and concepts relating to water/wastewater engineering and provides illustrative examples of the subject covered. To the extent possible, examples rely on practical field data and regulatory requirements have been integrated into the environmental design process. Each of the calculations provided herein is solved step-by-step in a streamlined manner that is intended to facilitate understanding. Examples (step-by-step solutions) range from calculations commonly used by operators to more complicated calculations required for research or design. For calculations provided herein using the US customary units, readers who use the International System may apply the conversion factors listed in Appendix E. Answers are also generally given in SI units for most of problems solved by the US customary units.

This book has been written for use by the following readers: students taking coursework relating to "Public Water Supply," "Waste-Water Engineering," or "Stream Sanitation"; practicing environmental (sanitary) engineers; regulatory officers responsible for the review and approval of engineering project proposals; operators, engineers, and managers of water and/or wastewater treatment plants; and other professionals, such as chemists and biologists, who need some knowledge of water/wastewater issues. This work will benefit all operators and managers of public water supply and of wastewater treatment plants, environmental design engineers, military environmental engineers, undergraduate and graduate students, regulatory officers, local public works engineers, lake managers, and environmentalists.

Advances and improvements in many fields are driven by competition or the need for increased profits. It may be fair to say, however, that advances and improvements in environmental engineering are driven instead by regulation. The US Environmental Protection Agency (US EPA) sets up maximum contaminant levels, which research and project designs must reach as a goal. The step-by-step solution examples provided in this book are guided by the integration of rules and regulations on every aspect of water and wastewater. The author has performed an extensive survey of literature on surface water and groundwater pertaining to environmental engineering and compiled them in this book. Rules and regulations are described as simply as possible, and practical examples are given.

The text includes calculations for surface water, groundwater, drinking water treatment, and wastewater engineering. Chapter 1 comprises calculations for river and stream waters. Stream sanitation had been studied for nearly 100 years. By mid-twentieth century, theoretical and empirical models for assessing waste-assimilating capacity of streams were well developed. Dissolved oxygen and biochemical oxygen demand in streams and rivers have been comprehensively illustrated in this book. Apportionment of stream users and pragmatic approaches for stream dissolved oxygen models also first appeared in this manual. From the 1950s through the 1980s, researchers focused extensively on wastewater treatment. In the 1970s, rotating biological contactors became a hot subject. Design criteria and examples for all of these are included in this volume. Some treatment and management technologies are no longer suitable in the United States. However, they are still of some use in developing countries.

Chapter 2 is a compilation of adopted methods and documented research. In the early 1980s, the US EPA published Guidelines for Diagnostic and Feasibility Study of Public Owned Lakes (Clean Lakes Program, or CLP). This was intended to be as a guideline for lake management. CLP and its calculation (evaluation) methods are presented for the first time in this volume. Hydrological, nutrient, and sediment budgets are presented for reservoir and lake waters. Techniques for classification of lake water quality and assessment of the lake trophic state index and lake use support are also presented.

Calculations for groundwater are given in Chapter 3. They include groundwater hydrology, flow in aquifers, pumping and its influence zone, setback zone, and soil remediation. Well setback zone is regulated by the state EPA. Determinations of setback zones are also included in the book. Well function for confined aquifers is presented in Appendix B.

Hydraulics for environmental engineering is included in Chapter 4. This chapter covers fluid (water) properties and definitions, hydrostatics, fundamental concepts of water flow in pipes, weirs, orifices, and in open channels, and flow measurements. Pipe networks for water supply distribution systems and hydraulics for water and wastewater treatment plants are also included.

Chapters 5 and 6 cover the unit process for drinking water and wastewater treatment, respectively. The US EPA developed design criteria and guidelines for almost all unit processes. These two chapters depict the integration of regulations (or standards) into water and wastewater design procedures. Drinking water regulations and membrane filtration are updated in Chapter 5. In addition, three new sections on pellet softening, disinfection by-products (DBP), and health risks, also are incorporated in Chapter 5. The DBP section provides concise information for drinking water professionals. Although the pellet softening process is not accepted in the United States, it has been successfully used in many other countries. It is believed that this is the first presentation of pellet softening in US environmental engineering books. Another new section of constructed wetlands is included in Chapter 6. These two chapters (5 and 6) are the heart of the book and provide the theoretical considerations of unit processes, traditional (or empirical) design concepts, and integrated regulatory requirements. Drinking water quality standards, wastewater effluent standards, and several new examples have also been added.

The current edition corrects certain computational, typographical, and grammatical errors found in the previous edition.

Charles C. C. Lee initiated the project of *Handbook of Environmental Engineering Calculations*. Gita Raman of ITC (India) did excellent editing of the final draft. The author also wishes to acknowledge Meiling Lin, for typing the corrected manuscript. Ben Movahed, President of WATEK Engineering, reviewed the section of membrane filtration. Alex Ya Ching Wu, Plant Manager of Cheng-Ching Lake Advanced Water Purification Plant in Taiwan, provided the operational manual for pellet softening. Mike Henebry of Illinois EPA reviewed the section of health risks. Jessica Moorman, Editor of *Water & Waste Digest*, provided 2006 drinking water regulatory updates. Thanks to Dr. Chuan-jui Lin, Dr. C. Eddie Tzeng, Nancy Simpson, Jau-hwan Tzeng, Heather Lin, Robert Greenlee, Luke Lin, Kevin Lin, and Lucy Lin for their assistance. Any reader suggestions and comments will be greatly appreciated.

SHUN DAR LIN
*Peoria, Illinois*

# Water and Wastewater Calculations Manual

Chapter

# 1

# Streams and Rivers

## 1 General

This chapter presents calculations on stream sanitation. The main portion covers the evaluation of water assimilative capacities of rivers or streams. The procedures include classical conceptual approaches and pragmatic approaches: the conceptual approaches use simulation models, whereas Butts and his coworkers (1973, 1974, 1981) of the Illinois State Water Survey use a pragmatic approach. Observed dissolved oxygen (DO) and biochemical oxygen demand (BOD) levels are measured at several sampling points along a stream reach. Both approaches are useful for developing or approving the design of wastewater treatment facilities that discharge into a stream.

In addition, biological factors such as algae, indicator bacteria, diversity index, and macroinvertebrate biotic index are also presented.

## 2 Point Source Dilution

Point source pollutants are commonly regulated by a deterministic model for an assumed design condition having a specific probability of occurrence. A simplistic dilution and/or balance equation can be written as

$$C_d = \frac{Q_u C_u + Q_e C_e}{Q_u + Q_e} \tag{1.1}$$

where $C_d$ = completely mixed constituent concentration downstream of the effluent, mg/L
$Q_u$ = stream flow upstream of the effluent, cubic feet per second, cfs
$C_u$ = constituent concentration of upstream flow, mg/L
$Q_e$ = flow of the effluent, cfs
$C_e$ = constituent concentration of the effluent, mg/L

Under the worst case, a 7-day, 10-year low flow is generally used for stream flow condition, for design purposes.

**Example:** A power plant pumps 27 cfs from a stream, with a flow of 186 cfs. The discharge of the plant's ash-pond is 26 cfs. The boron concentrations for upstream water and the effluent are 0.051 and 8.9 mg/L, respectively. Compute the boron concentration in the stream after completely mixing.

**solution:** By Eq. (1.1)

$$C_d = \frac{Q_u C_u + Q_e C_e}{Q_u + Q_e}$$

$$= \frac{(186 - 27)(0.051) + 26 \times 8.9}{(186 - 27) + 26}$$

$$= 1.29 \text{ (mg/L)}$$

## 3 Discharge Measurement

Discharge (flow rate) measurement is very important to provide the basic data required for river or stream water quality. The total discharge for a stream can be estimated by float method with wind and other surface effects, by die study, or by actual subsection flow measurement, depending on cost, time, manpower, local conditions, etc. The discharge in a stream cross section can be measured from a subsection by the following formula:

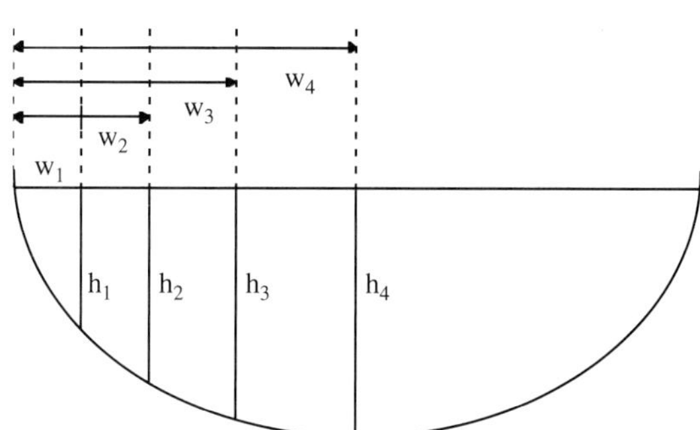


$$Q = \text{Sum (mean depth} \times \text{width} \times \text{mean velocity)}$$

$$Q = \sum_{n=1}^{n} \frac{1}{2}(h_n + h_{n-1})(w_n - w_{n-1}) \times \frac{1}{2}(v_n + v_{n-1}) \quad (1.2)$$

**TABLE 1.1 Velocity and Discharge Measurements**

| (1) Distance from 0, ft | (2) Depth, ft | (3) Velocity, ft/s | (4) Width, ft | (5) Mean depth, ft | (6) Mean velocity, ft/s | (7) = (4) × (5) × (6) Discharge, cfs |
|---|---|---|---|---|---|---|
| 0 | 0 | 0 | | | | |
| 2 | 1.1 | 0.52 | 2 | 0.55 | 0.26 | 0.3 |
| 4 | 1.9 | 0.84 | 2 | 1.50 | 0.68 | 2.0 |
| 7 | 2.7 | 1.46 | 3 | 2.30 | 1.15 | 7.9 |
| 10 | 3.6 | 2.64 | 3 | 3.15 | 2.05 | 19.4 |
| 14 | 4.5 | 4.28 | 4 | 4.05 | 3.46 | 56.1 |
| 18 | 5.5 | 6.16 | 4 | 5.00 | 5.22 | 104.4 |
| 23 | 6.6 | 8.30 | 5 | 6.05 | 7.23 | 349.9 |
| 29 | 6.9 | 8.88 | 6 | 6.75 | 8.59 | 302.3 |
| 35 | 6.5 | 8.15 | 6 | 6.70 | 7.52 | 302.3 |
| 40 | 6.2 | 7.08 | 5 | 6.35 | 6.62 | 210.2 |
| 44 | 5.5 | 5.96 | 4 | 5.85 | 6.52 | 152.2 |
| 48 | 4.3 | 4.20 | 4 | 4.90 | 5.08 | 99.6 |
| 50 | 3.2 | 2.22 | 2 | 3.75 | 3.21 | 24.1 |
| 52 | 2.2 | 1.54 | 2 | 2.70 | 1.88 | 10.2 |
| 54 | 1.2 | 0.75 | 2 | 1.45 | 1.15 | 3.3 |
| 55 | 0 | 0 | 1 | 0.35 | 0.38 | 0.1 |
| | | | | | | 1559.0* |

*The discharge is 1559 cfs.

If equal width $w$

$$Q = \sum_{n=1}^{n} \frac{w}{4}(h_n + h_{n-1})(v_n + v_{n-1}) \tag{1.2a}$$

where $Q$ = discharge, cfs
$w_n$ = $n$th distance from initial point 0, ft
$h_n$ = $n$th water depth, ft
$v_n$ = $n$th velocity, ft/s

Velocity $v$ is measured by a velocity meter, of which there are several types.

**Example:** Data obtained from the velocity measurement are listed in the first three columns of Table 1.1. Determine the flow rate at this cross section.

**solution:** Summarized field data and complete computations are shown in Table 1.1. The flow rate is 1559 cfs.

## 4 Time of Travel

The time of travel can be determined by dye study or by computation. The river time of travel and stream geometry characteristics can be computed using a volume displacement model. The time of travel is determined at any specific reach as the channel volume of the reach divided by the flow as follows:

$$t = \frac{V}{Q} \times \frac{1}{86{,}400} \tag{1.3}$$

where $t$ = time of travel at a stream reach, days
$V$ = stream reach volume, $ft^3$ or $m^3$
$Q$ = average stream flow in the reach, $ft^3/s$(cfs) or $m^3/s$
86,400 = a factor, s/d

**Example:** The cross-sectional areas at river miles 62.5, 63.0, 63.5, 64.0, 64.5, and 64.8 are, respectively, 271, 265, 263, 259, 258, and 260 $ft^2$ at a surface water elevation. The average flow in 34.8 cfs. Find the time of travel for a reach between river miles 62.5 and 64.8.

**solution:**

Step 1. Find average area in the reach

$$\begin{aligned}\text{Average area} &= \frac{1}{6}(271 + 265 + 263 + 259 + 258 + 260)\ \text{ft}^2 \\ &= 262.7\ \text{ft}^2\end{aligned}$$

Step 2. Find volume

$$\begin{aligned}\text{Distance of the reach} &= (64.8 - 62.5)\ \text{miles} \\ &= 2.3\ \text{miles} \times 5280\ \frac{\text{ft}}{\text{mile}} \\ &= 12{,}144\ \text{ft} \\ V &= 262.7\ \text{ft}^2 \times 12{,}144\ \text{ft} \\ &= 3{,}190{,}000\ \text{ft}^3\end{aligned}$$

Step 3. Find $t$

$$\begin{aligned}t &= \frac{V}{Q} \times \frac{1}{86{,}400} \\ &= \frac{3{,}190{,}000\ \text{ft}^3}{34.8\ \text{ft}^3/\text{s} \times 86{,}400\ \text{s/d}} \\ &= 1.06\ \text{days}\end{aligned}$$

## 5 Dissolved Oxygen and Water Temperature

Dissolved oxygen (DO) and water temperature are most commonly in situ monitored parameters for surface waters (rivers, streams, lakes, reservoirs, wetlands, oceans, etc.). DO concentration in milligrams per liter (mg/L) is a measurement of the amount of oxygen dissolved in water. It can be determined with a DO meter or by a chemical titration method.

The DO in water has an important impact on aquatic animals and plants. Most aquatic animals, such as fish, require oxygen in the water to survive. The two major sources of oxygen in water are from diffusion from the atmosphere across the water surface and the photosynthetic oxygen production from aquatic plants such as algae and macrophytes. Important factors that affect DO in water (Fig. 1.1) may include water temperature, aquatic plant photosynthetic activity, wind and wave mixing, organic contents of the water, and sediment oxygen demand.

Excessive growth of algae (bloom) or other aquatic plants may provide very high concentration of DO, so called supersaturation. On the other hand, oxygen deficiencies can occur when plant respiration depletes oxygen beyond the atmospheric diffusion rate. This can occur especially during the winter ice cover period and when intense decomposition of organic matter in the lake bottom sediment occurs during the summer. These oxygen deficiencies will result in fish being killed.

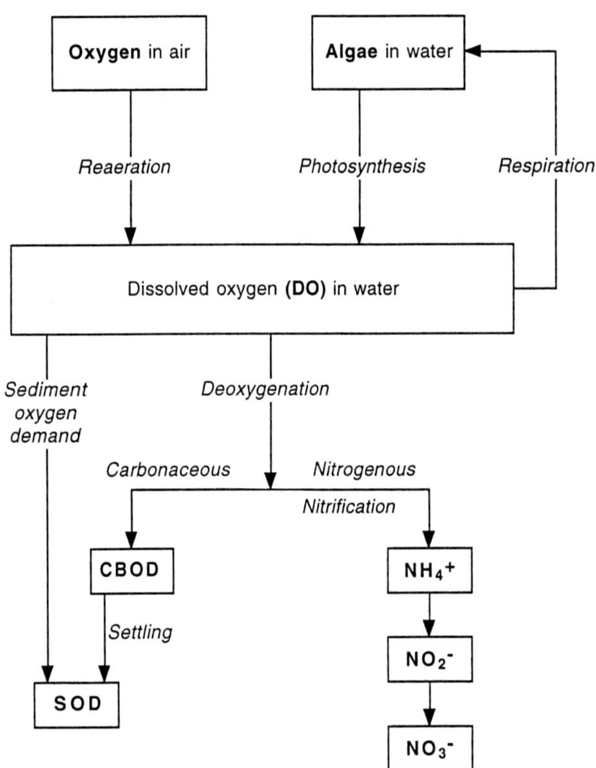


**Figure 1.1** Factors affecting dissolved oxygen concentration in water.

### 5.1 Dissolved oxygen saturation

DO saturation ($DO_{sat}$) values for various water temperatures can be computed using the American Society of Civil Engineers' formula (American Society of Civil Engineering Committee on Sanitary Engineering Research, 1960):

$$DO_{sat} = 14.652 - 0.41022T + 0.0079910T^2 - 0.000077774T^3 \quad (1.4)$$

where $DO_{sat}$ = dissolved oxygen saturation concentration, mg/L
$T$ = water temperature, °C

This formula represents saturation values for distilled water ($\beta = 1.0$) at sea level pressure. Water impurities can increase the saturation level ($\beta > 1.0$) or decrease the saturation level ($\beta < 1.0$), depending on the surfactant characteristics of the contaminant. For most cases, $\beta$ is assumed to be unity. The $DO_{sat}$ values calculated from the above formula are listed in Table 1.2 (example: $DO_{sat} = 8.79$ mg/L, when $T = 21.3$°C) for water temperatures ranging from zero to 30°C (American Society of Engineering Committee on Sanitary Engineering Research, 1960).

**Example 1:** Calculate DO saturation concentration for a water temperature at 0, 10, 20, and 30°C, assuming $\beta = 1.0$.

**solution:**

(a) At $T = 0$°C

$$DO_{sat} = 14.652 - 0 + 0 - 0$$
$$= 14.652 \text{ (mg/L)}$$

(b) At $T = 10$°C

$$DO_{sat} = 14.652 - 0.41022 \times 10 + 0.0079910 \times 10^2 - 0.000077774 \times 10^3$$
$$= 11.27 \text{ (mg/L)}$$

(c) At $T = 20$°C

$$DO_{sat} = 14.652 - 0.41022 \times 20 + 0.0079910 \times 20^2 - 0.000077774 \times 20^3$$
$$= 9.02 \text{ (mg/L)}$$

(d) At $T = 30$°C

$$DO_{sat} = 14.652 - 0.41022 \times 30 + 0.0079910 \times 30^2 - 0.000077774 \times 30^3$$
$$= 7.44 \text{ (mg/L)}$$

The DO saturation concentrations generated by the formula must be corrected for differences in air pressure caused by air temperature changes and

**TABLE 1.2 Dissolved Oxygen Saturation Values in mg/L**

| Temp, °C | 0.0 | 0.1 | 0.2 | 0.3 | 0.4 | 0.5 | 0.6 | 0.7 | 0.8 | 0.9 |
|---|---|---|---|---|---|---|---|---|---|---|
| 0 | 14.65 | 14.61 | 14.57 | 14.53 | 14.49 | 14.45 | 14.41 | 14.37 | 14.33 | 14.29 |
| 1 | 14.25 | 14.21 | 14.17 | 14.13 | 14.09 | 14.05 | 14.02 | 13.98 | 13.94 | 13.90 |
| 2 | 13.86 | 13.82 | 13.79 | 13.75 | 13.71 | 13.68 | 13.64 | 13.60 | 13.56 | 13.53 |
| 3 | 13.49 | 13.46 | 13.42 | 13.38 | 13.35 | 13.31 | 13.28 | 13.24 | 13.20 | 13.17 |
| 4 | 13.13 | 13.10 | 13.06 | 13.03 | 13.00 | 12.96 | 12.93 | 12.89 | 12.86 | 12.82 |
| 5 | 12.79 | 12.76 | 12.72 | 12.69 | 12.66 | 12.62 | 12.59 | 12.56 | 12.53 | 12.49 |
| 6 | 12.46 | 12.43 | 12.40 | 12.36 | 12.33 | 12.30 | 12.27 | 12.24 | 12.21 | 12.18 |
| 7 | 12.14 | 12.11 | 12.08 | 12.05 | 12.02 | 11.99 | 11.96 | 11.93 | 11.90 | 11.87 |
| 8 | 11.84 | 11.81 | 11.78 | 11.75 | 11.72 | 11.70 | 11.67 | 11.64 | 11.61 | 11.58 |
| 9 | 11.55 | 11.52 | 11.49 | 11.47 | 11.44 | 11.41 | 11.38 | 11.35 | 11.33 | 11.30 |
| 10 | 11.27 | 11.24 | 11.22 | 11.19 | 11.16 | 11.14 | 11.11 | 11.08 | 11.06 | 11.03 |
| 11 | 11.00 | 10.98 | 10.95 | 10.93 | 10.90 | 10.87 | 10.85 | 10.82 | 10.80 | 10.77 |
| 12 | 10.75 | 10.72 | 10.70 | 10.67 | 10.65 | 10.62 | 10.60 | 10.57 | 10.55 | 10.52 |
| 13 | 10.50 | 10.48 | 10.45 | 10.43 | 10.40 | 10.38 | 10.36 | 10.33 | 10.31 | 10.28 |
| 14 | 10.26 | 10.24 | 10.22 | 10.19 | 10.17 | 10.15 | 10.12 | 10.10 | 10.08 | 10.06 |
| 15 | 10.03 | 10.01 | 9.99 | 9.97 | 9.95 | 9.92 | 9.90 | 9.88 | 9.86 | 9.84 |
| 16 | 9.82 | 9.79 | 9.77 | 9.75 | 9.73 | 9.71 | 9.69 | 9.67 | 9.65 | 9.63 |
| 17 | 9.61 | 9.58 | 9.56 | 9.54 | 9.52 | 9.50 | 9.48 | 9.46 | 9.44 | 9.42 |
| 18 | 9.40 | 9.38 | 9.36 | 9.34 | 9.32 | 9.30 | 9.29 | 9.27 | 9.25 | 9.23 |
| 19 | 9.21 | 9.19 | 9.17 | 9.15 | 9.13 | 9.12 | 9.10 | 9.08 | 9.06 | 9.04 |
| 20 | 9.02 | 9.00 | 8.98 | 8.97 | 8.95 | 8.93 | 8.91 | 8.90 | 8.88 | 8.86 |
| 21 | 8.84 | 8.82 | 8.81 | 8.79 | 8.77 | 8.75 | 8.74 | 8.72 | 8.70 | 8.68 |
| 22 | 8.67 | 8.65 | 8.63 | 8.62 | 8.60 | 8.58 | 8.56 | 8.55 | 8.53 | 8.52 |
| 23 | 8.50 | 8.48 | 8.46 | 8.45 | 8.43 | 8.42 | 8.40 | 8.38 | 8.37 | 8.35 |
| 24 | 8.33 | 8.32 | 8.30 | 8.29 | 8.27 | 8.25 | 8.24 | 8.22 | 8.21 | 8.19 |
| 25 | 8.18 | 8.16 | 8.14 | 8.13 | 8.11 | 8.10 | 8.08 | 8.07 | 8.05 | 8.04 |
| 26 | 8.02 | 8.01 | 7.99 | 7.98 | 7.96 | 7.95 | 7.93 | 7.92 | 7.90 | 7.89 |
| 27 | 7.87 | 7.86 | 7.84 | 7.83 | 7.81 | 7.80 | 7.78 | 7.77 | 7.75 | 7.74 |
| 28 | 7.72 | 7.71 | 7.69 | 7.68 | 7.66 | 7.65 | 7.64 | 7.62 | 7.61 | 7.59 |
| 29 | 7.58 | 7.56 | 7.55 | 7.54 | 7.52 | 7.51 | 7.49 | 7.48 | 7.47 | 7.45 |
| 30 | 7.44 | 7.42 | 7.41 | 7.40 | 7.38 | 7.37 | 7.35 | 7.34 | 7.32 | 7.31 |

SOURCE: American Society of Civil Engineering Committee on Sanitary Engineering Research, 1960

for elevation above the mean sea level (MSL). The correction factor can be calculated as follows:

$$f = \frac{2116.8 - (0.08 - 0.000115A)E}{2116.8} \tag{1.5}$$

where $f$ = correction factor for above MSL
$A$ = air temperature, °C
$E$ = elevation of the site, feet above MSL

**Example 2:** Find the correction factor of $DO_{sat}$ value for water at 620 ft above the MSL and air temperature of 25°C. What is $DO_{sat}$ at a water temperature of 20°C?

**solution:**

Step 1. Using Eq. (1.5)

$$f = \frac{2116.8 - (0.08 - 0.000115A)E}{2116.8}$$

$$= \frac{2116.8 - (0.08 - 0.000115 \times 25)620}{2116.8}$$

$$= \frac{2116.8 - 47.8}{2116.8}$$

$$= 0.977$$

Step 2. Compute $DO_{sat}$

From Example 1, at $T = 20°C$

$$DO_{sat} = 9.02 \text{ (mg/L)}$$

With an elevation correction factor of 0.977

$$DO_{sat} = 9.02 \text{ mg/L} \times 0.977 = 8.81 \text{ (mg/L)}$$

## 5.2 Dissolved oxygen availability

Most regulatory agencies have standards for minimum DO concentrations in surface waters to support indigenous fish species in surface waters. In Illinois, for example, the Illinois Pollution Control Board stipulate that dissolved oxygen shall not be less than 6.0 mg/L during at least 16 h of any 24-h period, nor less than 5.0 mg/L at any time (IEPA, 1999).

The availability of dissolved oxygen in a flowing stream is highly variable due to several factors. Daily and seasonal variations in DO levels have been reported. The diurnal variations in DO are primarily induced by algal productivity. Seasonal variations are attributable to changes in temperature that affect DO saturation values. The ability of a stream to absorb or reabsorb oxygen from the atmosphere is affected by flow factors such as water depth and turbulence, and it is expressed in terms of the reaeration coefficient. Factors that may represent significant sources of oxygen use or oxygen depletion are biochemical oxygen demand (BOD) and sediment oxygen demand (SOD). BOD, including carbonaceous BOD (CBOD) and nitrogenous BOD (NBOD), may be the product of both naturally occurring oxygen use in the decomposition of organic material and oxygen depletion in the stabilization of effluents discharged from wastewater treatment plants (WTPs). The significance of any of these factors depends upon the specific stream conditions. One or all of these factors may be considered in the evaluation of oxygen use and availability.

## 6 Biochemical Oxygen Demand Analysis

Laboratory analysis for organic matter in water and wastewater includes testing for biochemical oxygen demand, chemical oxygen demand (COD), total organic carbon (TOC), and total oxygen demand (TOD). The BOD test is a biochemical test involving the use of microorganisms. The COD test is a chemical test. The TOC and TOD tests are instrumental tests.

The BOD determination is an empirical test that is widely used for measuring waste (loading to and from wastewater treatment plants), evaluating the organic removal efficiency of treatment processes, and assessing stream assimilative capacity. The BOD test measures: (1) the molecular oxygen consumed during a specific incubation period for the biochemical degradation of organic matter (CBOD); (2) oxygen used to oxidize inorganic material such as sulfide and ferrous iron; and (3) reduced forms of nitrogen (NBOD) with an inhibitor (trichloromethylpyridine). If an inhibiting chemical is not used, the oxygen demand measured is the sum of carbonaceous and nitrogenous demands, so-called total BOD or ultimate BOD.

The extent of oxidation of nitrogenous compounds during the 5-day incubation period depends upon the type and concentration of microorganisms that carry out biooxidation. The nitrifying bacteria usually are not present in raw or settleable primary sewage. These nitrifying organisms are present in sufficient numbers in biological (secondary) effluent. A secondary effluent can be used as "seeding" material for an NBOD test of other samples. Inhibition of nitrification is required for a CBOD test for secondary effluent samples, for samples seeded with secondary effluent, and for samples of polluted waters.

The result of the 5-day BOD test is recorded as carbonaceous biochemical oxygen demand, $CBOD_5$, when inhibiting nitrogenous oxygen demand. When nitrification is not inhibited, the result is reported as $BOD_5$ (incubation at 15°C for 5 days).

The BOD test procedures can be found in *Standard Methods for the Examination of Water and Wastewater* (APHA, AWWA, and WEF, 1995). When the dilution water is seeded, oxygen uptake (consumed) is assumed to be the same as the uptake in the seeded blank. The difference between the sample BOD and the blank BOD, corrected for the amount of seed used in the sample, is the true BOD. Formulas for calculation of BOD are as follows (APHA, AWWA, and WEF, 1995):

When dilution water is not seeded:

$$\text{BOD, mg/L} = \frac{D_1 - D_2}{P} \tag{1.6}$$

When dilution water is seeded:

$$\text{BOD, mg/L} = \frac{(D_i - D_e) - (B_i - B_e)f}{P} \tag{1.7}$$

where $D_1, D_i$ = DO of diluted sample immediately after preparation, mg/L
$D_2, D_e$ = DO of diluted sample after incubation at 20°C, mg/L
$P$ = decimal volumetric fraction of sample used; mL of sample/300 mL for Eq. (1.6)
$B_i$ = DO of seed control before incubation, mg/L
$B_e$ = DO of seed control after incubation, mg/L
$f$ = ratio of seed in diluted sample to seed in seed control
$P$ = percent seed in diluted sample/percent seed in seed control for Eq. (1.7)

If seed material is added directly to the sample and to control bottles:

$f$ = volume of seed in diluted sample/volume of seed in seed control

**Example 1:** For a BOD test, 75 mL of a river water sample is used in the 300 mL of BOD bottles without seeding with three duplications. The initial DO in three BOD bottles read 8.86, 8.88, and 8.83 mg/L, respectively. The DO levels after 5 days at 20°C incubation are 5.49, 5.65, and 5.53 mg/L, respectively. Find the 5-day BOD ($BOD_5$) for the river water.

**solution:**

Step 1. Determine average DO uptake

$$x = \frac{\Sigma(D_1 - D_2)}{3}$$

$$= \frac{[(8.86 - 5.49) + (8.88 - 5.65) + (8.83 - 5.53)]}{3}$$

$$= 3.30(\text{mg/L})$$

Step 2. Determine $P$

$$P = \frac{75}{300} = 0.25$$

Step 3. Compute $BOD_5$

$$BOD_5 = \frac{x}{P} = \frac{3.30 \text{ mg/L}}{0.25} = 13.2 \text{ (mg/L)}$$

**Example 2:** The wastewater is diluted by a factor of 1/20 using seeded control water. DO levels in the sample and control bottles are measured at 1-day intervals. The results are shown in Table 1.3. One milliliter of seed material is added directly to diluted and to control bottles. Find daily BOD values.

**solution:**

Step 1. Compute $f$ and $P$

$$f = \frac{1\ \text{mL}}{1\ \text{mL}} = 1.0$$

$$P = \frac{1}{20} = 0.05$$

Step 2. Find BODs using Eq. (1.7)

Day 1:

$$\begin{aligned} \text{BOD}_1 &= \frac{(D_i - D_e) - (B_i - B_e)f}{P} \\ &= \frac{(7.98 - 5.05) - (8.25 - 8.18)1}{0.05} \\ &= 57.2\ (\text{mg/L}) \end{aligned}$$

Similarly

Day 2:

$$\begin{aligned} \text{BOD}_2 &= \frac{(7.98 - 4.13) - (8.25 - 8.12)1}{0.05} \\ &= 74.4\ (\text{mg/L}) \end{aligned}$$

**TABLE 1.3 Change in Dissolved Oxygen and Biochemical Oxygen Demand with Time**

| | Dissolved oxygen, mg/L | | |
|---|---|---|---|
| Time, days | Diluted sample | Seeded control | BOD, mg/L |
| 0 | 7.98 | 8.25 | — |
| 1 | 5.05 | 8.18 | 57.2 |
| 2 | 4.13 | 8.12 | 74.4 |
| 3 | 3.42 | 8.07 | 87.6 |
| 4 | 2.95 | 8.03 | 96.2 |
| 5 | 2.60 | 7.99 | 102.4 |
| 6 | 2.32 | 7.96 | 107.4 |
| 7 | 2.11 | 7.93 | 111.0 |

For other days, BOD can be determined in the same manner. The results of BODs are also presented in the above table. It can be seen that $BOD_5$ for this wastewater is 102.4 mg/L.

## 7 Streeter–Phelps Oxygen Sag Formula

The method most widely used for assessing the oxygen resources in streams and rivers subjected to effluent discharges is the Streeter–Phelps oxygen sag formula that was developed for the use on the Ohio River in 1914. The well-known formula is defined as follows (Streeter and Phelps, 1925):

$$D_t = \frac{K_1 L_a}{K_2 - K_1}[e^{-K_1 t} - e^{-K_2 t}] + D_a e^{-K_2 t} \tag{1.8a}$$

or

$$D_t = \frac{k_1 L_a}{k_2 - k_1}[10^{-k_1 t} - 10^{k_2 t}] + D_a 10^{-k_2 t} \tag{1.8b}$$

where $D_t$ = DO saturation deficit downstream, mg/L or lb ($DO_{sat} - DO_a$) at time $t$
$t$ = time of travel from upstream to downstream, days
$D_a$ = initial DO saturation deficit of upstream water, mg/L or lb
$L_a$ = ultimate upstream BOD, mg/L
$e$ = base of natural logarithm, 2.7183
$K_1$ = deoxygenation coefficient to the base $e$, per day
$K_2$ = reoxygenation coefficient to the base $e$, per day
$k_1$ = deoxygenation coefficient to the base 10, per day
$k_2$ = reoxygenation coefficient to the base 10, per day

In the early days, $K_1$ or $K_2$ and $k_1$ or $k_2$ were used classically for values based on $e$ and 10, respectively. Unfortunately, in recent years, many authors have mixed the usage of $K$ and $k$. Readers should be aware of this. The logarithmic relationships between $k$ and $K$ are $K_1 = 2.3026k_1$ and $K_2 = 2.3026k_2$; or $k_1 = 0.4343K_1$ and $k_2 = 0.4343K_2$.

The Streeter–Phelps oxygen sag equation is based on two assumptions: (1) at any instant, the deoxygenation rate is directly proportional to the amount of oxidizable organic material present; and (2) the reoxygenation rate is directly proportional to the DO deficit. Mathematical expressions for assumptions (1) and (2) are

$$\frac{\mathrm{d}D}{\mathrm{d}t} = K_1(L_a - L_t) \tag{1.9}$$

and

$$\frac{dD}{dt} = -K_2 D \tag{1.10}$$

where $\frac{dD}{dt}$ = the net rate of change in the DO deficit, or the absolute change of DO deficit ($D$) over an increment of time $dt$ due to stream waste assimilative capacity affected by deoxygenation coefficient $K_1$ and due to an atmospheric exchange of oxygen at the air/water interface affected by the reaeration coefficient $K_2$

$L_a$ = ultimate upstream BOD, mg/L

$L_t$ = ultimate downstream BOD at any time $t$, mg/L

Combining the above two differential equations and integrating between the limits $D_a$, the initial upstream sampling point, and $t$, any time of flow below the initial point, yields the basic equation devised by Streeter and Phelps. This stimulated intensive research on BOD, reaction rates, and stream sanitation.

There are some shortcomings in the Streeter–Phelps equation. The two assumptions are likely to generate errors. It is assumed that: (1) wastes discharged to a receiving water are evenly distributed over the river's cross section; and (2) the wastes travel down the river as a plug flow without mixing along the axis of the river. These assumptions only apply within a reasonable distance downstream. Effluent discharge generally travels as a plume for some distance before mixing. In addition, it is assumed that oxygen is removed by microbial oxidation of the organic matter (BOD) and is replaced by reaeration from the surface. Some factors, such as the removal of BOD by sedimentation, conversion of suspended BOD to soluble BOD, sediment oxygen demand, and algal photosynthesis and respiration are not included.

The formula is a classic in sanitary engineering stream work. Its detailed analyses can be found in almost all general environmental engineering texts. Many modifications and adaptations of the basic equation have been devised and have been reported in the literature. Many researches have been carried out on BOD, $K_1$, and $K_2$ factors. Illustrations for oxygen sag formulas will be presented in the latter sections.

## 8 BOD Models and $K_1$ Computation

Under aerobic conditions, organic matter and some inorganics can be used by bacteria to create new cells, energy, carbon dioxide, and residue. The oxygen used to oxidize total organic material and all forms of nitrogen for 60 to 90 days is called the ultimate BOD (UBOD). It is common that measurements of oxygen consumed in a 5-day test period called

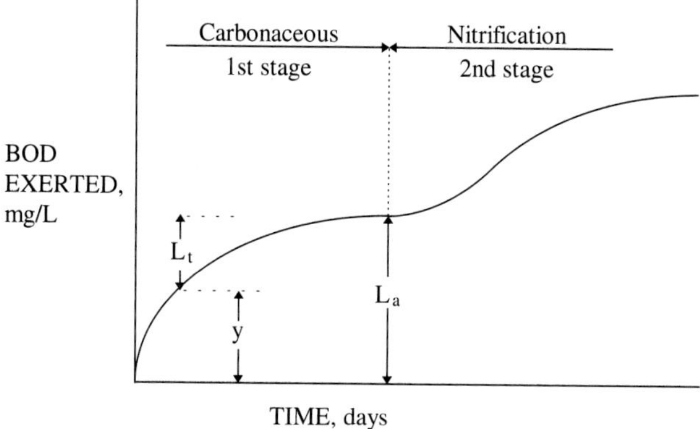


**Figure 1.2** BOD progressive curve.

5-day BOD or $BOD_5$ are practiced. The BOD progressive curve is shown in Fig. 1.2.

## 8.1 First-order reaction

Phelps law states that the rate of biochemical oxidation of organic matter is proportional to the remaining concentration of unoxidized substance. Phelps law can be expressed in differential form as follows (monomolecular or unimolecular chemical reaction):

$$-\frac{\mathrm{d}L_t}{\mathrm{d}t} = K_1 L_t$$

$$\frac{\mathrm{d}L_t}{L_t} = -K_1 \mathrm{d}t \qquad (1.11)$$

by integration

$$\int_{L_a}^{L_t} \frac{\mathrm{d}L_t}{L_t} = -K_1 \int_0^t \mathrm{d}t_1$$

$$\ln \frac{L_t}{L_a} = -K_1 t \qquad \text{or} \qquad \log \frac{L_t}{L_a} = -0.434 K_1 t = -k_1 t$$

$$\frac{L_t}{L_a} = e^{-K_1 t} \qquad \text{or} \qquad \frac{L_t}{L_a} = 10^{-k_1 t}$$

$$L_t = L_a \cdot e^{-K_1 t} \qquad \text{or} \qquad L_t = L_a \cdot 10^{-k_1 t} \qquad (1.12)$$

where $L_t$ = BOD remaining after time $t$ days, mg/L
$L_a$ = first-stage BOD, mg/L

$K_1$ = deoxygenation rate, based on $e$, $K_1 = 2.303k_1$, per day
$k_1$ = deoxygenation rate, based on 10,
$k_1 = 0.4343K_1(k_1 = 0.1$ at 20°C), per day
$e$ = base of natural logarithm, 2.7183

Oxygen demand exerted up to time $t$, $y$, is a first-order reaction (see Fig. 1.2):

$$y = L_a - L_t$$

$$y = L_a(1 - e^{-K_1 t}) \tag{1.13a}$$

or based on $\log_{10}$

$$y = L_a(1 - 10^{-k_1 t}) \tag{1.13b}$$

When a delay occurs in oxygen uptake at the onset of a BOD test, a lag-time factor $t_0$ should be included and Eqs. (1.13a) and (1.13b) become

$$y = L_a[1 - e^{-K_1(t-t_0)}] \tag{1.14a}$$

or

$$y = L_a[1 - 10^{-k_1(t-t_0)}] \tag{1.14b}$$

For the Upper Illinois Waterway study, many of the total and NBOD curves have an S-shaped configuration. The BOD in waters from pools often consists primarily of high-profile second-stage or NBOD, and the onset of the exertion of this NBOD is often delayed 1 or 2 days. The delayed NBOD and the total BOD (TBOD) curves, dominated by the NBOD fraction, often exhibit an S-shaped configuration. The general mathematical formula used to simulate the S-shaped curve is (Butts *et al.*, 1975)

$$y = L_a[1 - e^{-K_1(t-t_0)^m}] \tag{1.15}$$

where $m$ is a power factor, and the other terms are as previously defined.

Statistical results show that a power factor of 2.0 in Eq. (1.15) best represents the S-shaped BOD curve generated in the Lockport and Brandon Road areas of the waterway. Substituting $m = 2$ in Eq. (1.15) yields

$$y = L_a[1 - e^{-K_1(t-t_0)^2}] \tag{1.15a}$$

**Example 1:** Given $K_1 = 0.25$ per day, $BOD_5 = 6.85$ mg/L, for a river water sample. Find $L_a$ when $t_0 = 0$ days and $t_0 = 2$ days.

**solution:**

Step 1.

When $t_0 = 0$

$$y = L_a(1 - e^{-K_1 t})$$

$$6.85 = L_a(1 - e^{-0.25\times 5})$$

$$L_a = \frac{6.85}{1 - 0.286}$$

$$= 9.60 \text{ (mg/L)}$$

Another solution using $k_1$: Since

$$k_1 = 0.4343K_1 = 0.4343 \times 0.25 \text{ per day} = 0.109 \text{ per day}$$

$$y = L_a(1 - 10^{-k_1 t})$$

$$6.85 = L_a(1 - 10^{-0.109\times 5})$$

$$L_a = \frac{6.85}{1 - 0.286}$$

$$= 9.60 \text{ (mg/L)}$$

Step 2.

When $t_0 = 2$ days, using Butts *et al.*'s equation (Eq. (1.15a)):

$$y = L_a[1 - e^{-K_1(t-t_0)^2}]$$

or

$$L_a = \frac{y}{1 - e^{-K_1(t-t_0)^2}}$$

$$= \frac{6.85}{1 - e^{-0.25(5-2)^2}}$$

$$= \frac{6.85}{1 - e^{-2.25}}$$

$$= \frac{6.85}{1 - 0.105}$$

$$= 7.65 \text{ (mg/L)}$$

**Example 2:** Compute the portion of BOD remaining to the ultimate BOD ($1 - e^{-K_1 t}$ or $1 - 10^{-k_1 t}$) for $k_1 = 0.10$ (or $K_1 = 0.23$).

**TABLE 1.4 Relationship between $t$ and the Ultimate BOD ($1 - 10^{-k_1 t}$ or $1 - e^{-K_1 t}$)**

| $t$ | $1 - 10^{-k_1 t}$ or $1 - e^{-K_1 t}$ | $t$ | $1 - 10^{-k_1 t}$ or $1 - e^{-K_1 t}$ |
|---|---|---|---|
| 0.25 | 0.056 | 4.5 | 0.646 |
| 0.50 | 0.109 | 5.0 | 0.684 |
| 0.75 | 0.159 | 6.0 | 0.749 |
| 1.00 | 0.206 | 7.0 | 0.800 |
| 1.25 | 0.250 | 8.0 | 0.842 |
| 1.50 | 0.291 | 9.0 | 0.874 |
| 1.75 | 0.332 | 10.0 | 0.900 |
| 2.0 | 0.369 | 12.0 | 0.937 |
| 2.5 | 0.438 | 16.0 | 0.975 |
| 3.0 | 0.500 | 20.0 | 0.990 |
| 3.5 | 0.553 | 30.0 | 0.999 |
| 4.0 | 0.602 | $\infty$ | 1.0 |

**solution:** By Eq. (1.13b)

$$y = L_a(1 - 10^{-k_1 t})$$

$$\frac{y}{L_a} = 1 - 10^{-k_1 t} \quad \text{or} \quad 1 - e^{-K_1 t}$$

When $t = 0.25$ days

$$\frac{y}{L_a} = 1 - 10^{-0.10 \times 0.25} = 1 - 0.944 = 0.056$$

Similar calculations can be performed as above. The relationship between $t$ and $(1 - 10^{-k_1 t})$ is listed in Table 1.4.

### 8.2 Determination of deoxygenation rate and ultimate BOD

Biological decomposition of organic matter is a complex phenomenon. Laboratory BOD results do not necessarily fit actual stream conditions. BOD reaction rate is influenced by immediate demand, stream or river dynamic environment, nitrification, sludge deposit, and types and concentrations of microbes in the water. Therefore, laboratory BOD analyses and stream surveys are generally conducted for raw and treated wastewaters and river water to determine BOD reaction rate.

Many investigators have worked on developing and refining methods and formulas for use in evaluating the deoxygenation ($K_1$) and reaeration ($K_2$) constants and the ultimate BOD ($L_a$). There are several methods proposed to determine $K_1$ values. Unfortunately, $K_1$ values determined by different methods given by the same set of data have considerable variations. Reed–Theriault least-squares method published in 1927 (US Public Health Service, 1927) gives the most consistent results, but it is time consuming and tedious. Computation using a digital computer was developed by Gannon and Downs (1964).

In 1936, a simplified procedure, the so-called log-difference method of estimating the constants of the first-stage BOD curve, was presented by Fair (1936). The method is also mathematically sound, but is also difficult to solve.

Thomas (1937) followed Fair *et al.* (1941a, 1941b) and developed the "slope" method, which, for many years, was the most used procedure for calculating the constants of the BOD curve. Later, Thomas (1950) presented a graphic method for BOD curve constants. In the same year, Moore *et al.* (1950) developed the "moment" method that was simple, reliable, and accurate to analyze BOD data; this soon became the most used technique for computing the BOD constants.

Researchers found that $K_1$ varied considerably for different sources of wastewaters and questioned the accepted postulate that the 5-day BOD is proportional to the strength of the sewage. Orford and Ingram (1953) discussed the monomolecular equation as being inaccurate and unscientific in its relation to BOD. They proposed that the BOD curve could be expressed as a logarithmic function.

Tsivoglou (1958) proposed a "daily difference" method of BOD data solved by a semigraphical solution. A "rapid ratio" method can be solved using curves developed by Sheehy (1960). O'Connor (1966) modified the least-squares method using $BOD_5$.

This book describes Thomas's slope method, method of moments, logarithmic function, and rapid methods calculating $K_1$ (or $k_1$) and $L_a$.

**Slope method.** The slope method (Thomas, 1937) gives the BOD constants via the least-squares treatment of the basic form of the first-order reaction equation or

$$\frac{dy}{dt} = K_1(L_a - y) = K_1L_a - K_1y \tag{1.16}$$

where $dy$ = increase in BOD per unit time at time $t$
$K_1$ = deoxygenation constant, per day
$L_a$ = first stage ultimate BOD, mg/L
$y$ = BOD exerted in time $t$, mg/L

This differential equation (Eq. (1.16)) is linear between $dy/dt$ and $y$. Let $y' = dy/dt$ to be the rate of change of BOD and $n$ be the number of BOD measurements minus one. Two normal equations for finding $K_1$ and $L_a$ are

$$na + b\Sigma y - \Sigma y' = 0 \tag{1.17}$$

and

$$a\Sigma y + b\Sigma y^2 - \Sigma yy' = 0 \tag{1.18}$$

Solving Eqs. (1.17) and (1.18) yields values of $a$ and $b$, from which $K_1$ and $L_a$ can be determined directly by following relations:

$$K_1 = -b \tag{1.19}$$

and

$$L_a = -a/b \tag{1.20}$$

The calculations include first determinations of $y'$, $y'y$, and $y^2$ for each value of $y$. The summation of these gives the quantities of $\Sigma y'$, $\Sigma y'y$, and $\Sigma y^2$ which are used for the two normal equations. The values of the slopes are calculated from the given data of $y$ and $t$ as follows:

$$\frac{dy_i}{dt} = y'_i = \frac{(y_i - y_{i-1})\left(\frac{t_{i+1} - t_i}{t_i - t_{i-1}}\right) + (y_{i+1} - y_i)\left(\frac{t_i - t_{i-1}}{t_{i+1} - t_i}\right)}{t_{i+1} - t_{i-1}} \tag{1.21}$$

For the special case, when equal time increments $t_{i+1} - t_i = t_3 - t_2 = t_2 - t_1 = \Delta t$, $y'$ becomes

$$\frac{dy_i}{dt} = \frac{y_{i+1} - y_{i-1}}{2\Delta t} \quad \text{or} \quad \frac{y_{i+1} - y_{i-1}}{t_{i+1} - t_{i-1}} \tag{1.21a}$$

A minimum of six observations ($n > 6$) of $y$ and $t$ are usually required to give consistent results.

**Example 1:** For equal time increments, BOD data at temperature of 20°C, $t$ and $y$, are shown in Table 1.5. Find $K_1$ and $L_a$.

**solution:**

Step 1. Calculate $y'$, $y'y$, and $y^2$

Step 2. Determine $a$ and $b$

Writing normal equations (Eqs. (1.17) and (1.18)), $n = 9$

$$na + b\Sigma y - \Sigma y' = 0$$

$$9a + 865.8b - 89.6 = 0$$

$$a + 96.2b - 9.96 = 0 \tag{1}$$

**TABLE 1.5 Calculations for $y'$, $y'y$, and $y^2$ Values**

| $t$, day (1) | $y$ (2) | $y'$ (3) | $y'y$ (4) | $y^2$ (5) |
|---|---|---|---|---|
| 0 | 0 | | | |
| 1 | 56.2 | 37.2* | 2090.64 | 3158.44 |
| 2 | 74.4 | 15.7 | 1168.08 | 5535.36 |
| 3 | 87.6 | 10.9 | 954.84 | 7673.76 |
| 4 | 96.2 | 7.4 | 711.88 | 9254.44 |
| 5 | 102.4 | 5.6 | 573.44 | 10485.76 |
| 6 | 107.4 | 4.3 | 461.82 | 11534.76 |
| 7 | 111.0 | 3.3 | 366.30 | 12321.00 |
| 8 | 114.0 | 2.8 | 319.20 | 12996.00 |
| 9 | 116.6 | 2.4 | 279.84 | 13595.56 |
| 10 | 118.8 | | | |
| Σ | 865.8† | 89.6 | 6926.04 | 86555.08 |

$^*y'_1 = \frac{y_3 - y_1}{t_3 - t_1} = \frac{74.4 - 0}{2 - 0} = 37.2$

†Sum of first nine observations.

and

$$a\Sigma y + b\Sigma y^2 - \Sigma yy' = 0$$

$$865.8a + 86{,}555b - 6926 = 0$$

$$a + 99.97b - 8.0 = 0 \qquad (2)$$

Eq. (2) – Eq. (1)

$$3.77b + 1.96 = 0$$

$$b = -0.52$$

From Eq. (2)

$$a + 99.97(-0.52) - 8.0 = 0$$

$$a = 59.97$$

Step 3. Calculate $K_1$ and $L_a$ with Eqs. (1.19) and (1.20)

$$K_1 = -b = -(-0.52)$$

$$= 0.52 \text{ (per day)}$$

$$L_a = -a/b$$

$$= -59.97/(-0.52)$$

$$= 115.3 \text{ (mg/L)}$$

**Example 2:** For unequal time increments, observed BOD data, $t$ and $y$ are given in Table 1.6. Find $K_1$ and $L_a$.

**solution:**

Step 1. Calculate $\Delta t$, $\Delta y$, $y'$, $yy'$, and $y^2$; then complete Table 1.6

From Eq. (1.21) (see Table 1.6)

$$^{*}y'_i = \frac{(y_i - y_{i-1})\left(\frac{t_{i+1} - t_i}{t_i - t_{i-1}}\right) + (y_{i+1} - y_i)\left(\frac{t_i - t_{i-1}}{t_{i+1} - t_i}\right)}{t_{i+1} - t_{i-1}}$$

$$= \frac{(\Delta y_{i-1})\left(\frac{\Delta t_{i+1}}{\Delta t_{i-1}}\right) + (\Delta y_{i+1})\left(\frac{\Delta t_{i-1}}{\Delta t_{i+1}}\right)}{(\Delta t_{i-1}) + (\Delta t_{i+1})}$$

$$y'_1 = \frac{(28.8)\left(\frac{0.6}{0.4}\right) + (27.4)\left(\frac{0.4}{0.6}\right)}{0.4 + 0.6} = 61.47$$

**TABLE 1.6 Various *t* and *y* Values**

| $t$ | $\Delta t$ | $y$ | $\Delta y$ | $y'$ | $yy'$ | $y^2$ |
|---|---|---|---|---|---|---|
| 0 | | 0 | | | | |
| | 0.4 | | 28.8 | | | |
| 0.4 | | 28.8 | | 61.47* | 1770.24 | 829.44 |
| | 0.6 | | 27.4 | | | |
| 1 | | 56.2 | | 30.90 | 1736.75 | 3158.44 |
| | 0.5 | | 9.3 | | | |
| 1.5 | | 65.5 | | 19.48 | 1276.00 | 4290.25 |
| | 0.7 | | 14.5 | | | |
| 2.2 | | 80.0 | | 15.48 | 1238.48 | 6400.00 |
| | 0.8 | | 7.6 | | | |
| 3 | | 87.6 | | 9.10 | 797.16 | 7673.76 |
| | 1 | | 8.6 | | | |
| 4 | | 96.2 | | 7.40 | 711.88 | 9254.44 |
| | 1 | | 6.2 | | | |
| 5 | | 102.4 | | 5.57 | 570.03 | 10485.76 |
| | 2 | | 8.6 | | | |
| 7 | | 111.0 | | 3.55 | 394.05 | 12321.00 |
| | 2 | | 5.6 | | | |
| 9 | | 116.6 | | 2.43 | 282.95 | 13595.56 |
| | 3 | | 5.6 | | | |
| 12 | | 122.2 | | | | |
| Sum | | 744.3 | | 155.37 | 8777.53 | 68008.65 |

*See text for calculation of this value.

Step 2. Compute $a$ and $b$; while $n = 9$

$$na + b\Sigma y - \Sigma y' = 0$$

$$9a + 744.3b - 155.37 = 0$$

$$a + 82.7b - 17.26 = 0 \qquad (1)$$

and

$$a\Sigma y + b\Sigma y^2 - \Sigma yy' = 0$$

$$744.3a + 68{,}008.65b - 8777.53 = 0$$

$$a + 91.37b - 11.79 = 0 \qquad (2)$$

Eq. (2) – Eq. (1)

$$8.67b + 5.47 = 0$$

$$b = -0.63$$

with Eq. (2)

$$a + 91.37(-0.63) - 11.79 = 0$$

$$a = 69.35$$

Step 3. Determine $K_1$ and $L_a$

$$K_1 = -b = 0.63 \text{ (per day)}$$

$$L_a = -a/b$$

$$= -69.35/-0.63$$

$$= 110.1 \text{ (mg/L)}$$

**Moment method.** This method requires that BOD measurements must be a series of regularly spaced time intervals. Calculations are needed for the sum of the BOD values, $\Sigma y$, accumulated to the end of a series of time intervals and the sum of the product of time and observed BOD values, $\Sigma ty$, accumulated to the end of the time series.

The rate constant $K_1$ and the ultimate BOD $L_a$ can then be easily read from a prepared graph by entering values of $\Sigma y/\Sigma ty$ on the appropriate scale. Treatments of BOD data with and without lag phase will be different. The authors (Moore *et al.*, 1950) presented three graphs for

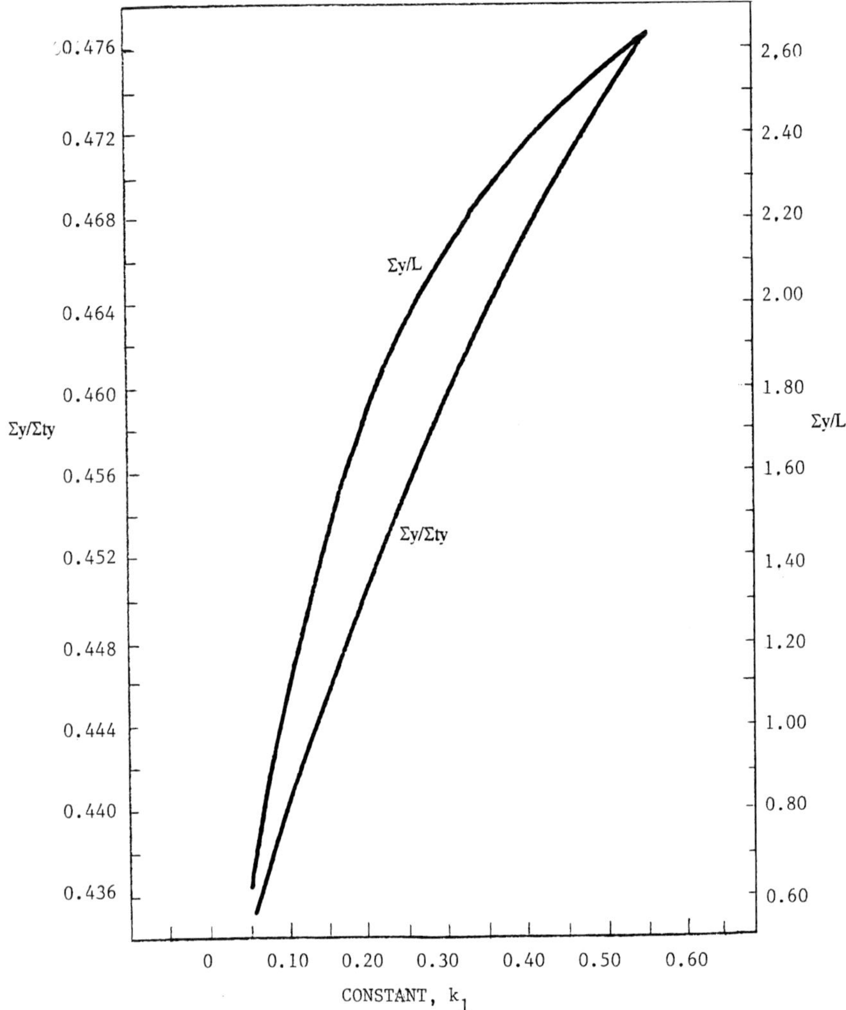


**Figure 1.3** $\Sigma y/L$ and $\Sigma y/\Sigma ty$ for various values of $k_1$ in a 3-day sequence.

3-, 5-, and 7-day sequences (Figs. 1.3, 1.4, and 1.5) with daily intervals for BOD value without lag phase. There is another chart presented for a 5-day sequence with lag phase (Fig 1.6).

**Example 1:** Use the BOD (without lag phase) on Example 1 of Thomas' slope method, find $K_1$ and $L_a$.

**solution:**

Step 1. Calculate $\Sigma y$ and $\Sigma ty$ (see Table 1.7)

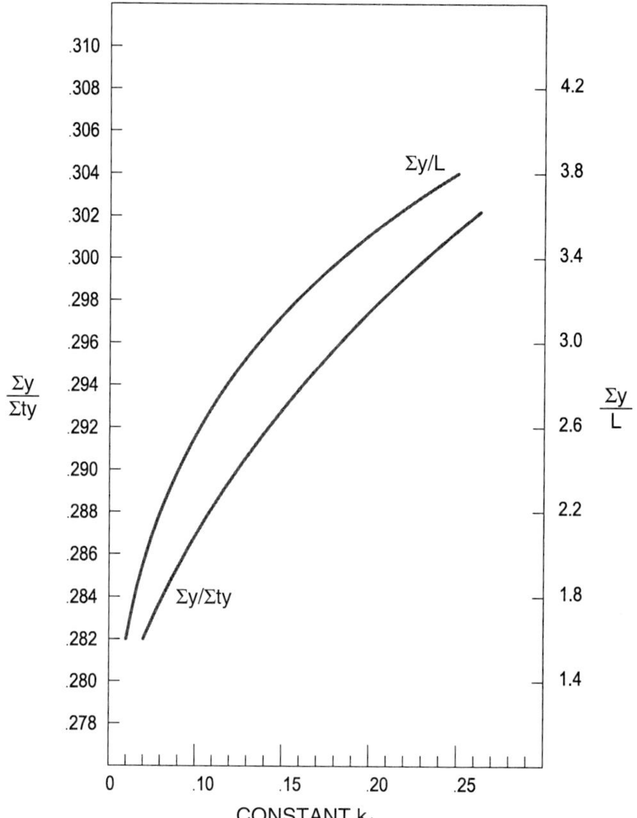


**Figure 1.4** $\Sigma y/L$ and $\Sigma y/\Sigma ty$ for various values of $k_1$ in a 5-day sequence.

Step 2. Compute $\Sigma y/\Sigma ty$

$$\Sigma y/\Sigma ty = 635.2/2786 = 0.228$$

Step 3. Find $K_1$ and $L_a$

On the 7-day time sequence graph (Fig. 1.5), enter the value 0.288 on the $\Sigma y/\Sigma ty$ scale, extend a horizontal line to the curve labeled $\Sigma y/\Sigma ty$, and from this point, follow a vertical line to the $k_1$ scale. A value of $k_1 = 0.264$.

$$K_1 = 2.3026 \times k_1 = 2.3026 \times 0.264$$

$$= 0.608 \text{ (per day)}$$

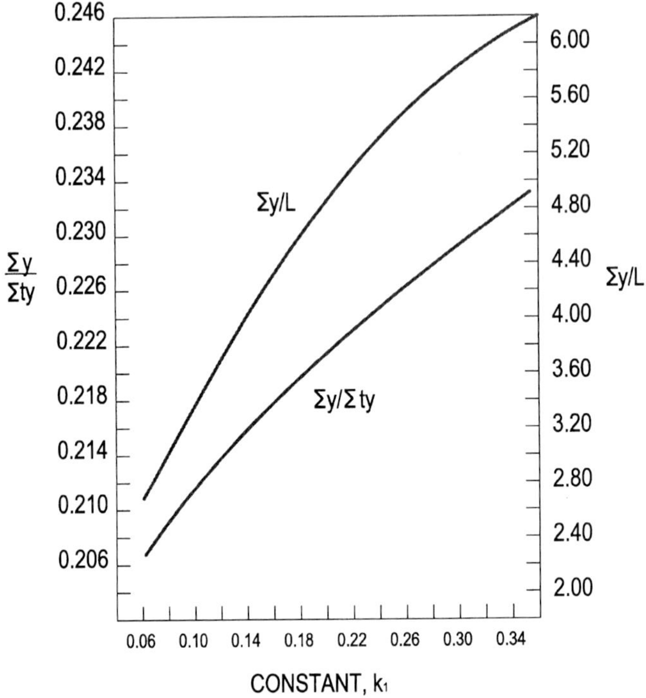


**Figure 1.5** $\Sigma y/L$ and $\Sigma y/\Sigma ty$ for various values of $k_1$ in a 7-day sequence.

Extend the same vertical line to the curve labeled $\Sigma y/L_a$, obtaining a value of 5.81. Since

$$L_a = \Sigma y/5.81 = 635.2/5.81$$

$$= 109.3 \text{ (mg/L)}$$

The technique of the moment method for analyzing a set of BOD data containing a lag phase is as follows:

1. Compute $\Sigma t$, $\Sigma y$, and $\Sigma ty$
2. Compute $t^2$ and $t^2y$ and take the sum of the values of each quantity as $\Sigma t^2$ and $\Sigma t^2y$
3. Compute the derived quantity:

$$\frac{\Sigma ty/\Sigma t - \Sigma y/n}{\Sigma(t^2y)/\Sigma t^2 - \Sigma y/n}$$

in which $n$ is the number of observations

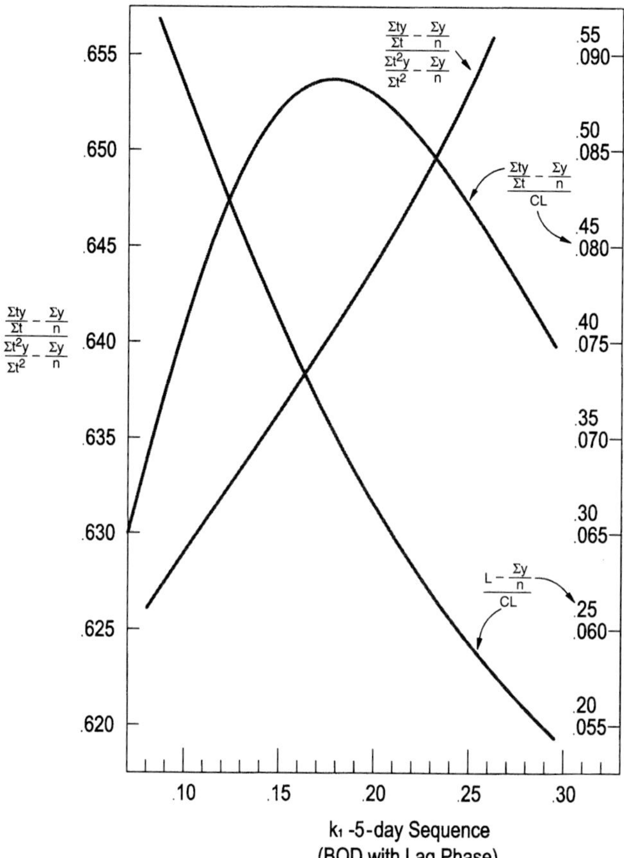


**Figure 1.6** Curves for BOD computation with lag phase, based on a 5-day time sequence.

4. Enter the above quantity on appropriate curve to find $k_1$
5. Project to other curves for values and solve equations for $C$ and $L_a$. The BOD equation with log phase is expressed as

$$y = L[1 - 10^{-k_1(t-t_0)}] \tag{1.14b}$$

$$y = L_a[1 - C10^{-k_1 t}] \tag{1.14c}$$

in which $t_0$ is the lag period and $C = 10^{k_1 t_0}$

**Example 2:** With a lag phase BOD data, BOD values are shown in Table 1.8. Find $K_1$ and $L_a$, and complete an equation of the curve of best fit for the BOD data.

**TABLE 1.7 Calculation of $\Sigma y$ and $\Sigma ty$**

| $t$ | $y$ | $ty$ |
|---|---|---|
| 1 | 56.2 | 56.2 |
| 2 | 74.4 | 148.8 |
| 3 | 87.6 | 262.8 |
| 4 | 96.2 | 384.8 |
| 5 | 102.4 | 512.0 |
| 6 | 107.4 | 644.4 |
| 7 | 111.0 | 777.0 |
| Sum | 635.2 | 2786.0 |

**solution:**

Step 1. Compute $\Sigma t$, $\Sigma y$, $\Sigma ty$, $\Sigma t^2$, and $\Sigma t^2 y$ (Table 1.8)

Step 2. Compute some quantities

$$\Sigma ty/\Sigma t = 1697/15 = 113.13$$

$$\Sigma y/n = 459/5 = 91.8$$

$$\Sigma t^2 y/\Sigma t^2 = 6883/55 = 125.14$$

and

$$\frac{\Sigma ty/\Sigma t - \Sigma y/n}{\Sigma t^2 y/\Sigma t^2 - \Sigma y/n} = \frac{113.13 - 91.8}{125.14 - 91.8} = 0.640$$

Step 3. Enter in Fig. 1.6 on the vertical axis labeled

$$\frac{\Sigma ty/\Sigma t - \Sigma y/n}{\Sigma t^2 y/\Sigma t^2 - \Sigma y/n}$$

with the value 0.640, and proceed horizontally to the diagonal straight line. Extend a vertical line to the axis labeled $k_1$ and read $k_1 = 0.173$ per day ($K_1 = 0.173$ per day $\times$ 2.3026 = 0.398 per day). Extend the vertical line to curve labeled

$$\frac{\Sigma ty/\Sigma t - \Sigma y/n}{CL}$$

**TABLE 1.8 Computations of $\Sigma t$, $\Sigma y$, $\Sigma ty$, $\Sigma t^2$, and $\Sigma t^2 y$**

| $t$ | $y$ | $ty$ | $t^2$ | $t^2 y$ |
|---|---|---|---|---|
| 1 | 12 | 12 | 1 | 12 |
| 2 | 74 | 148 | 4 | 296 |
| 3 | 101 | 303 | 9 | 909 |
| 4 | 126 | 504 | 16 | 2016 |
| 5 | 146 | 730 | 25 | 3650 |
| 15 | 459 | 1697 | 55 | 6883 |

and proceed horizontally from the intersection to the scale at the far right; read

$$\frac{\Sigma ty/\Sigma t - \Sigma y/n}{CL} = 0.0878$$

Step 4. Find $CL$

$$CL = \frac{113.13 - 91.8}{0.0878} = 242.94$$

Step 5. Find $L_a$, $L_a = L$ for this case

Continue to the same vertical line to the curve labeled

$$\frac{L - \Sigma y/n}{CL}$$

and read horizontally on the inside right-hand scale

$$\frac{L - \Sigma y/n}{CL} = 0.353$$

Then

$$L = L_a = 0.353(242.94) + 91.8$$
$$= 177.6 \text{ (mg/L)}$$

Also

$$C = \frac{CL}{L} = \frac{242.94}{177.6} = 1.368$$

and lag period

$$t_0 = \frac{1}{k_1}\log_{10}C$$
$$= \frac{1}{0.173}\log_{10}1.368$$
$$= 0.787 \text{ (days)}$$

Step 6. Write complete equation of the best fit for the data

$$y = 177.6[1 - 10^{-0.173(t-0.787)}]$$

or

$$y = 177.6[1 - e^{-0.398(t-0.787)}]$$

**Logarithmic formula.** Orford and Ingram (1953) reported that there is a relationship between the observed BOD from domestic sewage and the logarithm of the time of observation. If the BOD data are plotted against the logarithm of time, the resultant curve is approximately a straight line. The general equation is expressed as

$$y_t = m \log t + b \tag{1.22}$$

where $m$ is the slope of the line and $b$ is a constant (the intercept).

The general equation can be transformed by dividing each side by the 5-day $BOD_5$ intercept of the line to give

$$\frac{y_t}{s} = \frac{m}{s} \log t + \frac{b}{s}$$

$$\frac{y_t}{s} = M \log t + B \tag{1.22a}$$

or

$$y_t = s(M \log t + B) \tag{1.22b}$$

where $s$ = $BOD_5$ intercept of the line
$M$ = $m/s$, BOD rate parameter
$B$ = $b/s$, BOD rate parameter

For domestic sewage oxidation at 20°C, the straight line through the observed plotted points, when extrapolated to the $\log t$ axis, intercepts the $\log t$ axis at 0.333 days. The general equation is

$$y_t = s(0.85 \log t + 0.41) \tag{1.23a}$$

where 0.41 is the BOD rate parameter for domestic sewage.

For any observed BOD curve with different oxidation time, the above equation may be generalized as

$$y_t = S(\log at + 0.41) \tag{1.23b}$$

where $S$ = BOD intercept ($y$-axis) of the line at $5/a$ days
= 5-day BOD at the standardized domestic sewage oxidation rate, when $a = 1$
= strength factor
$a$ = $\log t$ at $x$-axis intercept of normal domestic sewage BOD curve divided by the $x$-axis intercept of
$= \dfrac{0.333}{x\text{-axis intercept}}$
= 1 for a standardized domestic sewage $BOD_5$ curve

The oxidation rate

$$\frac{dy}{dt} = \frac{0.85S}{t} = K_1(L_a - y_t) \tag{1.24a}$$

or

$$\frac{0.85S}{2.303t} = k_1(L_a - y_t) \tag{1.24b}$$

For the logarithmic formula method, you need to determine the two constants $S$ and $a$. Observed BOD data are plotted on semilogarithmic graph paper. Time in days is plotted on a logarithmic scale on the $x$-axis and percent of 5-day BOD on a regular scale on the $y$-axis. The straight line of best fix is drawn through the plotted points. The time value of the $x$-axis intercept is then read, $x_1$ (for a standardized domestic sewage sample, this is 0.333). The $a$ value can be calculated by $a = 0.333/x_1$.

**Example:** At 20°C, $S$ = 95 mg/L and $x_1$ = 0.222 days. What is the BOD equation and $K_1$ for the sample?

**solution:**

Step 1. Compute $a$

$$\begin{aligned} a &= 0.333/x_1 = 0.333/0.222 \\ &= 1.5 \end{aligned}$$

Step 2. Determine the simplified Eq. (1.23b)

$$\begin{aligned} y_t &= S(0.85 \log at + 0.41) \\ &= 95(0.85 \log 1.5t + 0.41) \end{aligned}$$

Step 3. Compute the oxidation rate at $t$ = 5 days, using Eq. (1.24a)

$$\begin{aligned} \frac{dy}{dt} &= \frac{0.85S}{t} = \frac{0.85 \times 95 \text{ mg/L}}{5 \text{ days}} \\ &= 16.15[\text{mg/(L} \cdot \text{d)}] \end{aligned}$$

Step 4. Determine $L_a$ when $t$ = 20 days from Step 2

$$\begin{aligned} L_a = y_{20} &= 95(0.85 \log(1.5 \times 20) + 0.41) \\ &= 158 \text{ (mg/L)} \end{aligned}$$

Step 5. Compute $K_1$

$$\frac{dy}{dt} = K_1(L_a - y_5)$$

$$K_1 = \frac{16.15 \text{ mg/(L}\cdot\text{d)}}{(158 - 95) \text{ mg/L}}$$

$$= 0.256 \text{ (per day)}$$

**Rapid methods.** Sheehy (1960) developed two rapid methods for solving first-order BOD equations. The first method, using the "BOD slide rule," is applicable to the ratios of observations on whole day intervals, from 1 to 8 days, to the 5-day BOD. The second method, the graphical method, can be used with whole or fractional day BOD values.

The BOD slide rule consists of scales A and B for multiplication and division. Scales C and D are the values of $1 - 10^{-k_1 t}$ and $10^{-k_1 t}$ plotted in relation to values of $k_1 t$ on the B scale. From given BOD values, the $k_1$ value is easily determined from the BOD slide rule. Unfortunately, the BOD slide rule is not on the market.

The graphical method is based on the same principles as the BOD slide rule method. Two figures are used to determine $k_1$ values directly. One figure contains $k_1$ values based on the ratios of BOD at time $t$ ($BOD_t$) to the 5-day BOD for $t$ less than 5 days. The other figure contains $k_1$ values based on the ratio of $BOD_t$ to $BOD_5$ at times $t$ greater than 5 days. The value of $k_1$ is determined easily from any one of these figures based on the ratio of $BOD_t$ to $BOD_5$.

### 8.3 Temperature effect on $K_1$

A general expression of the temperature effect on the deoxygenation coefficient (rate) is

$$\frac{K_{1a}}{K_{1b}} = \theta^{(T_a - T_b)} \qquad (1.25)$$

where $K_{1a}$ = reaction rate at temperature $T_a$, per day
$K_{1b}$ = reaction rate at temperature $T_b$, per day
$\theta$ = temperature coefficient

On the basis of experimental results over the usual range of river temperature, $\theta$ is accepted as 1.047. Therefore the BOD reaction rate at any $T$ in Celsius (working temperature), deviated from 20°C, is

$$K_{1(T)} = K_{1(20°\text{C})} \times 1.047^{(T-20)} \qquad (1.26)$$

or

$$k_{1(T)} = k_{1(20°C)} \times 1.047^{(T-20)} \qquad (1.26a)$$

Thus

$$L_{a(T)} = L_{a(20°C)}[1 + 0.02(T - 20)] \qquad (1.27)$$

or

$$L_{a(T)} = L_{a(20°C)}(0.6 + 0.2T) \qquad (1.28)$$

**Example 1:** A river water sample has $k_1 = 0.10$ (one base 10) and $L_a =$ 280 mg/L at 20°C. Find $k_1$ and $L_a$ at temperatures 14°C and 29°C.

**solution:**

Step 1.

Using Eq. (1.26a)

$$k_{1(T)} = k_{1(20°C)} \times 1.047^{(T-20)}$$

at 14°C

$$k_{1(14°C)} = k_{1(20°C)} \times 1.047^{(14-20)}$$

$$= 0.10 \times 1.047^{-6}$$

$$= 0.076 \text{ (per day)}$$

at 29°C

$$k_{1(29°C)} = k_{1(20°C)} \times 1.047^{(29-20)}$$

$$= 0.10 \times 1.047^{9}$$

$$= 0.15 \text{ (per day)}$$

Step 2. Find $L_{a(T)}$, using Eq. (1.28)

$$L_{a(T)} = L_{a(20°C)}(0.6 + 0.02T)$$

at 14°C

$$L_{a(14°C)} = L_{a(20°C)}(0.6 + 0.02T)$$

$$= 280(0.6 + 0.02 \times 14)$$

$$= 246 \text{ (mg/L)}$$

at 29°C

$$L_{a(29°C)} = 280(0.6 + 0.02 \times 29)$$
$$= 330 \text{ (mg/L)}$$

The ultimate BOD and $K_1$ values found in the laboratory at 20°C have to be adjusted to river temperatures using the above formulas. Three types of BOD can be determined: i.e. total, carbonaceous, and nitrogenous. TBOD (uninhibited) and CBOD (inhibited with trichloromethylpyridine for nitrification) are measured directly, while NBOD can be computed by subtracting CBOD values from TBOD values for given time elements.

**Example 2:** Tables 1.9 and 1.10 show typical long-term (20-day) BOD data for the Upper Illinois Waterway downstream of Lockport (Butts and Shackleford, 1992). The graphical plots of these BOD progressive curves are presented in Figs. 1.7 to 1.9. Explain what are their unique characteristics.

**TABLE 1.9 Biochemical Oxygen Demand at 20°C in the Upper Illinois Waterway**

| | Station sample: | Lockport 18 | | | Station sample: | Lockport 36 | |
|---|---|---|---|---|---|---|---|
| | Date: | 01/16/90 | | | Date: | 09/26/90 | |
| | pH: | 7.03 | | | pH: | 6.98 | |
| | Temp: | 16.05°C | | | Temp: | 19.10°C | |
| Time, days | TBOD, mg/L | CBOD, mg/L | NBOD, mg/L | Time, days | TBOD, mg/L | CBOD, mg/L | NBOD, mg/L |
| 0.89 | 1.09 | 0.98 | 0.11 | 0.72 | 0.19 | 0.18 | 1.01 |
| 1.88 | 1.99 | 1.73 | 0.25 | 1.69 | 0.66 | 0.18 | 0.48 |
| 2.87 | 2.74 | 2.06 | 0.68 | 2.65 | 0.99 | 0.92 | 0.07 |
| 3.87 | 3.39 | 2.25 | 1.14 | 3.65 | 1.38 | 1.25 | 0.13 |
| 4.87 | 4.60 | 3.08 | 1.52 | 4.72 | 1.81 | 1.25 | 0.56 |
| 5.62 | 5.48 | 3.45 | 2.04 | 5.80 | 2.74 | 1.62 | 1.12 |
| 6.58 | 7.41 | 4.17 | 3.24 | 6.75 | 3.51 | 2.09 | 1.43 |
| 7.62 | 9.87 | 4.83 | 5.04 | 7.79 | 4.17 | 2.53 | 1.64 |
| 8.57 | 12.62 | 5.28 | 7.34 | 8.67 | 4.65 | 2.92 | 1.74 |
| 9.62 | 16.45 | 6.15 | 10.30 | 9.73 | 5.59 | 2.92 | 2.67 |
| 10.94 | 21.55 | 6.62 | 14.94 | 10.69 | 6.22 | 3.08 | 3.14 |
| 11.73 | 25.21 | 6.96 | 18.25 | 11.74 | 6.95 | 3.08 | 3.88 |
| 12.58 | 29.66 | 7.26 | 22.40 | 12.66 | 7.28 | 3.12 | 4.16 |
| 13.62 | 34.19 | 7.90 | 26.29 | 13.69 | 7.79 | 3.56 | 4.23 |
| 14.67 | 35.68 | 8.14 | 27.55 | 14.66 | 7.79 | 3.56 | 4.23 |
| 15.67 | 36.43 | 8.38 | 28.05 | 15.65 | 8.36 | 3.56 | 4.80 |
| 16.58 | 37.08 | 8.75 | 28.33 | 16.64 | 8.56 | 3.57 | 4.99 |
| 17.81 | 37.45 | 8.92 | 28.53 | 17.76 | 8.80 | 3.82 | 4.99 |
| 18.80 | 37.64 | 9.16 | 28.47 | 17.67 | 8.80 | 3.82 | 4.98 |
| 19.90 | 38.19 | 9.39 | 28.79 | 19.66 | 8.92 | 3.87 | 5.05 |
| 20.60 | 38.56 | 9.55 | 29.01 | | | | |

**TABLE 1.10 Biochemical Oxygen Demand at 20°C in the Kankakee River**

| | Station sample: | Kankakee 11 | |
|---|---|---|---|
| | Date: | 08/13/90 | |
| | pH: | 8.39 | |
| | Temp: | 23.70°C | |
| **Time, days** | **TBOD, mg/L** | **CBOD, mg/L** | **NBOD, mg/L** |
| 0.78 | 0.81 | 0.44 | 0.37 |
| 2.79 | 1.76 | 1.13 | 0.63 |
| 3.52 | 2.20 | 1.26 | 0.93 |
| 6.03 | 2.42 | 1.62 | 0.79 |
| 6.77 | 2.83 | 1.94 | 0.89 |
| 7.77 | 3.20 | 2.22 | 0.98 |
| 8.76 | 3.54 | 2.49 | 1.06 |
| 9.49 | 3.83 | 2.81 | 1.02 |
| 10.76 | 4.19 | 3.09 | 1.10 |
| 13.01 | 4.85 | 3.76 | 1.10 |
| 13.43 | 5.08 | 3.94 | 1.14 |
| 14.75 | 5.24 | 4.02 | 1.22 |
| 15.48 | 5.57 | 4.42 | 1.15 |
| 17.73 | 6.64 | 5.30 | 1.17 |
| 19.96 | 7.03 | 5.85 | 1.19 |

**solution:**

These three sets of data have several unique characteristics and considerations. Figure 1.7 demonstrates that S-shaped NBOD, and TBOD curves exist at Lockport station (Illinois) that fit the mathematical formula represented by Eq. (1.8). These curves usually occur at Lockport during cold weather periods, but not always.

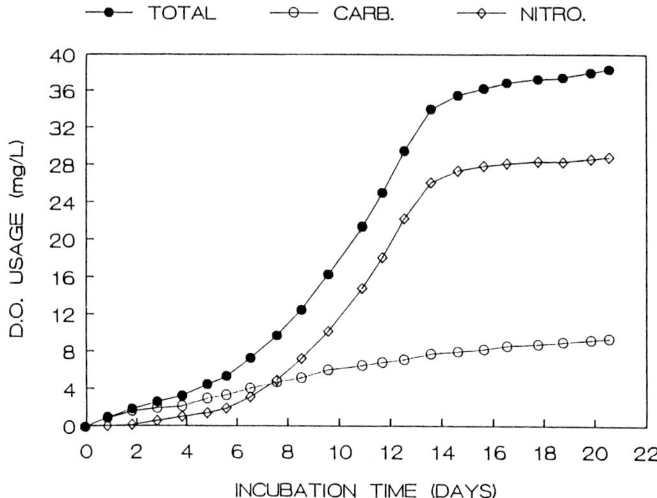


**Figure 1.7** BOD progressive curves for Lockport 18, January 16, 1990 (*Butts and Shackleford, 1992*).

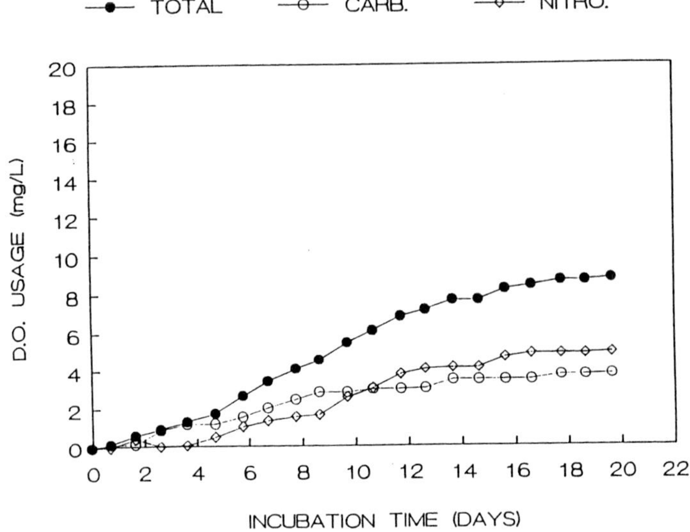


**Figure 1.8** BOD progressive curves for Lockport 36, September 26, 1990 *(Butts and Shackleford, 1992).*

Figure 1.8 illustrates Lockport warm weather BOD progression curves. In comparisons with Figs. 1.7 and 1.8, there are extreme differences between cold and warm weather BOD curves. The January $NBOD_{20}$ represents 75.3% of the $TBOD_{20}$, whereas the September $NBOD_{20}$ contains

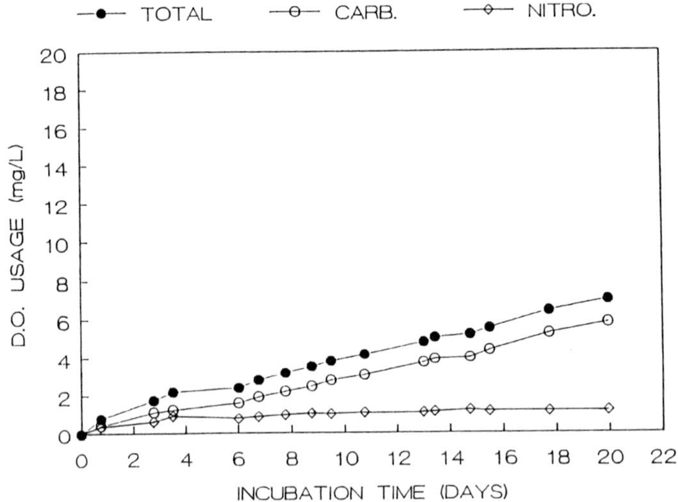


**Figure 1.9** BOD progressive curves for Kankakee 10, August 13, 1990 *(Butts and Shackleford, 1992).*

only 56.6% of the $TBOD_{20}$. Furthermore, the September $TBOD_{20}$ is only 23.3% as great as the January $TBOD_{20}$ (see Table 1.9).

Figure 1.9 shows most tributary (Kankakee River) BOD characteristics. The warm weather $TBOD_{20}$ levels for the tributary often approach those observed at Lockport station (7.03 mg/L versus 8.92 mg/L), but the fraction of $NBOD_{20}$ is much less (1.19 mg/L versus 5.05 mg/L) (Tables 1.9 and 1.10). Nevertheless, the $TBOD_{20}$ loads coming from the tributaries are usually much less than those originating from Lockport since the tributary flows are normally much lower.

### 8.4 Second-order reaction

In many cases, researchers stated that a better fit of BOD data can be obtained by using a second-order chemical reaction equation, i.e. the equation of a rotated rectangular hyperbola. A second-order chemical reaction is characterized by a rate of reaction dependent upon the concentration of two reactants. It is defined as (Young and Clark, 1965):

$$-\frac{\mathrm{d}C}{\mathrm{d}t} = KC^2 \qquad (1.29)$$

When applying this second-order reaction to BOD data, $K$ is a constant; $C$ becomes the initial substance concentration; $L_a$, minus the BOD; $y$, at any time, $t$; or

$$-\frac{\mathrm{d}(L_a - y)}{\mathrm{d}t} = K(L_a - y)^2 \qquad (1.30)$$

rearranging

$$\frac{\mathrm{d}(L_a - y)}{(L_a - y)^2} = -K\,\mathrm{d}t$$

Integrating yields

$$\int_{y=0}^{y=y} \frac{\mathrm{d}(L_a - y)}{(L_a - y)^2} = \int_{t=0}^{t=t} -K\,\mathrm{d}t$$

or

$$\frac{1}{L_a} - \frac{1}{L_a - y} = -Kt \qquad (1.31)$$

Multiplying each side of the equation by $L_a$ and rearranging,

$$1 - \frac{L_a}{L_a - y} = -KL_a t$$

$$\frac{L_a - y - L_a}{L_a - y} = -KL_a t$$

$$\frac{-y}{L_a - y} = -KL_a t$$

$$y = KL_a t(L_a - y)$$

$$y = KL_a^2 t - KL_a t y \qquad (1.32)$$

$$(1 + KL_a t)y = KL_a^2 t$$

$$y = \frac{KL_a^2}{1 + KL_a t} t$$

$$y = \frac{t}{\dfrac{1}{KL_a^2} + \dfrac{1}{L_a} t} \qquad (1.33)$$

or

$$y = \frac{t}{a + bt} \qquad (1.34)$$

where $a$ is $1/KL_a^2$ and $b$ represents $1/L_a$. The above equation is in the form of a second-order reaction equation for defining BOD data. The equation can be linearized in the form

$$\frac{t}{y} = a + bt \qquad (1.35)$$

in which $a$ and $b$ can be solved by a least-squares analysis. The simultaneous equations for the least-squares treatment are as follows:

$$\Sigma\left(a + bt - \frac{t}{y}\right) = 0$$

$$\Sigma\left(a + bt - \frac{t}{y}\right)t = 0$$

$$at + b\Sigma t - \frac{\Sigma t}{y} = 0$$

$$a\Sigma t + b\Sigma t^2 - \frac{\Sigma t^2}{y} = 0$$

$$a + b\left(\frac{\Sigma t}{t}\right) - \left(\frac{\Sigma t/y}{t}\right) = 0$$

$$a + b\left(\frac{\Sigma t^2}{\Sigma t}\right) - \left(\frac{\Sigma t^2/y}{\Sigma t}\right) = 0$$

To solve $a$ and $b$ using five data points, $t = 5$, $\Sigma t = 15$, and $\Sigma t^2 = 55$; then

$$a + b\left(\frac{15}{5}\right) - \left(\frac{\Sigma t/y}{5}\right) = 0$$

$$a + b\left(\frac{55}{15}\right) - \left(\frac{\Sigma t^2/y}{15}\right) = 0$$

Solving for $b$

$$\left(\frac{55}{15} - \frac{15}{5}\right)b - \left[\left(\frac{\Sigma t^2/y}{15}\right) - \left(\frac{\Sigma t/y}{5}\right)\right] = 0$$

$$\left(\frac{55 - 45}{15}\right)b - \frac{1}{15}\left(\frac{\Sigma t^2}{y} - \frac{3\Sigma t}{y}\right) = 0$$

$$10b = \frac{\Sigma t^2}{y} - \frac{3\Sigma t}{y}$$

or

$$b = 0.10\left(\frac{\Sigma t^2}{y} - \frac{3\Sigma t}{y}\right) \qquad (1.36)$$

Solving for $a$, by substituting $b$ in the equation

$$\begin{aligned} a &= \frac{1}{5}\left(\frac{\Sigma t}{y} - 15b\right) \\ &= \frac{1}{5}\left[\frac{\Sigma t}{y} - 15 \times 0.1\left(\frac{\Sigma t^2}{y} - \frac{3\Sigma t}{y}\right)\right] \\ &= \frac{1}{5}\left(\frac{5.5\Sigma t}{y} - \frac{1.5\Sigma t^2}{y}\right) \\ &= 1.1\left(\frac{\Sigma t}{y}\right) - 0.3\left(\frac{\Sigma t^2}{y}\right) \qquad (1.37) \end{aligned}$$

The velocity of reaction for the BOD curve is $y/t$. The initial reaction is $1/a$ [in mg/(L · d)] and is the maximum velocity of the BOD reaction denoted as $v_m$.

The authors (Young and Clark, 1965) claimed that a second-order equation has the same precision as a first-order equation at both 20°C and 35°C.

Dougal and Baumann (1967) modified the second-order equation for BOD predication as

$$y = \frac{t}{a + bt} = \frac{t/bt}{\frac{a}{bt} + 1} = \frac{1/b}{\frac{a}{bt} + 1} = \frac{L_a}{\frac{1}{K't} + 1} \tag{1.38}$$

where

$$L_a = 1/b \tag{1.39}$$

$$K' = b/a \tag{1.40}$$

This formula has a simple rate constant and an ultimate value of BOD. It can also be transformed into a linear function for regression analysis, permitting the coefficients $a$ and $b$ to be determined easily.

**Example:** The laboratory BOD data for wastewater were 135, 198, 216, 235, and 248 mg/L at days 1, 2, 3, 4, and 5, respectively. Find $K$, $L_a$, and $v_m$ for this wastewater.

**solution:**

Step 1. Construct a table for basic calculations (Table 1.11)

Step 2. Solve $a$ and $b$ using Eqs. (1.36) and (1.37)

$$\begin{aligned} b &= 0.10\left(\frac{\Sigma t^2}{y} - \frac{3\Sigma t}{y}\right) \\ &= 0.10(0.238167 - 3 \times 0.068579) \\ &= 0.003243 \end{aligned}$$

**TABLE 1.11 Data for Basic Calculations for Step 1**

| $t$ | $y$ | $t/y$ | $t^2/y$ |
|---|---|---|---|
| 1 | 135 | 0.007407 | 0.007407 |
| 2 | 198 | 0.010101 | 0.020202 |
| 3 | 216 | 0.013889 | 0.041667 |
| 4 | 235 | 0.017021 | 0.068085 |
| 5 | 248 | 0.020161 | 0.100806 |
| Sum | | 0.068579 | 0.238167 |

$$a = 1.1\left(\frac{\Sigma t}{y}\right) - 0.3\left(\frac{\Sigma t^2}{y}\right)$$
$$= 1.1 \times 0.068579 - 0.3 \times 0.238167$$
$$= 0.003987$$

Step 3. Calculate $L_a$, $K$, $v_m$, and $K'$ using Eqs. (1.38) to (1.40)

$$L_a = 1/b$$
$$= 1/0.003243$$
$$= 308 \text{ (mg/L)}$$
$$K = 1/aL_a^2$$
$$= 1/0.003987(308)^2$$
$$= 0.00264 \text{ [per mg/(L} \cdot \text{d)]}$$
$$v_m = 1/a$$
$$= 1/0.003987 \text{ [mg/(L} \cdot \text{d)]}$$
$$= 250.8 \text{ [mg/(L} \cdot \text{d)]}$$

or

$$= 10.5 \text{ mg/(L} \cdot \text{h)}$$
$$K' = b/a$$
$$= 0.003243/0.003987$$
$$= 0.813 \text{ (per day)}$$

## 9 Determination of Reaeration Rate Constant $K_2$

### 9.1 Basic conservation

For a stream deficient in DO but without BOD load, the classical formula is

$$\frac{dD}{dt} = -K_2 D \tag{1.41}$$

where $dD/dt$ is the absolute change in DO deficit $D$ over an increment of time $dt$ due to an atmospheric exchange of oxygen at air/water interface; $K_2$ is the reaeration coefficient, per day; and $D$ is DO deficit, mg/L.

Integrating the above equation from $t_1$ to $t_2$ gives

$$\int_{D_1}^{D_2} \frac{dD}{D} = -K_2 \int_{t_1}^{t_2} dt \qquad \text{or} \qquad \int_{D_a}^{D_t} \frac{dD}{D} = -K_2 \int_0^t dt$$

$$\ln \frac{D_2}{D_1} = -K_2(t_2 - t_1) \qquad \text{or} \qquad \ln \frac{D_t}{D_a} = -K_2 t$$

$$\ln \frac{D_2}{D_1} = -K_2 \Delta t \qquad \text{or} \qquad D_t = D_a e^{-K_2 t}$$

$$K_2 = -\frac{\ln \dfrac{D_2}{D_1}}{\Delta t} \tag{1.42}$$

The $K_2$ values are needed to correct for river temperature according to the equation.

$$K_{2(@T)} = K_{2(@20)}(1.02)^{T-20} \tag{1.43}$$

where $K_{2(@T)} = K_2$ value at any temperature $T$°C and $K_{2(@20)} = K_2$ value at 20°C.

**Example:** The DO deficits at upstream and downstream stations are 3.55 and 2.77 mg/L, respectively. The time of travel in this stream reach is 0.67 days. The mean water temperature is 26.5°C. What is the $K_2$ value for 20°C?

**solution:**

Step 1. Determine $K_2$ at 26.5°C

$$\begin{aligned} k_{2(@26.5)} &= -\frac{\ln \dfrac{D_2}{D_1}}{\Delta t} \\ &= -\left(\ln \frac{2.77}{3.55}\right) \Big/ 0.67 \\ &= -(-0.248)/0.67 \\ &= 0.37 \text{ (per day)} \end{aligned}$$

Step 2. Calculate $K_{2(@20)}$

$$K_{2(@T)} = K_{2(@20)}(1.02)^{T-20}$$

therefore

$$\begin{aligned} K_{2(@20)} &= 0.37/(1.02)^{26.5-20} = 0.37/1.137 \\ &= 0.33 \text{ (per day)} \end{aligned}$$

### 9.2 From BOD and oxygen sag constants

In most stream survey studies involving oxygen sag equations (discussed later), the value of the reoxygenation constant ($K_2$) is of utmost importance. Under different conditions, several methods for determining $K_2$ are listed below:

I. $K_2$ may be computed from the oxygen sag equation, if all other parameters are known; however, data must be adequate to support the conclusions. A trial-and-error procedure is generally used.

II. The amount of reaeration ($r_m$) in a reach (station A to station B) is equal to the BOD exerted ($L_{aA} - L_{aB}$) plus oxygen deficiency from station A to station B, ($D_A - D_B$). The relationship can be expressed as

$$r_m = (L_{aA} - L_{aB}) + (D_A - D_B) \tag{1.44}$$

$$K_2 = r_m / D_m \tag{1.45}$$

where $r_m$ is the amount of reaeration and $D_m$ is the mean (average) deficiency.

**Example:** Given $L_a$ for the upper and lower sampling stations are 24.6 and 15.8 mg/L, respectively; the DO concentration at these two stations are 5.35 and 5.83 mg/L, respectively; and the water temperature is 20°C; find the $K_2$ value for the river reach.

**solution:**

Step 1. Find $D_A$, $D_B$, and $D_m$ at 20°C

$$\begin{aligned} \text{DO}_{sat} \text{ at } 20°\text{C} &= 9.02 \text{ mg/L (Table 1.2)} \\ D_A &= 9.02 - 5.35 = 3.67 \text{ (mg/L)} \\ D_B &= 9.02 - 5.83 = 3.19 \text{ (mg/L)} \\ D_m &= (3.67 + 3.19)/2 = 3.43 \text{ (mg/L)} \end{aligned}$$

Step 2. Calculate $r_m$ and $K_2$

$$\begin{aligned} r_m &= (L_{aA} - L_{aB}) + (D_A - D_B) \\ &= (24.6 - 15.8) + (3.67 - 3.19) \\ &= 9.28 \text{ (mg/L)} \\ K_2 &= r_m/D_m = 9.28/3.43 \\ &= 2.71 \text{ (per day)} \end{aligned}$$

III. For the case, where DO = 0 mg/L for a short period of time and without anaerobic decomposition (O'Connor, 1958),

$$K_2 D_{max} = K_2 C_s \tag{1.46}$$

where $D_{max}$ is the maximum deficit, mg/L, and $C_s$ is the saturation DO concentration, mg/L. The maximum deficiency is equal to the DO saturation concentration and the oxygen transferred is oxygen utilized by organic matter.

Organic matter utilized $= L_{aA} - L_{aB}$, during the time of travel $t$; therefore

$$\text{rate of exertion} = (L_{aA} - L_{aB})/t = K_2 D_{max}$$

and

$$(L_{aA} - \mathrm{L}_{aB})/t = K_2 C_s$$

then

$$K_2 = (L_{aA} - L_{aB})/C_s t \tag{1.47}$$

**Example:** Given that the water temperature is 20°C; $L_{aA}$ and $L_{aB}$ are 18.3 and 13.7 mg/L, respectively; and the time of travel from station A to station B is 0.123 days. Compute $K_2$.

**solution:** At 20°C,

$$\begin{aligned} C_s &= 9.02 \text{ (mg/L)} \\ K_2 &= (L_{aA} - L_{aB})/C_s t \\ &= (18.3 - 13.7)/(9.02 \times 0.123) \\ &= 4.15 \text{ (per day)} \end{aligned}$$

IV. At the critical point in the river (O'Connor, 1958)

$$\frac{\mathrm{d}D}{\mathrm{d}t} = 0 \tag{1.48}$$

and

$$K_d L_c = K_2 D_c \tag{1.49}$$

$$K_2 = K_d L_c / D_c \tag{1.50}$$

where $K_d$ is the deoxygenation rate in stream conditions, $L_c$ is the first-stage ultimate BOD at critical point, and $D_c$ is the DO deficit at the critical point. These will be discussed in a later section.

V. Under steady-state conditions, at a sampling point (O'Connor, 1958)

$$\frac{\mathrm{d}D}{\mathrm{d}t} = 0 \tag{1.48}$$

or at this point

$$K_2 D = K_d L \tag{1.51}$$

The value of $D$ can be obtained from a field measurement. The value of $K_d L$ can be obtained by measuring oxygen uptake from a sample taken from a given point on a river. Thus, $K_2$ can be computed from the above equation.

### 9.3 Empirical formulas

The factors in the Streeter–Phelps equation has stimulated much research on the reaeration rate coefficient $K_2$. Considerable controversy exists as to the proper method of formulas to use. Several empirical and semi-empirical formulas have been developed to estimate $K_2$, almost all of which relate stream velocity and water depth to $K_2$, as first proposed by Streeter (1926). Three equations below are widely known and employed in stream studies:

$$K_2 = \frac{13.0V^{0.5}}{H^{1.5}} \quad \text{O'Connor and Dobbins (1958)} \tag{1.52}$$

$$K_2 = \frac{11.57V^{0.969}}{H^{1.673}} \quad \text{Churchill } et\ al. \text{ (1962)} \tag{1.53}$$

$$K_2 = \frac{7.63V}{H^{1.33}} \quad \text{Langbein and Durum (1967)} \tag{1.54}$$

where $K_2$ = reaeration rate coefficient, per day
$V$ = average velocity, ft/s (fps)
$H$ = average water depth, ft

On the basis of $K_2$-related physical aspects of a stream, O'Connor (1958) and Eckenfelder and O'Connor (1959) proposed $K_2$ formulas as follows:
For an isotropic turbulence (deep stream)

$$K_2 = \frac{(D_L V)^{0.5}}{2.3H^{1.5}} \tag{1.55}$$

For a nonisotropic stream

$$K_2 = \frac{480 D_L^{0.5} S^{0.25}}{H^{1.25}} \tag{1.56}$$

where $D_L$ = diffusivity of oxygen in water, ft$^2$/d
$H$ = average depth, ft

$S$ = slope
$V$ = velocity of flow, fps = $B\sqrt{(HS)}$ (nonisotropic if $B < 17$; isotropic if $B > 17$)

The O'Connor and Dobbins equation, Eq. (1.52), is based on a theory more general than the formula developed by several of the other investigators. The formula proposed by Churchill *et al.* (Eq. (1.53)) appears to be more restrictive in use. The workers from the US Geological Survey (Langbein and Durum, 1967) have analyzed and summarized the work of several investigators and concluded that the velocities and cross section information were the most applicable formulation of the reaeration factor.

**Example:** A stream has an average depth of 9.8 ft (3.0 m) and velocity of flow of 0.61 ft/s (0.20 m/s). What are the $K_2$ values determined by the first three empirical formulas in this section (Eqs. (1.52) to (1.54))? What is $K_2$ at 25°C of water temperature (use Eq. (1.54) only).

**solution:**

Step 1. Determine $K_2$ at 20°C

(a) by the O'Conner and Dobbins formula (Eq. (1.52))

$$\begin{aligned} K_2 &= \frac{13.0V^{0.5}}{H^{1.5}} = \frac{13.0 \times (0.61)^{0.5}}{(9.8)^{1.5}} \\ &= 0.33 \text{ (per day)} \end{aligned}$$

(b) by the Churchill *et al.* formula (Eq. (1.53))

$$K_2 = \frac{11.57 \times (0.61)^{0.969}}{(9.8)^{1.673}} = 0.16 \text{ (per day)}$$

(c) by the Langbein and Durum formula (Eq. (1.54))

$$K_2 = \frac{7.63 \times 0.61}{(9.8)^{1.33}} = 0.22 \text{ (per day)}$$

Step 2. For a temperature of 25°C

$$\begin{aligned} K_{2(@25)} &= K_{2(@20)}(1.02)^{T-20} \\ &= 0.22(1.02)^{25-20} \\ &= 0.24 \text{ (per day)} \end{aligned}$$

### 9.4 Stationary field monitoring procedure

Larson *et al.* (1994) employed physical, biochemical, and biological factors to analyze the $K_2$ value. Estimations of physical reaeration and the associated effects of algal photosynthetic oxygen production (primary productivity) are based on the schematic formulations as follows:

Physical aeration = ambient DO − light chamber DO + SOD

Algal productivity:

Gross = light chamber DO − dark chamber DO

Net = light chamber DO at end − light chamber DO at beginning

The amount of reaeration (REA) is computed with observed data from DataSonde as

$$\text{REA} = C_2 - C_1 - \text{POP} - \text{PAP} + \text{TBOD} + \text{SOD} \tag{1.57}$$

where REA = reaeration ($=D_2 - D_1 = \mathrm{d}D$), mg/L
$C_2$ and $C_1$ = observed DO concentrations in mg/L at $t_2$ and $t_1$, respectively
POP = net periphytonic (attached algae) oxygen production for the time period ($t_2 - t_1 = \Delta t$), mg/L
PAP = net planktonic algae (suspended algae) oxygen production for the time period $t_2 - t_1$, mg/L
TBOD = total biochemical oxygen demand (usage) for the time period $t_2 - t_1$, mg/L
SOD = net sediment oxygen demand for the time period $t_2 - t_1$, mg/L

The clear periphytonic chamber is not used if periphytonic productivity/respiration is deemed insignificant, i.e. POD = 0. The combined effect of TBOD – PAP is represented by the gross output of the light chamber, and the SOD is equal to the gross SOD chamber output less than dark chamber output.

The DO deficit $D$ in the reach is calculated as

$$D = \frac{S_1 + S_2}{2} - \frac{C_1 + C_2}{2} \tag{1.58}$$

where $S_1$ and $S_2$ are the DO saturation concentration in mg/L at $t_1$ and $t_2$, respectively, for the average water temperature ($T$) in the reach. The DO saturation formula is given in the preview section (Eq. (1.4)).

Since the Hydrolab's DataSondes logged data at hourly intervals, the DO changes attributable to physical aeration (or deaeration) are available for small time frames, which permits the following modification of the basic natural reaeration equation to be used to calculate $K_2$ values (Broeren *et al.*, 1991)

$$\sum_{i=1}^{24} (R_{i+1} - R_i) = -K_2 \sum_{i=1}^{24} \left( \frac{S_i + S_{i+1}}{2} - \frac{C_i + C_{i+1}}{2} \right) \tag{1.59}$$

where $R_i$ = an $i$th DO concentration from the physical aeration DO-used "mass diagram" curve
$R_{i+1}$ = a DO concentration 1 hour later than $R_i$ on the physical reaeration DO-used curve
$S_i$ = an $i$th DO saturation concentration
$S_{i+1}$ = a DO saturation concentration 1 hour later than $S_i$
$C_i$ = an $i$th-observed DO concentration
$C_{i+1}$ = an observed DO concentration 1 hour later than $C_i$

*Note*: All units are in mg/L.

Thus, the reaeration rate can be computed by

$$K_2 = \frac{\text{REA}}{D(t_2 - t_1)} \tag{1.60}$$

Both the algal productivity/respiration (*P*/*R*) rate and SOD are biologically associated factors which are normally expressed in terms of grams of oxygen per square meter per day (g/(m$^2 \cdot$ d)). Conversion for these areal rates to mg/L of DO usage for use in computing physical aeration (REA) is accomplished using the following formula (Butts *et al.*, 1975):

$$U = \frac{3.28Gt}{H} \tag{1.61}$$

where $U$ = DO usage in the river reach, mg/L
$G$ = SOD or *P*/*R* rate, g/(m$^2 \cdot$ d)
$t$ = time of travel through the reach, days
$H$ = average depth, ft

**Example:** Given that $C_1$ = 6.85 mg/L, $C_2$ = 7.33 mg/L, POP = 0, at $T$ = 20°C, gross DO output in light chamber = 6.88 mg/L, dark chamber DO output = 5.55 mg/L, gross SOD chamber output = 6.15 mg/L, and water temperature at beginning and end of field monitoring = 24.4 and 24.6°C, respectively, $t_2 - t_1$ = 2.50 days, calculate $K_2$ at 20°C.

**solution:**

Step 1. Determine REA

$$\text{TBOD} - \text{PAP} = \text{gross} = 6.88 \text{ mg/L} - 5.55 \text{ mg/L} = 1.33 \text{ mg/L}$$

$$\text{SOD} = 6.15 \text{ mg/L} - 5.55 \text{ mg/L} = 0.60 \text{ mg/L}$$

From Eq. (1.57)

$$\begin{aligned}\text{REA} &= C_2 - C_1 - \text{POP} + (-\text{PAP} + \text{TBOD}) + \text{SOD} \\ &= 7.33 - 6.85 - 0 + 1.33 + 0.60 \\ &= 2.41 \text{ mg/L}\end{aligned}$$

Step 2. Calculate $S_1$, $S_2$, and $D$

$S_1$ at $T = 24.4°C$

$$S_1 = 14.652 - 0.41022(24.4) + 0.007991(24.4)^2 - 0.00007777(24.4)^3$$
$$= 8.27 \text{ (mg/L) or from Table 1.2}$$

$S_2$ at $T = 24.6°C$, use same formula

$$S_2 = 8.24 \text{ mg/L}$$

*Note*: The elevation correction factor is ignored, for simplicity.

From Eq. (1.58)

$$D = \frac{S_1 + S_2}{2} - \frac{C_1 + C_2}{2}$$
$$= \frac{1}{2}(8.27 + 8.24 - 6.85 - 7.33)$$
$$= 1.17 \text{ (mg/L)}$$

Step 3. Compute $K_2$ with Eq. (1.60)

$$K_2 = \frac{\text{REA}}{D(t_2 - t_1)}$$
$$= \frac{2.41 \text{ mg/L}}{1.17 \text{ mg/L} \times 2.50 \text{ days}}$$
$$= 0.82 \text{ per day}$$

## 10 Sediment Oxygen Demand

Sediment oxygen demand (SOD) is measure of the oxygen demand characteristics of the bottom sediment which affects the dissolved oxygen resources of the overlying water. To measure SOD, a bottom sample is especially designed to entrap and seal a known quantity of water at the river bottom. Changes in DO concentrations (approximately 2 h, or until the DO usage curve is clearly defined) in the entrapped water are recorded by a DO probe fastened in the sampler (Butts, 1974). The test temperature should be recorded and factored for temperature correction at a specific temperature.

SOD curves can be plotted showing the accumulated DO used ($y$-axis) versus elapsed time ($x$-axis). SOD curves resemble first-order carbonaceous BOD curves to a great extent; however, first-order kinetics are not applicable to the Upper Illinois Waterway's data. For the most part, SOD is caused by bacteria reaching an "unlimited" food supply. Consequently, the oxidation rates are linear in nature.

The SOD rates, as taken from SOD curves, are in linear units of milligrams per liter per minute (mg/(L · min)) and can be converted into grams per square meter per day (g/(m$^2$ · d)) for practical application. A bottle sampler SOD rate can be formulated as (Butts, 1974; Butts and Shackleford, 1992):

$$\text{SOD} = \frac{1440\,SV}{10^3 A} \tag{1.62}$$

where SOD = sediment oxygen demand, g/(m$^2$ · d)
$S$ = slope of stabilized portion of the curve, mg/(L · min)
$V$ = volume of sampler, L
$A$ = bottom area of sampler, m$^2$

**Example:** An SOD sampler is a half-cylinder in shape (half-section of steel pipe). Its diameter and length are $1\frac{1}{6}$ ft (14 in) and 2.0 ft, respectively.

(1) Determine the relationship of SOD and slope of linear portion of usage curve.
(2) What is SOD in g/(m$^2$ · d) for $S$ = 0.022 mg/(L · min)?

**solution:**

Step 1. Find area of the SOD sampler

$$A = 1\frac{1}{6}\,(\text{ft}) \times 2.0\,(\text{ft}) \times 0.0929\,(\text{m}^2/\text{ft}^2)$$

$$= 0.217\ \text{m}^2$$

Step 2. Determine the volume of the sampler

$$V = \frac{1}{2}\pi r^2 l$$

$$V = \frac{1}{2} \times 3.14 \times (7/12\ \text{ft})^2 \times (1.0\ \text{ft}) \times (28.32\ \text{L/ft}^3)$$

$$= 30.27\ \text{L}$$

Step 3. Compute SOD with Eq. (1.62)

$$\text{SOD} = \frac{1440SV}{10^3 A}$$

$$= \frac{1400 \times S \times 30.27}{10^3 \times 0.217}$$

$$= 195S \qquad \text{ignoring the volume of two union connections}$$

*Note*: If two union connections are considered, the total volume of water contained within the sampler is 31.11 L. Thus, the Illinois State Water Survey's SOD formula is

$$\text{SOD} = 206.6S$$

Step 4. Compute SOD

$$\begin{aligned}\text{SOD} &= 206.6 \times 0.022 \\ &= 4.545\ [\text{g}/(\text{m}^2 \cdot \text{d})]\end{aligned}$$

### 10.1 Relationship of sediment characteristics and SOD

Based on extensive data collections by the Illinois State Water Survey, Butts (1974) proposed the relationship of SOD and the percentage of dried solids with volatile solids.

The prediction equation is

$$\text{SOD} = 6.5(\text{DS})^{-0.46}\,(\text{VS})^{0.38} \tag{1.63}$$

where SOD = sediment oxygen demand $\text{g}/(\text{m}^2 \cdot \text{d})$
DS = percent dried solids of the decanted sample by weight
VS = percent volatile solids by weight

**Example:** Given DS = 68% and VS = 8%, predict the SOD value of this sediment.

**solution:**

$$\begin{aligned}\text{SOD} &= 6.5(\text{DS})^{-0.46}(\text{VS})^{0.38} \\ &= \frac{6.5 \times (8)^{0.38}}{(68)^{0.46}} = \frac{6.5 \times 2.20}{6.965} \\ &= 2.05\ [\text{g}/(\text{m}^2 \cdot \text{d})]\end{aligned}$$

### 10.2 SOD versus DO

An SOD–DO relationship has been developed from the data given by McDonnell and Hall (1969) for 25-cm deep sludge. The formula is

$$\text{SOD}/\text{SOD}_1 = \text{DO}^{0.28} \tag{1.64}$$

where SOD = SOD at any DO level, $\text{g}/(\text{m}^2 \cdot \text{d})$
$\text{SOD}_1$ = SOD at a DO concentration of 1.0 mg/L, $\text{g}/(\text{m}^2 \cdot \text{d})$
DO = dissolved oxygen concentration, mg/L

This model can be used to estimate the SOD rate at various DO concentration in areas having very high benthic invertebrate populations.

**Example:** Given $SOD_1 = 2.4\ g/(m^2 \cdot d)$, find SOD at DO concentrations of 5.0 and 8.0 mg/L.

**solution:**

Step 1. For DO = 5.0 mg/L, with Eq. (1.64)

$$\begin{aligned} SOD_5 &= SOD_1 \times DO^{0.28} \\ &= 2.4 \times 5^{0.28} \\ &= 2.4 \times 1.57 \\ &= 3.77\ (mg/L) \end{aligned}$$

Step 2. For DO = 8.0 mg/L

$$\begin{aligned} SOD_8 &= 2.4 \times 8^{0.28} \\ &= 2.4 \times 1.79 \\ &= 4.30\ (mg/L) \end{aligned}$$

Undisturbed samples of river sediments are collected by the test laboratory to measure the oxygen uptake of the bottom sediment: the amount of oxygen consumed over the test period is calculated as a zero-order reaction (US EPA, 1997):

$$\frac{dC}{dt} = -\frac{SOD}{H} \tag{1.65}$$

where $dC/dt$ = rate change of oxygen concentration, g $O_2/(m^2 \cdot d)$
SOD = sediment oxygen demand, g $O_2/(m^2 \cdot d)$
$H$ = average river depth, m

## 11 Organic Sludge Deposits

Solids in wastewaters may be in the forms of settleable, flocculation or coagulation of colloids, and suspended. When these solids are being carried in a river or stream, there is always a possibility that the velocity of flow will drop to some value at which sedimentation will occur. The limiting velocity at which deposition will occur is probably about 0.5 to 0.6 ft/s (15 to 18 cm/s).

Deposition of organic solids results in a temporary reduction in the BOD load of the stream water. Almost as soon as organic matter is deposited, the deposits will start undergoing biological decomposition, which results in some reduction in the DO concentration of the water adjacent to the sediment material. As the deposited organic matter increases in volume, the rate of decomposition also increases. Ultimately, equilibrium will be established. Velz (1958) has shown that at equilibrium the rate of decomposition (exertion of a BOD) equals the rate of deposition.

In order to properly evaluate the effects of wastewater on the oxygen resources of a stream, it may be necessary to account for the contribution made by solids being deposited or by sediments being scoured and carried into suspension.

From field observations, Velz (1958) concluded that enriched sediments will deposit and accumulate at a stream velocity of 0.6 ft/s (18 cm/s) or less, and resuspension of sediments and scouring will occur at a flow velocity of 1.0 to 1.5 ft/s (30 to 45 cm/s). Velz (1958) has reported that the effect of sludge deposits can be expressed mathematically. The accumulation of sludge deposition ($L_d$) is

$$L_d = \frac{P_d}{2.303k}(1 - 10^{-kt}) \tag{1.66}$$

where $L_d$ = accumulation of BOD in area of deposition, lb/d
$P_d$ = BOD added, lb/d
$k$ = rate of oxidation of deposit on base 10
$t$ = time of accumulation, days

**Example:** Given (A) $k = 0.03$; $t = 2$ days
(B) $k = 0.03$; $t = 30$ days
(C) $k = 0.03$; $t = 50$ days

Find the relationship of $L_d$ and $P_d$.

**solution for A:** (with Eq. (1.66))

$$L_d = \frac{P_d}{2.303k}(1 - 10^{-kt})$$

$$= \frac{P_d}{2.303 \times 0.03}(1 - 10^{-0.03 \times 2})$$

$$= \frac{P_d}{0.0691}(0.13)$$

$$= 1.88P_d$$

This means that in 2 days, 188% of daily deposit ($P_d$) will have been accumulated in the deposit area, and 13% of $P_d$ will have been oxidized. The rate of utilization is about 6.9% per day of accumulated sludge BOD.

**solution for B:**

$$\begin{aligned}L_{\mathrm{d}} &= \frac{P_{\mathrm{d}}}{0.0691}(1 - 10^{-0.03\times 30}) \\ &= \frac{P_{\mathrm{d}}}{0.0691}(0.874) \\ &= 12.65P_{\mathrm{d}}\end{aligned}$$

In 30 days, 1265% of daily deposit will have been accumulated and the rate of utilization is about 87.4% of daily deposit.

**solution for C:**

$$\begin{aligned}L_{\mathrm{d}} &= \frac{P_{\mathrm{d}}}{0.0691}(1 - 10^{-0.03\times 50}) \\ &= \frac{P_{\mathrm{d}}}{0.0691}(0.97) \\ &= 14.03P_{\mathrm{d}}\end{aligned}$$

In 50 days, 1403% of daily deposit will have been accumulated and the rate of utilization is about 97% of daily deposit (98% in 55 days, not shown). This suggests that equilibrium is almost reached in approximately 50 days, if there is no disturbance from increased flow.

## 12 Photosynthesis and Respiration

Processes of photosynthesis and respiration by aquatic plants such as phytoplankton (algae), periphyton, and rooted aquatic plants (macrophytes) could significantly affect the DO concentrations in the water column. Plant photosynthesis consumes nutrients and carbon dioxide under the light, and produces oxygen. During dark conditions, and during respiration, oxygen is used. The daily average oxygen production and reduction due to photosynthesis and respiration can be expressed in the QUALZE model as (US EPA, 1997):

$$\frac{\mathrm{d}C}{\mathrm{d}t} = P - R \tag{1.67}$$

$$\frac{\mathrm{d}C}{\mathrm{d}t} = (\alpha_3\mu - \alpha_4\rho)Ag \tag{1.68}$$

where $\mathrm{d}C/\mathrm{d}t$ = rate of change of oxygen concentration, mg $O_2$/(L · d)
$P$ = average gross photosynthesis production, mg $O_2$/(L · d)
$R$ = average respiration, mg $O_2$/(L · d)
$\alpha_3$ = stoichiometric ratio of oxygen produced per unit of algae photosynthesis, mg/mg

$\alpha_4$ = stoichiometric ratio of oxygen uptake per unit of algae respired, mg/mg
$\mu$ = algal growth rate coefficient, per day
$\rho$ = algal respiration rate coefficient, per day
$Ag$ = algal mass concentration, mg/L

## 13 Natural Self-Purification in Streams

When decomposable organic waste is discharged into a stream, a series of physical, chemical, and biological reactions are initiated and thereafter the stream ultimately will be relieved of its pullutive burden. This process is so-called natural self-purification. A stream undergoing self-purification will exhibit continuously changing water quality characteristics throughout the reach of the stream.

Dissolved oxygen concentrations in water are perhaps the most important factor in determining the overall effect of decomposable organic matters in a stream. It is also necessary to maintain mandated DO levels in a stream. Therefore, the type and the degree of wastewater treatment necessary depend primarily on the condition and best usage of the receiving stream.

Recently the US Environmental Protection Agency has revised the water quality model QUALZE for total maximum daily loads to rivers and streams (US EPA, 1997). Interested readers may refer to an excellent example of total maximum daily load analysis in Appendix B of the US EPA technical guidance manual.

### 13.1 Oxygen sag curve

The dissolved oxygen balance in a stream which is receiving wastewater effluents can be formulated from a combination of the rate of oxygen utilization through BOD and oxygen transfer from the atmosphere into water. Many factors involved in this process are discussed in the previous sections. The oxygen sag curve (DO balance) is as a result of DO added minus DO removed. The oxygen balance curve or oxygen profile can be mathematically expressed (Streeter–Phelps, 1925) as previously discussed:

$$D_t = \frac{k_1 L_a}{k_2 - k_1}(10^{-k_1 t} - 10^{-k_2 t}) + D_a \times 10^{-k_2 t} \qquad (1.8b)$$

where $k_1$ and $k_2$ are, respectively, deoxygenation and reoxygenation rates to the base 10 which are popularly used. Since $k_1$ is determined under laboratory conditions, the rate of oxygen removed in a stream by oxidation may be different from that under laboratory conditions. Thus, a term $k_d$

is often substituted. Likewise, the rate of BOD removal in a stream may not equal the deoxygenation rate in a laboratory bottle and the oxidation rate in a stream, so the term $k_r$ is used to reflect this situation. Applying these modified terms, the oxygen sag equation becomes

$$D_t = \frac{k_d L_a}{k_2 - k_r}(10^{-k_r t} - 10^{-k_2 t}) + D_a \times 10^{-k_2 t} \tag{1.69}$$

If deposition occurs, $k_r$ will be greater than $k_1$; and if scour of sediment organic matter occurs, $k_r$ will be less then $k_1$.

Computation of organic waste load-capacity of streams or rivers may be carried out by using Eq. (1.69) (the following example), the Thomas method, the Churchill–Buckingham method, and other methods.

**Example:** Station 1 receives a secondary effluent from a city. BOD loading at station 2 is negligible. The results of two river samplings at stations 3 and 4 (temperature, flow rates, $BOD_5$, and DO) are adopted from Nemerow (1963) and are shown below. Construct an oxygen sag curve between stations 3 and 4.

| | | Flow Q | $BOD_5$ | | DO, mg/L | | DO deficit | |
|---|---|---|---|---|---|---|---|---|
| Station (1) | Temp °C (2) | cfs (3) | mg/L (4) | lb/d (5) | Measure (6) | Saturated (7) | mg/L (8) | lb/d (9) |
| 3 | **20.0** | **60.1** | **36.00** | 11662 | **4.1** | 9.02 | 4.92 | 1594 |
| | **17.0** | **54.0** | **10.35** | 3012 | **5.2** | 9.61 | 4.41 | 1284 |
| | | | *Mean* | *7337* | | | | *1439* |
| 4 | **20.0** | **51.7** | **21.2** | 5908 | **2.6** | 9.02 | 6.42 | 1789 |
| | **16.5** | **66.6** | **5.83** | 2093 | **2.8** | 9.71 | 6.91 | 2481 |
| | | | *Mean* | *4000* | | | | *2135* |

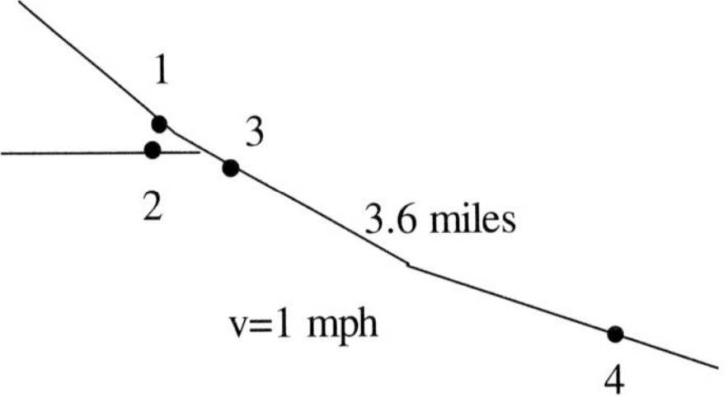


**solution:**

Step 1. Calculation for columns 5, 7, 8, and 9

Col. 5: lb/d = 5.39 × col. 3 × col. 4

Col. 7: taken from Table 1.2

Col. 8: mg/L = col. 7 − col. 6
Col. 9: lb/d = 5.39 × col. 3 × col. 8

*Note*: 5.39 lb/d = 1 cfs × 62.4 lb/ft$^3$ × 86,400 s/d × (1 mg/10$^6$ mg)

Step 2. Determine ultimate BOD at stations 3 ($L_{a3}$) and 4 ($L_{a4}$)

Under normal deoxygenation rates, factor of 1.46 is used as a multiplier to convert $BOD_5$ to the ultimate first-stage BOD.

$$L_{a3} = 7337 \text{ lb/d} \times 1.46 = 10{,}712 \text{ lb/d}$$

$$L_{a4} = 4000 \text{ lb/d} \times 1.46 = 5840 \text{ lb/d}$$

Step 3. Compute deoxygenation rate $k_d$

$$\Delta t = \frac{3.6 \text{ miles}}{1.0 \text{ mph} \times 24 \text{ h/d}} = 0.15 \text{ days}$$

$$k_d = \frac{1}{\Delta t} \log \frac{L_{a3}}{L_{a4}} = \frac{1}{0.15 \text{ days}} \log \frac{10{,}712}{5840}$$

$$= 1.76 \text{ (per day)}$$

Step 4. Compute average BOD load $\overline{L}$, average DO deficit $\overline{D}$, and the difference in DO deficit $\Delta D$ in the reach (Stations 3 and 4) from above

$$\overline{L} = \frac{1}{2}(L_{a3} + L_{a4}) = \frac{1}{2}(10{,}712 + 5840) \text{ lb/d}$$

$$= 8276 \text{ lb/d}$$

$$\overline{D} = \frac{1}{2}(1439 + 2135) \text{ lb/d}$$

$$= 1787 \text{ lb/d}$$

$$\Delta D = (1789 + 2481) - (1594 + 1284)$$

$$= 1392 \text{(lb/d)}$$

Step 5. Compute reaeration rate $k_2$

$$k_2 = k_d \frac{\overline{L}}{\overline{D}} - \frac{\Delta D}{2.303 \Delta t \overline{D}}$$

$$= 1.76 \times \frac{8276}{1787} - \frac{1392}{2.303 \times 0.15 \times 1787}$$

$$= 5.90 \text{ (per day)}$$

$$f = k_2/k_d = 5.9/1.76 = 3.35$$

Step 6. Plot the DO sag curve

Assuming the reaction rates $k_d$ ($k_d = k_r$) and $k_2$ remain constant in the reach (stations 3–4), the initial condition at station 3:

$$L_a = 10{,}712 \text{ lb/d}$$
$$D_a = 1439 \text{ lb/d}$$

Using Eq. (1.69)

$$D_t = \frac{k_d L_a}{k_2 - k_r}(10^{-k_r t} - 10^{-k_2 t}) + D_a \times 10^{-k_2 t}$$

When $t = 0.015$

$$D_{0.015} = \frac{1.76 \times 10{,}712}{5.9 - 1.76}(10^{-1.76\times0.015} - 10^{-5.9\times0.015}) + 1439 \times 10^{-5.9\times0.015}$$
$$= 1745 \text{ (lb/d)}$$

$D$ values at various $t$ can be computed in the same manner. We get $D$ values in lb/d as below. It was found that the critical point is around $t = 0.09$ days. The profile of DO deficit is depicted as below (Fig. 1.10):

$D_0 = 1439$    $D_{0.09} = 2245$
$D_{0.015} = 1745$    $D_{0.105} = 2227$
$D_{0.03} = 1960$    $D_{0.12} = 2190$
$D_{0.045} = 2104$    $D_{0.135} = 2137$
$D_{0.06} = 2192$    $D_{0.15} = 2073$
$D_{0.075} = 2235$

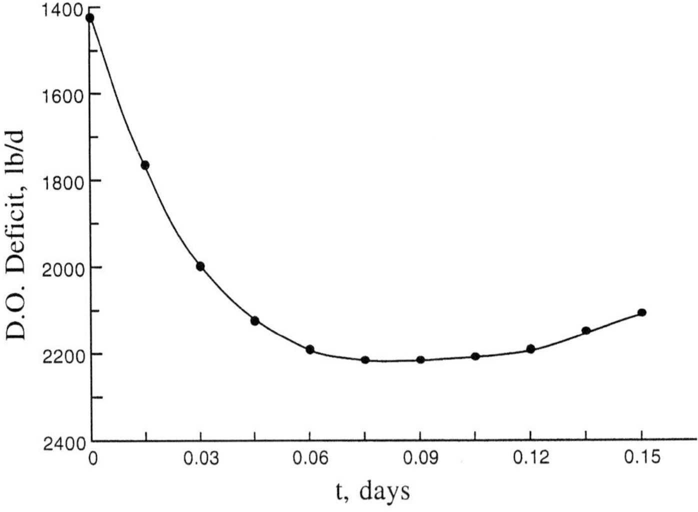


**Figure 1.10** Profile of DO deficit.

**Thomas method.** Thomas (1948) developed a useful simplification of the Streeter-Phelps equation for evaluating stream waste–assimilative capacity. His method presumes the computation of the stream deoxygenation coefficient ($k_1$) and reoxygenation coefficient ($k_2$) as defined in previous sections (also example). He developed a nomograph to calculate the DO deficit at any time, $t$ (DO profile), downstream from a source of pollution load. The nomograph (not shown, the interested reader should refer to his article) is plotted as $D/L_a$ versus $k_2t$ for various ratios of $k_2/k_1$. In most practical applications, this can be solved only by a tedious trial-and-error procedure. Before the nomograph is used, $k_1$, $k_2$, $D_a$, and $L_a$ must be computed (can be done as in the previous example). By means of a straightedge, a straight line (isopleth) is drawn, connecting the value of $D_a/L_a$ at the left of the point representing the reaeration constant × time of travel ($k_2t$) on the appropriate $k_2/k_1$ curve (a vertical line at $k_2t$ and intercept on the $k_2/k_1$ curve). The value $D_a/L_a$ is read at the intersection with the isopleth. Then, the value of the deficit at time $t(D_t)$ is obtained by multiplying $L_a$ with the interception value.

Nemerow (1963) claimed that he had applied the Thomas nomograph in many practical cases and found it very convenient, accurate, and timesaving. He presented a practical problem-solving example for an industrial discharge to a stream. He illustrated three different DO profiles along a 38-mile river: i.e. (1) under current loading and flow conditions; (2) imposition of an industrial load at the upstream station under current condition; and (3) imposition of industrial load under 5-year low-stream conditions.

**Churchill–Buckingham method.** Churchill and Buckingham (1956) found that the DO values in streams depend on only three variables: temperature, BOD, and stream flow. They used multiple linear correlation with three normal equations based on the principle of least squares:

$$\Sigma x_1 y = a x_1^2 + b x_1 x_2 + c x_1 x_3 \tag{1.70}$$

$$\Sigma x_2 y = a x_1 x_2 + b x_2^2 + c x_2 x_3 \tag{1.71}$$

$$\Sigma x_3 y = a x_1 x_3 + b x_2 x_3 + c x_3^2 \tag{1.72}$$

where $y$ = DO dropped in a reach, mg/L
$x_1$ = BOD at sag, mg/L
$x_2$ = temperature, °C
$x_3$ = flow, cfs (or MGD)
$a$, $b$, and $c$ = constants

By solving these three equations, the three constants $a$, $b$, and $c$ can be obtained. Then, the DO drop is expressed as

$$y = ax_1 + bx_2 + cx_3 + d \tag{1.73}$$

This method eliminates the often questionable and always cumbersome procedure for determining $k_1$, $k_2$, $k_d$, and $k_r$. Nemerow (1963) reported that this method provides a good correlation if each stream sample is collected during maximum or minimum conditions of one of the three variables. Only six samples are required in a study to produce practical and dependable results.

**Example:** Data obtained from six different days' stream surveys during medium- and low-flow periods showed that the DO sag occurred between stations 2 and 4. The results and computation are as shown in Table 1.12. Develop a multiple regression model for DO drop and BOD loading.

**solution:**

Step 1. Compute elements for least-squares method (see Table 1.13)

Step 2. Apply to three normal equations by using the values of the sum or average of each element

$$229.84a + 326.7b + 122.81c = 39.04$$
$$326.7a + 474.42b + 179.77c = 56.98$$
$$122.81a + 179.77b + 92.33c = 23.07$$

Solving these three equations by dividing by the coefficient of $a$:

$$a + 1.421b + 0.534c = 0.1699$$
$$a + 1.452b + 0.550c = 0.1744$$
$$a + 1.464b + 0.752c = 0.1879$$

**TABLE 1.12 DO Sag between Stations 2 and 4**

| Observed DO, mg/L | | DO drop, mg/L | BOD @ sag, mg/L | Temperature, °C | Flow, 1000 cfs |
|---|---|---|---|---|---|
| Station 2 | Station 4 | $y$ | $x_1$ | $x_2$ | $x_3$ |
| 7.8 | 5.7 | 2.1 | 6.8 | 11.0 | 13.21 |
| 8.2 | 5.9 | 2.3 | 12.0 | 16.5 | 11.88 |
| 6.1 | 4.8 | 1.9 | 14.8 | 21.2 | 9.21 |
| 6.0 | 3.2 | 2.8 | 20.2 | 23.4 | 6.67 |
| 5.5 | 2.3 | 3.2 | 18.9 | 28.3 | 5.76 |
| 6.2 | 2.9 | 3.3 | 14.3 | 25.6 | 8.71 |
| Sum | | 15.6 | 87.0 | 126.0 | 55.44 |
| Mean | | 2.6 | 14.5 | 21.0 | 9.24 |

**TABLE 1.13 Elements for Least-Squares Method**

| | $y^2$ | $yx_1$ | $yx_2$ | $yx_3$ | $x_1^2$ | $x_1x_2$ | $x_1x_3$ | $x_2^2$ | $x_2x_3$ | $x_3^2$ |
|---|---|---|---|---|---|---|---|---|---|---|
| | 4.41 | 14.28 | 23.10 | 27.74 | 46.24 | 74.8 | 89.83 | 121.00 | 145.31 | 174.50 |
| | 5.29 | 27.60 | 37.95 | 27.32 | 144.00 | 198.0 | 142.56 | 272.25 | 176.02 | 141.13 |
| | 3.61 | 28.12 | 40.28 | 17.50 | 219.04 | 313.76 | 136.31 | 449.44 | 195.25 | 84.82 |
| | 7.84 | 56.56 | 65.52 | 18.68 | 408.04 | 472.68 | 134.74 | 547.56 | 196.08 | 44.49 |
| | 10.24 | 60.48 | 90.56 | 18.43 | 357.21 | 534.87 | 108.86 | 800.89 | 163.01 | 33.18 |
| | 10.89 | 47.19 | 84.48 | 28.74 | 204.49 | 366.08 | 124.55 | 655.36 | 222.98 | 75.86 |
| Sum | 42.28 | 234.23 | 341.89 | 138.42 | 1379.02 | 1960.19 | 736.85 | 2846.50 | 1078.64 | 553.99 |
| Mean | 7.05 | 39.04 | 56.98 | 23.07 | 229.84 | 326.70 | 122.81 | 474.42 | 179.77 | 92.33 |

Subtracting one equation from the other:

$$0.043b + 0.218c = 0.0180$$

$$0.012b + 0.202c = 0.0135$$

Dividing each equation by the coefficient of $b$:

$$b + 5.07c = 0.4186$$

$$b + 16.83c = 1.125$$

Subtracting one equation from the other:

$$11.76c = 0.7064$$

$$c = 0.0601$$

Substituting $c$ in the above equation:

$$b + 5.07 \times 0.0601 = 0.4186$$

$$b = 0.1139$$

Similarly

$$\begin{aligned} a &= 0.1699 - 1.421 \times 0.1139 - 0.534 \times 0.0601 \\ &= -0.0241 \end{aligned}$$

Check on to the three normal equations with values of $a$, $b$, and $c$:

$$\begin{aligned} d &= Y - (ax_1 + bx_2 + cx_3) \\ &= 39.04 - (-229.84 \times 0.0241 + 326.7 \times 0.1139 + 122.81 \times 0.0601) \\ &= 0 \end{aligned}$$

The preceding computations yield the following DO drop:

$$Y = -0.0241x_1 + 0.1139x_2 + 0.0601x_3$$

Step 3. Compute the DO drop using the above model

The predicted and observed DO drop is shown in Table 1.14.

**TABLE 1.14 Predicted and Observed DO Drop**

| | Dissolved oxygen drop, mg/L | |
|---|---|---|
| Sampling date | Calculated | Observed |
| 1 | 1.93 | 2.1 |
| 2 | 2.35 | 2.3 |
| 3 | 2.61 | 1.9 |
| 4 | 2.62 | 2.8 |
| 5 | 3.07 | 3.2 |
| 6 | 3.04 | 3.3 |

It can be seen from Table 1.14 that the predicted versus the observed DO drop values are reasonably close.

Step 4. Compute the allowable BOD loading at the source of the pollution

The BOD equation can be derived from the same least-squares method by correlating the upstream BOD load (as $x_1$) with the water temperature (as $x_2$), flow rate (as $x_3$), and resulting BOD (as $z$) at the sag point in the stream. Similar to steps 1 and 2, the DO drop is replaced by BOD at station 4 (not shown). It will generate an equation for the allowable BOD load:

$$z = ax_1 + bx_2 + cx_3 + d$$

Then, applying the percent of wastewater treatment reduction, the BOD in the discharge effluent is calculated as $x_1$. Selecting the design temperature and low-flow data, one can then predict the BOD load (lb/d or mg/L) from the above equation. Also, applying $z$ mg/L in the DO drop equation gives the predicted value for DO drop at the sag.

## 13.2 Determination of $k_r$

The value $k_r$ can be determined for a given reach of stream by determining the BOD of water samples from the upper (A) and lower (B) ends of the section under consideration:

$$\log \frac{L_{aB}}{L_{aA}} = -k_r t \tag{1.74}$$

$$k_r = -\frac{1}{t}\log\frac{L_{aB}}{L_{aA}} = \frac{1}{t}\log\frac{L_{aA}}{L_{aB}} \tag{1.75a}$$

or

$$k_r = \frac{1}{t}\,(\log L_{aA} - \log L_{aB}) \tag{1.75b}$$

While the formula calls for first-stage BOD values, any consistent BOD values will give satisfactory results for $k_r$.

**Example 1:** The first-stage BOD values for river stations 3 and 4 are 34.6 and 24.8 mg/L, respectively. The time of travel between the two stations is 0.99 days. Find $k_r$ for the reach.

**solution:**

$$\begin{aligned} k_r &= \frac{1}{t}(\log L_{aA} - \log L_{aB}) \\ &= \frac{1}{0.99}(\log 34.6 - \log 24.8) \\ &= 0.15 \text{ (per day for base 10)} \\ K_r &= 2.3026k_r = 0.345 \text{ per day for log base } e \end{aligned}$$

**Example 2:** Given

$$\begin{aligned} L_{aA} &= 32.8 \text{ mg/L}, \qquad L_{aB} = 24.6 \text{ mg/L} \\ D_A &= 3.14 \text{ mg/L}, \qquad D_B = 2.58 \text{ mg/L} \\ t &= 1.25 \text{ days} \end{aligned}$$

Find $k_2$ and $k_r$

**solution:**

Step 1. Determine $k_2$ using Eqs. (1.44) and (1.45)

Reaeration

$$\begin{aligned} r_m &= (L_{aA} - L_{aB}) + (D_A - D_B) \\ &= (32.8 - 24.6) + (3.14 - 2.58) \\ &= 8.76 \text{ (mg/L)} \end{aligned}$$

Mean deficit

$$\begin{aligned} D_m &= (3.14 + 2.58)/2 = 2.86 \\ K_2 &= \frac{r_m}{D_m} = \frac{8.76}{2.86} = 3.06 \text{ (per day) or } k_2 = \frac{3.06}{2.303} = 1.33 \text{ (per day)} \end{aligned}$$

Step 2. Compute $k_r$

$$\begin{aligned} k_r &= \frac{1}{t}(\log L_{aA} - \log L_{aB}) \\ &= \frac{1}{1.25}(\log 32.8 - \log 24.6) \\ &= 0.1 \text{(per day)} \end{aligned}$$

### 13.3 Critical point on oxygen sag curve

In many cases, only the lowest point of the oxygen sag curve is of interest to engineers. The equation can be modified to give the critical value for DO deficiency ($D_c$) and the critical time ($t_c$) downstream at the critical point. At the critical point of the curve, the rate of deoxygenation equals the rate of reoxygenation:

$$\frac{dD}{dt} = K_1 L_t - K_2 D_c = 0 \tag{1.76}$$

Thus

$$K_1 L_t = K_2 D_c$$

$$D_c = \frac{K_1}{K_2} L_t \tag{1.77a}$$

or

$$D_c = \frac{K_d}{K_2} L_t \tag{1.77b}$$

where $L_t$ = BOD remaining.

Use $K_d$ if deoxygenation rate at critical point is different from $K_1$. Since

$$L_t = L_a \times e^{-K_r t_c}$$

then

$$D_c = \frac{K_d}{K_2} (L_a \times e^{-K_r t_c}) \tag{1.78a}$$

or

$$D_c = \frac{k_d}{k_2} (L_a \times 10^{-k_r t_c}) \tag{1.78b}$$

Let

$$f = \frac{k_2}{k_d} = \frac{K_2}{K_d} \tag{1.79}$$

or

$$D_c = \frac{1}{f} (L_a \times e^{-K_r t_c}) \tag{1.80}$$

and from Thomas (1948)

$$t_c = \frac{1}{k_2 - k_r} \log \frac{k_2}{k_r}\left[1 - \frac{D_a(k_2 - k_r)}{k_d L_a}\right] \tag{1.81}$$

then

$$t_c = \frac{1}{k_r(f-1)} \log\left\{f\left[1 - (f-1)\frac{D_a}{L_a}\right]\right\} \tag{1.82}$$

Substitute $t_c$ in $D_c$ formula (Thomas, 1948):

$$L_a = D_c\left(\frac{k_2}{k_r}\right)\left[1 + \frac{k_r}{k_2 - k_r}\left(1 - \frac{D_a}{\mathrm{D}_c}\right)^{0.418}\right] \tag{1.83}$$

or

$$\log L_a = \log D_c + \left[1 + \frac{k_r}{k_2 - k_r}\left(1 - \frac{D_a}{D_c}\right)^{0.418}\right] \log\left(\frac{k_2}{k_r}\right) \tag{1.84}$$

Thomas (1948) provided this formula which allows us to approximate $L_a$: the maximum BOD load that may be discharged into a stream without causing the DO concentration downstream to fall below a regulatory standard (violation).

**Example 1:** The following conditions are observed at station A. The water temperature is 24.3°C, with $k_1 = 0.06$, $k_2 = 0.24$, and $k_r = 0.19$. The stream flow is 880 cfs. DO and $L_a$ for river water is 6.55 and 5.86 mg/L, respectively. The state requirement for minimum DO is 5.0 mg/L. How much additional BOD ($Q$ = 110 cfs, DO = 2.22 mg/L) can be discharged into the stream and still maintain 5.0 mg/L DO at the flow stated?

**solution:**

Step 1. Calculate input data with total flow $Q = 880 + 110 = 990$ cfs; at $T$ = 24.3°C

Saturated DO from Table 1.2

$$\mathrm{DO_s} = 8.29 \text{ (mg/L)}$$

After mixing

$$\begin{aligned}\mathrm{DO_a} &= \frac{6.55 \times 880 + 2.22 \times 110}{990} \\ &= 6.07 \text{ (mg/L)}\end{aligned}$$

Deficit at station A

$$D_a = DO_s - DO_a = 8.29 \text{ mg/L} - 6.07 \text{ mg/L} = 2.22 \text{ (mg/L)}$$

Deficit at critical point

$$D_c = DO_s - DO_{min} = 8.29 \text{ mg/L} - 5.00 \text{ mg/L} = 3.29 \text{ (mg/L)}$$

Rates

$$k_d = k_r = 0.19 \text{ (per day)}$$

$$f = \frac{k_2}{k_1} = \frac{0.24}{0.16} = 1.5$$

$$f - 1 = 0.5$$

Step 2. Assume various values of $L_a$ and calculate resulting $D_c$

$$t_c = \frac{1}{k_r(f-1)} \log \left\{ f\left[1 - (f-1)\frac{D_a}{L_a}\right]\right\}$$

$$= \frac{1}{0.19(0.5)} \log \left\{ 1.5\left[1 - (0.5)\frac{2.22}{L_a}\right]\right\}$$

$$= 10.53 \log \left(1.5 - \frac{1.655}{L_a}\right)$$

Let $L_a = 10$ mg/L

$$t_c = 10.53 \log (1.5 - 0.1665) = 1.32 \text{ days}$$

$$D_c = \frac{k_d}{k_2}(L_a \times 10^{-k_r t_c})$$

$$= \frac{0.19}{0.24}(10 \times 10^{-0.19 \times 1.32})$$

$$= 0.792(10 \times 0.561)$$

$$= 4.45 \text{ mg/L}$$

Similarly, we can develop a table

| $L_a$, mg/L | $T_c$, days | $D_c$, mg/L |
|---|---|---|
| 10.00 | 1.32 | 4.45 |
| 9.00 | 1.25 | 4.13 |
| 8.00 | 1.17 | 3.80 |
| 7.00 | 1.06 | 3.48 |
| **6.40** | **0.98** | **3.29** |
| 6.00 | 0.92 | 3.17 |

Therefore, maximum $L_{a,\max} = 6.40$ mg/L

Step 3. Determine effluent BOD load ($Y_e$) that can be added

BOD that can be added = maximum load – existing load

$$110Y_e = 6.40 \times 990 - 5.86 \times 880$$
$$Y_e = 10.72 \text{ (mg/L)}$$

This means that the first-stage BOD of the effluent should be less than 10.72 mg/L.

**Example 2:** Given: At the upper station A of the stream reach, under standard conditions with temperature of 20°C, $BOD_5$ = 3800 lb/d, $k_1 = 0.14$, $k_2 = 0.25$, $k_r = 0.24$, and $k_d = k_r$. The stream temperature is 25.8°C with a velocity 0.22 mph. The flow in the reach (A → B) is 435 cfs (including effluent) with a distance of 4.8 miles. $DO_A$ = 6.78 mg/L, $DO_{min}$ = 6.00 mg/L. Find how much additional BOD can be added at station A and still maintain a satisfactory DO level at station B?

Step 1. Calculate $L_a$ at $T$ = 25.8°C

$$y = L_a(1 - 10^{-k_1 t})$$

When $t$ = 5 days, $T$ = 20°C (using loading unit of lb/d):

$$3800 = L_a(1 - 10^{-0.14\times 5})$$
$$L_a = 3800/0.80 = 4750 \text{(lb/d)} = L_{a(20)}$$

Convert to $T$ = 25.8°C:

$$L_{a(T)} = L_{a(20)}[1 + 0.02(T - 20)]$$
$$L_{a(25.8)} = 4750[1 + 0.02(25.8 - 20)]$$
$$= 5300 \text{(lb/d)}$$

Step 2. Change all constants to 25.8°C basis

At 25.8°C,

$$k_{1(T)} = k_{1(20)} \times 1.047^{T-20}$$
$$k_{1(25.8)} = 0.14 \times 1.047^{(25.8-20)}$$
$$= 0.18 \text{ (per day)}$$
$$k_{2(T)} = k_{2(20)} \times (1.02)^{T-20}$$
$$k_{2(25.8)} = 0.25 \times (1.02)^{25.8-20}$$
$$= 0.28 \text{ (per day)}$$
$$k_{r(25.8)} = 0.24 \times 1.047^{(25.8-20)}$$
$$= 0.31 \text{ (per day)}$$
$$k_{d(25.8)} = 0.31 \text{ (per day)}$$

Step 3. Calculate allowable deficit at station B

At 25.8°C,

$$\begin{aligned} DO_{sat} &= 8.05 \text{ mg/L} \\ DO_B &= 6.00 \text{ mg/L} \\ D_B &= 8.05 \text{ mg/L} - 6.00 \text{ mg/L} = 2.05 \text{ mg/L} \\ D_A &= 8.05 \text{ mg/L} - 6.78 \text{ mg/L} = 1.27 \text{ mg/L} \end{aligned}$$

Step 4. Determine the time of travel ($t$) in the reach

$$t = \frac{\text{distance}}{V} = \frac{4.8 \text{ miles}}{0.22 \text{ mph} \times 24 \text{ h/d}} = 0.90 \text{ days}$$

Step 5. Compute allowable $L_a$ at 25.8°C using Eq. (1.69)

$$D_t = \frac{k_d L_a}{k_2 - k_r}(10^{-k_r t} - 10^{-k_2 t}) + D_a \times 10^{-k_2 t}$$

Here $D_t = D_B$ and $D_a = D_A$

$$\begin{aligned} 2.05 &= \frac{0.31 L_a}{0.28 - 0.31}(10^{-0.31\times 0.9} - 10^{-0.28\times 0.9}) + 1.27 \times 10^{-0.28\times 0.9} \\ 2.05 &= (-10.33) L_a (0.526 - 0.560) + 0.71 \\ L_a &= 1.34/0.35 = 3.83 \text{ (mg/L)} \end{aligned}$$

Allowable load:

$$\begin{aligned} \text{lb/d} &= 5.39 \times L_a \cdot Q \\ &= 5.39(3.83 \text{ mg/L})(435 \text{ cfs}) \\ &= 8980 \end{aligned}$$

Step 6. Find load that can be added at station A

$$\begin{aligned} \text{Added} &= \text{allowable} - \text{existing} \\ &= 8980 \text{ lb/d} - 5300 \text{ lb/d} \\ &= 3680 \text{ (lb/d)} \end{aligned}$$

Step 7. Convert answer back to 5-day 20°C BOD

$$\begin{aligned} 3680 &= L_{a(20)}[1 + 0.02(25.8 - 20)] \\ L_{a(20)} &= 3297 \text{(lb/d)} \end{aligned}$$

**Example 3:**

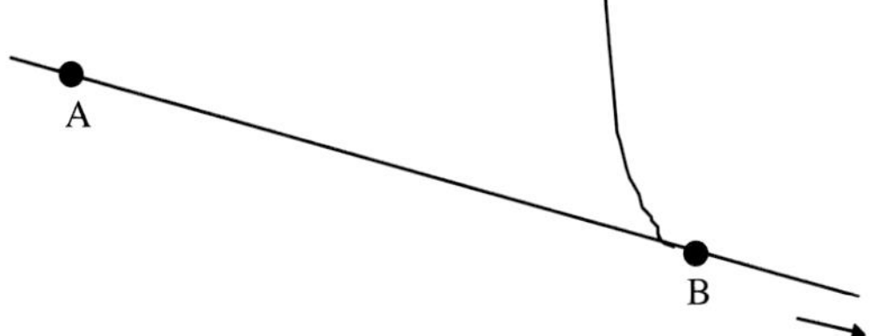


Given: The following data are obtained from upstream station A.

$k_1 = 0.14$ @ 20°C

$k_2 = 0.31$ @ 20°C

$k_r = 0.24$ @ 20°C

$k_d = k_r$

$DO_A = 5.82$ mg/L

Allowable $DO_{min} = 5.0$ mg/L

$BOD_5$ load at station A = 3240 lb/d

Stream water temperature = 26°C

Flow $Q = 188$ cfs

Velocity $V = 0.15$ mph

A–B distance = 1.26 miles

Find: Determine the BOD load that can be discharged at downstream station B.

**solution:**

Step 1. Compute DO deficits

At $T = 26$°C

$DO_s = 8.02$ mg/L and $D_a$ at station A

$$D_a = DO_s - DO_A = 8.02 \text{ mg/L} - 5.82 \text{ mg/L} = 2.20 \text{ mg/L}$$

Critical deficit at station B.

$$D_c = 8.02 - DO_{min} = 8.02 \text{ mg/L} - 5.0 \text{ mg/L} = 3.02 \text{ mg/L}$$

Step 2. Convert all constants to 26°C basis

At 26°C,

$$\begin{aligned} k_{1(26)} &= k_{1(20)} \times 1.047^{(26-20)} \\ &= 0.14 \times 1.047^6 \\ &= 0.18 \text{ (per day)} \\ k_{2(26)} &= k_{2(20)} \times (1.02)^{26-20} \\ &= 0.31 \times (1.02)^6 \\ &= 0.34 \text{ (per day)} \\ k_{r(26)} &= 0.24 \times 1.047^6 \\ &= 0.32 \text{ (per day)} \\ \frac{k_{2(26)}}{k_{r(26)}} &= \frac{0.34}{0.32} = 1.0625 \end{aligned}$$

Step 3. Calculate allowable BOD loading at station B at 26°C

From Eq. (1.84):

$$\log L_{aB} = \log D_c + \left[1 + \frac{k_r}{k_2 - k_r}\left(1 - \frac{D_a}{D_c}\right)^{0.418}\right]\log\left(\frac{k_2}{k_r}\right)$$

$$\log L_{aB} = \log 3.02 + \left[1 + \frac{0.32}{0.34 - 0.32}\left(1 - \frac{2.2}{3.02}\right)^{0.418}\right]\log 1.0625$$

$$= 0.480 + [1 + 16 \times (0.272)^{0.418}]0.0263$$

$$= 0.480 + 10.28 \times 0.0263$$

$$= 0.750$$

$$L_{aB} = 5.52\ (\text{mg/L})$$

Allowable BOD loading at station B:

$$\text{lb/d} = 5.39 \times L_{aB}\ (\text{mg/L}) \cdot Q\ (\text{cfs})$$

$$= 5.39 \times 5.52 \times 188$$

$$= 5594$$

Step 4. Compute ultimate BOD at station A at 26°C

$$L_{aA} = \text{BOD}_5\ \text{loading}/(1 - 10^{-k_1 t})$$

$$= 3240/(1 - 10^{-0.14\times5})$$

$$= 4050\ (\text{lb/d, at } 20°\text{C})$$

$$L_{aA(26)} = L_{aA}[1 + 0.02 \times (26 - 20)]$$

$$= 4050(1 + 0.12)$$

$$= 4536\ (\text{lb/d})$$

Step 5. Calculate BOD at 26°C from station A oxidized in 1.26 miles

Time of travel,

$$t = \frac{1.26\ \text{miles}}{0.15\ \text{mph} \times 24\ \text{h/d}} = 0.35\ \text{days}$$

$$y_B = L_{aA}(1 - 10^{-k_r t})$$

$$= 4536(1 - 10^{-0.32\times0.35})$$

$$= 4536 \times 0.227$$

$$= 1030\ (\text{lb/d})$$

Step 6. Calculate BOD remaining from station A at station B

$$L_t = 4536 \text{ lb/d} - 1030 \text{ lb/d} = 3506 \text{ (lb/d)}$$

Step 7. Additional BOD load that can be added in stream at station B (=$x$)

$$\begin{aligned} x_{(26)} &= \text{allowable BOD} - L_t \\ &= 5594 \text{ lb/d} - 3506 \text{ lb/d} \\ &= 2088 \text{(lb/d)} \end{aligned}$$

This is first-stage BOD at 26°C.

At 20°C,

$$x_{(26)} = x_{(20)}[1 + 0.02 \times (26 - 20)]$$

$$x_{(20)} = 2088/1.12 \text{ lb/d} = 1864 \text{ (lb/d)}$$

For $\text{BOD}_5$ at 20°C ($y_5$)

$$\begin{aligned} y_5 &= x_{20}\,(1 - 10^{-0.14\times 5}) \\ &= 1864\,(1 - 10^{-0.7}) \\ &= 1492 \text{ (lb/d)} \end{aligned}$$

**Example 4:** Given: Stations A and B are selected in the main stream, just below the confluences of tributaries. The flows are shown in the sketch below. Assume there is no significant increased flow between stations A and B, and the distance is 11.2 miles (18.0 km). The following information is derived from the laboratory results and stream field survey:

$k_1 = 0.18$ per day @ 20°C — stream water temperature = 25°C
$k_2 = 0.36$ per day @ 20°C — DO above station A = 5.95 mg/L
$k_r = 0.22$ per day @ 20°C — $L_a$ just above A = 8.870 lb/d @ 20°C
$k_d = k_r$ — $L_a$ added at A = 678 lb/d @ 20°C
Velocity, $V = 0.487$ mph — DO deficit in the tributary above A = 256 lb/d

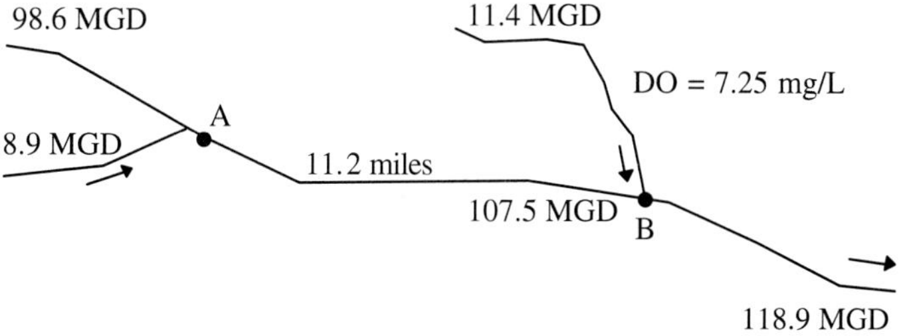


Find: Expected DO concentration just below station B.

**solution:**

Step 1. Calculate total DO deficit just below station A at 25°C

DO deficit above station A $D_A$ = 8.18 mg/L − 5.95 mg/L = 2.23 mg/L

$$\text{lb/d of DO deficit above A} = D_A\ (\text{mg/L}) \times Q\ (\text{MGD}) \times 8.34\ (\text{lb/gal})$$
$$= 2.23 \times 98.6 \times 8.34$$
$$= 1834$$

DO deficit in tributary above A = 256 lb/d

Total DO deficit below A = 1834 lb/d + 256 lb/d = 2090 lb/d

Step 2. Compute ultimate BOD loading at A at 25°C

At station A at 20°C, $L_{aA} = 8870 + 678 = 9548$ lb/d

At station A at 25°C, $L_{aA} = 9548[1 + 0.02(25 - 20)]$

$= 10{,}503$ lb/d

Step 3. Convert rate constants for 25°C

$$k_{1(25)} = k_{1(20)} \times 1.047^{(25-20)}$$
$$= 0.18 \times 1.258$$
$$= 0.23\ (\text{per day})$$
$$k_{2(25)} = k_{2(20)} \times 1.024^{25-20}$$
$$= 0.36 \times 1.126$$
$$= 0.40\ (\text{per day})$$
$$k_{r(25)} = k_{r(20)} \times 1.047^{25-20}$$
$$= 0.22 \times 1.258$$
$$= 0.28\ (\text{per day}) = k_{d(25)}$$

Step 4. Calculate DO deficit at station B from station A at 25°C

Time of travel $t = 11.2/(0.487 \times 24) = 0.958$ days

$$D_B = \frac{k_d L_a}{k_2 - k_r}(10^{-k_r t} - 10^{-k_2 t}) + D_a \times 10^{-k_2 t}$$
$$= \frac{0.28 \times 10{,}503}{0.40 - 0.28}(10^{-0.28\times0.958} - 10^{-0.4\times0.958}) + 2090 \times 10^{-0.4\times0.958}$$
$$= 24{,}507(0.5392 - 0.4138) + 2090 \times 0.4138$$
$$= 3073 - 865$$
$$= 2208(\text{lb/d})$$

Convert the DO deficit into concentration.

$$D_B = \text{amount, lb/d}/(8.34\ \text{lb/gal} \times \text{flow, MGD})$$
$$= 2208/(8.34 \times 107.5)$$
$$= 2.46\ (\text{mg/L})$$

This is DO deficiency at station B from BOD loading at station A.

Step 5. Calculate tributary loading above station B

$$\text{DO deficit} = 8.18\text{ mg/L} - 7.25\text{ mg/L} = 0.93\text{ mg/L}$$

$$\text{Amount of deficit} = 0.93 \times 8.34 \times 11.4 = 88\text{ (lb/d)}$$

Step 6. Compute total DO deficit just below station B

Total deficit

$$D_B = 2208\text{ lb/d} + 88\text{ lb/d} = 2296\text{ lb/d}$$

or

$$= 2296/(8.34 \times 118.9)$$
$$= 2.23\text{ (mg/L)}$$

Step 7. Determine DO concentration just below station B

$$\text{DO} = 8.18\text{ mg/L} - 2.23\text{ mg/L} = 5.95\text{ (mg/L)}$$

**Example 5:** A treated wastewater effluent from a community of 108,000 persons is to be discharged into a river which is not receiving any other significant wastewater discharge. Normally, domestic wastewater flow averages 80 gal (300 L) per capita per day. The 7-day, 10-year low flow of the river is 78.64 cfs. The highest temperature of the river water during the critical flow period is 26.0°C. The wastewater treatment plant is designed to produce an average carbonaceous 5-day BOD of 7.8 mg/L; an ammonia–nitrogen concentration of 2.3 mg/L, and DO for 2.0 mg/L. Average DO concentration in the river upstream of the discharge is 6.80 mg/L. After the mixing of the effluent with the river water, the carbonaceous deoxygenation rate coefficient ($K_{rC}$ or $K_C$) is estimated at 0.25 per day (base $e$) at 20°C and the nitrogenous deoxygenation coefficient ($K_{rN}$ or $K_N$) is 0.66 per day at 20°C. The lag time ($t_0$) is approximately 1.0 day. The river cross section is fairly constant with mean width of 30 ft (10 m) and mean depth of 4.5 ft (1.5 m). Compute DO deficits against time $t$.

**solution:**

Step 1. Determine total flow downstream $Q$ and $V$

$$\text{Effluent flow } Q_e = 80 \times 108{,}000\text{ gpd}$$

$$= 8.64 \times 10^6\text{ gpd} \times 1.54/10^6\text{ cfs/gpd}$$

$$= 13.31\text{ cfs}$$

$$\text{Upstream flow } Q_u = 78.64\text{ cfs}$$

$$\text{Downstream flow } Q_d = Q_e + Q_u = 13.31\text{ cfs} + 78.64\text{ cfs}$$

$$= 91.95\text{ cfs}$$

$$\text{Velocity } V = Q_d/A = 91.95\text{ ft}^2\text{/s } (30 \times 4.5)\text{ ft}^3$$

$$= 0.681\text{ (ft/s)}$$

Step 2. Determine reaeration rate constant $K_2$

The value of $K_2$ can be determined by several methods as mentioned previously. From the available data, the method of O'Connor and Dobbins (1958), Eq. (1.52) is used at 20°C:

$$\begin{aligned} K_2 &= 13.0V^{1/2}\mathrm{H}^{-3/2} \\ &= 13.0(0.681)^{1/2}\,(4.5)^{-3/2} \\ &= 1.12 \text{ (per day)} \end{aligned}$$

Step 3. Correct temperature factors for coefficients

$$\begin{aligned} K_{2(26)} &= 1.12 \text{ per day} \times 1.024^{26-20} = 1.29 \text{ (per day)} \\ K_{\mathrm{C}(26)} &= 0.25 \text{ per day} \times 1.047^{6} = 0.33 \text{ (per day)} \end{aligned}$$

Zanoni (1967) proposed correction factors for nitrogenous $K_{\mathrm{N}}$ in wastewater effluent at different temperatures as follows:

$$K_{\mathrm{N}(T)} = K_{\mathrm{N}(20)} \times 1.097^{T-20} \text{ for 10 to 22°C} \tag{1.85}$$

and

$$K_{\mathrm{N}(T)} = K_{\mathrm{N}(20)} \times 0.877^{T-22} \text{ for 22 to 30°C} \tag{1.86}$$

For 26°C,

$$\begin{aligned} K_{\mathrm{N}(26)} &= 0.66 \text{ per day} \times 0.877^{26-22} \\ &= 0.39 \text{ (per day)} \end{aligned}$$

Step 4. Compute ultimate carbonaceous BOD ($L_{\mathrm{aC}}$)

At 20°C,

$$\begin{aligned} K_{1\mathrm{C}} &= 0.25 \text{ per day} \\ \mathrm{BOD}_5 &= 7.8 \text{ mg/L} \\ L_{\mathrm{aC}} &= \mathrm{BOD}_5/(1 - e^{-K_{1\mathrm{C}}\times 5}) \\ &= 7.8 \text{ mg/L}/(1 - e^{-0.25\times 5}) \\ &= 10.93 (\text{mg/L}) \end{aligned}$$

At 26°C, using Eq. (1.28)

$$\begin{aligned} L_{\mathrm{aC}(26)} &= 10.93 \text{ mg/L}(0.6 + 0.02 \times 26) \\ &= 12.24 \text{ (mg/L)} \end{aligned}$$

Step 5. Compute ultimate nitrogenous oxygen demand ($L_{\mathrm{aN}}$)

Reduced nitrogen species ($NH_4^+$, $NO_3$, and $NO_2^-$) can be oxidized aerobically by nitrifying bacteria which can utilize carbon compounds but always require nitrogen as an energy source. The two-step nitrification can be expressed as

$$NH_4^+ + \frac{3}{2}O_2 \xrightarrow{\text{Nitrosomonas}} NO_2^- + 2H^+ + H_2O \tag{1.87}$$

and

$$NO_2^- + \frac{1}{2}O_2 \xrightarrow{\text{Nitrobacter}} NO_3^- \tag{1.88}$$

$$NH_4^+ + 2O_2 \rightarrow NO_3^- + 2H^+ + H_2O \tag{1.89}$$

Overall 2 moles of $O_2$ are required for each mole of ammonia; i.e. the N : $O_2$ ratio is 14 : 64 (or 1 : 4.57). Typical domestic wastewater contains 15 to 50 mg/L of total nitrogen. For this example, effluent $NH_3$ = 2.3 mg/L as N; therefore, the ultimate nitrogenous BOD is

$$L_{aN} = 2.3 \text{ mg/L} \times 4.57 = 10.51 \text{ (mg/L)}$$

Step 6. Calculate DO deficit immediately downstream of the wastewater load

$$\text{DO} = \frac{13.31 \times 2 + 78.64 \times 6.8}{13.31 + 78.64}$$

$$= 6.10 \text{ (mg/L)}$$

From Table 1.2, the DO saturation value at 26°C is 8.02 mg/L.

The initial DO deficit $D_a$ is

$$D_a = 8.02 \text{ mg/L} - 6.10 \text{ mg/L} = 1.92 \text{ mg/L}$$

Step 7. Compute $D_t$ and DO values at various times $t$

Using

$$D_t = \frac{K_C L_{aC}}{K_2 - K_C}(e^{-K_C t} - e^{-K_2 t}) + D_a e^{-K_2 t} + \frac{K_N L_{aN}}{K_2 - K_N}[e^{-K_N(t-t_0)} - e^{-K_2(t-t_0)}] \tag{1.90}$$

and at 26°C

$$K_2 = 1.29 \text{ per day}$$

$$K_C = 0.33 \text{ per day}$$

$$K_N = 0.39 \text{ per day}$$

$$L_{aC} = 12.24 \text{ mg/L}$$

$$L_{aN} = 10.51 \text{ mg/L}$$

$$D_a = 1.92 \text{ mg/L}$$

$$t_0 = 1.0 \text{ day}$$

Fot $t \leq 1.0$ day:

When $t = 0.1$ days

$$
\begin{aligned}
D_{0.1} &= \frac{0.33 \times 12.24}{1.29 - 0.33}(e^{-0.33\times0.1} - e^{-1.29\times0.1}) + 1.92e^{-1.29\times0.1} \\
&= 4.2075(0.9675 - 0.8790) + 1.6876 \\
&= 2.060 \text{ (mg/L)} \\
D_{0.2} &= 2.17 \text{ mg/L} \quad \text{and so on}
\end{aligned}
$$

For $t > 1.0$ day

$$
\begin{aligned}
D_{1.1} &= 4.2075(e^{-0.33\times1.1} - e^{-1.29\times1.1}) + 1.92e^{-1.29\times1.1} \\
&\quad + \frac{0.39 \times 10.51}{1.29 - 0.39}\left[e^{-0.39(1.1-1)} - e^{-1.29(1.1-1)}\right] \\
&= 4.2075(0.6956 - 0.2420) + 0.4646 \\
&\quad + 4.5543(0.9618 - 0.8790) \\
&= 2.75 \text{ (mg/L)} \\
D_{1.2} &= 3.04 \text{ mg/L} \quad \text{and so on}
\end{aligned}
$$

Step 8. Produce at table for DO sag

Table 1.15 gives the results of the above calculations for DO concentrations in the stream at various locations.

$$
\begin{aligned}
V &= 0.681 \text{ ft/s} = 0.681 \text{ ft/s} \times 3600 \text{ s/h} \times \frac{1 \text{ mile}}{5280 \text{ ft}} \\
&= 0.463 \text{ mph}
\end{aligned}
$$

$$
\begin{aligned}
\text{When } t &= 0.1 \text{ days} \\
\text{Distance} &= 0.463 \text{ mph} \times 24 \text{ h/d} \times 0.1 \text{ days} \\
&= 1.11 \text{ miles}
\end{aligned}
$$

Step 9. Explanation

From Table 1.15, it can be seen that minimum DO of 5.60 mg/L occurred at $t = 0.8$ days due to carbonaceous demand. After this location, the stream starts recovery. However, after $t = 1.0$ day, the stream DO is decreasing due to nitrogenous demand. At $t = 2.0$, it is the critical location (22.22 miles below the outfall) with a critical DO deficit of 3.83 mg/L and the DO level in the water is 4.19 mg/L.

**TABLE 1.15 DO Concentrations at Various Locations**

| $t$, days | Distance below outfall, miles | DO deficit, mg/L | Expected DO,* mg/L |
|---|---|---|---|
| 0.0 | 0.00 | 1.92 | 6.10 |
| 0.1 | 1.11 | 2.06 | 5.96 |
| 0.2 | 2.22 | 2.17 | 5.85 |
| 0.3 | 3.33 | 2.26 | 5.74 |
| 0.4 | 4.44 | 2.32 | 5.70 |
| 0.5 | 5.55 | 2.37 | 5.64 |
| 0.6 | 6.67 | 2.40 | 5.62 |
| 0.7 | 7.78 | 2.41 | 5.61 |
| **0.8** | **7.89** | **2.42** | **5.60** |
| 0.9 | 10.00 | 2.41 | 5.61 |
| 1.0 | 11.11 | 2.40 | 5.62 |
| 1.1 | 12.22 | 2.75 | 5.27 |
| 1.2 | 13.33 | 3.04 | 4.98 |
| 1.3 | 14.44 | 3.27 | 4.75 |
| 1.4 | 15.55 | 3.45 | 4.57 |
| 1.5 | 16.67 | 3.59 | 4.43 |
| 1.6 | 17.78 | 3.69 | 4.33 |
| 1.8 | 18.89 | 3.81 | 4.21 |
| **2.0** | **22.22** | **3.83** | **4.19** |
| 2.1 | 23.33 | 3.81 | 4.21 |
| 2.2 | 24.44 | 3.79 | 4.23 |

* Expected DO = saturated DO (8.02 mg/L) − DO deficit

### 13.4 Simplified oxygen sag computations

The simplified oxygen sag computation was suggested by Le Bosquet and Tsivoglou (1950). As shown earlier at the critical point

$$\frac{\mathrm{d}D}{\mathrm{d}t} = 0 \tag{1.48}$$

and

$$K_\mathrm{d} L_\mathrm{c} = K_2 D_\mathrm{c} \tag{1.49}$$

Let the lb/d of first-stage BOD be $C$

$$C(\mathrm{lb/d}) = \frac{L_\mathrm{a}(\mathrm{mg/L})\ Q\ (\mathrm{cfs}) \times 62.38\ (\mathrm{lb/cf}) \times 86{,}400\ (\mathrm{s/d})}{1{,}000{,}000}$$

$$= 5.39 L_\mathrm{a} Q$$

or

$$L_\mathrm{a} = \frac{C}{5.39Q} = \frac{C_1}{Q} \tag{1.91}$$

where $C_1$ = constant

If $D_a = 0$, then $D_c$ is a function of $L_a$, or

$$D = L_a C_2 = \frac{C_1}{Q} \times C_2 = \frac{C_3}{Q} \tag{1.92}$$

where $C_2$, $C_3$ = constants

$$DQ = C_3 \tag{1.93}$$

Since $K_d L_c = K_2 D_c$

$$D_c = \frac{K_d}{K_2} L_c = L_a C_2 = \frac{C_3}{Q} \tag{1.94}$$

Hence the minimum allowable DO concentration at the critical point ($DO_C$) is the DO saturation value ($S$, or $DO_{sat}$) minus $D_c$. It can be written as

$$DO_C = S - D_c \tag{1.95}$$

$$= S - C_3\left(\frac{1}{Q}\right) \tag{1.95a}$$

From the above equation, there is a linear relationship between $DO_C$ and $1/Q$. Therefore, a plot of $DO_C$ versus $1/Q$ will give a straight line, where $S$ is the $y$-intercept and $C_3$ is the slope of the line. Observed values from fieldwork can be analyzed by the method of least squares to determine the degree of correlation.

## 14 SOD of DO Usage

The SOD portion of DO usage is calculated using the following formula:

$$DO_{sod} = \frac{3.28Gt}{H} \tag{1.96}$$

where $DO_{sod}$ = oxygen used per reach, mg/L
$G$ = the SOD rate, $g/(m^2 \cdot d)$
$t$ = the retention time for the reach
$H$ = the average water depth in the reach, ft

All biological rates, including BOD and SOD, have to be corrected for temperature using the basic Arrhenius formula:

$$R_T = R_A(\theta^{T-A}) \tag{1.97}$$

where $R_T$ = biological oxygen usage rate at a temperature, $T\,°\mathrm{C}$
$R_A$ = biological oxygen usage rate at ambient or standard temperature (20°C as usual), $A°\mathrm{C}$
$\theta$ = proportionality constant, 1.047 for river or stream

**Example:** A stream reach has an average depth of 6.6 ft. Its SOD rate was determined as 3.86 g/(m$^2$ · d) at 24°C. The detention time for the reach is 0.645 days. What is the SOD portion of DO usage at the standard temperature of 20°C?

**solution:**

Step 1. Determine SOD rate $G$ at 20°C

$$G_{24} = G_{20}(1.047^{24-20})$$

$$G_{20} = G_{24}/(1.047)^4 = 3.86/1.2017 = 3.21[\mathrm{g/(m^2 \cdot d)}]$$

Step 2. Calculate $\mathrm{DO_{sod}}$

$$\begin{aligned}\mathrm{DO_{sod}} &= \frac{3.28Gt}{H} \\ &= \frac{3.28 \times 3.21 \times 0.645}{6.6} \\ &= 1.03\ (\mathrm{mg/L})\end{aligned}$$

## 15 Apportionment of Stream Users

In the United States, effluent standards for wastewater treatment plants are generally set for BOD, total suspended solids (TSS), and ammonia-nitrogen ($NH_3$-N) concentrations. Each plant often should meet standards of 10–10–12 (mg/L for BOD, TSS, and $NH_3$-N, respectively), right or wrong. It is not like classically considering the maximum use of stream-assimilative capacity. However, in some parts of the world, maximum usage of stream natural self-purification capacity may be a valuable tool for cost saving of wastewater treatment. Thus, some concepts of apportionment of self-purification capacity of a stream among different users are presented below.

The permissible load equation can be used to compute the BOD loading at a critical point $L_c$, as previously stated (Eq. (1.84); Thomas, 1948):

$$\log L_c = \log D_c + \left[1 + \frac{k_r}{k_2 - k_r}\left(1 - \frac{D_a}{D_c}\right)^{0.418}\right]\log\left(\frac{k_2}{k_r}\right) \qquad (1.98)$$

where $L_c$ = BOD load at the critical point
$D_c$ = DO deficit at the critical point
$D_a$ = DO deficit at upstream pollution point
$k_r$ = river BOD removal rate to the base 10, per day
$k_2$ = reoxygenation rate to the base 10, per day

$L_c$ also may be computed from limited BOD concentration and flow. Also, if city A and city B both have the adequate degree of treatment to meet the regulatory requirements, then

$$L_c = p\alpha L_A + \beta L_B \tag{1.99}$$

where $p$ = part or fraction of $L_A$ remaining at river point B
$\alpha$ = fraction of $L_A$ discharged into the stream
$\beta$ = fraction of $L_B$ discharged into the stream
$L_A$ = BOD load at point A, lb/d or mg/L
$L_B$ = BOD load at point B, lb/d or mg/L

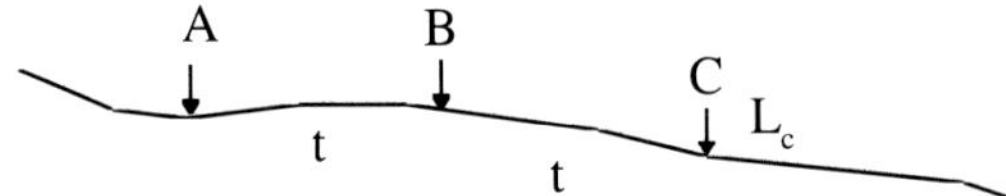


If $L_A = L_B = L$

$$L_c = p\alpha L + \beta L \tag{1.99a}$$

A numerical value of $L_c$ is given by Eq. (1.98) and a necessary relationship between $\alpha$, $\beta$, $p$, $L_A$, and $L_B$ is set up by Eq. (1.99). The value $p$ can be computed from

$$\log p = -k_r t$$

Since

$$\log \frac{Lt}{L_a} = -k_r t = \log \text{(fraction remaining)}$$

Therefore

$$\log p = -k_r t \tag{1.100}$$

or

$$p = 10^{-k_r t} \tag{1.100a}$$

The apportionment factors $\alpha$ and $\beta$ can be determined by the following methods.

### 15.1 Method 1

Assume: City A is further upstream than city B, with time of travel $t$ given, by nature, a larger degree of stream purification capacity. The time from city B to the lake inlet is also $t$. Since

$$y = L_a(1 - 10^{-k_r t})$$

$$\frac{y}{L_a} = 1 - 10^{-k_r t} = 1 - p \tag{1.101}$$

The proportion removed by the stream of BOD added by city A is

$$\frac{y}{L_A} = 1 - 10^{-k_r t \times 2} = 1 - 10^{-2k_r t} \tag{1.102}$$

$$= 1 - p^2 \tag{1.102a}$$

The proportion removed by the stream of BOD added by city B is

$$\frac{y}{L_B} = 1 - 10^{-k_r t} = 1 - p \tag{1.102b}$$

Then the percent amount removed BOD from cities A and B, given as $P_A$ and $P_B$, respectively, are

$$P_A = \frac{1 - p^2}{(1 - p^2) + (1 - p)} = \frac{(1 - p)(1 + p)}{(1 - p)(1 + p) + (1 - p)}$$

$$= \frac{(1 - p)(1 + p)}{(1 - p)(1 + p + 1)}$$

$$P_A = \frac{1 + p}{2 + p} \tag{1.103}$$

$$P_B = \frac{1 - p}{(1 - p^2) + (1 - p)} = \frac{(1 - p)}{(1 - p)(1 + p + 1)}$$

$$P_B = \frac{1}{2 + p} \tag{1.104}$$

Calculate the degree of BOD removal proportion:

$$\frac{\alpha}{\beta} = \frac{P_A}{P_B} = \frac{(1 + p)/(2 + p)}{1/(2 + p)} = \frac{1 + p}{1}$$

$$\alpha = (1 + p)\beta \tag{1.105}$$

Solving Eqs. (1.99a) and (1.105) simultaneously for $\alpha$ and $\beta$:

$$L_c = p\alpha L + \beta L$$

$$\frac{L_c}{L} = p\alpha + \beta$$

$$\frac{L_c}{L} = p(1 + p)\beta + \beta = (p + p^2 + 1)\beta$$

$$\beta = \frac{L_c}{L}\left(\frac{1}{1 + p + p^2}\right) \tag{1.106}$$

$$\alpha = (1 + p)\beta$$

$$= \frac{L_c}{L} \cdot \frac{1 + p}{1 + p + p^2} \tag{1.107}$$

The required wastewater treatment plant efficiencies for plants A and B are $1 - \alpha$ and $1 - \beta$, respectively. The ratio of cost for plant A to plant B is $(1 - \alpha)$ to $(1 - \beta)$.

**Example 1:** Assume all conditions are as in Method 1. The following data are available: $k_r = 0.16$ per day; $t = 0.25$ days; flow at point $C$, $Q_C = 58$ MGD $L = 6910$ lb/d; and regulation required $L_c \leq 10$ mg/L. Determine the apportionment of plants A and B, required plant BOD removal efficiencies, and their cost ratio.

**solution:**

Step 1. Compute allowable $L_c$

$$L_c = 10 \text{ mg/L} \times 8.34 \text{ lb/d MGD} \cdot \text{mg/L} \times 58 \text{ MGD}$$

$$= 4837 \text{ (lb/d)}$$

Step 2. Compute $p$

$$p = 10^{-k_r t} = 10^{-0.16 \times 0.25}$$

$$= 0.912$$

Step 3. Calculate $\alpha$ and $\beta$ by Eqs. (1.106) and (1.107)

$$\alpha = \frac{L_c}{L}\left(\frac{1 + p}{1 + p + p^2}\right) = \frac{4837}{6910}\left(\frac{1 + 0.912}{1 + 0.912 + 0.912^2}\right)$$

$$= 0.70 \times 0.697$$

$$= 0.49$$

$$\beta = \frac{L_c}{L}\left(\frac{1}{1 + p + p^2}\right) = 0.70 \times \frac{1}{2.744}$$

$$= 0.255$$

Step 4. Determine percent BOD removal required

For plant A:

$$1 - \alpha = 1 - 0.49 = 0.51 \quad \text{i.e. 51\% removal needed}$$

For plant B:

$$1 - \beta = 1 - 0.255 = 0.745 \quad \text{i.e. 74.5\% removal needed}$$

Step 5. Calculate treatment cost ratio

$$\frac{\text{Cost, plant } A}{\text{Cost, plant } B} = \frac{1 - \alpha}{1 - \beta} = \frac{0.51}{0.745} = \frac{0.68}{1} = \frac{1}{1.46}$$

## 15.2 Method 2

Determine a permissible BOD load for each city as if it were the only city using the river or stream. In other words, calculate maximum BOD load $L_A$ for plant A so that dissolved oxygen concentration is maintained above 5 mg/L (most state requirements), assuming that plant B does not exist. Similarly, also calculate maximum BOD load $L_B$ for plant B under the same condition. It should be noted that $L_B$ may be larger than $L_A$ and this cannot be the case in Method 1.

The BOD removed by the stream from that added

at point A: referring to Eq. (1.102a)

$$(1 - p^2)L_A$$

at point B: referring to Eq. (1.102b)

$$(1 - p)L_B$$

The percentage removed is
at point A,

$$P_A = \frac{(1 - p^2)L_A}{(1 - p^2)L_A + (1 - p)L_B} \tag{1.108}$$

at point B,

$$P_B = \frac{(1 - p)L_B}{(1 - p^2)L_A + (1 - p)L_B} \tag{1.109}$$

The degree of BOD removal is

$$\frac{\alpha}{\beta} = \frac{P_A}{P_B} \tag{1.105}$$

Solve Eqs. (1.99), (1.105), (1.108), and (1.109) for $\alpha$ and $\beta$. Also determine $(1-\alpha)$ and $(1-\beta)$ for the required wastewater treatment efficiencies and cost ratios as $(1-\alpha)/(1-\beta)$.

**Example 2:** All conditions and questions are the same as for Example 1, except using Method 2.

**solution:**

Step 1. Compute allowable $L_A$ and $L_B$

For point B, using Eq. (1.100)

$$\log p = \log\frac{L_t}{L_B} = -k_r t = -0.16 \times 0.25 = -0.04$$

$$p = \frac{L_t}{L_B} = 10^{-0.04} = 0.912$$

$$L_B = L_t/0.912 \ (L_t = L_c \text{ in this case})$$

$$= 4837/0.912$$

$$= 5304 \text{ lb/d}$$

For point A,

$$\frac{L_c}{L_A} = 10^{-0.16\times0.25\times2} = 10^{-0.08} = 0.8318$$

$$L_A = 4837 \text{ lb/d}/0.8318$$

$$= 5815 \text{ (lb/d)}$$

Step 2. Calculate $P_A$ and $P_B$, using Eq. (1.108)

$$P_A = \frac{(1-p^2)L_A}{(1-p^2)L_A + (1-p)L_B}$$

$$= \frac{(1-0.912^2)5815}{(1-0.912^2)5815 + (1-0.912)5304}$$

$$= \frac{978.4}{978.4 + 466.8}$$

$$= 0.677$$

$$P_B = 1 - 0.677 = 0.323$$

Step 3. Determine $\alpha$ and $\beta$

$$\alpha/\beta = P_A/P_B = 0.677/0.323 = 2.1/1$$

$$\alpha = 2.1\beta$$

$$L_c = p\alpha L_A + \beta L_B$$

$$4837 = 0.912 \times (1.1\beta) \times 5815 + \beta \times 5304$$

$$4837 = 11{,}137\beta + 5304\beta$$

$$\beta = 4837/16{,}441 = 0.29$$

$$\alpha = 2.1 \times 0.29 = 0.61$$

Step 4. Determine degree of BOD removal

For plant A

$1 - \alpha = 1 - 0.61 = 0.39$   i.e. 39% removal efficiency

For plant B

$1 - \beta = 1 - 0.29 = 0.71$   i.e. 71% removal efficiency

Step 5. Determine cost ratio

$$\frac{\text{Plant A}}{\text{Plant B}} = \frac{1 - \alpha}{1 - \beta} = \frac{0.39}{0.71} = \frac{1}{1.82} \text{ or } \frac{0.55}{1}$$

### 15.3 Method 3

The principle of this method is to load the river to the utmost so as to minimize the total amount of wastewater treatment. The proportion of $P_A$ and $P_B$ can be computed by either Method 1 or Method 2. The cost of treatment can be divided in proportion to population.

There are two cases of initial BOD loading:

**Case 1. $L_A > L$ ($L$ is the BOD load of plant A influent)**

- Let plant A discharge $L$ without a treatment.
- Have plant B treat an amount equal to $L - L_c$.
- City A is assessed for $P_A\,(L - L_c)$.
- City B is assessed for $P_B(L - L_c)$.

### Case 2. $L_A < L$

- Let plant A discharge $L_A$ and treat $L - L_A$.
- Have plant B treat an amount equal to $(L - L_c) - pL_A$.
- Then the total treatment will remove $(L - L_A) + (L - L_c - pL_A) = 2L - (1 + p)L_A - L_c$.
- City A is assessed for $P_A[2L - (1 + p)L_A - L_c]$.
- City B is assessed for $P_B[2L - (1 + p)L_A - L_c]$.

**Example 3:** Using data listed in Examples 1 and 2, determine the cost ratio assessed to plants A and B by Method 3.

**solution:**

Step 1. Select case

Since $L_A = 5815$ lb/d $< L = 6910$ lb/d
Case 2 will be applied

Step 2. Determine amounts needed to be treated

For plant A

$$6910 - 5815 = 1095 \text{ (lb/d) treated}$$

$$5815 = \text{lb/d discharged without treatment}$$

For plant B

$$L - L_c - pL_A = 6910 - 4837 - 0.912 \times 1095$$

$$= 1074 \text{ (lb/d)}$$

Total treatment = 1095 + 1074 = 2169 (lb/d)

Step 3. Compute cost assessments

For Example 2,

$$P_A = 0.677$$

$$P_B = 0.323$$

For plant A

$$\text{Cost} = 0.677 \times \text{treatment cost of 2169 lb/d}$$

For plant B

$$\text{Cost} = 0.323 \times \text{treatment cost of 2169 lb/d}$$

## 16 Velz Reaeration Curve (a Pragmatic Approach)

Previous sections discuss the conceptual approach of the relationships between BOD loading and DO in the stream. Another approach for determining the waste-assimilative capacity of a stream is a method first developed and used by Black and Phelps (1911) and later refined and used by Velz (1939). For this method, deoxygenation and reoxygenation computations are made separately, then added algebraically to obtain the net oxygen balance. This procedure can be expressed mathematically as

$$\mathrm{DO_{net}} = \mathrm{DO_a} + \mathrm{DO_{rea}} - \mathrm{DO_{used}} \tag{1.110}$$

where $\mathrm{DO_{net}}$ = dissolved oxygen at the end of a reach
$\mathrm{DO_a}$ = initial dissolved oxygen at beginning of a reach
$\mathrm{DO_{rea}}$ = dissolved oxygen absorbed from the atmosphere
$\mathrm{DO_{used}}$ = dissolved oxygen consumed biologically

All units are in lb/d or kg/d.

### 16.1 Dissolved oxygen used

The dissolved oxygen used (BOD) can be computed with only field-observed dissolved oxygen concentrations and the Velz reoxygenation curve. From the basic equation, the dissolved oxygen used in pounds per day can be expressed as:

$$\mathrm{DO_{used}} = (\mathrm{DO_a} - \mathrm{DO_{net}}) + \mathrm{DO_{rea}} \text{ (in lb/d)}$$

The terms $\mathrm{DO_a}$ and $\mathrm{DO_{net}}$ are the observed field dissolved oxygen values at the beginning and end of a subreach, respectively, and $\mathrm{DO_{rea}}$ is the reoxygenation in the subreach computed by the Velz curve. If the dissolved oxygen usage in the river approximates a first-order biological reaction, a plot of the summations of the $\mathrm{DO_{used}}$ versus time of travel can be fitted to the equation:

$$\mathrm{DO_{used}} = L_a[1 - \exp(-K_1 t)] \tag{1.111}$$

and $L_a$ and $K_1$ are determined experimentally in the same manner as for the Streeter–Phelps method. The fitting of the computed values to the equations must be done by trial and error, and a digital computer is utilized to facilitate this operation.

### 16.2 Reaeration

The reaeration term ($\mathrm{DO_{rea}}$) in the equation is the most difficult to determine. In their original work, Black and Phelps (1911) experimentally derived a reaeration equation based on principles of gas transfer of diffusion across a thin water layer in acquiescent or semiquiescent system. A modified but equivalent form of the original Black and Phelps equation is given by Gannon (1963):

$$R = 100\left(\frac{1 - B_0}{100}\right)(81.06)\left(e^{-K} + \frac{e^{-9K}}{9} + \frac{e^{-25K}}{25} + \cdots\right) \tag{1.112}$$

where $R$ = percent of saturation of DO absorbed per mix
$B_0$ = initial DO in percent of saturation
$K = \pi^2 am/4L^2$ in which $m$ is the mix or exposure time in hours, $L$ is the average depth in centimeters, and $a$ is the diffusion coefficient used by Velz

The diffusion coefficient $a$ was determined by Velz (1947) to vary with temperature according to the expression:

$$a_T = a_{20}(1.1^{T-20}) \tag{1.113}$$

where $a_T$ = diffusion coefficient at $T$°C
$a_{20}$ = diffusion coefficient at 20°C

when the depth is in feet,

$$a_{20} = 0.00153$$

when the depth is in centimeters,

$$a_{20} = 1.42$$

Although the theory upon which this reaeration equation was developed is for quiescent conditions, it is still applicable to moving, and even turbulent, streams. This can be explained in either of two ways (Phelps, 1944):

> (1) Turbulence actually decreases the effective depth through which diffusions operate. Thus, in a turbulent stream, mixing brings layers of saturated water from the surface into intimate contact with other less oxygenated layers from below. . . The actual extent of such mixing is difficult to envision, but it clearly depends upon the frequency with which surface layers are thus transported. In this change-of-depth concept, therefore, there is a time element involved.
>
> (2) The other and more practical concept is a pure time effect. It is assumed that the actual existing conditions of turbulence can be replaced by an equivalent condition composed of successive periods of perfect quiescence between which there is instantaneous and

complete mixing. The length of the quiescent period then becomes the time of exposure in the diffusion formula, and the total aeration per hour is the sum of the successive increments or, for practical purposes, the aeration period multiplied by the periods per hour.

Because the reaeration equation involves a series expansion in its solution, it is not readily solved by desk calculations. To facilitate the calculations, Velz (1939, 1947) published a slide-rule curve solution to the equation. This slide-rule curve is reprinted here as Fig. 1.11 (for its use, consult Velz, 1947). Velz's curve has been verified as accurate by two independent computer checks, one by Gannon (1963) and the other by Butts (1963).

Although the Velz curve is ingenious, it is somewhat cumbersome to use. Therefore, a nomograph (Butts and Schnepper, 1967) was developed for a quick, accurate desk solution of $R$ at zero initial DO($B_0$ in the Black and Phelps equation). The nomograph, presented in Fig. 1.12, was constructed on the premise that the exposure or mix time is characterized by the equation

$$M = (13.94)\log_e(H) - 7.45 \tag{1.114}$$

in which $M$ is the mix time in minutes and $H$ is the water depth in feet. This relationship was derived experimentally and reported by Gannon (1963) as valid for streams having average depths greater than 3 ft. For

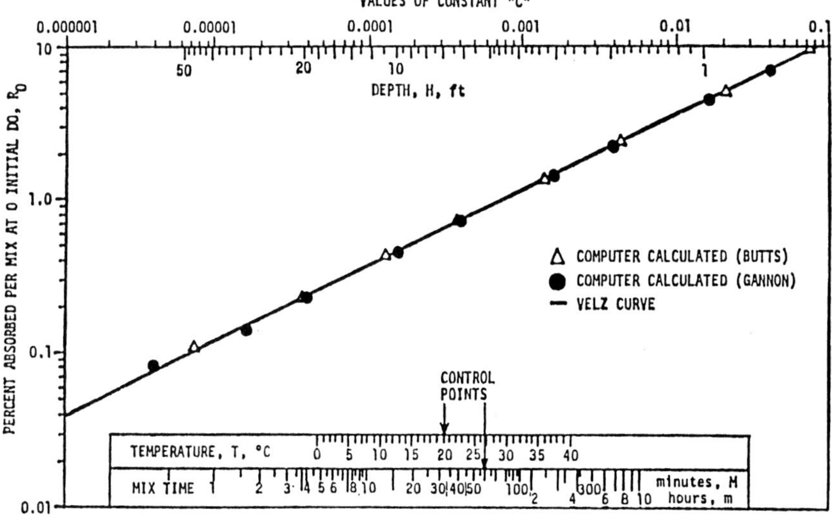


Figure 1.11 Standard reoxygeneration curve (*Butts et al., 1973*).

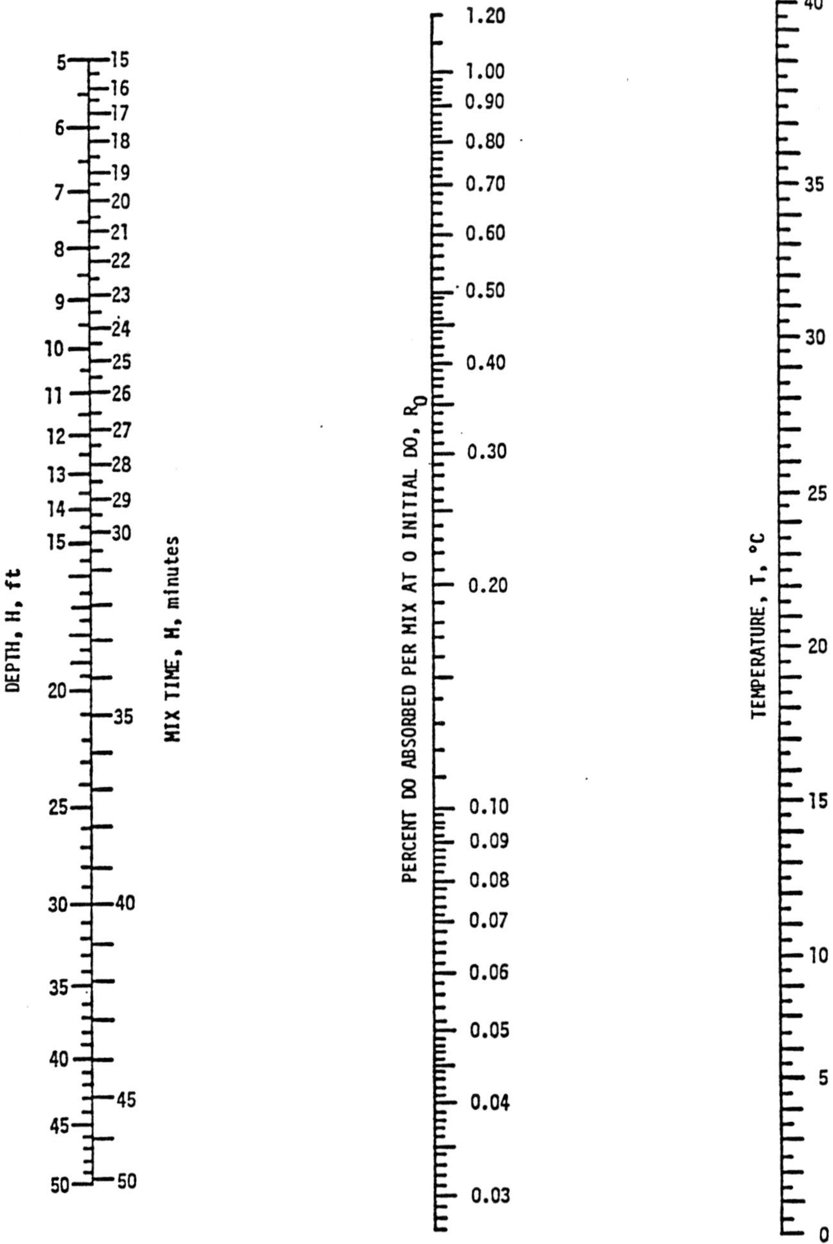


**Figure 1.12** Oxygen absorption nomograph (*Butts and Schnepper, 1967*).

an initial DO greater than zero, the nomograph value is multiplied by $(1 - B_0)$ expressed as a fraction because the relationship between $B_0$ and the equation is linear.

The use of the nomograph and the subsequent calculations required to compute the oxygen absorption are best explained by the example that follows:

**Example:** Given that $H = 12.3$ ft; initial DO of stream = 48% of saturation; $T = 16.2$°C; DO load at saturation at 16.2°C = 8800 lb; and time of travel in the stream reach = 0.12 days. Find the mix time ($M$) in minutes; the percent DO absorbed per mix at time zero initial DO ($R_0$); and the total amount of oxygen absorbed in the reach ($DO_{rea}$).

**solution:**

Step 1. Determine $M$ and $R_0$

From Fig. 1.12, connect 12.3 ft on the depth-scale with 16.2°C on the temperature scale. It can be read that:

$$M = 27.5 \text{ min on the mix timescale, and}$$
$$R_0 = 0.19\% \text{ on the } R_0\text{-scale}$$

Step 2. Compute $DO_{rea}$

Since $R_0$ is the percent of the saturated DO absorbed per mix when the initial DO is at 100% deficit (zero DO), the oxygen absorbed per mix at 100% deficit will be the product of $R_0$ and the DO load at saturation. For this example, lb/mix per 100% deficit is

$$8800 \text{ lb} \times \frac{0.19}{100} = 16.72 \text{ (lb/mix per 100\% deficit)}$$

However, the actual DO absorbed (at 48% saturation) per mix is only

$$16.72 \times [1 - (48/100)] = 8.7 \text{(lb/mix per 100\% deficit)}$$

Because the amount absorbed is directly proportional to the deficit, the number of mixes per reach must be determined:

The time of travel $\quad t = 0.12$ days

$$= 0.12 \text{ days} \times 1440 \text{ min/d}$$

$$= 173 \text{ min}$$

The mix time

$$M = 27.5 \text{ min}$$

Consequently, the number of mixes in the reach = 173/27.5 = 6.28.

Reaeration in the reach is therefore equal to the product of the DO absorbed per mix × the number of mixes per reach. $DO_{rea}$ is computed according to the expression:

$$DO_{rea} = \left(1 - \frac{\%\ \text{DO saturation}}{100}\right)\left(\frac{R_0}{100}\right)\left(\frac{t}{M}\right)(\text{DO saturation load}) \quad (1.115)$$

$$= \left(1 - \frac{48}{100}\right)\left(\frac{0.19}{100}\right)\left(\frac{173}{27.5}\right)(8800)$$

$$= 8.7 \times 6.28$$

$$= 54.6\ \text{(lb)}$$

## 17 Stream DO Model (a Pragmatic Approach)

Butts and his coworkers (Butts *et al.*, 1970, 1973, 1974, 1975, 1981; Butts, 1974; Butts and Shackleford, 1992) of the Illinois State Water Survey expanded Velz's oxygen balance equation. For the pragmatic approach to evaluate the BOD/DO relationship in a flowing stream, it is formulated as follows:

$$DO_n = DO_a - DO_u + DO_r + DO_x \quad (1.116)$$

where $DO_n$ = net dissolved oxygen at the end of a stream reach
$DO_a$ = initial dissolved oxygen at the beginning of a reach
$DO_u$ = dissolved oxygen consumed biologically within a reach
$DO_r$ = dissolved oxygen derived from natural reaeration within a reach
$DO_x$ = dissolved oxygen derived from channel dams, tributaries, etc.

Details of computation procedures for each component of the above equation have been outlined in several reports mentioned above. Basically the above equation uses the same basic concepts as employed in the application of the Streeter–Phelps expression and expands the Velz oxygen balance equation by adding $DO_x$. The influences of dams and tributaries are discussed below.

### 17.1 Influence of a dam

Low head channel dams and large navigation dams are very important factors when assessing the oxygen balance in a stream. There are certain disadvantages from a water quality standpoint associated with such structures but their reaeration potential is significant during overflow, particularly if a low DO concentration exists in the upstream pooled waters. This source of DO replenishment must be considered.

Procedures for estimating reaeration at channel dams and weirs have been developed by researchers in England (Gameson, 1957; Gameson *et al.*, 1958; Barrett *et al.*, 1960; Grindrod, 1962). The methods are easily applied and give satisfactory results. Little has been done in developing similar procedures for large navigation and power dams. Preul and Holler (1969) investigated larger structures.

Often in small sluggish streams, more oxygen may be absorbed by water overflowing a channel dam than in a long reach between dams. However, if the same reach were free flowing, this might not be the case; i.e. if the dams were absent, the reaeration in the same stretch of river could conceivably be greater than that provided by overflow at a dam. If water is saturated with oxygen, no uptake occurs at the dam overflow. If it is supersaturated, oxygen will be lost during dam overflow. Water, at a given percent deficit, will gain oxygen.

The basic channel dam reaeration formula takes the general form

$$r = 1 + 0.11qb(1 + 0.46T)h \tag{1.117}$$

where $r$ = dissolved oxygen deficit ratio at temperature $T$
$q$ = water quality correction factor
$b$ = weir correction factor
$T$ = water temperature, °C
$h$ = height through which the water falls, ft

The deficit dissolved oxygen ratio is defined by the expression

$$r = (C_s - C_A)/(C_s - C_B) = D_A/D_B \tag{1.118}$$

where $C_A$ = dissolved oxygen concentration upstream of the dam, mg/L
$C_B$ = dissolved oxygen concentration downstream of the dam, mg/L
$C_s$ = dissolved oxygen saturation concentration, mg/L
$D_A$ = dissolved oxygen deficit upstream of the dam, mg/L
$D_B$ = dissolved oxygen deficit downstream of the dam, mg/L

Although Eqs. (1.117) and (1.118) are rather simplistic and do not include all potential parameters which could affect the reaeration of water overflowing a channel dam, they have been found to be quite reliable in predicting the change in oxygen content of water passing over a dam or weir (Barrett *et al.*, 1960). The degree of accuracy in using the equations is dependent upon the estimate of factors $q$ and $b$.

For assigning values for $q$, three generalized classifications of water have been developed from field observations. They are $q = 1.25$ for clean or slightly polluted water; $q = 1.0$ for moderately polluted water; and $q = 0.8$ for grossly polluted water. A slightly polluted water is one in which no noticeable deterioration of water quality exists from sewage discharges; a moderately polluted stream is one which receives a significant quantity of sewage effluent; and a grossly polluted stream is one in which noxious conditions exist.

For estimating the value of $b$, the geometrical shape of the dam is taken into consideration. This factor is a function of the ratio of weir coefficients $W$ of various geometrical designs to that of a free weir where

$$W = (r - 1)/h \tag{1.119}$$

Weir coefficients have been established for a number of spillway types, and Gameson (1957) in his original work has suggested assigning $b$ values as follows:

| *Spillway type* | $b$ |
|---|---|
| Free | 1.0 |
| Step | 1.3 |
| Slope (ogee) | 0.58 |
| Slopping channel | 0.17 |

For special situations, engineering judgment is required. For example, a number of channel dams in Illinois are fitted with flashboards during the summer. In effect, this creates a free fall in combination with some other configuration. A value of $b$, say 0.75, could be used for a flashboard installation on top of an ogee spillway (Butts *et al.*, 1973). This combination would certainly justify a value less than 1.0, that for a free weir, because the energy dissipation of the water flowing over the flashboards onto the curved ogee surface would not be as great as that for a flat surface such as usually exists below free and step weirs.

Aeration at a spillway takes place in three phases: (1) during the fall; (2) at the apron from splashing; and (3) from the diffusion of oxygen due to entrained air bubbles. Gameson (1957) has found that little aeration occurs in the fall. Most DO uptake occurs at the apron; consequently, the greater the energy dissipation at the apron, the greater the aeration. Therefore, if an ogee spillway is designed with an energy dissipater, such as a hydraulic jump, the value of $b$ should increase accordingly.

**Example:** Given $C_A = 3.85$ mg/L, $T = 24.8°C$, $h = 6.6$ ft (1.0 m), $q = 0.9$, and $b = 0.30$ (because of hydraulic jump use 0.30).

Find DO concentration below the dam $C_B$.

**solution:**

Step 1. Using dam reaeration equation to find $r$

$$\begin{aligned} r &= 1 + 0.11qb(1 + 0.046T)h \\ &= 1 + 0.11 \times 0.9 \times 0.30(1 + 0.046 \times 24.8) \times 6.6 \\ &= 1.42 \end{aligned}$$

Step 2. Determine DO saturation value

From Table 1.2, or by calculation, for $C_s$ at 24.8°C this is

$$C_s = 8.21 \text{ mg/L}$$

Step 3. Compute $C_B$

Since

$$r = (C_s - C_A)/(C_s - C_B)$$

therefore

$$\begin{aligned} C_B &= C_s - (C_s - C_A)/r \\ &= 8.21 - (8.21 - 3.85)/1.42 \\ &= 8.21 - 3.07 \\ &= 5.14 \text{ (mg/L)} \end{aligned}$$

### 17.2 Influence of tributaries

Tributary sources of DO are often an important contribution in deriving a DO balance in stream waters. These sources may be tributary streams or outfalls of wastewater treatment plants. A tributary contribution of $DO_x$ can be computed based on mass balance basis of DO, ammonia, and BOD values from tributaries. The downstream effect of any DO input is determined by mass balance computations: in terms of pounds per day the tributary load can simply add to the mainstream load occurring above the confluence. The following two examples are used to demonstrate the influence of tributary sources of DO. Example 1 involves a tributary stream. Example 2 involves the design of an outfall structure to achieve a minimum DO at the point of discharge.

**Example 1:** Given: Tributary flow $Q_1 = 123$ cfs
Tributary DO $= 6.7$ mg/L
Mainstream flow $Q_2 = 448$ cfs
Mainstream DO $= 5.2$ mg/L

Find: DO concentration and DO load at the confluence.

**solution:**

Step 1. Determine DO concentration

Assuming there is a complete mix immediately below the confluence, the DO concentration downstream on the confluence can be determined by mass balance.

$$\text{DO} = \frac{Q_1 \times \text{DO}_1 + Q_2 \times \text{DO}_2}{Q_1 + Q_2}$$

$$= (123 \times 6.7 + 448 \times 5.2)/(123 + 448)$$

$$= 5.5 \text{ (mg/L)}$$

Step 2. Compute DO loadings

Since

$$\text{DO load} = (\text{DO, mg/L}) \times (Q, \text{cfs})$$

$$= \text{DO} \times Q\left(\frac{\text{mg}}{\text{L}} \times \text{cfs} \times 28.32 \frac{\text{L/s}}{\text{cfs}}\right)$$

$$= 28.32 \times \text{DO} \times Q\left(\frac{\text{mg}}{\text{s}} \times \frac{1 \text{ lb}}{454{,}000 \text{ mg}} \times \frac{86{,}400 \text{ s}}{\text{d}}\right)$$

$$= 5.39 \times \text{DO} \times Q \text{ lb/d}$$

$$\text{or} = 5.39 \text{ lb/(d} \cdot \text{mg/L} \cdot \text{cfs)} \times \text{DO mg/L} \times \text{Q cfs}$$

$$\text{Tributary load} = 5.39 \times 6.7 \times 123$$

$$= 4442 \text{ lb/d}$$

*Note*: The factor 5.39 is the same as that in Eq. (1.91).

Similarly,

$$\text{Mainstream load} = 5.39 \times 5.2 \times 448$$

$$= 12{,}557 \text{ (lb/d)}$$

$$\text{Total DO load} = 4442 \text{ lb/d} + 12{,}557 \text{ lb/d}$$

$$= 16{,}999 \text{ (lb/d)}$$

**Example 2:** The difference in elevation between the outfall crest and the 7-day, 10-year low flow is 4.0 ft (1.22 m). Utilizing this head difference, determine if a free-falling two-step weir at the outfall will insure a minimum DO of 5.0 mg/L at the point of discharge. The flow of wastewater effluent and the stream are, respectively, 6.19 and 14.47 cfs (175 and 410 L/s, or 4 and 10 MGD). DO for the effluent and stream are 1.85 and 6.40 mg/L, respectively. The temperature of the effluent is 18.9°C. The stream is less than moderately polluted. Determine the stream DO at the point of effluent discharge and the oxygen balance.

**solution:**

Step 1. The water quality correction factor is selected as $q = 1.1$

$$b = 1.3 \text{ for free-fall step-weir}$$

$$C_s = 9.23 \text{ mg/L, when } T = 18.9°\text{C from Table 1.2}$$

$$C_A = 1.85 \text{ mg/L as effluent DO}$$

$$h = 4.0 \text{ ft}$$

Step 2. Determine $r$ with the channel dam reaeration model

$$\begin{aligned} r &= 1 + 0.11qb(1 + 0.046T)h \\ &= 1 + 0.11 \times 1.1 \times 1.3(1 + 0.046 \times 18.9) \times 4 \\ &= 2.18 \end{aligned}$$

Step 3. Compute $C_B$

$$\begin{aligned} r &= (C_s - C_A)/(C_s - C_B) \\ C_B &= C_s - (C_s - C_A)/r \\ &= 9.23 - (9.23 - 1.85)/2.18 \\ &= 5.85 \text{ (mg/L)} \end{aligned}$$

*Note*: $C_B$ is the effluent DO as it reaches the stream.

Step 4. Compute the resultant DO in the stream

Assuming there is a complete mixing with the stream water, the resultant DO at the point of discharge is

$$\begin{aligned} \text{DO} &= \frac{5.85 \times 6.19 + 6.40 \times 14.47}{6.19 + 14.47} \\ &= 6.23 \text{ (mg/L)} \end{aligned}$$

Step 5. Calculate oxygen mass balance

$$\begin{aligned} \text{DO load from effluent} &= 5.39 \times \text{DO} \times Q \\ &= 5.39 \times 5.85 \times 6.19 \\ &= 195 \text{ (lb/d)} \end{aligned}$$

DO load in stream above the point of effluent discharge

$$\begin{aligned} &= 5.39 \times \text{DO} \times Q \\ &= 5.39 \times 6.40 \times 14.47 \\ &= 499 \text{ (lb/d)} \end{aligned}$$

The total amount of oxygen below the point of discharge is 694 (195 + 499) lb/d.

### 17.3 DO used

The term $DO_{used}$ ($DO_u$) in Eq. (1.111) represents the oxygen consumed biologically within a stream reach. This term can be determined by three methods (Butts *et al.*, 1973): (1) observed DO concentrations in conjunction with reaeration estimate; (2) bottle BOD and deoxygenation rate determinations of river water samples; and (3) long-term bottle BOD progression evaluation of a wastewater effluent. The method used will probably be dictated by the existing data or the resources available for collecting usable data.

The term $DO_u$ includes DO usage due to carbonaceous and nitrogenous BOD and to SOD, as stated previously. The ratio of DO contribution by algal photosynthesis to DO consumption by algal respiration is assumed to be unity, although it can handle values greater or less than one when derived on a diurnal basis. For a series of stream reaches, each incremental $DO_u$ value is added and the accumulated sums, with the corresponding time of travel in the stream, are fitted to the first-order exponential expression

$$y = L_a(1 - e^{-K_d t}) \tag{1.120}$$

where $y$ = oxygen demand exerted ($DO_u$)
$L_a$ = ultimate oxygen demand, including carbonaceous and nitrogenous
$K_d$ = in-stream deoxygenation coefficient to the base $e$, per day
$t$ = time of travel, days

*Note*: The coefficient $K_d$ is comparable to the composite of the terms $K_C$ and $K_N$ previously defined in the discussion regarding the conceptual approach to waste-assimilative analysis.

The use of field DO values in estimating the waste-degradation characteristics of a stream has certain advantages: the need for laboratory BOD tests is eliminated which saves time and cost; the reliability of the results should be better since the measurement of DO is far more precise and accurate than the BOD test; also, stream measured DO concentrations take into account the in situ oxygen demand in the stream, which includes both dissolved and benthic demand, whereas laboratory BOD results generally reflect only the oxygen demand exerted by dissolved matter.

**Estimations of $K_d$ and $L_a$.** The data obtained from steam survey, a summation of $DO_u$ values and corresponding time of travel $t$ at an observed water temperature, have to be computed along the reach of a stream. Nowadays, for determinations of $K_d$ and $L_a$, one can simplify the calculation by using a computer or a programmable calculator. However, the following example illustrates the use of the Thomas slope method for determining $K_d$ and $L_a$.

## 17.4 Procedures of pragmatic approach

The steps of the pragmatic approach involved in estimating the waste-assimilative capacity of a stream based upon a pragmatic approach can be summarized as follows (Butts *et al.*, 1973):

1. Develop a full understanding of the stream length, its channel geometry, water stage and flow patterns, and the general hydrologic features of the watershed.
2. Determine the 7-day, 10-year flow of the stream and select a design water temperature.
3. Define the location of all dams and their physical features; define also the location of all tributary flows and relevant data regarding them.
4. Divide the stream into reaches consistent with significant changes in cross sections and determine the volumes and average depth in each reach.
5. At the beginning and end of each reach, during low-flow conditions and summer temperatures, undertake a series of field determinations for at least water temperature and DO concentrations and, if desired, collect water samples for BOD determinations.
6. Compute the time of travel within each reach at stream flows observed during the time of sampling as well as that during 7-day, 10-year low flow.
7. From the observed DO values, flow, and time of travel, compute $DO_a - DO_n$, as demonstrated in Table 1.17 (see later).
8. Select DO saturation values from Table 1.2 for observed stream temperature conditions, and compute the natural reaeration for each reach using Figs. 1.12 and 1.13 in conjunction with appropriate equations for finding the mix time $M$ and the percent absorption at 100% deficit $R_0$. They are determined as the same manner of the example in Section 16.2. Keep in mind the need to make adjustments in accordance with weir and mass balance formulas where dams and tributaries are encountered.
9. Calculate, by summation, the $DO_u$ for each reach as demonstrated in Table 1.18 (see later).
10. From an array of the $DO_u$ versus $t$ data, determine $L_a$ and $K_d$, preferably by the methods of Read–Theriault, steepest descent, or least squares. For a graphical solution, the Thomas slope method is satisfactory. Adjust the values for $L_a$ and $K_d$ for the selected design water temperature by the use of Eqs. (1.26) and (1.28).

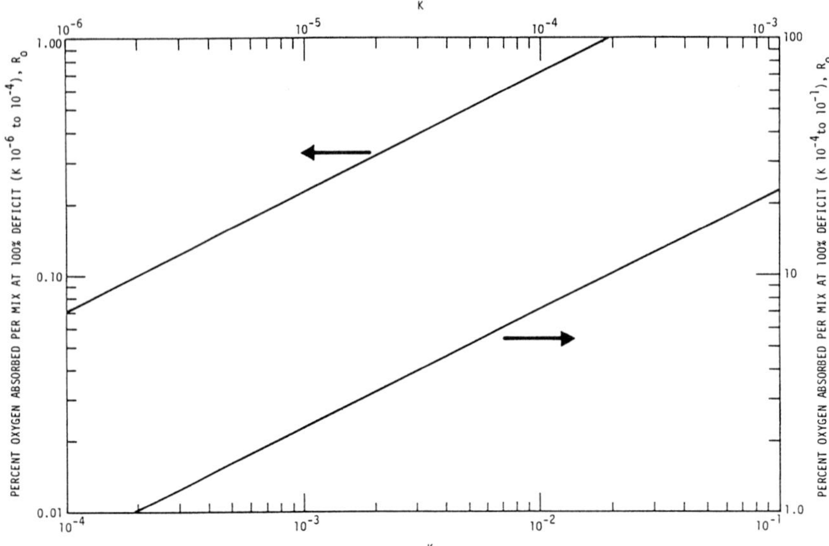


**Figure 1.13** Modified Velz curve (*Butts et al., 1973*).

11. Apply the removal efficiency anticipated to the computed ultimate oxygen demand $L_a$ and develop the required expression $DO_u = L_a(1 - e^{-Kt})$.
12. From the values developed in step 6 for time of travel and depth at 7-day, 10-year low flow, use the observed DO just upstream of the discharge point ($D_a$ at the beginning of the reach) as a starting-point, and follow the computation.
13. Note whether or not the removal efficiency selected will permit predicted DO concentrations in the stream compatible with water quality standards.

**Example:** A long reach of a moderately sized river is investigated. There is an overloaded secondary wastewater treatment plant discharging its effluent immediately above sampling station 1 (upstream). The plant, correctly, has BOD removal efficiency of 65%. A total of 12 sampling stations are established in reaches of the study river. Downstream stations 3 and 4 are above and below an ogee channel dam, respectively. Stations 10 and 11 are immediately above and below the confluence of the main stream and a tributary which has relatively clean water. The 7-day, 10-year low flow of the main receiving stream and the tributary are 660 and 96 cfs, respectively. The design water temperature is selected as 27°C.

A stream field survey for flow, water temperature, DO, average depth, and time of travel is conducted during summer low-flow conditions. During the field survey, the receiving stream in the vicinity of the effluent discharge was 822 cfs and the tributary flow was 111 cfs. The DO values are 5.96 and 6.2 mg/L, respectively. Data obtained from field measurements and subsequent computations for DO values are presented in Tables 1.16, 1.17, and 1.18. The DO levels in the stream are not less than 5.0 mg/L, as required.

*Question*: Determine the DO concentrations along the reach of the stream at design temperature and design flow conditions, if the efficiency of sewage treatment plant is updated to at least 90% BOD removal.

*Answer*: The solution is approached in two sessions: (1) the deoxygenation rate coefficients $K_d$ and $L_a$ should be determined under the existing conditions; this involves steps 1 to 5; (2) a predictive profile of DO concentrations will be calculated under 7-day, 10-year low flow and 27°C design conditions (steps 6 and 7).

**solution:**

Step 1. Compute DO values

In Table 1.16, data are obtained from fieldwork and subsequent DO computations for other purposes. The term $DO_a - DO_n$ (lb/d) will be used in the equation for Table 1.18. A plot of DO measured and time of travel $t$ can be made (not shown). It suggests that a profound DO sag exists in the stream below the major pollution source of effluent discharge.

Step 2. Compute the natural reaeration, $DO_r$

In Table 1.17, values of saturated DO ($DO_s$) at the specified water temperature are obtained from Table 1.2. The values of $M$ and $R_0$ are taken from Figs. 1.12 and 1.13 on the basis of the physical dimension of the stream listed in Table 1.16 or calculated from Eq. (1.114) ($M = 13.94 \text{ In } H - 745$).

Step 3. Calculate DO used, $DO_u$

The $DO_u$ is calculated from equation $DO_u = (DO_a - DO_n) + DO_r + DO_x$. Without consideration of $DO_x$, the values of $DO_u$ for each reach of the stream are given in Table 1.18. It should be noted that $DO_x$ does not apply yet.

Step 4. Determine $K_{d(23)}$ and $L_{d(23)}$ for $T = 23°C$ (during the field survey)

Using the Thomas (1950) slope method, the factor $(\Sigma t/\Sigma DO_u)^{1/3}$ is plotted on arithmetic graph paper against $\Sigma t$ as shown in Fig. 1.14. A line of best fit is then drawn, often neglecting the first two points. The ordinate intercept $b$ is read as 0.0338; and the slope of the line $S$ is computed as 0.00176 [i.e. (0.0426 − 0.0338)/5]. The determine $K_d$ and $L_a$ at 23°C.

$$K_{d(23)} = 6S/b = 6 \times 0.00176/0.0338 = 0.31 \text{ (per day)}$$

$$L_{a(23)} = 1/(K_d \cdot b^3) = 1/(0.31 \times 0.0338^3)$$

$$= 83{,}500 \text{ (lb/d)}$$

**TABLE 1.16 Field Data and DO Computations**

| Station | Flow, cfs at station | Flow, cfs mean | $DO_a$ observed† mg/L | $DO_a$ observed† lb/d* | $DO_a - DO_n$, lb/d | DO averaged mg/L | DO averaged lb/d |
|---|---|---|---|---|---|---|---|
| 1 | 822 | | 5.96 | 26406 | | | |
| 2 | 830 | 826 | 3.44 | 15390 | 11016 | 4.70 | 20925 |
| 3 | 836 | 833 | 2.40 | 10814 | 4575 | 2.92 | 13110 |
| Dam | | | | | | | |
| 4 | 836 | | 3.96 | 17844 | | | |
| 5 | 844 | 840 | 3.90 | 17741 | 103 | 3.93 | 17793 |
| 6 | 855 | 849 | 3.70 | 17051 | 690 | 3.80 | 17389 |
| 7 | 871 | 863 | 3.60 | 16900 | 151 | 3.65 | 16978 |
| 8 | 880 | 875 | 3.90 | 18498 | −1598 | 3.75 | 17686 |
| 9 | 880 | 880 | 4.04 | 19163 | −665 | 3.97 | 18831 |
| 10 | 888 | 884 | 4.44 | 21251 | −2288 | 4.27 | 20202 |
| Tributary | | | | | | | |
| 11 | 999 | | 5.18 | 27892 | −6641 | 4.81 | 24474 |
| 12 | 1010 | 1005 | 5.28 | 28743 | −851 | 5.23 | 28330 |

*DO (lb/d) = 5.39 × flow (cfs) × DO (mg/L).
†$DO_a$ at the end of a reach is the $DO_n$ of that reach.

The curve is expressed as

$$y = 83{,}500(1 - e^{-0.31t})$$

Step 5. Compute $K_{d(27)}$ and $L_{a(27)}$ for design temperature 27°C

Using Eq. (1.26)

$$K_{d(27)} = K_{d(20)} \times 1.047^{27-20}$$

$$K_{d(23)} = K_{d(20)} \times 1.047^{23-20}$$

$$K_{d(27)} = K_{d(23)} \times 1.047^7/1.047^3$$

$$= 0.31 \times 1.3792/1.1477$$

$$= 0.37 \text{ (per day)}$$

Similarly, using Eq. (1.28)

$$L_{a(T)} = L_{a(20)} \times (0.6 + 0.02T)$$

$$L_{a(27)} = L_{a(23)} \times (0.6 + 0.02 \times 27)/(0.6 + 0.02 \times 23)$$

$$= 83{,}500 \times 1.14/1.06$$

$$= 89{,}800 \text{ (lb/d)}$$

Next step goes to design phase.

**TABLE 1.17 Computations of Natural Reaeration**

| Station | Temp, °C | $DO_s$ mg/L | Mean $D_s$, mg/L | Mean $D^{\dagger}_a$, mg/L | Depth h, ft | $M$, min | $t'$, min | $1 - \frac{DO_a}{DO_s}$ | $\frac{R_0}{100}$ | $\frac{t'}{M}$ | $5.39Q \times DO_s$ lb/d | $DO_r$,* lb/d |
|---|---|---|---|---|---|---|---|---|---|---|---|---|
| 1 | 22.8 | 8.53 | | | | | | | | | | |
| 2 | 22.9 | 8.52 | 8.53 | 4.70 | 6.6 | 18.8 | 846.7 | 0.449 | 0.0042 | 45.0 | 37977 | 3225 |
| 3 | 23.0 | 8.50 | 8.51 | 2.92 | 6.8 | 19.4 | 1201.0 | 0.657 | 0.0041 | 61.9 | 38209 | 6370 |
| 4 | 23.0 | 8.50 | | | | | | | | | | |
| 5 | 23.0 | 8.50 | 8.50 | 3.93 | 2.9 | 4.4 | 396.0 | 0.538 | 0.0046 | 90.0 | 38485 | 8566 |
| 6 | 23.1 | 8.48 | 8.49 | 3.80 | 3.8 | 11.2 | 792.0 | 0.552 | 0.0052 | 70.7 | 38851 | 7892 |
| 7 | 23.1 | 8.48 | 8.48 | 3.65 | 5.2 | 15.5 | 1591.2 | 0.57 | 0.0049 | 102.7 | 39445 | 11301 |
| 8 | 23.2 | 8.46 | 8.47 | 3.75 | 6.1 | 17.6 | 1630.0 | 0.557 | 0.0045 | 92.6 | 39946 | 9277 |
| 9 | 23.2 | 8.46 | 8.46 | 3.97 | 6.3 | 18.2 | 1118.9 | 0.531 | 0.0044 | 61.5 | 40127 | 5761 |
| 10 | 23.3 | 8.45 | 8.46 | 4.24 | 6.5 | 18.6 | 911.5 | 0.499 | 0.0043 | 49.0 | 40310 | 4237 |
| 11 | 23.1 | 8.48 | 8.47 | 4.81 | | | | 0.432 | | | | |
| 12 | 23.1 | 8.48 | 8.48 | 5.23 | 7.2 | 19.9 | 656.6 | 0.383 | 0.0040 | 33.0 | 45936 | 2324 |

*$DO_r = \left(1 - \frac{DO_a}{DO_s}\right)\left(\frac{R_0}{100}\right)\left(\frac{t'}{M}\right)\left(5.39Q \times DO_s\right)$ and it is calculated as the same manner of Example in Section 16.2.

†Mean $D_a$ is from Table 1.16.

**TABLE 1.18 Calculation of DO Used**

| Station | $t$, days | $\Sigma t$, days | $DO_a - DO_n$, lb/d | $DO_r$, lb/d | $DO_u$,* lb/d | $\Sigma DO_u$, lb/d | $(\Sigma t/\Sigma DO_u)^{1/3}$ |
|---|---|---|---|---|---|---|---|
| 1 | | | | | | | |
| 2 | 0.588 | 0.588 | 11016 | 3225 | 14241 | 14241 | 0.03456 |
| 3 | 0.834 | 1.422 | 4575 | 6370 | 10945 | 25186 | 0.03836 |
| 4 | | | | | | 25186 | |
| 5 | 0.275 | 1.697 | 103 | 8566 | 8669 | 33855 | 0.03687 |
| 6 | 0.550 | 2.247 | 690 | 7892 | 8582 | 42437 | 0.03755 |
| 7 | 1.105 | 3.352 | 151 | 11301 | 11452 | 53889 | 0.03962 |
| 8 | 1.132 | 4.484 | −1598 | 9277 | 7679 | 61568 | 0.04176 |
| 9 | 0.777 | 5.261 | −665 | 5761 | 5096 | 66664 | 0.04289 |
| 10 | 0.633 | 5.894 | −2888 | 4237 | 1349 | 68013 | 0.04425 |
| 11 | | | | | | 68013 | |
| 12 | 0.456 | 6.350 | −851 | 2324 | 1473 | 69486 | 0.04504 |

*$DO_u = DO_a - DO_n + DO_r$

Step 6. Calculate the design BOD loading after plant improvement

Since the current overloaded wastewater treatment plant only removes 65% of the incoming BOD load, the raw wastewater BOD load is

$$89{,}800 \text{ lb/d}/(1 - 0.65) = 256{,}600(\text{lb/d})$$

The expanded (updated) activated-sludge process may be expected to remove 90% to 95% of the BOD load from the raw wastewater. For the safe side, 90% removal is selected for the design purpose. Therefore, the design load $L_a$ at 27°C to the stream will be 25,660 lb/d (256,600 × 0.1).

Step 7. Gather new input data for DO profile

In order to develop the predictive profile for DO concentration in the receiving stream after the expanded secondary wastewater treatment plant is functioning, the following design factors are obtained from field survey data and the results computed:

| | | | |
|---|---|---|---|
| Stream flow | = 660 cfs | At the dam | $q = 1.1$ |
| Tributary flow | = 96 cfs | | $b = 0.58$ |
| Tributary DO | = 6.2 mg/L | | $h = 6.1$ ft |
| DO at station 1 | = 5.90 mg/L | | $K_{d(27)} = 0.37$ per day |
| Water temperature | = 27°C | | $L_{a(27)} = 25{,}660$ lb/d |

$H$ and $t$ are available from cross-sectional data.

Step 8. Perform DO profile computations

Computations are essentially the same as previous steps with some minor modifications (Table 1.19).

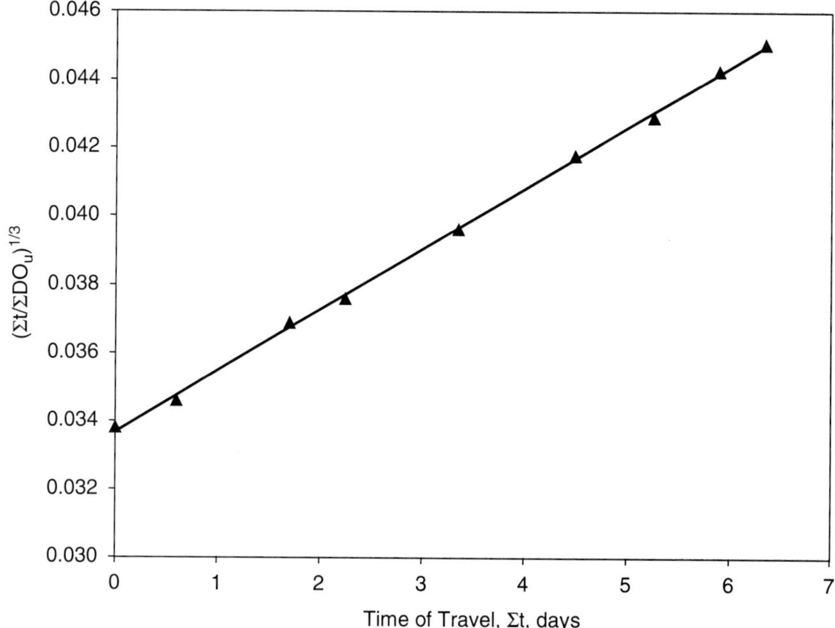


**Figure 1.14** Plot of $(\Sigma t/\Sigma DO_u)^{1/3}$ versus $\Sigma t$.

**TABLE 1.19 DO Profile Computation at 27°C and a 7-Day, 10-Year Low Flow**

| (1) | (2) Flow, cfs | (3) | (4) | (5) | (6) | (7) Total | (8) Subreach |
|---|---|---|---|---|---|---|---|
| Station | at station | mean | $t$, days | $\Sigma t$, days | $1 - e^{-K_d \Sigma t}$ | $DO_u$, lb/d | $DO_u$, lb/d |
| 1 | 660 | | | | | 0 | |
| 2 | 666 | 663 | 0.750 | 0.750 | 0.242 | 6218 | 6218 |
| 3 | 670 | 668 | 1.076 | 1.826 | 0.491 | 12603 | 6385 |
| Dam | | | | | | | |
| 4 | 670 | | | | | | |
| 5 | 674 | 672 | 0.345 | 2.171 | 0.551 | 14168 | 1565 |
| 6 | 682 | 678 | 0.626 | 2.797 | 0.645 | 16544 | 2376 |
| 7 | 696 | 689 | 1.313 | 4.110 | 0.781 | 20052 | 3508 |
| 8 | 702 | 699 | 1.370 | 5.48 | 0.868 | 22282 | 2230 |
| 9 | 704 | 703 | 0.938 | 6.418 | 0.907 | 23272 | 990 |
| 10 | 708 | 706 | 0.845 | 7.263 | 0.932 | 23913 | 941 |
| Tributary | 96 | | | | | | |
| 11 | 804 | | | | | | |
| 12 | 810 | 806 | 0.608 | 7.871 | 0.946 | 24265 | 351 |

**TABLE 1.19 (contd.)**

| Station | (9) $DO_a - DO_u$, lb/d | (10) Mean $DO_a$, lb/d | (11) $DO_s$ mg/L | (12) $DO_s$ lb/d | (13) $1 - \frac{DO_a}{DO_s}$ | (14) Mean depth $H$, ft | (15) $M$, min |
|---|---|---|---|---|---|---|---|
| 1 | | | | 21084 | | | |
| 2 | 14866 | 17975 | 7.87 | 28124 | 0.361 | 5.9 | 17.4 |
| 3 | 11977 | 15170 | 7.87 | 28336 | 0.465 | 6.1 | 17.5 |
| Dam | | | | | | | |
| 4 | | | | | | | |
| 5 | 21584 | 22366 | 7.87 | 28506 | 0.215 | 2.5 | 4.1 |
| 6 | 23895 | 25083 | 7.87 | 28760 | 0.128 | 3.4 | 9.6 |
| 7 | 22908 | 24662 | 7.87 | 29227 | 0.165 | 4.6 | 13.8 |
| 8 | 24556 | 25671 | 7.87 | 29651 | 0.134 | 5.5 | 16.2 |
| 9 | 26367 | 26862 | 7.87 | 29821 | 0.099 | 5.7 | 16.6 |
| 10 | 27093 | 27404 | 7.87 | 29948 | 0.085 | 6.0 | 17.4 |
| Tributary | | | | | | | |
| 11 | | | | | | | |
| 12 | 30924 | 31100 | 7.87 | 34190 | 0.090 | 6.5 | 18.6 |

**TABLE 1.19 (concluded)**

| Station | (16) $t' = 1440t$, min | (17) $t'/M$ | (18) $R_0/100$ | (19) $DO_r$, lb/d | (20) $DO_n$ (or $DO_a$) lb/d | (21) | Remark |
|---|---|---|---|---|---|---|---|
| 1 | | | | | 21084 | 5.90 | |
| 2 | 1080 | 62.1 | 0.00555 | 3696 | 18362 | 5.14 | |
| 3 | 1549 | 88.5 | 0.00545 | 6353 | 18330 | 5.09 | Above dam |
| Dam | | | | | | | |
| 4 | | | | | 23148 | 6.41 | Below dam |
| 5 | 497 | 121.2 | 0.00630 | 4687 | 26271 | 7.25 | |
| 6 | 901 | 93.9 | 0.00730 | 2521 | 26416 | 7.23 | |
| 7 | 1891 | 137.0 | 0.00620 | 3878 | 26786 | 7.21 | |
| 8 | 1973 | 121.8 | 0.00578 | 2802 | 27357 | 7.26 | |
| 9 | 1351 | 81.4 | 0.00568 | 1367 | 27734 | 7.32 | |
| 10 | 1217 | 69.9 | 0.00550 | 975 | 28067 | 7.38 | Above trib. |
| Tributary | | | | | | | |
| 11 | | | | | 31276 | | Below trib. |
| 12 | 876 | 47.1 | 0.00521 | 788 | 31682 | 7.29 | |

*Notes*: Col 7 = $L_a$ × Col 6 = 25,660 × Col 6
Col $8_n$ = Col $7_n$ – Col $7_{n-1}$
Col $9_n$ = Col $20_{n-1}$ –Col $8_n$
Col $10_n$ = (Col $20_{n-1}$ + Col $9_n$)/2
Col 11 = $DO_s$ at $T$ of 27°C
Col 12 = 5.39 × Col 3 × Col 11
Col 15 = Calculated or from monogram Fig. 1.12
Col 18 = Calculated or from monogram Fig. 1.12
Col 19 = Col 12 × Col 13 × Col 17 × Col 18
Col 20 = initial value is 663 × 5.90 × 5.39 and thereafter = col 9 + col 19
Col 21 = Col 20/Col 3/5.39
trib. = tributary

For columns 9, 10, 13, 19, 20, and 21, calculate iteratively from station by station from upstream downward. The other columns can be calculated independently. At station 4 (below the dam), the dam aeration ratio is given as Eqs. (1.117) and (1.118)

$$r = 1 + 0.11qb(1 + 0.046T)h$$

$$= 1 + 0.11 \times 1.1 \times 0.58(1 + 0.046 \times 27) \times 6.1$$

$$= 1.960$$

$$r = (C_s - C_A)/(C_s - C_B)$$

$$1.96 = (7.87 - 5.09)/(7.87 - C_B)$$

$$C_B = -2.78/1.96 + 7.87 = 6.45 \text{ (mg/L)}$$

This DO value of 6.45 mg/L is inserted into column 21 at station 4. The corresponding $DO_n$ (in lb/d) is computed by multiplying $5.39Q$ to obtain 23,148 lb/d, which is inserted into column 20 at station 4.

At station 11, immediately below the tributary, DO contribution is 3208 lb/d (i.e. $5.39 \times 6.2 \times 96$). This value plus 28,068 lb/d (31,276 lb/d) is inserted into column 20. Then computations continue with iterative procedures for station 12.

It should be noted that the parameters $H$, $M$, $t'$, $t'/M$, and $R/100$ can be computed earlier (i.e. in earlier columns). Therefore, the other DO computations will be in the latter part of Table 1.19. It seems easier for observation. Also, these steps of computation can be easily programmed using a computer.

The results of computations for solving equation, $DO_n = DO_a - DO_u + DO_r + DO_x$ are given in Table 1.19, and the DO concentrations in column 21 of Table 1.19 are the predictive DO values for 7-day, 10-year low flow with secondary wastewater treatment. All DO values are above the regulatory standard of 5.0 mg/L, owing to expanded wastewater treatment and dam aeration.

## 18 Biological Factors

### 18.1 Algae

Algae are most commonly used to assess the extent that primary productivity (algal activity) affects the DO resources of surface waters. Many factors affect the distribution, density, and species composition of algae in natural waters. These include the physical characteristics of the water, length of storage, temperature, chemical composition, in situ reproduction and elimination, floods, nutrients, human activities, trace elements, and seasonal cycles.

Of the many methods suggested for defining the structure of a biological community, the most widely used procedure has been the diversity index. A biological diversity index provides a means of evaluating the richness of species within a biological community using a mathematical computation. Although different formulas have been used, the one determined by the information theory formula, the Shannon–Weiner diversity index, is widely used for algal communities or communities of other organisms. The formula is (Shannon and Weaver, 1949):

$$D = -\sum_{i=1}^{m} P_i \log_2 P_i \tag{1.121}$$

where $D$ = diversity index
$P_i = n_i/N$
$n_i$ = the number (density) of the $i$th genera or species
$N$ = total number (density) of all organisms of the sample
$i = 1, 2, \ldots, m$
$m$ = the number of genera or species
$\log_2 P_i = 1.44 \ln P_i$

or

$$D = -1.44\sum_{i-1}^{m} (n_i/N) \ln(n_i/N) \tag{1.122}$$

The index $D$ has a minimum value, when a community consisting solely of one ($m = 1$) species has no diversity or richness and takes on a value of unity. As the number of species increases and as long as each species is relatively equal in number, the diversity index increases numerically. It reaches a maximum value when $m = N$.

**Example:** Seven species of algae are present in a stream water sample. The numbers of seven species per mL are 32, 688, 138, 98, 1320, 424, and 248, respectively. Compute the diversity index of this algal community.

**solution:** Using formula (Eq. (1.122))

$$D = -1.44\sum_{i-1}^{m} (n_i/N) \ln(n_i/N)$$

and construct a table of computations (Table 1.20). Answer: $D = 2.15$

### 18.2 Indicator bacteria

Pathogenic bacteria, pathogenic protozoan cysts, and viruses have been isolated from wastewaters and natural waters. The sources of these pathogens are the feces of humans and of wild and domestic animals. Identification and enumeration of these disease-causing organisms in water and wastewater are not recommended because no single technique

**TABLE 1.20 Values for Eq. (2.122)**

| $i$ | $n_i$ | $n_i/N$ | $-1.44 \ln (n_i/N)$ | $-1.44(n_i/N) \ln (n_i/N)$ |
|---|---|---|---|---|
| 1 | 32 | 0.0109 | 6.513 | 0.071 |
| 2 | 688 | 0.2334 | 2.095 | 0.489 |
| 3 | 138 | 0.0468 | 4.409 | 0.206 |
| 4 | 98 | 0.0332 | 4.902 | 0.163 |
| 5 | 1320 | 0.4478 | 1.157 | 0.518 |
| 6 | 424 | 0.1438 | 2.792 | 0.402 |
| 7 | 248 | 0.0841 | 3.565 | 0.300 |
| Σ | 2948($N$) | 1.0000 | | 2.149 |

is currently available to isolate and identify all the pathogens. In fact, concentrations of these pathogens are generally low in water and wastewater. In addition, the methods for identification and enumeration of pathogens are labor-intensive and expensive.

Instead of direct isolation and enumeration of pathogens, total coliform (TC) has long been used as an indicator of pathogen contamination of a water that poses a public health risk. Fecal coliform (FC), which is more fecal-specific, has been adopted as a standard indicator of contamination in natural waters in Illinois, Indiana, and many other states. Both TC and FC are used in standards for drinking-water and natural waters. Fecal streptococcus (FS) is used as a pollution indicator in Europe. FC/FS ratios have been employed for identifying pollution sources in the United States. Fecal streptococci are present in the intestines of warm-blooded animals and of insects, and they are present in the environment (water, soil, and vegetation) for long periods of time. *Escherichia coli* bacteria have also been used as an indicator.

**Calculation of bacterial density.** The determination of indicator bacteria, total coliform, fecal coliform, and fecal streptococcus or enterococcus is very important for assessing the quality of natural waters, drinking-waters, and wastewaters. The procedures for enumeration of these organisms are presented elsewhere (APHA *et al.*, 1995). Examinations of indicator bacterial density in water and in wastewater are generally performed by using a series of four-decimal dilutions per sample, with 3 to 10, usually 5, tubes for each dilution. Various special broths are used for the presumptive, confirmation, and complete tests for each of the bacteria TC, FC, and FS, at specified incubation temperatures and periods.

**MPN method.** Caliform density is estimated in terms of the most probable number (MPN). The multiple-tube fermentation procedure is often called an MPN procedure. The MPN values for a variety of inoculation series are listed in Table 1.21. These values are based on a series of five tubes for three dilutions. MPN values are commonly determined using

**TABLE 1.21 MPN Index for Various Combinations of Positive Results When Five Tubes Are Used per Dilution (10, 1.0, 0.1 mL)**

| Combination of positives | MPN index 100 mL | Combination of positives | MPN index 100 mL |
|---|---|---|---|
| | | 4-2-0 | 22 |
| 0-0-0 | <2 | 4-2-1 | 26 |
| 0-0-1 | 2 | 4-3-0 | 27 |
| 0-1-0 | 2 | 4-3-1 | 33 |
| 0-2-0 | 4 | 4-4-0 | 34 |
| | | 5-0-0 | 23 |
| 1-0-0 | 2 | 5-0-1 | 30 |
| 1-0-1 | 4 | 5-0-2 | 40 |
| 1-1-0 | 4 | 5-1-0 | 30 |
| 1-1-1 | 6 | 5-1-1 | 50 |
| 1-2-0 | 6 | 5-1-2 | 60 |
| 2-0-0 | 4 | 5-2-0 | 50 |
| 2-0-1 | 7 | 5-2-1 | 70 |
| 2-1-0 | 7 | 5-2-2 | 90 |
| 2-1-1 | 9 | 5-3-0 | 80 |
| 2-2-0 | 9 | 5-3-1 | 110 |
| 2-3-0 | 12 | 5-3-2 | 140 |
| 3-0-0 | 8 | 5-3-3 | 170 |
| 3-0-1 | 11 | 5-4-0 | 130 |
| 3-1-0 | 11 | 5-4-1 | 170 |
| 3-1-1 | 14 | 5-4-2 | 220 |
| 3-2-0 | 14 | 5-4-3 | 280 |
| 3-2-1 | 17 | 5-4-4 | 350 |
| | | 5-5-0 | 240 |
| 4-0-0 | 13 | 5-5-1 | 300 |
| 4-0-1 | 17 | 5-5-2 | 500 |
| 4-1-0 | 17 | 5-5-3 | 900 |
| 4-1-1 | 21 | 5-5-4 | 1600 |
| 4-1-2 | 26 | 5-5-5 | ≥1600 |

SOURCE: APHA, AWWA, and WEF (1995)

Table 1.21 and are expressed as MPN/100 mL. Only two significant figures are used for the reporting purpose. Other methods are based on five 20 mL or ten 10 mL of the water sample used.

Table 1.21 presents MPN index for combinations of positive and negative results when five 10-mL, five 1-mL, and five 0.1-mL volumes of sample are inoculated. When the series of decimal dilution is different from that in the table, select the MPN value from Table 1.21 for the combination of positive tubes and compute the MPN index using the following formula:

$$\frac{\text{MPN}}{100\text{ mL}} = \text{MPN value from table} \times \frac{10}{\text{largest volume tested in dilution series used}} \tag{1.123}$$

For the MPN index for combinations not appearing in Table 1.21 or for other combinations of dilutions or numbers of tubes used, the MPN can be estimated by the Thomas equation:

$$\frac{\text{MPN}}{100\text{ mL}} = \frac{\text{No. of positive tubes} \times 100}{\sqrt{(\text{mL sample in negative tubes})(\text{mL sample in all tubes})}} \tag{1.124}$$

Although the MPN tables and calculations are described for use in the coliform test, they are equally applicable for determining the MPN of any other organisms for which suitable test media are available.

**Example 1:** Estimate the MPN index for the following six samples.

**solution:**

| Sample | Positive/five tubes, mL used | | | | Combination of positive | MPN index/ 100 mL |
|---|---|---|---|---|---|---|
| | 1 | 0.1 | 0.01 | 0.001 | | |
| A-raw | 5/5 | **5/5** | **3/5** | **1/5** | 5-3-1 | 11000 |
| $B \times 10^{-3}$ | 5/5 | **5/5** | **3/5** | **1/5** | 5-3-1 | 11000000 |
| C | **5/5** | **3/5** | **2/5** | 0/5 | 5-3-2 | 1400 |
| D | **5/5** | **3/5** | **1/5** | 1/5 | 5-3-2 | 1400 |
| E | **4/5** | **3/5** | **1/5** | 0/5 | 4-3-1 | 330 |
| F | **0/5** | **1/5** | **0/5** | 0/5 | 0-1-0 | 20 |

Step 1. For sample A

For 5-3-1; from Table 1.21, MPN = 110. Using Eq. (1,123)

$$\text{For sample: MPN} = 110 \times \frac{10}{0.1} = 11{,}000$$

Step 2. For sample B

Same as sample A; however, it is a polluted water and a $10^{-3}$ dilution is used. Thus,

$$\text{MPN} = 11{,}000 \times 10^3 = 11{,}000{,}000$$

Step 3. For sample C

From Table 1.21, MPN = 140 for 5-3-2 combination

$$\text{For sample: MPN} = 140 \times \frac{10}{1} = 1400$$

Step 4. For sample D

Since one is positive at 0.001 mL used, add 1 positive into 0.01 dilution. Therefore, the combination of positive is 5-3-2.

Thus, MPN = 1400

Step 5. For sample E with a 4-3-1 combination

$$\text{MPN} = 33 \times \frac{10}{1} = 330$$

Step 6. For sample F, 0-1-0 positive

$$\text{MPN} = 2 \times \frac{10}{1} = 20$$

*Note*: Unit for MPN is the number per 100 mL sample.

**Example 2:** Calculate the MPN value by the Thomas equation with data given in Example 1.

**solution:**

Step 1. For sample A
Number of positive tubes for the last two dilutions = 3 + 1 = 4
mL sample in negative tubes = 2 × 0.01 + 4 × 0.001 = 0.024
mL sample in all tubes = 5 × 0.01 + 5 × 0.001 = 0.055

With the Thomas equation:

$$\frac{\text{MPN}}{100\text{ mL}} = \frac{\text{No. of positive tubes} \times 100}{\sqrt{(\text{mL sample in negative tubes})(\text{mL sample in all tubes})}}$$

$$= \frac{4 \times 100}{\sqrt{0.024 \times 0.055}}$$

$$= 11{,}000 \text{ (for 10, 1, and 0.1 mL)}$$

Step 2. For sample B

As Step 1, in addition multiply by $10^3$

$$\frac{\text{MPN}}{100\text{ mL}} = 11{,}000 \times 10^3 = 11{,}000{,}000$$

Step 3. For sample C

$$\frac{\text{MPN}}{100\text{ mL}} = \frac{(3 + 2) \times 100}{\sqrt{0.23 \times 0.55}}$$

$$= 1406$$

$$= 1400$$

*Note*: Use only two significant figures for bacterial count.

Step 4. For sample D

The answer is exactly the same as for sample C

$$\frac{\text{MPN}}{100\text{ mL}} = 1400$$

Step 5. For sample E

$$\frac{\text{MPN}}{100\text{ mL}} = \frac{(4 + 3 + 1) \times 100}{\sqrt{1.24 \times 5.55}}$$

$$= 305$$

$$= 310$$

Step 6. For sample F

$$\frac{\text{MPN}}{100\text{ mL}} = \frac{100}{\sqrt{(5.45 \times 5.55)}}$$

$$= 18$$

**Example 3:** The results of positive tubes of 4 tubes with five dilutions (100, 10, 1, 0.1, and 0.01 mL) are 4/4, 3/4, 1/4, 1/4, and 0/4. Estimate the MPN value of the sample.

**solution:**

Step 1. Adjust the combination of positive tubes
It will be 4-3-2 at 100, 10, and 1 mL dilutions.

Step 2. Find MPN by the Thomas formula:

$$\frac{\text{MPN}}{100\text{ mL}} = \frac{(3 + 2) \times 100}{\sqrt{(10 + 2 \times 1) \times 4(10 + 1)}}$$

$$= 22$$

**Membrane filter method.** The membrane filtration (MF) method is the most widely used method. Calculations of MF results for these indicators or other organisms are presented here (Illinois EPA, 1987):

Step 1. Select the membrane filter with the number of colonies in the acceptable range
The acceptable range for TC is 20 to 80 TC colonies and no more than 200 colonies of all types per membrane. Sample quantities producing MF counts of 20 to 60 colonies of FC or FS are desired.

Step 2. Compute count per 100 mL, according to the general formula:

$$\frac{\text{Colonies}}{100\ \text{mL}} = \frac{\text{No. colonies counted} \times 100}{\text{Volume of sample, mL}}$$

**Example 1:** Counts with the acceptable limits:
Assume that filtration of volumes 75, 25, 10, 3, and 1 mL produced FC colony counts of 210, 89, 35, 11, and 5, respectively. What is the FC density for the sample?

**solution:**

The MF with 35 FC colonies is the best MF for counting.
Thus, the FC density is

$$\frac{\text{FC}}{100\ \text{mL}} = \frac{35 \times 100}{10} = 350$$

which will be recorded as 350 FC/100 mL.

It should be noted that an analyst would not count the colonies on all five filters. After inspection of five membrane filters, the analyst would select the MF(s) with 20 to 60 FC colonies and then limit the actual counting to such a membrane. If there are acceptable counts on replicate filters, count colonies on each MF independently and determine final reporting densities. Then compute the arithmetic mean of these densities to obtain the final recorded density.

**Example 2:** For duplicate samples:
A duplicate filtration of 2.0 mL of water sample gives plate counts of 48 and 53 TC colonies. Compute the recorded value.

**solution:**

Step 1. Determine FS densities for two membranes independently, with the general formula:

$$\frac{\text{TC}}{100\ \text{mL}} = \frac{48 \times 100}{2} = 2400$$

and

$$\frac{\text{TC}}{100\text{ mL}} = \frac{53 \times 100}{2} = 2650$$

Step 2. Taking the average

$$\frac{2400 + 2650}{2} = 2525 \text{ [ reporting 2500A]}$$

The FS density for the water sample is reported as 2500 A/100 mL. Code A represents two or more filter counts within acceptable range and same dilution.

*Note*: Use two significant figures for all reported bacteria densities.

If more than one dilution is used, calculate the acceptable range results to final reporting densities separately; then average for final recorded value.

**Example 3:** More than one dilution:
Volumes of 1.00, 0.30, 0.10, 0.03, and 0.01 mL generate TC colony counts of TNTC (too numerous to count), 210, 78, 33, and 6. What is the TC density in the water?

**solution:**

Step 1. In this case, only two volumes, 0.10 and 0.03 mL, produce TC colonies in the acceptable limit. Calculate each MF density, separately,

$$\frac{\text{TC}}{100\text{ mL}} = \frac{78 \times 100}{0.10} = 78{,}000$$

and

$$\frac{\text{TC}}{100\text{ mL}} = \frac{33 \times 100}{0.03} = 110{,}000$$

Step 2. Compute the arithmetic mean of these two densities to obtain the final recorded TC density

$$\frac{\text{TC}}{100\text{ mL}} = \frac{78{,}000 + 110{,}000}{2} = 94{,}000$$

which would be recorded as 94,000 TC per 100 mL C. Code C stands for the calculated value from two or more filter colony counts within acceptable range but different dilution.

If all MF colony counts are below the acceptable range, all counts should be added together. Also, all mL volumes should be added together.

**Example 4:** All results below the desirable range:
Volumes of 25, 10, 4, and 1 mL produced FC colony counts of 18, 10, 3, and 0, respectively. Calculate the FC density for the water sample.

**solution:**

$$\frac{\text{FC}}{100\text{ mL}} = \frac{(18 + 10 + 3 + 0) \times 100}{25 + 10 + 4 + 1}$$

$$= 77.5$$

which would be recorded as 77 FC per 100 mL B, where code B represents results based on colony counts outside the acceptable range.

If colony counts from all membranes are zero, there is no actual calculation possible, even as an estimated result. If there had been one colony on the membrane representing the largest filtration volume, the density is estimated from this filter only.

If all MF counts are above the acceptable limit, calculate the density as shown in Example 4.

**Example 5:** Filtration volumes 40, 15, 6, and 2 mL produced FC colony counts of 1, 0, 0, and 0, respectively. Estimate the FC density for the sample.

**solution:**

$$\frac{\text{FC}}{100\text{ mL}} = \frac{1 \times 100}{40} = 2.5$$

which would be recorded as 2.5 FC per 100 mL K1, where K1 means less than 2.5/100 mL using one filter to estimate the value.

**Example 6:** Membrane filtration volumes of 0.5, 1.0, 3.0, and 10 mL produced FC counts of 62, 106, 245, and TNTC, respectively. Compute the FC density.

**solution:**

$$\frac{\text{FC}}{100\text{ mL}} = \frac{(62 + 106 + 245) \times 100}{0.5 + 1 + 3} = 9180 \text{ (reporting 9200B)}$$

which would be reported as 9200 FC per 100 mL B.

If bacteria colonies on the membrane are too numerous to count, use a count of 200 colonies for the membrane filter with the smallest filtration volume. The number at which a filter is considered TNTC is when the colony count exceeds 200 or when the colonies are too indistinct for accurate counting.

**Example 7:** Given that filtration volumes of 2, 8, and 20 mL produce TC colony counts of 244, TNTC, and TNTC, respectively, what is the TC density in the sample?

**solution:**

Using the smallest filtration volume, 2 mL in this case, and TC for 200 per filter, gives a better representative value. The estimated TC density is

$$\frac{\text{TC}}{100\text{ mL}} = \frac{200 \times 100}{2} = 10{,}000 \text{ L}$$

Code L, for bacterial data sheet description, has to be changed to *greater than*; i.e. all colony counts greater than acceptable range, calculated as when smallest filtration volume had 200 counts.

If some MF colony counts are below and above the acceptable range and there is a TNTC, use the useable counts and disregard any TNTCs and its filtration volume.

**Example 8:** Assume that filtration volumes 0.5, 2, 5, and 15 mL of a sample produced FC counts of 5, 28, 67, and TNTC. What is the FC density?

**solution:**

The counts of 5 and TNTC with their volume are not included for calculation.

$$\frac{\text{FC}}{100\text{ mL}} = \frac{(28 + 67) \times 100}{2 + 5} = 1357$$

$$= [\text{reporting } 1400 \text{ B}]$$

**Bacterial standards.** Most regulatory agencies have stipulated standards based on the bacterial density for waters. For example, in Illinois, the Illinois Department of Public Health (IDPH) has promulgated the indicator bacteria standards for recreational-use waters as follows:

- A beach will be posted "Warning—Swim At Your Own Risk" when bacterial counts exceed 1000 TC per 100 mL or 100 FC per 100 mL.
- A beach will be closed when bacterial densities exceed 5000 TC per 100 mL or 500 FC per 100 mL in water samples collected on two consecutive days.

The Illinois Pollution Control Board (IEPA, 1999) has adopted rules regarding FC limits for general-use water quality standards applicable to lakes and streams. The rules of Section 302.209 are

a. During the months May through October, based on a minimum of five samples taken over not more than a 30-day period, FCs (STORET number 31616) shall not exceed a geometric mean of 200 per100 mL, nor shall more than 10% of the samples during any 30-day period exceed 400 per 100 mL in protected waters. Protected waters are defined as waters that, due to natural characteristics, aesthetic value, or environmental significance, are deserving of protection from pathogenic organisms. Protected waters must meet one or both of the following conditions:

(1) They presently support or have the physical characteristics to support primary contact.
(2) They flow through or adjacent to parks or residential areas.

**Example:** FC densities were determined weekly for the intake water from a river. The results are listed in Table 1.22. Calculate the 30-day moving geometric mean of FC densities and complete the table.

**solution:**

Step 1. Check the minimum sampling requirement: Five samples were collected in less than a 30-day period

Step 2. Calculate geometric means ($M_g$) from each consecutive five samples

$$M_g = \sqrt[5]{42 \times 20 \times 300 \times 50 \times 19} = 47$$

**Bacterial die-off in streams.** There have been many studies on the die-off rate of bacteria in streams. Most of the work suggests that Chick's law, one of the first mathematics formulations to describe die-off curves, remains quite applicable for estimating the survival of pathogens and nonpathogens of special interest in stream sanitation investigations. Chick's law is

$$N = N_0 10^{-kt} \tag{1.125}$$

or

$$\log N/N_0 = -kt \tag{1.126}$$

where $N$ = bacterial density at time $t$, days
$N_0$ = bacterial density at time 0
$k$ = die-off or death rate, per day

**TABLE 1.22 Fecal Coliform Densities**

| | FC density/100 mL | |
|---|---|---|
| Date, 1997 | Observed | Moving geometric mean |
| 5 May | 42 | |
| 12 May | 20 | |
| 19 May | 300 | |
| 26 May | 50 | |
| 2 June | 19 | 47 |
| 9 June | 20 | 41 |
| 16 June | 250 | 68 |
| 23 June | 3000 | 110 |
| 30 June | 160 | 140 |
| 7 July | 240 | 220 |
| 14 July | 20 | 220 |
| 21 July | 130 | 200 |
| 28 July | 180 | 110 |

Before Chick's law is applied, the river data are generally transformed to bacterial population equivalents (BPE) in the manner proposed by Kittrell and Furfari (1963). The following expressions were used for TC and FC data:

$$\text{BPE} = Q(\text{cfs}) \times \text{TC/100 mL} \times 6.1 \times 10^{-5} \tag{1.127}$$

$$\text{BPE} = Q(\text{cfs}) \times \text{FC/100 mL} \times 6.1 \times 0.964 \times 10^{-5} \tag{1.128}$$

Here the FC : TC ratio is assumed to be 0.964, as reported by Geldreich (1967).

Average stream flows and geometric means for each stream sampling station are used for BPE calculation. The relationship of BPE versus time of travel are plotted on semilog paper for both TC and FC. Reasonably good straight lines of fit can be developed. The decay or death rates for TC and FC can be estimated from the charts.

**Example:** TC density in the effluent is 9800 per 100 mL. Its densities are determined to be 5700, 2300, and 190 TC/100 mL, respectively, after the time of travel for 8, 24, and 60 h. Determine the die-off rate of TC in the stream.

**solution:**

Step 1. Calculate $k$ for each stream reach using Eq. (1.126); $t = 1/3$, 1.0, and 2.5 days

$$-kt = \log(N/N_0)$$

$$-k(1/3) = \log(5700/9800) = -0.235$$

$$k = 0.706 \text{ (per day)}$$

Similarly

$$-k(1) = \log(2300/9800) = -0.629$$

$$k = 0.629 \text{ (per day)}$$

Also

$$-k(2.5) = \log(190/9800) = -1.712$$

$$k = 0.685 \text{ (per day)}$$

The average

$$k = 0.673 \text{(per day)}$$

*Note*: This problem also can be solved by graphic method. Plot the data on semilogarithmic paper and draw a straight line. The slope of the line is the die-off rate.

### 18.3 Macroinvertebrate biotic index

The macroinvertebrate biotic index (MBI) was developed to provide a rapid stream-quality assessment. The MBI is calculated at each stream station as a tool to assess the degree and extent of wastewater discharge impacts. The MBI is an average of tolerance rating weighted by macroinvertebrate abundance, and is calculated from the formula (IEPA, 1987):

$$\text{MBI} = \sum_{i=1}^{n} (n_i t_i)/N \tag{1.129}$$

where $n_i$ = number of individuals in each taxon $i$
$t_i$ = tolerance rating assigned to that taxon $i$
$N$ = total number of individuals in the sediment sample

Most macroinvertebrate taxa known to occur in Illinois have been assigned a pollution tolerance rating, ranging from 0 to 11 based on literature and field studies. A 0 is assigned to taxa found only in unaltered streams of high water quality, and an 11 is assigned to taxa known to occur in severely polluted or disturbed streams. Intermediate ratings are assigned to taxa that occur in streams with intermediate degrees of pollution or disturbance. Appendix A presents a list of these tolerance ratings for each taxon.

**Example:** The numbers of macroinvertebrates in a river sediment sample are 72, 29, 14, 14, 144, 445, 100, and 29 organisms/m$^2$ for *Corbicula*, *Perlesta placida*, *Stenonema*, *Caenis*, *Cheumatopsyche*, *Chironomidae*, *Stenelmis*, and *Tubificidae*, respectively. The tolerance values for these organisms are, respectively, 4, 4, 4, 6, 6, 6, 7, and 10 (Appendix A). Compute MBI for this sample.

**solution:**

$$n = 8$$

$$N = 72 + 29 + 14 + 14 + 144 + 445 + 100 + 29 = 847$$

$$\begin{aligned}\text{MBI} &= \sum_{n=1}^{8} (n_i t_i)/N \\ &= \frac{1}{847}(72 \times 4 + 29 \times 4 + 14 \times 4 + 14 \times 6 + 144 \times 6 + 445 \\ &\quad \times 6 + 100 \times 7 + 29 \times 10) \\ &= 5.983 \\ &\cong 6.0\end{aligned}$$

## References


American Public Health Association (APHA), American Water Works Association (AWWA), and Water Environment Federation (WEF). 1995. *Standard methods for the examination of water and wastewater*, 19th edn. Washington, DC: American Public Health Association.

American Society of Civil Engineering Committee on Sanitary Engineering Research. 1960. Solubility of atmospheric oxygen in water. *J. Sanitary Eng. Div.* **86**(7): 41–53.

Barrett, M. J., Gameson, A. L. H. and Ogden, C. G. 1960. Aeration studies at four weir systems. *Water and Water Eng.* **64**: 407.

Black, W. M. and Phelps, E. B. 1911. *Location of sewer outlets and discharge of sewage in New York Harbor*. New York City Board of Estimate, March 23.

Broeren, S. M., Butts, T. A. and Singh, K. P. 1991. *Incorporation of dissolved oxygen in aquatic habitat assessment for the upper Sangamon River*. Contract Report 513. Champaign: Illinois State Water Survey.

Butts, T. A. 1963. *Dissolved oxygen profile of Iowa River*. Masters Thesis. Iowa City: University of Iowa.

Butts, T. A. 1974. *Measurement of sediment oxygen demand characteristics of the Upper Illinois Waterway*. Report of Investigation 76. Urbana: Illinois State Water Survey.

Butts, T. A. and Schnepper, D. H. 1967. Oxygen absorption in streams. *Water and Sewage Works* **144**(10): 385–386.

Butts, T. A. and Shackleford, D. B. 1992. *Reduction in peak flows and improvements in water quality in the Illinois Waterway downstream of Lockport due to implementation of Phase I and II of TARP*, Vol. 2: Water Quality. Contract Report 526. Urbana: Illinois State Water Survey.

Butts, T. A., Evans, R. L. and Stall, J. B. 1974. *A waste allocation study of selected streams in Illinois*. Contract Report prepared for Illinois EPA. Urbana: Illinois State Water Survey.

Butts, T. A., Evans, R. L. and Lin, S. D. 1975. *Water quality features of the Upper Illinois Waterway*. Report of Investigation 79. Urbana: Illinois State Water Survey.

Butts, T. A., Kothandaraman, V. and Evans, R. L. 1973. *Practical considerations for assessing waste assimilative capacity of Illinois streams*. Circular 110. Urbana: Illinois State Water Survey.

Butts, T. A., Roseboom, D., Hill, T., Lin, S. D., Beuscher, D., Twait, R. and Evans, R. L. 1981. *Water quality assessment and waste assimilative analysis of the LaGrange Pool, Illinois River*. Contract Report 260. Urbana: Illinois State Water Survey.

Butts, T. A., Schnepper, D. H. and Evans, R. L. 1970. *Dissolved oxygen resources and waste assimilative capacity of a LaGrange pool, Illinois River*. Report of Investigation 64. Urbana: Illinois State Water Survey.

Churchill, M. A. and Buckingham, R. A. 1956. Statistical method for analysis of stream purification capacity. *Sewage and Industrial Wastes* **28**(4): 517–537.

Churchill, M. A., Elmore, R. L. and Buckingham, R. A. 1962. The prediction of stream reaeration rates. *J. Sanitary Eng. Div.* **88**(7): 1–46.

Dougal, M. D. and Baumann, E. R. 1967. Mathematical models for expressing the biochemical oxygen demand in water quality studies. *Proc. 3rd Ann. American Water Resources Conference*, pp. 242–253.

Eckenfelder, W. and O'Conner, D. J. 1959. Stream analysis biooxidation of organic wastes—theory and design. Civil Engineering Department, Manhattan College, New York.

Fair, G. M. 1936. The log-difference method of estimating the constants of the first stage biochemical oxygen-demand curve. *Sewage Works J.* **8**(3): 430.

Fair, G. M., Moore, W. E. and Thomas, Jr., H. A. 1941a. The natural purification of river muds and pollutional sediments. *Sewage Works J.* **13**(2): 270–307.

Fair, G. M., Moore, W. E. and Thomas, Jr., H. A., 1941b. The natural purification of river muds and pollutional sediments. *Sewage Works J.* **13**(4): 756–778.

Gameson, A. L. H. 1957. Weirs and the aeration of rivers. *J. Inst. Water Eng.* **11**(6): 477.

Gameson, A. L. H., Vandyke, K. G. and Ogden, C. G. 1958. The effect of temperature on aeration at weirs. *Water and Water Eng.* **62:** 489.

Gannon, J. J. 1963. *River BOD abnormalities.* Final Report USPHS Grants RG-6905, WP-187(C1) and WP-187(C2). Ann Arbor, Michigan: University of Michigan School of Public Health.

Gannon, J. J. and Downs, T. D. 1964. Professional Paper Department of Environmental Health. Ann Arbor, Michigan: University of Michigan.

Geldreich, E. E. 1967. Fecal coliform concepts in stream pollution. *Water and Sewage Works* **144**(11): R98–R110.

Grindrod, J. 1962. British research on aeration at weirs. *Water and Sewage Works* **109**(10): 395.

Illinois Environmental Protection Agency. 1987. *Quality Assurance and Field Method Manual.* Springfield, Illinois: IEPA.

Illinois Environmental Protection Agency. 1999. *Title 35: Environmental Protection, Subtitle C: Water pollution, Chapter I: Pollution Control Board.* Springfield, Illinois: IEPA.

Kittrel, F. W. and Furfari, S. A. 1963. Observation of coliform bacteria in streams. *J. Water Pollut. Control Fed.* **35:** 1361.

Langbein, W. B. and Durum, W. H. 1967. Aeration capacity of streams. Circular 542. Washington, DC: US Geological Survey, 67.

Larson, R. S., Butts, T. A. and Singh, K. P. 1994. *Water quality and habitat suitability assessment: Sangamon River between Decatur and Petersburg.* Contract Report 571. Champaign: Illinois State Water Survey.

Le Bosquet, M. and Tsivoglou, E. C. 1950. Simplified dissolved oxygen computations. *Sewage and Industrial Wastes* **22:** 1054–1061.

McDonnell, A. J. and Hall, S. D. 1969. Effect of environmental factors on benthal oxygen uptake. *J. Water Pollut. Control Fed. Research Supplement, Part 2* **41:** R353–R363.

Moore, E. W., Thomas, H. A. and Snow, W. B. 1950. Simplified method for analysis of BOD data. *Sewage and Industrial Wastes* **22**(10): 1343.

Nemerow, N. L. 1963. *Theories and practices of industrial waste treatment.* Reading, Massachusetts: Addison-Wesley.

O'Connor, D. J. 1958. The measurement and calculation of stream reaeration ratio. In: *Oxygen relation in streams.* Washington, DC: US Department of Health, Education, and Welfare, pp. 35–42.

O'Connor, D. J. 1966. *Stream and estuarine analysis.* Summer Institute in Water Pollution Control, Training Manual. New York: Manhattan College.

O'Connor, D. J. and Dobbins, W. E. 1958. The mechanics of reaeration in natural streams. *Trans. Am. Soc. Civil Engineers* **123:** 641–666.

Orford, H. E. and Ingram, W. T. 1953. Deoxygenation of sewage. *Sewage and Industrial Waste* **25**(4):419–434.

Perry, R. H. 1959. *Engineering manual*, 2nd edn. New York: McGraw-Hill.

Phelps, E. B. 1944. *Stream sanitation.* New York: John Wiley.

Preul, H. C. and Holler, A. G. 1969. *Reaeration through low days in the Ohio River.* Proceedings of the 24th Purdue University Industrial Waste Conference. Lafayette, Indiana: Purdue University.

Shannon, C. E. and Weaver, W. 1949. *The mathematical theory of communication.* Urbana: University of Illinois Press, p. 125.

Sheehy, J. P. 1960. Rapid methods for solving monomolecular equations. *J. Water Pollut. Control Fed.* **32**(6): 646–652.

Streeter, H. W. 1926. The rate of atmospheric reaeration of sewage polluted streams. *Trans. Amer. Soc. Civil Eng.* **89**: 1351–1364.

Streeter, H. W. and Phelps, E. B. 1925. *A study of the pollution and natural purification of the Ohio River*. Cincinati: US Public Health Service, Bulletin No. 146.

Thomas, H. A., Jr. 1937. The 'slope' method of evaluating the constants of the first-stage biochemical oxygen demand curve. *Sewage Works J.* **9**(3): 425.

Thomas, H. A. 1948. Pollution load capacity of streams. *Water and Sewage Works* **95**(11): 409.

Thomas, H. A., Jr. 1950. Graphical determination of BOD curve constants. *Water and Sewage Works* **97**(3): 123–124.

Tsivoglou, E. C. 1958. *Oxygen relationships in streams*. Robert A. Taft Sanitary Engineering Center, Technical Report W-58-2. Cincinnati, Ohio.

US Environmental Protection Agency. 1997. *Technical guidance manual for developing total maximum daily loads, book 2: streams and rivers. Part 1: Biological oxygen demand/dissolved oxygen and nutrients/eutrophication*. EPA 823-B-97-002. Washington, DC: US EPA.

US Public Health Service. 1927. *The oxygen demand of polluted waters*. Public Health Bulletin No. 172. Washington, DC: US Public Health Service.

Velz, C. J. 1958. Significance of organic sludge deposits. In: Taft, R. A. (ed.), *Oxygen relationships in streams*. Sanitary Engineering Center Technical Report W58-2. Cincinnati, Ohio: US Department of Health, Education, and Welfare.

Velz, C. J. 1939. Deoxygenation and reoxygenation. *Trans. Amer. Soc. Civil Eng.* **104:** 560–572.

Wetzel, R. G. 1975. *Limnology*. Philadelphia, Pennsylvania: Saunders.

Young, J. C. and Clark, J. W. 1965. Second order equation for BOD. *J. Sanitary Eng. Div., ASCE* **91**(SA): 43–57.

Zanoni, A. E. 1967. *Effluent deoxygenation at different temperatures*. Civil Engineering Department Report 100-5A. Milwaukee, Wisconsin: Marquette University.

Chapter

# 2

# Lakes and Reservoirs



This chapter includes mainly lake morphometry, evaporation, and the Clean Lakes Program (CLP). Since most lake management programs in the United States are based on the CLP, the CLP is discussed in detail. Regulatory requirements and standardization of research and application are provided with a focus on the Phase 1, diagnostic/feasibility study.

## 1 Lakes and Impoundment Impairments

Lakes are extremely complex systems whose conditions are a function of physical, chemical, and biological (the presence and predominance of the various plants and organisms found in the lake) factors. Lakes inherently function as traps or sinks for pollutants from tributaries, watersheds (drainage basins) or from atmospheric deposits and precipitation.

Like streams, lakes are most often impaired by agricultural activities (main sources in the United States), hydrologic/habitat modification (stream channelization), and point pollution sources. These activities contribute to nutrient and sediment loads, suspended solids, and organic matter, and subsequently cause overgrowth of aquatic plants. The resulting decline in water quality limits recreation, impairs other beneficial uses, and shortens the expected life span of a lake.

Common lake problems are eutrophication, siltation, shoreline erosion, algal bloom, bad taste and/or odor, excessive growth of aquatic vegetation, toxic chemicals, and bacterial contamination. Eutrophication, or aging, the process by which a lake becomes enriched with nutrients, is caused primarily by point and nonpoint pollution sources from human activities. Some man-made lakes and impoundments may be untrophic from their birth. These problems impact esthetic and practical uses of the lake. For example, the growth of planktonic algae in water supply impoundments may cause taste and odor problems, shortened filter runs, increased chlorine demand, increased turbidity, and, for some facilities, increased trihalomethane precursors. The effects ultimately lead to increased water treatment costs and, in some instances, even to abandonment of the lake as a public water supply source.

Lakes and reservoirs are sensitive to pollution inputs because they flush out their contents relatively slowly. Even under natural conditions, lakes and reservoirs undergo eutrophication, an aging process caused by the inputs of organic matters and siltation.

## 2 Lake Morphometry

Lake morphometric data can be calculated from either a recent hydrographic map or a pre-impoundment topographic map of the basin. In general, pre-impoundment maps may be obtainable from the design engineering firm or local health or environmental government agencies. If the map is too old and there is evidence of significant siltation, sections of the lake, such as the upper end and coves with inflowing streams, will need to be remapped.

In some cases, it is necessary to create a new lake map. The procedures include the following:

1. The outline of the shoreline is drawn either from aerial photographs or from United States Geological Survey (USGS) 7.5-minute topographic maps.
2. The water depth for transacts between known points on the shoreline of the lake is measured by a graphing sonar.
3. The sonar strip chart of the transacts is interpreted and drawn on to an enlarged copy of the lake outline.
4. Contours may be drawn by hand or with a computer on the map.
5. All maps are then either digitized or scanned and entered into a geographic information system (GIS).
6. Coverages are then converted into sea level elevations by locating the map on a 7.5-minute USGS topographic map and digitizing reference points both on the map and on the quadrangle for known sea level elevations.
7. Depths are assigned to the contour. The GIS gives the length of each contour and areas between adjacent contour lines.

From this data, surface area, maximum depth of the lake, and shoreline length can be computed. Lake volume ($V$), shoreline development index (SDI) or shoreline configuration ratio, and mean depth ($\overline{D}$) can be calculated by the formula in Wetzel (1975):

1. Volume

$$V = \sum_{i=0}^{n} \frac{h}{3}(A_i + A_{i+1} + \sqrt{A_i \times A_{i+1}}) \tag{2.1}$$

where $V$ = volume, ft$^3$, acre · ft, or m$^3$
$h$ = depth of the stratum, ft or m
$i$ = number of depth stratum
$A_i$ = area at depth $i$, ft$^2$, acre, or m$^2$

2. Shoreline development index

$$\text{SDI} = \frac{L}{2\sqrt{\pi \times A_0}} \tag{2.2}$$

where $L$ = length of shoreline, miles or m
$A_0$ = surface area of lake, acre, ft$^2$, or m$^2$

3. Mean depth

$$\overline{D} = \frac{V}{A_0} \tag{2.3}$$

where $\overline{D}$ = mean depth, ft or m
$V$ = volume of lake, ft$^3$, acre · ft, or m$^3$
$A_0$ = surface area, ft$^2$, acre, or m$^2$

Other morphometric information also can be calculated by the following formulas:

4. Bottom slope

$$S = \frac{\overline{D}}{D_m} \tag{2.4}$$

where $S$ = bottom slope
$\overline{D}$ = mean depth, ft or m
$D_m$ = maximum depth, ft or m

5. Volume development ratio, $V_d$ (Cole, 1979)

$$V_d = 3 \times \frac{\overline{D}}{D_m} \tag{2.5}$$

6. Water retention time

$$\text{RT} = \frac{\text{storage capacity, acre·ft or m}^3}{\text{annual runoff, acre·ft/yr or m}^3\text{/yr}} \tag{2.6}$$

where RT = retention time, year

7. Ratio of drainage area to lake capacity $R$

$$R = \frac{\text{drainage area, acre or m}^2}{\text{storage capacity, acre·ft or m}^3} \tag{2.7}$$

**Example:** A reservoir has a shoreline length of 9.80 miles. Its surface area is 568 acres. Its maximum depth is 10.0 ft. The areas for each foot depth are 480, 422, 334, 276, 205, 143, 111, 79, 30, and 1 acres. Annual rainfall is 38.6 in. The watershed drainage is 11,620 acres. Calculate morphometric data with the formulas described above.

**solution:** Compute the following parameters:

1. The volume of the lake

$$V = \sum_{i=0}^{10} \frac{h}{3}(A_i + A_{i+1} + \sqrt{A_i \times A_{i+1}})$$

$$= \frac{1}{3}[(568 + 480 + \sqrt{568 \times 480}) + (480 + 422 + \sqrt{480 \times 422})$$

$$+ (422 + 324 + \sqrt{422 \times 324}) + \cdots + (30 + 1 + \sqrt{30 \times 1})]$$

$$= \frac{1}{3}[7062]$$

$$= 2354(\text{acre}\cdot\text{ft})$$

$$= 2{,}902{,}000\,\text{m}^3$$

2. Shoreline development index or shoreline configuration ratio

$$A_0 = 568 \text{ acres} = 568 \text{ acres} \times \frac{1 \text{ sq. miles}}{640 \text{ acres}} = 0.8875 \text{ sq. miles}$$

$$\text{SDI} = \frac{L}{2\sqrt{\pi \times A_0}}$$

$$= \frac{9.80 \text{ miles}}{2\sqrt{3.14 \times 0.8875 \text{ sq. miles}}}$$

$$= \frac{9.80}{3.34}$$

$$= 2.93$$

3. Mean depth

$$\bar{D} = \frac{V}{A_0}$$

$$= \frac{2354 \text{ acre}\cdot\text{ft}}{568 \text{ acres}}$$

$$= 4.14 \text{ ft}$$

4. Bottom slope

$$S = \frac{\bar{D}}{D_\text{m}} = \frac{4.13 \text{ ft}}{10.0 \text{ ft}} = 0.41$$

5. Volume development ratio

$$V_\text{d} = 3 \times \frac{\bar{D}}{D_\text{m}} = 3 \times 0.41 = 1.23 \text{ (or 1.23:1)}$$

6. Water retention time

$$\text{Storage capacity } V = 2354 \text{ acre} \cdot \text{ft}$$

$$\begin{aligned}
\text{Annual runoff} &= 38.6 \text{ in/yr} \times 11{,}620 \text{ acres} \\
&= 38.6 \text{ in/yr} \times \frac{1 \text{ ft}}{12 \text{ in}} \times 11{,}620 \text{ acres} \\
&= 37{,}378 \text{ acre}\cdot\text{ft/yr}
\end{aligned}$$

$$\begin{aligned}
\text{RT} &= \frac{\text{storage capacity}}{\text{annual runoff}} \\
&= \frac{2354 \text{ acre}\cdot\text{ft}}{37{,}378 \text{ acre}\cdot\text{ft/yr}} \\
&= 0.063 \text{ years}
\end{aligned}$$

7. Ratio of drainage area to lake capacity

$$\begin{aligned}
R &= \frac{\text{drainage area}}{\text{storage capacity}} \\
&= \frac{11{,}620 \text{ acres}}{2354 \text{ acre}\cdot\text{ft}} \\
&= \frac{4.94 \text{ acres}}{1 \text{ acre}\cdot\text{ft}} \quad \text{or} = 4.94 \text{ acres/(acre}\cdot\text{ft)}
\end{aligned}$$

## 3 Water Quality Models

Lakes and reservoirs are usually multipurpose, serving municipal and industrial water supplies, recreation, hydroelectric power, flood control, irrigation, drainage, and/or agriculture, due to the importance of protecting these natural resources. Water quality involves the physical, chemical, and biological integrity of water resources. Water quality standards promulgated by the regulatory agencies define the water quality goals for protection of water resources in watershed management.

Modeling the water quality in lakes and reservoirs is very different from that in rivers (Chapter 1) or estuaries. A variety of models are based on some of physical, chemical, and biological parameters and/or combinations. Physical models deal with temperature, dissolved oxygen (DO), energy budget diffusion, mixing, vertical and horizontal aspects, seasonal cycles, and meteorological data setting. Chemical models involve mass balance and toxic substances. Biological models deal with nutrients and the food chain, biological growth, perdition, oxygen balance, and tropical conditions, etc.

A detailed modeling of lake water quality is given in a book by Chapra and Reckhow (1983). Clark *et al.* (1977), Tchobanoglous and Schroeder (1985), and James (1993) also present modeling water quality in lakes and reservoirs. However, most of those models are not used for lake management practices.

For construction of a new lake or reservoir, it is required that the owner submit the prediction of water quality in the future new lake to the US Army Corps of Engineers. For example, Borah *et al.* (1997) used the US Army Corps of Engineers' HEC-5Q model (US Army Corps of Engineers, 1986, 1989) for a central Illinois lake and two proposed new lakes. The model was calibrated and verified on the existing water supply lake used monitored data in the draft years of 1986 and 1988. Eight water quality constituents were simulated. Calibration, verification, and predictions were made only for water temperature, DO, nitrate nitrogen, and phosphate phosphorus. The calibrated and verified model was used to predict water surface elevations and constituent concentrations in the three lakes, individually and in combination, with meteorological and estimated flow data for a 20-month period from May 1953 through December 1954, the most severe drought of record. The HEC-5Q model provided a useful tool in the water quality evaluation of the three lakes. The HEC-5Q model was developed by the Hydraulic Engineering Center in Davis, California, for simulation of flood control and conservation systems with water quality analysis. The details of these models are beyond the scope of this book.

The major portion of this chapter will cover evaporation and lake management programs in the Clean Lakes Program. Because these subjects are general, they are not usually included in environmental engineering textbooks.

## 4 Evaporation

Evaporation converts water in its liquid or solid state into water vapor which mixes with the atmosphere. The rate of evaporation is controlled by the availability of energy at the evaporating surface, and the ease with which water vapor can diffuse into the atmosphere.

Shuttleworth (1993) presented a detailed description with modeling of evaporation and transpiration from soil surfaces and crops. Modelings include fundamental and empirical equations. Measurements of evaporation and methods for estimating evaporation are also given.

Evaporation of water from a lake surface uses energy provided by the sun to heat the water. The rate of evaporation is controlled primarily by the water temperature, air temperature, and the level of moisture saturation in the air. Knowledge of evaporative processes is important in understanding how water losses through evaporation from a lake or reservoir are determined. Evaporation increases the storage requirement and decreases the yield of lakes and reservoirs.

Determination of evaporation from a lake surface can be modeled by the water budget method, the mass transfer method, and the energy budget method (Robert and Stall, 1967). Many empirical equations have been proposed elsewhere (Fair *et al.*, 1966, Linsley and Franzinni, 1964).

### 4.1 Water budget method

The water budget method for lake evaporation depends on an accurate measurement of the inflow and outflow of the lake. The simple mathematical calculation shows that the change in storage equals the input minus output. It is expressed as

$$\Delta S = P + R + GI - GO - E - T - O \tag{2.8}$$

where $\Delta S$ = change in lake storage, mm (or in)
$P$ = precipitation, mm
$R$ = surface runoff or inflow, mm
$GI$ = groundwater inflow, mm
$GO$ = groundwater outflow, mm
$E$ = evaporation, mm
$T$ = transpiration, mm
$O$ = surface water release, mm

For the case of a lake with little vegetation and negligible groundwater inflow and outflow, lake evaporation can be estimated by

$$E = P + R - O \pm \Delta S \tag{2.9}$$

The water budget method has been used successfully to estimate lake evaporation in some areas.

### 4.2 Energy budget method

The principal elements in energy budget of the lake evaporation are shown in Fig. 2.1. The law of conservation of energy suggests that the total energy reaching the lake must be equal to the total energy leaving the lake plus the increase in internal energy of the lake. The energy budget in Fig. 2.1 can be expressed as (US Geological Survey, 1954)

$$Q_e = Q_s - Q_r - Q_b - Q_h - Q_\theta \pm Q_v \tag{2.10}$$

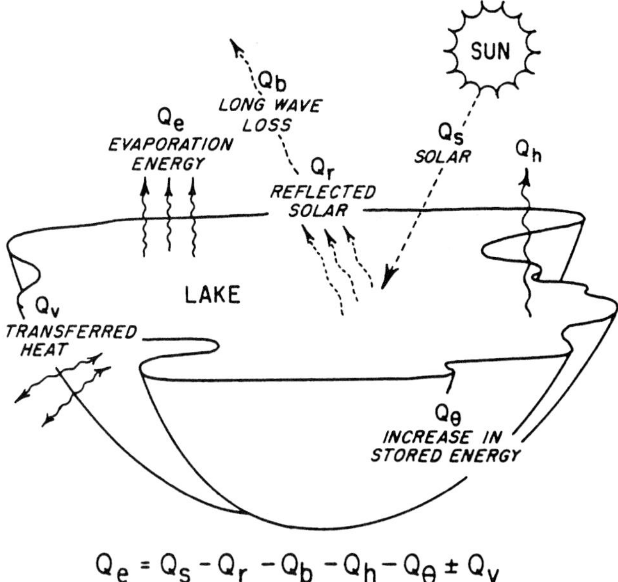


$$Q_e = Q_s - Q_r - Q_b - Q_h - Q_\theta \pm Q_v$$

**Figure 2.1** Elements of energy budget method for determining lake evaporation (*Robert and Stall, 1967*).

where $Q_e$ = energy available for evaporation
$Q_s$ = solar radiation energy
$Q_r$ = reflected solar radiation
$Q_b$ = net long-wave radiation
$Q_h$ = energy transferred from the lake to the atmosphere
$Q_\theta$ = increase in energy stored in the lake
$Q_v$ = energy transferred into or from the lake bed

The energy budget shown in Fig. 2.1 can be applied to evaporation from a class A pan (illustrated in Fig. 2.2). The pan energy budget (Fig. 2.2) contains the same elements as the lake budget (Fig. 2.1) except that heat is lost through the side and the bottom of the pan. Equation (2.10) is also applied to pan energy budget calculation.

A pioneering and comprehensive research project centered on techniques for measuring evaporation was conducted at Lake Hefner, Oklahoma, in 1950–51 with the cooperation of five federal agencies (US Geological Survey, 1954). The energy budget data from a Weather Bureau class A evaporation pan was used to evaluate the study. Growing out of that project was the US Weather Bureau nomograph, a four-gradient diagram (Fig. 2.3) for general use for lake evaporation (Kohler *et al.*, 1959). Subsequently, the formula for a mathematical solution of the nomograph was adapted from computer use (Lamoreux, 1962).

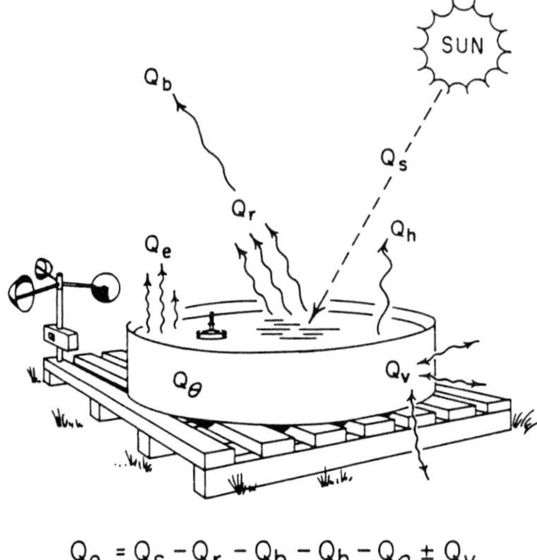


$$Q_e = Q_s - Q_r - Q_b - Q_h - Q_\theta \pm Q_v$$

**Figure 2.2** Energy budget method applied to class A pan (*Robert and Stall, 1967*).

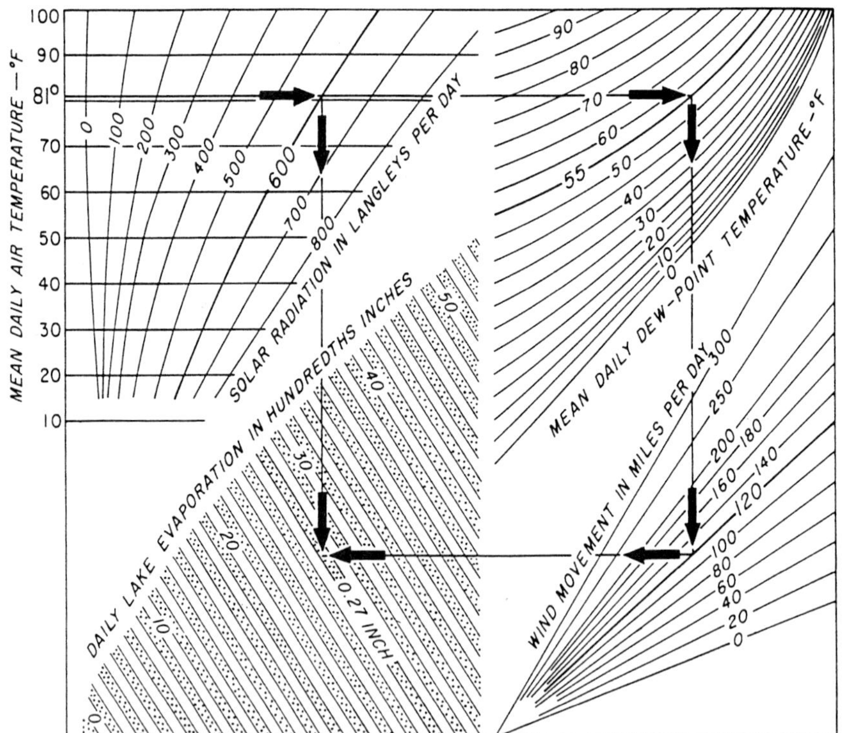


**Figure 2.3** Lake evaporation nomograph (*Kohler et al., 1959*).

These procedures allow lake and pan evaporation to be computed from four items of climate data, i.e. air temperature, dew point temperature, wind movement, and solar radiation. Since long-term climate data are mostly available, these techniques permit extended and refined evaporation determination.

Although the nomograph (Fig. 2.3) was developed for the use with daily values, it is also usable with monthly average values and gives reasonably reliable results with annual average values. The four-quadrant diagram can be used to determine a lake's evaporation rate as illustrated in Fig. 2.3. For example, at the upper left quadrant, a horizontal line passes from the mean daily air temperature 81°F to the first intersection at the solar radiation 600-langley curve, and projects toward the right to intersection of the 55°F mean daily dew point curve. At this stop, a vertical line runs downward to the line of 120-miles per day wind movement from where a horizontal line projects toward to the left to intersect a vertical line from the first intersection. The average daily lake evaporation, 0.27/in, is read at this point.

The following two equations for evaporation can be computed by computer (Lamoreux, 1962) as exemplified by the Weather Bureau computational procedures for the United States.

For pan evaporation:

$$\begin{aligned}E_{\mathrm{p}} = \{&\exp[(T_{\mathrm{a}} - 212)(0.1024 - 0.01066 \ln R)] - 0.0001 \\ &+ 0.025(e_{\mathrm{s}} - e_{\mathrm{a}})^{0.88}(0.37 + 0.0041\,U_{\mathrm{p}})\} \times \{0.025 + (T_{\mathrm{a}} + 398.36)^{-2} \\ &\times 4.7988 \times 10^{10} \exp[-7482.6/(T_{\mathrm{a}} + 398.36)]\}^{-1}\end{aligned} \tag{2.11}$$

For lake evaporation, the expression is

$$\begin{aligned}E_{\mathrm{L}} = \{&\exp[(T_{\mathrm{a}} - 212)(0.1024 - 0.01066 \ln R)] - 0.0001 + 0.0105(e_{\mathrm{s}} - e_{\mathrm{a}})^{0.88} \\ &\times (0.37 + 0.0041 U_{\mathrm{p}})\} \times \{0.015 + (T_{\mathrm{a}} + 398.36)^{-2}\, 6.8554 \times 10^{10} \\ &\times \exp[-7482.6/(T_{\mathrm{a}} + 398.36)]\}^{-1}\end{aligned} \tag{2.12}$$

where $E_{\mathrm{p}}$ = pan evaporation, in
$E_{\mathrm{L}}$ = lake evaporation, in
$T_{\mathrm{a}}$ = air temperature, °F
$e_{\mathrm{a}}$ = vapor pressure, inches of mercury at temperature $T_{\mathrm{a}}$
$e_{\mathrm{s}}$ = vapor pressure, inches of mercury at temperature $T_{\mathrm{d}}$
$T_{\mathrm{d}}$ = dew point temperature, °F
$R$ = solar radiation, langleys per day
$U_{\mathrm{p}}$ = wind movement, miles per day

Pan evaporation is used widely to estimate lake evaporation in the United States. The lake evaporation $E_{\mathrm{L}}$ is usually computed for yearly time periods. Its relationship with the pan evaporation $E_p$ can be expressed as

$$E_L = p_c E_p \tag{2.13}$$

where $p_c$ is the pan coefficient. The pan coefficient on an annual basis has been reported as 0.65 to 0.82 by Kohler *et al.* (1955).

Robert and Stall (1967) used Eqs. (2.9) and (2.10) for 17 stations in and near Illinois over the May–October season for a 16-year period to develop the general magnitudes and variability of lake evaporation in Illinois.

Evaporation loss is a major factor in the net yield of a lake or reservoir. The net evaporation loss from a lake is the difference between a maximum expected gross lake evaporation and the minimum expected precipitation on the lake surface for various recurrence intervals and for critical periods having various durations. This approach assumes maximum evaporation and minimum precipitation would occur simultaneously. The following example illustrates the calculation of net draft rate for a drought recurrence interval of 40 years.

**Example:** Calculate the net yield or net draft for the 40-year recurrence interval for a reservoir (333 acres, 135 ha). Given: drainage area = 15.7 miles$^2$, unit reservoir capacity (RC) = 2.03 in, draft = 1.83 MGD, $E$ = 31 in/acre.

**solution:**

Step 1. Compute reservoir capacity in million gallons (Mgal)

$$\begin{aligned}
\text{RC} &= 2.03 \text{ in} \times 15.7 \text{ miles}^2 \\
&= 2.03 \text{ in} \times (1 \text{ ft}/12 \text{ in}) \times 15.7 \text{ miles}^2 \times (640 \text{ acres/miles}^2) \\
&= 1700 \text{ acre} \cdot \text{ft} \\
&= 1700 \text{ acre} \cdot \text{ft} \times 43{,}560 \text{ ft}^2/\text{acre} \times 7.48 \text{ gal/ft}^3 \\
&= 554 \text{ Mgal} = 2{,}097{,}000 \text{ m}^3
\end{aligned}$$

*Note:* 1 Mgal = 3785 m$^3$

Step 2. Compute the total gross draft, the duration of critical drawdown period for 40-year recurrence interval is 16 months

$$\begin{aligned}
16 \text{ months} &= 365 \text{ days}/12 \times 16 = 486 \text{ days} \\
\text{Draft} &= 1.83 \text{ MGD} \times 486 \text{ days} \\
&= 888 \text{ Mgal}
\end{aligned}$$

Step 3. Compute the inflow to the reservoir (total gross draft minus RC)

$$\begin{aligned}
\text{Inflow} &= \text{draft} - \text{RC} = 888 \text{ Mgal} - 554 \text{ Mgal} \\
&= 334 \text{ Mgal}
\end{aligned}$$

Step 4. Compute the evaporation loss $E$

Effective evaporation surface area of the lake is 333 acres $\times$ 0.65 = 216 acres

$$\begin{aligned} E &= 31 \text{ in} \times (1 \text{ ft}/12 \text{ in}) \times 216 \text{ acres} \\ &= 558 \text{ acre} \cdot \text{ft} \\ &= 182 \text{ Mgal} \end{aligned}$$

Step 5. Compute the net usable reservoir capacity (the total reservoir capacity minus the evaporation loss)

$$\begin{aligned} \text{RC} - E &= 554 \text{ Mgal} - 182 \text{ Mgal} \\ &= 372 \text{ Mgal} \end{aligned}$$

Step 6. Compute the total net draft which the reservoir can furnish (the net usable reservoir capacity plus the inflow)

$$\begin{aligned} \text{Net draft} &= 372 \text{ Mgal} + 334 \text{ Mgal} \\ &= 706 \text{ Mgal} \end{aligned}$$

Step 7. Compute the net draft rate, or the net yield, which the reservoir can furnish (the total net draft divided by total days in the critical period)

$$\begin{aligned} \text{Net draft rate} &= 706 \text{ Mgal}/486 \text{ days} \\ &= 1.45 \text{ MGD} \\ &= 5490 \text{ m}^3/\text{d} \end{aligned}$$

*Note:* 1 MGD = 3785 $\text{m}^3$/d

## 5 The Clean Lakes Program

In the 1960s, many municipal wastewater treatment plants were constructed or upgraded (point source pollution control) in the United States. However, the nation's water quality in lakes and reservoirs has not been improved due to nonpoint sources of pollution. In 1972, the US Congress amended Section 314 of the Federal Water Pollution Control Act (Public Law 92-500) to control the nation's pollution sources and to restore its freshwater lakes. In 1977, the Clean Lakes Program (CLP) was established pursuant to Section 314 of the Clean Water Act. Under the CLP, publicly owned lakes can apply and receive financial assistance from the US Environmental Protection Agency (US EPA) to conduct diagnostic studies of the lakes and to develop feasible pollution control measures and water quality enhancement techniques for lakes. This is the so-called Phase 1 diagnostic/feasibility (D/F) study.

The *objectives* of the CLP are to (US EPA, 1980):

1. classify publicly owned freshwater lakes according to trophic conditions;
2. conduct diagnostic studies of specific publicly owned lakes, and develop feasible pollution control and restoration programs for them; and
3. implement lake restoration and pollution control projects.

### 5.1 Types of funds

The CLP operates through four types of financial assistance by cooperative agreements. They are

State lake classification survey
Phase 1—diagnostic/feasibility study
Phase 2—implementation
Phase 3—post-restoration monitoring

Funding is typically provided yearly for the state lake classification survey, but on a long-term basis for Phases 1, 2, and 3. Phase 1 funds are used to investigate the existing or potential causes of decline in the water quality of a publicly owned lake; to evaluate possible solutions to existing or anticipated pollution problems; and to develop and recommend the most "feasible" courses of action to restore or preserve the quality of the lake. Activities typically associated with sample collection, sample analyses, purchase of needed equipment, information gathering, and report development are eligible for reimbursement. During Phase 1, total project costs cannot exceed $100,000. Fifty percent of this funding will come from the CLP, while the rest must come from nonfederal or local sources.

### 5.2 Eligibility for financial assistance

Only states are eligible for CLP financial assitance, and cooperative agreements will be awarded to state agencies. A state may in turn make funds available to a substate agency or agencies for all or any portion of a specific project. The D/F study is generally carried out by a contracted organization, such as research institutes or consulting engineers.

For nearly two decades, 46% of the US lake acres were assessed for their water qualities and impairments (US EPA, 1994). Since the early 1990s, federal funds for CLP were terminated due to the budget cuts. Some states, such as Illinois, continued the CLP with state funds.

## 5.3 State lake classification survey

In general, most states use state lake classification survey funding to operate three types of lake survey activities. These include the volunteer lake monitoring program (VLMP), the ambient lake monitoring program (ALMP), and lake water quality assessment (LWQA) grant. All three programs are partially supported by a Section 314 Federal CLP LWQA grant. State funds matched equally with federal grant funds are used to improve the quantity and quality of lake information reported in the annual 305(b) report to the US Congress.

**Volunteer lake monitoring program.** The VLMP is a statewide cooperative program which volunteers to monitor lake conditions twice a month from May through October and transmit the collected data to the state agency (EPA or similar agency) for analysis and report preparation. In Illinois, the VLMP was initiated by Illinois Environmental Protection Agency (IEPA) in 1981 (IEPA, 1984). The volunteers can be personnel employed by the lake owner (a water treatment plant, Department of Conservation or other state agency, city, lake association, industry, etc.), or they can be local citizens. (The lakes monitored are not limited to publicly owned lakes.) Volunteers must have a boat and an anchor in order to perform the sampling. Volunteers receive a report prepared by the state EPA, which evaluates their sampling results.

Volunteers measure Secchi disc transparencies and total depths at three sites in the lake. For reservoirs, site 1 is generally located at the deepest spot (near the dam); site 2 is at midlake or in a major arm; and site 3 is in the headwater, a major arm, or the tributary confluence. The data are recorded on standard forms. In addition, the volunteers also complete a field observation form each time the lake is sampled and record the number that best describes lake conditions during sampling for each site on the lake. Observations include color of the water, the amount of sediment suspended in the water, visible suspended algae, submerged or floating aquatic weeds, weeds near the shore, miscellaneous substances, odors, cloudiness, precipitation, waves, air temperature at the lake, lake water levels, recreational usage, and lake management done since the last sampling (IEPA, 1983).

**Ambient lake monitoring program.** The ALMP is also a statewide regular water quality monitoring program. Water quality samples are collected and analyzed annually by IEPA personnel on selected lakes throughout the state. The major objectives of the ALMP are to (IEPA, 1992):

- characterize and define trends in the condition of significant lakes in the state;
- diagnose lake problems, determine causes/sources of problems, and provide a basis for identifying alternative solutions;

- evaluate progress and success of pollution control/restoration programs;
- judge effectiveness of applied protection/management measures and determine applicability and transferability to other lakes;
- revise and update the lake classification system; and
- meet the requirements of Section 314 CLP regulation and/or grant agreements.

The ALMP provides a much more comprehensive set of chemical analyses on the collected water samples than does the VLMP.

In Illinois, the ALMP was initiated in 1977. Water samples are analyzed for temperature, dissolved oxygen, Secchi disc transparency, alkalinity, conductivity, turbidity, total and volatile suspended solids, dissolved and total phosphorus, and nitrogen (ammonia, nitrite/nitrate, and total kjeldahl). In addition, 9 metals and 11 organics analyses are performed on water and sediment samples.

**Lake water quality assessment.** The LWQA grant is focused on non-routinely monitored lakes to gather basic data on in-lake water quality and sediment quality. This information increases the efficiency, effectiveness, and quality of the state EPA's lake data management and 305(b) report.

LWQA fieldwork involves two major tasks: namely, collection of lake assessment information and in-lake water and sediment sampling.

Lake assessment information collected includes lake identification and location, morphometric data, public access, designated uses and impairments, lake and shoreline usages, watershed drainage area usage, water quality and problems, status of fisheries, causes and sources of impairment (if any), past lake protection and management techniques, and lake maps.

In general, only one sampling trip per lake is made in order to evaluate a maximum number of lakes with the limited resources. The water and sediment samples are collected in summer by contracted agencies and analyzed by the state EPA laboratory. Chemical analyses for water and sediment samples are the same as in the ALMP.

## 5.4 Phase 1: Diagnostic/feasibility study

A diagnostic/feasibility study is a two-part study to determine a lake's current condition and to develop plans for its restoration and management. Protocol for the D/F study of a lake (Lin, 1994) is a digest from the *CLP Guidance Manual* (US EPA, 1980).

For a Phase 1 D/F study, the following activities should be carried out and completed within 3 years (US EPA, 1980):

1. Development of a detailed work plan
2. Study of the natural characteristics of the lake and watershed
3. Study of social, economic, and recreational characteristics of the lake and watershed
4. Lake monitoring
5. Watershed monitoring
6. Data analysis
7. Development and evaluation of restoration alternatives
8. Selection and further development of watershed management plans
9. Projection of benefits
10. Environmental evaluation
11. Public participation
12. Public hearings (when appropriate)
13. Report production

Activities 1 through 6 are part of the diagnostic study; the other activities pertain to the feasibility study.

**Diagnostic study.** The diagnostic study is devoted to data gathering and analysis. It involves collecting sufficient limnological, morphological, demographic, and socioeconomic information about the lake and its watershed.

The study of lake and watershed natural characteristics, and most of the social, economic, and recreational information of the lake region can often be accomplished by obtaining and analyzing secondary data, i.e. data already available from other sources.

Baseline limnological data include a review of historical data and 1 year of current limnological data. Lake monitoring is expensive. Monthly and bimonthly samples are required for assessing physical and chemical characteristics. At least three sites are chosen for each lake, as mentioned in the discussion of the ALMP (Section 5.3). In addition, biological parameters, such as chlorophylls, phytoplankton, zooplankton, aquatic macrophytes, indicator bacteria, benthic macroinvertebrate, and fish surveys, must be assessed at different specified frequencies. At least one surficial and/or core sediment sample and water sample at each site must be collected for heavy metals and organic compounds analyses. Water quality analyses of tributaries and groundwater may also be included, depending on each lake's situation.

Watershed monitoring is done by land-use stream monitoring manually or automatically to determine the nutrient, sediment, and hydraulic budget for a lake.

Primary and secondary data collected during the diagnostic study are analyzed to provide the basic information for the feasibility portion of the Phase 1 study. Data analysis involves:

1. inventory of point source pollutant discharges;
2. watershed land use and nonpoint nutrient/solids loading;
3. analyses of lake data—water and sediment quality;
4. analyses of stream and groundwater data;
5. calculating the hydrologic budget of the watershed and lake;
6. calculating the nutrient budget of the lake;
7. assessing biological resources and ecological relationships—lake fauna, terrestrial vegetation, and animal life;
8. determining the loading reductions necessary to achieve water quality objectives.

The following analyses of lake data are typically required: (1) identification of the limiting nutrient based on the ratio of total nitrogen to total phosphorus; (2) determination of the trophic state index based on total phosphorus, chlorophyll *a*, Secchi depth, and primary productivity; (3) calculation of the fecal coliform to fecal streptococcus ratios to identify causes of pollution in the watershed; and (4) evaluation of the sediment quality for the purpose of dredging operations.

A description of the biological resources and ecological relationships is included in the diagnostic study based on information gathered from secondary sources. This description generally covers lake flora (terrestrial vegetation, such as forest, prairie, and marsh) and fauna (mammals, reptiles and amphibians, and birds).

**Feasibility study.** The feasibility study involves developing alternative management programs based on the results of the diagnostic study. First, existing lake quality problems and their causes should be identified and analyzed. Problems include turbid water, eroding shorelines, sedimentation and shallow water depths, low water levels, low dissolved oxygen levels, excessive nutrients, algal bloom, unbalanced aquatic vegetative growth (excessive growth of macrophytes), nonnative exotic species, degraded or unbalanced fishery and aquatic communities, bad taste and odor, acidity, toxic chemicals, agricultural runoff, bacterial contamination, poor lake esthetics, user conflicts, and negative human impacts.

Once the lake problems have been defined, a preliminary list of corrective alternatives needs to be established and discussed. Commonly adopted pollution control and lake restoration techniques can be found in the literature (US EPA, 1980, 1990). The corrective measures include in-lake restoration and/or best management practices in the watershed. Each alternative must be evaluated in terms of cost; reliability; technical feasibility; energy consumption; and social, economic, and environmental impacts. For each feasible alternative, a detailed description of measures to be taken, a quantitative analysis of pollution control effectiveness, and expected lake water quality improvement must be provided.

There are usually many good methods of achieving specific objectives or benefits. The Phase 1 report should document the various alternatives by describing their relative strengths and weaknesses and showing how the one chosen is superior. Final selection should be discussed in a public meeting.

Once an alternative has been selected, work can proceed on developing other program details required as part of Phase 1. These include developing the Phase 2 monitoring program, schedule and budget, and source of nonfederal funds; defining the relationship of Phase 2 plans to other programs; developing an operation and maintenance plan; and obtaining required permits. In other words, the feasibility study portion of the Phase 1 report actually constitutes the proposal for Phase 2 study. The report also must include the projection of the project benefits, environmental evaluations, and public participation.

**Final report.** The final report will have to follow strictly the format and protocol stipulated in the *Clean Lakes Program Guidance Manual.* Requirements for diagnostic feasibility studies and environmental evaluations are at a minimum, but not limited, as follows:

1. Lake identification and location
2. Geologic and soils description of drainage basins
   a. Geologic description
   b. Groundwater hydrology
   c. Topography
   d. Soils
3. Description of public access
4. Description of size and economic structure of potential user population

5. Summary of historical lake uses
6. Population segments adversely affected by lake degradation
7. Comparison of lake uses to uses of other lakes in the region
8. Inventory of point-source pollution discharges
9. Land uses and nonpoint pollutant loadings
10. Baseline and current limnological data
    a. Summary analysis and discussion of historical baseline limnological data
    b. Presentation, analysis, and discussion of 1 year of current limnological data
    c. Tropic condition of lake
    d. Limiting algal nutrient
    e. Hydraulic budget for lake
    f. Nutrient sediment budget
11. Biological resources and ecological relationships
12. Pollution control and restoration procedures
13. Benefits expected from restoration
14. Phase II monitoring program
15. Schedule and budget
16. Source of matching funds
17. Relationship to other pollution control programs
18. Public participation summary
19. Operation and maintenance plan
20. Copies of permits or pending applications
21. Environmental evaluation
    a. Displacement of people
    b. Defacement of residential area
    c. Changes in land use patterns
    d. Impacts on prime agricultural land
    e. Impacts on parkland, other public land and scenic resources
    f. Impacts on historic, architectural, archaeological, or cultural resources
    g. Long range increases in energy demand
    h. Changes in ambient air quality or noise levels
    i. Adverse effects of chemical treatments
    j. Compliance with executive order 11988 on floodplain management
    k. Dredging and other channel bed or shoreline modifications

l. Adverse effects on wetlands and related resources
m. Feasible alternatives to proposed project
n. Other necessary mitigative measures requirements

### 5.5 Phase 2: Implementation

A Phase 1 D/F study identifies the major problems in a lake and recommends a management plan. Under the CLP, if funds are available, the plan can then be implemented and intensive monitoring of the lake and tributaries conducted either by the state EPA or a contracted agency.

The physical, chemical, and biological parameters monitored and the frequencies of monitoring are similar to those for Phase 1 D/F study (or more specific, depending on the objectives). One year of water quality monitoring is required for post-implementation study. Non-federal cost-share (50%) funds are usually provided by state and local landowners. A comprehensive monitoring program should be conducted to obtain post-implementation data for comparison with pre-implementation (Phase 1) data.

The implementation program may include keeping point- and nonpoint-source pollutants from entering a lake, implementing in-lake restoration measures to improve lake water quality, a monitoring program, environmental impact and cost evaluation, and public participation.

### 5.6 Phase 3: Post-restoration monitoring

The CLP has recently begun setting up a system for managing and evaluating lake project data, designed to derive benefit from past projects. Quantitative scientific data on long-term project effectiveness will become increasingly available through post-restoration monitoring studies being conducted under Phase 3 CLP grants instituted in 1989.

Several years after Phase 2 implementation, a lake may be selected for Phase 3 post-restoration monitoring with matching funds. For Phase 3 monitoring, the water quality parameters monitored and the frequency of sampling in the lake and tributaries are similar to that for Phases 1 and 2. However, the period of monitoring for Phase 3 is generally 3 years with a total project period of 4 to 5 years. The purpose is to build upon an extensive database of information gathered under Phases 1 and 2 investigations.

Data gathered from Phase 1 are compared with Phases 2 and 3 databases to determine the long-term effectiveness of watershed protection measures and in-lake management techniques implemented during and since Phase 2 completion.

### 5.7 Watershed protection approach

The watershed protection approach (WPA) is an integrated strategy for more effective restoration and protection of aquatic ecosystems and human health; i.e. drinking water supplies and fish consumption (US EPA, 1993). The WPA focuses on hydrologically defined drainage basins ("watersheds") rather than on areas arbitrarily defined by political boundaries. Local decisions on the scale of geographic units consider many factors including the ecological structure of the basin, the hydrologic factors of groundwaters, the economic lake uses, the type and scope of pollution problems, and the level of resources available for protection and restoration projects (US EPA, 1991, 1993).

The WPA has three major principles (US EPA, 1991, 1993):

1. Problem identification—identify the primary risk to human health and ecosystem within the watershed;
2. Stakeholder involvement—involve all parties most likely to be concerned or most able to take action for solution;
3. Integrated actions—take corrective actions in a manner that provides solutions and evaluates results.

The WPA is not a new program, and any watershed planning is not mandated by federal law. It is a flexible framework to achieve maximum efficiency and effect by collaborative activities. Everyone—individual citizens, the public and private sectors—can benefit from a WPA.

The US EPA's goal for the WPA is to maintain and improve the health and integrity of aquatic ecosystems using comprehensive approaches that focus resources on the major problems facing these systems within the watershed (US EPA, 1993).

For more than two decades, the CLP has emphasized using the watershed protection approach. A long-standing program policy gives greater consideration to applicants who propose restoration and protection techniques that control pollutants at the source through watershed-wide management rather than dealing with symptoms in the lake.

### 5.8 In-lake monitoring

In Phase 1, 2, and 3 studies, in-lake water quality parameters that need to be monitored generally include water temperature, dissolved oxygen, turbidity, Secchi transparency, solids (total suspended and volatile), conductivity, pH, alkalinity, nitrogen (ammonia, nitrite/nitrate, and total kjeldahl), phosphorus (total and dissolved), total and fecal coliforms,

**TABLE 2.1 Illinois General Use Water Quality Standards**

1. Dissolved oxygen: $> 5$ mg/L at any time
   $> 6$ mg/L at 16 h/24 h
2. Temperature: $< 30$°C
3. Total dissolved solids: 100 mg/L
   (conductivity: 1700 $\mu\Omega^{-1}$/cm)
4. Chloride: $< 500$ mg/L for protection of aquatic life
5. pH: 6.5–9.0, except for natural causes
   (alkalinity in Illinois lakes 20–200 mg/L as $CaCO_3$)
6. $NH_3$-N: $< 1.5$ mg/L at 20°C (68°F) and pH of 8.0
   $< 15$ mg/L under no conditions
7. $NO_3$-N: $< 10$ mg/L in all waters
   $< 1.0$ mg/L in public water supply
8. Total P: $< 0.05$ mg/L in any lake $> 8.1$ ha (20 acres)

SOURCE: IEPA, 1987

fecal streptococci, algae, macrophytes, and macroinvertebrates. Except for the last two parameters, samples are collected bimonthly (April to October) and monthly (other months) for at least 1 year. At a minimum, three stations (at the deepest, the middle, and upper locations of the lake) are generally monitored (Lin *et al.*, 1996, 1998).

**Lake water quality standards and criteria.** The comprehensive information and data collected during Phase 1, 2, or 3 study are evaluated with state water quality standards and state lake assessment criteria. Most states in the United States have similar state standards and assessment criteria. For example, Illinois generally uses the water quality standards applied to lake water listed in Table 2.1; and Illinois EPA's lake assessment criteria are presented in Table 2.2.

The data obtained from Phase 1, 2, and 3 of the study can be plotted from temporal variations for each water quality parameter, as shown in the example given in Fig. 2.4. Similarly, the historical data also can be plotted in the same manner for comparison purposes.

Indicator bacteria (total and fecal coliforms, and fecal streptococcus) densities in lakes and reservoirs should be evaluated in the same manner as that given for rivers and streams in Chapter 1. The density and moving geometric mean of FC are usually stipulated by the state's

**TABLE 2.2 Illinois EPA Lake Assessment Criteria**

| Parameter | Minimal | Slight | Moderate | High | Indication |
|---|---|---|---|---|---|
| Secchi depth | > 79 | 49–79 | 18–48 | < 18 | lake use impairment |
| TSS, mg/L | < 5 | 5–15 | 15–25 | > 25 | lake use impairment |
| Turbidity, NTU | < 3 | 3–7 | 7–15 | > 15 | suspended sediment |
| COD, mg/L | < 10 | 10–20 | 20–30 | > 30 | organic enrichment |

SOURCE: IEPA, 1987

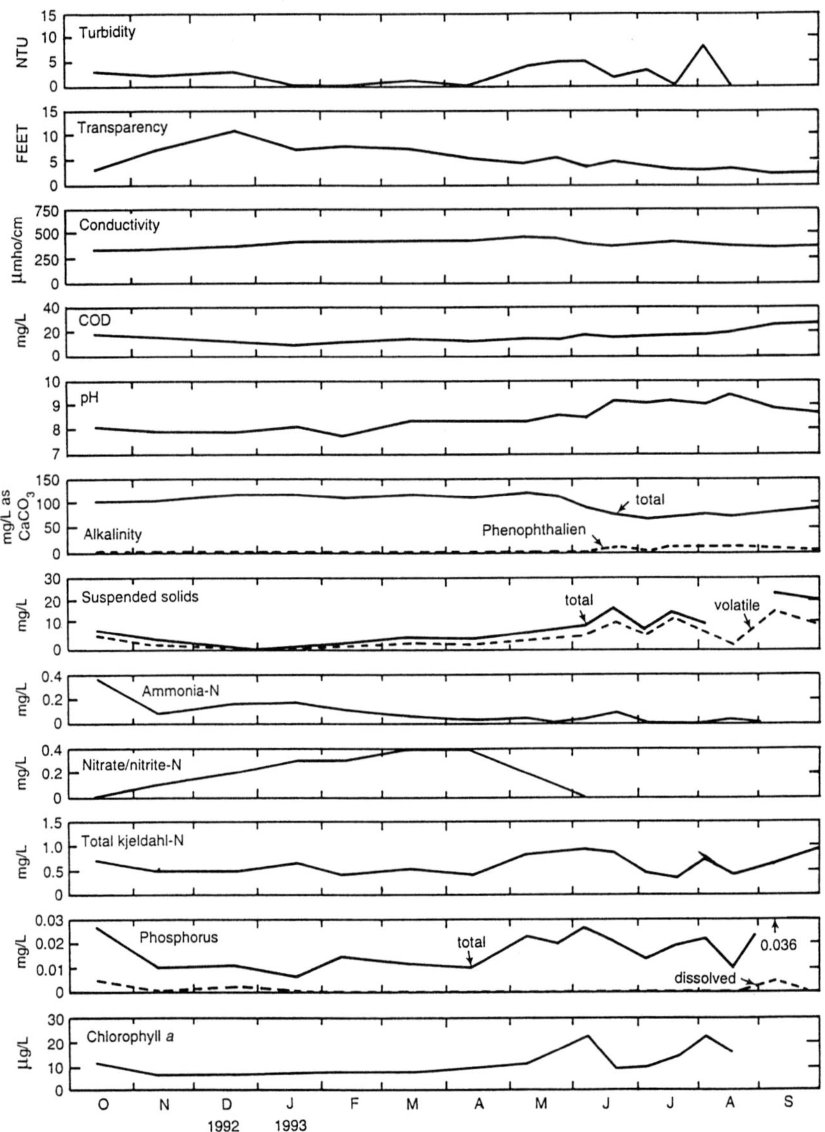


**Figure 2.4** Temporal variations of surface water characteristics at RHA-2 (*Lin et al., 1996*).

bacterial quality standards. The ratio of FC/FS is used to determine the source of pollution.

**Temperature and dissolved oxygen.** Water temperature and dissolved oxygen in lake waters are usually measured at 1 ft (30 cm) and 2 ft (60 cm) intervals. The obtained data are used to calculate values of the percent

DO saturation using Eq. (1.4) or from Table 1.2. The plotted data is shown in Fig. 2.5: observed vertical DO and temperature profiles on selected dates at a station. Data from long-term observations can also be depicted in Fig. 2.6 which shows the DO isopleths and isothermal plots for a near-bottom (2 ft above the bottom station).

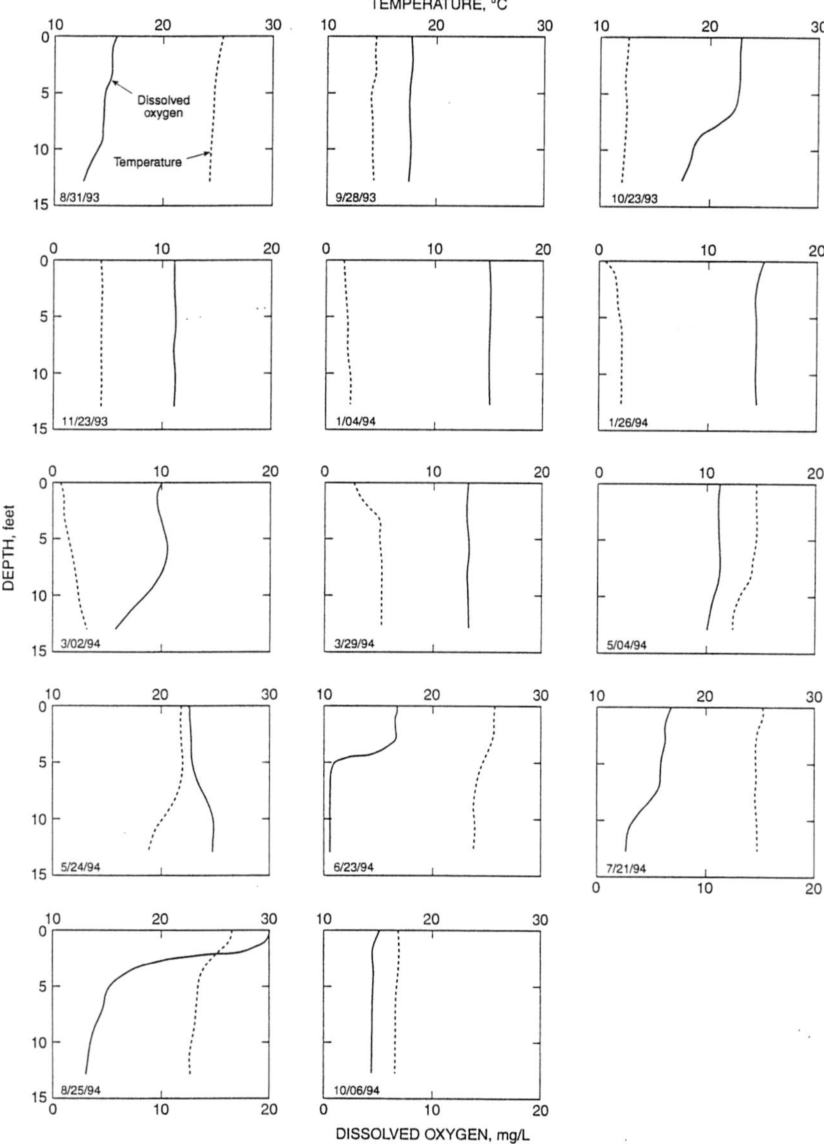


**Figure 2.5** Temperature and dissolved oxygen profiles at station 2 (*Lin and Raman, 1997*).

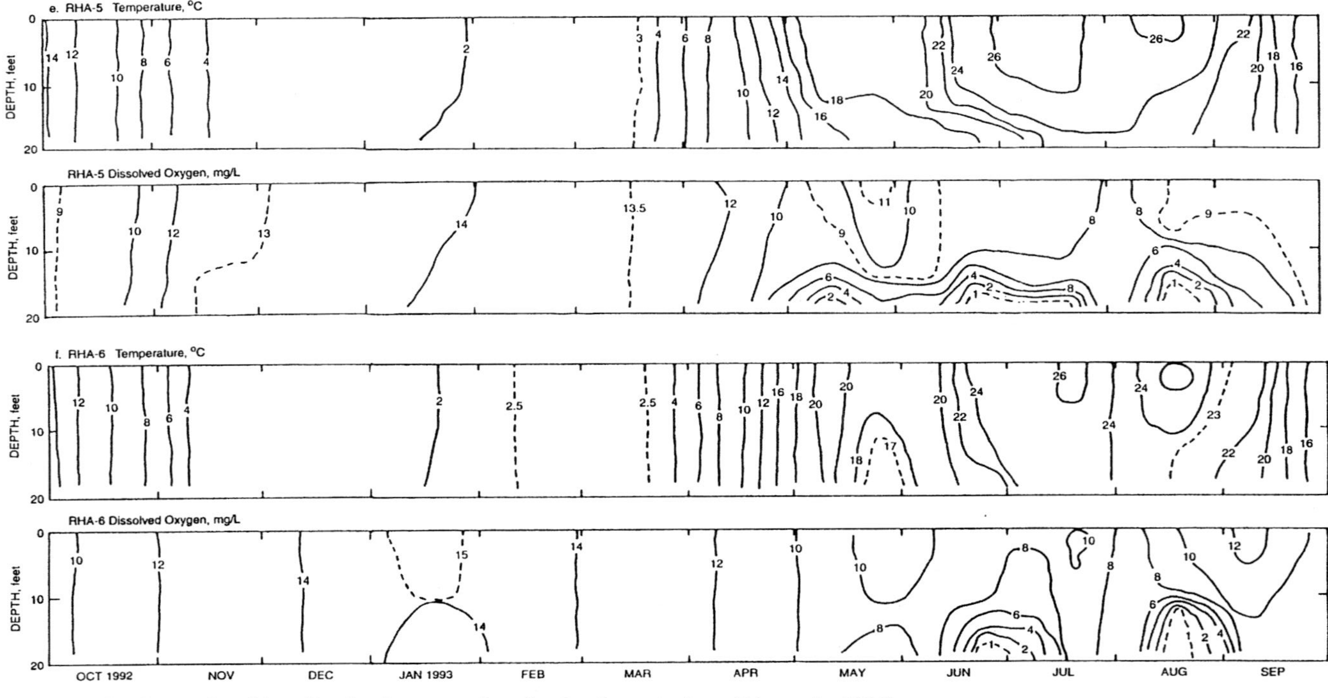


**Figure 2.6** Isothermal and iso-dissolved oxygen plots for the deep stations (*Lin et al., 1996*).

**Algae.** Most algae are microscopic, free-floating plants. Some are filamentous and attached ones. Algae are generally classified into four major types: blue-greens, greens, diatoms, and flagellates. Through their photosynthesis processes, algae use energy from sunlight and carbon dioxide from bicarbonate sources, converting it to organic matter and oxygen (Fig. 2.7). The removal of carbon dioxide from the water results in an increase in pH and a decrease in alkalinity. Algae are important sources of dissolved oxygen in water. Phytoplanktonic algae form at the base of the aquatic food web and provide the primary source of food for fish and other aquatic insects and animals.

Excessive growth (blooms) of algae may cause problems such as bad taste and odor, increased color and turbidity, decreased filter run at a water treatment plant, unsightly surface scums and esthetic problems, and even oxygen depletion after die-off. Blue-green algae tend to cause the worst problems. To prevent such proliferation (upset the ecological balance), lake and watershed managers often try to reduce the amounts of nutrients entering lake waters, which act like fertilizers in promoting algal growth. Copper sulfate is frequently used for algal control, and sometimes it is applied on a routine basis during summer months, whether or not it is actually needed. One should try to identify the

$$\underset{\text{Carbon dioxide}}{CO_2} + \underset{\text{Water}}{H_2O} \xrightarrow[\text{Enzymes}]{\text{Sunlight}} \underset{\text{Organic matter}}{(CH_2O)_n} + \underset{\text{Oxygen}}{O_2}$$

(a) **The photosynthetic process**

$$CO_2 + H_2O \rightleftharpoons H_2CO_3 \rightleftharpoons HCO_3^- + H^+$$

$$HCO_3^- \rightleftharpoons CO_3^= + H^+$$

$$CO_3^= + H_2O \rightleftharpoons HCO_3^- + OH^-$$

(b) **Carbonate equilibria**

Removed by photosynthesis

$$2HCO_3^- \rightleftharpoons CO_2 + CO_3^= + H_2O$$

$$Ca^{++} + CO_3^= \rightarrow CaCO_3 \downarrow \text{ (Marl deposits)}$$

(c) **Chemical processes that occur during photosynthesis**

**Figure 2.7** Photosynthesis and its chemical processes (*Illinois State Water Survey, 1989*).

causes of the particular problems and then take corrective measures selectively rather than on a shot-in-the-dark basis.

Copper sulfate ($CuSO_4 \cdot 5H_2O$) was applied in an Illinois impoundment at an average rate of 22 pounds per acre (lb/acre); in some lakes, as much as 80 lb/acre was used (Illinois State Water Survey, 1989). Frequently, much more copper sulfate is applied than necessary. Researchers have shown that 5.4 lb of copper sulfate per acre of lake surface (6.0 kg/ha) is sufficient to control problem-causing blue-green algae in waters with high alkalinity (>40 mg/L as $CaCO_3$). The amount is equivalent to the rate of 1 mg/L of copper sulfate for the top 2 ft (60 cm) of the lake surface (Illinois State Water Survey, 1989). The literature suggests that a concentration of 0.05 to 0.10 mg/L as $Cu^{2+}$ is effective in controlling blue-green algae in pure cultures under laboratory conditions.

**Example 1:** What is the equivalent concentration of $Cu^{2+}$ of 1 mg/L of copper sulfate in water?

**solution:**

$$\text{MW of } CuSO_4 \cdot 5H_2O = 63.5 + 32 + 16 \times 4 + 5(2 + 16)$$

$$= 249.5$$

$$\text{MW of } Cu^{2+} = 63.5$$

$$\frac{Cu^{2+}, \text{ mg/L}}{CuSO_4 \cdot 5H_2O, \text{ mg/L}} = \frac{63.5}{249.5} = 0.255$$

$$Cu^{2+} = 0.255 \times 1 \text{ mg/L}$$

$$= 0.255 \text{ mg/L}$$

**Example 2:** Since 0.05 to 0.10 mg/L of $Cu^{2+}$ is needed to control blue-green algae, what is the theoretical concentration expressed as $CuSO_4 \cdot 5H_2O$?

**solution:**

$$\text{Copper sulfate} = 0.05 \text{ mg/L} \times \frac{249.5}{63.5}$$

$$\cong 0.20 \text{ mg/L}$$

Answer: This is equivalent to 0.20 to 0.4 mg/L as $CuSO_4 \cdot 5H_2O$.

*Note:* For field application, however, a concentration of 1.0mg/L as $CuSO_4 \cdot 5H_2O$ is generally suggested.

**Example 3:** For a lake with 50 acres (20.2 ha), 270 lb (122 kg) of copper sulfate is applied. Compute the application rate in mg/L on the basis of the top 2 ft of lake surface.

**solution:**

Step 1. Compute the volume ($V$) of the top 2 ft

$$\begin{aligned} V &= 2\text{ ft} \times 50\text{ acres} \\ &= 2\text{ ft} \times 0.3048\text{ m/ft} \times 50\text{ acres} \times 4047\text{ m}^2\text{/acre} \\ &= 123.350\text{ m}^3 \\ &= 123.3 \times 10^6\text{ L} \end{aligned}$$

Step 2. Convert weight ($W$) in pounds to milligrams

$$\begin{aligned} W &= 270\text{ lb} \times 453{,}600\text{ mg/lb} \\ &= 122.5 \times 10^6\text{ mg} \end{aligned}$$

Step 3. Compute copper sulfate application rate

$$\begin{aligned} \text{Rate} &= \frac{W}{V} = \frac{122.5 \times 10^6\text{ mg}}{123.3 \times 10^6\text{ L}} \\ &= 0.994\text{ mg/L} \\ &\cong 1.0\text{ mg/L} \end{aligned}$$

**Secchi disc transparency.** The Secchi disc is named after its Italian inventor Pietro Angelo Secchi, and is a black and white round plate used to measure water clarity. Values of Secchi disc transparency are used to classify a lake's trophic state.

Secchi disc visibility is a measure of a lake's water transparency, which suggests the depth of light penetration into a body of water (its ability to allow sunlight to penetrate). Even though Secchi disc transparency is not an actual quantitative indication of light transmission, it provides an index for comparing similar bodies of water or the same body of water at different times. Since changes in water color and turbidity in deep lakes are generally caused by aquatic flora and fauna, transparency is related to these entities. The euphotic zone or region of a lake where enough sunlight penetrates to allow photosynthetic production of oxygen by algae and aquatic plants is taken as two to three times the Secchi disc depth (US EPA, 1980).

Suspended algae, microscopic aquatic animals, suspended matter (silt, clay, and organic matter), and water color are factors that interfere with light penetration into the water column and reduce Secchi disc transparency. Combined with other field observations, Secchi disc readings may furnish information on (1) suitable habitat for fish and other aquatic life; (2) the lake's water quality and esthetics; (3) the state of the lake's nutrient enrichment; and (4) problems with and potential solutions for the lake's water quality and recreational use impairment.

**Phosphorus.** The term total phosphorus (TP) represents all forms of phosphorus in water, both particulate and dissolved forms, and includes three chemical types: reactive, acid-hydrolyzed, and organic. Dissolved phosphorus (DP) is the soluble form of TP (filterable through a 0.45-μm filter).

Phosphorus as phosphate may occur in surface water or groundwater as a result of leaching from minerals or ores, natural processes of degradation, or agricultural drainage. Phosphorus is an essential nutrient for plant and animal growth and, like nitrogen, it passes through cycles of decomposition and photosynthesis.

Because phosphorus is essential to the plant growth process, it has become the focus of attention in the entire eutrophication issue. With phosphorus being singled out as probably the most limiting nutrient and the one most easily controlled by removal techniques, various facets of phosphorus chemistry and biology have been extensively studied in the natural environment. Any condition which approaches or exceeds the limits of tolerance is said to be a limiting condition or a limiting factor.

In any ecosystem, the two aspects of interest for phosphorus dynamics are phosphorus concentration and phophorus flux (concentration × flow rate) as functions of time and distance. The concentration alone indicates the possible limitation that this nutrient can place on vegetative growth in the water. Phosphorus flux is a measure of the phosphorus transport rate at any point in flowing water.

Unlike nitrate-nitrogen, phosphorus applied to the land as a fertilizer is held tightly to the soil. Most of the phosphorus carried into streams and lakes from runoff over cropland will be in the particulate form adsorbed to soil particles. On the other hand, the major portion of phosphate-phosphorus emitted from municipal sewer systems is in a dissolved form. This is also true of phosphorus generated from anaerobic degradation of organic matter in the lake bottom. Consequently, the form of phosphorus, namely, particulate or dissolved, is indicative of its source to a certain extent. Other sources of dissolved phosphorus in the lake water may include the decomposition of aquatic plants and animals. Dissolved phosphorus is readily available for algae and macrophyte growth. However, the DP concentration can vary widely over short periods of time as plants take up and release this nutrient. Therefore, TP in lake water is the more commonly used indicator of a lake's nutrient status.

From his experience with Wisconsin lakes, Sawyer (1952) concluded that aquatic blooms are likely to develop in lakes during summer months when concentrations of inorganic nitrogen and inorganic phosphorus exceed 0.3 and 0.01mg/L, respectively. These critical levels for nitrogen and phosphorus concentrations have been accepted and widely quoted in scientific literature.

To prevent biological nuisance, the IEPA (1990) stipulates, "Phosphorus as P shall not exceed a concentration of 0.05 mg/L in any reservoir or lake with a surface area of 8.1 ha (20 acres) or more or in any stream at the point where it enters any reservoir or lake."

**Chlorophyll.** All green plants contain chlorophyll *a*, which constitutes approximately 1% to 2% of the dry weight of planktonic algae (APHA *et al.*, 1992). Other pigments that occur in phytoplankton include chlorophyll *b* and *c*, xanthophylls, phycobilius, and carotenes. The important chlorophyll degration products in water are the chlorophyllides, pheophorbides, and pheophytines. The concentration of photosynthetic pigments is used extensively to estimate phytoplanktonic biomass. The presence or absence of the various photosynthetic pigments is used, among other features, to identify the major algal groups present in the water body.

Chlorophyll *a* is a primary photosynthetic pigment in all oxygen-evolving photosynthetic organisms. Extraction and quantification of chlorophyll *a* can be used to estimate biomass or the standing crop of planktonic algae present in a body of water. Other algae pigments, particularly chlorophyll *b* and *c*, can give information on the type of algae present. Blue-green algae (Cyanophyta) contain only chlorophyll *a*, while both the green algae (Chlorophyta) and the euglenoids (Euglenophyta) contain chlorophyll *a* and *c*. Chlorophyll *a* and *c* are also present in the diatoms, yellow-green and yellow-brown algae (Chrysophyta), as well as dinoflagellates (Pyrrhophyta). These accessory pigments can be used to identify the types of algae present in a lake. Pheophytin *a* results from the breakdown of chlorophyll *a*, and a large amount indicates a stressed algal population or a recent algal die-off. Because direct microscopic examination of water samples is used to identify and enumerate the type and concentrations of algae present in the water samples, the indirect method (chlorophyll analyses) of making such assessments may not be employed.

Nutrient in lake water will impact the aquatic community in the surface water during summer. Dillon and Rigler (1974) used data from North American lakes to drive the relationship between spring phosphorus and chlorophyll *a* concentration in summer as follows:

$$\mathrm{CHL} = 0.0731\ (\mathrm{TP})^{1.449} \tag{2.14}$$

where CHL = summer average chlorophyll *a* concentration at lake surface water, $\mathrm{mg/m^3}$

TP = spring average total phosophorus concentration at lake surface water, $\mathrm{mg/m^3}$

**Example:** Estimate the average chlorophyll *a* concentration in a North American lake during the summer if the average spring total phosphorus is 0.108 mg/L.

**solution:**

$$\begin{aligned} \text{TP} &= 0.108 \text{ mg/L} \\ &= 0.108 \text{ mg/L} \times 1000 \text{ L/m}^3 \\ &= 108 \text{ mg/m}^3 \end{aligned}$$

Estimate summer CHL using Eq. (2.14)

$$\begin{aligned} \text{CHL} &= 0.0731\,(\text{TP})^{1.449} \\ &= 0.0731\,(108)^{1.449} \text{ mg/m}^3 \\ &= 64.6 \text{ mg/m}^3 \end{aligned}$$

### 5.9 Trophic state index

Eutrophication is a normal process that affects every body of water from its time of formation (Walker, 1981a, 1981b). As a lake ages, the degree of enrichment from nutrient materials increases. In general, the lake traps a portion of the nutrients originating in the surrounding drainage basin. Precipitation, dry fallout, and groundwater inflow are the other contributing sources.

A wide variety of indices of lake trophic conditions have been proposed in the literature. These indices have been based on Secchi disc transparency; nutrient concentrations; hypolimnetic oxygen depletion; and biological parameters, including chlorophyll *a*, species abundance, and diversity. In its *Clean Lake Program Guidance Manual*, the US EPA (1980) suggests the use of four parameters as trophic indicators: Secchi disc transparency, chlorophyll *a*, surface water total phosphorus, and total organic carbon.

In addition, the lake trophic state index (TSI) developed by Carlson (1977) on the basis of Secchi disc transparency, chlorophyll *a*, and surface water total phosphorus can be used to calculate a lake's trophic state. The TSI can be calculated from Secchi disc transparency (SD) in meters (m), chlorophyll *a* (CHL) in micrograms per liter (μg/L), and total phosphorus (TP) in μg/L as follows:

$$\text{on the basis of SD, TSI} = 60 - 14.4 \ln(\text{SD}) \tag{2.15}$$

$$\text{on the basis of CHL, TSI} = 9.81 \ln(\text{CHL}) + 30.6 \tag{2.16}$$

$$\text{on the basis of TP, TSI} = 14.42 \ln(\text{TP}) + 4.15 \tag{2.17}$$

**TABLE 2.3 Quantitative Definition of a Lake Trophic State**

| Trophic state | Secchi disc transparency in | Secchi disc transparency m | Chlorophyll *a* (μg/L) | Total phosphorus, lake surface (μg/L) | TSI |
|---|---|---|---|---|---|
| Oligotrophic | >157 | >4.0 | <2.6 | <12 | <40 |
| Mesotrophic | 79–157 | 2.0–4.0 | 2.6–7.2 | 12–24 | 40–50 |
| Eutrophic | 20–79 | 0.5–2.0 | 7.2–55.5 | 24–96 | 50–70 |
| Hypereutrophic | <20 | <0.5 | >55.5 | >96 | >70 |

The index is based on the amount of algal biomass in surface water, using a scale of 0 to 100. Each increment of 10 in the TSI represents a theoretical doubling of biomass in the lake. The advantages and disadvantages of using the TSI were discussed by Hudson *et al.* (1992). The accuracy of Carlson's index is often diminished by water coloration or suspended solids other than algae. Applying TSI classification to lakes that are dominated by rooted aquatic plants may indicate less eutrophication than actually exists.

Lakes are generally classified by limnologists into one of three trophic states: oligotrophic, mesotrophic, or eutrophic (Table 2.3). Oligotrophic lakes are known for their clean and cold waters and lack of aquatic weeds or algae, due to low nutrient levels. There are few oligotrophic lakes in the Midwest. At the other extreme, eutrophic lakes are high in nutrient levels and are likely to be very productive in terms of weed growth and algal blooms. Eutrophic lakes can support large fish populations, but the fish tend to be rougher species that can better tolerate depleted levels of DO. Mesotrophic lakes are in an intermediate stage between oligotrophic and eutrophic. The great majority of Midwestern lakes are eutrophic. A hypereutrophic lake is one that has undergone extreme eutrophication to the point of having developed undesirable esthetic qualities (e.g. odors, algal mats, and fish kills) and water-use limitations (e.g. extremely dense growths of vegetation). The natural aging process causes all lakes to progress to the eutrophic condition over time, but this eutrophication process can be accelerated by certain land uses in the contributing watershed (e.g. agricultural activities, application of lawn fertilizers, and erosion from construction sites, unprotected areas, and streams). Given enough time, a lake will grow shallower and will eventually fill in with trapped sediments and decayed organic matter, such that it becomes a shallow marsh or emergent wetland.

**Example 1:** Lake monitoring data shows that the Secchi disc transparency is 77 in; total phosphorus, 31μg/L; and chlorophyll *a* 3.4 μg/L. Calculate TSI values using Eqs. (2.15) to (2.17).

**solution:**

Step 1.

Using Eq. (2.15)

$$\text{SD} = 77\ \text{in} \times 2.54\ \text{cm/in} \times 0.01\ \text{m/cm} = 1.956\ \text{m}$$

$$\text{TSI} = 60 - 14.4\ \ln\ (\text{SD}) = 60 - 14.4\ \ln\ (1.956)$$
$$= 50.3$$

Step 2.

Using Eq. (2.16)

$$\text{TSI} = 9.81\ \ln\ (\text{CHL}) + 30.6 = 9.81\ \ln\ (2.4) + 30.6$$
$$= 42.6$$

Step 3.

Using Eq. (2.17)

$$\text{TSI} = 14.42\ \ln\ (\text{TP}) + 4.15 = 14.42\ \ln\ (31) + 4.15$$
$$= 53.7$$

**Example 2:** One-year lake monitoring data for a Secchi disc transparency, total phosphorus, and chlorophyll *a* in a southern Illinois lake (at the deepest station) are listed in Table 2.4. Determine the trophic condition of the lake.

**solution:**

Step 1. Calculate TSI values:

As in Example 1, the TSI values for the lake are calculated using Eqs. (2.15) to (2.17) based on SD, TP, and chlorophyll *a* concentration of each sample. The TSI values are included in Table 2.4.

Step 2. Determine trophic stage

a. Calculate average TSI based on each parameter.
b. Classify trophic state based on the average TSI and the criteria listed in Table 2.3.
c. Calculate overall mean TSI for the lake.
d. Determine overall trophic state. For this example, the lake is classified as *eutrophic*. The classifications are slightly different if based on each of the three water quality parameters.

## 5.10 Lake use support analysis

**Definition.** An analysis of a lake's use support can be carried out employing a methodology (similar to Federal and other states) developed by the

**TABLE 2.4 Trophic State Index and Trophic State of an Illinois Lake**

| | Secchi disc trans | | Total phosphorus | | Chlorophyll *a* | |
|---|---|---|---|---|---|---|
| Date | in | TSI | μg/L | TSI | μg/L | TSI |
| 10/26/95 | 77 | 50.3 | 31 | 53.7 | 3.4 | 42.6 |
| 11/20/95 | 104 | 46.0 | 33 | 54.6 | 4.7 | 45.8 |
| 12/12/95 | 133 | 42.5 | 21 | 48.1 | 3 | 41.4 |
| 1/8/96 | 98 | 46.9 | 29 | 52.7 | | |
| 2/14/96 | 120 | 44.0 | 26 | 51.1 | 2.1 | 37.9 |
| 3/11/96 | 48 | 57.1 | 22 | 48.7 | 6.2 | 48.5 |
| 4/15/96 | 43 | 58.7 | 49 | 60.3 | 10.4 | 53.6 |
| 5/21/96 | 52 | 56.0 | 37 | 56.2 | 12.4 | 55.3 |
| 6/6/96 | 51 | 56.3 | 36 | 55.8 | 7.1 | 49.8 |
| 6/18/96 | 60 | 53.9 | 30 | 53.2 | 14.6 | 56.9 |
| 7/1/96 | 60 | 53.9 | 21 | 48.1 | 7.7 | 50.6 |
| 7/16/96 | 42 | 59.1 | 38 | 56.6 | 12.4 | 55.3 |
| 8/6/96 | 72 | 51.3 | 19 | 46.6 | 6 | 48.2 |
| 8/20/96 | 48 | 57.1 | 20 | 47.3 | 1.7 | 35.8 |
| 9/6/96 | 54 | 55.4 | 32 | 54.1 | 9.8 | 53.0 |
| 9/23/96 | 69 | 51.9 | 22 | 48.9 | 11.3 | 54.4 |
| 10/23/96 | 42 | 59.1 | 42 | 58.0 | | |
| Mean | | 52.9 | | 52.6 | | 48.6 |
| Trophic state | | eutrophic | | eutrophic | | mesotrophic |
| Overall mean | | 51.4 | | | | |
| Overall trophic state | | eutrophic | | | | |

SOURCE: Bogner *et al.* (1997)

IEPA (1994). The degree of use support identified for each designated use indicates the ability of the lake to (1) support a variety of high-quality recreational activities, such as boating, sport fishing, swimming, and esthetic enjoyment; (2) support healthy aquatic life and sport fish populations; and (3) provide adequate, long-term quality and quantity of water for public or industrial water supply (if applicable). Determination of a lake's use support is based upon the state's water quality standards as described in Subtitle C of Title 35 of the State of Illinois Administrative Code (IEPA, 1990). Each of four established use-designation categories (including general use, public and food processing water supply, Lake Michigan, and secondary contact and indigenous aquatic life) has a specific set of water quality standards.

For the lake uses assessment, the general use standards—primarily the 0.05 mg/L TP standard—should be used. The TP standard has been established for the protection of aquatic life, primary-contact (e.g. swimming), fish consumption, and secondary-contact (e.g. boating) recreation, agriculture, and industrial uses. In addition, lake use support is based in part on the amount of sediment, macrophytes, and algae in the lake and how these might impair designated lake uses. The following is a summary of the various classifications of use impairment:

- *Full* = full support of designated uses, with minimal impairment.
- *Full/threatened* = full support of designated uses, with indications of declining water quality or evidence of existing use impairment.
- *Partial/minor* = partial support of designated uses, with slight impairment.
- *Partial/moderate* = partial support of designated uses, with moderate impairment, only Partial is used after the year of 2000.
- *Nonsupport* = no support of designated uses, with severe impairment.

Lakes that fully support designated uses may still exhibit some impairment, or have slight-to-moderate amounts of sediment, macrophytes, or algae in a portion of the lake (e.g. headwaters or shoreline); however, most of the lake acreage shows minimal impairment of the aquatic community and uses. *It is important to emphasize that if a lake is rated as not fully supporting designated uses, it does not necessarily mean that the lake cannot be used for those purposes or that a health hazard exists.* Rather, it indicates impairment in the ability of significant portions of the lake waters to support either a variety of quality recreational experiences or a balanced sport fishery. Since most lakes are multiple-use water bodies, a lake can fully support one designated use (e.g. aquatic life) but exhibit impairment of another (e.g. swimming).

Lakes that partially support designated uses have a designated use that is slightly to moderately impaired in a portion of the lake (e.g. swimming impaired by excessive aquatic macrophytes or algae, or boating impaired by sediment accumulation). So-called nonsupport lakes have a designated use that is severely impaired in a substantial portion of the lake (e.g. a large portion of the lake has so much sediment that boat ramps are virtually inaccessible, boating is nearly impossible, and fisheries are degraded). However, in other parts of the same nonsupport lake (e.g. near a dam), the identical use may be supported. *Again, nonsupport does not necessarily mean that a lake cannot support any uses, that it is a public health hazard, or that its use is prohibited.*

Lake-use support and level of attainment shall be determined for aquatic life, recreation, swimming, drinking water supply, fish consumption, secondary contact, and overall lake use, using methodologies described in the IEPA's *Illinois Water Quality Report 1994–1995* (IEPA, 1996).

The primary criterion in the aquatic life use assessment is an aquatic life use impairment index (ALI), while in the recreation use assessment the primary criterion is a recreation use impairment index (RUI). While both indices combine ratings for TSI (Carlson, 1977) and degree of use impairment from sediment and aquatic macrophytes, each index is specifically designed for the assessed use. ALI and RUI relate directly to the TP standard of 0.05mg/L. If a lake water sample is found to have a TP concentration at or below the standard, the lake is given a "full support" designation. The aquatic life use rating reflects the degree of attainment of the "fishable goal" of the Clean Water Act, whereas the recreation use rating reflects the degree to which pleasure boating, canoeing, and esthetic enjoyment may be obtained at an individual lake.

The assessment of swimming use for primary-contact recreation was based on available data using two criteria: (1) Secchi disc transparency depth data and (2) Carlson's TSI. The swimming use rating reflects the degree of attainment of the "swimmable goal" of the Clean Water Act. If a lake is rated "nonsupport" for swimming, it does not mean that the lake cannot be used or that health hazards exist. It indicates that swimming may be less desirable than at those lakes assessed as fully or partially supporting swimming.

Finally, in addition to assessing individual aquatic life, recreation, and swimming uses, and drinking water supply, the overall use support of the lake is also assessed. The overall use support methodology aggregates the use support attained for each of the individual lake uses assessed. Values assigned to each use-support attainment category are summed and averaged, and then used to assign an overall lake use attainment value for the lake.

**Designated uses assessment.** Multiple lakes designated are assessed for aquatic life, recreation, drinking water supply, swimming, fish consumption, and overall use. Specific criteria for determining attainment of these designated lake uses are described below. The degree of use support attainment is described as full, full/threatened, partial/minor impairment, partial/moderate impairment, or nonsupport.

**Aquatic life.** An aquatic life use impairment index (ALI) which combines ratings for trophic state index and the amount of use impairments from aquatic macrophytes and sediment is used as the primary criteria for assessing aquatic life lake use (Table 2.5). The higher the ALI number, the more impaired the lake. Specific criteria used for each level of aquatic life use support attainment are presented in Table 2.6.

**Recreation.** A recreation use impairment index (RUI), which combines TSI and the amount of use impairments from aquatic life and from sediment is utilized as the primary criteria for assessing recreation lake use (Table 2.7). Lake uses include pleasure boating, canoeing, skiing, sailing, esthetic enjoyment, and fishing. The higher the RUI number, the more impaired the lake. Specific criteria used for each level of attaining recreation use support are listed in Table 2.8.

**TABLE 2.5 Aquatic Life Use Impairment Index (ALI)**

| Evaluation factor | Parameter | Weighting criteria | Points |
|---|---|---|---|
| 1. Mean trophic state index (Carson, 1977) | Mean TSI value between 30 and 100 | a. TSI < 60<br>b. 60 ≤ TSI < 85<br>c. 85 ≤ TSI < 90<br>d. 90 < TSI | a. 40<br>b. 50<br>c. 60<br>d. 70 |
| 2. Macrophyte impairment | Percent of lake surface area covered by weeds, or amount of weeds recorded on form | a. 15 ≤ % < 40; or minimal (1)<br>b. 10 ≤ % < 15 & < 40 ≤ % < 50; or slight (2)<br>c. 5 ≤ % < 10 & 50 ≤ % < 70; or moderate (3)<br>d. % < 5 & 70 ≤ %; or substantial (4) | a. 0<br>b. 5<br>c. 10<br>d. 15 |
| 3. Sediment impairment | Concentration of nonvolatile suspended solids (NVSS); or amount of sediment value reported on form | a. NVSS < 12; or minimal (1)<br>b. 12 ≤ NVSS < 15; or slight (2)<br>c. 15 ≤ NVSS < 20; or moderate (3)<br>d. 20 ≤ NVSS; or substantial (4) | a. 0<br>b. 5<br>c. 10<br>d. 15 |

SOURCE: Modified from Illinois EPA (1996)

**Swimming.** The assessment criteria for swimming use is based primarily on the Secchi disc transparency depth and on the fecal coliform (FC) density-percent that exceed the 200 FC/100mL standard. If FC data are not available, a TSI is calculated and used to make the assessment. The degree of swimming use support attainment is presented in Table 2.9.

**TABLE 2.6 Assessment Criteria for Aquatic Life and Overall Use in Illinois Lakes**

| Degree of use support | Criteria |
|---|---|
| Full | a. Total ALI points are < 75<br>b. Direct field observations of minimal aquatic life impairment |
| Full/threatened | a. Total ALI point are < 75 and evidence of a declined water quality trend exists<br>b. Specific knowledge of existing or potential threats to aquatic life impairment |
| Partial/minor | a. 75 ≤ total ALI points < 85<br>b. Direct field observations of slight aquatic life impairment |
| Partial/moderate | a. 85 ≤ total ALI points < 95<br>b. Direct field observations of moderate aquatic life impairment |
| Nonsupport | a. Total ALI points ≥ 95<br>b. Direct field observation of substantial aquatic life impairment |

SOURCE: Modified form Illinois EPA (1996)

**TABLE 2.7 Recreation Use Impairment Index (RUI)**

| Evaluation factor | Parameter | Weighting criteria | Points |
|---|---|---|---|
| 1. Mean trophic state index (Carlson, 1977) | Mean TSI value (30–110) | a. Actual TSI value | Actual TSI value |
| 2. Macrophyte impairment | Percent of lake surface area covered by weeds; or amount of weeds value reported on form | a. % < 5; or minimal (1)<br>b. 5 < % < 15; or slight (2)<br>c. 15 < % < 25; or moderate (3)<br>d. 25 < %; or substantial (4) | a. 0<br>b. 5<br>c. 10<br>d. 15 |
| 3. Sediment impairment | Concentration of nonvolatile suspended solid (NVSS); or amount of sediment value reported on form | a. NVSS % < 3; or minimal (1)<br>b. 3 ≤ NVSS < 7; or slight (2)<br>c. 7 ≤ NVSS < 15; or moderate (3)<br>d. 15 ≤ %; or substantial (4) | a. 0<br>b. 5<br>c. 10<br>d. 15 |

SOURCE: Modified form Illinois EPA (1996)

**Drinking water supply.** Drinking water supply use assessment for a lake is determined on the basis of water supply advisories or closure issued through state regulatory public water supply programs. For example, in Illinois, the primary criteria used is the length of time (greater than 30 days) nitrate and/or atrazine concentration exceeding the public

**TABLE 2.8 Assessment Criteria for Recreation Use in Illinois Lakes**

| Degree of use support | Criteria |
|---|---|
| Full | a. Total RUI points are less than 60<br>b. Direct field observation of minor recreation impairment |
| Full/threatened | a. Total RUI points are < 60, and evidence of a decline in water quality trend exists<br>b. Specific knowledge or potential threats to recreation impairment |
| Partial/minor moderate | a. 60 ≤ total RUI points < 75<br>b. Direct field observation of slight impairment |
| Partial/moderate | a. 75 ≤ total RUI points < 90<br>b. Direct field observation of moderate recreation impairment |
| Nonsupport | a. Total RUI points ≥ 90<br>b. Direct field observation of substantial recreation impairment |

SOURCE: Illinois EPA (1996)

**TABLE 2.9 Assessment Criteria for Swimming Use in Illinois Lakes**

| Degree of use support | Criteria |
|---|---|
| Full | a. No Secchi depths are $<$ 24 in<br>b. 10% $\geq$ fecal coliforms (FC) samples exceed the standard<br>c. TSI $\leq$ 50 |
| Partial/minor | a. $\leq$ 50% of Secchi depths were $<$ 24 in<br>b. 10% $<$ FC $\leq$ 25% exceed the standard<br>c. TSI $\leq$ 65 |
| Partial/moderate | a. 50–100% of Secchi depths were $<$ 24 in<br>b. 10% $<$ FC sample $\leq$ 25% exceed the standard<br>c. TSI $\leq$ 75 |
| Nonsupport | a. 100% of Secchi depths were $<$ 24 in<br>b. 25% $<$ FC sample exceed the standard<br>c. TSI $>$ 75 |

SOURCE: Modified from Illinois EPA (1996)

water supply standards of 10mg/L and 3 μg/L, respectively. Other problems which affect the quality of finished waters, such as chemical or oil spills or severe taste and odor problems requiring immediate attention, are also included in assessing use support. Specific criteria used for assessing the drinking water supply and the degree of use support are shown in Table 2.10.

**Fish consumption.** The assessment of fish consumption use is based on fish tissue data and resulting sport fish advisories generated by the

**TABLE 2.10 Assessment Criteria for Drinking Water Supply in Illinois Lakes**

| Degree of use support | Criteria |
|---|---|
| Full | No drinking water closures or advisories in effect during reporting period; no treatment necessary beyond "reasonable levels" (copper sulfate may occasionally be applied for algae/taste and odor control). |
| Partial/minor | One or more drinking water supply advisory(ies) lasting 30 days or less; or problems not requiring closure or advisories but adversely affecting treatment costs and quality of treated water, such as taste and odor problems, color, excessive turbidity, high dissolved solids, pollutants requiring activated carbon filters. |
| Partial/moderate | One or more drinking water supply advisories lasting more than 30 d/yr. |
| Nonsupport | One or more drinking water supply closures per year. |

SOURCE: Illinois EPA (1996)

**TABLE 2.11 Assessment Criteria for Fish Consumption Use in Illinois Lakes**

| Degree of use support | Criteria |
|---|---|
| Full | No fish advisories or bans are in effect. |
| Partial/moderate | Restricted consumption fish advisory or ban in effect for general population or a subpopulation that could be at potentially greater risk (pregnant women, children). Restricted consumption is defined as limits on the number or meals or size of meals consumed per unit time for one or more fish species. In Illinois, this is equivalent to a Group II advisory. |
| Nonsupport | No consumption fish advisory or ban in effect for general population for one or more fish species; commercial fishing ban in effect. In Illinois, this is equivalent to a Group III advisory. |

SOURCE: Illinois EPA (1996)

state fish contaminant monitoring program. The degree of fish consumption use support attainment can be found in Table 2.11.

**Overall use.** After assessing individual lake uses, the overall use support of a lake can be determined. The overall use support methodology aggregates the use support attained for each of the individual lake uses assessed (i.e. aquatic life, recreation, swimming, drinking water supply, and fish consumption). The aggregation is achieved by averaging individual use attainments for a lake. For instance, individual uses meeting full, full/threatened, partial/minor, partial/moderate, or nonsupport are assigned values from five (5) to one (1), respectively. The values assigned to each individual use are subsequently summed and averaged. The average value is rounded down to the next whole number, which is then applied to assign an overall lake use attainment. Full support attainment is assigned to an average value of 5; full/threatened, 4; partial/minor, 3; partial/moderate, 2; and nonsupport, 1.

**Example:** The mean TSI is determined by averaging 18 months' SD-TSI, TP-TSI, and CHL-TSI values at station 1-surface (deepest station) and station 2-surface (midlake station) of a central Illinois lake, for values of 54.3 and 58.9, respectively. All Secchi disc readings were greater than 24 in. The mean of nonvolatile suspended solids concentrations observed during 1996–97 at stations 1-surface and 2-surface are 5 and 7 mg/L, respectively. Estimated macrophyte impairment and other observed data are given in Table 2.12. Determine support of designated uses in the lake based on Illinois lake-use support assessment criteria.

**solution:** Solve for station 1 (Steps 1 to 5); then station 2 (Step 6) can be solved in the same manner

**Table 2.12 Use Support Assessment for Otter Lake, 1996–97**

| | Station 1 | | Station 2 | |
|---|---|---|---|---|
| | value | ALI points* | value | ALI points |
| *I. Aquatic life use* | | | | |
| 1. Mean trophic state index | 54.3 | 40 | 58.9 | 40 |
| 2. Macrophyte impairment | < 5% | 10 | < 5% | 10 |
| 3. Mean NVSS | 5 mg/L | 0 | 7 mg/L | 5 |
| Total points: | | 50 | | 55 |
| Criteria points: | | < 75 | | < 75 |
| *Use support*: | | *Full* | | *Full* |
| | Value | RUI points* | Value | RUI points |
| *II. Recreation use* | | | | |
| 1. Mean trophic state index | 54.3 | 54 | 58.9 | 59 |
| 2. Macrophyte impairment | < 5% | 0 | < 5% | 0 |
| 3. Mean NVSS | 5 mg/L | 5 | 7 mg/L | 5 |
| Total points: | | 59 | | 64 |
| Criteria points: | | < 60 | | $65 \leq R \leq 75$ |
| *Use support*: | | *Full* | | *Partial* |
| | Value | Degree of use support | Value | Degree of use support |
| *III. Swimming use* | | | | |
| 1. Secchi depth < 24 in | 0% | Full | 0% | Full |
| 2. Fecal coliform > 200/100 mL | 0% | Full | 0% | Full |
| 3. Mean trophic state index | 54.3 | Full | 58.9 | Full |
| *Use support*: | | *Full* | | *Full* |
| *IV. Drinking water supply* | | *Full* | | *Full* |
| *V. Overall use* | 5 | | 4 | |
| *Use support*: | | *Full* | | *Full/threatened* |

*ALI, aquatic life use impairment index; RUI, recreation use impairment index.

Step 1. Assess aquatic life use

1. Find ALI points for TSI
   Since mean TSI = 54.3, TSI points = 40 is obtained from Table 2.5, section 1.

2. Determine macrophyte impairment (MI) point of ALI points
   The estimated macrophyte impairment is 5% of the lake surface area. From section 2 of Table 2.5, the MI point is 10.

3. Determine mean nonvolatile suspended solids (NVSS) of ALI point
   The NVSS = 5 mg/L. From section 3 of Table 2.5, we read NVSS point = 0.

4. Calculate total ALI points

$$\begin{aligned}\text{Total ALI points} &= \text{TSI points} + \text{MI points} + \text{NVSS points}\\ &= 40 + 10 + 0\\ &= 50\end{aligned}$$

5. Determine the degree of aquatic life index use support
   Since ALI points = 50 (i.e. < 75) of the critical value, and there are potential threats to aquatic life impairment, the degree of use support is *full* from Table 2.6.

Step 2. Assess recreation use

1. Determine the mean TSI of RUI point
   Since mean TSI value is = 54.3, then RUI points for TSI = 54, from section 1 of Table 2.7.

2. Determine macrophyte impairment of RUI point
   Since < 5% of the lake's surface area is covered by weeds, from section 2 of Table 2.7, the RUI point for MI = 0.

3. Determine sediment impairment
   The mean NVSS is 5 mg/L. From section 3 of Table 2.7, for sediment impairment (NVSS), RUI = 5.

4. Compute total RUI points

$$\begin{aligned}\text{Total RUI points} &= \text{TSI value} + \text{MI points} + \text{NVSS points}\\ &= 54 + 0 + 5\\ &= 59\end{aligned}$$

5. Determine the degree of recreation use support
   Total RUI points < 60, therefore the degree of recreation use is considered as *Full* from Table 2.8.

Step 3. Assess swimming use

1. Based on Secchi disc transparency
   Since no Secchi depth is less than 24 in, from Table 2.9, the degree of use support is classified as *Full.*

2. Fecal coliform criteria
   No fecal coliform density is determined

3. Based on trophic state index
   Mean TSI = 54.3. From Table 2.9, it can be classified as *Full.*

4. Determine swimming use support
   Based on the above use analysis, it is assessed as *Full* for swimming use support.

Step 4. Assess drinking water supply use

There was no drinking water supply closure or advisories during the 1996–97 study period. Also no chemicals were applied to the lake. It is considered as *Full* use support for drinking water supply.

Step 5. Assess overall use for station 1

On the basis of overall use criteria, the assigned score values for each individual use are as follows:

| | |
|---|---|
| Aquatic life use | 5 |
| Recreation use | 5 |
| Swimming use | 5 |
| Drinking water use | 5 |
| Total | 20 |

The average value is 5. Then the overall use for station 1 is attained as *Full* (Table 2.12).

Step 6. Assess for station 2

Most assessments are as for station 1 except the following:

1. Recreation use:
   From Table 2.12, we obtain RUI points = 64 which is in the critical range 60 ≤; RUI < 75; therefore, the degree of research on use is considered as partial/minor from Table 2.8.

2. The assigned scores for overall use are

| | |
|---|---|
| Aquatic life use | 5 |
| Recreation use | 3 |
| Swimming use | 5 |
| Drinking water supply | 5 |
| Total | 18 |

The average is 4.5, which is rounded down to the next whole number, 4. Then, the overall use for station 2 is attained as *full/threatened.*

### 5.11 Lake budgets (fluxes)

Calculation of lake budgets (fluxes) usually includes the hydrologic budget, nutrients (nitrogen and phosphorus) budgets, and the sediment budget. These data should be generated in Phases I, II, and III studies.

**Hydrologic budget.** A lake's hydrologic budget is normally quantified on the basis of the height of water spread over the entire lake surface area entering or leaving the lake in a year period. The hydrologic budget of a lake is determined by the general formula:

$$\text{Storage change} = \text{inflows} - \text{outflows} \tag{2.18}$$

or

$$\Delta S = P + I + U - E - O - R \tag{2.18a}$$

(Equations. (2.18) and (2.18a) are essentially the same.)

**TABLE 2.13 Summary of Hydrologic Analysis for Vienna Correction Center Lake, October 1995–September 1996**

| Date | Storage change (acre · ft) | Direct preci- pitation (acre · ft) | Ground- water inflow outflow (−) (acre · ft) | Surface inflow (acre · ft) | Monthly evapo- ration (acre · ft) | Spillway discharge (acre · ft) | Water supply with- drawal (acre · ft) |
|---|---|---|---|---|---|---|---|
| 1995 | | | | | | | |
| October | −56 | 9 | 7 | 3 | 14 | 0 | 61 |
| November | −14 | 16 | −13 | 40 | 7 | 0 | 49 |
| December | −13 | 14 | −2 | 31 | 4 | 0 | 52 |
| 1996 | | | | | | | |
| January | 40 | 19 | 18 | 63 | 4 | 0 | 55 |
| February | −11 | 4 | 24 | 19 | 6 | 0 | 51 |
| March | 153 | 27 | 36 | 154 | 12 | 0 | 52 |
| April | 74 | 34 | 31 | 221 | 20 | 142 | 50 |
| May | −8 | 33 | 24 | 262 | 28 | 244 | 55 |
| June | −37 | 21 | −3 | 76 | 30 | 42 | 59 |
| July | −55 | 30 | −6 | 11 | 33 | 0 | 58 |
| August | −76 | 3 | 4 | 4 | 29 | 0 | 59 |
| September | −36 | 34 | −18 | 27 | 21 | 0 | 58 |
| Annual | −39 | 244 | 102 | 911 | 208 | 428 | 658 |

In general, inflow to the lake includes direct precipitation ($P$), watershed surface runoff ($I$), subsurface groundwater inflow through lake bottom ($U$), and pumped input, if any. Outflows include lake surface evaporation ($E$), discharge through surface outlet ($O$), outflow through lake bottom (groundwater recharge, $R$), and pumped outflow for water supply use, if any.

The storage term is positive if the water level increases in the period and negative if it decreases. The unit of the storage can be, simply, in (or mm) or acre · ft (or ha · cm). The hydrologic budget is used to compute nutrient budgets and the sediment budget and in selecting and designing pollution control and lake restoration alternatives.

**Example:** Table 2.13 illustrates the hydrologic analysis from Vienna Correctional Center Lake, a water supply lake (Bogner *et al.*, 1997). Date needed for evaluating various parameters (Eq. (2.8)) to develop a hydrologic budget for the lake were collected for a 1-year period (October 1995 to September 1996). Table 2.13 presents monthly and annual results of this monitoring. Data sources and methods are explained as below:

**solution:**

Step 1. Determine lake storage change

Lake storage change was determined on the basis of direct measurement of the lake level during the study period. Lake-level data were collected by automatic water-level recorder at 15-min intervals and recorded at 6-h intervals or less. On the basis of water-level record frequency, changes in storage were estimated by multiplying the periodic change in lake storage from the water-level recorder by the lake surface area ($A$) to determine net inflow or outflow volume.

Step 2. Compute direct precipitation

Direct precipitation was obtained from the precipitation record at the University of Illinois' Dixon Spring Experiment Station. The volume of direct precipitation input to the lake was determined by multiplying the precipitation depth by $A$.

Step 3. Compute spillway discharge $Q$

The general spillway rating equation was used as below:

$$Q = 3.1\, LH^{1.5} = 3.1(120)(H^{1.5})$$

Where $L$ = spillway length, ft
$H$ = the height or water level exceeding the spillway

Step 4. Estimate evaporation

Evaporation was estimated using average monthly values for Carbondale in Lake Evaporation in Illinois (Robert and Stall, 1967). Monthly evaporation rates were reduced to daily rates by computing an average daily value for each month. The daily lake surface evaporation volume was determined for the study period by multiplying the daily average evaporation depth by $A$.

Step 5. Obtain water supply withdraw rate

The daily water supply withdraw rates were taken directly from the monthly reports of the water treatment plant.

Step 6. Estimate groundwater inflow and surface inflow

Groundwater inflow and surface water inflow could not be estimated from direct measurements and instead were determined on the basis of a series of sorting steps:

a. If the daily spillway discharge was zero and no rainfall occurred during the preceding 3 days, all inflow was attributed to seepage from the groundwater system.
b. If there was precipitation during the preceding 3-day period, or if there was discharge over the spillway, the groundwater input was not determined in Step 6a. In this case, a moving average was used for the groundwater parameter based on the Step 6a values determined for the preceding 5-day and following 10-day periods.
c. If the daily balance indicated an outflow from the lake, it was attributed to seepage into the groundwater inflow/outflow to the lake; the surface water inflow to the lake was determined to be any remaining inflow volume needed to achieve a daily balance.

**TABLE 2.14 Annual Summary of the Hydrologic Budget for Vienna Correction Center Lake, October 1995–Septmber 1996**

| Source | Inflow volume (acre · ft) | Outflow volume (acre · ft) | Inflow (%) | Outflow (%) |
|---|---|---|---|---|
| Storage change | 37.1 | | 2.9 | |
| Direct precipitation | 243.8 | | 18.9 | |
| Surface inflow | 910.2 | | 70.4 | |
| Groundwater inflow | 100.6 | | 7.8 | |
| Spillway discharge | | 426.9 | | 33.0 |
| Evaporation | | 207.8 | | 16.1 |
| Water supply withdrawal | | 656.9 | | 50.9 |
| Totals | 1291.7 | 1291.7 | 100.0 | 100.0 |

*Note:* Blank spaces—not applicable; 1 acre · ft = 1233 $m^3$.

Step 7. Summary

Table 2.14 summarizes the hydrologic budget for the 1-year monitoring period. The inflows and outflows listed in Table 2.14 are accurate within the limits of the analysis. During the study period, 18.9, 70.4, 7.8, and 2.9% of the inflow volume to the lake were, respectively, direct precipitation on the lake surface, watershed surface runoff, groundwater inflow, and decrease in storage. Outflow volume was 50.9% water supply withdrawal, 33.0% spillway overflow, and 16.1% evaporation.

**Nutrient and sediment budgets.** Although nitrogen and phosphorus are not the only nutrients required for algal growth, they are generally considered to be the two main nutrients involved in lake eutrophication. Despite the controversy over the role of carbon as a limiting nutrient, the vast majority of researchers regard phosphorus as the most frequently limiting nutrient in lakes.

Several factors have complicated attempts to quantify the relationship between lake trophic status and measured concentrations of nutrients in lake waters. For example, measured inorganic nutrient concentrations do not denote nutrient availability but merely represent what is left over by the lake production process. A certain fraction of the nutrients (particularly phosphorus) becomes refractory while passing through successive biological cycles. In addition, numerous morphometric and chemical factors affect the availability of nutrients in lakes. Factors such as mean depth, basin shape, and detention time affect the amount of nutrients a lake can absorb without creating nuisance conditions. Nutrient budget calculations represent the first step in quantifying the dependence of lake water quality on the nutrient supply. It is often essential to quantify nutrients from various sources for effective management and eutrophication control.

A potential source of nitrogen and phosphorus for lakes is watershed drainage, which can include agricultural runoff, urban runoff, swamp and forest runoff, domestic and industrial waste discharges, septic tank discharges from lakeshore developments, precipitation on the lake surface, dry fallout (i.e. leaves, dust, seeds, and pollen), groundwater influxes, nitrogen fixation, sediment recycling, and aquatic bird and animal wastes. Potential sink can include outlet losses, fish catches, aquatic plant removal, denitrification, groundwater recharge, and sediment losses.

The sources of nutrients considered for a lake are tributary inputs from both gaged and ungaged streams, direct precipitation on the lake surface, and internal nutrient recycling from bottom sediments under anaerobic conditions. The discharge of nutrients from the lake through spillway is the only readily quantifiable sink.

The flow weighted-average method of computing nutrient transport by the tributary are generally used in estimating the suspended sediments, phosphorus, and nitrogen loads delivered by a tributary during normal flow conditions. Each individual measurement of nitrogen and phosphorus concentrations in a tributary sample is used with the mean flow values for the period represented by that sample to compute the nutrient transport for the given period. The total amount of any specific nutrient transported by the tributary is given by the expression (Kothandaraman and Evans, 1983; Lin and Raman, 1997):

$$T\,\text{lb} = 5.394 \sum q_i c_i n_i \tag{2.19a}$$

$$T\,\text{kg} = 2.446 \sum q_i c_i n_i \tag{2.19b}$$

where $T$ = total amount of nutrient (nitrogen or phosphorus) or TSS, lb or kg

$q_i$ = average daily flow in cfs for the period represented by the $i$th sample

$c_i$ = concentration of nutrient, mg/L

$n_i$ = number of days in the period represented by the $i$th sample

A similar algorithm with appropriate constant (0.0255) can be used for determining the sediment and nutrient transport during storm events. For each storm event, $n_i$ is the interval of time represented by the $i$th sample and $q_i$ is the instantaneous flow in cfs for the period represented by the $i$th sample. The summation is carried out for all the samples collected in a tributary during each storm event. An automatic sampler can be used to take stormwater samples.

The level of nutrient or sediment input is expressed either as a concentration (mg/L) of pollutant or as mass loading per unit of land area per unit time (kg/ha · yr). There is no single correct way to express the quantity of nutrient input to a lake. To analyze nutrient inputs to a lake, the appropriate averaging time is usually 1 year, since the approximate hydraulic residence time in a lake is of this order of magnitude and the concentration value is a long-term average.

## 5.12 Soil loss rate

Specific soil loss rate from the watershed (for agricultural land, large construction sites, and other land uses of open land) due to rainfall can be estimated through the universal soil loss equation or USLE (Wischmeier and Smith, 1965):

$$A = R \times LS \times K \times C \times P \qquad (2.20)$$

where $A$ = average soil loss rate, tons/(acre · yr)
$R$ = rainfall factor
$L$ = slope length, ft
$S$ = slope steepness, %
$LS$ = length-slope factor or topographic factor (not $L$ times $S$)
$K$ = soil erodibility
$C$ = cropping and management factor
$P$ = conservation practice factor

The slope steepness, slope length, cropping factors, and erodibility of each soil type cropping factor can be determined for various land uses in consultation with the local (county) USDA Soil and Water Conservation Service.

The $R \times P$ factor value is assigned as 135 for Illinois agricultural cropland and 180 for other land uses. Wischmeier and Smith (1965) and Wischmeier *et al.* (1971) are useful references. The value of $P$ is one for some CRP (Conservation Reserve Program) implementations after 1985.

Based on the soil information compiled in the watershed or subwatersheds (for example, Table 2.15), the soil loss rates can be computed. The soil loss for each soil type for each subwatershed is obtained by multiplying the rate and soil acreage. The total soil loss for the watershed is the sum of soil loss in all subwatersheds expressed as tons per year. Excluding the lake surface area, the mean soil erosion rate for the watershed is estimated in terms of tons/(acre · yr).

**TABLE 2.15 Soil Classifications in the Nashville City Reservoir Watershed**

| Soil type | Soil name | Slope | Area (acres) |
|---|---|---|---|
| 2 | Cisne silt loam | – | 38.0 |
| 3A | Hoyleton silt loam | 0–2 | 14.9 |
| 3B | Hoyleton silt loam | 2–5 | 11.9 |
| 5C2 | Blair silt loam, eroded | 5–10 | 36.1 |
| 5C3 | Blair silt loam, severely eroded | 5–10 | 31.0 |
| 5D3 | Blair silt loam, severely eroded | 10–15 | 9.8 |
| 8D2 | Hickory silt loam | 10–15 | 6.2 |
| 12 | Wynoose silt loam | – | 186.0 |
| 13A | Bluford silt loam | 0–2 | 208.4 |
| 13B | Bluford silt loam | 2–5 | 73.7 |
| 13B2 | Bluford silt loam, eroded | 2–5 | 17.5 |
| 14B | Ava silt loam | 2–5 | 14.2 |
| 48 | Ebbert silt loam | – | 7.5 |
| 912A | Dartstadt-Hoyleton complex | 0–2 | 61.4 |
| 912B2 | Dartstadt-Hoyleton complex, eroded | 2–5 | 164.7 |
| 991 | Huey-Cisne complex | – | 32.6 |
| 1334 | Birds silt loam | – | 10.9 |
| 3415 | Orian silt loam, wet | – | 17.5 |
| Water | | | 42.0 |
| Total | | | 984.3 |

*Note*: – means flat, using 0.5% for calculation
SOURCES: USDA, 1998; J. Quinn, personal communication, 2000

The USLE accounts for a series of factors that are the most significant influences on the erosion of soil by precipitation. The USLE is a calculation of in-field soil losses and does not account for deposition from the field to the stream or lake. Redeposition within the field or drainage system is accounted for by a sediment delivery ratio that defines the proportion of the upstream soil losses that actually pass through the stream then to the lake.

The RUSLE (Revised Universal Soil Loss Equation) is an updated version of the USLE and was adopted recently by the USDA and many states. The original USLE has been retained in RUSLE; however it has been put into a computer program to facilitate calculations, the technology for factor evaluation has been altered, and new data has been introduced to evaluate each factor under more specific conditions. There are three tables for $LS$ value determinations.

For the following example, taking the sum of annual loss for each soil type, the total annual soil loss in the Nashville City Reservoir watershed (984.3 acres) is 2.953 tons (Table 2.16). This analysis shows that the average annual soil loss in the watershed is 3.5 tons/(acre · yr).

**Example:** Estimate the watershed soil loss rate for the Nashville City Reservoir watershed in Washington County, Illinois (southern). Use USLE instead of RUSLE, since new factors are not yet available.

**TABLE 2.16 Summary of Soil Losses in the Nashville City Reservoir Watershed Due to Precipitation Runoff**

| Soil type | Areal coverage (acres) | *K* | *L* (ft) | *S* (%) | *LS* | *C* | Annual soil loss rate (tons/acre) | Total annual loss (tons) |
|---|---|---|---|---|---|---|---|---|
| 2 | 38.0 | 0.37 | 175 | 0.5 | 0.11 | 0.37 | 3.0 | 114 |
| 3A | 14.9 | 0.32 | 175 | 1.1 | 0.24 | 0.20 | 3.1 | 46 |
| 3B | 11.9 | 0.32 | 125 | 3.5 | 0.37 | 0.13 | 3.1 | 37 |
| 5C2 | 36.1 | 0.37 | 100 | 7.5 | 0.91 | 0.06 | 4.0 | 146 |
| 5C3 | 31.0 | 0.37 | 100 | 7.0 | 0.82 | 0.05 | 3.0 | 94 |
| 5D3 | 9.8 | 0.37 | 75 | 12.5 | 1.67 | 0.02 | 2.5 | 24 |
| 8D2 | 6.2 | 0.37 | 75 | 12.5 | 1.67 | 0.04 | 4.9 | 31 |
| 12 | 186.0 | 0.43 | 175 | 0.5 | 0.11 | 0.32 | 3.0 | 563 |
| 13A | 208.4 | 0.43 | 175 | 1.1 | 0.24 | 0.15 | 3.1 | 645 |
| 13B | 73.7 | 0.43 | 125 | 3.5 | 0.37 | 0.09 | 2.9 | 211 |
| 13B2 | 17.5 | 0.43 | 100 | 4.0 | 0.40 | 0.09 | 3.1 | 54 |
| 14B | 14.2 | 0.43 | 125 | 3.5 | 0.37 | 0.13 | 4.1 | 59 |
| 48 | 7.5 | 0.32 | 175 | 0.5 | 0.11 | 0.71 | 5.0 | 37 |
| 912A | 61.4 | 0.37 | 175 | 1.1 | 0.24 | 0.17 | 3.0 | 185 |
| 912B2 | 164.7 | 0.37 | 125 | 3.5 | 0.37 | 0.11 | 3.0 | 496 |
| 991 | 32.6 | 0.43 | 175 | 0.5 | 0.11 | 0.21 | 2.0 | 65 |
| 1334 | 10.9 | 0.43 | 175 | 0.5 | 0.11 | 0.57 | 5.4 | 59 |
| 3415 | 17.5 | 0.37 | 175 | 0.5 | 0.11 | 0.61 | 5.0 | 87 |
| Water | 42.0 | | | | | | | |
| Total | 984.3 | | | | | | | 2953 |
| Average | | 0.38 | | | | | 3.5 | |

*Note*: Blank spaces—not applicable
SOURCE: Lin, 2001

The following values are given for each of these factors for Nashville City Reservoir:

- The rainfall factor ($R$) is set to 200, the value applicable for southern Illinois.
- The soil classifications for the watershed and related information are presented in Table 2.15.
- The soil erodibility factor ($K$), the slope length ($L$), slope steepness ($S$), and the cropping factor ($C$) are provided in Table 2.16 by the Washington County Soil and Water Conservation District.
- The conservation factor ($P$) is one.

**solution:** (Assistance in developing these estimates can be obtained through the county conservation district or directly from the Soil and Water Conservation District of US Department of Agriculture. Both are in the same office.)

Step 1. Determine watershed boundary

The watershed boundary of the lake can be determined from a 7.5-minute US Geological Survey topographic map and digitized into a GIS coverage.

Step 2. Obtain the values for, $K$, $L$, $S$, and $C$

Obtain "Soil Survey of Washington County, Illinois" from the USDA office. The soil types, values of $K$ and $S$, and soil classifications, etc. are listed in the book. Values for $L$ and $C$ can be obtained by consultation with USDA or the county's Soil and Water Conservation District personnel. These values for each type of soil are presented in Table 2.16.

Step 3. Measure the area coverage for each type of soil

The area coverage for each type of soil in the watershed can be directly measured from "Soil Survey of Washington County, Illinois." The area coverage for each type of soil is listed in Table 2.15.

Step 4. Find $LS$ value

The $LS$ value can be determined from tables of topographic factor based on $L$, $S$, $R$, and land use; and then listed in Table 2.16 for calculations.

Step 5. Calculate annual soil loss rate for a type of soil

For soil type 2: Using Eq. (2.20), $A = R \times K \times LS \times C \times P$

$$= 200 \times 0.37 \times 0.11 \times 0.37 \times 1.0$$

$$= 3.0 \text{ (tons/acre} \cdot \text{yr) (see Table 2.16)}$$

Step 6. Calculate total annual soil loss for each type of soil

The total annual soil loss for each type of soil is calculated from annual soil loss rates for that soil type multiplying by the area coverage, such as

For soil type 2: Total soil loss $= 3.0 \text{ tons/acre} \cdot \text{yr} \times 38 \text{ acres}$

$$= 114 \text{ tons/yr}$$

Step 7. The total annual soil loss for other soil types can be determined in the same manner as above and listed in Table 2.16.

Step 8. Calculate the total annual soil loss (2.953 tons from Table 2.16) in the watershed

## References

American Public Health Association, American Water Works Association and Water Environment Federation. 1992. *Standard methods for the examination of water and wastewater,* 18th edn. Washington, DC: APHA.

Bogner, W. C., Lin, S. D., Hullinger, D. L. and Raman, R. K. 1997. *Diagnostic—feasibility study of Vienna Correctional Center Lake Johnson County, Illinois*. Contract report 619. Champaign: Illinois State Water Survey.

Borah, D. K., Raman, R. K., Lin, S. D., Knapp, H. V. and Soong, T. W. D. 1997. *Water quality evaluations for Lake Springfield and proposed Hunter Lake and proposed Lick Creek Reservoir*. Contract Report 621. Champaign: Illinois State Water Survey.

Carlson, R. E. 1977. A trophic state index for lakes. *Limnology and Oceanography* **22**(2): 361–369.

Chapra, S. C. and Reckhow, K. H. 1983. *Engineering approaches for lake management. Vol. 2: Mechanistic modeling. Woburn*, Massachusetts: Butterworth.

Clark, J. W., Viessman, Jr., W. and Hammer, M. J. 1977. *Water supply and pollution control*, 3rd edn. New York: Dun-Donnelley.

Cole, G. A. 1979. *Textbook of limnology*, 2nd edn. St. Louis, Missouri: Mosby.

Dillon, P. J. and Rigler, F. H. 1974. The phosphorus-chlorophyll relationship in lakes. *Limnology Oceanography* **19**(5): 767–773.

Fair, G. M., Geyer, J. C. and Okun, D. A. 1966. *Water and Wastewater Engineering, Vol. 1: Water supply and wastewater removal*. New York: John Wiley.

Hudson, H. L., Kirschner, R. J. and Clark, J. J. 1992. *Clean Lakes Program, Phase 1: Diagnostic/feasibility study of McCullom Lake, McHenry County, Illinois*. Chicago: Northeastern Illinois Planning Commission.

Illinois Environmental Protection Agency 1983. *Illinois Water Quality Report 1990–1991*. IEPA/WPC/92-055. Springfield, Illinois: IEPA.

Illinois Environmental Protection Agency 1984. *Volunteer Lake Monitoring Program—Report for 1983 Wolf Lake/Cook Co*. IEPA/WPC/84-015. Springfield, Illinois: IEPA.

Illinois Environmental Protection Agency 1990. *Title 35: Environmental protection, Subtitle C: Water pollution*. State of Illinois, Rules and Regulations, Springfield, Illinois: IEPA.

Illinois Environmental Protection Agency 1992. *Illinois Water Quality Report 1990–1991*. IEPA/WPC/92-055. Springfield, Illinois: IEPA.

Illinois Environmental Protection Agency 1996. *Illinois Water Quality Report 1994–1995*. Vol. 1. Springfield, Illinois: Bureau of Water, IEPA.

Illinois State Water Survey. 1989. *Using copper sulfate to control algae in water supply impoundment*. Misc. Publ. 111. Champaign: Illinois State Water Survey.

James, A. 1993. Modeling water quality in lakes and reservoirs. In: James, A. (ed.), *An introduction to water quality modeling*, 2nd edn. Chichester: John Wiley.

Kohler, M. A., Nordenson, T. J. and Fox, W. E. 1955. *Evaporation from pan and lakes*. US Weather Bureau Research Paper 38. Washington, DC.

Kohler, M. A., Nordenson, T. J. and Baker, D. R. 1959. Evaporation maps for the United States. US Weather Bureau Technical Paper 37. Washington, DC.

Kothandaraman, V. and Evans, R. L. 1983. *Diagnostic-feasibility study of Lake Le-Aqua-Na.*, Contract Report 313. Champaign: Illinois State Water Survey.

Lamoreux, W. M. 1962. Modern evaporation formulae adapted to computer use. *Monthly Weather Review*, January.

Lin, S. D. 1994. Protocol for diagnostic/feasibility study of a lake. *Proc. Aquatech Asia '94*, November 22–24, 1994, Singapore, pp. 165–176.

Lin, S. D. and Raman, R. K. 1997. *Phase III, Post-restoration monitoring of Lake Le-Aqua-na*. Contract Report 610. Champaign: Illinois State Water Survey.

Lin, S. D. *et al.* 1996. *Diagnostic feasibility study of Wolf Lake, Cook County, Illinois, and Lake County, Indiana*. Contract Report 604. Champaign: Illinois State Water Survey.

Lin, S. D., Bogner, W. C. and Raman, R. K. 1998. *Diagnostic-feasibility study of Otter Lake, Macoupin County, Illinois*. Contract Report 652. Champaign: Illinois State Water Survey.

Lin, S. D. 2001. *Diagnostic-feasibility study of Nashville City Reservoir, Washington County, Illinois*. Contract Report (draft). Champaign: Illinois State Water Survey.

Linsley, R. K. and Franzinni, J. B. 1964. *Water resources engineering*. New York: McGraw-Hill.

Reckhow, K. H. and Chapra, S. C. 1983. *Engineering approaches for lake management. Vol. 1: Data analysis and empirical modeling*. Woburn, Massachusetts: Butterworth.

Robert, W. J. and Stall, J. B. 1967. *Lake evaporation in Illinois*. Report of Investigation 57. Urbana: Illinois State Water Survey.

Sawyer, C. N. 1952. Some aspects of phosphate in relation to lake fertilization. *Sewage and Industrial Wastes* **24**(6): 768–776.

Shuttleworth, W. J. 1993. Evaporation. In: Maidment, D. R. (ed.) *Handbook of hydrology*. New York: McGraw-Hill.

Tchobanoglous, G. and Schroeder, E. D. 1985. *Water quality*. Reading, Massachusetts: Addison-Wesley.

US Army Corps of Engineers. 1986. *HEC-5: Simulation of flood control and conservation systems: Appendix on Water Quality Analysis*. Davis, California: Hydrologic Engineering Center, US Army Corps of Engineers.

US Army Corps of Engineers. 1989. *HEC-5: Simulation of flood control and conservation systems: Exhibit 8 of user's manual: input description*. Davis, California: Hydrologic Engineering Center, US Army Corps of Engineers.

US Department of Agriculture. 1998. *Soil survey of Washington County, Illinois*. Washington, DC: USDA.

US Environmental Protection Agency. 1980. *Clean Lakes Program Guidance Manual*. USEPA-440/5-81-003. Washington, DC: USEPA.

US Environmental Protection Agency. 1990. *The lake and reservoir restoration guidance manual*, 2nd edn. USEPA-440/4-90-006. Washington, DC: USEPA.

US Environmental Protection Agency. 1991. *The watershed protection approach—an overview*. USEPA/503/9-92/002. Washington, DC: USEPA.

US Environmental Protection Agency. 1993. *The watershed protection approach—annual report 1992*. USEPA840-S-93-001. Washington, DC: USEPA.

US Environmental Protection Agency. 1994. *The quality of our nation's water: 1992*. USEPA841-S-94-002. Washington, DC: USEPA.

US Geological Survey. 1954. *Water loss investigations*: *Lake Hefner Studies*. Technical Report, Professional paper 269.

Walker, Jr., W. W. 1981a. *Empirical methods for predicting eutrophication in impoundments. Part 1. Phase I: data base development*. Environmental Engineers. Tech. Report E-81-9, Concord, Massachusetts.

Walker, Jr., W. W. 1981b. *Empirical methods for predicting eutrophication in impoundments. Part 2. Phase II: model testing*. Environmental Engineers Tech. Report E-81-9, Concord, Massachusetts.

Wetzel, R. G. 1975. *Limnology*. Philadelphia, Pennsylvania: Saunders.

Wischmeier, W. H. and Smith, D. D. 1965. *Predicting rainfall—erosion losses from cropland east of the Rocky Mountains*. Agriculture Handbook 282. Washington, DC: US Department of Agriculture.

Wischmeier, W. H., Johnson, C. B. and Cross, B. V. 1971. A soil-erodibility monograph for farmland and construction sites. *J. Soil and Water Conservation* **26**(5): 189–193.

Chapter

# 3

# Groundwater

# 1 Definition

## 1.1 Groundwater and aquifer

Groundwater is subsurface water which occurs beneath the earth's surface. In a hydraulic water cycle, groundwater comes from surface waters (precipitation, lake, reservoir, river, sea, etc.) and percolates into the ground beneath the water table. The groundwater table is the surface of the groundwater exposed to an atmospheric pressure beneath the ground surface (the surface of the saturated zone). A water table may fluctuate in elevation.

An aquifer is an underground water-saturated stratum or formation that can yield usable amounts of water to a well. There are two different types of aquifers based on physical characteristics. If the saturated zone is sandwiched between layers of impermeable material and the groundwater is under pressure, it is called a confined aquifer (Fig. 3.1). If there is no impermeable layer immediately above the saturated zone, that is called an unconfined aquifer. In an unconfined aquifer, the top of the saturated zone is the water table defined as above.

Aquifers are replenished by water infiltrated through the earth above from the upland area. The area replenishing groundwater is called the recharge area. In reverse, if the groundwater flows to the lower land area, such as lakes, streams, or wetlands, it is called discharge.

On the earth, approximately 3% of the total water is freshwater. Of this, groundwater comprises 95%, surface water 3.5%, and soil moisture 1.5%. Out of all the freshwater on earth, only 0.36% is readily available to use (Leopold, 1974).

Groundwater is an important source of water supply. Fifty-three percent of the population of the United States receives its water supply from groundwater or underground sources (US EPA, 1994). Groundwater is also a major source of industrial uses (cooling, water supply, etc.) and agricultural uses (irrigation and livestock). The quantity of groundwater available is an important value. The so-called safe yield of an aquifer is the practicable rate of withdrawing water from it perennially. Such a safe amount does not exist, however.

The quantity of groundwater is also affected by water engineering. For decades and centuries, through improper disposal of wastes (solid, liquid, and gaseous) to the environment and subsurface areas, many groundwaters have become contaminated. Major sources of contaminants are possibly from landfill leachate, industrial wastes, agricultural chemicals, wastewater effluents, oil and gasoline (underground tanks, animal wastes, acid-mine drainage, road salts, hazardous wastes spillage, household and land chemicals, etc.). Illegal dumping of wastes and toxic chemicals to waterways and to lands are major environmental problems in many countries. Water quality problem is grown in the United States and in global.

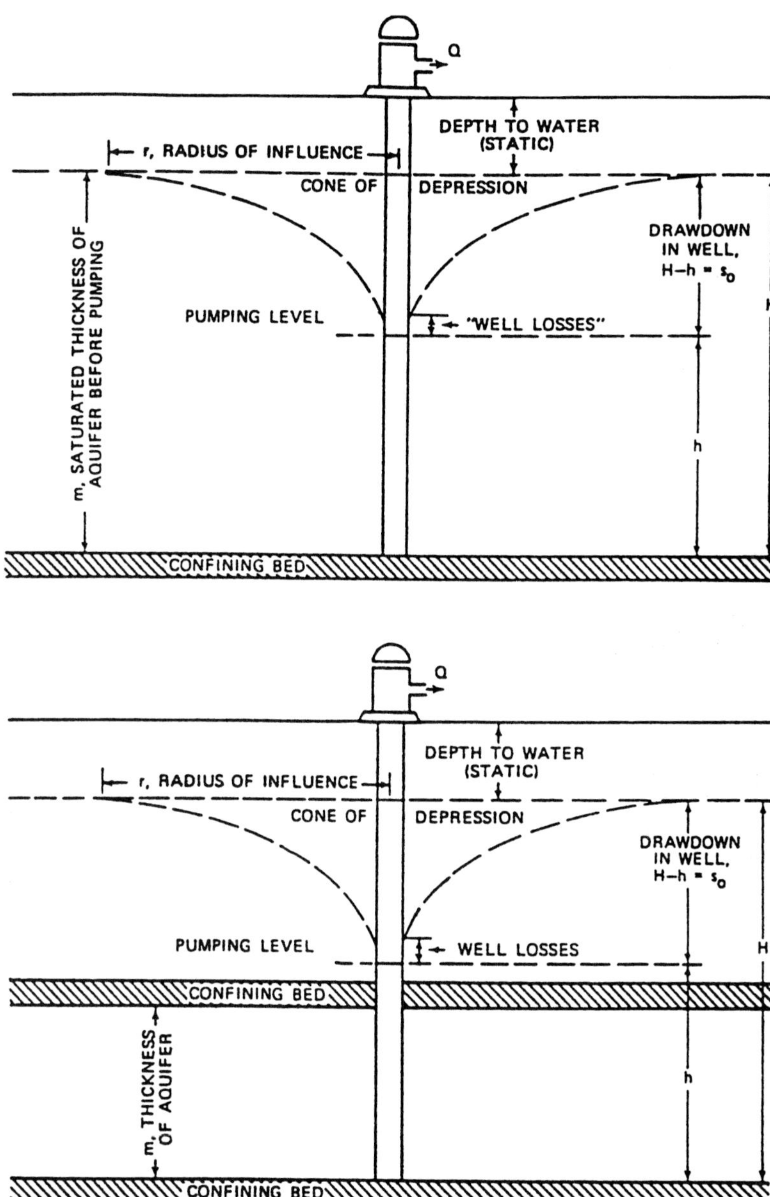


**Figure 3.1** Drawdown, cone of depression, and radius of influence in unconfined and confined aquifers (*Illinois EPA, 1995*).

Efforts to protect the quantity and quality of groundwater have been made by cooperation between all government agencies, interested parties, and researchers. Most states are responsible for research, education, establishment of minimum setback zones for public and private water supply wells, and contamination survey and remediation.

## 1.2 Zones of influence and capture

The withdrawal of groundwater by a pumping well causes a lowering of the water level. Referring to Fig. 3.2, the difference between water levels during nonpumping and pumping is called drawdown. The pattern of drawdown around a single pumping well resembles a cone. The area

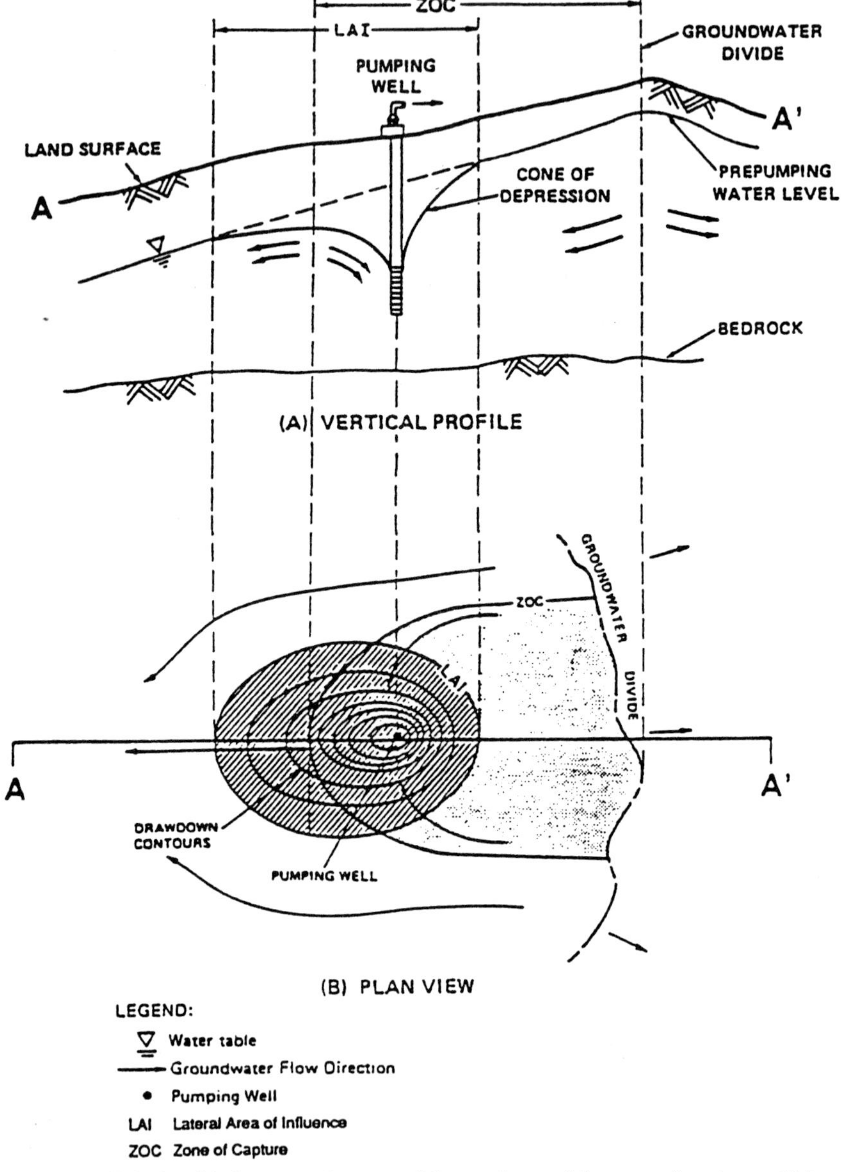


**Figure 3.2** Relationship between the cone of depression and the zone of capture within a regional flow field (*Illinois EPA, 1995*).

affected by the pumping well is called the cone of depression, the radius of influence, or lateral area of influence (LAI). Within the LAI, the flow velocity continuously increases as it flows toward the well due to gradually increased slope or hydraulic gradient.

As a well pumps groundwater to the surface, the groundwater withdrawn from around the well is replaced by water stored within the aquifer. All water overlaid on the nonpumping potentiometric surface will eventually be pulled into the well. This area of water entering the area of influence of the well is called the zone of capture (ZOC), zone of contribution, or capture zone (Fig. 3.2). The ZOC generally extends upgradient from the pumping well to the edge of the aquifer or to a groundwater divide. The ZOC is usually asymmetrical. It is important to identify the ZOC because any pollution will be drawn toward the well, subsequently contaminating the water supply.

A ZOC is usually referred to the time of travel as a time-related capture zone. For instance, a "10-year" time-related capture zone is the area within which the water at the edge of the zone will reach the well within 10-years. The state primacy has established setback zones for wells which will be discussed later.

**Example 1:** Groundwater flows into a well at 4 in/d (10 cm/d) and is 365 ft (111 m) from the well. Estimate the capture time.

**solution:**

$$\text{Time} = 365 \text{ ft}/(4/12 \text{ ft/d})$$
$$= 1095 \text{ days}$$

or

$$= 3\text{-year time-related capture zone}$$

**Example 2:** Calculate the distance from a well to the edge of the 3-year capture zone if the groundwater flow is 2 ft/d (0.61 m/d).

**solution:**

$$\text{Distance} = 2 \text{ ft/d} \times 3 \text{ years} \times 365 \text{ d/yr}$$
$$= 2190 \text{ ft}$$

or

$$= 668 \text{ m}$$

## 1.3 Wells

Wells are classified according to their uses by the US Environmental Protection Agency (EPA). A public well is defined as having a minimum of 15 service connections or serving 25 persons at least 60 d/yr. Community public wells serve residents the year round. Noncommunity public wells serve nonresidential populations at places such as schools, factories, hotels, restaurants, and campgrounds. Wells that do not meet the definition of public wells are classified as either semiprivate or private.

Semiprivate wells serve more than one single-family dwelling, but fewer than 15 connections or serving 25 persons. Private wells serve an owner-occupied single-family dwelling.

Wells are also classified into types based on construction. The most common types of wells are drilled and bored dependent on the aquifer to be tapped, and the needs and economic conditions of the users (Babbitt *et al.*, 1959; Forest and Olshansky, 1993).

## 2 Hydrogeologic Parameters

There are three critical aquifer parameters: porosity, specific yield (or storativity for confined aquifers), and hydraulic conductivity (including anisotropy). These parameters required relatively sophisticated field and laboratory procedures for accurate measurement.

Porosity and specific yield/storativity express the aquifer storage properties. Hydraulic conductivity (permeability) and transmissivity describe the groundwater transmitting properties.

### 2.1 Aquifer porosity

The porosity of soil or fissured rock is defined as the ratio of void volume to total volume (Wanielista, 1990; Bedient and Huber, 1992):

$$n = \frac{V_{\mathrm{v}}}{V} = \frac{(V - V_{\mathrm{s}})}{V} = 1 - \frac{V_{\mathrm{s}}}{V} \tag{3.1}$$

where $n$ = porosity

$V_{\mathrm{v}}$ = volume of voids within the soil

$V$ = total volume of sample (soil)

$V_{\mathrm{s}}$ = volume of the solids within the soil

= dry weight of sample/specific weight

The ratio of the voids to the solids within the soil is called the void ratio $e$ expressed as

$$e = V_{\mathrm{v}}/V_{\mathrm{s}} \tag{3.2}$$

Then the relationship between void ratio and porosity is

$$e = \frac{n}{1 - n} \tag{3.3}$$

or

$$n = \frac{e}{1 + e} \tag{3.4}$$

The porosity may range from a small fraction to about 0.90. Typical values of porosity are 0.2 to 0.4 for sands and gravels depending on the grain size, size of distribution, and the degree of compaction; 0.1 to 0.2 for sandstone; and 0.01 to 0.1 for shale and limestone depending on the texture and size of the fissures (Hammer, 1986).

When groundwater withdraws from an aquifer and the water table is lowered, some water is still retained in the voids. This is called the specific retention. The quantity drained out is called the specific yield. The specific yield for alluvial sand and gravel is of the order of 90% to 95%.

**Example:** If the porosity of sands and gravels in an aquifer is 0.38 and the specific yield is 92%, how much water can be drained per cubic meter of aquifer?

**solution:**

$$\begin{aligned}\text{Volume} &= 0.38 \times 0.92 \times 1\ \text{m}^3 \\ &= 0.35\ \text{m}^3\end{aligned}$$

### 2.2 Storativity

The term storativity ($S$) is the quantity of water that an aquifer will release from storage or take into storage per unit of its surface area per unit change in land. In unconfined aquifers, the storativity is in practice equal to the specific yield. For confined aquifers, storability is between 0.005 and 0.00005, with leaky confined aquifers falling in the high end of this range (US EPA, 1994). The smaller storativity of confined aquifers, the larger the pressure change throughout a wide area to obtain a sufficient supply from a well. However, this is not the case for unconfined aquifers due to gravity drainage.

### 2.3 Transmissivity

Transmissivity describes the capacity of an aquifer to transmit a water. It is the product of hydraulic conductivity (permeability) and the aquifer's saturated thickness:

$$T = Kb \tag{3.5}$$

where $T$ = transmissivity of an aquifer, gpd/ft or $\text{m}^3/(\text{d} \cdot \text{m})$
$K$ = permeability, gpd/ft$^2$ or $\text{m}^3/(\text{d} \cdot \text{m}^2)$
$b$ = thickness of aquifer, ft or m

A rough estimation of $T$ is by multiplying specific capacity by 2000 (US EPA, 1994).

**Example:** If the aquifer's thickness is 50 ft, estimate the permeability of the aquifer using data in the example of the specific capacity (in Section 2.7).

**solution:**

$$T = 2000 \times \text{specific capacity} = 2000 \times 15 \text{ gpm/ft}$$
$$= 30{,}000 \text{ gpm/ft}$$

Rearranging Eq. (3.5)

$$K = T/b = (30{,}000 \text{ gpm/ft})/50 \text{ ft}$$
$$= 600 \text{ gpm/ft}^2$$

## 2.4 Flow nets

Many groundwater systems are two or three dimensional. Darcy's law was first derived in a one dimensional equation. Using Darcy's law can establish a set of streamlines and equipotential lines to develop a two dimensional flow net. The details of this concept are discussed elsewhere in the text (Bedient *et al.,* 1994).

A flow net is constructed by flow lines that intersect the equipotential lines or contour lines at a right angle. Equipotential lines are developed based on the observed water levels in wells penetrating an isotropic aquifer. Flow lines are then drawn orthogonally to indicate the flow direction.

Referring to Fig. 3.3, the horizontal flow within a segment in a flow net can be determined by the following equation (US EPA, 1994):

$$q_a = T_a \,\Delta H_a W_a / L_a \tag{3.6}$$

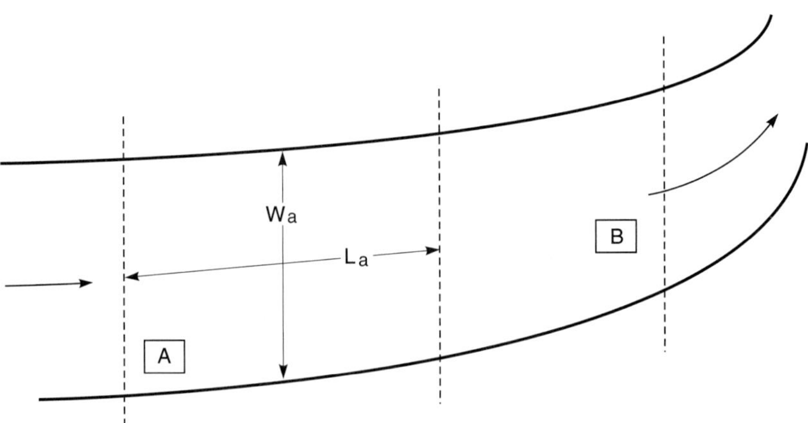


**Figure 3.3** Distribution of transmissivity in flow nets.

where $q_a$ = groundwater flow in segment A, $m^3/d$ or $ft^3/d$
$T_a$ = transmissivity in segment A, $m^3/d$ or $ft^3/d$
$\Delta H_a$ = drop in groundwater level across segment A, m or ft
$W_a$ = average width of segment A, m or ft
$L_a$ = average length of segment A, m or ft

The flow in the next segment, B, is similarly computed as Eq. (3.7)

$$q_b = T_b \, \Delta H_b W_b / L_b \tag{3.7}$$

Assuming that there is no flow added between segments A and B by recharge (or that recharge is insignificant), then

$$q_b = q_a$$

or

$$T_b \, \Delta H_b W_b / L_b = T_a \Delta H_a W_a / L_a$$

solving $T_b$ computation of which allows $T_b$ from $T_a$

$$T_b = T_a \, (L_b \Delta H_a W_a / L_a \Delta H_b W_b) \tag{3.8}$$

Measurement or estimation of transmissivity ($T$) for one segment allows the computation of variations in $T$ upgradient and downgradient. If variations in aquifer thickness are known, or can be estimated for different segments, variation in hydraulic conductivity can also be calculated as

$$K = T/b \tag{3.9}$$

where $K$ = hydraulic conductivity, m/d or ft/d
$T$ = transmissivity, $m^2/d$ or $ft^2/d$
$b$ = aquifer thickness, m or ft

Equation (3.9) is essentially the same as Eq. (3.5).

## 2.5 Darcy's law

The flow movement of water through the ground is entirely different from the flow in pipes and in an open channel. The flow of fluids through porous materials is governed by Darcy's law. It states that the flow velocity of fluid through a porous medium is proportional to the hydraulic gradient (referring to Fig. 3.4):

$$v = Ki \tag{3.10a}$$

or

$$v = K \frac{(h_1 + z_1) - (h_2 + z_2)}{L} \tag{3.10b}$$

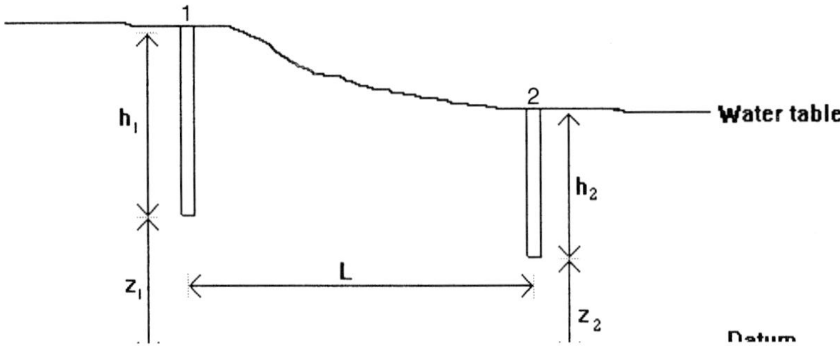


**Figure 3.4** Groundwater flow (one dimensional).

where $v$ = Darcy velocity of flow, mm/s or ft/s
$K$ = hydraulic conductivity of medium or coefficient of permeability, mm/s or ft/s
$i$ = hydraulic gradient, mm or ft/ft
$h_1$, $h_2$ = pressure heads at points 1 and 2, m or ft
$z_1$, $z_2$ = elevation heads at points 1 and 2, m or ft
$L$ = distance between points (piezometers) 1 and 2, m or ft

The pore velocity $v_p$ is equal to the Darcy velocity divided by porosity as follows:

$$v_p = v/n \tag{3.11}$$

Darcy's law is applied only in the laminar flow region. Groundwater flow may be considered as laminar when the Reynolds number is less than unity (Rose, 1949). The Reynolds number can be expressed as

$$\mathbf{R} = \frac{VD}{\nu} \tag{3.12}$$

where $\mathbf{R}$ = Reynolds number
$V$ = velocity, m/s or ft/s
$D$ = mean grain diameter, mm or in
$\nu$ = kinematic viscosity, $m^2$/d or $ft^2$/d

**Example 1:** Determine the Reynolds number when the groundwater temperature is 10°C (from Table 4.1a, $\nu = 1.31 \times 10^{-6}$ $m^2$/s); the velocity of flow is 0.6 m/d (2 ft/d); and the mean grain diameter is 2.0 mm (0.079 in).

**solution:**

$$V = 0.6\ \text{m/d} = (0.6\ \text{m/d})/(86{,}400\ \text{s/d})$$
$$= 6.94 \times 10^{-6}\ \text{m/s}$$
$$D = 2\ \text{mm} = 0.002\ \text{m}$$
$$\mathbf{R} = \frac{VD}{\nu} = \frac{6.94 \times 10^{-6}\text{m/s} \times 0.002\ \text{m}}{1.31 \times 10^{-6}\text{m}^2/\text{s}}$$
$$= 0.011$$

*Note*: It is a laminar flow ($\mathbf{R} < 1$).

**Example 2:** If the mean grain diameter is 0.12 in (3.0 mm), its porosity is 40%, and the groundwater temperature is 50°F (10°C). Determine the Reynolds number for a flow 5 ft (1.52 m) from the centerline of the well with a 4400 gpm in a confined aquifer depth of 3.3 ft (1 m) thick.

**solution:**

Step 1. Find flow velocity ($V$)

$$Q = 4400\ \text{gpm} \times 0.002228\ \text{cfs/gpm}$$
$$Q = 9.80\ \text{cfs}$$
$$n = 0.4$$
$$r = 5\ \text{ft}$$
$$h = 3.3\ \text{ft}$$

Since

$$Q = nAV = n(2\pi rh)V$$
$$V = Q/n(2\pi rh)$$
$$= 9.8\ \text{cfs}/(0.4 \times 2 \times 3.14 \times 5\ \text{ft} \times 3.3\ \text{ft})$$
$$= 0.236\ \text{fps}$$

Step 2. Compute $\mathbf{R}$

$$\nu = 1.41 \times 10^{-5}\ \text{ft}^2/\text{s (see Table 4.1b at 50°F)}$$
$$D = 0.12\ \text{in} = 0.12\ \text{in}/(12\ \text{in/ft}) = 0.01\ \text{ft}$$
$$\mathbf{R} = \frac{VD}{\nu} = \frac{0.236\ \text{fps} \times 0.01\ \text{ft}}{1.41 \times 10^{-5}\ \text{ft}^2/\text{s}}$$
$$= 167$$

**Example 3:** The slope of a groundwater table is 3.6 m per 1000 m. The coefficient of permeability of coarse sand is 0.51 cm/s (0.2 in/s). Estimate the flow velocity and the discharge rate through this aquifer of coarse sand 430 m (1410 ft) wide and 22 m (72 ft) thick.

**solution:**

Step 1. Determine the velocity of flow, $v$, using Eq. (3.10a)

$$
\begin{aligned}
i &= 3.6\ \text{m}/1000\ \text{m} = 0.0036 \\
v &= Ki = 0.51\ \text{cm/s}(0.0036) \\
&= 0.00184\ \text{cm/s}\ (86{,}400\ \text{s/d})(0.01\ \text{m/cm}) \\
&= 1.59\ \text{m/d} \\
&= 5.21\ \text{ft/d}
\end{aligned}
$$

Step 2. Compute discharge

$$
\begin{aligned}
Q &= vA = 1.59\ \text{m/d} \times 430\ \text{m} \times 22\ \text{m} \\
&= 15{,}040\ \text{m}^3/\text{d} \\
&= 15{,}040\ \text{m}^3/\text{d} \times 264.17\ \text{gal/m}^3 \\
&= 3.97\ \text{MGD (million gallons per day)}
\end{aligned}
$$

**Example 4:** The difference in water level between two wells 1.6 miles (2.57 km) apart is 36 ft (11 m), and the hydraulic conductivity of the media is 400 gpd/ft$^2$ (16,300 L/(d · m$^2$)). The depth of the media (aquifer) is 39 ft (12 m). Estimate the quantity of groundwater flow moving through the cross section of the aquifer.

**solution:**

$$
\begin{aligned}
i &= 36\ \text{ft}/1.6\ \text{miles} = 36\ \text{ft}/(1.6\ \text{miles} \times 5280\ \text{ft/miles}) \\
&= 0.00426 \\
A &= 1.6 \times 5280\ \text{ft} \times 39\ \text{ft} \\
&= 329{,}500\ \text{ft}^2
\end{aligned}
$$

then

$$
\begin{aligned}
Q &= KiA = 400\ \text{gpd/ft}^2 \times 0.00426 \times 329{,}500\ \text{ft}^2 \\
&= 561{,}400\ \text{gpd} \\
&= 0.561\ \text{MGD} \\
&= 2{,}123\ \text{m}^3/\text{d}
\end{aligned}
$$

**Example 5:** If the water moves from the upper to the lower lake through the ground. The following data are given:

difference in elevation $\Delta h$ = 25 m (82 ft)
length of low path $L$ = 1500 m (4920 ft)
cross-sectional area of flow $A$ = 120 m$^2$ (1290 ft$^2$)
hydraulic conductivity $K$ = 0.15 cm/s
porosity of media $n$ = 0.25

Estimate the time flow between the two lakes.

**solution:**

Step 1. Determine the Darcy velocity $v$

$$v = Ki = K\Delta h/L = 0.0015 \text{ m/s } (25 \text{ m}/1500 \text{ m})$$
$$= 2.5 \times 10^{-5} \text{m/s}$$

Step 2. Calculate pore velocity

$$v_p = v/n = (2.5 \times 10^{-5} \text{ m/s})/0.25$$
$$= 1.0 \times 10^{-4} \text{ m/s}$$

Step 3. Compute the time of travel $t$

$$t = L/v_p = 1500 \text{ m}/(1 \times 10^{-4} \text{ m/s})$$
$$= 1.5 \times 10^{7} \text{ s} \times (1 \text{ day}/86{,}400 \text{ s})$$
$$= 173.6 \text{ days}$$

## 2.6 Permeability

The terms permeability ($P$) and hydraulic conductivity ($K$) are often used interchangeably. Both are measurements of water moving through the soil or an aquifer under saturated conditions. The hydraulic conductivity, defined by Nielsen (1991), is the quantity of water that will flow through a unit cross-sectional area of a porous media per unit of time under a hydraulic gradient of 1.0 (measured at right angles to the direction of flow) at a specified temperature.

**Laboratory measurement of permeability.** Permeability can be determined using permeameters in the laboratory. Rearranging Eq. (3.5) and $Q = vA$, for constant head permeameter, the permeability is

$$K = \frac{LQ}{H\pi R^2} \tag{3.13}$$

where $K$ = permeability, m/d or ft/d
$L$ = height of sample (media), m or ft
$Q$ = flow rate at outlet, $m^3$/d or $ft^3$/d
$H$ = head loss, m or ft
$R$ = radius of sample column

The permeability measure from the falling-head permeameter is

$$K = \left(\frac{L}{t}\right)\left(\frac{r}{R}\right)^2 \ln\left(\frac{h_1}{h_2}\right) \tag{3.14}$$

where $r$ = radius of standpipe, m or ft
$h_1$ = height of water column at beginning, m or ft
$h_2$ = height of the water column at the end, m or ft
$t$ = time interval between beginning and end, day
Other parameters are the same as Eq. (3.13).

Groundwater flows through permeable materials, such as sand, gravel, and sandstone, and is blocked by less permeable material, such as clay. Few materials are completely impermeable in nature. Even solid bedrock has fine cracks, so groundwater can flow through. Groundwater recharge occurs when surface water infiltrates the soil faster than it is evaporated, used by plants, or stored as soil moisture.

**Field measurement of permeability.** Ideal steady-state flow of groundwater is under the conditions of uniform pump withdrawal, a stable drawdown curve, laminar and horizontal uniform flow, a flow velocity proportional to the tangent of the hydraulic gradient, and a homogeneous aquifer. Assuming these ideal conditions, the well flow is a function of the coefficient of permeability, the shape of the drawdown curve, and the thickness of the aquifer. For an unconfined aquifer the well discharge can be expressed as an equilibrium equation (Steel and McGhee, 1979; Hammer and Mackichan, 1981):

$$Q = \pi K \frac{H^2 - h_{\mathrm{w}}^2}{\ln (r/r_{\mathrm{w}})} \tag{3.15}$$

where $Q$ = well discharge, L/s or gpm
$\pi$ = 3.14
$K$ = coefficient of permeability, mm/s or fps
$H$ = saturated thickness of aquifer before pumping, m or ft (see Fig. 3.1)
$h_{\mathrm{w}}$ = depth of water in the well while pumping, m or ft
= $h$ + well losses in Fig. 3.1
$r$ = radius of influence, m or ft
$r_{\mathrm{w}}$ = radius of well, m or ft

Also under ideal conditions, the well discharge from a confined aquifer can be calculated as

$$Q = 2\pi K m \frac{H - h_{\mathrm{w}}}{\ln (r/r_{\mathrm{w}})} \tag{3.16}$$

where $m$ is the thickness of the aquifer, m or ft. Other parameters are the same as Eq. (3.15). Values of $Q$, $H$, and $r$ may be assumed or measured from field well tests, with two observation wells, often establishing a steady-state condition for continuous pumping for a long period. The coefficient of permeability can be calculated by rearranging Eqs. (3.14) and (3.15). The $K$ value of an unconfined aquifer is also computed by the equation:

$$K = \frac{Q \ln (r_2/r_1)}{\pi(h_2^2 - h_1^2)} \tag{3.17}$$

and for a confined aquifer:

$$K = \frac{Q \ln (r_2/r_1)}{2m\pi(h_2 - h_1)} \tag{3.18}$$

where $h_1, h_2$ = depth of water in observation wells 1 and 2, m or ft
$r_1, r_2$ = centerline distance from the well and observation wells 1 and 2, respectively, m or ft

**Example 1:** A well is pumped to equilibrium at 4600 gpm (0.29 $m^3$/s) in an unconfined aquifer. The drawdown in the observation well at 100 ft (30.5 m) away from the pumped well is 10.5 ft (3.2 m) and at 500 ft (152 m) away is 2.8 ft (0.85 m). The water table is 50.5 ft (15.4 m). Determine the coefficient of permeability.

**solution:**

$$h_1 = 50.5 \text{ ft} - 10.5 \text{ ft} = 40.0 \text{ ft}$$
$$h_2 = 50.5 \text{ ft} - 2.8 \text{ ft} = 47.7 \text{ ft}$$
$$r_1 = 100 \text{ ft}$$
$$r_2 = 500 \text{ ft}$$
$$Q = 4600 \text{ gpm} = 4600 \text{ gmp} \times 0.002228 \text{ cfs/gpm}$$
$$= 10.25 \text{ cfs}$$

Using Eq. (3.17)

$$K = \frac{Q\ln(r_2/r_1)}{\pi(h_2^2 - h_1^2)}$$
$$= \frac{(10.25 \text{ cfs}) \ln (500 \text{ ft}/100 \text{ ft})}{3.14[(47.7 \text{ ft})^2 - (40 \text{ ft})^2]}$$
$$= 0.00778 \text{ ft/s}$$

or

$$= 0.00237 \text{ m/s}$$

**Example 2:** Referring to Fig. 3.1, a well with a diameter of 0.46 m (1.5 ft) is in a confined aquifer which has a uniform thickness of 16.5 m (54.1 ft). The depth of the top impermeable bed to the ground surface is 45.7 m (150 ft). Field pumping tests are carried out with two observation wells to determine the coefficient of permeability of the aquifer. The distances between the test well and observation wells 1 and 2 are 10.0 and 30.2 m (32.8 and 99.0 ft), respectively. Before pumping, the initial piezometric surface in the test well and the observation wells are 10.4 m (34.1 ft) below the ground surface. After pumping at a discharge rate of 0.29 $m^3/s$ (4600 gpm) for a few days, the water levels in the wells are stabilized with the following drawdowns: 8.6 m (28.2 ft) in the test well, 5.5 m (18.0 ft) in the observation well 1, and 3.2 m (10.5 ft) in the observation well 2. Compute (a) the coefficient of permeability and transmissivity of the aquifer and (b) the well discharge with the drawdown in the well 10 m (32.8 ft) above the impermeable bed if the radius of influence ($r$) is 246 m (807 ft) and ignoring head losses.

**solution:**

Step 1. Find $K$

Let a datum be the top of the aquifer, then

$$
\begin{aligned}
H &= 45.7 \text{ m} - 10.4 \text{ m} = 35.3 \text{ m} \\
h_w &= H - 8.6 \text{ m} = 35.3 \text{ m} - 8.6 \text{ m} = 26.7 \text{ m} \\
h_1 &= 35.3 \text{ m} - 5.5 \text{ m} = 29.8 \text{ m} \\
h_2 &= 35.3 \text{ m} - 3.2 \text{ m} = 32.1 \text{ m} \\
m &= 16.5 \text{ m}
\end{aligned}
$$

Using Eq. (3.18)

$$
\begin{aligned}
K &= \frac{Q \ln (r_2/r_1)}{2\pi m(h_2 - h_1)} \\
&= \frac{0.29 \text{ m}^3\text{/s} \ln (30.2 \text{ m}/10 \text{ m})}{2 \times 3.14 \times 16.5 \text{ m} \times (32.1 \text{ m} - 29.8 \text{ m})} \\
&= 0.00134 \text{ m/s} = 115.8 \text{ m/d}
\end{aligned}
$$

or $$= 0.00441 \text{ ft/s}$$

Step 2. Find $T$ using Eq. (3.5)

$$
\begin{aligned}
T = Kb &= 115.8 \text{ m/d} \times 16.5 \text{ m} \\
&= 1910 \text{ m}^2\text{/d}
\end{aligned}
$$

Step 3. Estimate well discharge

$$
\begin{aligned}
h_w &= 10 \text{ m}, \qquad r_w = 0.46 \text{ m} \\
H &= 35.3 \text{ m}, \qquad r = 246 \text{ m}
\end{aligned}
$$

Using Eq. (3.16)

$$Q = 2\pi Km \frac{H - h_w}{\ln (r/r_w)}$$

$= 2 \times 3.14 \times 0.00134$ m/s $\times$ 16.5 m $\times$ (35.3 −10) m/ln(264/0.46)

$= 0.553$ m$^3$/s

or $= 8765$ gpm

## 2.7 Specific capacity

The permeability can be roughly estimated by a simple field well test. The difference between the static water level prior to any pumping and the level to which the water drops during pumping is called drawdown (Fig. 3.2). The discharge (pumping) rate divided by the drawdown is the specific capacity. The specific capacity gives the quantity of water produced from the well per unit depth (ft or m) of drawdown. It is calculated by

$$\text{Specific capacity} = Q/wd \tag{3.19}$$

where $Q$ = discharge rate, gpm or m$^3$/s
$wd$ = well drawdown, ft or m

**Example:** The static water elevation is at 572 ft (174.3 m) before pumping. After a prolonged normal well pumping rate of 120 gpm (7.6 L/s), the water level is at 564 ft (171.9 m). Calculate the specific capacity of the well.

**solution:**

Specific capacity $= Q/wd =$ 120 gpm/(572 − 564)ft

$=$ 15 gpm/ft

or Specific capacity $=$ (7.6 L/s)/(174.3 − 171.9)m

$=$ 3.17 L/s · m

## 3 Steady Flows in Aquifers

Referring to Fig. 3.3, if $z_1 = z_2$, for an unconfined aquifer,

$$Q = KA\,dh/dL \tag{3.20}$$

and let the unit width flow be $q$, then

$$q = Kh\,dh/dL$$
$$qdL = Kh\,dh \tag{3.21}$$

By integration:

$$q\int_0^L dL = K\int_{h_2}^{h_1} h\,dh$$

$$qL = (K/2)(h_1^2 - h_2^2) \tag{3.22}$$

$$q = \frac{K(h_1^2 - h_2^2)}{2L}$$

This is the so-called Dupuit equation.

For a confined aquifer, it is a linear equation

$$q = \frac{KD(h_1 - h_2)}{L} \tag{3.23}$$

where $q$ = unit width flow, $m^2/d$ or $ft^2/d$
$K$ = coefficient of permeability, m/d or ft/d
$h_1, h_2$ = piezometric head at locations 1 and 2, m or ft
$L$ = length of aquifer between piezometric measurements, m or ft
$D$ = thickness of aquifer, m of ft

**Example:** Two rivers are located 1800 m (5900 ft) apart and fully penetrate an aquifer. The water elevation of the rivers are 48.5 m (159 ft) and 45.6 m (150 ft) above the impermeable bed. The hydraulic conductivity of the aquifer is 0.57 m/d. Estimate the daily discharge per meter of width between the two rivers, neglecting recharge.

**solution:**

This case can be considered as an unconfined aquifer:

$$K = 0.57 \text{ m/d}$$
$$h_1 = 48.5 \text{ m}$$
$$h_2 = 45.6 \text{ m}$$
$$L = 1800 \text{ m}$$

Using Eq. (3.22) the Dupuit equation:

$$q = \frac{K(h_1^2 - h_2^2)}{2L} = \frac{0.57 \text{ m/d}[(48.5 \text{ m})^2 - (45.6 \text{ m})^2]}{2 \times 1800 \text{ m}}$$

$$= 0.0432 \text{ m}^2/\text{d}$$

## 4 Anisotropic Aquifers

Most real geologic formations tend to have more than one direction for the movement of water due to the nature of the material and its orientation. Sometimes, a soil formation may have a hydraulic conductivity (permeability) in the horizontal direction, $K_x$, radically different from that in the vertical direction, $K_z$. This phenomenon ($K_x \neq K_z$), is called anisotropy. When hydraulic conductivities are the same in all directions, ($K_x = K_z$), the aquifer is called isotropic. In typical alluvial deposits, $K_x$ is greater than $K_z$. For a two-layered aquifer of different hydraulic conductivities and different thicknesses, applying Darcy's law to horizontal flow can be expressed as

$$K_x = \frac{K_1 z_1 + K_2 z_2}{z_1 + z_2} \tag{3.25}$$

or, in general form

$$K_x = \frac{\Sigma K_i z_i}{\Sigma z_i} \tag{3.26}$$

where $K_i$ = hydraulic conductivity in layer $i$, mm/s or fps
$z_i$ = aquifer thickness of layer $i$, m or ft

For a vertical groundwater flow through two layers, let $q_z$ be the flow per unit horizontal area in each layer. The following relationship exists:

$$dh_1 + dh_2 = \left(\frac{z_1}{K_1} + \frac{z_2}{K_2}\right) q_z \tag{3.27}$$

Since

$$(dh + dh_2)k_z = (z_1 + z_2)q_z$$

then

$$dh_1 + dh_2 = \left(\frac{z_1 + z_2}{K_2}\right) q_z \tag{3.28}$$

where $K_z$ is the hydraulic conductivity for the entire aquifer. Comparison of Eqs. (3.27) and (3.28) yields

$$\frac{z_1 + z_2}{K_z} = \frac{z_1}{K_1} + \frac{z_2}{K_2}$$

$$K_z = \frac{z_1 + z_2}{z_1/K_1 + z_2/K_2} \tag{3.29}$$

or, in general form

$$K_z = \frac{\Sigma z_i}{\Sigma z_i/K_i} \tag{3.30}$$

The ratios of $K_x$ to $K_z$ for alluvium are usually between 2 and 10.

## 5 Unsteady (Nonequilibrium) Flows

Equilibrium equations described in the previous sections usually overestimate hydraulic conductivity and transmissivity. In practical situations, equilibrium usually takes a long time to reach. Theis (1935) originated the equation relations for the flow of groundwater into wells and was then improved on by other investigators (Jacob, 1940, 1947; Wenzel 1942; Cooper and Jacob, 1946).

Three mathematical/graphical methods are commonly used for estimations of transmissivity and storativity for nonequilibrium flow conditions. They are the Theis method, the Cooper and Jacob (straight-line) method, and the distance-drawdown method.

### 5.1 Theis method

The nonequilibrium equation proposed by Theis (1935) for the ideal aquifer is

$$d = \frac{Q}{4\pi T}\int_m^\infty \frac{e^{-u}}{u}du = \frac{Q}{4\pi T}W(u) \tag{3.31}$$

and

$$u = \frac{r^2 S}{4Tt} \tag{3.32}$$

where $d$ = drawdown at a point in the vicinity of a well pumped at a constant rate, ft or m

$Q$ = discharge of the well, gpm or m$^3$/s

$T$ = Transmissibility, ft$^2$/s or m$^2$/s

$r$ = distance from pumped well to the observation well, ft or m

$S$ = coefficient of storage of aquifer

$t$ = time of the well pumped, min

$W(u)$ = well function of $u$

The integral of the Theis equation is written as $W(u)$, and is the exponential integral (or well function) which can be expanded as a series:

$$W(u) = -0.5772 - \ln u + u - \frac{u^2}{2 \cdot 2!} + \frac{u^3}{3 \cdot 3!} - \frac{u^4}{4 \cdot 4!} + \cdots \quad (3.33a)$$

$$= -0.5772 - \ln u + u - u^2/4 + u^3/18 - u^4/96 + \cdots$$

$$+ (-1)^{n-1}\frac{u^n}{n \cdot n!} \quad (3.33b)$$

Values of $W(u)$ for various values of $u$ are listed in Appendix B, which is a complete table by Wenzel (1942) and modified from Illinois EPA (1990).

If the coefficient of transmissibility $T$ and the coefficient of storage $S$ are known, the drawdown $d$ can be calculated for any time and at any point on the cone of depression including the pumped well. Obtaining these coefficients would be extremely laborious and is seldom completely satisfied for field conditions. The complete solution of the Theis equation requires a graphical method of two equations (Eqs. (3.31) and (3.32)) with four unknowns.

Rearranging as:

$$d = \frac{Q}{4\pi T} W(u) \quad (3.31)$$

and

$$\frac{r^2}{t} = \frac{4T}{S}u \quad (3.34)$$

Theis (1935) first suggested plotting $W(u)$ on log-log paper, called a *type curve*. The values of $S$ and $T$ may be determined from a series of drawdown observations on a well with known times. Also, prepare another plot, of values of $d$ against $r^2/t$ on transparent log-log paper, with the same scale as the other figure. The two plots (Fig. 3.5) are superimposed so that a match point can be obtained in the region of which the curves nearly coincide when their coordinate axes are parallel. The coordinates of the match point are marked on both curves. Thus, values of $u$, $W(u)$, $d$, and $r^2/t$ can be obtained. Substituting these values into Eqs. (3.31) and (3.32), values of $T$ and $S$ can be calculated.

**Example:** An artesian well is pumped at a rate of 0.055 $m^3/s$ for 60 h. Observations of drawdown are recorded and listed below as a function of time at an observation hole 90 m away. Estimate the transmissivity and storativity using the Theis method.

| Time $t$, min | Drawdown $d$, m | $r^2/t$, $m^2$/min |
|---|---|---|
| 1 | 0.12 | 8100 |
| 2 | 0.21 | 4050 |
| 3 | 0.33 | 2700 |
| 4 | 0.39 | 2025 |
| 5 | 0.46 | 1620 |
| 6 | 0.5 | 1350 |
| 10 | 0.61 | 810 |
| 15 | 0.75 | 540 |
| 20 | 0.83 | 405 |
| 30 | 0.92 | 270 |
| 40 | 1.02 | 203 |
| 50 | 1.07 | 162 |
| 60 | 1.11 | 135 |
| 80 | 1.2 | 101 |
| 90 | 1.24 | 90 |
| 100 | 1.28 | 81 |
| 200 | 1.4 | 40.5 |
| 300 | 1.55 | 27 |
| 600 | 1.73 | 13.5 |
| 900 | 1.9 | 10 |

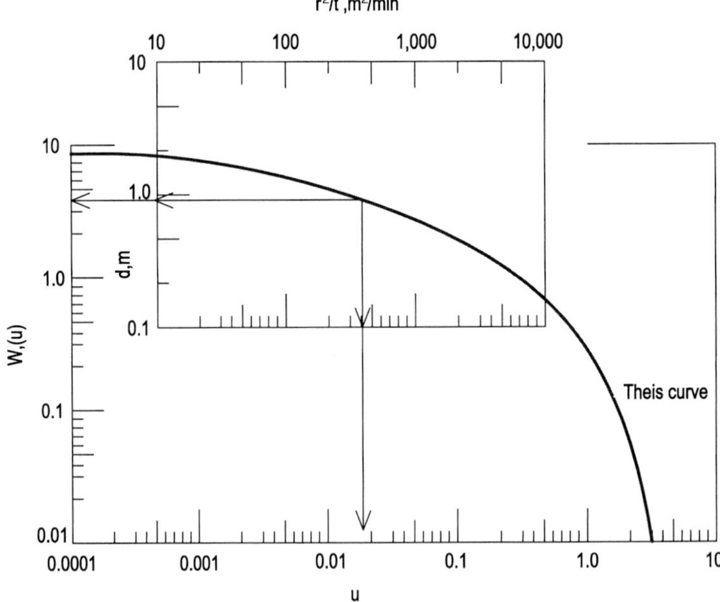


**Figure 3.5** Observed data and Theis type curve.

**solution:**

Step 1. Calculate $r^2/t$ and construct a table with $t$ and $d$

Step 2. Plot the observed data $d$ versus $r^2/t$ on log-log (transparency) paper in Fig. 3.5

Step 3. Plot a Theis type curve $W(u)$ versus $u$ on log-log paper (Fig. 3.5) using the data listed in Appendix B (Illinois EPA, 1988)

Step 4. Select a match point on the superimposed plot of the observed data on the type curve

Step 5. From the plots, the coordinates of the match points on the two curves are

$$\text{Observed data: } r^2/t = 320 \text{ m}^2/\text{min}$$

$$d = 0.90 \text{ m}$$

$$\text{Type curve: } W(u) = 3.5$$

$$u = 0.018$$

Step 6. Substitute the above values to estimate $T$ and $S$

Using Eq. (3.31)

$$d = \frac{Q}{4\pi T} W(u)$$

$$T = \frac{QW(u)}{4\pi d} = \frac{0.055 \text{ m}^3/\text{s} \times 3.5}{4\pi \times 0.90 \text{ m}}$$

$$= 0.017 \text{ m}^2/\text{s}$$

Using Eq. (3.34)

$$\frac{r^2}{t} = \frac{4Tu}{S}$$

$$S = \frac{4Tu}{r^2 t} = \frac{4 \times 0.017 \text{ m}^2/\text{s} \times 0.018}{320 \text{ m}^2/\text{min} \times (1 \text{ min}/60 \text{ s})}$$

$$= 2.30 \times 10^{-4}$$

## 5.2 Cooper–Jacob method

Cooper and Jacob (1946) modified the nonequilibrium equation. It is noted that the parameter $u$ in Eq. (3.32) becomes very small for large values of $t$ and small values of $r$. The infinite series for small $u$, $W(u)$, can be approximated by

$$W(u) = -0.5772 - \ln u = -0.5772 - \ln \frac{r^2 S}{4Tt} \tag{3.35}$$

Then

$$d = \frac{Q}{4\pi T} W(u) = \frac{Q}{4\pi T}\left(-0.5772 - \ln \frac{r^2 S}{4Tt}\right) \tag{3.36}$$

Further rearrangement and conversion to decimal logarithms yields

$$d = \frac{2.303Q}{4\pi T} \log \frac{2.25Tt}{r^2 S} \tag{3.37}$$

Therefore, the drawdown is to be a linear function of log $t$. A plot of $d$ versus the logarithm of $t$ forms a straight line with the slope $Q/4\pi T$ and an intercept at $d = 0$, when $t = t_0$, yielding

$$0 = \frac{2.3Q}{4\pi T} \log \frac{2.25Tt_0}{r^2 S} \tag{3.38}$$

Since $\log(1) = 0$

$$1 = \frac{2.25Tt_0}{r^2 S}$$

$$S = \frac{2.25Tt_0}{r^2} \tag{3.39}$$

If the slope is measured over one log cycle of time, the slope will equal the change in drawdown $\Delta d$, and Eq. (3.37) becomes

$$\Delta d = \frac{2.303Q}{4\pi T}$$

then

$$T = \frac{2.303Q}{4\pi \Delta d} \tag{3.40}$$

The Cooper and Jacob modified method solves for $S$ and $T$ when values of $u$ are less than 0.01. The method is not applicable to periods immediately after pumping starts. Generally, 12 h or more of pumping are required.

**Example:** Using the given data in the above example (by the Theis method) with the Cooper and Jacob method, estimate the transmissivity and storativity of a confined aquifer.

**solution:**

Step 1. Determine $t_0$ and $\Delta d$

Values of drawdown ($d$) and time ($t$) are plotted on semilog paper with the $t$ in the logarithmic scale as shown in Fig. 3.6. A best-fit straight line is drawn through the observed data. The intercept of the $t$-axis is 0.98 min. The slope of the line $\Delta d$ is measured over 1 log cycle of $t$ from the figure. We obtain:

$$t_0 = 0.98 \text{ min}$$
$$= 58.8 \text{ s}$$

and for a cycle ($t$ = 10 to 100 min)

$$\Delta d = 0.62 \text{ m}$$

Step 2. Compute $T$ and $S$

Using Eq. (3.40)

$$T = \frac{2.303Q}{4\pi\Delta d} = \frac{2.303 \times 0.055 \text{ m}^3\text{/s}}{4 \times 3.14 \times 0.62 \text{ m}}$$
$$= 0.016 \text{ m}^2\text{/s}$$

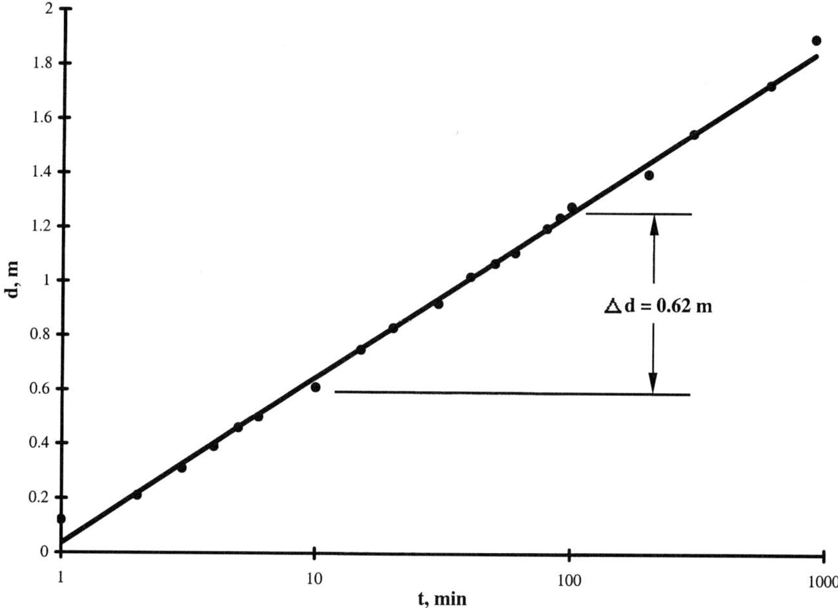


**Figure 3.6** Relationship of drawdown and the time of pumping.

Using Eq. (3.39)

$$S = \frac{2.25Tt_0}{r^2} = \frac{2.25 \times 0.016\ \mathrm{m^2/s} \times 58.8\ \mathrm{s}}{(90\ \mathrm{m})^2}$$

$$= 2.61 \times 10^{-4}$$

### 5.3 Distance-drawdown method

The distance-drawdown method is a modification of the Cooper and Jacob method and is applied to obtain quick information about the aquifer characteristics while the pumping test is in progress. The method needs simultaneous observations of drawdown in three or more observation wells. The aquifer properties can be determined from pumping tests by the following equations (Watson and Burnett, 1993):

$$T = \frac{0.366Q}{\Delta(h_0 - h)} \quad \text{for SI units} \tag{3.41}$$

$$T = \frac{528Q}{\Delta(h_0 - h)} \quad \text{for English units} \tag{3.42}$$

and

$$S = \frac{2.25Tt}{r_0^2} \quad \text{for SI units} \tag{3.43}$$

$$S = \frac{Tt}{4790r_0^2} \quad \text{for English units} \tag{3.44}$$

where $T$ = transmissivity, $\mathrm{m^2/d}$ or gpd/ft
$Q$ = normal discharge rate, $\mathrm{m^3/d}$ or gpm
$\Delta(h_0 - h)$ = drawdown per log cycle of distance, m or ft
$S$ = storativity, unitless
$t$ = time since pumping when the simultaneous readings are taken in all observation wells, days or min
$r_0$ = intercept of the straight-line plot with the zero-drawdown axis, m or ft

The distance-drawdown method involves the following procedures:

1. Plot the distance and drawdown data on semilog paper; drawdown on the arithmetic scale and the distance on the logarithmic scale.
2. Read the drawdown per log cycle in the same manner for the Cooper and Jacob method: this gives the value of $\Delta(h_0 - h)$.

3. Draw a best-fit straight line.
4. Extend the line to the zero-drawdown and read the value of the intercept, $r_0$.

**Example:** A water supply well is pumping at a constant discharge rate of 1000 $m^3$/d (11,000 gpm). It happens that there are five observation wells available. After pumping for 3 h, the drawdown at each observation well is recorded as below. Estimate transmissivity and storativity of the aquifer using the distance-drawdown method.

| Distance | | |
|---|---|---|
| m | ft | Drawdown, m |
| 3 | 10 | 3.22 |
| 7.6 | 25 | 2.21 |
| 20 | 66 | 1.42 |
| 50 | 164 | 0.63 |
| 70 | 230 | 0.28 |

**solution:**

Step 1. Plot the distance-drawdown data on semilog paper as shown in Fig. 3.7

Step 2. Draw a best-fit straight line over the observed data and extend the line to the $x$-axis

Step 3. Read the drawdown value for one log cycle

From Fig. 3.7, for the distance for a cycle from 4 to 40 m, the value of the drawdown, then $\Delta(h_0 - h)$ is read as 2.0 m (2.8 m – 0.8 m).

Step 4. Read the intercept on the $X$ axis for $r_0$

$$r_0 = 93 \text{ m}$$

Step 5. Determine $T$ and $S$ by Eqs. (3.41) and (3.43)

$$T = \frac{0.366Q}{\Delta(h_0 - h)} = \frac{0.366 \times 1000 \text{ m}^3/\text{d}}{2.0 \text{ m}}$$

$$= 183 \text{ m}^2/\text{d}$$

$$\text{Time of pumping: } t = 3 \text{ h}/24 \text{ h/d}$$

$$= 0.125 \text{ days}$$

$$S = \frac{2.25Tt}{r_0^2} = \frac{2.25 \times 183 \text{ m}^2/\text{d} \times 0.125 \text{ days}}{(93 \text{ m})^2}$$

$$= 0.006$$

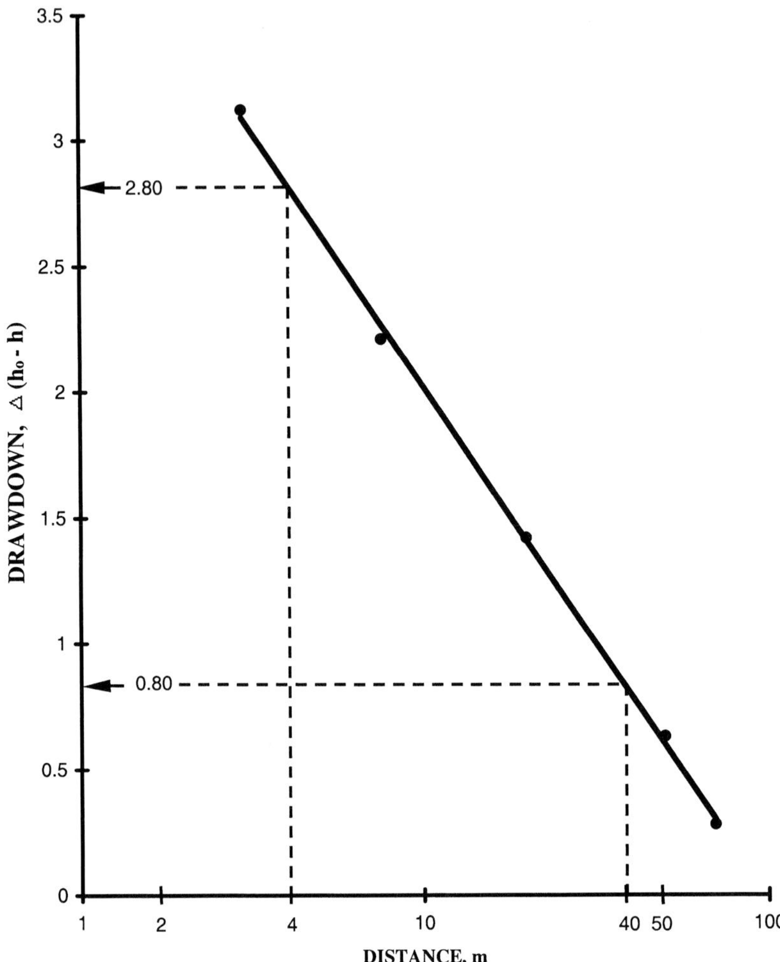


**Figure 3.7** Plot of observed distance-drawdown data.

### 5.4 Slug tests

In the preceding sections, the transmissivity $T$ and storativity $S$ of the aquifer and permeability $K$ of the soil are determined by boring one or two more observation wells. Slug tests use only a single well for the determination of those values by careful evaluation of the drawdown curve and information of screen geometry. The tests involve either raising or lowering the water level in the well and measuring the return to a static water level as a function of time.

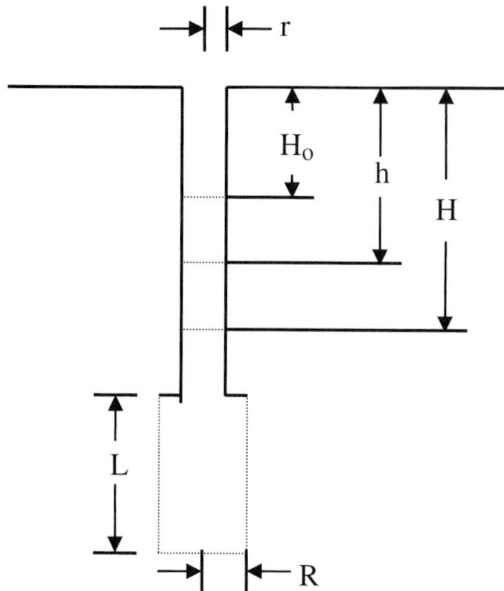


**Figure 3.8** Hvorslev slug test.

A typical test procedure requires introducing an object to record the volume (the slug) of the well. The Hvorslev (1951) method using a piezometer in an confined aquifer is widely used in practice due to it being quick and inexpensive. The procedures of conducting and analyzing the Hvorslev test (Fig. 3.8) are as follows:

1. Record the instantaneously raised (or lowered) water level to the static water level as $H_0$.
2. Read subsequently changing water levels with time as $h$. Thus $h = H_0$ at $t = 0$.
3. Measure the final raised head as $H$ at infinite time.
4. There is a relationship as

$$\frac{H - h}{H - H_0} = e^{-t/T_0} \tag{3.45}$$

where

$$T_0 = \frac{\pi r^2}{Fk} \tag{3.46}$$

$= $ Hvorslev-defined basic time lag

$F = $ shape factor

5. Calculate ratios $(H - h)/(H - H_0)$ as recovery.
6. Plot on semilogarithmic paper $(H - h)/(H - H_0)$ on the logarithmic scale versus time on the arithmetic scale.
7. Find $T_0$ at recovery equals 0.37 (37% of the initial change caused by the slug).
8. For piezometer intake length divided by radius $(L/R)$ greater than 8, Hvorslev evaluated the shape factor $F$ and proposed an equation for hydraulic conductivity $K$ as

$$K = \frac{r^2 \ln(L/R)}{2LT_0} \tag{3.47}$$

where $K$ = hydraulic conductivity, m/d or ft/d
$r$ = radius of the well casing, cm or in
$R$ = radius of the well screen, cm or in
$L$ = length of the well screen, cm or in
$T_0$ = time required for water level to reach 37% of the initial change, s

Other slug test methods have been developed for confined aquifers (Cooper *et al.*, 1967; Papadopoulous *et al.*, 1973; Bouwer and Rice, 1976). These methods are similar to Theis's type curves in that a curve-matching method is used to determine $T$ and $S$ for a given aquifer. A family of type curves $H_t/H_0$ versus $Tr/r_c^2$ were published for five values of the variable, (defined as $(r_s^2/r_c^2)\ S$, to estimate transmissivity, storativity, and hydraulic conductivity.

The Bouwer and Rice (1976) slug test method is most commonly used for estimating hydraulic conductivity in groundwater. Although the method was originally developed for unconfined aquifers, it can also be applied for confined or stratified aquifers if the top of the screen is some distance below the upper confined layer. The following formula is used to compute hydraulic conductivity:

$$K = \frac{r^2 \ln (R/r_w)}{2L} \frac{1}{t} \ln \frac{y_0}{y_t} \tag{3.48}$$

where $K$ = hydraulic conductivity, cm/s
$r$ = radius of casing, cm
$y_0, y_t$ = vertical difference in water levels between inside and outside the well at time $t = 0$, and $t = t$, m
$R$ = effective radius distance over which $y$ is dissipated, cm

$r_w$ = radius distance of undisturbed portion of aquifer from well centerline (usually $r$ plus thickness of gravel).
$L$ = length of screen, m
$t$ = time, s

**Example 1:** The internal diameters of the well casing and well screen are 10 cm (4 in) and 15 cm (6 in), respectively. The length of the well screen is 2 m (6.6 ft). The static water level measured from the top of the casing is 2.50 m (8.2 ft). A slug test is conducted and pumped to lower the water level to 3.05 m (10 ft). The time-drawdown $h$ in the unconfined aquifer is recorded every 3 s as shown in the following table. Determine the hydraulic conductivity of the aquifer by the Hvorslev method.

**solution:**

Step 1. Calculate $(h - H_0)/(H - H_0)$

Given: $H_0 = 2.50$ m, $H = 3.05$ m
Then $H - H_0 = 3.05\text{ m} - 2.50\text{ m} = 0.55$ m.

| Time $t$, s | $h$, m | $h - H_0$, m | $(h - H_0)/0.55$ |
|---|---|---|---|
| 0 | 3.05 | 0.55 | 1.00 |
| 3 | 2.96 | 0.46 | 0.84 |
| 6 | 2.89 | 0.39 | 0.71 |
| 9 | 2.82 | 0.32 | 0.58 |
| 12 | 2.78 | 0.28 | 0.51 |
| 15 | 2.73 | 0.23 | 0.42 |
| 18 | 2.69 | 0.19 | 0.34 |
| 21 | 2.65 | 0.15 | 0.27 |
| 24 | 2.62 | 0.12 | 0.22 |
| 27 | 2.61 | 0.10 | 0.18 |
| 30 | 2.59 | 0.09 | 0.16 |

Step 2. Plot $t$ versus $(h - H_0)/(H - H_0)$ on semilog paper as shown in Fig. 3.9, and draw a best-fit straight line

Step 3. Find $T_0$

From Fig. 3.9, read 0.37 on the $(h - H_0)/(H - H_0)$ scale and note the time for the water level to reach 37% of the initial change $T_0$ caused by the slug. This is expressed as $T_0$. In this case $T_0 = 16.2$ s

Step 4. Determine the $L/R$ ratio

$$R = 15\text{ cm}/2 = 7.5\text{ cm}$$

$$L/R = 200\text{ cm}/7.5\text{ cm} = 26.7 > 8$$

Thus Eq. (3.47) can be applied

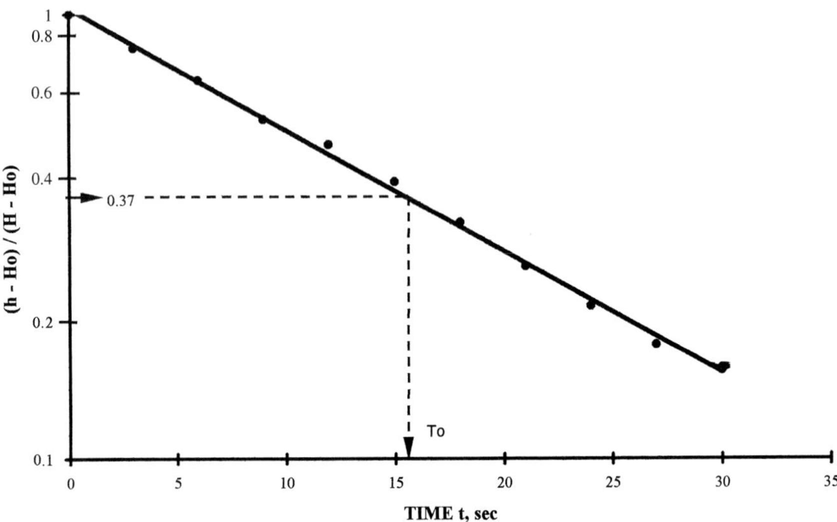


**Figure 3.9** Plot of Hvorslev slug test results.

Step 5. Find $K$ by Eq. (3.47)

$$r = 10 \text{ cm}/2 = 5 \text{ cm}$$

$$\begin{aligned} K &= \frac{r^2 \ln (L/R)}{2LT_0} \\ &= (5 \text{ cm})^2 \ln 26.7/(2 \times 200 \text{ cm} \times 16.2 \text{ s}) \\ &= 0.0127 \text{ cm/s} \\ &= 0.0127 \text{ cm/s} \times (1 \text{ m}/100 \text{ cm}) \times 86{,}400 \text{ s/d} \\ &= 11.0 \text{ m/d} \\ &= 36 \text{ ft/d} \end{aligned}$$

**Example 2:** A screened, cased well penetrates a confined aquifer with gravel pack 3.0-cm thickness around the well. The radius of casing is 5.0 cm and the screen is 1.2 m long. A slug of water is injected and water level raised by $m$. The effective radial distance over which $y$ is dissipated is 12 cm. Estimate hydraulic conductivity for the aquifer. The change of water level with time is as follows:

| $t$, s | $y_t$, cm | $t$,s | $y_t$, cm |
|---|---|---|---|
| 1 | 30 | 10 | 4.0 |
| 2 | 24 | 13 | 2.0 |
| 3 | 17 | 16 | 1.1 |
| 4 | 14 | 20 | 0.6 |
| 5 | 12 | 30 | 0.2 |
| 6 | 9.6 | 40 | 0.1 |
| 8 | 5.5 | | |

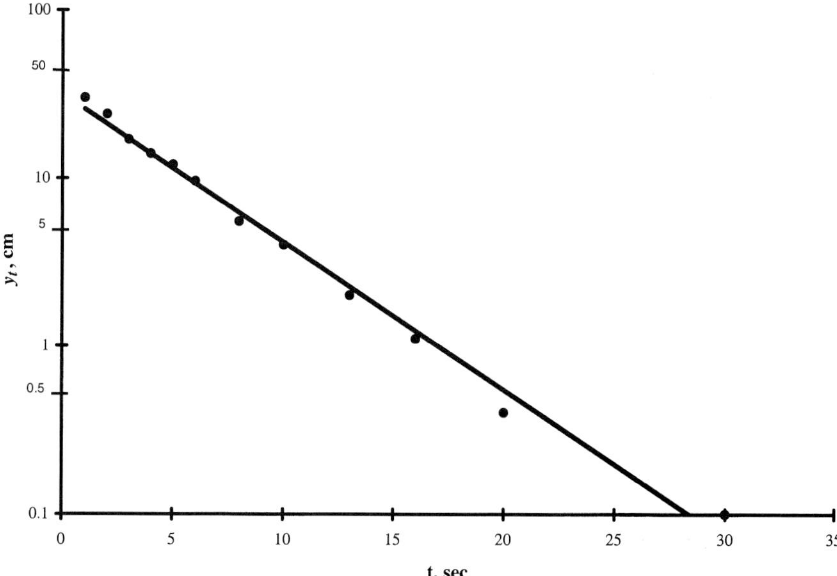


**Figure 3.10** Plot of $y_t$ versus $t$.

**solution:**

Step 1. Plot values of $y$ versus $t$ on semilog paper as shown in Fig. 3.10

Draw a best-fit straight line. The line from $y_0 = 36$ cm to $y_t = 0.1$ cm covers 2.5 log cycles. The time increment between the two points is 26 s.

Step 2. Determine $K$ by Eq. (3.48), the Bouwer and Rice equation

$$r = 5 \text{ cm}$$

$$R = 12 \text{ cm}$$

$$r_w = 5 \text{ cm} + 3 \text{ cm} = 8 \text{ cm}$$

$$K = \frac{r^2 \ln (R/r_w)}{2L} \frac{1}{t} \ln \frac{y_0}{y_t}$$

$$= \frac{(5 \text{ cm})^2 \ln(12 \text{ cm}/8 \text{ cm})}{2(120 \text{ cm})} \frac{1}{26 \text{ s}} \ln \frac{36 \text{ cm}}{0.1 \text{ cm}}$$

$$= 9.56 \times 10^{-3} \text{ cm/s}$$

## 6 Groundwater Contamination

### 6.1 Sources of contamination

There are various sources of groundwater contamination and various types of contaminants. Underground storage tanks, agricultural activity, municipal landfills, abandoned hazardous waste sites, and septic tanks are the major threats to groundwater. Other sources may be from industrial spill, injection wells, land application, illegal dumps (big problem in many countries), road salt, saltwater intrusion, oil and gas wells, mining and mine drainage, municipal wastewater effluents, surface impounded waste material stockpiles, pipelines, radioactive waste disposal, and transportation accidents.

Large quantities of organic compounds are manufactured and used by industries, agriculture, and municipalities. These man-made organic compounds are of most concern. The inorganic compounds occur in nature and may come from natural sources as well as human activities. Metals from mining, industries, agriculture, and fossil fuels also may cause groundwater contamination.

Types of contaminants are classified by chemical group. They are metals (arsenic, lead, mercury, etc.), volatile organic compounds (VOCs) (gasoline, oil, paint thinner), pesticides and herbicides, and radionuclides (radium, radon). The most frequently reported groundwater contaminants are the VOCs (Voelker, 1984; Rehfeldt *et al.*, 1992).

Volatile organic compounds are made of carbon, hydrogen, and oxygen and have a vapor pressure less than one atmosphere. Compounds such as gasoline or dry cleaning fluid evaporate when left open to the air. They are easily dissolved in water. There are numerous incidences of VOCs contamination caused by leaking underground storage tanks. Groundwater contaminated by VOCs poses cancer risks to humans either by ingestion of drinking water or inhalation.

## 6.2 Contaminant transport pathways

The major contaminant transport mechanisms in groundwater are advection, diffusion, dispersion, adsorption, chemical reaction, and biodegradation. Advection is the movement of contaminant(s) with the flowing groundwater at the seepage velocity in the pore space and is expressed as Darcy's law:

$$v_x = \frac{K}{n}\frac{dh}{L} \tag{3.49}$$

where $v_x$ = seepage velocity, m/s or pps
$K$ = hydraulic conductivity, m/s or fps
$n$ = porosity
$dh$ = pressure head, m or ft
$L$ = distance, m or ft

Equation (3.49) is similar to Eq. (3.10b).

Diffusion is a molecular-scale mass transport process that moves solutes from an area of higher concentration to an area of lower concentration. Diffusion is expressed by Fick's law:

$$F_x = -D_{\mathrm{d}} \frac{dc}{dx} \tag{3.50}$$

where $F_x$ = mass flux, mg/m$^2$ · s or lb/ft$^2$ · s
$D_{\mathrm{d}}$ = diffusion coefficient, m$^2$/s or ft$^2$/s
$dc/dx$ = concentration gradient, mg/(m$^3$ · m) or lb/(ft$^3$ · ft)

Diffusive transport can occur at low or zero flow velocities. In a tight soil or clay, typical values of $D_{\mathrm{d}}$ range from 1 to $2 \times 10^9$ m$^2$/s at 25°C (Bedient *et al.* 1994). However, typical dispersion coefficients in groundwater are several orders of magnitude greater than that in clay.

Dispersion is a mixing process caused by velocity variations in porous media. Mass transport due to dispersion can occur parallel and normal to the direction of flow with two-dimensional spreading.

Sorption is the interaction of a contaminant with a solid. It can be divided into adsorption and absorption. An excess concentration of contaminants at the surfaces of solids is called adsorption. Absorption refers to the penetration of the contaminants into the solids.

Biodegradation is a biochemical process that transforms contaminants (certain organics) into simple carbon dioxide and water by microorganisms. It can occur in aerobic and anaerobic conditions. Anaerobic biodegradation may include fermentation, denitrification, iron reduction, sulfate reduction, and methane production.

Excellent and complete coverage of contaminant transport mechanisms is presented by Bedient *et al.* (1994). Theories and examples are covered for mass transport, transport in groundwater by advection, diffusion, dispersion, sorption, chemical reaction, and bio-degradation.

Some example problems of contaminant transport are also given by Tchobanoglous and Schroeder (1985). Mathematical models that analyze complex contaminant pathways in groundwater are also discussed elsewhere (Canter and Knox, 1986; Willis and Yeh, 1987; Canter *et al.*, 1988; Mackay and Riley, 1993; Smith and Wheatcraft, 1993; Watson and Burnett, 1993; James, 1993; Gupta, 1997).

The transport of contaminants in groundwater involves adsorption, advection, diffusion, dispersion, interface mass transfer, biochemical transformations, and chemical reactions. On the basis of mass balance, the general equation describing the transport of a dissolved contaminant through an isotropic aquifer under steady-state flow conditions can be mathematically expressed as (Gupta, 1997).

$$\frac{\partial c}{\partial t} + \frac{\rho_b}{\rho}\frac{\partial S}{\partial t} + ck_l + Sk_s\frac{\rho_b}{\varphi}$$
$$= D_x\frac{\partial^2 c}{\partial x^2} + D_y\frac{\partial^2 c}{\partial y^2} + D_z\frac{\partial^2 c}{\partial z^2} - V_x\frac{\partial c}{\partial x} - V_y\frac{\partial c}{\partial y} - V_z\frac{\partial c}{\partial z} \quad (3.51)$$

where
$c$ = solute (contaminant in liquid phase) concentration, g/m$^3$
$S$ = concentration in solid phase as mass of contaminant per unit mass of dry soil, g/g
$t$ = time, day
$\rho_b$ = bulk density of soil, kg/m$^3_x$
$\varphi$ = effective porosity
$k_l$, $k_s$ = first-order decay rate in the liquid and soil phases, respectively, day$^{-1}$
$x$, $y$, $z$ = Cartesian coordinates, m
$D_x$, $D_y$, $D_z$ = directional hydrodynamic dispersion coefficients, m$^2$/d
$V_x$, $V_y$, $V_z$ = directional seepage velocity components, m/d

There are two unknowns ($c$ and $S$) in Eq. (3.51). Assuming a linear adsorption isotherm of the form

$$S = K_d c \quad (3.52)$$

where $K_d$ = distribution coefficient due to chemical reactions and biological degradation, and substituting Eq. (3.52) into Eq. (3.51), we obtain

$$R\frac{\partial c}{\partial t} + kc = D_x\frac{\partial^2 c}{\partial x^2} + D_y\frac{\partial^2 c}{\partial y^2} + D_z\frac{\partial^2 c}{\partial z^2} - V_x\frac{\partial c}{\partial x} - V_y\frac{\partial c}{\partial y} - V_z\frac{\partial c}{\partial z} \quad (3.53)$$

where $R = 1 + k_d\,(\rho_b/\varphi)$
= retardation factor which slows the movement of solute due to adsorption
$k = k_1 + k_s K_d(\rho_b/\varphi)$
= overall first-order decay rate, day$^{-1}$

The general equation under steady-state flow conditions in the $x$ direction, we modify from Eq. (3.53).

### 6.3 Underground storage tank

There are 3 to 5 million underground storage tanks (USTs) in the United States. It is estimated that 3% to 10% of these tanks and their associated piping systems may be leaking (US EPA 1987). The majority of

USTs contain petroleum products (gasoline and other petroleum products). When the UST leaks and is left unattended, and subsequently contaminates subsurface soils, surface and groundwater monitoring and corrective actions are required.

The migration or transport pathways of contaminants from the UST depend on the quantity released, the physical properties of the contaminant, and characteristics of the soil particles.

When a liquid contaminant is leaked from a UST below the ground surface, it percolates downward to the unconfined groundwater surface. If the soil characteristics and contaminant properties are known, it can be estimated whether the contaminant will reach the groundwater. For hydrocarbons in the unsaturated (vadose) zone, it can be estimated by the equation (US EPA 1987):

$$D = \frac{R_v V}{A} \tag{3.54}$$

where $D$ = maximum depth of penetration, m
$R_v$ = a coefficient of retention capacity of the soil and the viscosity of the product (contaminant)
$V$ = volume of infiltrating hydrocarbon, $m^3$
$A$ = are of spell, $m^2$

The typical values for $R_v$ are as follows (US EPA, 1987):

| Soil | Gasoline | Kerosene | Light fuel oil |
|---|---|---|---|
| Coarse gravel | 400 | 200 | 100 |
| Gravel to coarse sand | 250 | 125 | 62 |
| Coarse to medium sand | 130 | 66 | 33 |
| Medium to fine sand | 80 | 40 | 20 |
| Fine sand to silt | 50 | 25 | 12 |

Retention (attenuative) capacities for hydrocarbons vary approximately from 5 L/$m^3$ in coarse gravel to more than 40 L/$m^3$ in silts. Leaked gasoline can travel 25 ft (7.6 m) through unsaturated, permeable, alluvial, or glacial sediments in a few hours, or at most a few days, which is extremely site specific.

The major movement of nonaqueous-phase liquids that are less dense than water (such as gasoline and other petroleum products) in the capillary zone (between unsaturated and saturated zones) is lateral. The plume will increase thickness (vertical plane) and width depending on leakage rates and the site's physical conditions. The characteristic shape of the flow is the so-called "oil package." Subsequently, it will plug the pores of the soil or be diluted and may be washed out into the water table.

The transport of miscible or dissolved substances in saturated zones follows the general direction of groundwater flow. The transport pathways can be applied to the models (laws) of advection and dispersion. The dispersion may include molecular diffusion, microscopic dispersion, and macroscopic dispersion.

Immiscible substances with a specific gravity of less than 1.0 (lighter than water) are usually found only in the shallow part of the saturated zone. The transport rate depends on the groundwater gradient and the viscosity of the substance.

Immiscible substances with a specific gravity of more than 1.0 (denser than water) move downward through the saturated zone. A dense immiscible substance poses a greater danger in terms of migration potential than less dense substances due to its deeper penetration into the saturated zone. When the quantity of released contaminant exceeds the retention capacity of the unsaturated and saturated zones, the denser nonaqueous-phase liquid continues its downward migration until it reaches an impermeable boundary. A liquid substance leaking from a UST enters the vapor phase in the unsaturated zone according to its specific vapor pressure. The higher the vapor pressure of the substance, the more it evaporates. The contaminant in the vapor phase moves by advection and by diffusion. Vapor moves primarily in a horizontal direction depending on the slope of the water table and the location of the impermeable bedrock. If the vapor, less dense than air, migrates in a vertical direction, it may accumulate in sewer lines, basements, and such areas.

### 6.4 Groundwater treatment

Once a leaking UST is observed, corrective action should be carried out, such as tank removal, abandonment, rehabilitation, removal/excavation of soil and sediment, on-site and/or off-site treatment and disposal of contaminants, product and groundwater recovery, groundwater treatment, etc.

Selection of groundwater treatment depends on the contaminants to be removed. Gasoline and volatile organic compounds can be removed by air stripping and stream stripping processes. Activated carbon adsorption, biological treatment, and granular media filtration can be used for removal of gasoline and other organics. Nonvolatile organics are removable by oxidation/reduction processes. Inorganic chemicals can be treated by coagulation/sedimentation, neutralization, dissolved air flotation, granular media filtration, ion exchange, resin adsorption, and reverse osmosis.

These treatment processes are discussed in detail in Chapter 5, Public Water Supply, and in Chapter 6, Wastewater.

## 7 Setback Zones

Section 1428 of the Safe Drinking Water Act (SDWA) requires each state to submit a wellhead protection program to the US Environmental Protection Agency (EPA). For example, Illinois EPA promulgated the Illinois Groundwater Protection Act (IGPA) in 1991. The Act assigned the responsibilities of existing state agencies and established groundwater classifications. On the basis of classification, different water quality standards, monitoring, and remedial requirements can be applied. Classifications are based on the PCB levels which may be associated with hazardous wastes.

Groundwater used as a public water supply source is called potable groundwater. This requires the highest degree of protection with the most stringent standards. The groundwater quality standards for potable groundwater are generally equal to the US EPA's maximum contamination levels applicable at the tap pursuant to the SDWA. The rationale is that potable groundwater should be safe for drinking water supply without treatment.

The state primacy agency establishes a comprehensive program for the protection of groundwater. Through interagency cooperation, local groundwater protection programs can help to prevent unexpected and costly water supply systems. An Illinois community experienced a leaking gasoline underground tank, operated by the city-owned garage. It contaminated one well and threatened to contaminate the entire well field. The city has spent more than $300,000 in an attempt to replace the water supply.

Some parts of the groundwater protection programs, such as minimum and maximum setback zones for wellhead protection, are used to protect public and private drinking water supplies from potential sources of groundwater contamination. Each community well must have a setback zone to restrict land use near the well. The setback zone provides a buffer between the well and potential contamination sources and routes. It will give time for cleanup efforts of contaminated groundwater or to obtain an alternative water supply source before the existing groundwater source becomes unfit for use.

A minimum setback zone is mandatory for each public well. Siting for new potential primary or secondary pollution sources or potential routes is prohibited within the setback zone. In Illinois, the minimum setback zone is a 200-ft (61-m) radius area around the wellhead for every water supply well. For some vulnerable aquifers, the zone may be 400 ft in radius.

The maximum setback zone is a second level of protection from pollution. It prohibits the siting of new potential primacy pollution sources within the area outside the minimum setback zone up to 1000 ft (305 m)

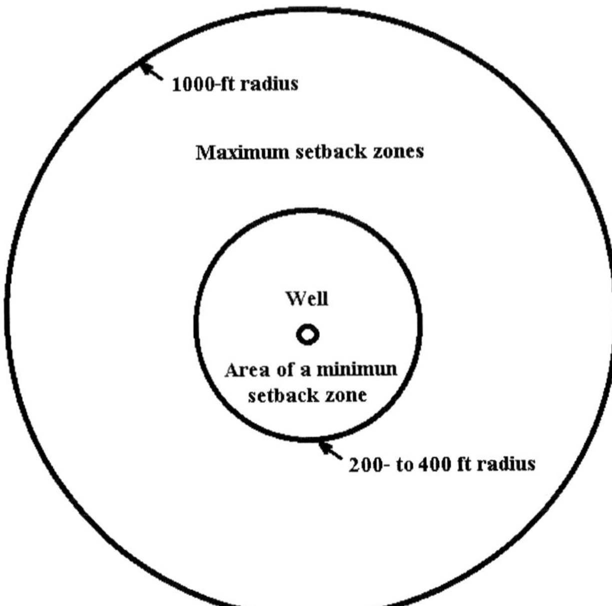


**Figure 3.11** Maximum and minimum setback zones.

from a wellhead in Illinois (Fig. 3.11). Maximum setback zones allow the well owners, county or municipal government, and state to regulate land use beyond the minimum setback zone.

The establishment of a maximum setback zone is voluntary. A request to determine the technical adequacy of a maximum setback zone determination must first be submitted to the state by a municipality or county. Counties and municipalities served by community water supply wells are empowered to enact maximum setback zone ordinances. If the community water supply wells are investor or privately owned, a county or municipality served by that well can submit an application on behalf of the owner.

### 7.1 Lateral area of influence

As described in the previous section, the lateral radius of influence (LRI) is the horizontal distance from the center of the well to the outer limit of the cone of depression (Figs. 3.1 and 3.12). It is the distance from the well to where there is no draw of groundwater (no reduction in water level). The lateral area of influence (LAI) outlines the extent of the cone of depression on the land surface as shown in the hatched area in Fig. 3.12.

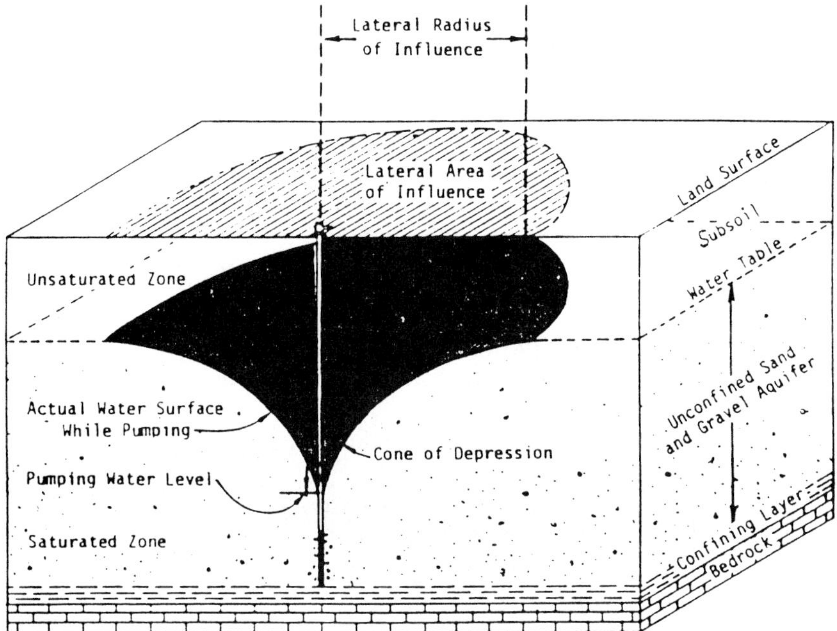


**Figure 3.12** Lateral area of influence and the cone of depression under normal operation conditions (*Illinois EPA, 1990*).

The LAI in a confined aquifer is generally 4000 times greater than the LAI in an unconfined aquifer (Illinois EPA, 1990). If a pollutant is introduced within the LAI, it will reach the well faster than other water replenishing the well. The slope of the water table steepens toward the well within the LAI. Therefore, it is extremely important to identify and protect the LAI. The LAI is used to establish a maximum setback zone.

## 7.2 Determination of lateral radius of influence

There are three types of determination methods for the LRI. They are the direct measurement method, the use of the Theis equation or volumetric flow equation, and the use of a curve-matching technique with the Theis equation. The third method involves the interpretation of pump test data from observation wells within the minimum setback zone using a curve-matching technique and the Theis equation to determine the transmissivity and storativity of the aquifer. Once these aquifer constants have been obtained, the Theis equation is then used to compute the LRI of the well. It should be noted that when the observation well or piezometer measurement was made outside of the minimum setback zone, the determination could have been conducted by the direct measurement method.

Volumetric flow equations adopted by the Illinois EPA (1990) are as follows. For unconfined unconsolidated or unconfined nonfractured bedrock aquifers, the LRI can be calculated by

$$r = \sqrt{\frac{Qt}{4524nH}} \tag{3.55}$$

where $r$ = lateral radius of influence, ft
$Q$ = daily well flow under normal operational conditions, cfs
$t$ = time that the well is pumped under normal operational conditions, min
$H$ = open interval or length of well screen, ft
$n$ = aquifer porosity, use following data unless more information is available:

| | |
|---|---|
| Sand | 0.21 |
| Gravel | 0.19 |
| Sand and gravel | 0.15 |
| Sandstone | 0.06 |
| Limestone primary dolomites | 0.18 |
| Secondary dolomites | 0.18 |

If pump test data is available for an unconfined/confined unconsolidated, or nonfractured bedrock aquifer, the LRI can be calculated by

$$r = \sqrt{\frac{uTt}{2693S}} \tag{3.56}$$

where $r$ = lateral radius of influence, ft
$T$ = aquifer transmissivity, gpd/ft
$t$ = time that well is pumped under normal operational conditions, min
$S$ = aquifer storativity or specific yield, dimensionless
$u$ = a dimensionless parameter related to the well function $W(u)$

and

$$W(u) = \frac{T(h_0 - h)}{114.6Q} \tag{3.57}$$

where $W(u)$ = well function
$h_0 - h$ = drawdown in the piezometer or observation well, ft
$Q$ = production well pumping rate under normal operational conditions, gpm

**Direct measurement method.** The direct measurement method involves direct measurement of drawdown in an observation well piezometer or in an another production well. Both the observation well and another production well should be located beyond the minimum setback zone of the production well and also be within the LAI.

**Example:** A steel tape is used to measure the water level of the observation well located 700 ft from a drinking water well. The pump test results show that under the normal operational conditions, 0.58 ft of drawdown compared to the original nonpumping water level is recorded. Does this make the well eligible for applying maximum setback zone?

**Answer:** Yes. A municipality or county may qualify to apply for a maximum setback zone protection when the water level measurements show that the drawdown is greater than 0.01 ft and the observation well is located beyond (700 ft) the minimum setback zone (200 to 400 ft). It is also not necessary to estimate the LRI of the drinking water well.

**Theis equation method.** The Theis equation (Eqs. (3.56) and (3.57)) can be used to estimate the LRI of a well if the aquifer parameters, such as transmissivity, storativity, and hydraulic conductivity, are available or can be determined. The Theis equation is more useful if the observation well is located within the minimum setback zone when the direct measurement method cannot be applied. Equations (3.52) and (3.53) can be used if pump test data are available for an unconfined/ confined, unconsolidated or nonfractured bedrock aquifer (Illinois EPA,1990).

**Example:** An aquifer test was conducted in a central Illinois county by the Illinois State Water Survey (Walton 1992). The drawdown data from one of the observations located 22 ft from the production well was analyzed to determine the average aquifer constants as follows:

Transmissivity: $T$ = 340,000 gpd/ft of drawdown

Storativity: $S$ = 0.09

The pumping rate under normal operational conditions was 1100 gpm. The volume of the daily pumpage was 1,265,000 gal. Is this aquifer eligible for a maximum setback zone protection?

**solution:**

Step 1. Determine the time of pumping, $t$

$$t = \frac{v}{Q} = \frac{1{,}265{,}000 \text{ gal}}{1100 \text{ gal/min}}$$

$$= 1150 \text{ min}$$

Step 2. The criteria used to represent a minimum level of drawdown $(h_0 - h)$ is assumed to be

$$h_0 - h = 0.01 \text{ ft}$$

Step 3. Calculate $W(u)$ using Eq. (3.57)

$$W(u) = \frac{T(h_0 - h)}{114.6Q}$$

$$= \frac{340{,}000 \text{ gpm/ft} \times 0.01 \text{ ft}}{114.6 \times 1100 \text{ gpm}}$$

$$= 0.027$$

Step 4. Find $u$ associated with a $W(u) = 0.027$ in Appendix B

$$u = 2.43$$

Step 5. Compute the lateral radius of influence $r$ using Eq. (3.56)

$$r = \sqrt{\frac{uTt}{2693S}}$$

$$= \sqrt{\frac{2.43 \times 340{,}000 \text{ gpd/ft} \times 1150 \text{ min}}{2693 \times 0.09}}$$

$$= 1980 \text{ ft}$$

**Answer:** Yes. The level of drawdown was estimated to be 0.01 ft at a distance of 1980 ft from the drinking water wellhead. Since $r$ is greater than the minimum setback distance of 400 ft from the wellhead, the well is eligible for a maximum setback zone.

### 7.3 The use of volumetric flow equation

The volumetric flow equation is a cost-effective estimation for the radius of influence and is used for wells that pump continuously. It is an alternative procedure where aquifer constants are not available and where the Theis equation cannot be used. The volumetric flow equation

(Eq. (3.55)) may be utilized for wells in unconfined unconsolidated or unconfined nonfractured bedrock aquifers.

**Example:** A well in Peoria, Illinois (Illinois EPA, 1995), was constructed in a sand and gravel deposit which extends from 40 to 130 ft below the land surface. The static water level in the aquifer is 65 ft below the land surface. Because the static water level is below the top of the aquifer, this would be considered an unconfined aquifer. The length of the well screen is 40 ft. Well operation records indicate that the pump is operated continuously for 30 days every other month. The normal discharge rate is 1600 gpm (0.101 $m^3$/s). Is the well eligible for a maximum setback zone protection?

**solution**

Step 1. Convert discharge rate from gpm to $ft^3$/d

$$\begin{aligned} Q &= 1600 \text{ gpm} \\ &= 1600 \frac{\text{gal}}{\text{min}} \times \frac{1 \text{ ft}^3}{7.48 \text{ gal}} \times \frac{1440 \text{ min}}{1 \text{ day}} \\ &= 308{,}021 \text{ ft}^3/\text{d} \\ &= 8{,}723 \text{ m}^3/\text{d} \end{aligned}$$

Step 2. Calculate $t$ in min

$$\begin{aligned} t &= 30 \text{ days} \\ &= 30 \text{ days} \times 1440 \text{ min/d} \\ &= 43{,}200 \text{ min} \end{aligned}$$

Step 3. Find the aquifer porosity $n$

$$n = 0.15 \quad \text{for sand and gravel}$$

Step 4. Calculate $r$ using Eq. (3.55)

$$\begin{aligned} H &= 40 \text{ ft} \\ r &= \sqrt{\frac{Qt}{4524 n\, H}} \\ &= \sqrt{\frac{308{,}021 \text{ ft}^3/\text{d} \times 43{,}200 \text{ min}}{4524 \times 0.15 \times 40 \text{ ft}}} \\ &= 700 \text{ ft} \\ &= 213.4 \text{ m} \end{aligned}$$

Yes, the well is eligible for maximum setback zone.

*Note*: The volumetric flow equation is very simple and economical.

## References

Babbitt, H. E., Doland, J. J. and Cleasby, J. L. 1959. *Water supply engineering*, 6th edn. New York: McGraw-Hill.

Bedient, P. B. and Huber, W. C. 1992. *Hydrology and floodplain analysis*, 2nd edn. Reading, Massachusetts: Addison-Wesley.

Bedient, P. B., Rifai, H. S. and Newell, C. J. 1994. *Ground water contamination transport and remediation*. Englewood Cliffs, New Jersey: Prentice Hall.

Bouwer, H. and Rice, R. C. 1976. A slug test for determining hydraulic conductivity of unconfined aquifers with completely or partially penetrating wills. *Water Resources Res.* **12**: 423–428.

Canter, L. W. and Knox, R. C. 1986. *Ground water pollution control*. Chelsea, Michigan: Lewis.

Canter, L. W., Knox, R. C. and Fairchild, D. M. 1988. *Gound water quality protection*. Chelsea, Michigan: Lewis.

Cooper, H. H., Jr., Bredehoeft, J. D. and Papadopoulos, I. S. 1967. Response of a finite-diameter well to an instantaneous charge of water. *Water Resources Res.* **3**: 263–269.

Cooper, H. H. Jr. and Jacob, C. E. 1946. A generalized graphical method for evaluating formation constants and summarizing well field history. *Trans. Amer. Geophys. Union* **27**: 526–534.

Forrest, C. W. and Olshansky, R. 1993. Groundwater protection by local government. University of Illinois at Urbana-Champaign. Illinois.

Gupta, A. D. 1997. Groundwater and the environment. In: Biswas, A. K. (ed.) *Water resources*. New York: McGraw-Hill.

Hammer, M. J. 1986. *Water and wastewater technology*. New York: John Wiley.

Hammer, M. J. and Mackichan, K. A. 1981. *Hydrology and quality of water resources*. New York: John Wiley.

Hvorslev, M. J. 1951. Time lag and soil permeability in groundwater observations. U.S. Army Corps of Engineers Waterways Experiment Station, Bulletin 36, Vicksburg, Mississippi.

Illinois Environmental Protection Agency (IEPA). 1987. *Quality assurance and field method manual*. Springfield: IEPA.

Illinois Environmental Protection Agency (IEPA). 1988. *A primer regarding certain provisions of the Illinois groundwater protection act*. Springfield: IEPA.

Illinois Environmental Protection Agency (IEPA). 1990. *Maximum setback zone workbooks*. Springfield: IEPA.

Illinois Environmental Protection Agency (IEPA) 1994. *Illinois water quality report* 1992–1993. Springfield: IEPA.

Illinois Environmental Protection Agency (IEPA). 1995. *Guidance document for groundwater protection needs assessments*. Springfield: IEPA.

Illinois Environmental Protection Agency (IEPA). 1996. *Illinois Water Quality Report* 1994–1995. Vol. I. Springfield, Illinois: Bureau of Water, IEPA.

Jacob, C. E. 1947. Drawdown test to determine effective radius of artesian well. *Trans. Amer. Soc. Civil Engs.* **112**(5): 1047–1070.

James, A. (ed.) 1993. *Introduction to water quality modelling*, 2nd edn. Chichester: John Wiley.

Leopold, L. B. 1974. *Water: A primer*. New York: W. H. Freeman.

Mackay, R. and Riley, M. S. 1993. Groundwater quality modeling. In: James, A. (ed.) *An introduction to water quality modeling*, 2nd edn. New York: John Wiley.

Nielsen, D. M. (ed.). 1991. *Practical handbook of groundwater monitoring*. Chelsea, Michigan: Lewis.

Papadopoulous, I. S., Bredehoeft, J. D. and Cooper, H. H. 1973. On the analysis of slug test data. *Water Resources Res.* **9**(4): 1087–1089.

Rehfeldt, K. R., Raman, R. K., Lin, S. D. and Broms, R. E. 1992. *Assessment of the proposed discharge of ground water to surface waters of the American Bottoms area of southwestern Illinois.* Contract report 539. Champaign: Illinois State Water Survey.

Rose, H. E. 1949. On the resistance coefficient-Reynolds number relationship for fluid flow through a bed of granular materials. *Proc. Inst. Mech. Engrs. (London)* **154**: 160.

Smith, L. and Wheatcraft, S. W. 1993. In: Maidment, D. R. (ed.) *Handbook of hydrology.* New York: McGraw-Hill.

Steel, E. W. and McGhee, T. J. 1979. *Water supply and sewerage,* 5th edn. New York: McGraw-Hill.

Tchobanoglous, G. and Schroeder, E. D. 1985. *Water quality, characteristics, modeling, modification.* Reading, Massachusetts: Addison-Wesley.

Theis, C. V. 1935. The relation between the lowering of the piezometric surface and the rate and duration of discharge of a well using ground storage. *Trans. Am. Geophys. Union* **16**: 519–524.

US Environmental Protection Agency (EPA). 1987. Underground storage tank corrective action technologies. EPA/625/6-87-015. Cincinnati, Ohio: US EPA.

US Environmental Protection Agency (EPA). 1994. Handbook: Groundwater and wellhead protection, EPA/625/R-94/001. Washington, DC: US EPA.

Voelker, D. C. 1984. *Quality of water in the alluvial aquifer, American Bottoms, East St. Louis, Illinois.* US Geological Survey Water Resources Investigation Report 84–4180. Chicago, Illinois.

Walton, W. C. 1962. *Selected analytical methods for well and aquifer evaluation.* Illinois State Water Survey. Bulletin 49, Urbana, Illinois.

Wanielista, M. 1990. *Hydrology and water quality control.* New York: John Wiley.

Watson, I., and Burnett, A. D. 1993. *Hydrology: An environmental approach.* Cambridge, Ft. Lauderdale, Florida: Buchanan Books.

Wenzel, L. K. 1942. *Methods for determining the permeability of water-bearing materials.* US Geological Survey. Water Supply Paper 887, p. 88.

Willis. R., and Yeh, W. W-G. 1987. *Groundwater systems planning & management.* Englewood Cliffs, New Jersey: Prentice-Hall.

Chapter

# 4

# Fundamental and Treatment Plant Hydraulics

# 1 Definitions and Fluid Properties

## 1.1 Weight and mass

The weight ($W$) of an object, in the International System of Units (SI), is defined as the product of its mass ($m$, in grams, kilograms, etc.) and the gravitational acceleration ($g = 9.81$ m/s$^2$ on the earth's surface) by Newton's second law of motion: $F = ma$. The weight is expressed as

$$W = mg \tag{4.1a}$$

The unit of weight is kg · m/s$^2$ and is usually expressed as newton (N).

In the SI system, 1N is defined as the force needed to accelerate 1kg of mass at a rate of 1m/s$^2$. Therefore

$$\begin{aligned} 1\text{N} &= 1 \text{ kg} \cdot \text{m/s}^2 \\ &= 1 \times 10^3\text{g} \cdot 10^2\text{cm/s}^2 \\ &= 10^5 \text{ g} \cdot \text{cm/s}^2 \\ &= 10^5 \text{ dyn} \end{aligned}$$

In the US customary units, mass is expressed in slugs. One slug is defined as the mass of an object which needs one pound of force to accelerate to one ft/s$^2$, i.e.

$$m = \frac{\text{W(lb)}}{g\,(\text{ft/s}^2)} = \frac{w}{g} \text{ (in slugs)} \tag{4.1b}$$

**Example:** What is the value of the gravitational acceleration ($g$) in the US customary units?

**solution:**

$$\begin{aligned} g &= 9.81 \text{ m/s}^2 = 9.81 \text{ m/s}^2 \times 3.28 \text{ ft/m} \\ &= 32.174 \text{ ft/s}^2 \end{aligned}$$

*Note*: This is commonly used as $g = 32.2$ ft/s$^2$.

## 1.2 Specific weight

The specific weight (weight per unit volume) of a fluid such as water, $\gamma$, is defined by the product of the density ($\rho$) and the gravitational acceleration ($g$), i.e.

$$\begin{aligned} \gamma &= \rho g \text{ (in kg/m}^3 \cdot \text{m/s}^2) \\ &= \rho g \text{ (in N/m}^3) \end{aligned} \tag{4.2}$$

Water at 4°C reaches its maximum density of 1000 kg/m$^3$ or 1.000 g/cm$^3$. The ratio of the specific weight of any liquid to that of water at 4°C is called the specific gravity of that liquid.

**Example:** What is the unit weight of water at 4°C in terms of N/m$^3$ and dyn/cm$^3$?

**solution:**

Step 1. In N/m$^3$, $\rho = 1000$ kg/m$^3$

$$\gamma = \rho g = 1000 \text{ kg/m}^3 \times 9.81 \text{ m/s}^2$$
$$= 9810 \text{ N/m}^3$$

Step 2. In dyn/cm$^3$, $\rho = 1.000$ g/cm$^3$

$$\gamma = 1 \text{ g/cm}^3 \times 981 \text{ cm/s}^2$$
$$= 981 \text{ (g} \cdot \text{cm/s}^2\text{)/cm}^3$$
$$= 981 \text{ dyn/cm}^3$$

### 1.3 Pressure

Pressure ($P$) is the force ($F$) applied to or distributed over a surface area ($A$) as a measure of force per unit area:

$$P = \frac{F}{A} \tag{4.3a}$$

In the SI system, the unit of pressure can be expressed as barye, bar, N/m$^2$, or pascal. One barye equals one dyne per square centimeter (dyn/cm$^2$). The bar is one megabarye, $10^6$ dynes per square centimeter ($10^6$ barye). One pascal equals one newton per square meter (N/m$^2$). In the US customary units, the unit of pressure is expressed as pounds per square inch, lb/in$^2$ (psi), or lb/ft$^2$, etc.

Pressure is also a measure of the height ($h$) of the column of mercury, water, or other liquid which it supports. The pressure at the liquid surface is atmospheric pressure ($P_a$). The pressure at any point in the liquid is the absolute pressure ($P_{ab}$) at that point. The absolute pressure is measured with respect to zero pressure. Thus, $P_{ab}$ can be written as

$$P_{ab} = \gamma h + P_a \tag{4.3b}$$

where $\gamma$ = specific weight of water or other liquid.

In engineering work, gage pressure is more commonly used. The gage pressure scale is designed on the basis that atmospheric pressure ($P_a$) is zero. Therefore, the gage pressure becomes:

$$P = \gamma h \tag{4.3c}$$

and pressure head

$$h = \frac{P}{\gamma} \tag{4.3d}$$

According to Pascal's law, the pressure exerted at any point on a confined liquid is transmitted undiminished in all directions.

**Example:** Under normal conditions, the atmospheric pressure at sea level is approximately 760-mm height of mercury. Convert it in terms of m of water, N/m$^2$, pascals, bars, and psi,

**solution:**

Step 1. Solve for m of water

Let $\gamma_1$ and $\gamma_2$ be specific weights of water and mercury, respectively, and $h_1$ and $h_2$ be column heights of water and mercury, respectively; then approximately

$$\gamma_1 = 1.00, \quad \gamma_2 = 13.5936$$

From Eq. (4.3c):

$$\begin{aligned} P &= \gamma_1 h_1 = \gamma_2 h_2 \\ h_1 &= \frac{\gamma_2}{\gamma_1} h_2 = \frac{13.5936}{1} \times 760 \text{ mm} = 10{,}331 \text{ mm} \\ &= 10.33 \text{ m (of water)} \end{aligned}$$

Step 2. Solve for N/m$^2$

Since $\gamma_1 = 9810$ N/m$^3$

$$P = \gamma_1 h_1 = 9810 \text{ N/m}^3 \times 10.33 \text{ m} = 101{,}337 \text{ N/m}^2 = 1.013 \times 10^5 \text{ N/m}^2$$

Step 3. Solve for pascals

Since 1 pascal (Pa) = 1 N/m$^2$, from Step 2:

$$\begin{aligned} P &= 1.013 \times 10^5 \text{ N/m}^2 \\ &= 1.013 \times 10^5 \text{ Pa} \end{aligned}$$

*Note*: 1 atm approximately equals $10^5$ N/m$^2$ or $10^5$ Pa.

Step 4. Solve for bars

Since approximately $\gamma_1 = 981$ dyn/cm$^3$

$$P = \gamma_1 h_1 = 981 \text{ dyn/cm}^3 \times 1033 \text{ cm}$$
$$= 1{,}013{,}400 \text{ dyn/cm}^2$$
$$= 1.013 \times 10^6 \text{ dyn/cm}^2 \times \frac{1 \text{ bar}}{1 \times 10^6 \text{ dyn/cm}^2}$$
$$= 1.013 \text{ bars}$$

*Note*: 1 atm pressure is approximately equivalent to 1 bar.

Step 5. Solve for lb/in$^2$ (psi)

From Step 1,

$$h_1 = 10.33 \text{ m} = 10.33 \text{ m} \times 3.28 \text{ ft/m}$$
$$= 33.88 \text{ ft of water}$$
$$P = 33.88 \text{ ft} \times 1(\text{lb/in}^2)/2.31 \text{ ft}$$
$$= 14.7 \text{ lb/in}^2$$

## 1.4 Viscosity of water

All liquids possess a definite resistance to change of their forms, and many solids show a gradual yielding to force (shear stress) tending to change their forms. Newton's law of viscosity states that, for a given rate of angular deformation of liquid, the shear stress is directly proportional to the viscosity. It can be expressed as

$$\tau = \mu \frac{du}{dy} \tag{4.4a}$$

where $\tau$ = shear stress
$\mu$ = proportionality factor, viscosity
$du/dy$ = velocity gradient
$u$ = angular velocity
$y$ = depth of fluid

The angular velocity and shear stress change with $y$.

The viscosity can be expressed as absolute (dynamic) viscosity or kinematic viscosity. From Eq. (4.4a), it may be rewritten as

$$\mu = \frac{\tau}{du/dy} \tag{4.4b}$$

The viscosity $\mu$ is frequently referred to as absolute viscosity, or mass per unit length and time. In the SI system, the absolute viscosity $\mu$ is expressed as poise or in dyne · second per square centimeter. One poise is defined as the tangent shear force ($\tau$) per unit area (dyn/cm$^2$) required to maintain unit difference in velocity (1 cm/s) between two parallel planes separated by 1 cm of fluid. It can be written as

$$\begin{aligned} 1 \text{ poise} &= 1 \text{ dyn}\cdot\text{s/cm}^2 \\ &= 1 \text{ dyn}\cdot\text{s/cm}^2 \times \frac{1\text{g}\cdot\text{cm/s}^2}{1 \text{ dyn}} \\ &= 1 \text{ g/cm}\cdot\text{s} \end{aligned}$$

or

$$\begin{aligned} 1 \text{ poise} &= 1 \text{ dyn}\cdot\text{s/cm}^2 \\ &= 1 \text{ dyn}\cdot\text{s/cm}^2 \times \frac{1\text{ N}}{10^5 \text{ dyn}} \Big/ \frac{1 \text{ m}^2}{10^4 \text{ cm}^2} \\ &= 0.1 \text{ N}\cdot\text{s/m}^2 \end{aligned}$$

Also

$$1 \text{ poise} = 100 \text{ centipoise}$$

The usual measure of absolute viscosity of fluid and gas is the centipoise. The viscosity of fluids is temperature dependent. The viscosity of water at room temperature (20°C) is one centipoise. The viscosity of air at 20°C is approximately 0.018 centipoise.

The British system unit of viscosity is 1 lb · s/ft$^2$ or 1 slug/ft · s.

**Example 1:** How much in British units is 1 poise absolute viscosity?

**solution:**

$$\begin{aligned} 1 \text{ poise} &= 0.1 \text{ N}\cdot\text{s/m}^2 \\ &= 0.1 \text{ N} \times \frac{1 \text{ lb}}{4.448 \text{ N}} \cdot \text{s/m}^2 \Big/ \frac{(3.28 \text{ ft})^2}{\text{m}^2} \\ &= 0.02248 \text{ lb}\cdot\text{s}/10.758 \text{ ft}^2 \\ &= 0.00209 \text{ lb}\cdot\text{s/ft}^2 \end{aligned}$$

On the other hand:

$$\begin{aligned} 1 \text{ lb} \cdot \text{s/ft}^2 &= 479 \text{ poise} \\ &= 47.9 \text{ N} \cdot \text{s/m}^2 \end{aligned}$$

The kinematic viscosity $\nu$ is the ratio of viscosity (absolute) to mass density $\rho$. It can be written as

$$\nu = \mu/\rho \tag{4.5}$$

The dimension of $\nu$ is length squared per unit time. In the SI system, the unit of kinematic viscosity is the stoke. One stoke is defined as 1 $cm^2/s$. However, the standard measure is the centistoke ($=10^{-2}$ stoke or $10^{-2}$ $cm^2/s$). Kinematic viscosity offers many applications, such as the Reynolds number $\mathbf{R} = VD/\nu$. For water, the absolute viscosity and kinematic viscosity are essentially the same, especially when the temperature is less than 10°C. The properties of water in SI units and British units are respectively given in Tables 4.1a and 4.1b.

**TABLE 4.1a Physical Properties of Water—SI Units**

| Temperature $T$, °C | Specific gravity | Specific weight $\gamma$, $N/m^3$ | Absolute viscosity $\mu$, $N \cdot s/m^2$ | Kinematic viscosity $\nu$, $m^2/s$ | Surface tension $\sigma$, $N/m^2$ | Vapor pressure $P_v$, $N/m^2$ |
|---|---|---|---|---|---|---|
| 0 | 0.9999 | 9805 | 0.00179 | $1.795 \times 10^{-6}$ | 0.0756 | 608 |
| 4 | 1.0000 | 9806 | 0.00157 | $1.568 \times 10^{-6}$ | 0.0750 | 809 |
| 10 | 0.9997 | 9804 | 0.00131 | $1.310 \times 10^{-6}$ | 0.0743 | 1226 |
| 15 | 0.9990 | 9798 | 0.00113 | $1.131 \times 10^{-6}$ | 0.0735 | 1762 |
| 21 | 0.9980 | 9787 | 0.00098 | $0.984 \times 10^{-6}$ | 0.0727 | 2504 |
| 27 | 0.9966 | 9774 | 0.00086 | $0.864 \times 10^{-6}$ | 0.0718 | 3495 |
| 38 | 0.9931 | 9739 | 0.00068 | $0.687 \times 10^{-6}$ | 0.0700 | 6512 |
| 93 | 0.9630 | 9444 | 0.00030 | $0.371 \times 10^{-6}$ | 0.0601 | 79,002 |

SOURCE: Brater *et al.* (1996)

**TABLE 4.1b Physical Properties of Water—British Units**

| Temperature $T$, °F | Density $\rho$, $slug/ft^3$ | Specific weight $\gamma$, $lb/ft^3$ | Absolute viscosity $\mu \times 10^5$ $lb \cdot s/ft^2$ | Kinematic viscosity $\nu \times 10^5$ $ft^2/s$ | Surface tension $\sigma$, lb/ft | Vapor pressure $P_v$, psia |
|---|---|---|---|---|---|---|
| 32 | 1940 | 62.42 | 3.746 | 1.931 | 0.00518 | 0.09 |
| 40 | 1938 | 62.43 | 3.229 | 1.664 | 0.00514 | 0.12 |
| 50 | 1936 | 62.41 | 2.735 | 1.410 | 0.00509 | 0.18 |
| 60 | 1934 | 62.37 | 2.359 | 1.217 | 0.00504 | 0.26 |
| 70 | 1931 | 62.30 | 2.050 | 1.059 | 0.00498 | 0.36 |
| 80 | 1927 | 62.22 | 1.799 | 0.930 | 0.00492 | 0.51 |
| 90 | 1923 | 62.11 | 1.595 | 0.826 | 0.00486 | 0.70 |
| 100 | 1918 | 62.00 | 1.424 | 0.739 | 0.00480 | 0.95 |
| 120 | 1908 | 61.71 | 1.168 | 0.609 | | |
| 140 | 1896 | 61.38 | 0.981 | 0.514 | | |
| 160 | 1890 | 61.00 | 0.838 | 0.442 | | |
| 180 | 1883 | 60.58 | 0.726 | 0.385 | | |
| 200 | 1868 | 60.12 | 0.637 | 0.341 | | |
| 212 | 1860 | 59.83 | 0.593 | 0.319 | | |

SOURCES: Benefield *et al.* (1984) and Metcalf & Eddy, Inc. (1972)

**Example 2:** At 21°C, water has an absolute viscosity of 0.00982 poise and a specific gravity of 0.998. Compute (a) the absolute (N · s/m$^2$) and kinematic viscosity in SI units and (b) the same quantities in British units.

**solution:**

Step 1.

For (a), at 21°C (69.8°F)

$$\mu = 0.00982 \text{ poise} = 0.00982 \text{ poise} \times \frac{0.1 \text{ N}\cdot\text{s/m}^2}{1 \text{ poise}}$$

$$= 0.000982 \text{ N}\cdot\text{s/m}^2 \quad \text{or}$$

$$= 0.000982, \text{ N}\cdot\text{s/m}^2 \times \frac{1 \text{ kg}\cdot\text{m/s}^2}{1 \text{ N}}$$

$$= 0.000982 \text{ kg/m}\cdot\text{s}$$

$$\nu = \frac{\mu}{\rho} = \frac{0.000982 \text{ kg/m}\cdot\text{s}}{0.998 \times 1000 \text{ kg/m}^3}$$

$$= 0.984 \times 10^{-6} \text{m}^2\text{/s}$$

Step 2.

For (b), from Step 1 and Example 1

$$\mu = 0.000982 \text{ N}\cdot\text{s/m}^2 \times \frac{1 \text{ slug/ft}\cdot\text{s}}{47.9 \text{ N}\cdot\text{s/m}^2}$$

$$= 2.05 \times 10^{-5} \text{ slug/ft}\cdot\text{s}$$

$$\nu = 0.984 \times 10^{-6} \text{ m}^2\text{/s} \times \left(\frac{3.28 \text{ ft}}{1 \text{ m}}\right)^2$$

$$\nu = 1.059 \times 10^{-5} \text{ ft}^2\text{/s}$$

### 1.5 Perfect gas

A perfect gas is a gas that satisfies the perfect gas laws, such as the Boyle–Mariotte law and the Charles–Gay-Lussac law. It has internal energy as a function of temperature only. It also has specific heats with values independent of temperature. The normal volume of a perfect gas is 22.4136 liters/mole (commonly quoted as 22.4 L/mol).

Boyle's law states that at a constant temperature, the volume of a given quantity of any gas varies inversely as the pressure applied to the gas. For a perfect gas, changing from pressure $P_1$ and volume $V_1$ to $P_2$ and $V_2$ at constant temperature, the following law exists:

$$P_1V_1 = P_2V_2 \tag{4.6}$$

According to the Boyle–Mariotte law for perfect gases, the product of pressure $P$ and volume $V$ is constant in an isothermal process.

$$PV = nRT \tag{4.7}$$

or

$$P = \frac{1}{V}RT = \rho RT$$

where $P$ = pressure, pascal ($P_a$) or lb/ft$^2$
$V$ = volume, m$^3$ or ft$^3$
$n$ = number of moles
$R$ = gas constant
$T$ = absolute temperature, K = °C + 273, or °R = °F + 459.6
$\rho$ = density, kg/m$^3$ or slug/ft$^3$

On a mole (M) basis, a pound mole (or kg mole) is the number of pounds (or kg) mass of gas equal to its molecular weight. The product $MR$ is called the universal gas constant and depends on the units used. It can be

$$MR = 1545 \text{ ft}\cdot\text{lb/lb}\cdot°\text{R}$$

The gas constant $R$ is determined as

$$R = \frac{1545}{M} \text{ ft}\cdot\text{lb/lb}\cdot°\text{R} \tag{4.8a}$$

or, in slug units

$$R = \frac{1545 \times 32.2}{M} \text{ ft}\cdot\text{lb/slug}\cdot°\text{R} \tag{4.8b}$$

For SI units

$$R = \frac{8312}{M} n \text{ N/kg}\cdot\text{K} \tag{4.8c}$$

**Example:** For carbon dioxide with a molecular weight of 44 at a pressure of 12.0 psia (pounds per square inch absolute) and at a temperature of 70°F (21°C), compute $R$ and its density.

**solution:**

Step 1. Determine $R$ from Eq. (4.8b)

$$R = \frac{1545 \times 32.2}{M} \text{ ft} \cdot \text{lb/slug} \cdot {}^\circ\text{R}$$
$$= \frac{1545 \times 32.2}{44}$$
$$= 1130 \text{ ft} \cdot \text{lb/slug} \cdot {}^\circ\text{R}$$

Step 2. Determine density $\rho$

$$\rho = \frac{P}{RT} = \frac{12 \text{ lb/in}^2 \cdot 144 \text{ in}^2/\text{ft}^2}{(1130 \text{ ft} \cdot \text{lb/slug} \cdot {}^\circ\text{R})(460 + 70^\circ\text{R})}$$
$$= 0.00289 \text{ slug/ft}^3$$

Based on empirical generation at constant pressure in a gaseous system (perfect gas), when the temperature varies, the volume of gas will vary approximately in the same proportion. If the volume is exactly one mole of gas at 0°C (273.15 K) and at atmosphere pressure, then, for ideal gases, these are the so-called standard temperature and pressure, STP. The volume of ideal gas at STP can be calculated from Eq. (4.7) (with $R = 0.08206$ L · atm/K · mol):

$$V = \frac{nRT}{P} = \frac{(1 \text{ mole})(0.08206 \text{ L} \cdot \text{atm/mol} \cdot \text{K})(273.15 \text{ K})}{1 \text{ atm}}$$
$$= 22.41 \text{ L}$$

Thus, according to Avogadro's hypothesis and the ideal-gas equation, one mole of any gas will occupy 22.41 L at STP.

## 2 Water Flow in Pipes

### 2.1 Fluid pressure

Water and wastewater professionals frequently encounter some fundamentals of hydraulics, such as pressure, static head, pump head, velocity of flow, and discharge rate. The total force acting on a certain entire space, commonly expressed as the force acting on unit area, is called intensity of pressure, or simply pressure, $p$. The US customary of units generally uses the pound per square inch (psi) for unit pressure. This quantity is also rather loosely referred to simply as pounds pressure; i.e. "20 pounds pressure" means 20 psi of pressure. The International System of Units uses the kg/cm$^2$ (pascal) or g/cm$^2$.

To be technically correct, the pressure is so many pounds or kilograms more than that exerted by the atmosphere (760 mm of mercury). However, the atmospheric pressure is ignored in most cases, since it is applied to everything and acts uniformly in all directions.

### 2.2 Head

The term *head* is frequently used, such as in energy head, velocity head, pressure head, elevation head, friction head, pump head, and loss of head (head loss). All heads can be expressed in the dimension of length, i.e. ft × lb/lb = ft, or m × kg/kg = m, etc.

The pump head equals the ft · lb (m · kg) of energy put into each pound (kg) of water passing through the pump. This will be discussed later in the section on pumps.

Pressure drop causes loss of head and may be due to change of velocity, change of elevation, or friction loss. Hydraulic head loss may occur at lateral entrances and is caused by hydraulic components such as valves, bends, control points, sharp-crested weirs, and orifices. These types of head loss have been extensively discussed in textbooks and handbooks of hydraulics. Bend losses and head losses due to dividing and combining flows are discussed in detail by James M. Montgomery, Consulting Engineers, Inc. (1985).

**Velocity head.** The kinetic energy (KE) of water with mass $m$ is its capacity to do work by reason of its velocity $V$ and mass and is expressed as $\frac{1}{2}mV^2$.

For a pipe with mean flow velocity $V$ (m/s or ft/s) and pipe cross-sectional area $A$ (cm$^2$ or sq. in), the total mass of water flowing through the cross section in unit time is $m = \rho VA$, where $\rho$ is the fluid density. Thus the total kinetic energy for a pipe flow is

$$\text{KE} = \frac{1}{2}mV^2 = \frac{1}{2}(\rho VA)V^2 = \frac{1}{2}\rho AV^3 \tag{4.9}$$

The total weight of fluid $W = mg = \rho AVg$, where $g$ is the gravitational acceleration. It is commonly expressed in terms of energy in a unit weight of fluid. The kinetic energy in unit weight of fluid is

$$\frac{\text{KE}}{W} = \frac{\frac{1}{2}\rho AV^3}{\rho gAV} = \frac{V^2}{2g} \tag{4.10}$$

This is the so-called velocity head, i.e. the height of the fluid column.

**Example 1:** Twenty-two pounds of water are moving at a velocity of 2 ft/s. What are the kinetic energy and velocity head?

**solution:**

Step 1.

$$\begin{aligned}\text{KE} &= \tfrac{1}{2}mV^2 = \tfrac{1}{2} \times 22 \text{ lb} \times (2 \text{ ft/s})^2 \\ &= 44 \text{ lb} \cdot \text{ft}^2/\text{s}^2 = 44 \text{ slug} \cdot \text{ft/s (since 1 slug} = 1 \text{ lb} \cdot \text{ft/s}^2)\end{aligned}$$

Step 2.

$$h_v = \frac{V^2}{2g} = \frac{(2\ \text{ft/s})^2}{2 \times 32.2\ \text{ft/s}^2} = 0.062\ \text{ft}$$

**Example 2:** Ten kilograms of water are moving with a velocity of 0.61 m/s. What are the kinetic energy and velocity head?

**solution:**

Step 1.

$$\text{KE} = \tfrac{1}{2}mV^2 = \tfrac{1}{2}(10\ \text{kg})(0.61\ \text{m/s})^2$$

$$= 1.86\ \text{kg}\cdot\text{m}^2/\text{s}^2 = 1.86\ \text{m}\cdot\text{kg}\cdot\text{m/s}^2$$

$$= 1.86\ \text{N}\cdot\text{m, since } 1\ \text{N} = 1\ \text{kg}\cdot\text{m/s}^2$$

Step 2.

$$h_v = \frac{V^2}{2g} = \frac{(0.61\ \text{m/s}^2)}{2 \times 9.81\ \text{m/s}^2}$$

$$= 0.019\ \text{m}$$

$$= 0.019\ \text{m} \times 3.28\ \text{ft/m}$$

$$= 0.062\ \text{ft}$$

*Note*: Examples 1 and 2 are essentially the same.

**Pressure head.** The pressure energy (PE) is a measure of work done by the pressure force on the fluid mass and is expressed as

$$\text{PE} = pAV \tag{4.11}$$

where $P$ = pressure at a cross section
$A$ = pipe cross-sectional area, $\text{cm}^2$ or $\text{in}^2$
$V$ = mean velocity

The pressure head ($h$) is the pressure energy in unit weight of fluid. The pressure is expressed in terms of the height of the fluid column $h$. The pressure head is

$$h = \frac{\text{PE}}{W} = \frac{pAV}{\rho gAV} = \frac{p}{\rho g} = \frac{p}{\gamma} = \frac{\text{pressure}}{\text{sp}\cdot\text{wt.}} \tag{4.12}$$

where $\gamma$ is the weight per unit volume of fluid or its specific weight in $\text{N/m}^3$ or $\text{lb/ft}^3$. The general expression of unit pressure is

$$P = \gamma h \tag{4.13}$$

and the pressure head is $h_p = p/\gamma$ in ft or m.

**Example:** At the bottom of a water storage tank, the pressure is 31.2 lb/in$^2$. What is the pressure head?

**solution:**

Step 1. Convert the pressure to lb/ft$^2$

$$P = 31.2 \text{ lb/in}^2 = 31.2 \text{ lb/in}^2 \times 144 \text{ in}^2\text{/ft}^2$$
$$= 31.2 \times 144 \text{ lb/ft}^2$$

Step 2. Determine $h_p$; $\gamma = 62.4$ lb/ft$^3$ for water

$$h_p = \frac{p}{\gamma} = \frac{31.2 \times 144 \text{ lb/ft}^2}{62.4 \text{ lb/ft}^3}$$
$$= 72 \text{ ft}$$

**Elevation head.** The elevation energy (EE) of a fluid mass is simply the weight multiplied by the height above a reference plane. It is the work to raise this mass $W$ to elevation $h$ and can be written as

$$\text{EE} = Wz \tag{4.13a}$$

The elevation head, $z$, is EE divided by the total weight $W$ of fluid:

$$\text{Elevation head} = \frac{\text{EE}}{W} = \frac{Wz}{W} = z \tag{4.13b}$$

**Example:** (1) Ten kilograms and (2) 1 lb of water are at 50 ft above the earth's surface. What are their elevation heads?

**solution:**

Step 1.

For 10 kg

$$h_e = \frac{Wh}{W} = \frac{10 \text{ kg} \times 50 \text{ ft}}{10 \text{ kg}}$$
$$= 50 \text{ ft}$$

Step 2.

For 1 lb

$$h_e = \frac{Wh}{W} = \frac{1 \text{ lb} \times 50 \text{ ft}}{1 \text{ lb}} = 50 \text{ ft}$$

Both have 50 ft of head.

**Bernoulli equation.** The total energy ($H$) at a particular section of water in a pipe is the algebraic sum of the kinetic head, pressure head, and elevation head. For sections 1 and 2, they can be expressed as

$$H_1 = \frac{V_1^2}{2g} + \frac{p_1}{\gamma} + z_1 \tag{4.14}$$

and

$$H_2 = \frac{V_2^2}{2g} + \frac{p_2}{\gamma} + z_2 \tag{4.15}$$

If water is flowing from section 1 to section 2, friction loss ($h_f$) is the major loss. The energy relationship between the two sections can be expressed as the Bernoulli equation

$$\frac{V_1^2}{2g} + \frac{p_2}{\gamma} + z_1 = \frac{V_2^2}{2g} + \frac{p_2}{\gamma} + z_2 = h_f \tag{4.16}$$

where $h_f$ is the friction head. This is also called the continuity equation.

**Example:** A nozzle of 12-cm diameter is located near the bottom of a storage water tank. The water surface is 3.05 m (10 ft) above the nozzle. Determine (a) the velocity of effluent from the nozzle and (b) the discharge.

**solution:**

Step 1. (a) Determine $V_2$

Let point 1 be at the water surface and point 2 at the center of the nozzle. By the Bernoulli equation

$$\frac{V_1^2}{2g} + \frac{P_1}{\gamma} + z_1 = \frac{V_2^2}{2g} + \frac{P_2}{\gamma} + z_2$$

Since $P_1 = P_2 = 0$, and $V_1^2 = 0$

$$\frac{V_2^2}{2g} = z_1 - z_2 = H = 3.05 \text{ m}$$

$$V_2 = \sqrt{2gH} = \sqrt{2 \times 9.806 \times 3.05}$$
$$= 7.73 \text{ m/s}$$

Step 2. (b) Solve for flow

$$Q = A_2V_2 = \pi r^2 V_2 = 3.14\,(0.06 \text{ m})^2 \times 7.73 \text{ m/s}$$
$$= 0.087 \text{ m}^3\text{/s}$$

**Friction head.** Friction head ($h_f$) equals the loss of energy by each unit weight of water or other liquid through friction in the length of a pipe, in which the energy is converted into heat. The values of friction heads are usually obtained from the manufacturer's tables.

**Darcy–Weisback equation.** The Darcy–Weisback formula can be calculated from the friction head:

$$h_f = f\left(\frac{L}{D}\right)\frac{V^2}{2g} \tag{4.17}$$

where $h_f$ = head of friction loss, cm or ft
$f$ = friction factor, dimensionless
$L$ = length of pipeline, cm or ft
$D$ = diameter of pipe, cm or ft
$\frac{V^2}{2g}$ = velocity head, cm or ft
$V$ = average velocity of flow, cm/s or ft/s

**Example 1:** Rewrite the Darcy–Weisback formula for $h_f$ in terms of flow rate $Q$ instead of velocity $V$.

**solution:**

Step 1. Convert $V$ to $Q$

In a circular pipe,

$$A = \pi(D/2)^2 = \pi D^2/4 = 0.785D^2$$

The volumetric flow rate may be expressed in terms of velocity and area ($A$) as

$$Q = VA$$

then

$$V = \frac{Q}{A} = 4Q/\pi D^2$$

$$V^2 = 16Q^2/\pi^2 D^4$$

Step 2. Substitute in the formula

$$h_f = f\left(\frac{L}{D}\right)\frac{V^2}{2g} = f\left(\frac{L}{D}\right)\frac{16Q^2}{2g\pi^2 D^4}$$

$$= f\left(\frac{L}{D^5}\right)\frac{8Q^2}{\pi^2 g} \tag{4.18}$$

**TABLE 4.2 Values of Equivalent Roughness *e* for New Commercial Pipes**

| Type of pipe | *e*, in |
|---|---|
| Asphalted cast iron | 0.0048 |
| Cast iron | 0.0102 |
| Concrete | 0.01–0.1 |
| Drawn tubing | 0.00006 |
| Galvanized iron | 0.006 |
| PVC | 0.00084 |
| Riveted steel | 0.04–0.4 |
| Steel and wrought iron | 0.0018 |
| Wood stave | 0.007 |
| | 0.04 |

SOURCE: Benefield *et al.* (1984)

The head of loss due to friction may be determined in three ways:

1. The friction factor $f$ depends upon the velocity of flow and the diameter of the pipe. The value of $f$ may be obtained from a table in many hydraulic textbooks and handbooks, or from the Moody chart (Fig. 4.1) of Reynolds number $\mathbf{R}$ versus $f$ for various grades and sizes of pipe $\mathbf{R} = DV/\nu$, where $\nu$ is the kinematic viscosity of the fluid.
2. The friction loss of head $h_f$ per 1000 ft of pipe may be determined from the Hazen-Williams formula for pipe flow. A constant $C$ for a particular pipe, diameter of pipe, and either the velocity or the quantity of flow should be known. A nomograph chart is most commonly used for solution by the Hazen-Williams formula.
3. Another empirical formula, the Manning equation, is also a popular formula for determining head loss due to free flow.

The friction factor $f$ is a function of Reynolds number $\mathbf{R}$ and the relative roughness of the pipe wall $e/D$. The value of $e$ (Table 4.2), the roughness of the pipe wall (equivalent roughness), is usually determined from experiment.

For laminar pipe flow ($\mathbf{R} < 2000$) $f$ is independent of surface roughness of the pipe. The $f$ value can be determined from

$$f = 64/\mathbf{R} \tag{4.19}$$

When $\mathbf{R} > 2000$ then the relative roughness will affect the $f$ value. The $e/D$ values can be found from the manufacturer or any textbook. The relationship between $f$, $\mathbf{R}$, and $e/D$ are summarized in graphical expression as the Moody diagram (Fig. 4.1).

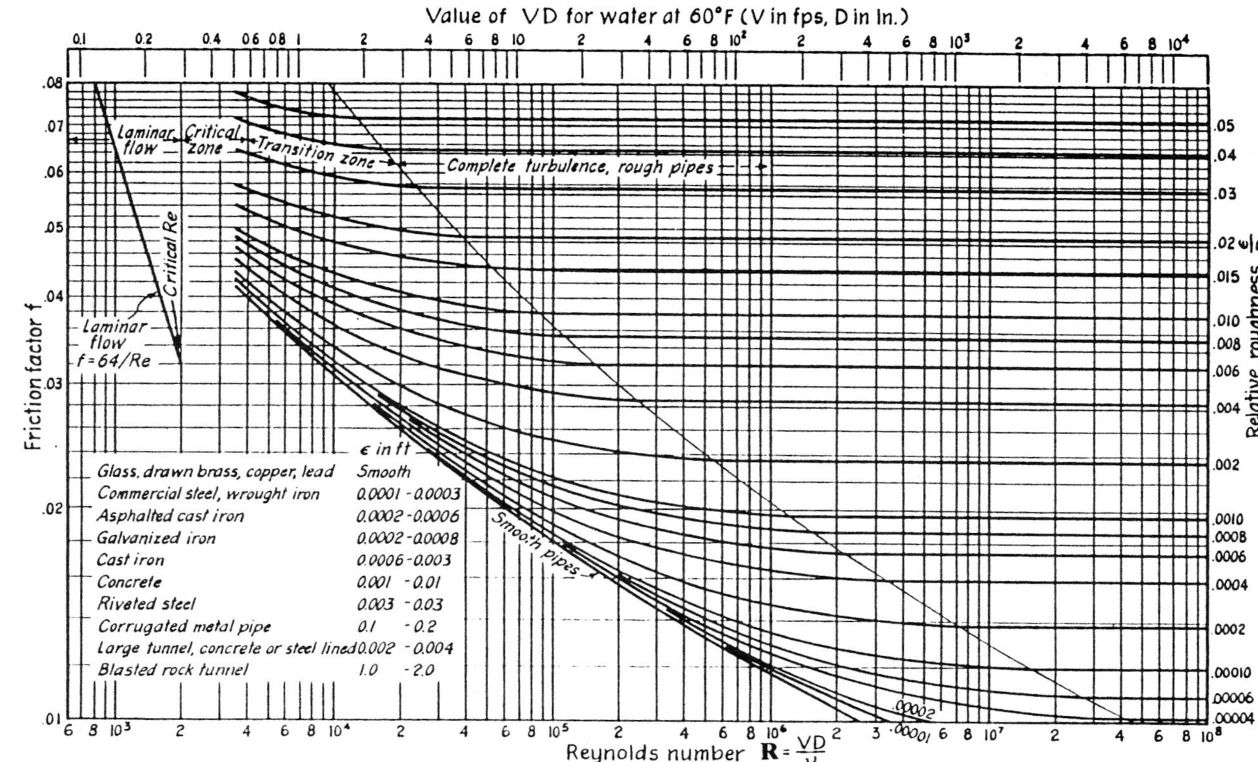


**Figure 4.1** Moody diagram (*Metcalf and Eddy, Inc., Wastewater Engineering Collection and Pumping of Wastewater, Copyright 1990, McGraw-Hill, New York, reproduced with permission of McGraw-Hill).*

**Example 2:** A pumping station has three pumps with 1 MGD, 2 MGD, and 4 MGD capacities. Each pumps water from a river at an elevation of 588 ft above sea level to a reservoir at an elevation of 636 ft, through a cast iron pipe of 24 in. diameter and 2600 ft long. The Reynolds number is 1600. Calculate the total effective head supplied by each pump and any combinations of pumping.

**solution:**

Step 1. Write Bernoulli's equation

Let stations 1 and 2 be river pumping site and discharge site at reservoir, respective; $h_p$ is effective head applied by pump in feet.

$$\frac{V_1^2}{2g} + \frac{p_1}{\gamma} + z_1 + h_p = \frac{V_2^2}{2g} + \frac{p_2}{\gamma} + z_2 + h_f$$

Since $V_1 = V_2$,

$p_1$ is not given (usually negative), assumed zero

$p_2$ = zero, to atmosphere

then

$$z_1 + h_p = z_2 = h_f$$

$$\begin{aligned} h_p &= (z_2 - z_1) + h_f \\ &= (636 - 588) + h_f \\ &= 48 + h_f \end{aligned}$$

Step 2. Compute flow velocity ($V$) for 1 MGD pump

$$A = \pi r^2 = 3.14(1\text{ ft})^2 = 3.14\text{ ft}^2$$

$$\begin{aligned} Q &= 1\text{ MGD} \\ &= \frac{10^6\text{ gal} \times 0.1337\text{ ft}^3\text{/gal}}{1\text{ day} \times 24 \times 60 \times 60\text{ s/d}} \\ &= 1.547\text{ ft}^3\text{/s (cfs)} \end{aligned}$$

$$\begin{aligned} V &= \frac{Q}{A} = \frac{1.547\text{ cfs}}{3.14\text{ ft}^2} \\ &= 0.493\text{ ft/s} \end{aligned}$$

Step 3. Find $f$, $h_f$ and $h_p$

From Eqs. (4.19) and (4.17),

$$f = 64/\mathbf{R} = 64/1600 = 0.04$$

$$\begin{aligned} h_f &= f\left(\frac{L}{D}\right)\frac{V^2}{2g} \\ &= 0.04\left(\frac{2600}{2}\right)\frac{(0.493)^2}{2 \times 32.2} \end{aligned}$$

$$= 0.20 \text{ ft}$$

$$h_p = 48 \text{ ft} + 0.2 \text{ ft} = 48.2 \text{ ft}$$

Step 4. Similar calculations for the 2-MGD pump

$$V = 0.493 \text{ ft/s} \times 2 = 0.986 \text{ ft/s}$$

$$h_f = 0.2 \text{ ft} \times 2^2 = 0.8 \text{ ft}$$

$$h_p = 48 \text{ ft} + 0.8 \text{ ft} = 48.8 \text{ ft}$$

Step 5. For the 4-MGD pump

$$V = 0.493 \text{ ft/s} \times 4 = 1.972 \text{ ft}^2\text{/s}$$

$$h_f = 0.2 \text{ ft} \times 4^2 = 3.2 \text{ ft}$$

$$h_p = 48 \text{ ft} + 3.2 \text{ ft} = 51.2 \text{ ft}$$

Step 6. For the 3 MGD: using the 1- and 2-MGD capacity pumps

$$h_f = (0.2 + 0.8) \text{ ft} = 1.0 \text{ ft}$$

$$h_p = (48.2 + 48.8) \text{ ft} = 97 \text{ ft}$$

Step 7. For the 5 MGD: using the 1- and 4-MGD capacity pumps

$$h_f = (0.2 + 3.2) \text{ ft} = 3.4 \text{ ft}$$

$$h_p = (48.2 + 51.2) \text{ ft} = 99.4 \text{ ft}$$

Step 8. For the 6 MGD: using the 2- and 4-MGD capacity pumps

$$h_f = (0.8 + 3.2) \text{ ft} = 4.0 \text{ ft}$$

$$h_p = (48.8 + 51.2) \text{ ft} = 100.0 \text{ ft}$$

Step 9. For the 7 MGD: using all 3 pumps simultaneously

$$h_f = (0.2 + 0.8 + 3.2) \text{ ft} = 4.2 \text{ ft}$$

$$h_p = (48.2 + 48.8 + 51.2) \text{ ft} = 148.2 \text{ ft}$$

**Hazen–Williams equation.** Due to the difficulty of using the Darcy–Weisback equation for pipe flow, engineers continue to make use of an exponential equation with empirical methods for determining friction losses in pipe flows. Among these the empirical formula of the Hazen–Williams equation is most widely used to express flow relations in pressure conduits, while the Manning equation is used for flow relations in free-flow conduits and in pipes partially full. The Hazen-Williams equation, originally developed for the British measurement system, has the following form:

$$V = 1.318\, CR^{0.63} S^{0.54} \qquad (4.20a)$$

where $V$ = average velocity of pipe flow, ft/s
$C$ = coefficient of roughness (see Table 4.3)
$R$ = hydraulic radius, ft
$S$ = slope of the energy gradient line or head loss per unit length of the pipe ($S = h_f/L$)

The hydraulic radius $R$ is defined as the water cross-sectional area $A$ divided by the wetted perimeter $P$. For a circular pressure pipe, if $D$ is the diameter of the pipe, then $R$ is

$$R = \frac{A}{P} = \frac{\pi D^2/4}{\pi D} = \frac{D}{4} \tag{4.21}$$

The Hazen–Williams equation was developed for water flow in large pipes ($D \geq 2$ in, 5 cm) with a moderate range of velocity ($V \leq 10$ ft/s or 3 m/s). The coefficient of roughness $C$ values range from 140 for very smooth (new), straight pipe to 90 or 80 for old, unlined tuberculated pipe. The value of 100 is used for the average conditioned pipe. It is not a function of the flow condition (i.e. Reynolds number **R**). The major limitation of this equation is that the viscosity and temperature are considered. A nomograph (Fig. 4.2) can be used to solve the Hazen-Williams equation.

**TABLE 4.3 Hazen–Williams Coefficient of Roughness *C* for Various Types of Pipe**

| Pipe material | *C* value |
|---|---|
| Brass | 130–140 |
| Brick sewer | 100 |
| Cast iron | |
| tar coated | 130 |
| new, unlined | 130 |
| cement lined | 130–150 |
| uncertain | 60–110 |
| Cement–asbestos | 140 |
| Concrete | 130–140 |
| Copper | 130–140 |
| Fire hose (rubber lined) | 135 |
| Galvanized iron | 120 |
| Glass | 140 |
| Lead | 130–140 |
| Plastic | 140–150 |
| Steel | |
| coal tar enamel lined | 145–150 |
| corrugated | 60 |
| new unlined | 140–150 |
| riveted | 110 |
| Tin | 130 |
| Vitrified clay | 110–140 |
| Wood stave | 110–120 |

SOURCES: Perry (1967), Hwang (1981), and Benefield *et al.* (1984)

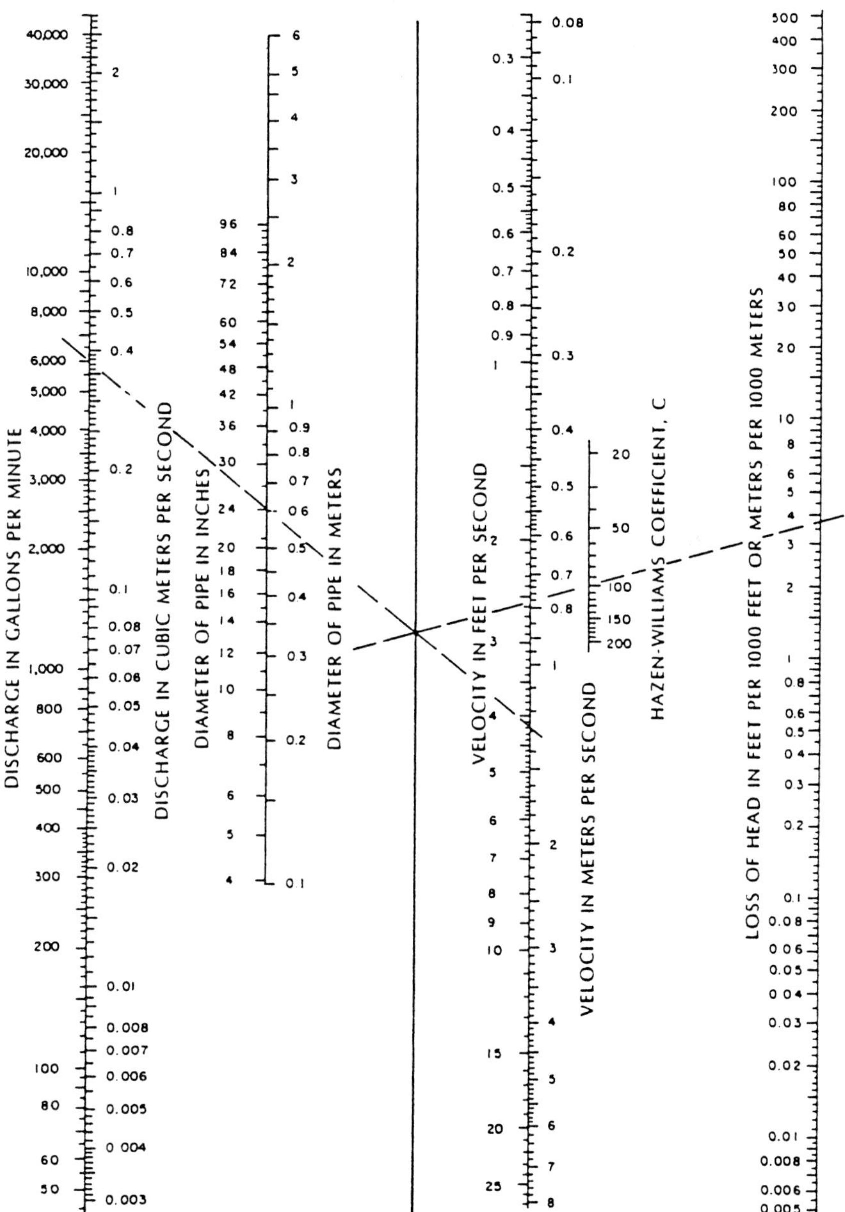


**Figure 4.2** Nomograph for Hazen–Williams formula.

In the SI system, the Hazen–Williams equation is written as

$$V = 0.85\,CR^{0.63}S^{0.54} \tag{4.20b}$$

The units are $R$ in m, $S$ in m/1000 m, $V$ in m/s, and $D$ in m, and discharge $Q$ in $m^3$/s (from chart).

**Manning equation.** Another popular empirical equation is the Manning equation, developed by the Irish engineer Robert Manning in 1889, and employed extensively for open channel flows. It is also commonly used for pipe free-flow. The Manning equation is

$$V = \frac{1.486}{n}R^{2/3}S^{1/2} \tag{4.22a}$$

where $V$ = average velocity of flow, ft/s
$n$ = Manning's coefficient of roughness (see Table 4.4)
$R$ = hydraulic radius (ft) (same as in the Hazen–Williams equation)
$S$ = slope of the hydraulic gradient, ft/ft

**TABLE 4.4 Manning's Roughness Coefficient *n* for Various Types of Pipe**

| Type of pipe | $n$ |
|---|---|
| Brick | |
| open channel | 0.014–0.017 |
| pipe with cement mortar | 0.012–0.017 |
| brass copper or glass pipe | 0.009–0.013 |
| Cast iron | |
| pipe uncoated | 0.013 |
| tuberculated | 0.015–0.035 |
| Cement mortar surface | 0.011–0.015 |
| Concrete | |
| open channel | 0.013–0.022 |
| pipe | 0.010–0.015 |
| Common clay drain tile | 0.011–0.017 |
| Fiberglass | 0.013 |
| Galvanized iron | 0.012–0.017 |
| Gravel open channel | 0.014–0.033 |
| Plastic pipe (smooth) | 0.011–0.015 |
| Rock open channel | 0.035–0.045 |
| Steel pipe | 0.011 |
| Vitrified clay | |
| pipes | 0.011–0.015 |
| liner plates | 0.017–0.017 |
| Wood, laminated | 0.015–0.017 |
| Wood stave | 0.010–0.013 |
| Wrought iron | 0.012–0.017 |

SOURCES: Perry (1967), Hwang (1981), and ASCE & WEF (1992)

This equation can be easily solved with the nomograph shown in Fig. 4.3. Figure 4.4 is used for partially filled circular pipes with varying Manning's $n$ and depth.

The Manning equation for SI units is

$$V = \frac{1}{n} R^{2/3} S^{1/2} \tag{4.22b}$$

where $V$ = mean velocity, m/s
$n$ = same as above
$R$ = hydraulic radius, m
$S$ = slope of the hydraulic gradient, m/m

**Example 1:** Assume the sewer line grade gives a sewer velocity of 0.6 m/s with a half full sewer flow. The slope of the line is 0.0081. What is the diameter of the uncoated cast iron sewer line?

**solution:**

Step 1. Using the Manning equation to solve $R$

$$V = \frac{1}{n} R^{2/3} S^{1/2}$$
$$R^{2/3} = Vn/S^{1/2}$$

From Table 4.4, $n = 0.013$

$$\begin{aligned} R^{2/3} &= Vn/S^{1/2} \\ &= 0.6 \times 0.013/0.0081^{1/2} \\ &= 0.0864 \\ R &= 0.0254 \end{aligned}$$

Step 2. Compute diameter of sewer line using Eq. (4.21)

$$\begin{aligned} D = 4R &= 4 \times 0.0254 \\ &= 0.102 \text{ (m)} \\ &= 4 \text{ in} \end{aligned}$$

**Example 2:** Determine the energy loss over 1600 ft in a new 24-in cast iron pipe when the water temperature is 60°F and the flow rate is 6.25 cfs using (1) the Darcy–Weisback equation, (2) the Hazen–Williams equation, and (3) the Manning equation.

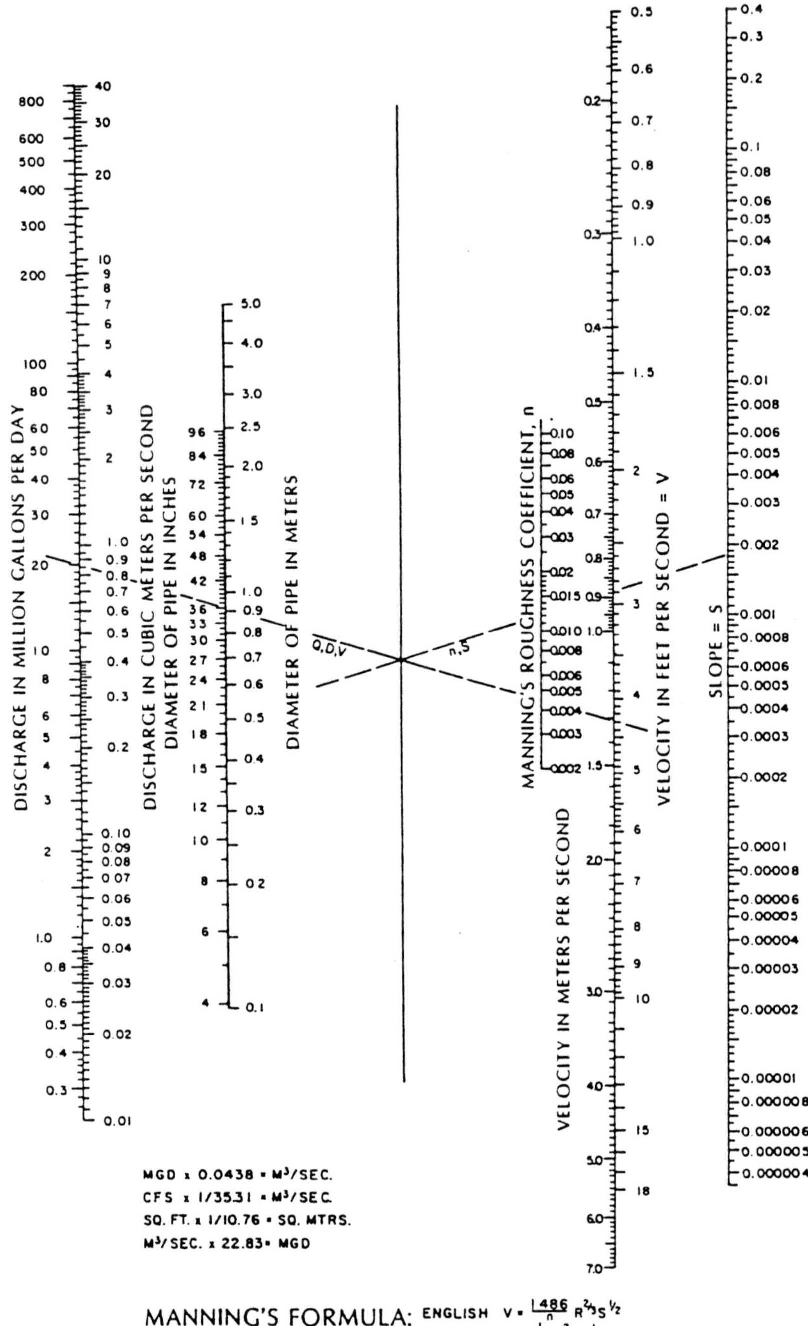


**Figure 4.3** Nomograph for Manning formula.

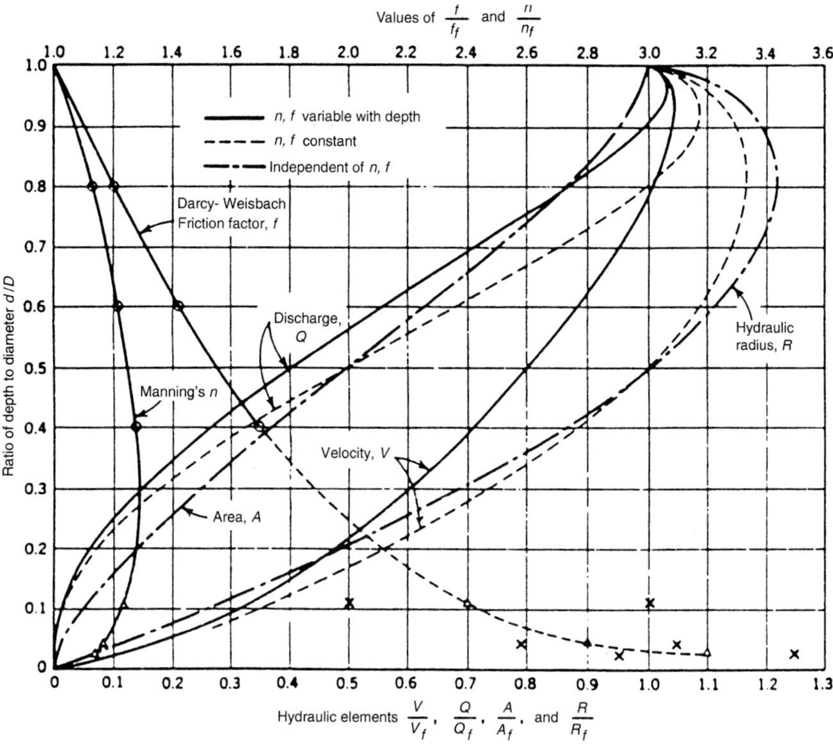


**Figure 4.4** Hydraulic elements graph for circular sewers *(Metcalf and Eddy, Inc., Wastewater Engineering: Collection and Pumping of Wastewater, Copyright, 1990, McGraw-Hill, New York, reproduced with permission of McGraw-Hill).*

**solution:**

Step 1. By the Darcy–Weisback equation

From Table 4.2, $e = 0.0102$ in

(a) Determine $e/D$, $D = 24$ in $= 2$ ft

$$\frac{e}{D} = \frac{0.0102}{24} = 0.000425$$

(b) Compute velocity $V$, $r = 1$ ft

$$V = \frac{Q}{A} = \frac{6.25\,\text{ft}^3/\text{s}}{\pi(1\,\text{ft})^2}$$
$$= 1.99\,\text{ft/s}$$

(c) Compute Reynolds number **R**

From Table 4.1b, $\nu = 1.217 \times 10^{-5} \text{ft}^2/\text{s}$ for 60°F

$$\mathbf{R} = \frac{VD}{\nu} = \frac{1.99 \text{ ft/s} \times 2 \text{ ft}}{1.217 \times 10^{-5} \text{ ft}^2/\text{s}}$$
$$= 3.27 \times 10^5$$

(d) Determine $f$ value from the Moody diagram (Fig. 4.1)

$$f = 0.018$$

(e) Compute loss of head $h_f$

$$h_f = f\left(\frac{L}{D}\right)\frac{V^2}{2g} = 0.018\left(\frac{1600}{2}\right)\frac{(1.99)^2}{2 \times 32.2}$$
$$= 0.885 \text{ (ft)}$$

Step 2. By the Hazen–Williams equation

(a) Find $C$

From Table 4.3, for new unlined cast pipe

$$C = 130$$

(b) Solve $S$

Eq. (4.21): $R = D/4 = 2/4 = 0.5$

Eq. (4.20a): $V = 1.318CR^{0.63}S^{0.54}$

Rearranging

$$S = \left[\frac{V}{1.318CR^{0.63}}\right]^{1.852}$$
$$= \left[\frac{1.99}{1.318 \times 130 \times (0.5)^{0.63}}\right]^{1.852}$$
$$= 0.000586 \text{ ft/ft}$$

(c) Compute $h_f$

$$h_f = S \times L = 0.000586 \text{ ft/ft} \times 1600 \text{ ft}$$
$$= 0.94 \text{ ft}$$

Step 3. By the Manning equation

(a) Find $n$

From Table 4.4, for unlined cast iron pipe

$$n = 0.013$$

(b) Determine slope of the energy line $S$

$$\text{From Eq. (4.21)}, R = D/4 = 2/4 = 0.5$$

$$\text{From Eq. (4.22a)}, V = \frac{1.486}{n} R^{2/3} S^{1/2}$$

Rearranging

$$S = \left(\frac{Vn}{1.486R^{2/3}}\right)^2 = \left[\frac{1.99 \times 0.013}{1.486 \times (0.5)^{2/3}}\right]^2 = 0.000764 \text{ (ft/ft)}$$

(c) Compute $h_f$

$$h_f = S \times L = 0.000764 \text{ ft/ft} \times 1600 \text{ ft}$$
$$= 1.22 \text{ ft}$$

**Example 3:** The drainage area of the watershed to a storm sewer is 10,000 acres (4047 ha). Thirty-eight percent of the drainage area has a maximum runoff of 0.6 $ft^3/s \cdot acre$ (0.0069 $m^3/s \cdot ha$), and the rest of the area has a runoff of 0.4 $ft^3/s \cdot acre$ (0.0046 $m^3/s$. ha). Determine the size of pipe needed to carry the storm flow with a grade of 0.12% and $n = 0.011$.

**solution:**

Step 1. Calculate the total runoff $Q$

$$Q = 0.6 \text{ cfs/acre} \times 0.38 \times 10{,}000 \text{ acre} + 0.4 \text{ ft}^3/(\text{s} \cdot \text{acre}) \times 0.62 \times 10{,}000 \text{ acre}$$
$$= 4760 \text{ ft}^3/\text{s}$$

Step 2. Determine diameter $D$ of the pipe by the Manning formula

$$A = \frac{\pi}{4} D^2 = 0.785D^2$$

$$R = D/4$$

$$Q = AV = A\frac{1.486}{n} R^{2/3} S^{1/2}$$

$$4760 = 0.785D^2 \times \frac{1.486}{0.011} \left(\frac{D}{4}\right)^{2/3} (0.0012)^{1/2}$$

$$3265 = D^{8/3}$$

$$D = 20.78 \text{ (ft)}$$

Minor head losses due to hydraulic devices such as sharp-crested orifices, weirs, valves, bend, construction, enlargement, discharge, branching, etc. have been extensively covered in textbooks and handbooks elsewhere. Basically they can be calculated from empirical formulas.

## 2.3 Pipeline systems

There are three major types of compound piping system: pipes connected in series, in parallel, and branching. Pipes in parallel occur in a pipe network when two or more paths are available for water flowing between two points. Pipe branching occurs when water can flow to or from a junction of three or more pipes from independent outlets or sources. On some occasions, two or more sizes of pipes may be connected in series between two reservoirs. Figure 4.5 illustrates all three cases.

The flow through a pipeline consisting of three or more pipes connected in various ways can be analyzed on the basis of head loss concept with two basic conditions. First, at a junction, water flows in and out should be the same. Second, all pipes meeting at the junction have the same water pressure at the junction. The general procedures are to apply the Bernoulli energy equation to determine the energy line and hydraulic gradient mainly from friction head loss equations. In summary, using Fig. 4.5:

1. For pipelines in series

   Energy: $H = h_{f1} + h_{f2}$

   Continuity: $Q = Q_1 = Q_2$

   Method: (a) Assume $h_f$

   (b) Calculate $Q_1$ and $Q_2$

   (c) If $Q_1 = Q_2$, the solution is correct

   (d) If $Q_1 \neq Q_2$, repeat step (a)

2. For pipelines in parallel

   Energy: $H = h_{f1} = h_{f2}$

   Continuity: $Q = Q_1 + Q_2$

   Method: Parallel problem can be solved directly for $Q_1$ and $Q_2$

3. For branched pipelines

   Energy: $h_{f1} = z_1 - H_J$

   $h_{f2} = z_2 - H_J$

   $h_{f3} = H_J - z_3$

   where $H_J$ is the energy head at junction J.

   Continuity: $Q_3 = Q_1 + Q_2$

   Method: (a) Assume $H_J$

   (b) Calculate $Q_1$, $Q_2$, and $Q_3$

   (c) If $Q_1 + Q_2 = Q_3$, the solution is correct

   (d) If $Q_1 + Q_2 \neq Q_3$, repeat step (a)

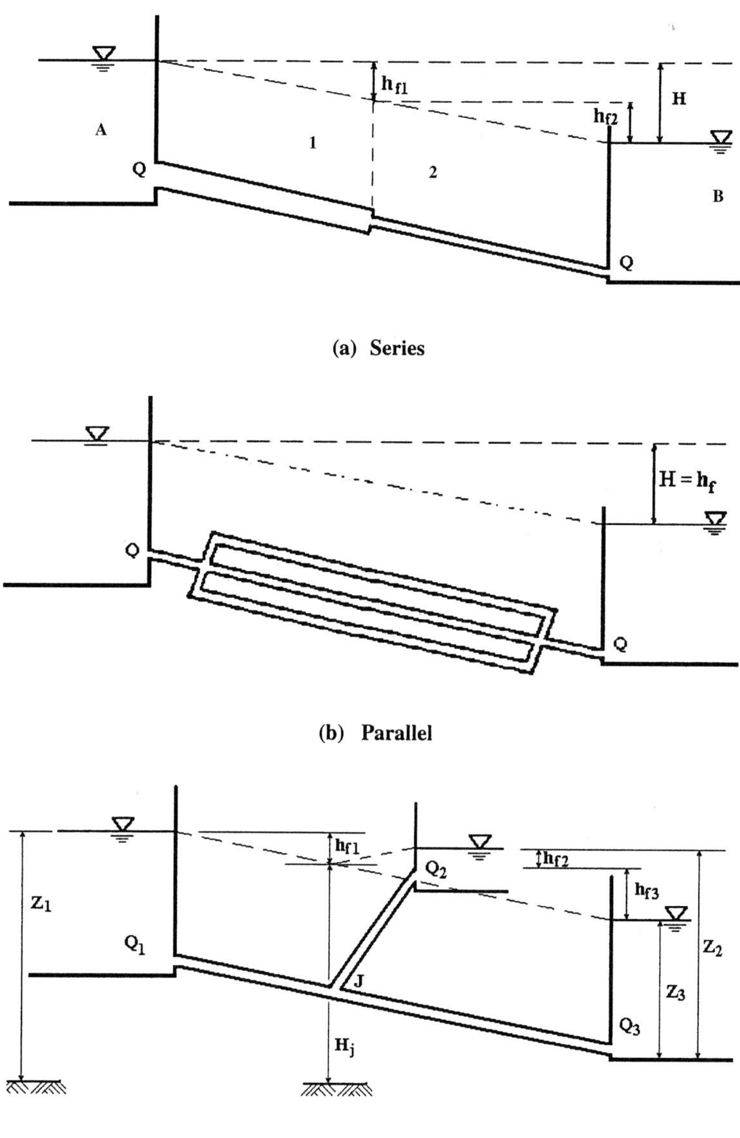


**Figure 4.5** Pipeline system for two reservoirs.

## Pipelines in series

**Example:** As shown in Fig. 4.5a, two concrete pipes are connected in series between two reservoirs A and B. The diameters of the upstream and downstream pipes are 0.6 m ($D_1 = 2$ ft) and 0.45 m ($D_2 = 1.5$ ft); their lengths are 300 m (1000 ft) and 150 m (500 ft), respectively. The flow rate of 15°C water from reservoir A to reservoir B is 0.4 $m^3/s$. Also given are coefficient of entrance $K_e = 0.5$, coefficient of contraction $K_c = 0.13$, and coefficient of discharge $K_d = 1.0$. Determine the elevation of the water surface of reservoir B when the elevation of the water surface of reservoir A is 100 m.

**solution:**

Step 1. Find the relative roughness $e/D$, from Fig. 4.1 for a circular concrete pipe, assume $e = 0.003$ ft $= 0.0009$ m

then

$$e/D_1 = 0.0009/0.60 = 0.0015$$
$$e/D_2 = 0.0009/0.45 = 0.0020$$

Step 2. Determine velocities ($V$) and Reynolds number ($\mathbf{R}$)

$$V_1 = \frac{Q}{A_1} = \frac{0.4 \text{ m}^3\text{/s}}{\frac{\pi}{4}(0.6\ m)^2} = 1.41 \text{ m/s}$$

$$V_2 = \frac{Q}{A_2} = \frac{0.4}{\frac{\pi}{4}(0.45)^2} = 2.52 \text{ m/s}$$

From Table 4.1a, $\nu = 1.131 \times 10^{-6}\text{m}^2\text{/s}$ for $T = 15°\text{C}$

$$\mathbf{R}_1 = \frac{V_1 D_1}{\nu} = \frac{1.41 \text{ m/s} \times 0.6 \text{ m}}{1.131 \times 10^{-6}\text{m}^2\text{/s}} = 7.48 \times 10^5$$

$$\mathbf{R}_2 = \frac{V_2 D_2}{\nu} = \frac{2.51 \times 0.45}{1.131 \times 10^{-6}} = 9.99 \times 10^5$$

Step 3. Find $f$ value from Moody chart (Fig. 4.1) corresponding to $\mathbf{R}$ and $e/D$ values

$$f_1 = 0.022$$
$$f_2 = 0.024$$

Step 4. Determine total energy loss for water flow from A to B

$$h_e = K_e \frac{V_1^2}{2g} = 0.5 \frac{V_1^2}{2g}$$

$$h_{f1} = f_1 \frac{L_1}{D_1} \frac{V_1^2}{2g} = 0.022 \times \frac{300}{0.6} \frac{V_1^2}{2g} = 11 \frac{V_1^2}{2g}$$

$$h_c = K_c \frac{V_2^2}{2g} = 0.13 \frac{V_2^2}{2g}$$

$$h_{f2} = f_2 \frac{L_2}{D_2} \frac{V_2^2}{2g} = 0.024 \frac{150}{0.45} \frac{V_2^2}{2g} = 8 \frac{V_2^2}{2g}$$

$$h_d = K_d \frac{V_2^2}{2g} = 1 \times \frac{V_2^2}{2g}$$

Step 5. Calculate total energy head $H$

$$\begin{aligned} H &= h_e + h_{f1} + h_c + h_{f2} + h_d \\ &= (0.5 + 11)\frac{V_1^2}{2g} + (0.13 + 8 + 1)\frac{V_2^2}{2g} \\ &= 11.5 \times \frac{(1.41)^2}{2 \times 9.81} + 9.13\frac{(2.52)^2}{2 \times 9.81} \\ &= 4.12 \text{ (m)} \end{aligned}$$

Step 6. Calculate the elevation of the surface of reservoir B

$$\begin{aligned} \text{Ele.} &= 100 \text{ m} - H = 100 \text{ m} - 4.12 \text{ m} \\ &= 95.88 \text{ m} \end{aligned}$$

## Pipelines in parallel

**Example:** Circular pipelines 1, 2, and 3, each 1000 ft (300 m) long and of 6, 8, and 12 in (0.3 m) diameter, respectively, carry water from reservoir A and join pipeline 4 to reservoir B. Pipeline 4 is 2 ft (0.6 m) in diameter and 500 ft (150 m) long. Determine the percentages of flow passing pipelines 1, 2, and 3 using the Hazen–Williams equation. The $C$ value for pipeline 4 is 110 and, for the other three, 100.

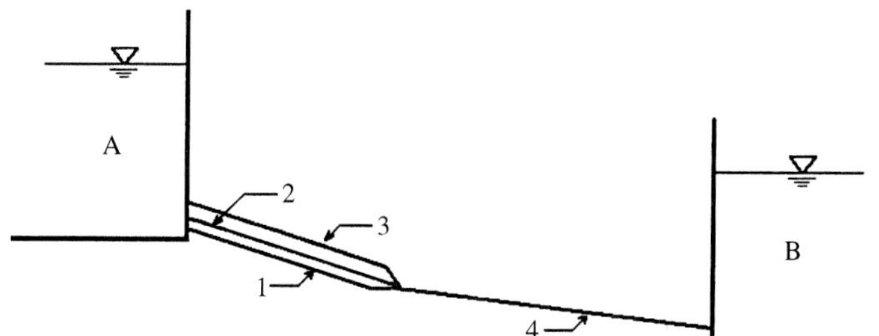


**solution:**

Step 1. Calculate velocity in pipelines 1, 2, and 3

Since no elevation and flow are given, any slope of the energy line is OK. Assuming the head loss in the three smaller pipelines is 10 ft per 1000 ft of pipeline:

$$S = 10/1000 = 0.01$$

For a circular pipe, the hydraulic radius is

$$R = \frac{D}{4}$$

then, from the Hazen–Williams equation, Eq. (4.20a)

$$V_1 = 1.318\, CR^{0.63}S^{0.54} = 1.318 \times 100 \times \left(\frac{0.5}{4}\right)^{0.63}(0.01)^{0.54}$$

$$= 10.96 \times \left(\frac{0.5}{4}\right)^{0.63}$$

$$= 2.96 \text{ (ft/s)}$$

$$V_2 = 10.96\left(\frac{0.667}{4}\right)^{0.63} = 3.54 \text{ (ft/s)}$$

$$V_3 = 10.96\left(\frac{1}{4}\right)^{0.63} = 4.58 \text{ (ft/s)}$$

Step 2. Determine the total flow $Q_4$

$$Q_1 = A_1V_1 = \frac{\pi}{4}D_1^2V_1 = 0.785(0.5)^2 \times 2.96$$

$$= 0.58 \text{ (ft}^3\text{/s)}$$

$$Q_2 = A_2V_2 = 0.785(0.667)^2 \times 3.54$$

$$= 1.24 \text{ (ft}^3\text{/s)}$$

$$Q_3 = A_3V_3 = 0.785(1)^2 \times 4.58$$

$$= 3.60 \text{ (ft}^3\text{/s)}$$

$$Q_4 = Q_1 + Q_2 + Q_3 = 1.58 + 1.24 + 3.60$$

$$= 5.42 \text{ (ft}^3\text{/s)}$$

Step 3. Determine percentage of flow in pipes 1, 2, and 3

For pipe 1 $= (Q_1/Q_4) \times 100\% = 0.58 \times 100/5.42$
$= 10.7\%$

pipe 2 $= 1.24 \times 100\%/5.42 = 22.9\%$

pipe 3 $= 3.60 \times 100\%/5.42 = 66.4\%$

## 2.4 Distribution networks

Most waterline or sewer distribution networks are complexes of looping and branching pipelines (Fig. 4.6). The solution methods described for the analysis of pipelines in series, parallel, and branched systems are

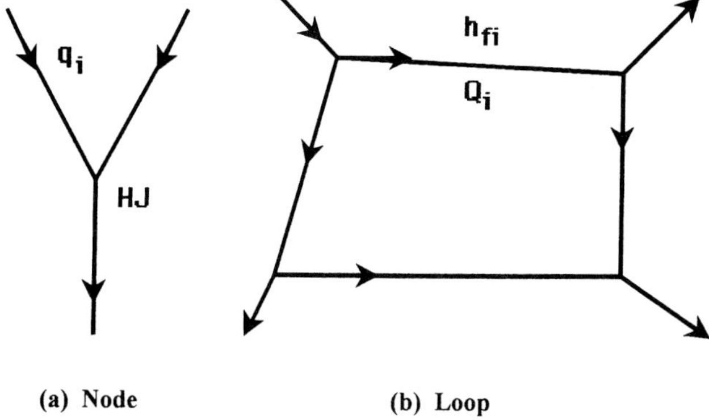


(a) Node (b) Loop

**Figure 4.6** Two types of pipe network.

not suitable for the more complex cases of networks. A trial and error procedure should be used. The most widely used is the loop method originally proposed by Hardy Cross in 1936. Another method, the nodal method, was proposed by Cornish in 1939 (Chadwick and Morfett, 1986).

For analysis flows in a network of pipes, the following conditions must be satisfied:

1. The algebraic sum of the pressure drops around each circuit must be zero.
2. Continuity must be satisfied at all junctions; i.e. inflow equals outflow at each junction.
3. Energy loss must be the same for all paths of water.

Using the continuity equation at a node, this gives

$$\sum_{i=1}^{m} q_i = 0 \tag{4.23}$$

where $m$ is the number of pipes joined at the node. The sign conventionally used for flow into a joint is positive, and outflow is negative.

Applying the energy equation to a loop, we get

$$\sum_{i=1}^{n} h_{\mathrm{f}i} = 0 \tag{4.24}$$

where $n$ is the number of pipes in a loop. The sign for flow ($q_i$) and head loss ($h_{\mathrm{f}i}$) is conventionally positive when clockwise.

Since friction loss is a function of flow

$$h_{\mathrm{f}i} = \phi(q_1) \tag{4.25}$$

Equations (4.23) and (4.24) will generate a set of simultaneous non-linear equations. An iterative solution is needed.

**Method of equivalent pipes.** In a complex pipe system, small loops within the system are replaced by single hydraulically equivalent pipes. This method can also be used for determination of diameter and length of a replacement pipe; i.e. one that will produce the same head loss as the old one.

**Example 1:** Assume $Q_{AB} = 0.042\ m^3/s\ (1.5\ ft^3/s)$, $Q_{BC} = 0.028\ m^3/s\ (1.0\ ft^3/s)$, and a Hazen–Williams coefficient $C = 100$. Determine an equivalent pipe for the network of the following figure.

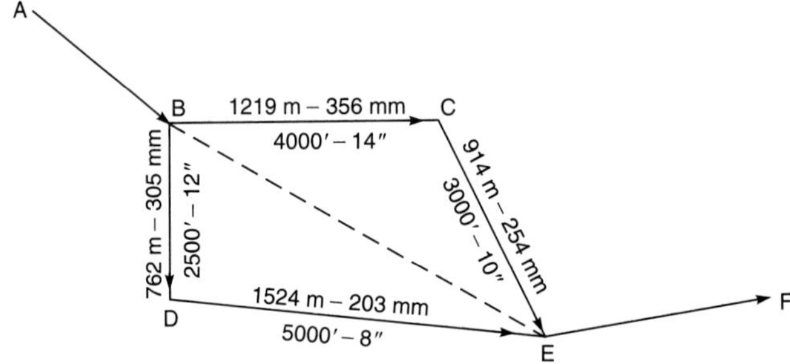


**solution:**

Step 1. For line BCE, $Q = 1\ ft^3/s$. From the Hazen–Williams chart (Fig. 4.2) with $C = 100$, to obtain loss of head in ft per 1000 ft, then compute the total loss of head for the pipe length by proportion.

(a) Pipe BC, 4000 ft, 14 in $\phi$

$$h_{BC} = \frac{0.44}{1000} \times 4000\ \text{ft} = 1.76\ \text{ft}$$

(b) Pipe CE, 3000 ft, 10 in $\phi$

$$h_{CE} = \frac{2.40}{1000} \times 3000\ \text{ft} = 7.20\ \text{ft}$$

(c) Total

$$h_{BCE} = 1.76\ \text{ft} + 7.20\ \text{ft} = 8.96\ \text{ft}$$

(d) Equivalent length of 10-in pipe: $1000\ \text{ft} \times \dfrac{8.96}{2.4} = 3733\ \text{ft}$

Step 2. For line BDE, $Q = 1.5 - 1.0 = 0.5\ ft^3/s$ or $0.014\ m^3/s$

(a) Pipe BD, 2500 ft, 12 in $\phi$

$$h_{\mathrm{BD}} = \frac{0.29}{1000} \times 2500 \text{ ft} = 0.725 \text{ ft}$$

(b) Pipe DE, 5000 ft, 8 in $\phi$

$$h_{\mathrm{DE}} = \frac{1.80}{1000} \times 5000 \text{ ft} = 9.0 \text{ ft}$$

(c) Total

$$h_{\mathrm{BDE}} = h_{\mathrm{BD}} + h_{\mathrm{DE}} = 9.725 \text{ ft}$$

(d) Equivalent length of 8-in pipe: $1000 \text{ ft} \times \dfrac{9.725}{1.80} = 5403 \text{ ft}$

Step 3. Determine equivalent line BE from results of Steps 1 and 2, assuming $h_{\mathrm{BE}} = 8.96$ ft

(a) Line BCE, 3733 ft, 10 in $\phi$

$$S = \frac{8.96}{3733} = 2.4/1000 = 24\%$$

$$Q_{\mathrm{BCD}} = 1.0 \text{ ft}^3/\text{s}$$

(b) Line BDE, 5403 ft, 8 in $\phi$

$$S = \frac{8.96}{5403} = 1.66/1000$$

From Fig. 4.2 $\quad Q_{\mathrm{BDE}} = 0.475 \text{ ft}^3/\text{s} = 0.013 \text{ m}^3/\text{s}$

(c) Total

$$Q_{\mathrm{BE}} = 1 \text{ ft}^3/\text{s} + 0.475 \text{ ft}^3/\text{s} = 1.475 \text{ ft}^3/\text{s} = 0.0417 \text{ m}^3/\text{s}$$

(d) Using $Q = 1.475$ ft$^3$/s, select 12-in pipe to find equivalent length From chart

$$S = 1.95$$

Since assumed total head loss $H_{\mathrm{BE}} = 8.96$

$$L = \frac{8.96}{1.95} \times 1000 \approx 4600 \text{ (ft)}$$

*Answer*: Equivalent length of 4600 ft of 12-in pipe.

**Example 2:** The sewer pipeline system below shows pipe diameter sizes ($\phi$), grades, pipe section numbers, and flow direction. Assume there is no surcharge and full flow in each of sections 1, 2, 3, and 4. All pipes are fiberglass with $n = 0.013$. A, B, C, and D represent inspection holes. Determine

(a) the flow rate and minimum commercial pipe size for section AB;
(b) the discharge, sewage depth, and velocity in section BC; and
(c) the slope required to maintain full flow in section CD.

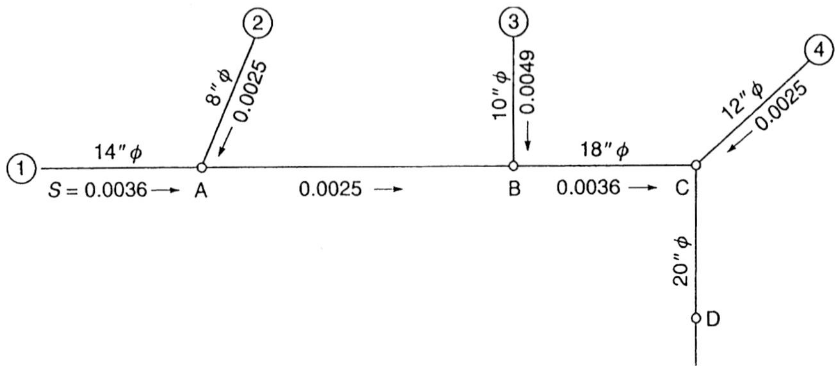


**solution:**

Step 1. Compute cross-sectional areas and hydraulic radius

| Section | Diameter, $D$ (ft) | $A$ (ft) | $R = D/4$ (ft) |
|---|---|---|---|
| 1 | 1.167 | 1.068 | 0.292 |
| 2 | 0.667 | 0.349 | 0.167 |
| 3 | 0.833 | 0.545 | 0.208 |
| 4 | 1.0 | 0.785 | 0.250 |
| BC | 1.5 | 1.766 | 0.375 |
| CD | 1.667 | 2.180 | 0.417 |

Step 2. Compute flows with Manning formula

$$Q = A\frac{1.486}{n}R^{2/3}S^{1/2}$$

$$Q_1 = 1.068 \times \frac{1.486}{0.013}(0.292)^{2/3}(0.0036)^{1/2}$$

$$= 3.22\ (\text{ft}^3/\text{s})$$

$$Q_2 = 0.349 \times 114.3(0.167)^{2/3}(0.0025)^{1/2}$$

$$= 0.60\ (\text{ft}^3/\text{s})$$

$$Q_{AB} = Q_1 + Q_2 = 3.22 + 0.60$$

$$= 3.82\ (\text{ft}^3/\text{s})$$

Step 3. Compute $D_{AB}$

$$Q_{AB} = 3.82 = 0.785D^2 \times 114.3\left(\frac{D}{4}\right)^{2/3} \times (0.0025)^{1/2}$$

$$D^{8/3} = 2.146$$

$$D = 1.33 \text{ ft}$$

$$= 15.9 \text{ in}$$

Use 18-in commercially available pipe for answer (a).

Step 4. Determine $Q_3$, $Q_{BC}$, $Q_{BC}/Q_f$

$$Q_3 = 0.545 \times 114.3 \times (0.208)^{2/3} (0.0049)^{1/2}$$

$$= 1.53 \text{ (ft}^3\text{/s)}$$

$$Q_{BC} = Q_{AB} + Q_3 = 3.82 \text{ ft}^3\text{/s} + 1.53 \text{ ft}^3\text{/s}$$

$$= 5.35 \text{ ft}^3\text{/s}$$

At section BC: velocity for full flow

$$V_f = 114.3 \times (0.375)^{2/3} (0.0036)^{1/2}$$

$$= 3.57 \text{ (ft/s)}$$

$$Q_f = AV_f = 1.766 \times 3.57$$

$$= 6.30 \text{ (ft}^3\text{/s)}$$

Percentage of actual flow

$$\frac{Q_{BC}}{Q_f} = \frac{5.35 \text{ ft}^3\text{/s}}{6.30 \text{ ft}^3\text{/s}} = 0.85$$

Step 5. Determine sewage depth $d$ and $V$ for section BC from Fig. 4.4

We obtain $d/D = 0.70$

Then

$$d = 0.70 \times 18 \text{ in} = 12.6 \text{ in}$$
$$= 1.05 \text{ ft}$$

and

$$V/V_f = 1.13$$

Then

$$V = 1.13 \times V_f = 1.13 \times 3.57 \text{ ft/s}$$
$$= 4.03 \text{ ft/s}$$

Step 6. Determine slope $S$ of section CD

$$Q_4 = 0.785 \times 114.3 \times (0.25)^{2/3}(0.0025)^{1/2}$$
$$= 1.78 \text{ (ft}^3\text{/s)}$$

$$Q_{\mathrm{CD}} = Q_4 + Q_{\mathrm{BC}} = 1.78\ \mathrm{ft^3/s} + 5.35\ \mathrm{ft^3/s}$$
$$= 7.13\ \mathrm{ft^3/s}$$

Section CD if flowing full:

$$7.13 = 2.18 \times 114.3(0.417)^{2/3} S^{1/2}$$
$$S^{1/2} = 0.05127$$
$$S = 0.00263$$

**Hardy–Cross method.** The Hardy–Cross method (1936) of network analysis is a loop method which eliminates the head losses from Eqs. (4.24) and (4.25) and generates a set of discharge equations. The basics of the method are:

1. Assume a value for $Q_i$ for each pipe to satisfy $\Sigma Q_i = 0$.
2. Compute friction losses $h_{\mathrm{fi}}$ from $Q_i$; find $S$ from Hazen–Williams equation.
3. If $\Sigma h_{\mathrm{fi}} = 0$, the solution is correct.
4. If $\Sigma_{\mathrm{fi}} \neq 0$, apply a correction factor $\Delta Q$ to all $Q_i$, then repeat step 1.
5. A reasonable value of $\Delta Q$ is given by Chadwick and Morfett (1986)

$$\Delta Q = \frac{\sum h_{\mathrm{fi}}}{2\sum h_{\mathrm{fi}}/Q_i} \tag{4.26}$$

This trial and error procedure solved by digital computer program is available in many textbooks on hydraulics (Hwang 1981, Streeter and Wylie 1975).

**The nodal method.** The basic concept of the nodal method consists of the elimination of discharges from Eqs. (4.23) and (4.25) to generate a set of head loss equations. This method may be used for loops or branches when the external heads are known and the heads within the networks are needed. The procedure of the nodal method is as follows:

1. Assume values of the head loss $H_{\mathrm{j}}$ at each junction.
2. Compute $Q_i$ from $H_{\mathrm{j}}$.
3. If $\Sigma Q_i = 0$, the solution is correct.
4. If $\Sigma Q_i \neq 0$, adjust a correction factor $\Delta H$ to $H_{\mathrm{j}}$, then repeat step 2.
5. The head correction factor is

$$\Delta H = \frac{2\sum Q_i}{\sum Q_i/h_{\mathrm{fi}}} \tag{4.27}$$

### 2.5 Sludge flow

To estimate every loss in a pipe carrying sludge, the Hazen–Williams equation with a modified $C$ value and a graphic method based on field experience are commonly used. The modified $C$ values for various total solids contents are as follows (Brisbin 1957):

| Total solids (%) | 0 | 2 | 4 | 6 | 8.5 | 10 |
|---|---|---|---|---|---|---|
| $C$ | 100 | 81 | 61 | 45 | 32 | 25 |

### 2.6 Dividing-flow manifolds and multiport diffusers

The related subjects of dividing-flow manifolds and multiport diffusers are discussed in detail by Benefield *et al.* (1984). They present basic theories and excellent design examples for these two subjects. In addition, there is a design example of hydraulic analysis for all unit processes of a wastewater treatment plant.

## 3 Pumps

### 3.1 Types of pump

The centrifugal pump and the displacement pump are most commonly used for water and wastewater works. Centrifugal pumps have a rotating impeller which imparts energy to the water. Displacement pumps are often of the reciprocating type in which a piston draws water or slurry into a closed chamber and then expels it under pressure. Reciprocating pumps are widely used to transport sludge in wastewater treatment plants. Air-lift pumps, jet pumps, and hydraulic rams are also used in special applications.

### 3.2 Pump performance

The Bernoulli equation may be applied to determine the total dynamic head on the pump. The energy equation expressing the head between the suction(s) and discharge(d) nozzles of the pumps is as follows:

$$H = \frac{P_\mathrm{d}}{\gamma} + \frac{V_\mathrm{d}^2}{2g} + z_\mathrm{d} - \left(\frac{P_\mathrm{s}}{\gamma} + \frac{V_\mathrm{s}^2}{2g} + z_\mathrm{s}\right)$$

where $H$ = total dynamic head, m or ft

$P_\mathrm{d}, P_\mathrm{s}$ = gage pressure at discharge and suction, respectively, $\mathrm{N/m^2}$ or $\mathrm{lb/in^2}$

$\gamma$ = specific weight of water, $\mathrm{N/m^3}$ or $\mathrm{lb/ft^3}$

$V_\mathrm{d}, V_\mathrm{s}$ = velocity in discharge and suction nozzles, respectively, m/s or ft/s

$g$ = gravitational acceleration, 9.81 $\mathrm{m/s^2}$ or 32.2 $\mathrm{ft/s^2}$

$z_\mathrm{d}, z_\mathrm{s}$ = elevation of discharge and suction gage above the datum, m or ft

The power $P_w$ required to pump water is a function of the flow $Q$ and the total head $H$ and can be written as

$$P_w = jQH \tag{4.28a}$$

where $P_w$ = water power, kW (m · m$^3$/min) or hp (ft · gal/min)
$j$ = constant, at 20°C
= 0.163 (for SI units)
= $2.525 \times 10^{-4}$ (for British units)
$Q$ = discharge, m$^3$/min or gpm
$H$ = total head, m or ft

**Example 1:** Calculate the water power for a pump system to deliver 3.785 m$^3$/min (1000 gpm) against a total system head of 30.48 m (100 ft) at a temperature of 20°C.

**solution:**

$$\begin{aligned} P_w &= jQH \\ &= 0.163 \times 3.785 \text{ m}^3/\text{min} \times 30.48 \text{ m} \\ &= 18.8 \text{ kW} \end{aligned}$$

or $= 25.3$ hp ($P_o = rQH$ for SI units)

The power output ($P_o = QrH$ for SI units) of a pump is the work done to lift the water to a higher elevation per unit time. The theoretical power output or horsepower (hp) required is a function of a known discharge and total pump lift. For US customary units, it may be written as

$$p_o(\text{in hp}) = \frac{Q\gamma H}{550} \tag{4.28b}$$

For SI units: $P_o$ (in N · m/s or Watt) $= Q\gamma H$

where $Q$ = discharge, ft$^3$/s (cfs) or m$^3$/s
$\gamma$ = specific weight of water or liquid, lb/ft$^3$ or N/m$^3$
$H$ = total lift, ft or m
550 = conversion factor from ft · lb/s to hp

**Example 2:** Determine how many watts equal one horsepower.

**solution:**

$$\begin{aligned} 1\text{hp} &= 550 \text{ ft} \cdot \text{lb/s} \\ &= 550 \text{ ft} \times 0.305 \text{ m/ft} \cdot 1 \text{ lb} \times 4.448 \text{ N/(lb} \cdot \text{s)} \end{aligned}$$

$$= 746 \text{ N} \cdot \text{m/s (J/s = watt)}$$

$$= 746 \text{ watts (W) or } 0.746 \text{ kilowatts (kW)}$$

Conversely, 1 kW = 1.341 hp

The actual horsepower needed is determined by dividing the theoretical horsepower by the efficiency of the pump and driving unit. The efficiencies for centrifugal pumps normally range from 50% to 85%. The efficiency increases with the size and capacity of the pump.

The design of a pump should consider total dynamic head which includes differences in elevations. The horsepower which should be delivered to a pump is determined by dividing Eq. (4.28b) by the pump efficiency, as follows:

$$\text{hp} = \frac{\gamma QH}{550 \times \text{efficiency}} \tag{4.28c}$$

The efficiency of a pump ($e_p$) is defined as the ratio of the power output ($P_o = \gamma QH/550$) to the input power of the pump ($P_i = \omega\tau$). It can be written as:

for US customary units

$$e_p = \frac{P_o}{P_i} = \frac{\gamma QH}{h_p \times 550} \tag{4.28d}$$

for SI units

$$e_p = \frac{\gamma QH}{P_i} = \frac{\gamma QH}{\omega\tau} \tag{4.28e}$$

where $e_p$ = efficiency of the pump, %
$\gamma$ = specific weight of water, kN/m$^3$ or lb/ft$^3$
$Q$ = capacity, m$^3$/s or ft$^3$/s
$h_p$ = brake horsepower
550 = conversion factor for horsepower to ft · lb/s
$\omega$ = angular velocity of the turbo-hydraulic pump
$\tau$ = torque applied to the pump by a motor

The efficiency of the motor ($e_m$) is defined as the ratio of the power applied to the pump by the motor ($P_i$) to the power input to the motor ($P_m$), i.e.

$$e_m = \frac{P_i}{P_m} \tag{4.29}$$

The overall efficiency of a pump system ($e$) is combined as

$$e = e_{\mathrm{p}} e_{\mathrm{m}} = \frac{P_{\mathrm{o}}}{P_{\mathrm{i}}} \times \frac{P_{\mathrm{i}}}{P_{\mathrm{m}}} = \frac{P_{\mathrm{o}}}{P_{\mathrm{m}}} \tag{4.30}$$

**Example 3:** A water treatment plant pumps its raw water from a reservoir next to the plant. The intake is 12 ft below the lake water surface at elevation 588 ft. The lake water is pumped to the plant influent at elevation 611 ft. Assume the suction head loss for the pump is 10 ft and the loss of head in the discharge line is 7 ft. The overall pump effiency is 72%. The plant serves 44,000 persons. The average water consumption is 200 gal per capita per day (gpcpd). Compute the horsepower output of the motor.

**solution:**

Step 1. Determine discharge $Q$

$$\begin{aligned} Q &= 44{,}000 \times 200 \text{ gpcpd} = 8.8 \text{ Mgal/d (MGD)} \\ &= 8.8 \text{ Mgal/d} \times 1.547 \text{ (ft}^3\text{/s)/(Mgal/d)} \\ &= 13.6 \text{ ft}^3\text{/s} \end{aligned}$$

Step 2. Calculate effective head $H$

$$H = (611 - 588) + 10 + 7 = 40 \text{ (ft)}$$

Step 3. Compute overall horsepower output

Using Eq. (4.28c),

$$\begin{aligned} \text{hp} &= \frac{Q\gamma H}{550 e_{\mathrm{p}}} = \frac{13.6 \text{ ft}^3\text{/s} \times 62.4 \text{ lb/ft}^3 \times 40 \text{ ft}}{550 \text{ ft} \cdot \text{lb/hp} \times 0.72} \\ &= 85.7 \text{ hp} \end{aligned}$$

**Example 4:** A water pump discharges at a rate of 0.438 m$^3$/s (10 Mgal/d). The diameters of suction and discharge nozzles are 35 cm (14 in) and 30 cm (12 in), respectively. The reading of the suction gage located 0.3m (1 ft) above the pump centerline is 11 kN/m$^2$ (1.6 lb/in$^2$). The reading of the discharge gage located at the pump centerline is 117 kN/m$^2$ (17.0 lb/in$^2$). Assume the pump efficiency is 80% and the motor efficiency is 93%; water temperature is 13°C. Find (a) the power input needed by the pump and (b) the power input to the motor.

**solution:**

Step 1. Write the energy equation

$$H = \left(\frac{P_d}{\gamma} + \frac{V_{\mathrm{d}}^2}{2g} + z_{\mathrm{d}}\right) - \left(\frac{P_{\mathrm{s}}}{\gamma} + \frac{V_{\mathrm{s}}^2}{2g} + z_{\mathrm{s}}\right)$$

Step 2. Calculate each term in the above equation

At $T = 13°C$, $\gamma = 9800\ \text{N/m}^3$, refer to table 4.1a

$$\frac{P_d}{\gamma} = \frac{117{,}000\ \text{N/m}^2}{9800\ \text{N/m}^3} = 11.94\ \text{m}$$

$$V_d = \frac{Q_d}{A_d} = \frac{0.438\ \text{m}^3/\text{s}}{(\pi/4)(0.30\ \text{m})^2} = 6.20\ \text{m/s}$$

$$\frac{V_d^2}{2g} = \frac{(6.20\ \text{m/s})^2}{2(9.81\ \text{m/s}^2)} = 1.96\ \text{m}$$

Let $z_d = 0$ be the datum at the pump centerline

$$\frac{P_s}{\gamma} = \frac{11{,}000\ \text{N/m}^2}{9800\ \text{N/m}^3} = 1.12\ \text{m}$$

$$V_s = \frac{0.438\ \text{m}^3/\text{s}}{(\pi/4)(0.35\ \text{m})^2} = 4.55\ \text{m/s}$$

$$\frac{V_s^2}{2g} = \frac{(4.55\ \text{m/s})^2}{2(9.81\ \text{m/s}^2)} = 1.06\ \text{m}$$

$$z_s = +0.30\ \text{m}$$

Step 3. Calculate the total head $H$

$$H = [11.94 + 1.96 + 0 - (1.12 + 1.06 + 0.30)]\ \text{m}$$
$$= 11.42\ \text{m or } 34.47\ \text{ft}$$

Step 4. Compute power input $P_i$ by Eq. (4.28e) for question (a)

$$P_i = \frac{P_q}{e_p} = \frac{\gamma QH}{e_p} = \frac{(9.8\ \text{kN})(0.438\ \text{m}^3/\text{s})(11.42\ \text{m})}{0.80}$$
$$= 61.27\ \text{kW}$$
$$= 61.27\ \text{kW} \times \frac{1.341\ \text{hp}}{1\ \text{kW}}$$
$$= 82.2\ \text{hp}$$

Step 5. Compute power input to the motor ($P_m$) for question (b)

$$P_m = \frac{P_i}{e_m} = \frac{61.27\ \text{kW}}{0.93}$$
$$= 65.88\ \text{kW}$$

or

$$= 88.4\ \text{hp}$$

### 3.3 Cost of pumping

The cost of pumping through a pipeline is a function of head loss, flow rate, power cost, and the total efficiency of the pump system. It can be expressed as (ductile Iron Pipe Research Association, 1997)

$$\text{CP} = 1.65\, HQ\$/E \tag{4.31}$$

where CP = pumping cost, \$/(yr · 1000 ft) (based on 24 h/d operation)
$H$ = head loss, ft/1000 ft
$Q$ = flow, gal/min
\$ = unit cost of electricity, \$/kWh
$E$ = total efficiency of pump system, %

Velocity is related to flow by the following equation

$$V = \frac{Q}{2.448d^2} \tag{4.32a}$$

where $V$ = velocity, ft/s
$Q$ = flow, gal/min
$d$ = actual inside diameter, in

Head loss is determined by the Hazen–Williams formula

$$H = 1000\left(\frac{V}{0.115\, Cd^{0.63}}\right)^{1.852} \tag{4.32b}$$

where $C$ = a coefficient, and other symbols are as above.

**Example:** Water is pumped at a rate of 6300 gal/min (0.40 $m^3$/s) through a 24-in pipeline of 10,000 ft (3048 m) length. The actual inside diameters for ductile iron pipe and PVC pipe are respectively 24.95 and 22.76 in. Assume the unit power cost is \$0.058/kWh; the total efficiency of pump system is 75%; the pump is operated 24 h per day. Estimate: (a) the cost of pumping for each kind of pipeline, (b) the present value of the difference of pumping cost, assuming 50-year design pipe life ($n$), 6.6% annual rate ($r$) of return on the initial investment, and 3.5% annual inflation rate of power costs.

**solution:**

Step 1. Compute the velocity of flow for each pipeline.

Let $V_d$ and $V_p$ represent velocity for ductile and PVC pipe, respectively, using Eq. (4.32a)

$$V_d = \frac{Q}{2.448\, d^2} = \frac{6300 \text{ gal/min}}{2.448(24.95 \text{ in})^2}$$

$$= 4.13 \text{ ft/s}$$

$$V_p = \frac{6300}{2.448(22.76)^2}$$

$$= 4.96 \text{ ft/s}$$

Step 2. Compute head losses ($H_d$ and $H_p$) for each pipe flow using Eq. (4.32b).

Coefficients $C$ for ductile and PVC pipes are 140 and 150, respectively.

$$H_d = 1000\left(\frac{V_d}{0.115Cd^{0.63}}\right)^{1.852}$$

$$= 1000\left(\frac{4.13}{0.115 \times 140 \times 24.95^{0.63}}\right)^{1.852}$$

$$= 1.89 \text{ (ft/1000 ft)}$$

$$H_p = 1000\left(\frac{4.96}{0.115 \times 150 \times 22.76^{0.63}}\right)^{1.852}$$

$$= 2.59 \text{ (ft/1000 ft)}$$

Step 3. Compute the costs of pumping, $CP_d$ and $CP_p$

$$CP_d = 1.65\, H_d Q\$/E = 1.65 \times 1.89 \times 6300 \times 0.058/0.75$$

$$= 1519 \text{ (\$/1000 ft/yr)}$$

$$CP_p = 1.65 \times 2.59 \times 6300 \times 0.058/0.75$$

$$= 2082 \text{ (\$/1000 ft/yr)}$$

Step 4. Compute the difference of total cost for 10,000 ft annually ($A$)

$$A = (2082 - 1519) \text{ \$/(1000 ft/yr)} \times (10{,}000 \text{ ft})$$

$$= 5630 \text{ \$/yr}$$

Step 5. Compute the present worth (PW) of $A$ adjusting for inflation using the appropriate equation below

When $g = r$

$$\text{PW} = An \tag{4.33a}$$

When $g \neq r$

$$\text{PW} = A\left[\frac{(1+i)^n - 1}{i(1+i)^n}\right] \tag{4.33b}$$

$$i = \frac{r - g}{1 + g} \tag{4.33c}$$

where PW = present worth of annual difference in pumping cost, \$
$A$ = annual difference in pumping cost, \$
$i$ = effective annual investment rate accounting for inflation, %
$n$ = design life of pipe, yr
$g$ = inflation (growth) rate of power cost, %
$r$ = annual rate of return on the initial investment, %

In this example, $g \neq n$

$$i = \frac{r - g}{1 + g} = \frac{0.066 - 0.035}{1 + 0.035} = 0.030$$

$$\text{PW} = A\left[\frac{(1 + i)^n - 1}{i(1 + i)^n}\right] = 5630\left[\frac{(1 + 0.03)^{50} - 1}{0.03(1 + 0.03)^{50}}\right]$$

$$= 144{,}855(\$)$$

## 4 Water Flow in Open Channels

### 4.1 Che'zy equation for uniform flow

In 1769, the French engineer Antoine Che'zy proposed an equation for uniform open channel flow in which the average velocity is a function of the hydraulic radius and the energy gradient. It can be written as

$$V = C\sqrt{RS} \tag{4.34}$$

where $V$ = average velocity, ft/s
$C$ = Che'zy discharge coefficient, $\text{ft}^{1/2}$/s
$R$ = hydraulic radius, ft
= cross-sectional area $A(\text{ft}^2)$ divided by the wetted perimeter P (ft)
= $A/P$
$S$ = energy gradient (slope of the bed, slope of surface water for uniform flow)
= head loss ($h_f$) over the length of channel ($L$) divided by L
= $h_f/L$

The value of $C$ can be determined from

$$C = \sqrt{\frac{8g}{f}} \tag{4.35a}$$

where $g$ = gravity constant, ft/s
$f$ = Darcy–Weisback friction factor

**Example:** A trapezoidal open channel has a bottom width of 20 ft and side slopes of inclination 1:1.5. Its friction factor $f = 0.056$. The depth of the channel is 4.0 ft. The channel slope is 0.025. Compute the flow rate of the channel.

**solution:**

Step 1. Determine $C$ value from Eq. (4.34)

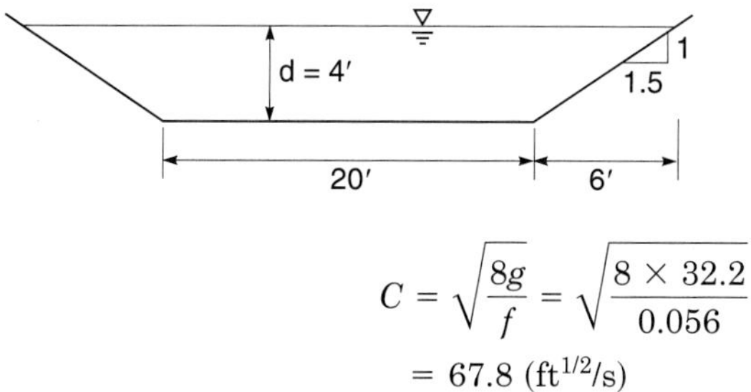


$$C = \sqrt{\frac{8g}{f}} = \sqrt{\frac{8 \times 32.2}{0.056}}$$
$$= 67.8\ (\text{ft}^{1/2}/\text{s})$$

Step 2. Compute $A$, $P$, and $R$

Width of water surface = 20 ft + 2 × 6 ft = 32 ft

$$A = \tfrac{1}{2}(20 + 32)\ \text{ft} \times 4\ \text{ft}$$
$$= 104\ \text{ft}^2$$
$$P = 2\ \text{ft} \times \sqrt{4^2 + 6^2} + 20\ \text{ft}$$
$$= 34.42\ \text{ft}$$
$$R = A/P = 104\ \text{ft}^2/34.42\ \text{ft}$$
$$= 3.02\ \text{ft}$$

Step 3. Determine the flow rate $Q$

$$Q = AV = AC\sqrt{RS}$$
$$= 104\ \text{ft}^2 \times 67.8\ \text{ft}^{1/2}/\text{s} \times \sqrt{3.02\ \text{ft} \times 0.025}$$
$$= 1937\ \text{ft}^3/\text{s}$$

## 4.2 Manning equation for uniform flow

As in the previous section, Eq. (4.22) was proposed by Robert Manning in 1889. The well-known Manning formula for uniform flow in open channel of nonpressure pipe is

$$V = \frac{1.486}{n} R^{2/3} S^{1/2} \quad \text{(for British units)} \tag{4.22a}$$

$$V = \frac{1}{n} R^{2/3} S^{1/2} \quad \text{(for SI units)} \tag{4.22b}$$

The flow rate (discharge) $Q$ can be determined by

$$Q = AV = A\frac{1.486}{n} R^{2/3} S^{1/2} \tag{4.36}$$

All symbols are the same as in the Che'zy equation. The Manning roughness coefficient ($n$) is related to the Darcy–Weisback friction factor as follows:

$$n = 0.093 f^{1/2} R^{1/6} \tag{4.37}$$

Manning also derived the relationship of $n$ (in s/ft$^{1/3}$) to the Che'zy coefficient $C$ by the equation

$$C = \frac{1}{n} R^{1/6} \tag{4.35b}$$

Typical values of $n$ for various types of channel surface are shown in Table 4.4.

**Example 1:** A 10-ft wide ($w$) rectangular source-water channel has a flow rate of 980 ft$^3$/s at a uniform depth ($d$) of 3.3 ft. Assume $n = 0.016$ s/ft$^{1/3}$ for concrete. (a) Compute the slope of the channel. (b) Determine the discharge if the normal depth of the water is 4.5 ft.

**solution:**

Step 1. Determine $A$, $P$, and $R$ for question (a)

$$A = wd = 10 \text{ ft} \times 3.3 \text{ ft} = 33 \text{ ft}^2$$

$$P = 2d + w = 2 \times 3.3 \text{ ft} + 10 \text{ ft} = 16.6 \text{ ft}$$

$$R = A/P = 33 \text{ ft}^2/16.6 \text{ ft} = 1.988 \text{ ft}$$

Step 2. Solve $S$ by the Manning formula

$$Q = A \frac{1.486}{n} R^{2/3} S^{1/2}$$

Rewrite

$$S = \left(\frac{Qn}{1.486 A R^{2/3}}\right)^2 = \left(\frac{980 \text{cfs} \times 0.016 \text{s/ft}^{1/3}}{1.486 \times 33 \text{ ft}^2 \times (1.988 \text{ ft})^{2/3}}\right)^2 = 0.041$$

Answer for (a), $S = 0.041$

Step 3. For (b), determine new $A$, $P$, and $R$

$$A = wd = 10 \text{ ft} \times 4.5 \text{ ft} = 45 \text{ ft}^2$$

$$P = 2d + w = 2(4.5 \text{ ft}) + 10 \text{ ft} = 19 \text{ ft}$$

$$R = A/P = 45 \text{ ft}^2/19 \text{ ft} = 2.368 \text{ ft}$$

Step 4. Calculate $Q$ for answer of question (b)

$$Q = A \frac{1.486}{n} R^{2/3} S^{1/2}$$

$$= 45 \times \frac{1.486}{0.016}(2.368)^{2/3}(0.041)^{1/2}$$
$$= 1500 \text{ (ft}^3\text{/s)}$$

**Example 2:** A rock trapezoidal channel has bottom width 5 ft (1.5 m), water depth 3 ft (0.9 m), side slope 2:1, $n = 0.044$, 5 ft wide. The channel bottom has 0.16% grade. Two equal-size concrete pipes will carry the flow downstream. Determine the size of pipes for the same grade and velocity.

**solution:**

Step 1. Determine the flow $Q$

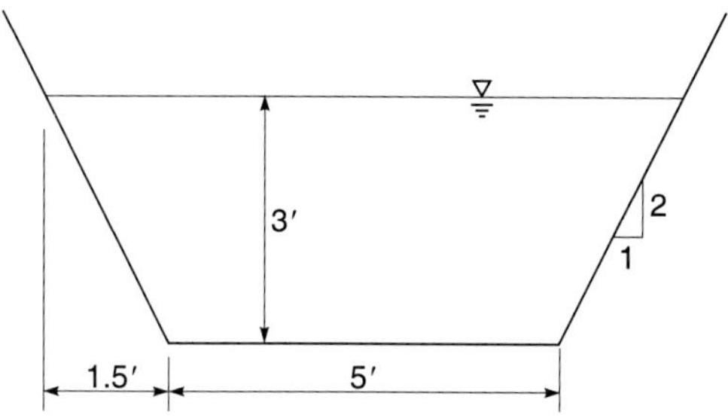


$$A = \frac{5 + 8}{2}\text{ft} \times 3 \text{ ft}$$
$$= 19.5 \text{ ft}^2$$
$$R = 5 \text{ ft} + 2\sqrt{1.5^2 + 3^2} \text{ ft}$$
$$= 11.71 \text{ ft}$$

By the Manning formula, Eq. (4.36)

$$Q = A\frac{1.486}{n}R^{2/3}S^{1/2}$$
$$= 19.5 \times \frac{1.486}{0.044}(11.71)^{2/3}(0.0016)^{1/2}$$
$$= 135.8 \text{ (ft}^3\text{/s)}$$

Step 2. Determine diameter of a pipe $D$

For a circular pipe flowing full, a pipe carries one-half of the total flow

$$R = \frac{D}{4}$$
$$\frac{135.8}{2} = \frac{\pi D^2}{4} \times \frac{1.486}{0.013}\left(\frac{D}{4}\right)^{2/3}(0.0016)^{1/2}$$
$$67.9 = 1.424\, D^{8/3}$$

$$D = 4.26 \text{ (ft)}$$
$$= 51.1 \text{ in}$$

*Note*: Use 48-in pipe, although the answer is slightly over 48 in.

Step 3. Check velocity of flow

Cross-sectional area of pipe $A_p$

$$A_p = \frac{\pi(4 \text{ ft})^2}{4} = 12.56 \text{ ft}^2$$

$$V = \frac{Q}{A_p} = \frac{67.9 \text{ ft}^3/\text{s}}{12.56 \text{ ft}^2}$$
$$= 5.4 \text{ ft/s}$$

This velocity is between 2 and 10 ft/s; it is thus suitable for a storm sewer.

## 4.3 Partially filled conduit

The conditions of partially filled conduit are frequently encountered in environmental engineering, particularly in the case of sewer lines. In a conduit flowing partly full, the fluid is at atm pressure and the flow is the same as in an open channel. The Manning equation (Eq. (4.22)) is applied.

A schematic pipe cross section is shown in Fig. 4.7. The angle $\theta$, flow area $A$, wetted perimeter $P$ and hydraulic radius $R$ can be determined by the following equations:

For angle

$$\cos\frac{\theta}{2} = \frac{\overline{BC}}{\overline{AC}} = \frac{r - d}{r} = 1 - \frac{d}{r} = 1 - \frac{2d}{D} \tag{4.38a}$$

$$\theta = 2\cos^{-1}\left(1 - \frac{2d}{D}\right) \tag{4.38b}$$

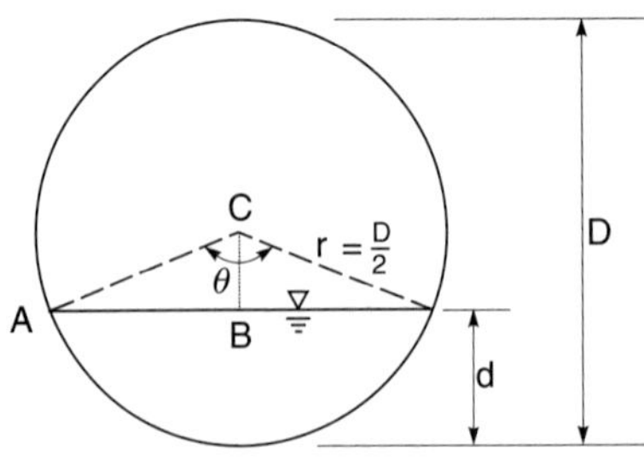


**Figure 4.7** Flow in partially filled circular pipe.

Area of triangle ABC, $a$

$$a = \frac{1}{2}\overline{AB}\,\overline{BC} = \frac{1}{2} r \sin\frac{\theta}{2} r \cos\frac{\theta}{2} = \frac{1}{2} r^2 \frac{\sin\theta}{2}$$

$$= \frac{1}{2}\frac{D^2}{4}\frac{\sin\theta}{2}$$

Flow area $A$

$$A = \frac{\pi D^2}{4}\frac{\theta}{360} - 2a$$

$$= \frac{\pi D^2}{4}\frac{\theta}{360} - 2\left(\frac{1}{2}\frac{D^2}{4}\frac{\sin\theta}{2}\right)$$

$$A = \frac{D^2}{4}\left(\frac{\pi\theta}{360} - \frac{\sin\theta}{2}\right) \qquad (4.39)$$

For wetted perimeter $P$

$$P = \pi D \frac{\theta}{360} \qquad (4.40)$$

For hydraulic radius $R$

$$R = A/P$$

Thus we can mathematically calculate the flow area $A$, the wetted perimeter $P$, and the hydraulic radius $R$. In practice, for a circular conduit, a chart is generally used which is available in hydraulic textbooks and handbooks (Chow, 1959; Morris and Wiggert, 1972; Zipparo and Hasen, 1993; Horvath, 1994).

**Example 1:** Assume a 24-in diameter sewer concrete pipe ($n = 0.012$) is placed on a slope of 2.5 in 1000. The depth of the sewer flow is 10 in. What is the average velocity and the discharge? Will this grade produce a self-cleansing velocity for the sanitary sewer?

**solution:**

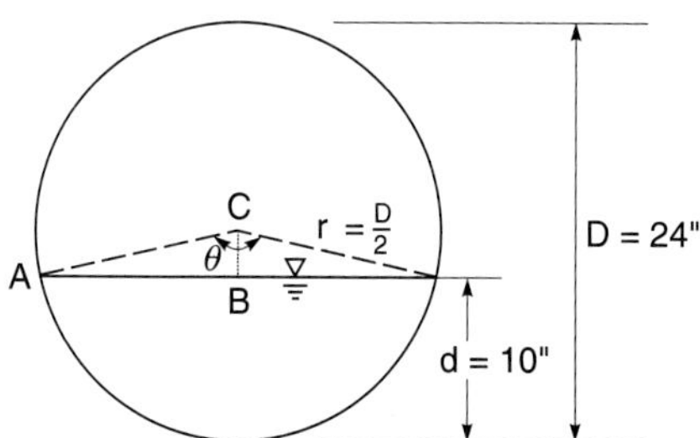

Step 1. Determine $\theta$ from (Eq. (4.38a), referring to Fig. 4.7 and to the figure above

$$\cos\frac{\theta}{2} = 1 - \frac{2d}{D} = 1 - \frac{2 \times 10}{24}$$

$$= 0.1667$$

$$\frac{\theta}{2} = \cos^{-1}(0.1667) = 80.4^\circ$$

$$\theta = 160.8^\circ$$

Step 2. Compute flow area $A$ from Eq. (4.39)

$$D = 24\,\text{in} = 2\ \text{ft}$$

$$A = \frac{D^2}{4}\left(\pi\frac{\theta}{360} - \frac{\sin\theta}{2}\right)$$

$$= \frac{2^2}{4}\left(3.14 \times \frac{160.8}{360} - \frac{1}{2}\sin 160.8^\circ\right)$$

$$= 1.40 - 0.16$$

$$= 1.24\ (\text{ft}^2)$$

Step 3. Compute $P$ from Eq. (4.40)

$$P = \pi D\frac{\theta}{360} = 3.14 \times 2\ \text{ft} \times \frac{160.8}{360}$$

$$= 2.805\ \text{ft}$$

Step 4. Calculate $R$

$$R = A/P = 1.24\ \text{ft}^2/2.805\ \text{ft}$$

$$= 0.442\ \text{ft}$$

Step 5. Determine $V$ from Eq. (4.22a)

$$V = \frac{1.486}{n}R^{2/3}S^{1/2}$$

$$= \frac{1.486}{0.012}\,(0.442)^{2/3}(2.5/1000)^{1/2}$$

$$= 123.83 \times 0.58 \times 0.05$$

$$= 3.591\ (\text{ft/s})$$

Step 6. Determine $Q$

$$Q = AV = 1.24 \text{ ft}^2 \times 3.591 \text{ ft/s}$$
$$= 4.453 \text{ ft}^3\text{/s}$$

Step 7. It will produce self-cleaning since $V > 2.0$ ft/s

**Example 2:** A concrete circular sewer has a slope of 1 m in 400 m. (a) What diameter is required to carry 0.1 $m^3$/s (3.5 $ft^3$/s) when flowing six-tenths full? (b) What is the velocity of flow? (c) Is this a self-cleansing velocity?

**solution:** These questions are frequently encountered in sewer design engineering. In the partially filled conduit, the wastewater is at atm pressure and the flow is the same as in an open channel, which can be determined with the Manning equation (Eq. (4.22)). Since this is inconvenient for mathematical calculation, a chart (Fig. 4.4) is commonly used for calculating area $A$, hydraulic radius $R$, and flow $Q$ for actual values (partly filled), as opposed to full-flow values.

Step 1. Find full-flow rate $Q_f$ (subscript "f" is for full flow)

Given:

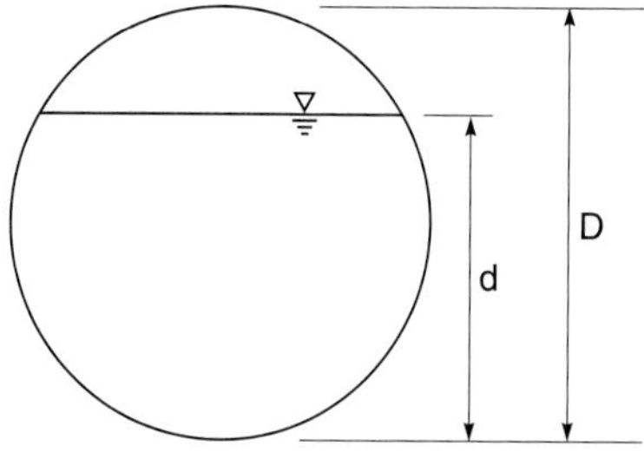


$$\frac{d}{D} = 0.60$$

From Fig. 4.4

$$\frac{Q}{Q_f} = 0.68$$

$$Q_f = \frac{Q}{0.68} = \frac{0.1 \text{ m}^3\text{/s}}{0.68}$$
$$= 0.147 \text{ m}^3\text{/s}$$

Step 2. Determine diameter of pipe $D$ from the Manning formula

For full flow

$$A_f = \frac{\pi}{4}D^2$$

$$R_f = \frac{D}{4}$$

For concrete $n = 0.013$
Slope $S = 1/400 = 0.0025$
From Eq. (4.22b):

$$V = \frac{1}{n} R^{2/3} S^{1/2}$$

$$Q = AV = A\frac{1}{n} R^{2/3} S^{1/2}$$

$$= \frac{\pi}{4} D^2 \times \frac{1}{0.013} \times \left(\frac{D}{4}\right)^{2/3} \times (0.0025)^{1/2}$$

$$0.147 = 1.20\ D^{8/3}$$

Answer (a)

$$D = 0.455\ \text{(m)}$$
$$= 1.5\ \text{ft}$$

Step 3. Calculate $V_f$

$$V_f = \frac{1}{0.013}\left(\frac{0.455}{4}\right)^{2/3} (0.0025)^{1/2}$$

$$= 0.903\ \text{(m/s)}$$

$$= 2.96\ \text{ft/s}$$

From Fig. 4.4

$$\frac{V}{V_f} = 1.07$$

Answer (b)

$$V = 1.07\ V_f = 1.07 \times 0.903\ \text{m/s}$$
$$= 0.966\ \text{m/s}$$
$$= 3.17\ \text{ft/s}$$

Step 4. For answer (c)

Since

$$V = 3.27\ \text{ft/s} > 2.0\ \text{ft/s}$$

it will provide self-cleansing for the sanitary sewer.

**Example 3:** A 12-in (0.3 m) sewer line is laid on a slope of 0.0036 with $n$ value of 0.012 and flow rate of 2.0 ft$^3$/s (0.057 m$^3$/s). What are the depth of flow and velocity?

**solution:**

Step 1. Calculate flow rate in full, $Q_f$

From Eq. (4.22a)

$$D = 1 \text{ ft}$$

$$Q_f = AV = \frac{\pi}{4}D^2 \frac{1.486}{0.012}\left(\frac{D}{4}\right)^{2/3}(0.0036)^{1/2}$$

$$= 0.785 \times 123.8 \times 0.397 \times 0.06$$

$$= 2.32 \text{ (ft}^3\text{/s)}$$

$$V_f = 2.96 \text{ ft/s}$$

Step 2. Determine depth of flow $d$

$$\frac{Q}{Q_f} = \frac{2.0}{2.32} = 0.862$$

From Fig. 4.4

$$\frac{d}{D} = 0.72$$

$$d = 0.72 \times 1 \text{ ft}$$

$$= 0.72 \text{ ft}$$

Step 3. Determine velocity of flow $V$

From chart (Fig. 4.4)

$$\frac{V}{V_f} = 1.13$$

$$\text{V} = 1.13 \times 2.96 \text{ ft/s}$$

$$= 3.34 \text{ ft/s}$$

or

$$= 1.0 \text{ m/s}$$

## 4.4 Self-cleansing velocity

The settling of suspended mater in sanitary sewers is of great concern to environmental engineers. If the flow velocity and turbulent motion are sufficient, this may prevent deposition and resuspend the sediment and move it along with the flow. The velocity sufficient to prevent

deposits in a sewer is called the self-cleansing velocity. The self-cleansing velocity in a pipe flowing full is (ASCE and WEF 1992):
for SI units

$$V = \frac{R^{1/6}}{n}[B(s - 1)D_p]^{1/2}$$

$$= \left[\frac{8B}{f}g(s - 1)D_p\right]^{1/2} \tag{4.41a}$$

for British units

$$V = \frac{1.486R^{1/6}}{n}[B(s - 1)D_p]^{1/2}$$

$$= \left[\frac{8B}{f}g(s - 1)D_p\right]^{1/2} \tag{4.41b}$$

where $V$ = velocity, m/s or ft/s
$R$ = hydraulic radius, m or ft
$n$ = Manning's coefficient of roughness
$B$ = dimensionless constant
= 0.04 to start motion
= 0.8 for adequate self-cleansing
$s$ = specific gravity of the particle
$D_p$ = diameter of the particle
$f$ = friction factor, dimensionless
$g$ = gravitational acceleration
= 9.81 m/s$^2$ or 32.2 ft/s$^2$

Sewers flowing between 50% and 80% full need not be placed on steeper grades to be as self-cleansing as sewers flowing full. The reason is that velocity and discharge are functions of attractive force which depends on the friction coefficient and flow velocity (Fair *et al.* 1966). Figure 4.8 presents the hydraulic elements of circular sewers that possess equal self-cleansing effect.

Using Fig. 4.8, the slope for a given degree of self-cleansing of partly full pipes can be determined. Applying Eq. (4.41) with the Manning equation (Eq. (4.22)) or Fig. 4.3, a pipe to carry a design full discharge $Q_f$ at a velocity $V_f$ that moves a particle of size $D_p$ can be selected. This same particle will be moved by a lesser flow rate between $Q_f$ and some lower discharge $Q_s$.

Figure 4.8 suggests that any flow ratio $Q/Q_f$ that causes the depth ratio $d/D$ to be larger than 0.5 requires no increase in slope because $S_s$ is less than $S_f$. For smaller flows, the invert slope must be increased to $S_s$ to avoid a decrease in self-cleansing.

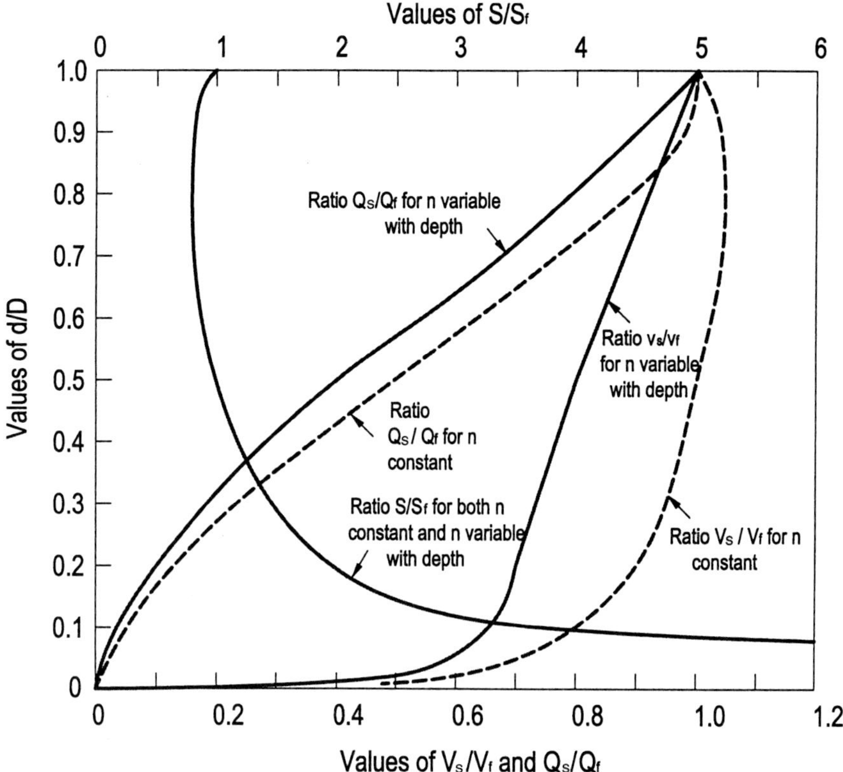


**Figure 4.8** Hydraulic elements of circular sewers that possess equal self-cleansing properties at all depths.

**Example:** A 10 in (25 cm) sewer is to discharge 0.353 ft$^3$/s (0.01m$^3$/s) at a self-cleansing velocity. When the sewer is flowing full, its velocity is 3 ft/s (0.9m/s). Determine the depth and velocity of flow and the required sewer line slope. Assume $N = n = 0.013$.

**solution:**

Step 1. Determine the flow $Q_f$ and slope $S_f$ during full flow

$$Q_f = \frac{\pi r^2}{4} V_f = 0.785\,(10/12\ \text{ft})^2 \times 3.0\ \text{ft/s}$$

$$= 1.64\ (\text{ft}^3/\text{s})$$

Using the Manning formula

$$Q_f = \frac{1.486}{n} R^{2/3} S_f^{1/2}$$

$$1.64 = \frac{1.486}{0.013} \left(\frac{0.833}{4}\right)^{2/3} S_f^{1/2}$$

Rearranging

$$S_f^{1/2} = 0.04084$$
$$S_f = 0.00167$$
$$= 1.67‰$$

Step 2. Determine depth $d$, velocity $V_s$, and slope $S$ for self-cleansing

From Fig. 4.8
For $N = n$ and $Q_s/Q_f = 0.353/1.64 = 0.215$, we obtain

$$d/D = 0.28$$
$$V_s/V_f = 0.95$$

and

$$S/S_f = 1.50$$

Then

$$d = 0.33D = 0.28 \times 10 \text{ in} = 2.8 \text{ in}$$
$$V_s = 0.95\ V = 0.95 \times 3 \text{ ft/s} = 2.85 \text{ ft/s}$$
$$S = 1.5\ S_f = 1.5 \times 1.67‰ = 2.50‰$$

## 4.5 Specific energy

For a channel with small slope (Fig. 4.9) the total energy head at any section may be generally expressed by the general Bernoulli equation as

$$E = \frac{V^2}{2g} + \frac{P}{\gamma} + z \tag{4.15}$$

where $z$ is the elevation of the bed. For any stream line in the section, $P/\gamma + z = D$ (the water depth at the section).

When the channel bottom is chosen as the datum ($z = 0$), the total head or total energy $E$ is called the specific energy ($H_e$). The specific energy at any section in an open channel is equal to the sum of the velocity head (kinetic energy) and water depth (potential energy) at the section. It is written as

$$H_e = \frac{V^2}{2g} + D \tag{4.42a}$$

Since $V = Q/A$ = flow/area, then

$$H_e = \frac{Q^2}{2gA^2} + D \tag{4.42b}$$

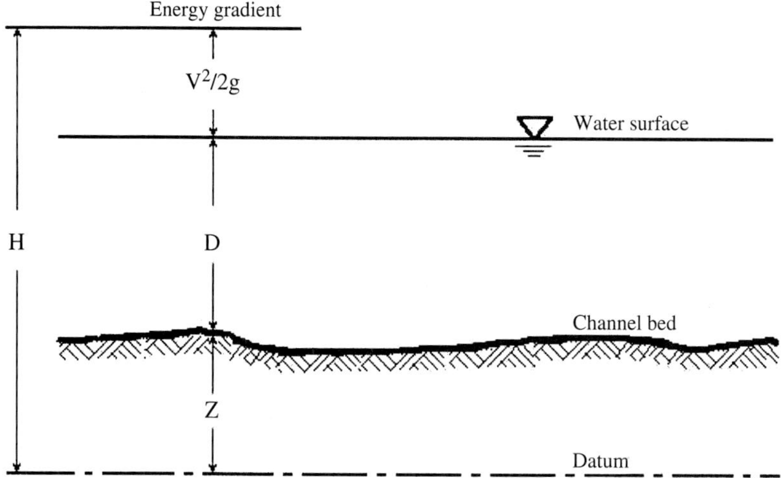


**Figure 4.9** Energy of open channel flow.

For flow rate in a rectangular channel, area $A$ is

$$A = WD \tag{4.43}$$

where $W$ is the width of the channel.

## 4.6 Critical depth

Given the discharge ($Q$) and the cross-sectional area ($A$) at a particular section, Eq. (4.42a) may be rewritten for specific energy expressed in terms of discharge

$$H_e = \frac{Q^2}{2gA^2} + D \tag{4.42b}$$

If the discharge is held constant, specific energy varies with $A$ and $D$ (Fig. 4.10). At the critical state the specific energy of the flow is at a minimum value. The velocity and depth associated with the critical state are called the critical velocity and critical depth, respectively. At

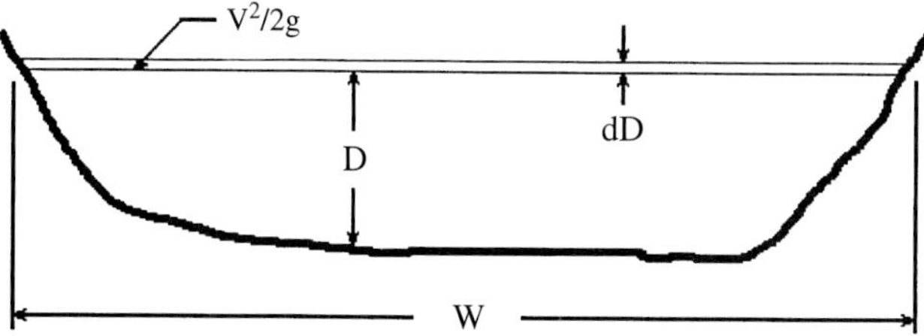


**Figure 4.10** Channel cross section.

the critical state, the first derivative of the specific energy with respect to the water depth should be zero:

$$\frac{dE}{dD} = \frac{d}{dD}\left(\frac{Q^2}{2gA^2} + D\right) = 0 \tag{4.44}$$

$$\frac{dE}{dD} = -\frac{Q^2}{gA^3}\frac{dA}{dD} + 1 = 0$$

Since near the free surface

$$dA = WdD$$

then

$$W = \frac{dA}{dD}$$

where $W$ is the top width of the channel section (Fig. 4.10). For open channel flow, $D_m = A/W$ is known as the hydraulic depth (or mean depth, $D_m$) for rectangular channels.

Hence for nonrectangular channels

$$-\frac{Q^2 W}{gA^3} + 1 = 0 \tag{4.45}$$

or

$$Q^2 W = gA^3 \tag{4.45a}$$

Then the general expression for flow at the critical depth is

$$Q = \sqrt{gA^3/W} = A\sqrt{gA/W}$$

$$Q = A\sqrt{gD_m} \tag{4.46}$$

and the critical velocity $V_c$ is

$$V_c = \sqrt{gA/W} = \sqrt{gD_m} \tag{4.47}$$

$$D_m = \frac{Q^2}{gA^2} \tag{4.48}$$

$$H_m = D_c + \frac{D_m}{2} \tag{4.49}$$

where $H_m$ is the minimum specific energy.

For the special case of a rectangular channel, the critical depth

$$D_c = \left(\frac{Q^2}{gW^2}\right)^{1/3} = \left(\frac{q^2}{g}\right)^{1/3} \tag{4.50}$$

where $q$ = discharge per unit width of the channel, $m^3/s/m$.

For rectangular channels, the hydraulic depth is equal to the depth of the flow. Equation (4.45) can be simplified as

$$\frac{Q^2}{A^2}\frac{W}{Ag} = 1$$

$$\frac{V^2}{gD} = 1$$

$$\frac{V}{\sqrt{gD}} = 1 \tag{4.51}$$

The quantity $V/\sqrt{gD}$ is dimensionless. It is the so-called Froude number, $F$. The Froude number is a very important factor for open channel flow. It relates three states of flow as follows:

| $F$ | Velocity | Flow state |
|---|---|---|
| 1 | $V = \sqrt{gD}$ | critical flow ($V$ = speed of surface wave) |
| $<1$ | $V < \sqrt{gD}$ | subcritical ($V <$ speed of surface wave) |
| $>1$ | $V > \sqrt{gD}$ | supercritical ($V >$ speed of surface wave) |

**Example:** Determine the critical depth of a channel when the water is flowing in a trapezoidal channel with a base width of 1.5 m (5.0 ft) and side slope of 2 on 1. The flow is 4.0 $m^3/s$ (63,400 gpm).

**solution:** Let $h$ = the critical depth of the channel

Then base = 1.5 m

$$W = 1.5\text{ m} + 2 \times (1/2\ h)\text{ m} = (1.5 + h)\text{ m}$$

$$A = \tfrac{1}{2} \times [1.5\text{ m} + (1.5 + h)\text{ m}] \times h\text{ m} = (1.5\,h + 0.5\,h^2)\text{ m}^2$$

$$Q = 4.0\text{ m}^3/\text{s}$$

$$g = 9.806\text{ m/s}^2$$

Solve $h$ using Eq. (4.45a)

$$Q^2W = gA^3$$

$$(4.0)^2 \times (1.5 + h) = 9.806 \times (1.5h + 0.5h^2)^3$$

$$(1.5h + 0.5h^2)^3/(1.5 + h) = 1.63$$

by trial $h$ = 0.818 m
= 2.68 ft

## 4.7 Hydraulic jump

When water flows in an open channel, an abrupt decrease in velocity may occur due to a sudden increase of water depth in the downstream direction. This is called hydraulic jump and may be a natural phenomenon. It occurs when the depth in an open channel increases from a depth less than the critical depth $D_c$ to one greater than $D_c$. Under these conditions, the water must pass through a hydraulic jump, as shown in Fig. 4.11.

The balance between the hydraulic forces $P_1$ and $P_2$, represented by the two triangles and the momentum flux through sections 1 and 2, per unit width of the channel, can be expressed as

$$P_1 - P_2 = \rho q\,(V_2 - V_1) \tag{4.52}$$

where $q$ is the flow rate per unit width of the channel. Substituting the following quantities in the above equation

$$P_1 = \frac{\gamma}{2}D_1^2; \quad P_2 = \frac{\gamma}{2}D_2^2$$

$$V_1 = \frac{q}{D_1}; \quad V_2 = \frac{q}{D_2}$$

we get

$$\frac{\gamma}{2}\,(D_1^2 - D_2^2) = \rho q\left(\frac{q}{D_2} - \frac{q}{D_1}\right)$$

$$\frac{1}{2}(D_1 + D_2)(D_1 - D_2) = \frac{\rho}{\gamma}\,q^2\left(\frac{D_1 - D_2}{D_1 D_2}\right)$$

since

$$\frac{\rho}{\gamma} = \frac{1}{g}$$

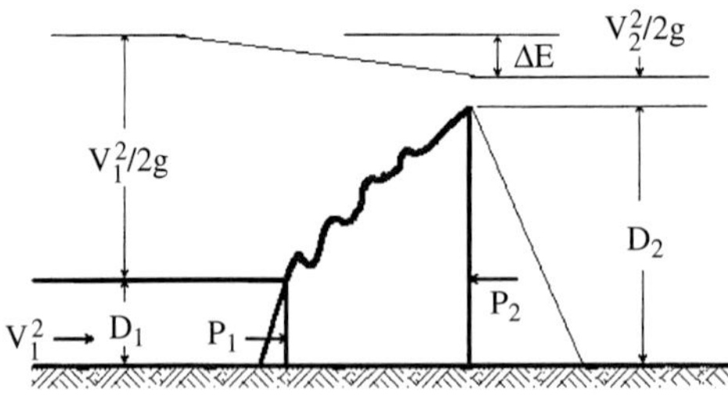


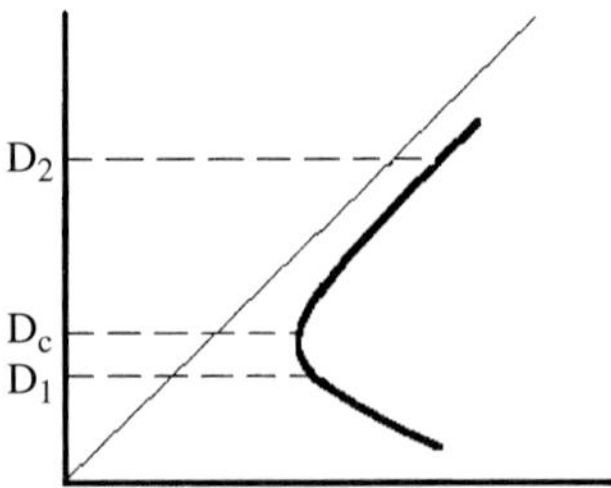


**Figure 4.11** Hydraulic jump.

then

$$\frac{q^2}{g} = D_1 D_2 \left(\frac{D_1 + D_2}{2}\right) \tag{4.53}$$

This equation may be rearranged into a more convenient from as follows:

$$\frac{D_2}{D_1} = \left(\frac{1}{4} + 2F_1^2\right)^{1/2} - \frac{1}{2} \tag{4.54}$$

where $F_1$ is the upstream Froude number and is expressed by

$$F_1 = \frac{V_1}{\sqrt{gD_1}} \tag{4.55}$$

If the hydraulic jump is to occur, $F_1$ must be greater than one; i.e. the upstream flow must be supercritical. The energy dissipated ($\Delta E$) in a hydraulic jump in a rectangular channel may be calculated by the equation

$$\Delta E = E_1 - E_2 = \frac{(D_2 - D_1)^3}{4D_1 D_2} \tag{4.56}$$

**Example 1:** A rectangular channel, 4 m wide and with 0.5 m water depth, is discharging 12 $m^3/s$ flow before entering a hydraulic jump. Determine the critical depth and the downstream water depth.

**solution:**

Step 1. Calculate discharge per unit width $q$

$$q = \frac{Q}{W} = \frac{12 \text{ m}^3/s}{4 \text{ m}} = 3(\text{m}^3/\text{m} \cdot \text{s or m}^2/\text{s})$$

Step 2. Calculate critical depth $D_c$ from Eq. (4.50)

$$D_c = \left(\frac{q^2}{g}\right)^{1/3} = \left(\frac{3 \text{ m}^2/\text{s} \times 3 \text{ m}^2/\text{s}}{9.81 \text{ m/s}^2}\right)^{1/3}$$

$$= 0.972 \text{ m}$$

Step 3. Determine upstream velocity $V_1$

$$V_1 = \frac{q}{D_1} = \frac{3 \text{ m}^2/\text{s}}{0.5 \text{ m}} = 6.0 \text{ m/s}$$

Step 4. Determine the upstream Froude number

$$F_1 = \frac{V_1}{\sqrt{gD_1}} = \frac{6.0}{\sqrt{9.81 \times 0.5}} = 2.71$$

Step 5. Compute downstream water depth $D_2$ from Eq. (4.54)

$$D_2 = D_1\left[\left(\frac{1}{4} + 2F_1^2\right)^{1/2} - \frac{1}{2}\right]$$

$$= 0.5\left[\left(\frac{1}{4} + 2 \times 2.71^2\right)^{1/2} - \frac{1}{2}\right]$$

$$= 1.68(\mathrm{m})$$

**Example 2:** A rectangular channel 9 ft (3 m) wide carries 355 ft³/s (10.0 m³/s) of water with a water depth of 2 ft (0.6 m). Is a hydraulic jump possible? If so, what will be the depth of water after the jump and what will be the horsepower loss through the jump?

**solution:**

Step 1. Compute the average velocity in the channel

$$V_1 = \frac{Q}{A_1} = \frac{355 \text{ ft}^3/\text{s}}{9 \text{ ft} \times 2 \text{ ft}}$$

$$= 19.72 \text{ ft/s}$$

Step 2. Compute $F_1$ from Eq. (4.55)

$$F_1 = \frac{V_1}{\sqrt{gD_1}} = \frac{19.72 \text{ ft/s}}{\sqrt{32.2 \text{ ft/s}^2 \times 2 \text{ ft}}}$$

$$= 2.46$$

Since $F_1 > 1$, the flow is supercritical and a hydraulic jump is possible.

Step 3. Compute the depth $D_2$ after the hydraulic jump

$$D_2 = D_1\left[(0.25 + 2F_1^2)^{1/2} - 0.5\right]$$

$$= 2\left[(0.25 + 2 \times 2.46^2)^{1/2} - 0.5\right]$$

$$= 6.03 \text{ ft}$$

Step 4. Compute velocity $V_2$ after jump

$$V_2 = \frac{Q}{A_2} = \frac{355 \text{ ft}^3/\text{s}}{9 \text{ ft} \times 6.03 \text{ ft}}$$

$$= 6.54 \text{ ft/s}$$

Step 5. Compute total energy loss $\Delta E$ (or $\Delta h$)

$$E_1 = D_1 + \frac{V_1^2}{2g} = 2 + \frac{19.72^2}{2 \times 32.2} = 8.04 \text{ (ft)}$$

$$E_2 = D_2 + \frac{V_2^2}{2g} = 6.03 + \frac{6.54^2}{2 \times 32.2} = 6.69 \text{ (ft)}$$

$$\Delta E = E_1 - E_2 = 8.04 \text{ ft} - 6.69 \text{ ft} = 1.35 \text{ ft}$$

Step 6. Compute horsepower (hp) loss

$$\text{Loss} = \frac{\Delta E \gamma Q}{550} = \frac{1.35 \text{ ft} \times 62.4 \text{ lb/ft}^3 \times 355 \text{ ft}^3\text{/s}}{550 \text{ ft} \cdot \text{lb/hp}}$$

$$= 54.4 \text{ hp}$$

## 5 Flow Measurements

Flow can be measured by velocity methods and direct discharge methods. The measurement flow velocity can be carried out by a current meter, Pitot tube, U-tube, dye study, or salt velocity. Discharge is the product of measured mean velocity and cross-sectional area. Direct discharge methods include volumetric gravimeter, Venturi meter, pipe orifice meter, standardized nozzle meter, weirs, orifices, gates, Parshall flumes, etc. Detail flow measurements in orifices, gates, tubes, weirs, pipes, and in open channals are discussed by Brater *et al.* (1996).

### 5.1 Velocity measurement in open channel

The mean velocity of a stream or a channel can be measured with a current meter. A variety of current meters is commercially available. An example of discharge calculation with known mean velocity in sub-cross section is presented in Chapter 1.

### 5.2 Velocity measurement in pipe flow

**Pitot tube.** A Pitot tube is bent to measure velocity due to the pressure difference between the two sides of the tube in a flow system. The flow velocity can be determined from

$$V = \sqrt{2g\Delta h} \tag{4.57}$$

where $V$ = velocity, m/s or ft/s

$g$ = gravitational acceleration, 9.81 m/s$^2$ or 32.2 ft/s$^2$

$\Delta h$ = height of the fluid column in the manometer or a different height of immersible liquid such as mercury, m or ft

**Example:** The height difference of the Pitot tube is 5.1 cm (2 in). The specific weight of the indicator fluid (mercury) is 13.55. What is the flow velocity of the water?

**solution 1:** For SI units

Step 1. Determine the water column equivalent to $\Delta h$

$$\Delta h = 5.1 \text{ cm} \times 13.55 = 69.1 \text{ cm} = 0.691 \text{ m of water}$$

Step 2. Determine velocity

$$V = \sqrt{2g\Delta h} = \sqrt{2 \times 9.81 \text{ m/s}^2 \times 0.691 \text{ m}}$$
$$= 3.68 \text{ m/s}$$

**solution 2:** For British units

Step 1.

$$\Delta h = 2/12 \text{ ft} \times 13.55 = 2.258 \text{ ft}$$

Step 2.

$$V = \sqrt{2g\Delta h} = \sqrt{2 \times 32.2 \text{ ft/s}^2 \times 2.258 \text{ ft}}$$
$$= 12.06 \text{ ft/s}$$
$$= 12.06 \text{ ft/s} \times 0.304 \text{ m/ft}$$
$$= 3.68 \text{ m/s}$$

### 5.3 Discharge measurement of water flow in pipes

Direct collection of volume (or weight) of water discharged from a pipe divided by time of collection is the simplest and most reliable method. However, in most cases it cannot be done by this method. A change of pressure head is related to a change in flow velocity caused by a sudden change of pipe cross-sectional geometry. Venturi meters, nozzle meters, and orifice meters use this concept.

**Venturi meter.** A Venturi meter is a machine-cased section of pipe with a narrow throat. The device, in a short cylindrical section, consists of an entrance cone and a diffuser cone which expands to full pipe diameter. Two piezometric openings are installed at the entrance (section 1) and at the throat (section 2). When the water passes through the throat, the velocity increases and the pressure decreases. The decrease of pressure is directly related to the flow. Using the Bernoulli equation at sections 1 and 2, neglecting friction head loss, it can be seen that

$$\frac{V_1^2}{2g} + \frac{P_1}{\gamma} = z_1 = \frac{V_2^2}{2g} + \frac{P_2}{\gamma} + z_2 \tag{4.16}$$

For continuity flow between sections 1 and 2, $Q_1 = Q_2$

$$A_1V_1 = A_2V_2$$

Solving the above two equations, we get

$$Q = \frac{A_1A_2\sqrt{2g[(h_1 - h_2) + (z_1 - z_2)]}}{\sqrt{A_1^2 - A_2^2}}$$

$$= \frac{A_1A_2\sqrt{2g(H + Z)}}{\sqrt{A_1^2 - A_2^2}}$$

$$= C_dA_1\sqrt{2g(H + Z)} \tag{4.58a}$$

where $Q$ = discharge, $m^3/s$ or $ft^3/s$
$A_1, A_2$ = cross-sectional areas at pipe and throat, respectively, $m^2$ or $ft^2$
$g$ = gravity acceleration, 9.81 $m/s^2$ or 32.2 $ft/s^2$
$H = h_1 - h_2$ = pressure drop in Venturi tube, m or ft
$Z = z_1 - z_2$ = difference of elevation head, m or ft
$C_d$ = coefficient of discharge

For a Venturi meter installed in a horizontal position, $Z = 0$

$$Q = \frac{A_1A_2\sqrt{2gH}}{\sqrt{A_1^2 - A_2^2}} = C_dA_1\sqrt{2gH} \tag{4.58b}$$

where

$$C_d = \frac{A_2}{\sqrt{A_1^2 - A_2^2}} = \frac{1}{\sqrt{\left(\frac{A_1}{A_2}\right)^2 - 1}} \tag{4.59}$$

**Example:** A 6-cm throat Venturi meter is installed in an 18-cm diameter horizontal water pipe. The reading of the differential manometer is 18.6 cm of mercury column (sp gr = 13.55). What is the flow rate in the pipe?

**solution:**

Step 1. Determine $A_1/A_2$

$$A_1 = \pi(18/2)^2 = 254.4\ \text{cm}^2 = 0.2544\ \text{m}^2$$
$$A_2 = \pi(6/2)^2 = 28.3\ \text{cm}^2 = 0.0283\ \text{m}^2$$
$$\frac{A_1}{A_2} = \frac{254.4}{28.3} = 8.99$$

Step 2. Calculate $C_d$

$$C_d = 1/\sqrt{\left(\frac{A_1}{A_2}\right)^2 - 1} = 1/\sqrt{(8.99)^2 - 1}$$

$$= 0.112$$

Step 3. Calculate $H$

$$H = \gamma y = 13.55 \times \frac{18.6}{100}\ \text{m}$$

$$= 2.52\ \text{m}$$

Step 4. Calculate $Q$

$$Q = C_d A_1 \sqrt{2gH} = 0.112 \times 0.2544 \times \sqrt{2 \times 9.81 \times 2.52}$$

$$= 0.20\ (\text{m}^3/\text{s})$$

**Nozzle meter and orifice meter.** The nozzle meter and the orifice meter are based on the same principles as the Venturi meter. However, nozzle and orifice meters encounter a significant loss of head due to a nozzle or orifice installed in the pipe. The coefficient ($C_v$) for the nozzle meter or orifice meter is added to the discharge equation for the Venturi meter, and needs to be determined by on-site calibration. The discharge for the nozzle meter and orifice meter is

$$Q = C_v C_d A_1 \sqrt{2g(H + Z)} \tag{4.60}$$

or, for horizontal installation

$$Q = C_v C_d A_1 \sqrt{2gH} \tag{4.61}$$

The nozzle geometry has been standardized by the American Society of Mechanical Engineers (ASME) and the International Standards Association. The nozzle coefficient $C_v$ is a function of Reynolds number ($\mathbf{R} = V_2 d_2/\nu$) and ratio of diameters ($d_2/d_1$). A nomograph (Fig. 4.12) is available. Values of $C_v$ range from 0.96 to 0.995 for $\mathbf{R}$ ranging from $5 \times 10^4$ to $5 \times 10^6$.

**Example:** Determine the discharge of a 30-cm diameter water pipe. An ASME nozzle of 12-cm throat diameter is installed. The attached differential manometer reads 24.6 cm of mercury column. The water temperature in the pipe is 20°C.

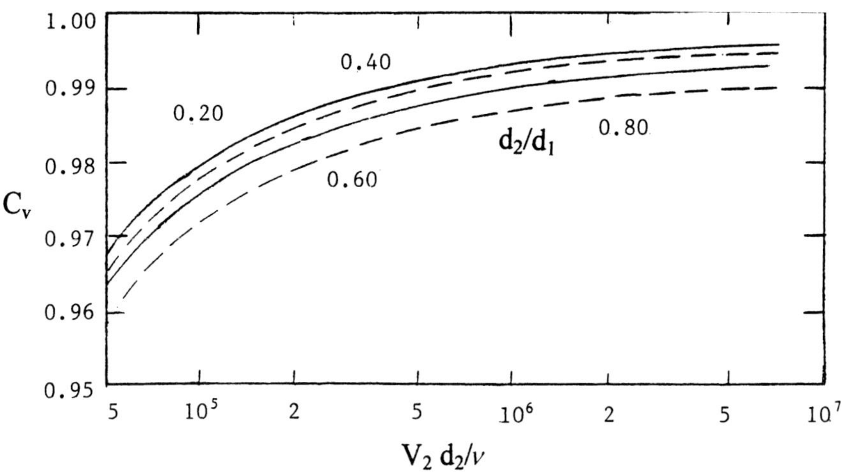


**Figure 4.12** ASME nozzle coefficients.

**solution:**

Step 1. Determine ratio $A_1 : A_2$

$$A_1 = \frac{\pi}{4}(30\ \text{cm})^2 = 707\ \text{cm}^2 = 0.0707\ \text{m}^2$$

$$A_2 = \frac{\pi}{1}(12\ \text{cm})^2 = 113\ \text{cm}^2 = 0.0113\ \text{m}^2$$

$$\frac{A_1}{A_2} = \frac{30^2}{12^2} = 6.25$$

Step 2. Calculate $C_d$

$$C_d = 1/\sqrt{(A_1/A_2)^2 - 1} = 1/\sqrt{(6.25)^2 - 1}$$
$$= 0.162$$

Step 3. Calculate $H$

$$\gamma = 13.55 \text{ at } 20°\text{C}$$
$$H = \gamma y = 13.55 \times 24.6\ \text{cm} = 333\ \text{cm}$$
$$= 3.33\ \text{m}$$

Step 4. Estimate discharge $Q$

Assuming $C_v = 0.98$

$$Q = C_v C_d A_1 \sqrt{2gH}$$
$$= 0.98 \times 0.162 \times 0.0707\ \sqrt{2 \times 9.81 \times 3.33}$$
$$= 0.091(\text{m}^3/\text{s})$$

This value needs to be verified by checking the corresponding Reynolds number **R** of the nozzle.

Step 5. Calculate **R**

$$V_2 = Q/A_2 = 0.091 \text{ m}^3\text{/s}/0.0113 \text{ m}^2$$
$$= 8.053 \text{ m/s}$$

The kinematic viscosity at 20°C is (Table 4.1a)

$$\nu = 1.007 \times 10^{-6} \text{ m}^2\text{/s}$$

then

$$\mathbf{R} = \frac{V_2 d_2}{\nu} = \frac{8.053 \text{ m/s} \times 0.12 \text{ m}}{1.007 \times 10^{-6} \text{ m}^2\text{/s}}$$
$$= 9.6 \times 10^5$$

Using this **R** value, with $d_1/d_2 = 0.4$, the chart in Fig. 4.12 gives

$$C_v = 0.977$$

Step 6. Correct $Q$ value

$$Q = \frac{0.977298}{0.98} \times 0.091 \text{ m}^3\text{/s}$$
$$= 0.0907 \text{ m}^3\text{/s}$$

### 5.4 Discharge measurements

**Orifices.** The discharge of an orifice is generally expressed as

$$Q = C_d A \sqrt{2gH} \tag{4.62}$$

where $Q$ = flow rate
$C_d$ = coefficient of discharge
$A$ = area of orifice
$g$ = gravity acceleration
$H$ = water height above center of orifice
All units are as Eq. (4.58a)

**Example:** The water levels upstream and downstream are 4.0 and 1.2 m, respectively. The rectangular orifice has a sharp-edged opening 1 m high and 1.2 m wide. Calculate the discharge, assuming $C_d = 0.60$.

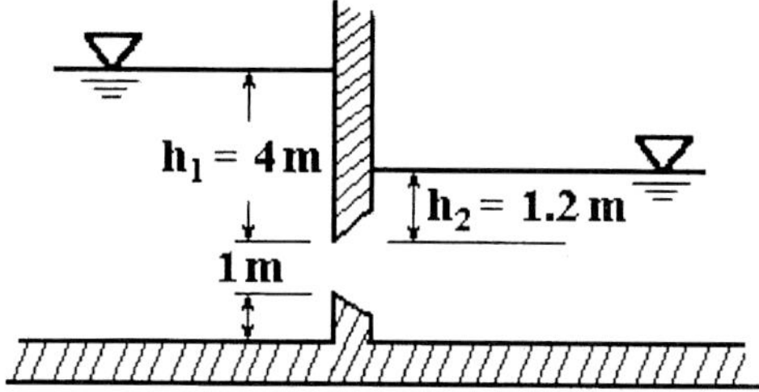


**solution:**

Step 1. Determine area $A$

$$A = 1 \text{ m} \times 1.2 \text{ m} = 1.2 \text{ m}^2$$

Step 2. Determine discharge $Q$

$$\begin{aligned} Q &= C_d A \sqrt{2g(h_1 - h_2)} \\ &= 0.60 \times 1.2 \text{ m}^2 \sqrt{2 \times 9.81 \text{ m/s}^2 \times (4 - 1.2) \text{ m}} \\ &= 5.34 \text{ m}^3/\text{s} \end{aligned}$$

**Weirs.** The general discharge equation for a rectangular, horizontal weir is

$$Q = C_d L H^{3/2} \tag{4.63a}$$

where $Q$ = flow rate, $\text{ft}^3/\text{s}$ or $\text{m}^3/\text{s}$
$L$ = weir length, ft or m
$H$ = head on weir, ft or m
= the water surface above the weir crest
$C_d$ = coefficient of discharge

In the US customary units, let $y$ be the weir height in feet.

$$C_d = 3.32 + 0.40\frac{H}{y} \quad \text{(US customary units)} \tag{4.64a}$$

$$C_d = 1.78 + 0.24\frac{H}{y} \quad \text{(SI units)} \tag{4.64b}$$

The theoretical equation for a rectangular streamlined weir is (ASCE & WEF, 1992)

$$Q = \frac{2}{3} L \sqrt{\frac{2}{3} g H^3} \tag{4.63b}$$

where $g$ is the gravitational acceleration rate. This formula reduces to

$$Q = 3.09\ LH^{3/2} \text{ (for U.S. customary units)} \tag{4.63c}$$

or

$$Q = 1.705\ LH^{3/2} \text{ (for SI units)} \tag{4.63d}$$

**Example 1:** A rectangular flow control tank has an outflow rectangular weir 1.5 m (5 ft) in length. The inflow to the box is 0.283 $m^3/s$ (10 $ft^3/s$). The crest of the weir is located 1.2 m above the bottom of the tank. Find the depth of water in the tank.

**solution:**

Step 1. Determine head on weir $H$

Using Eq. (4.63d)

$$Q = 1.705\ LH^{3/2}$$

$$0.283 = 1.705(1.5)H^{3/2}$$

$$H^{3/2} = 0.111$$

$$H = 0.23 \text{ (m)}$$

Step 2. Calculate water depth $D$

$$D = 1.2 \text{ m} + H = 1.2 \text{ m} + 0.23 \text{ m}$$

$$= 1.43 \text{ m}$$

**Example 2:** In a rectangular channel 4 m (13 ft) high and 12 m (40 ft) wide, a sharp-edged rectangular weir 1.2 m (4 ft) high without end contraction will be installed. The flow of the channel is 0.34 $m^3/s$ (12 $ft^3/s$). Determine the length of the weir to keep the head on the weir 0.15 m (0.5 ft).

**solution:**

Step 1. Compute velocity of approach at channel, $V$

$$V = \frac{Q}{A} = \frac{0.34 \text{ m}^3/\text{s}}{12 \times 4} = 0.007 \text{ m/s (negligible)}$$

Head due to velocity of approach

$$h = \frac{V^2}{2g} \text{ (negligible)}$$

Step 2. Use weir formula (Eq. (4.63d))

(a) Without velocity of approach

$$Q = 1.705\ LH^{3/2}$$

$$0.34 = 1.705\ LH^{3/2}$$

$$0.34 = 1.705\, L\, (0.15)^{1.5}$$
$$L = 3.43\,(\text{m})$$
$$= 11.2 \text{ ft}$$

(b) Including velocity of approach

$$Q = 1.705\, L\left[(0.15 + h)^{1.5} - h^{1.5}\right]$$
$$L = 3.4 \text{ m}$$

For freely discharging rectangular weirs (sharp-crested weirs), the Francis equation is most commonly used to determine the flow rate. The Francis equation is

$$Q = 3.33\, LH^{3/2} \tag{4.63e}$$

where $Q$ = flow rate, $\text{ft}^3/\text{s}$
$L$ = weir length, ft
$H$ = head on weir, ft
= the water surface above the weir crest

For constracted rectangular weirs

$$Q = 3.33\,(L - 0.1nH)H^{3/2} \quad \text{(US customary units)} \tag{4.63f}$$

$$Q = 1.705\,(L - 0.1\,nH)\,H^{3/2} \quad \text{(SI units)} \tag{4.63g}$$

where $n$ = number of end constractions.

**Example 3:** Conditions are the same as in Example 2, except that the weir has two-end constractions. Compute the width of the weir.

**solution:**

Step 1. For constracted weir, without velocity of approach

$$Q = 1.705\,(L - 0.2H)H^{3/2}$$
$$0.34 = 1.705\,(L - 0.2 \times 0.15) \times (0.15)^{1.5}$$
$$L = 3.40\ (\text{m})$$
$$= 11.1 \text{ ft}$$

Step 2. Discharge including velocity of approach

$$Q = 1.075\,(L - 0.2H)\left[(H + h)^{3/2} - h^{3/2}\right]$$

Since velocity head is negligible

$$L = 3.4 \text{ m}$$

**Example 4:** Estimate the discharge of a weir 4 ft long where the head on the weir is 3 in.

**solution:**

By the Francis equation

$$\begin{aligned} Q &= 3.33\ LH^{3/2} \\ &= 3.33 \times 4 \times (3/12)^{3/2} \\ &= 1.66\ (\text{ft}^3/\text{s}) \end{aligned}$$

For the triangular weir and V-notch weir, the flow is expressed as

$$Q = C_d\left(\tan\frac{\theta}{2}\right)H^{5/2} \tag{4.65}$$

where $\theta$ = weir angle
$C_d$ = discharge coefficient, calibrated in place
= 1.34 for SI units
= 2.50 for US customary units
Different values have been proposed

The triangular weir is commonly used for measuring small flow rates. Several different notch angles are available. However, the 90° V-notch weir is the one most commonly used. The discharge for 90° V-notch with free flow is

$$Q = 2.5\ H^{2.5} \tag{4.66}$$

where $\theta$ = discharge, ft$^3$/s
$H$ = head on the weir, ft

Detailed discussion of flow equations for SI units for various types of weir is presented by Brater *et al.* (1996).

**Example 5:** A rectangular control tank has an outflow rectangular weir 8 ft (2.4 m) in length. The crest of the weir is 5 ft (1.5 m) above the tank bottom. The inflow from a pipe to the tank is 10 ft$^3$/s (0.283 m$^3$/s). Estimate the water depth in the tank, using the Francis equation.

**solution:**

$$Q = 3.33\ LH^{3/2}$$

Rewrite equation for $H$

$$\begin{aligned} H &= \left(\frac{Q}{3.33L}\right)^{2/3} = \left(\frac{10}{3.33 \times 8}\right)^{2/8} \\ &= 0.52\ (\text{ft}) \end{aligned}$$

**Example 6:** A rectangular channel 6 ft (1.8 m) wide has a sharp-crested weir across its whole width. The weir height is 4 ft (1.2 m) and the head is 1 ft (0.3 m). Determine the discharge.

**solution:**

Step 1. Determine $C_d$ using Eq. (4.64)

$$C_d = 3.22 + 0.40\frac{H}{y} = 3.22 + 0.40 \times \frac{1}{4}$$
$$= 3.32$$

Step 2. Compute discharge $Q$

$$Q = C_d LH^{3/2}$$
$$= 3.32 \times 6 \times (1)^{3/2}$$
$$= 19.92\ (\text{ft}^3/\text{s})$$

**Example 7:** A circular sedimentation basin has a diameter of 53 ft (16m). The inflow from the center of the basin is 10 MGD (0.438 $m^3$/s). A circular effluent weir with 90° V-notches located at 0.5 ft (0.15 m) intervals is installed 1.5 ft (0.45 m) inside the basin wall. Determine the water depth on each notch and the elevation of the bottom of the V-notch if the water surface of the basin is at 560.00 ft above mean sea level (MSL).

**solution:**

Step 1. Determine the number of V-notches

Diameter of the weir

$$d = 53\ \text{ft} - 2 \times 1.5\ \text{ft} = 50\ \text{ft}$$

Weir length

$$l = \pi \text{d} = 3.14 \times 50\ \text{ft} = 157\ \text{ft}$$

No. of V-notches

$$n = 157\ \text{ft} \times \frac{2\ \text{notches}}{\text{ft}}$$
$$= 314\ \text{notches}$$

Step 2. Compute the discharge per notch

$$Q = \frac{10{,}000{,}000\ \text{gal/d}}{314\ \text{notches}} = 31{,}847\,(\text{gal/d})/\text{notch}$$

$$= 31{,}847\frac{\text{gal/d}}{\text{notch}} \times \frac{1\ \text{ft}^3}{7.48\ \text{gal}} \times \frac{1\,\text{day}}{86{,}400\ \text{s}}$$

$$= 0.0493\ (\text{ft}^3/\text{s})\,\text{notch}$$

Step 3. Compute the head of each notch from Eq. (4.66)

$$Q = 2.5\,H^{2.5}$$

$$H = (Q/2.5)^{1/2.5} = (Q/2.5)^{0.4} = (0.0493/2.5)^{0.4}$$

$$= 0.21\ \text{(ft)}$$

Step 4. Determine the elevation of the bottom of the V-notch

$$\text{Elevation} = 560.00\ \text{ft} - 0.21\ \text{ft}$$

$$= 559.79\ \text{ft MSL}$$

**Parshall flume.** The Parshall flume was developed by R. L. Parshall in 1920 for the British measurement system. It is widely used for measuring the flow rate of an open channel. It consists of a converging section, a throat section, and a diverging section. The bottom of the throat section is inclined downward and the bottom of the diverging section is inclined upward. The dimensions of the Parshall flume are specified in the British system, not the metric system. The geometry creates a critical depth to occur near the beginning of the throat section and also generates a back water curve that allows the depth $H_a$ to be measured from observation well a. There is a second measuring point $H_b$ located at the downstream end of the throat section. The ratio of $H_b$ to $H_a$ is defined as the submergence of the flume.

Discharge is a function of $H_a$. The discharge equation of a Parshall flume is determined by its throat width, which ranges from 3 in to 50 ft. The relationship between the empirical discharge and the gage reading $H_a$ for various sizes of flume is given below:

| Throat width $W$, ft | Discharge equation, ft$^3$/s | Flow capacity, ft$^3$/s |
|---|---|---|
| 3 in (0.25 ft) | $Q = 0.992 H_a^{1.547}$ | 0.03–1.9 |
| 6 in (0.5 ft) | $Q = 2.06\,H_a^{1.58}$ | 0.05–3.9 |
| 9 in (0.75 ft) | $Q = 3.07\,H_a^{1.53}$ | 0.09–8.9 |
| 1–8 ft | $Q = 4\,\mathrm{W} H_a^{1.522 W^{0.26}}$ | up to 140 |
| 10–50 ft | $Q = (3.6875\,W + 2.5)\,H_a^{1.6}$ | |

SOURCE: R. L. Parshall (1926)

When the ratio of $H_a$ to $H_b$ exceeds the following values:

0.5 for flumes of 1, 2, and 3 in width,

0.6 for flumes of 6 and 9 in width,

0.7 for flumes of 1 to 8 ft width,

0.8 for flumes of 8 to 50 ft width.

The flume is said to be submerged. When a flume is submerged, the actual discharge is less than that determined by the above equations. The diagram presented in Figs. 4.13 and 4.14 can be used to determine

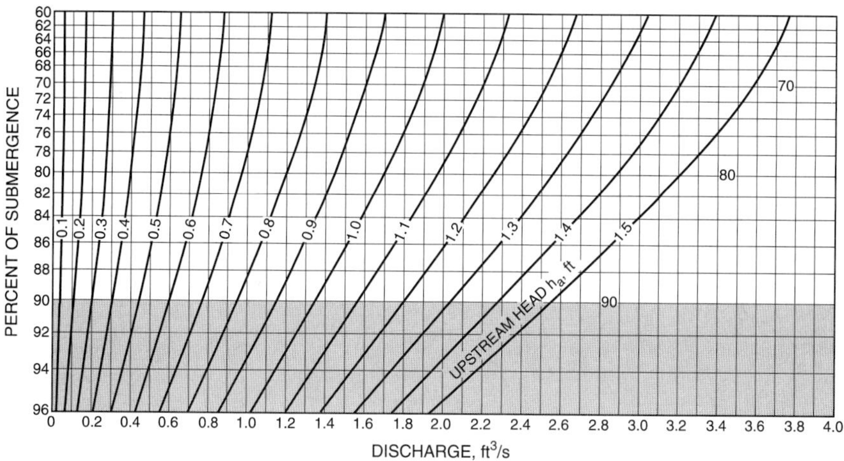


**Figure 4.13** Diagram for determining rate of submerged flow through a 6-in Parshall flume (*US Department of the Interior, 1997*).

discharges for submerged Parshall flumes of 6 and 9 in width, respectively. Figure 4.15 shows the discharge correction for flumes with 1 to 8 ft throat width and with various percentages of submergence. The correction for the 1-ft flume can be applicable to larger flumes by multiplying by a correction factor of 1.0, 1.4, 1.8, 2.4, 3.1, 4.3, and 5.4 for flume sizes of 1, 1.5, 2, 3, 4, 6, and 8 ft width, respectively. Figure 4.16 is used for determining the correction to be subtracted from the free-flow value for a 10-ft Parshall flume. For larger sizes, the correction equals the value from the diagram multiplied by the size factor $M$. Figures. 4.13–4.16 are improved curves from earlier developments (Parshall 1932, 1950).

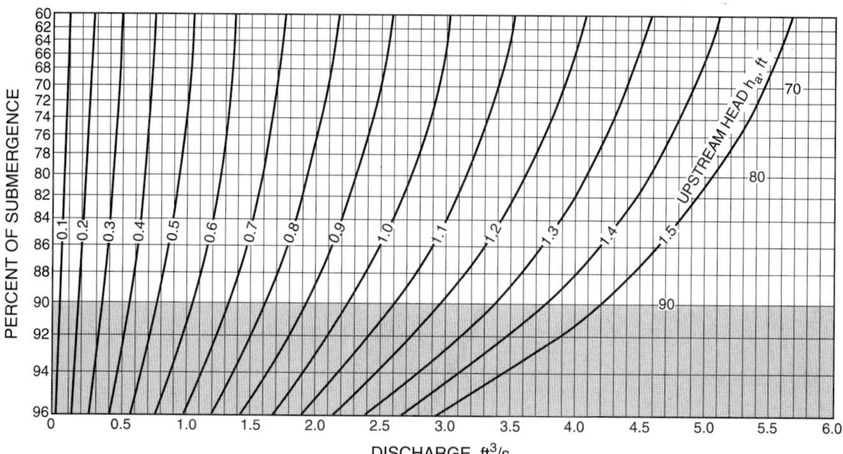


**Figure 4.14** Diagram for determining rate of submerged flow through a 9-in Parshall flume (*US Department of the Interior, 1997*).

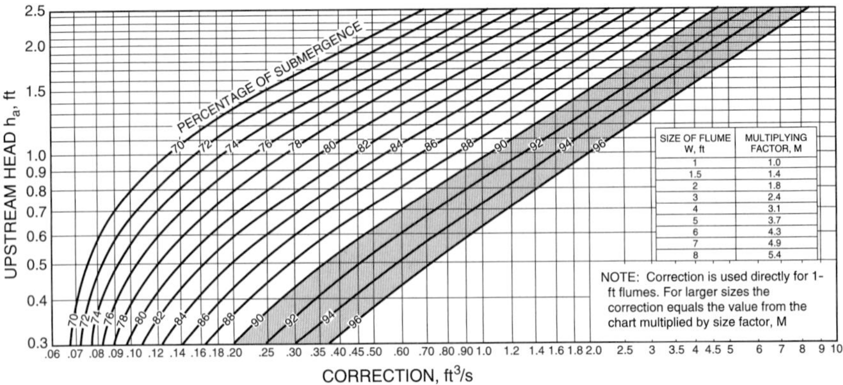


**Figure 4.15** Diagram for determining correction to be subtracted from the free discharge to obtain rate of submerged flow through 1- to 8-ft Parshall flumes (*US Department of the Interior, 1997*).

**Example 1:** Calculate the discharge through a 9-in Parshall flume for the gage reading $H_a$ of 1.1 ft when (a) the flume has free flow, and (b) the flume is operating at 72% of submergence.

**solution:**

Step 1. Calculate $Q$ for free flow for (a)

$$Q = 3.07H_a^{1.53} = 3.07\,(1.1)^{1.53}$$
$$= 3.55\ (\text{ft}^3/\text{s})$$

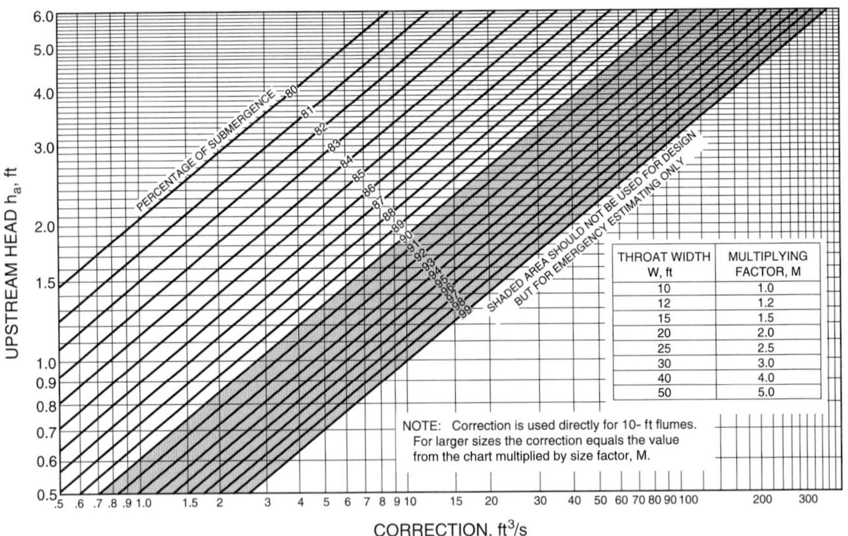


**Figure 4.16** Diagram for determining correction to be subtracted from discharge flow to obtain rate of submerged flow through 10-to 50- ft Parshall flumes (*US Department of the Interior, 1997*).

Step 2. Determine discharge with 0.72 submergence for (b)

From Fig. 4.14, enter the plot at $H_b/H_a = 0.72$ and proceed horizontally to the line representing $H_a = 1.1$. Drop downward and read a discharge of

$$Q_s = 3.33 \text{ ft}^3/\text{s}$$

**Example 2:** Determine the discharge of a 2-ft Parshall flume with gage readings $H_a$=1.55 ft and $H_b$ = 1.14 ft.

**solution:**

Step 1. Compute $H_b/H_a$ ratio

$$\frac{H_b}{H_a} = \frac{1.14 \text{ ft}}{1.55 \text{ ft}} = 0.74 \text{ or } 74\%$$

This ratio exceeds 0.7, therefore the flume is submerged.

Step 2. Compute the free discharge of the 2-ft flume

$$\begin{aligned} Q &= 4WH_a^{1.522W^{0.26}} \\ &= 4(2)(1.55)^{1.522 \times 2^{0.26}} \\ &= 8 \times 2.223 \\ &= 17.78 \text{ (ft}^3\text{/s)} \end{aligned}$$

Step 3. Determine the correction discharge for a 1-ft flume from Fig. 4.15, enter the plot at a value of $H_a$=1.55 ft and proceed horizontally to the line representing the submergence curve of 74 %. Then drop vertically downward and read a discharge correction for the 1-ft flume of

$$Q_{c1} = 0.40 \text{ ft}^3/\text{s}$$

Step 4. Compute discharge correction for the 2-ft flume

$$\begin{aligned} Q_{c2} &= 1.8 \times 0.40 \text{ ft}^3/\text{s} \quad \text{(factor M = 1.8, Fig. 4.15)} \\ &= 0.72 \text{ ft}^3/\text{s} \end{aligned}$$

Step 5. Compute the actual discharge

$$\begin{aligned} Q &= Q \text{ (at Step 2} - \text{at Step 4)} \\ &= 17.78 \text{ ft}^3/\text{s} - 0.72 \text{ ft}^3/\text{s} \\ &= 17.06 \text{ ft}^3/\text{s} \end{aligned}$$

## References


American Society of Civil Engineers (ASCE) and Water Environment Federation (WEF). 1992. *Design and construction of urban stormwater management systems.* New York: ASCE & WEF.

Benefield, L. D., Judkins, J. F. Jr and Parr, A. D. 1984. *Treatment plant hydraulics for environmental engineers.* Englewood Cliffs, New Jersey: Prentice-Hall.

Brater, E. F., King, H. W., Lindell, J. E. and Wei, C. Y. 1996. *Handbook of hydraulics* 7th edn. New York: McGraw-Hill.

Brisbin, S. G. 1957. Flow of concentrated raw sewage sludges in pipes. *J. Sanitary Eng. Div., ASCE* **83:** 201.

Chadwick, A. J. and Morfett, J. C. 1986. *Hydraulics in civil engineering.* London: Allen & Unwin.

Chow, V. T. 1959. *Open-channel hydraulics,* New York: McGraw-Hill.

Cross, H. 1936. Analysis of flow in networks of conduits or conductors. University of Illinois Bulletin no. 286. Champaign-Urbana: University of Illinois.

Ductile Iron Pipe Research Association. 1997. *Ductile Iron Pipe News,* fall/winter 1997.

Fair, G. M., Geyer, J. C. and Okun, D. A. 1966. *Water and wastewater engineering,* Vol. 1: *Water supply and wastewater removal.* New York: John Wiley.

Fetter, C. W. 1994. *Applied hydrogeology,* 3rd edn. Upper Saddle River, New Jersey: Prentice-Hall.

Horvath, I. 1994. *Hydraulics in water and waste-water treatment technology.* Chichester: John Wiley.

Hwang, N. H. C. 1981. *Fundamentals of hydraulic engineering systems.* Englewood Cliffs, New Jersey: Prentice-Hall.

Metcalf & Eddy, Inc. 1972. *Wastewater engineering: Collection, treatment, disposal.* New York: McGraw-Hill.

Metcalf & Eddy, Inc. 1990. *Wastewater engineering: Collection and pumping of wastewater.* New York: McGraw-Hill.

James M. Montgomery, Consulting Engineers, Inc. 1985. *Water treatment principles and design.* New York: John Wiley.

Morris, H. M. and Wiggert, J. M. 1972. *Applied hydraulics in engineering,* 2nd edn. New York: John Wiley.

Parshall, R. L. 1926. The improved Venturi flume. *Trans. Amer. Soc. Civil Engineers* **89**: 841–851.

Parshall, R. L. 1932. *Parshall flume of large size.* Colorado Agricultural Experiment Station, Bulletin 386.

Parshall, R. L. 1950. *Measuring water in irrigation channels with Parshall flumes and small weirs.* US Soil Conservation Service, Circular 843.

Perry, R. H. 1967. *Engineering manual: A practical reference of data and methods in architectural, chemical, civil, electrical, mechanical, and nuclear engineering.* New York: McGraw-Hill.

Streeter, V. L. and Wylie, E. B. 1975. *Fluid mechanics,* 6th edn. New York: McGraw-Hill.

US Department of the Interior, Bureau of Reclamation. 1997. *Water measurement manual,* 3rd edn. Denver, Colorado: US Department of the Interior.

Williams, G. S. and Hazen, A. 1933. *Hydraulic tables,* 3rd edn. New York: John Wiley.

Zipparro, V. J. and Hasen, H. 1993. *Davis handbook of applied hydraulics.* New York: McGraw-Hill.

Chapter

# 5

# Public Water Supply

## 1 Sources and Quantity of Water

The sources of a water supply may include rainwater, surface waters, and groundwater. During the water cycle, rainwater recharges the surface waters and groundwater. River, stream, natural pond, impoundment, reservoir, and lake waters are major surface water sources. Some communities use groundwater, such as galleries, wells, aquifers, and springs as their water supply sources.

Water supplies may be drawn from a single source or from a number of different ones. The water from multiple sources could be mixed before distribution or separately distributed. Any new source water has to be approved by the federal, state, and related authorities. The quantity of water needed varies with season, geography, size and type of community, and culture. A water supply system may provide for domestic, industrial and commercial, public services, fire demands, unaccounted losses as well as farm uses. The design flow for a water supply system is discussed in the sections of water requirements and fire demands. The design engineers should examine the availabilities of water sources and quantities in the area. The prediction of the nature population should be made for design purpose. Based on long-term meteorological data, the amount of water stored in the lake or reservoirs can be estimated.

Runoff refers to the precipitation that reaches a stream or river. Theoretically, every unit volume of water passing the observation station should be measured and the sum of all these units passing in a certain period of time would be the total runoff. However, it is not available due to cost. Observations should be carried out at reasonably close intervals so that the observed flow over a long period of time is the sum of flows for the individual shorter periods of observation and not the sum of the observed rates of flow multiplied by the total period.

**Example 1:** Records of observations during a month (30 days) period show that a flow rate 2.3 cfs (0.065 $m^3$/s) for 8 days, 3.0 cfs (0.085 $m^3$/s) for 10 days, 56.5 cfs (16.0 $m^3$/s) for 1 day, 12.5 cfs (0.354 $m^3$/s) for 2 days, 5.3 cfs (0.150 $m^3$/s) for 6 days, and 2.65 cfs (0.075 $m^3$/s) for 3 days. What is the mean flow rate?

**solution:**

$$Q = \frac{2.3 \times 8 + 3.0 \times 10 + 56.5 \times 1 + 12.5 \times 2 + 5.3 \times 6 + 2.65 \times 3}{30}$$

$$= 5.66 \text{ ft}^3/\text{s}$$

$$= 0.160 \text{ m}^3/\text{s}$$

**Example 2:** If the mean annual rainfall is 81 cm (32 in). A horizontal projected roof area is 300 $m^2$ (3230 $ft^2$). Make a rough estimate of how much water can be caught.

**solution:**

Step 1. Calculate annual gross yield, $V_y$

$$V_y = 300 \text{ m}^2 \times 0.81 \text{ m/yr} = 243 \text{ m}^3/\text{yr}$$

$$= 0.666 \text{ m}^3/d$$

Step 2. Estimate net yield $V_n$

According to Fair *et al.* (1966), $V_n = \frac{2}{3} V_y$

$$V_n = \frac{2}{3} V_y = \frac{2}{3}(243 \text{ m}^3/\text{yr}) = 162 \text{ m}^3/\text{yr}$$

$$= 0.444 \text{ m}^3/\text{d}$$

Step 3. Estimate the water that can be stored and then be used, $V_u$

$$V_u = 0.5V_n = 0.5(162\ \text{m}^3/\text{yr}) = 81\ \text{m}^3/\text{yr}$$

or

$$= 2860\ \text{ft}^3/\text{yr}$$

$$= 0.5(0.444\ \text{m}^3/\text{d}) = 0.222\ \text{m}^3/\text{d}$$

or

$$= 7.83\ \text{ft}^3/\text{d}$$

**Example 3:** Determine the rainfall-runoff relationship.

Using a straight line by method of average. The given values are as follows:

| Year | Rainfall, in | Runoff, cfs/miles$^2$ |
|---|---|---|
| 1988 | 15.9 | 26.5 |
| 1989 | 19.4 | 31.2 |
| 1990 | 23.9 | 38.6 |
| 1991 | 21.0 | 32.5 |
| 1992 | 24.8 | 37.4 |
| 1993 | 26.8 | 40.2 |
| 1994 | 25.4 | 38.7 |
| 1995 | 24.3 | 36.4 |
| 1996 | 27.3 | 39.6 |
| 1997 | 22.7 | 34.5 |

**solution:**

Step 1. Write a straight line equation as $y = mx + b$ in which m is the slope of the line, and $b$ is the intercept on the $Y$ axis.

Step 2. Write a equation for each year in which the independent variable is $X$ (rainfall) and the dependent variable is $Y$ (runoff). Group the equations, in two groups, and alternate the years. Then total each group.

| | | | |
|---|---|---|---|
| 1988 | $26.5 = 15.9\ m + b$ | 1989 | $31.2 = 19.4\ m + b$ |
| 1990 | $38.6 = 23.9\ m + b$ | 1991 | $32.5 = 21.0\ m + b$ |
| 1992 | $37.4 = 24.8\ m + b$ | 1993 | $40.2 = 26.8\ m + b$ |
| 1994 | $38.7 = 25.4\ m + b$ | 1995 | $36.4 = 24.3\ m + b$ |
| 1996 | $39.6 = 27.3\ m + b$ | 1997 | $34.5 = 22.7\ m + b$ |
| | $180.8 = 117.3\ m + 5b$ | | $174.8 = 114.2\ m + 5b$ |

Step 3. Solve the total of each group of equations simultaneously for $m$ and $b$

$$\begin{aligned} 180.8 &= 117.3m + 5b \\ -\ (174.8 &= 114.2m + 5b) \\ \hline 6.0 &= 3.1m \\ m &= 1.94 \end{aligned}$$

Step 4. Substituting for $m$ and solving for $b$

$$180.8 = 117.3(1.94) + 5b$$
$$5b = -46.76$$
$$b = -9.35$$

Step 5. The equation of straight line of best fit is

$$Y = 1.94X - 9.35$$

where $X$ = rainfall, in
$Y$ = runoff, cfs/miles$^2$

**Example 4:** A watershed has a drainage area of 1000 ha (2470 acres). The annual rainfall is 927 mm (36.5 in). The expected evaporation loss is 292 mm (11.5 in) per year. The estimated loss to groundwater is 89 mm (3.5 in) annually. Estimate the amount of water that can be stored in a lake and how many people can be served, assuming 200 L(c · d) is needed.

**solution:**

Step 1. Using a mass balance

$R$ (rainfall excess) = $P$ (precipitation) − $E$ (evaporation) − $G$ (loss to groundwater)

$$R = 927 \text{ mm} - 292 \text{ mm} - 89 \text{ mm}$$
$$= 546 \text{ mm}$$

Step 2. Convert $R$ from mm to m$^3$ (volume) and L

$$R = 564\,\text{mm} \times \frac{1\,\text{m}}{1000\,\text{mm}} \times 1000\,\text{ha} \times \frac{10{,}000\,\text{m}^2}{1\,\text{ha}}$$
$$= 5.46 \times 10^6\,\text{m}^3$$
$$= 5.46 \times 10^6\,\text{m}^3 \times 10^3\,\text{L/1}\,\text{m}^3$$
$$= 5.46 \times 10^9\,\text{L}$$

Step 3. Compute the people that can be served

$$\text{Annual usage per capita} = 200\ \text{L/(c} \cdot \text{d)} \times 365\ \text{days}$$
$$= 7.3 \times 10^4\ \text{L/c}$$
$$\text{No. of people served} = \frac{5.46 \times 10^9\ \text{L}}{7.3 \times 10^4\ \text{L/c}}$$
$$= 74{,}800\ \text{capita}$$

## 2 Population Estimates

Prior to the design of a water treatment plant, it is necessary to forecast the future population of the communities to be served. The plant should be sufficient generally for 25 to 30 years. It is difficult to estimate the population growth due to economic and social factors involved. However, a few methods have been used for forecasting population. They include the arithmetic method and uniform percentage growth rate method (Clark and Viessman, 1966; Steel and McGhee, 1979; Viessman and Hammer, 1993). The first three methods are short-term (<10 years) forecasting.

## 2.1 Arithmetic method

This method of forecasting is based upon the hypothesis that the rate of increase is constant. It may be expressed as follows:

$$\frac{dp}{dt} = k_a \tag{5.1}$$

where $p$ = population
$t$ = time, year
$k_a$ = arithmetic growth rate constant

Rearrange and integrate the above equation, $p_1$ and $p_2$ are the populations at time $t_1$ and $t_2$, respectively.

$$\int_{p_1}^{p_2} dp = \int_{t_1}^{t_2} k_a dt$$

We get

$$p_2 - p_1 = k_a(t_2 - t_1)$$

$$k_a = \frac{p_2 - p_1}{t_2 - t_1} = \frac{\Delta p}{\Delta t} \tag{5.2a}$$

or

$$p_t = p_0 + k_a t \tag{5.2b}$$

where $p_t$ = population at future time
$p_0$ = present population, usually use $p_2$ (recent censused)

## 2.2 Constant percentage growth rate method

The hypothesis of constant percentage or geometric growth rate assumes that the rate increase is proportional to population. It can be written as

$$\frac{dp}{dt} = k_p\, p \tag{5.3a}$$

Integrating this equation yields

$$\ln p_2 - \ln p_1 = k_p(t_2 - t_1)$$

$$k_p = \frac{\ln p_2 - \ln p_1}{t_2 - t_1} \tag{5.3b}$$

The geometric estimate of population is given by

$$\ln p = \ln p_2 + k_p (t - t_2) \tag{5.3c}$$

## 2.3 Declining growth method

This is a decreasing rate of increase on the basis that the growth rate is a function of its population deficit. Mathematically it is given as

$$\frac{dp}{dt} = k_d(p_s - p) \tag{5.4a}$$

where $p_s$ = saturation population, assume value

Integration of the above equation gives

$$\int_{p_1}^{p_2} \frac{dp}{p_s - p} = k_d \int_{t_1}^{t_2} dt$$

$$-\ln\frac{p_s - p_2}{p_s - p_1} = k_d(t_2 - t_1)$$

Rearranging

$$k_d = -\frac{1}{t_2 - t_1} \ln \frac{p_s - p_2}{p_s - p_1} \tag{5.4b}$$

The future population $P$ is

$$P = P_0 + (P_s - P_0)(1 - e^{-k_d t}) \tag{5.4c}$$

where $P_0$ = population of the base year

## 2.4 Logistic curve method

The logistic curve-fitting method is used for modeling population trends with an S-shape for large population center, or nations for long-term population predictions. The logistic curve form is

$$P = \frac{P_s}{1 + e^{a + b\Delta t}} \tag{5.5a}$$

where $P_s$ = saturation population

$a, b$ = constants

They are

$$P_s = \frac{2p_0p_1p_2 - p_1^2(p_0 + p_2)}{p_0p_2 - p_1^2} \tag{5.5b}$$

$$a = \ln \frac{p_s - p_0}{p_0} \tag{5.6}$$

$$b = \frac{1}{n} \ln \frac{p_0(p_s - p_1)}{p_1(p_s - p_0)} \tag{5.7}$$

where $n$ = time interval between successive censuses

Substitution of these values in Eq. (5.5a) gives the estimation of future population on $P$ for any period $\Delta t$ beyond the base year corresponding to $P_0$.

**Example:** A mid size city recorded populations of 113,000 and 129,000 in the April 1980 and April 1990 census, respectively. Estimate the population in January 1999 by comparing (a) arithmetic method, (b) constant percentage method, and (c) declining growth method.

**solution:**

Step 1. Solve with the arithmetic method

Let $t_1$ and $t_2$ for April 1980 and April 1990, respectively

$$\Delta t = t_2 - t_1 = 10 \text{ years}$$

Using Eq. (5.2 a) $\quad k_a = \dfrac{p_2 - p_1}{t_2 - t_1} = \dfrac{129{,}000 - 113{,}000}{10} = 1600$

Predict $p_t$ for January 1999 from $t_2$, using Eq. (5.2b)

$$\begin{aligned} t &= 8.75 \text{ years} \\ p_t &= p_2 + k_a t \\ &= 129{,}000 + 1600 \times 8.75 \\ &= 143{,}000 \end{aligned}$$

Step 2. Solve with constant percentage method, using Eq. (5.3b)

$$\begin{aligned} k_p &= \frac{\ln p_2 - \ln p_1}{t_2 - t_1} = \frac{\ln 129{,}000 - \ln 113{,}000}{10} \\ &= 0.013243 \end{aligned}$$

Then using Eq. (5.3c) $\quad$

$$\begin{aligned} \ln P &= \ln P_2 + k_p (t - t_2) \\ &= \ln 129{,}000 + 0.013243 \times 8.75 \\ &= 11.8834 \\ p &= 144{,}800 \end{aligned}$$

Step 3. Solve with declining growth method
Assuming

$$p_s = 200{,}000$$

Using Eq. (5.4b)

$$k_d = -\frac{1}{t_2 - t_1}\ln\frac{p_s - p_2}{p_s - p_1}$$

$$= -\frac{1}{10}\ln\frac{200{,}000 - 129{,}000}{200{,}000 - 113{,}000}$$

$$= 0.02032$$

From Eq. (5.4c)

$$\begin{aligned} P &= P_0 + (P_s - P_0)(1 - e^{-k_d t}) \\ &= 129{,}000 + (200{,}000 - 129{,}000)(1 - e^{-0.02032 \times 8.75}) \\ &= 129{,}000 + 71{,}000 \times 0.163 \\ &= 140{,}600 \end{aligned}$$

## 3 Water Requirements

The uses of water include domestic, commercial and industrial, public services such as fire fighting and public buildings, and unaccounted pipeline system losses and leakage. The average usage in the United States for the above four categories are 220, 260, 30, and 90L per capita per day (L/(c · d)), respectively (Tchobanoglous and Schroeder 1985). These correspond to 58, 69, 8, and 24 gal/(c · d), respectively. Total municipal water use averages 600 L/(c · d) or 160 gal/(c · d) in the United States.

The maximum daily water use ranges from about 120% to 400% of the average daily use with a mean of about 180%. Maximum hourly use is about 150% to 12,000% of the annual average daily flow; and 250% to 270% are typically used in design.

### 3.1 Fire demand

Fire demand of water is often the determining factor in the design of mains. Distribution is a short-term, small quantity but with a large flow rate. According to uniform fire code, the minimum fire flow requirement for a one- and two-family dwelling shall be 1000 gal per min (gpm). For the water demand for fire fighting based on downtown business districts and high-value area for communities of 200,000 people or less, the National Board of Fire Underwriters (1974) recommended the following fire flow rate and population relationship:

$$Q = 3.86\sqrt{p}\,(1 - 0.01\sqrt{p}) \quad \text{(SI units)} \tag{5.8a}$$

$$Q = 1020\sqrt{p}\,(1 - 0.01\sqrt{p}) \quad \text{(US customary units)} \tag{5.8b}$$

where $Q$ = discharge, $m^3$/min or gal/min (gpm)
$P$ = population in thousands

The required flow rate for fire fighting must be available in addition to the coincident maximum daily flow rate. The duration during the required fire flow must be available for 4 to 10 h. National Board of Fire Underwriters recommends providing for a 10-h fire in towns exceeding 2500 in population.

The Insurance Services Office Guide (International Fire Service Training Association, 1993) for determination of required fire flow recommends the formula

$$F = 18\,C\sqrt{A} \quad \text{(US customary units)} \tag{5.9a}$$

$$F = 320\,C\sqrt{A} \quad \text{(SI units)} \tag{5.9b}$$

where $F$ = required fire flow, gpm or $m^3$/d
$C$ = coefficient related to the type of construction
$A$ = total floor area, $ft^2$ or $m^2$

| $C$ value | Construction | Maximum flow, gpm ($m^3$/d) |
|---|---|---|
| 1.5 | wood frame | 8000 (43,600) |
| 1.0 | ordinary | 8000 (43,600) |
| 0.9 | heavy timber type building | |
| 0.8 | noncombustible | 6000 (32,700) |
| 0.6 | fire-resistant | 6000 (32,700) |

**Example 1:** A 4-story building of heavy timber type building of 715 $m^2$ (7700 $ft^2$) of ground area. Calculate the water fire requirement.

**solution:** Using Eq. (5.9b)

$$\begin{aligned} F &= 320\,C\sqrt{A} \\ &= 320 \times 0.9\sqrt{4 \times 715} \\ &= 15{,}400 \text{ (m}^3\text{/d)} \\ \text{or} \quad &= 2800 \text{ (gpm)} \end{aligned}$$

**Example 2:** A 5-story building of ordinary construction of 7700 $ft^2$ (715 $m^2$) of ground area communicating with a 3-story building of ordinary construction of 9500 $ft^2$ (880 $m^2$) ground area. Compute the required fire flow.

**solution:** Using Eq. (5.9a)

Total area $A = 5 \times 7700 + 3 \times 9500 = 67{,}000 \text{ ft}^2$

$$F = 18 \times 1.0\sqrt{67{,}000}$$

$$= 4{,}660 \text{ (gpm)}$$

$$\cong 4{,}750 \text{ gpm rounded to nearest 250 gpm for design purpose}$$

or

$$= 25{,}900 \text{ (m}^3\text{/d)}$$

**Example 3:** Assuming a high-value residential area of 100 ha (247 acres) has a housing density of 10 houses/ha with 4 persons per household, determine the peak water demand, including fire, in this residential area.

**solution:**

Step 1. Estimate population $P$

$$P = \text{(4 capita/house) (10 houses/ha) (100 ha)}$$

$$= 4000 \text{ capita}$$

Step 2. Estimate average daily flow $Q_a$

$$Q_a = \text{residential + public service + unaccounted}$$

$$= (220 + 30 + 90)$$

$$= 340 \text{ (L/(c} \cdot \text{d))}$$

Step 3. Estimate maximum daily flow $Q_{md}$ for the whole area

Using the basis of $Q_{md}$ is 180% of $Q_a$

$$Q_{md} = (340 \text{ L/(c} \cdot \text{d))(1.8) (4000 c)}$$

$$= 2{,}448{,}000 \text{ L/d}$$

$$\cong 2400 \text{ m}^3\text{/d}$$

Step 4. Estimate the fire demand, using Eq. (5,8a)

$$Q_f = 3.86\sqrt{p}(1 - 0.01\sqrt{p}) \text{ m}^3\text{/min}$$

$$= 3.86\sqrt{4}(1 - 0.01\sqrt{4})$$

$$= 7.57 \text{ m}^3\text{/min}$$

$$= 7.57 \text{ m}^3\text{/min} \times 60 \text{ min/h} \times 10 \text{ h/d}$$

$$= 4540 \text{ m}^3\text{/d}$$

Step 5. Estimate total water demand $Q$

$$Q = Q_{md} + Q_f$$

$$= 2400 \text{ m}^3\text{/d} + 4540 \text{ m}^3\text{/d}$$

$$= 6940 \text{ m}^3\text{/d}$$

*Note*: In this area, fire demand is a control factor. It is measured to compare $Q$ and peak daily demand.

Step 6. Check with maximum hourly demand $Q_{mh}$

The $Q_{mh}$ is assumed to be 250% of average daily demand.

$$Q_{mh} = 2400 \text{ m}^3/\text{d} \times 2.5 = 6000 \text{ m}^3/\text{d}$$

Step 7. Compare $Q$ versus $Q_{mh}$

$$Q = 6940 \text{ m}^3/\text{d} > Q_{mh} = 6000 \text{ m}^3/\text{d}$$

Use $Q = 6940 \text{ m}^3/\text{d}$ for the main pipe to this residential area.

**Example 4:** Estimate the municipal water demands for a city of 225,000 persons.

**solution:**

Step 1. Estimate the average daily demand $Q_{avg}$

$$\begin{aligned} Q_{avg} &= 600 \text{ L/(c·d)} \times 225{,}000 \text{ c} \\ &= 135{,}000{,}000 \text{ L/d} \\ &= 1.35 \times 10^5 \text{ m}^3/\text{d} \\ &= 25.7 \text{ MGD} \end{aligned}$$

Step 2. Estimate the maximum daily demand $Q_{md}$ ($f = 1.8$)

$$\begin{aligned} Q_{md} &= 1.35 \times 10^5 \text{ m}^3/\text{d} \times 1.8 \\ &= 2.43 \times 10^5 \text{ m}^3/\text{d} \\ &= 64.2 \text{ MGD} \end{aligned}$$

Step 3. Calculate the fire demand $Q_f$, using Eq. (5.8a)

$$\begin{aligned} Q_f &= 3.86\sqrt{p}\,(1 - 0.01\sqrt{p}) \text{ m}^3/\text{min} \\ &= 3.86\sqrt{225}\,(1 - 0.01\sqrt{225})\,(\text{m}^3/\text{min}) \\ &= 49.215 \text{ m}^3/\text{min} \end{aligned}$$

For 10-h duration of daily rate

$$Q_f = 49.215 \text{ m}^3/\text{min} \times 60 \text{ min/h} \times 10 \text{ h/d} = 0.30 \times 10^5 \text{ m}^3/\text{d}$$

Step 4. Sum of $Q_{md}$ and $Q_f$ (fire occurs coincident to peak flow)

$$\begin{aligned} Q_{md} + Q_f &= (2.43 + 0.30) \times 10^5 \text{ m}^3/\text{d} \\ &= 2.73 \times 10^5 \text{ m}^3/\text{d} \\ &= 72.1 \text{ MGD} \end{aligned}$$

Step 5. Estimate the maximum hourly demand $Q_{mh}$ ($f = 2.7$)

$$\begin{aligned} Q_{mh} &= f\,Q_{avg} \\ &= 2.7 \times 1.35 \times 10^5 \text{ m}^3/\text{d} \\ &= 3.645 \times 10^5 \text{ m}^3/\text{d} \\ &= 96.3 \text{ MGD} \end{aligned}$$

Step 6. Compare Steps 4 and 5

$$Q_{mh} = 3.645 \times 10^5 \text{ m}^3/\text{d} > Q_{md} + Q_f = 2.73 \times 10^5 \text{ m}^3/\text{d}$$

Use $Q_{mh} = 3.65 \times 10^5$ m$^3$/d (96.4 MGD) to design the plant's storage capacity.

**Fire flow tests.** Fire flow tests involve discharging water at a known flow rate from one or more hydrants and examining the corresponding pressure drop from the main through another nearby hydrant. The discharge from a hydrant nozzle can be determined as follows (Hammer, 1975):

$$Q = 29.8\,C\,d^2\sqrt{p} \tag{5.10}$$

where $Q$ = hydrant discharge, gpm
$C$ = coefficient, normally 0.9
$d$ = diameter of outlet, in
$p$ = pitot gage reading, psi

The computed discharge at a specified residual pressure can be computed as

$$\frac{Q_p}{Q_f} = \left(\frac{\Delta H_p}{\Delta H_f}\right)^{0.54} \tag{5.11a}$$

or

$$Q_p = Q_f\left(\frac{\Delta H_p}{\Delta H_f}\right)^{0.54} \tag{5.11b}$$

where $Q_p$ = computed flow rate at the specified residual pressure, gpm
$Q_f$ = total discharge during fire flow test, gpm
$\Delta H_p$ = pressure drop from beginning to special pressure, psi
$\Delta H_f$ = pressure drop during fire flow test, psi

**Example:** All four hydrant numbers 2, 4, 6, and 8, as shown in the following figure, have the same nozzle size of 2.5 in and discharge at the same rate. The pitot tube pressure at each hydrant during the test is 25 psi. At this discharge the residual pressure at hydrant #5 dropped from the original

100 to 65 psi. Compute the flow rate at a residual pressure of 30 psi based on the test.

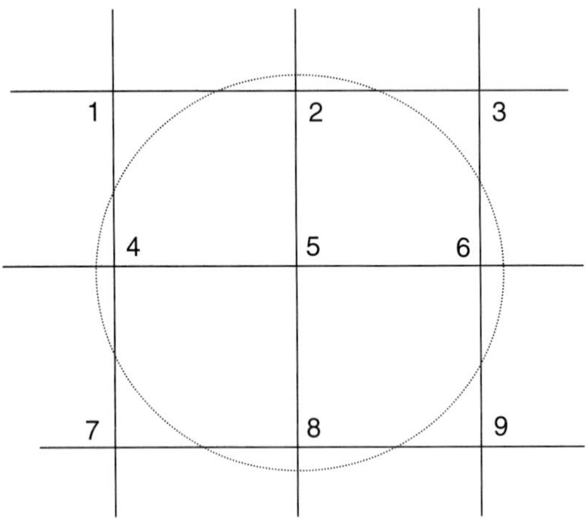


**solution:**

Step 1. Determine total discharge during the test

Using Eq. (5.10)

$$
\begin{aligned}
Q_f &= 4 \times 29.8\, C d^2 \sqrt{p} = 4 \times 29.8 \times 0.9 \times (2.5)^2 \times \sqrt{25} \\
&= 3353 \text{ (gpm)}
\end{aligned}
$$

Step 2. Compute $Q_p$ at 30 psi by using Eq. (5.11b)

$$
\begin{aligned}
Q_p &= Q_f \left(\frac{\Delta H_p}{\Delta H_f}\right)^{0.54} = 3353 \times \left(\frac{100 - 30}{100 - 65}\right)^{0.54} \\
&= 4875 \text{ (gpm)}
\end{aligned}
$$

### 3.2 Leakage test

After a new main is installed, a leakage test is usually required. The testing is generally carried out for a pipe length not exceeding 300 m (1000 ft). The pipe is filled with water and pressurized (50% above normal operation pressure) for at least 30 min. The leakage limit recommended by the American Water Work Association is determined as:

$$L = (ND\sqrt{p})/326 \qquad \text{for SI units} \qquad (5.12a)$$

$$L = (ND\sqrt{P})/1850 \qquad \text{for English units} \qquad (5.12b)$$

where $L$ = allowable leakage, L/(mm diameter · km · h) or gal/(in · mi · h)
$N$ = number of joints in the test line
$D$ = normal diameter of the pipe, mm or in
$P$ = average test pressure

Leakage allowed in a new main is generally specified in the design. It ranges from 5.6 to 23.2 L/mm diameter per km per day (60 to 250 gal/in diameter/miles/d). Nevertheless, recently, some water companies are not allowing for any leakage in a new main due to the use of better sealers.

## 4 Regulations for Water Quality

Water quality is a term used to describe the physical, chemical, and biological characteristics of water with respect to its suitability for a particular use. In the United States, all federal drinking water standards are set under the Safe Drinking Water Act (SDWA).

### 4.1 Safe drinking water act

The Safe Drinking Water Act was originally passed by the Congress in 1974. The SDWA gives the US Environmental Protection Agency (EPA) the authority to set drinking water standards to protect public health by regulating public drinking water supply. It was amended in 1986 (interim) and 1996.

There are two categories of federal drinking water standards: National Primary Drinking Water Regulations (Table 5.1) and National Secondary Drinking Water Regulations (Table 5.2). The Maximum Contaminant Level Goals (MCLGs) for chemicals suspected or known to cause cancer in humans are set to zero. If a chemical is carcinogenic and a safe dose is determined, the MCLG is set at a level above zero that is safe. For microbial concentrations (*Giardia, Legionella,* TC, FC, *E. coli,* and *ciruses*) the MCLGs are set at zero.

The SDWA regulates every public water supply in the United States (160,000 plants) with some variances and does not apply to water systems that serve fewer than 25 individuals. The SDWA established a multiple barrier protection for drinking water. It starts watershed management to protect drinking water as well as its water sources (lakes, reservoirs, rivers, streams, springs, and groundwater). It also requires assessing and protecting collection systems, properly treated by qualified operators, the integrity of the distribution systems, monitoring for the regulated and unregulated contaminants, providing information available to the public on the drinking water quality, and submitting the required operational reports to the state EPA. The state EPA is responsible for enforcement, technical assistances, granting variances, and even financial assistance

**TABLE 5.1 National Primary Drinking Water Regulations**

| Contaminant | MCLG[1] (mg/L)[2] | MCL or TT[1] (mg/L)[2] | Potential health effects from ingestion of water | Sources of contaminant in drinking water |
|---|---|---|---|---|
| **Microorganisms** | | | | |
| Cryptosporidium | zero | TT [3] | Gastrointestinal illness (e.g. diarrhea, vomiting, cramps) | Human and fecal animal waste |
| *Giardia lamblia* | zero | TT[3] | Gastrointestinal illness (e.g. diarrhea, vomiting, cramps) | Human and animal fecal waste |
| Heterotrophic plate count | n/a | TT[3] | HPC has no health effects; it is an analytic method used to measure the variety of bacteria that are common in water. The lower the concentration of bacteria in drinking water, the better maintained the water system is | HPC measures a range of bacteria that are naturally present in the environment |
| *Legionella* | zero | TT[3] | Legionnaire's disease, a type of pneumonia | Found naturally in water; multiplies in heating systems |
| Total coliforms (including fecal coliform and *E. Coli*) | zero | 5.0%[4] | Not a health threat in itself; it is used to indicate whether other potentially harmful bacteria may be present[5] | Coliforms are naturally present in the environment; as well as feces; fecal coliforms and *E. coli* only come from human and animal fecal waste |
| Turbidity | n/a | TT[3] | Turbidity is a measure of the cloudiness of water. It is used to indicate water quality and filtration effectiveness (e.g. whether disease-causing organisms are present). Higher turbidity levels are often associated with higher levels of disease-causing microorganisms such as viruses, parasites, and some bacteria. These organisms can cause symptoms such as nausea, cramps, diarrhea, and associated headaches | Soil runoff |
| Viruses (enteric) | zero | TT[3] | Gastrointestinal illness (e.g. diarrhea, vomiting, cramps) | Human and animal fecal waste |

| Contaminant | MCLG[1] (mg/L)[2] | MCL or TT[1] (mg/L)[2] | Potential health effects from ingestion of water | Sources of contaminant in drinking water |
|---|---|---|---|---|
| **Disinfection Byproducts** | | | | |
| Bromate | zero | 0.010 | Increased risk of cancer | By-product of drinking water disinfection |
| Chlorite | 0.8 | 1.0 | Anemia; infants and young children: nervous system effects | By-product of drinking water disinfection |
| Haloacetic acids (HAA5) | n/a[6] | 0.060 | Increased risk of cancer | By-product of drinking water disinfection |
| Total Trihalomethanes (TTHMs) | none[7] — n/a[6] | 0.10 — 0.080 | Liver, kidney, or central nervous system problems; increased risk of cancer | By-product of drinking water disinfection |

| Contaminant | MRDLG[1] (mg/L)[2] | MRDL[1] (mg/L)[2] | Potential health effects from ingestion of water | Sources of contaminant in drinking water |
|---|---|---|---|---|
| **Disinfectants** | | | | |
| Chloramines (as $Cl_2$) | MRDLG=4[1] | MRDL=4.0[1] | Eye/nose irritation; stomach discomfort, anemia | Water additive used to control microbes |
| Chlorine (as $Cl_2$) | MRDLG=4[1] | MRDL=4.0[1] | Eye/nose irritation; stomach discomfort | Water additive used to control microbes |
| Chlorine dioxide (as $ClO_2$) | MRDLG=0.8[1] | MRDL=0.8[1] | Anemia; infants and young children: nervous system effects | Water additive used to control microbes |

| Contaminant | MCLG[1] (mg/L)[2] | MCL or TT[1] (mg/L)[2] | Potential health effects from ingestion of water | Sources of contaminant in drinking water |
|---|---|---|---|---|
| **Inorganic Chemicals** | | | | |
| Antimony | 0.006 | 0.006 | Increase in blood cholesterol; decrease in blood sugar | Discharge from petroleum refineries; fire retardants; ceramics; electronics; solder |
| Arsenic | 0[7] | 0.010 as of 01/23/06 | Skin damage or problems with circulatory systems, and may have increased risk of getting cancer | Erosion of natural deposits; runoff from orchards, runoff from glass, and electronics-production wastes |

*(Continued)*

TABLE 5.1 (*continued*)

| Contaminant | MCLG[1] (mg/L)[2] | MCL or TT[1] (mg/L)[2] | Potential health effects from ingestion of water | Sources of contaminant in drinking water |
|---|---|---|---|---|
| Asbestos (fiber >10 $\mu$m) | 7 million fibers per liter | 7 MFL | Increased risk of developing benign intestinal polyps | Decay of asbestos cement in water mains; erosion of natural deposits |
| Barium | 2 | 2 | Increase in blood pressure | Discharge of drilling wastes; discharge from metal refineries; erosion of natural deposits |
| Beryllium | 0.004 | 0.004 | Intestinal lesions | Discharge from metal refineries and coal-burning factories; discharge from electrical, aerospace, and defense industries |
| Cadmium | 0.005 | 0.005 | Kidney damage | Corrosion of galvanized pipes; erosion of natural deposits; discharge from metal refineries; runoff from waste batteries and paints |
| Chromium (total) | 0.1 | 0.1 | Allergic dermatitis | Discharge from steel and pulp mills; erosion of natural deposits |
| Copper | 1.3 | TT[8]; Action Level=1.3 | Short term exposure: Gastrointestinal distress<br>Long-term exposure: Liver or kidney damage<br>People with Wilson's disease should consult their personal doctor if the amount of copper in their water exceeds the action level | Corrosion of household plumbing systems; erosion of natural deposits |
| Cyanide (as free cyanide) | 0.2 | 0.2 | Nerve damage or thyroid problems | Discharge from steel/metal factories; discharge from plastic and fertilizer factories |

| Contaminant | MCLG[1] (mg/L)[2] | MCL or TT[1] (mg/L)[2] | Potential health effects from ingestion of water | Sources of contaminant in drinking water |
|---|---|---|---|---|
| Fluoride | 4.0 | 4.0 | Bone disease (pain and tenderness of the bones); Children may get mottled teeth | Water additive which promotes strong teeth; erosion of natural deposits; discharge from fertilizer and aluminum factories |
| Lead | zero | TT[8]; Action Level =0.015 | Infants and children: Delays in physical or mental development; children could show slight deficits in attention span and learning abilities<br>Adults: Kidney problems; high blood pressure | Corrosion of household plumbing systems; erosion of natural deposits |
| Mercury (inorganic) | 0.002 | 0.002 | Kidney damage | Erosion of natural deposits; discharge from refineries and factories; runoff from landfills and croplands |
| Nitrate (measured as nitrogen) | 10 | 10 | Infants below the age of 6 months who drink water containing nitrate in excess of the MCL could become seriously ill and, if untreated, may die. Symptoms include shortness of breath and blue-baby syndrome | Runoff from fertilizer use; leaching from septic tanks, sewage; erosion of natural deposits |
| Nitrite (measured as nitrogen) | 1 | 1 | Infants below the age of 6 months who drink water containing nitrite in excess of the MCL could become seriously ill and, if untreated, may die. Symptoms include shortness of breath and blue-baby syndrome | Runoff from fertilizer use; leaching from septic tanks, sewage; erosion of natural deposits |
| Selenium | 0.05 | 0.05 | Hair or fingernail loss; numbness in fingers or toes; circulatory problems | Discharge from petroleum refineries; erosion of natural deposits; discharge from mines |

*(Continued)*

TABLE 5.1 *(continued)*

| Contaminant | MCLG[1] (mg/L)[2] | MCL or TT[1] (mg/L)[2] | Potential health effects from ingestion of water | Sources of contaminant in drinking water |
|---|---|---|---|---|
| Thallium | 0.0005 | 0.002 | Hair loss; changes in blood; kidney, intestine, or liver problems | Leaching from ore-processing sites; discharge from electronics, glass, and drug factories |
| **Organic Chemicals** | | | | |
| Acrylamide | zero | TT[9] | Nervous system or blood problems; increased risk of cancer | Added to water during sewage/ wastewater treatment |
| Alachlor | zero | 0.002 | Eye, liver, kidney, or spleen problems; anemia; increased risk of cancer | Runoff from herbicide used on row crops |
| Atrazine | 0.003 | 0.003 | Cardiovascular system or reproductive problems | Runoff from herbicide used on row crops |
| Benzene | zero | 0.005 | Anemia; decrease in blood platelets; increased risk of cancer | Discharge from factories; leaching from gas storage tanks and landfills |
| Benzo(a)pyrene (PAHs) | zero | 0.0002 | Reproductive difficulties; increased risk of cancer | Leaching from linings of water storage tanks and distribution lines |
| Carbofuran | 0.04 | 0.04 | Problems with blood, nervous system, or reproductive system | Leaching of soil fumigant used on rice and alfalfa |
| Carbon tetrachloride | zero | 0.005 | Liver problems; increased risk of cancer | Discharge from chemical plants and other industrial activities |
| Chlordane | zero | 0.002 | Liver or nervous system problems; increased risk of cancer | Residue of banned termiticide |
| Chlorobenzene | 0.1 | 0.1 | Liver or kidney problems | Discharge from chemical and agricultural chemical factories |
| 2,4-D | 0.07 | 0.07 | Kidney, liver, or adrenal gland problems | Runoff from herbicide used on row crops |

| Contaminant | MCLG[1] (mg/L)[2] | MCL or TT[1] (mg/L)[2] | Potential health effects from ingestion of water | Sources of contaminant in drinking water |
|---|---|---|---|---|
| Dalapon | 0.2 | 0.2 | Minor kidney changes | Runoff from herbicide used on rights of way |
| 1,2-Dibromo-3-chloropropane (DBCP) | zero | 0.0002 | Reproductive difficulties; increased risk of cancer | Runoff/leaching from soil fumigant used on soybeans, cotton, pineapples, and orchards |
| o-Dichlorobenzene | 0.6 | 0.6 | Liver, kidney, or circulatory system problems | Discharge from industrial chemical factories |
| p-Dichlorobenzene | 0.075 | 0.075 | Anemia; liver, kidney, or spleen damage; changes in blood | Discharge from industrial chemical factories |
| 1,2-Dichloroethane | zero | 0.005 | Increased risk of cancer | Discharge from industrial chemical factories |
| 1,1-Dichloroethylene | 0.007 | 0.007 | Liver problems | Discharge from industrial chemical factories |
| cis-1,2-Dichloroethylene | 0.07 | 0.07 | Liver problems | Discharge from industrial chemical factories |
| trans-1,2-Dichloroethylene | 0.1 | 0.1 | Liver problems | Discharge from industrial chemical factories |
| Dichloromethane | zero | 0.005 | Liver problems; increased risk of cancer | Discharge from drug and chemical factories |
| 1,2-Dichloropropane | zero | 0.005 | Increased risk of cancer | Discharge from industrial chemical factories |
| Di(2-ethylhexyl) adipate | 0.4 | 0.4 | Weight loss, liver problems, or possible reproductive difficulties | Discharge from chemical factories |
| Di(2-ethylhexyl) phthalate | zero | 0.006 | Reproductive difficulties; liver problems; increased risk of cancer | Discharge from rubber and chemical factories |
| Dinoseb | 0.007 | 0.007 | Reproductive difficulties | Runoff from herbicide used on soybeans and vegetables |

*(Continued)*

**TABLE 5.1** ***(continued)***

| Contaminant | MCLG[1] (mg/L)[2] | MCL or TT[1] (mg/L)[2] | Potential health effects from ingestion of water | Sources of contaminant in drinking water |
|---|---|---|---|---|
| Dioxin (2,3,7,8-TCDD) | zero | 0.00000003 | Reproductive difficulties; increased risk of cancer | Emissions from waste incineration and other combustion; discharge from chemical factories |
| Diquat | 0.02 | 0.02 | Cataracts | Runoff from herbicide use |
| Endothall | 0.1 | 0.1 | Stomach and intestinal problems | Runoff from herbicide use |
| Endrin | 0.002 | 0.002 | Liver problems | Residue of banned insecticide |
| Epichlorohydrin | zero | TT[9] | Increased cancer risk, and over a long period of time, stomach problems | Discharge from industrial chemical factories; an impurity of some water treatment chemicals |
| Ethylbenzene | 0.7 | 0.7 | Liver or kidneys problems | Discharge from petroleum refineries |
| Ethylene dibromide | zero | 0.00005 | Problems with liver, stomach, reproductive system, or kidneys; increased risk of cancer | Discharge from petroleum refineries |
| Glyphosate | 0.7 | 0.7 | Kidney problems; reproductive difficulties | Runoff from herbicide use |
| Heptachlor | zero | 0.0004 | Liver damage; increased risk of cancer | Residue of banned termiticide |
| Heptachlor epoxide | zero | 0.0002 | Liver damage; increased risk of cancer | Breakdown of heptachlor |
| Hexachlorobenzene | zero | 0.001 | Liver or kidney problems; reproductive difficulties; increased risk of cancer | Discharge from metal refineries and agricultural chemical factories |
| Hexachlorocyclopentadiene | 0.05 | 0.05 | Kidney or stomach problems | Discharge from chemical factories |
| Lindane | 0.0002 | 0.0002 | Liver or kidney problems | Runoff/leaching from insecticide used on cattle, lumber, gardens |

| Contaminant | MCLG[1] (mg/L)[2] | MCL or TT[1] (mg/L)[2] | Potential health effects from ingestion of water | Sources of contaminant in drinking water |
|---|---|---|---|---|
| Methoxychlor | 0.04 | 0.04 | Reproductive difficulties | Runoff/leaching from insecticide used on fruits, vegetables, alfalfa, livestock |
| Oxamyl (Vydate) | 0.2 | 0.2 | Slight nervous system effects | Runoff/leaching from insecticide used on apples, potatoes, and tomatoes |
| Polychlorinated biphenyls (PCBs) | zero | 0.0005 | Skin changes; thymus gland problems; immune deficiencies; reproductive or nervous system difficulties; increased risk of cancer | Runoff from landfills; discharge of waste chemicals |
| Pentachlorophenol | zero | 0.001 | Liver or kidney problems; increased cancer risk | Discharge from wood preserving factories |
| Picloram | 0.5 | 0.5 | Liver problems | Herbicide runoff |
| Simazine | 0.004 | 0.004 | Problems with blood | Herbicide runoff |
| Styrene | 0.1 | 0.1 | Liver, kidney, or circulatory system problems | Discharge from rubber and plastic factories; leaching from landfills |
| Tetrachloroethylene | zero | 0.005 | Liver problems; increased risk of cancer | Discharge from factories and dry cleaners |
| Toluene | 1 | 1 | Nervous system, kidney, or liver problems | Discharge from petroleum factories |
| Toxaphene | zero | 0.003 | Kidney, liver, or thyroid problems; increased risk of cancer | Runoff/leaching from insecticide used on cotton and cattle |
| 2,4,5-TP (Silvex) | 0.05 | 0.05 | Liver problems | Residue of banned herbicide |
| 1,2,4-Trichlorobenzene | 0.07 | 0.07 | Changes in adrenal glands | Discharge from textile finishing factories |
| 1,1,1-Trichloroethane | 0.20 | 0.2 | Liver, nervous system, or circulatory problems | Discharge from metal degreasing sites and other factories |
| 1,1,2-Trichloroethane | 0.003 | 0.005 | Liver, kidney, or immune system problems | Discharge from industrial chemical factories |

*(Continued)*

**TABLE 5.1 (*continued*)**

| Contaminant | MCLG[1] (mg/L)[2] | MCL or TT[1] (mg/L)[2] | Potential health effects from ingestion of water | Sources of contaminant in drinking water |
|---|---|---|---|---|
| Trichloroethylene | zero | 0.005 | Liver problems; increased risk of cancer | Discharge from metal degreasing sites and other factories |
| Vinyl chloride | zero | 0.002 | Increased risk of cancer | Leaching from PVC pipes; discharge from plastic factories |
| Xylenes (total) | 10 | 10 | Nervous system damage | Discharge from petroleum factories; discharge from chemical factories |
| **Radionuclides** | | | | |
| Alpha particles | none[7] — zero | 15 picocuries per Liter (pCi/L) | Increased risk of cancer | Erosion of natural deposits of certain minerals that are radioactive and may emit a form of radiation known as alpha radiation |
| Beta particles and photon emitters | none[7] — zero | 4 millirems per year | Increased risk of cancer | Decay of natural and man-made deposits of certain minerals that are radioactive and may emit forms of radiation known as photons and beta radiation |
| Radium 226 and Radium 228 (combined) | none[7] — zero | 5 pCi/L | Increased risk of cancer | Erosion of natural deposits |
| Uranium | zero | 30 $\mu$g/L as of 12/08/03 | Increased risk of cancer, kidney toxicity | Erosion of natural deposits |

NOTES

[1] Definitions:

Maximum Contaminant Level (MCL)—The highest level of a contaminant that is allowed in drinking water. MCLs are set as close to MCLGs as feasible using the best available treatment technology and taking cost into consideration. MCLs are enforceable standards.

Maximum Contaminant Level Goal (MCLG)—The level of a contaminant in drinking water below which there is no known or expected risk to health. MCLGs allow for a margin of safety and are nonenforceable public health goals.

Maximum Residual Disinfectant Level (MRDL)—The highest level of a disinfectant allowed in drinking water. There is convincing evidence that addition of a disinfectant is necessary for control of microbial contaminants.

Maximum Residual Disinfectant Level Goal (MRDLG)—The level of a drinking water disinfectant below which there is no known or expected risk to health. MRDLGs do not reflect the benefits of the use of disinfectants to control microbial contaminants.

Treatment Technique—A required process intended to reduce the level of a contaminant in drinking water.

[2] Units are in milligrams per liter (mg/L) unless otherwise noted. Milligrams per liter are equivalent to parts per million.

[3] EPA's surface water treatment rules require systems using surface water or groundwater under the direct influence of surface water to: (1) disinfect their water, and (2) filter their water or meet criteria for avoiding filtration so that the following contaminants are controlled at the following levels:

- Cryptosporidium: (as of 1/1/02 for systems serving >10,000 and 1/14/05 for systems serving <10,000) 99% removal.
- *Giardia lamblia:* 99.9% removal/inactivation.
- Viruses: 99.99% removal/inactivation.
- *Legionella*: No limit, but EPA believes that if *Giardia* and viruses are removed/inactivated, *Legionella* will also be controlled.
- Turbidity: At no time can turbidity (cloudiness of water) go above 5 nephelolometric turbidity units (NTU); systems that filter must ensure that the turbidity go no higher than 1 NTU (0.5 NTU for conventional or direct filtration) in at least 95% of the daily samples in any month. As of January 1, 2002, turbidity may never exceed 1 NTU, and must not exceed 0.3 NTU in 95% of daily samples in any month.
- HPC: No more than 500 bacterial colonies per milliliter.
- Long Term 1 Enhanced Surface Water Treatment (Effective Date: January 14, 2005): Surface water systems or (GWUDI) systems serving fewer than 10,000 people must comply with the applicable Long Term 1 Enhanced Surface Water Treatment Rule provisions (e.g. turbidity standards, individual filter monitoring, Cryptosporidium removal requirements, updated watershed control requirements for unfiltered systems).
- Filter Backwash Recycling: The Filter Backwash Recycling Rule requires systems that recycle to return specific recycle flows through all processes of the system's existing conventional or direct filtration system or at an alternate location approved by the state.

[4] More than 5.0% samples total coliform-positive in a month. (For water systems that collect fewer than 40 routine samples per month, no more than one sample can be total coliform-positive per month.) Every sample that has total coliform must be analyzed for either fecal coliforms or *E. coli* if two consecutive TC-positive samples, and one is also positive for *E.coli* fecal coliforms, system has an acute MCL violation.

[5] Fecal coliform and *E. coli* are bacteria whose presence indicates that the water may be contaminated with human or animal wastes. Disease-causing microbes (pathogens) in these wastes can cause diarrhea, cramps, nausea, headaches, or other symptoms. These pathogens may pose a special health risk for infants, young children, and people with severely compromised immune systems.

[6] Although there is no collective MCLG for this contaminant group, there are individual MCLGs for some of the individual contaminants:

- Trihalomethanes: Bromodichloromethane (zero); bromoform (zero); dibromochloromethane (0.06 mg/L). Chloroform is regulated with this group but has no MCLG.
- Haloacetic acids: Dichloroacetic acid (zero); trichloroacetic acid (0.3 mg/L). Monochloroacetic acid, bromoacetic acid, and dibromoacetic acid are regulated with this group but have no MCLGs.

[7] MCLGs were not established before the 1986 Amendments to the Safe Drinking Water Act. Therefore, there is no MCLG for this contaminant.

[8] Lead and copper are regulated by a treatment technique that requires systems to control the corrosiveness of their water. If more than 10% of tap water samples exceed the action level, water systems must take additional steps. For copper, the action level is 1.3 mg/L, and for lead is 0.015 mg/L.

[9] Each water system must certify, in writing, to the state (using third-party or manufacturer's certification) that when acrylamide and epichlorohydrin are used in drinking water systems, the combination (or product) of dose and monomer level does not exceed the levels specified, as follows:

- Acrylamide = 0.05% dosed at 1 mg/L (or equivalent)
- Epichlorohydrin = 0.01% dosed at 20 mg/L (or equivalent)

SOURCE: www.epa.gov/OGWDW/regs.html, 2007

**TABLE 5.2 EPA's Three–Category Approach for Establishing MCLGs**

| Category | Evidence of carcinogenicity via ingestion | MCLG setting approach |
|---|---|---|
| I | Strong evidence considering weight of evidence, pharmacokinetics, and exposure | Zero |
| II | Limited evidence considering weight of evidence, pharmacokinetics, and exposure | RfD approach with added safety margin or $10^{-5}$ to $10^{-6}$ cancer risk range |
| III | Inadequate or no animal evidence | RfD approach |

for treatment process improvements. The SDWA Amendments prohibit the uses of lead pipes, lead solder, and lead flux.

Under the SDWA, the National Drinking Water Contaminant Candidate List (CCL) originally published on March 2, 1998 that listed 60 contaminants (50 chemicals or chemical groups and 10 microbial contaminants). The SDWA also requires the US EPA to review each primary standard at least once every 6 years for revision if needed. On April 2, 2004, the US EPA listed 50 contaminants (41 chemicals or chemical groups plus 9 microbials) as a draft CCL2.

**How MCLGs are developed.** MCLGs are set at concentration levels at which no known or anticipated adverse health effects would occur, allowing for an adequate margin of safety. Establishment of a specific MCLG depends on the evidence of carcinogenicity from drinking water exposure or the US EPA's reference dose (RfD), which is calculated for each specific contaminant. The drinking water standards are developed as follows (Federal Register, 1991).

The RfD is an estimate, with an uncertainty spanning perhaps an order of magnitude of a daily exposure to the human population (including sensitive subgroups) that is likely to be without an appreciable risk of deleterious health effects during a lifetime. The RfD is derived from a no- or lowest-observed-adverse-effect level (called a NOAEL or LOAEL, respectively) that has been identified from a subchronic or chronic scientific study of humans or animals. The NOAEL or LOAEL is then divided by the uncertainty factor to derive the RfD.

The use of an uncertainty factor is important in the derivation of the RfD. The US EPA has established certain guidelines (shown below) to determine which uncertainty factor should be used:

10—Valid experimental results for appropriate duration of human exposure.

100—Human data not available. Extrapolation from valid long-term animal studies.

1000—Human data not available. Extrapolation from animal studies of less than chronic exposure.

1 to 10—Additional safety factor for use of a LOAEL instead of a NOAEL.

Other—Other uncertainty factors are used according to scientific judgment when justified.

In general, an uncertainly factor is calculated to consider intra- and interspecies variations, limited or incomplete data, use of subchronic studies, significance of the adverse effect, and the pharmacokinetic factors.

From the RfD, a drinking water equivalent level (DWEL) is calculated by multiplying the RfD by an assumed adult body weight (generally 70 kg) and then dividing by an average daily water consumption of 2 L/d. The DWEL assumes the total daily exposure to a substance is from drinking water exposure. The MCLG is determined by multiplying the DWEL by the percentage of the total daily exposure contributed by drinking water, called the relative source contribution. Generally, EPA assumes that the relative source contribution from drinking water is 20% of the total exposure, unless other exposure data for the chemical are available. The calculation below expresses the derivation of the MCLG:

$$\mathrm{RfD} = \frac{\text{NOAEL or LOAEL}}{\text{uncertainty factor (UF)}} = \text{mg/kg body weight/d} \qquad (5.13a)$$

$$\mathrm{DWEL} = \frac{\mathrm{RfD} \times \text{body weight, kg}}{\text{daily water consumption (WC), L/d}} = \text{mg/L} \qquad \text{or}$$

$$\mathrm{DWEL} = \frac{\text{NOAEL or LOAEL} \times \text{body weight, kg}}{(\mathrm{UF}) \times (\text{WC, L/d})} = \text{mg/L} \qquad (5.13b)$$

$$\mathrm{MCLG} = \mathrm{DWEL} \times \text{drinking water contribution} \qquad (5.13c)$$

For chemicals suspected as carcinogens, the assessment for nonthreshold toxicants consists of the weight of evidence of carcinogenicity in humans, using bioassays in animals and human epidemiological studies as well as information that provides indirect evidence (i.e. mutagenicity and other short-term test results). The objectives of the assessment are (1) to determine the level or strength of evidence that the substance is a human or animal carcinogen and (2) to provide an upper-bound estimate of the possible risk of human exposure to the substance in drinking water. A summary of US EPA's carcinogen classification scheme is:

Group A—Human carcinogen based on sufficient evidence for epidemiological studies.

Group B1—Probable human carcinogen based on at least limited evidence of carcinogenicity to humans.

Group B2—Probable human carcinogen based on a combination of sufficient evidence in animals and inadequate data in humans.

Group C—Possible human carcinogen based on limited evidence of carcinogenicity in animals in the absence of human data.

Group D—Not classifiable due to lack of data or inadequate evidence of carcinogenicity from animal data.

Group E—No evidence of carcinogenicity for humans (no evidence for carcinogenicity in at lease two adequate animal tests in different species or in both epidemiological and animal studies).

Establishing the MCLG for a chemical is generally accomplished in one of three ways depending upon its categorization (Table 5.2). The starting point in EPA's analysis is the EPA's cancer classification (i.e. A, B, C, D, or E). Each chemical is analyzed for evidence.

EPA's policy is to set MCLGs for Category I contaminants at zero. The MCLG for Category II contaminants is calculated by using the RfD/DWEL with an added margin of safety to account for cancer effects, or is based on a cancer risk range of $10^{-5}$ to $10^{-6}$ when noncancer data are inadequate for deriving an RfD. The MCLG for Category III contaminants is calculated using the RfD/DWEL approach.

The MCLG for Category I contaminants is set at zero because it is assumed, in the absence of other data, that there is no known threshold. Category I contaminants are those contaminants for which EPA has determined that there is strong evidence of carcinogenicity from drinking water ingestion. If there is no additional information to consider on potential cancer risk from drinking water ingestion, chemicals classified as A or B carcinogens are placed in Category I.

Category II contaminants include those contaminants for which EPA has determined that there is limited evidence, pharmacokinetics, and exposure. If there is no additional information to consider on potential cancer risks from drinking water ingestion, chemicals classified by the Agency as Group C carcinogens are placed in Category II. For Category II contaminants, two approaches are used to set the MCLGs—either (1) setting the goal based upon noncarcinogenic endpoints (the RfD) and then applying an additional uncertainty (safety) factor of up to 10 or (2) setting the goals based upon a normal lifetime cancer risk calculation in the range of $10^{-5}$ to $10^{-6}$, using a conservative calculation model. The first approach is generally used. However, the second is used when valid noncarcinogenicity data are not available and adequate experimental data are available to quantify the cancer risk. EPA is currently evaluating the approach to establishing MCLGs for Category II contaminants.

Category III contaminants include those contaminants for which there is inadequate evidence of carcinogenicity via ingestion. If there is no additional information to consider, contaminants classified as Group D or E carcinogens are placed in Category III. For these contaminants, the MCLG is established using the RfD approach.

**Maximum contamination level or treatment technique.** Once the MCLG is defined, the US EPA sets an enforceable standard. In most cases, it is as maximum contamination level (MCL). The MCL is the maximum permissible concentration of a contaminant in water that is delivered to any user of a public water system. The MCL is set as close to the MCLG as feasible that the contaminant level may be achieved with the use of the best available technology, treatment techniques, and other means of corrections and improvement.

If there is no reliable method that is technically and economically feasible to measure a contaminant at particularly low levels, a TT (treatment technique) is set rather than an MCL. The TT is an enforceable procedure or level of technical performance that public water treatment systems must follow to ensure control of a contaminant.

The US EPA must conduct a cost-benefit analysis, after MCL or TT is determined. Based on the results of the cost-benefit analysis, the US EPA may justify any variances and exemptions from standards for some water utilities.

The SDWA required the US EPA to review each primary standard at least once every 6 years for revision if needed. The primary standards went onto effect 6 years after they were finalized.

### 4. 2 Two major rules updated in 2006

Over the past 8 years, the US EPA has worked collaboratively with stakeholders to develop regulations that will provide a balance between the needs to disinfect drinking water and to protect citizens from potentially harmful contaminants. Two of the most recent major rules finalized and published in 2006 are:

1. The Stage 2 Disinfectants and Disinfection By-products Rule (Stage 2 DBP rule or DBPR, Stage 2).
2. The Long Term 2 Enhanced Surface Water Treatment Rule (LT2 rule or LT2ESWTR).

More information about the US EPA drinking water regulations can be found at www.epa.gov/OGWDW/regs.html. This site, developed by the EPA's Office of Ground Water and Drinking Water, lists current drinking water rules, proposed rules, the code of federal regulations, drinking water standards, and a wealth of other information, such as fact sheets, guidance manuals, and more. Moorman (2006) summarized information below from various EPA documents helping to understand the immediate requirements and the overall purposes of these two new rules.

*Cryptosporidium* has caused serious disease outbreak and are resistant to traditional water disinfection practices. Consumption of water that contains *Cryptosporidium* can cause gastrointestinal illness that can be fatal for people with weakened immune systems. In addition, water chlorination can react with naturally occurring organic matters to form disinfection by-products (DBPs), which also pose health concerns.

To respond to these new threats, amendments to the SDWA in 1996 require the US EPA to develop rules to balance the risks between pathogens and DBPs. As the first phase of this requirement, the Stage 1 Disinfectants and Disinfection By-products Rule (Stage 1 DBP rule) was promulgated in December 1998 to reduce exposure to DBPs for customers of community water systems and nontransient noncommunity systems, including those serving fewer than 10,000 people, that add a disinfectant to the drinking water during any part of the treatment process.

The Interim Enhanced Surface Water Treatment Rule (IESWTR) was also promulgated in December 1998, but this rule was made to improve control of microbial contaminants, particularly *Cryptosporidium,* in systems using surface water, or groundwater under the direct influence (GWUDI) of surface water, that serve 10,000 or more persons. As the small-system counterpart to the IESWTR, the Long Term 1 Enhanced Surface Water Treatment Rule was finalized in January 2002 and strengthens control of microbial contaminants for systems serving fewer than 10,000 people.

**The Stage 2 DBP rule.** The Stage 2 DBP rule building upon the Stage 1 DBP rule was published in the Federal Register on January 4, 2006. This rule is intended to reduce potential cancer and reproductive and developmental health risks from DBPs in drinking water through more stringent methods for determining compliance.

**Requirements.** Under the Stage 2 DBP rule, community and nontransient noncommunity water systems that add and/or deliver water that is treated with a primary or residual disinfectant other than ultraviolet light must conduct an evaluation of their distribution systems, called an Initial Distribution System Evaluation (IDSE), to identify locations with high DBP concentrations. These locations must then be used by the systems as the sampling sites for Stage 2 DBP rule compliance monitoring. The rule requires two steps for the IDSE. First, the system must develop a plan of how they will approach the IDSE. There are four options: a very small system waiver, 40/30 certification, standard monitoring program (SMP), and system specific study (SSS). This plan must be submitted to the state or primary agency with proposed DBP monitoring sites. After receiving approval, the system must conduct the IDSE with 1 year of DBP monitoring if the standard option is used.

Compliance with the MCLs for trihalomethanes (THMs) and five haloacetic acids (HAA5) will be calculated as an average at each compliance-monitoring location (instead of as a system-wide average as in previous rules). This approach is referred to as the locational running annual average (LRAA).

The Stage 2 DBP rule also establishes operational evaluation levels. A system that exceeds this level must conduct an operational evaluation (a review of operational practices) to determine ways to reduce DBP concentrations. This provides an early warning of possible future MCL violations, and allows the system to take proactive steps to remain in compliance.

The Stage 2 DBP rule and the LT2 rule are the second phase of rules required by the Congress to help balance the risks between pathogens and DBPs. These rules strengthen protection against microbial contaminants, especially *Cryptosporidium,* and at the same time, reduce potential health risks of DBPs.

**Time frame.** The time frame of the Stage 2 DBP rule is a multiyear process for water systems to determine where higher levels of DBPs occur in their distribution systems. These locations will become the new DBP monitoring sites, and corrective action will be taken if DBP levels are above the MCL. These actions could include small operational changes up to major facility construction. Depending on system size and the extent of needed construction, systems will begin the first year of compliance monitoring between 2012 and 2015, and must be in compliance with the Stage 2 DBP rule MCLs at the end of a full year of monitoring (Table 5.3).

**Costs.** The Stage 2 DBP rule will result in increased costs to public water utilities and states. Although the rule applies to nearly 75,000 systems, only a small portion will be required to make treatment changes. The average cost of the rule is estimated at $79 million annually (using a 3% discount rate), according to the US EPA.

**The LT2 rule.** The LT2 rule was published in the Federal Register on June 5, 2006. This rule is intended to improve public health by reducing waterborne outbreaks due to *Cryptosporidium* and other pathogens in drinking water.

**Requirements.** The LT2 rule requires large public water systems (serving at least 10,000 people) that are supplied by surface water or GWUDI of surface water to conduct 2 years of monthly sampling for *Cryptosporidium* (and *E. coli* and turbidity for filtered systems). To

**TABLE 5.3 Timeframes for the Stage 2 DBP Rule and the Long Term 2 Enhanced Surface Water Treatment Rule**

| DBPR, Stage 2 | Year | LT2ESWTR |
|---|---|---|
| Proposed rules | 2004 | |
| | 2005 | |
| Final rules | 2006 | |
| IDSE[*] Plan to Primary Agency: | | ←A – 7/1/06, *Crypto* sample schedule due |
| Person served[†] A –10/1/06 → | | ←A – 10/1/06, Monitoring starts |
| B – 4/1/07 → | 2007 | ← B – 1/1/07, *Crypto* sample schedule due |
| C – 10/1/07 → | | ← B – 4/1/07, Monitoring starts |
| D – 4/1/08 → | 2008 | ← C – 1/1/08, *Crypto* sample schedule due |
| | | ← C – 4/1/08, Monitoring starts |
| IDSE Report due: A – 1/1/09 → | 2009 | ← D – 7/1/09, *Crypto* sample schedule due |
| B – 7/1/09 → | | ← D – 10/1/09, Monitoring starts |
| C – 1/1/10 → | 2010 | |
| D – 7/1/10 → | | |
| | 2011 | |
| Stage 2 LRAA starts (80/60 μg/L) at sampling site for: | | Additional *Cryptosporidium* treatment required if > 0.75 oocysts/L |
| | 2012 | |
| A – 4/1/12 → | | ← A – 4/1/12 |
| B – 10/1/12 → | | ← B – 10/1/12 |
| C – 10/1/13 → | 2013 | ← C – 10/1/13 |
| | 2014 | |
| D – 10/1/14 → | | ← D – 10/1/14 |
| | 2015 | |

[*]Other abbreviations—See the text
[†]Person served – A: ≥100,000; B: 50,000–99,999; C: 10,000–49,999; D: <10,000

reduce the cost of these monitoring, small water systems will monitor for *E. coli* for 1 year only because it is less expensive to analyze. The systems will only monitor for *Cryptosporidium* if their *E. coli* results exceed specified concentrations.

After monitoring is complete, filtered water systems will be classified in one of four treatment categories (bins) based on their monitoring results. The EPA expects the majority of systems to be placed in the lowest bin, and require no additional treatment. Systems placed in the higher bins will be required to provide additional treatment to reduce *Cryptosporidium* levels. Water systems in the higher bins and systems that do not filter their water will choose additional treatment options from a "microbial toolbox" of treatment and management processes.

Because of contamination risks, the LT2 rule also addresses uncovered finished water storage facilities, such as reservoirs. The rule requires systems that store treated water in uncovered facilities to either cover the facility or treat discharge to inactivate 4-log virus, 3-log *Giardia lamblia,* and 2-log *Cryptosporidium.*

The LT2 rule also establishes disinfection benchmarking, in which systems are required to review their current level of microbial treatment before making significant changes in disinfection practices. This will help systems maintain protection against microbial pathogens as they strive to reduce DBPs under the Stage 2 DBP rule.

**Time frame.** System-monitoring start dates correspond to system size. Approximately 3 months before the monitoring start date, systems must submit a sampling schedule and sampling location description. Systems serving at least 100,000 people should begin monitoring in October 2006. Those serving 50,000 to 99,000 people will begin monitoring in April 2007. Systems serving 10,000 to 49,000 people will begin monitoring in April 2008. The smallest systems (serving less than 10,000 people) will begin monitoring for *E. coli* in October 2008. If needed, additional *Cryptosporidium* treatment will start in April 2012 through October 2014 depending on the size of the system (Table 5.3.).

When these 2 years of monitoring are complete and bin classification is determined, systems will have approximately 3 years to implement any additional treatment requirements. Systems will then conduct a second round of monitoring 6 years after completing the initial round to determine if source water conditions have changed significantly.

**Costs.** The LT2 rule also will result in increased costs to public water utilities and states. The US EPA estimates that the average annualized present-value costs of the LT2 rule range from $92 to $133 million (using a 3% discount rate). Public water systems will be responsible for nearly all of the total cost (99%), and states will pick up the remaining 1%. The LT2 rule will also result in increased costs for households, with the US EPA estimating an average annual household cost of $1.67 to $2.59.

**Remark.** Singer (2006) stated that accordingly, all utilities should modify their monitoring programs to measure all nine bromine- and chlorine-containing HAA species. Even though they are only required (sum of five HAA) species, they will have a much better understanding of their system and the interrelationships among source water quality, treatment practices, and DBP formation if all nine species are measured. Furthermore, it is that in future versions of the DBP regulations—and most assuredly, there will be future versions—US EPA will regulate all nine bromine- and chlorine-containing THMs.

### 4.3 Compliance with standards

The US EPA has established a maximum contaminant level (MCL) of drinking water for a number of biological, physical, and chemical (inorganics and organics) parameters, and radioactivity (Table 5.1, National Primary Drinking Water Regulations). All the primary standards are based on health effects to the customers, and they are mandatory standards. The National Secondary Drinking Water Regulations (Table 5.2) pertain to those contaminants such as taste, odor, and color and some chemicals that may adversely affect the aesthetic quality of drinking water. They are intended as guidelines for the states. The states may establish higher or lower values as appropriate to their particular circumstances to adequately protect public health and welfare.

The design engineer should be thoroughly familiar with and integrate the various regulatory requirements into a water treatment design. In addition to the federal regulations stated above, the other rules and regulations pertaining to drinking water are the Surface Water Treatment Rule (SWTR), the revised Total Coliform Rule, the Lead and Copper Rule, consumer confidence report, Interim Enhanced SWTR, and the Groundwater Rule. It is necessary to obtain a permit from the state regulatory agency. Meetings with the representatives of the necessary regulatory agencies and related parties are recommended. The state review of the designs is based on considerations of minimum design requirements or standards, local conditions, and good engineering practice and experience. The most widely used standards are the Recommended Standards for Water Works promulgated by the Great Lakes-Upper Mississippi River Board of State Sanitary Engineers (GLUMRB) that are called the Ten-State Standards. This board includes representatives from the states of Illinois, Indiana, Iowa, Michigan, Minnesota, Missouri, New York, Ohio, Pennsylvania, and Wisconsin. Recently, the Province of Ontario (Canada) has joined the board. Many other states have used identical or similar versions of the Ten-State Standards.

The operational personnel at the drinking waterworks also have to be familiar and compliant with the regulations (especially the two rules updated in 2006) related to the system. They have to take more aggressive management and technical approaches to produce drinking water that meets all standards and requirements. For these safeguards to their water supply, consumers are expected to pay more than the current very low price.

Maximum contaminant level goal (MCLG) for water quality is set by the US Environmental Protection Agency (EPA). For agents in drinking water not considered to have carcinogenic potential, i.e. no effect levels for chromic and/or lifetime period of exposure including a margin of safety, are commonly referred to as reference dosages (RfDs).

Reference dosages are the exposure levels estimated to be without significant risk to humans when received daily over a lifetime. They are traditionally reported as mg/(kg·d). For MCLG purposes, however, no effect level in drinking water is measured by mg/L which has been in terms of the drinking water equivalent level (DWEL). The DWEL is calculated as (Cotruvo and Vogt, 1990)

$$\text{DWEL} = \frac{(\text{NOAEL})(70)}{(\text{UF})(2)} \tag{5.13}$$

where NOAEL = no observed adverse effect level, mg/(kg·d)
(70) = assumed weight of an adult, kg
UF = uncertainty factor (usually 10, 100, or 1000)
(2) = assumed quantity of water consumed by an adult, L/d

When sufficient data is available on the relative contribution from other sources, i.e. from food and air, the MCLG can be determined as follows:

$$\text{MCLG} = \text{RfD} - \text{contribution from (food + air)}$$

In fact, comprehensive data on the contributions from food and air are generally lacking. Therefore, in this case, MCLG is determined as MCLG = DWEL. The drinking water contribution often used in the absence of specific data for food and air is 20% of RfD. This effectively provides an additional safety factor of 5.

**Example:** The MCLG for nitrate–nitrogen is 10 mg/L. Assume UF = 100. What is the "no observed adverse effect level for the drinking water?"

**solution:** Using Eq. (5.13)

$$\text{DWEL} = \frac{(\text{NOAEL}) \times 70}{(\text{UF}) \times 2}$$

since

$$\text{DWEL} = \text{MCLG} = 10 \text{ mg/L}$$

$$10 \text{ mg/L} = \frac{(\text{NOAEL, mg/kg}\cdot\text{d})) \times (70 \text{ kg})}{100 \times 2 \text{ L/d}}$$

$$\text{NOAEL} = \frac{2000}{70}$$

$$= 28.6 \text{ (mg/(kg}\cdot\text{d))}$$

### 4.4 Atrazine

Atrazine is a widely used organic herbicide due to its being inexpensive compared to other chemicals. Often, excessive applications are used. When soil and climatic conditions are favorable, atrazine may be transported to the drinking water sources by runoff or by leaching into groundwater.

Atrazine has been shown to affect offspring of rats and the hearts of dogs. The US EPA has set the drinking water standard for atrazine concentration to 0.003 mg/L (or 3 μg/L) to protect the public against the risk of the adverse health effects. When atrazine or any other herbicide or pesticide is detected in surfacewater or groundwater, it is not meant as a human health risk. Treated (finished) water meets the EPA limits with little or no health risk and is considered safe with respect to atrazine.

**Example:** The lifetime health advisory level of 0.003 mg/L for atrazine is set at 1000 times below the amount that causes no adverse health effects in laboratory animals. The no observed adverse effect level is 0.086 mg/(kg · d). How is the 0.003 mg/L of atrazine in water derived?

**solution:** As previous example,

Using Eq. (5.13)

$$\begin{aligned} \text{DWEL} &= \frac{(\text{NOAEL})(70)}{(\text{UF})(2)} \\ &= \frac{(0.086\ \text{mg/(kg}\cdot\text{d)})(70\ \text{kg})}{(1000)(2\ \text{L/d})} \\ &= 0.003\ \text{mg/L} \end{aligned}$$

## 5 Water Treatment Processes

As the raw surface water comes to the treatment plant, physical screening is the first step to remove coarse material and debris. Thereafter, following the basic treatment process of clarification, it would include coagulation, flocculation, and sedimentation prior to filtration, then disinfection (mostly by the use of chlorination). With a good quality source, the conventional treatment processes may be modified by removing the sedimentation process and to just have the coagulation and flocculation processes followed by filtration. This treatment process scheme is called direct filtration.

Groundwater is generally better quality; however, it is typically associated with high hardness, iron, and manganese content. Aeration or air stripping is required to remove volatile compounds in groundwater. Lime softening is also necessary to remove the impurities and recarbonation is used to neutralize excess lime and to lower the pH value. Ion exchange processes can also be employed for the softening of water and the removal of other impurities, if low in iron, manganese, particulates, and organics. For a groundwater source high in ion and manganese, but with acceptable hardness, the aeration process and/or chemical oxidation can be followed by filtration to remove these compounds.

The reverse osmosis process is used to remove chemical constituents and salts in water. It offers the promise of conversion of saltwater to freshwater. Other membrane processes, such as microfiltration, ultrafiltration, nanofilteration, and electrodialysis, are also applied in the water industry.

## 6 Aeration and Air Stripping

Aeration has been used to remove trace volatile organic compounds (VOCs) in water either from surface water or groundwater sources. It has also been employed to transfer a substance, such as oxygen, from air or a gas phase into water in a process called gas adsorption or oxidation (to oxidize iron and/or manganese). Aeration also provides the escape of dissolved gases, such as carbon dioxide and hydrogen sulfide.

Air stripping has also been utilized effectively to remove ammonia gas from wastewater and to remove volatile tastes and other such substances in water.

The solubility of gases which do not react chemically with the solvent and the partial pressure of the gases at a given temperature can be expressed as Henry's law:

$$p = Hx \tag{5.14}$$

where $x$ = solubility of a gas in the solution phase
$H$ = Henry's constant
$p$ = partial pressure of a gas over the solution

In terms of the partial pressure, Dalton's law states that the total pressure ($P_t$) of a mixture of gases is just the sum of the pressures that each gas ($P_i$) would exert if it were present alone:

$$P_t = P_1 + P_2 + P_3 + \cdots + P_j \tag{5.15}$$

Since

$$PV = nRT$$

$$P = n\left(\frac{RT}{V}\right)$$

$$P_t = \frac{RT}{V}(n_1 + n_2 + \cdots n_j)$$

$$P_1 = \frac{n_1 P_t}{n_1 + n_2 + \cdots n_j} \tag{5.16}$$

Combining Henry's law and Dalton's law, we get

$$Y_i = \frac{H_i x_i}{P_t} \tag{5.17}$$

where $Y_i$ = mole fraction of $i$th gas in air
$x_i$ = mole fraction of $i$th gas in fluid (water)
$H_i$ = Henry's constant for $i$th gas
$P_t$ = total pressure, atm

The greater the Henry's constant, the more easily a compound can be removed from a solution. Generally, increasing temperature would increase the partial pressure of a component in the gas phase. The Henry's constant and temperature relationship is (James M. Montgomery Consulting Engineering, 1985; American Society of Civil Engineers and American Water Works Association, 1990):

$$\log H = \frac{-\Delta H}{RT} + J \tag{5.18}$$

where $H$ = Henry's constant
$\Delta H$ = heat absorbed in evaporation of 1 mole of gas from solution at constant temperature and pressure, kcal/kmol
$R$ = universal gas constant = 1.9872 kcal/kmol · °k
= 0.082057 atm · L/mol · k
$T$ = temperature, Kelvin
$J$ = empirical constant

Values of $H$, $\Delta H$, and $J$ of some gases are given in Table 5.4.

**Example 1:** At 20°C the partial pressure (saturated) of chloroform $CHC1_3$ is 18 mm of mercury in a storage tank. Determine the equilibrium concentration of chloroform in water assuming that gas and liquid phases are ideal.

**solution:**

Step 1. Determine mole fraction $(x)$ of $CHC1_3$

From Table 5.4, at 20°C and a total pressure of 1 atm, the Henry's constant for chloroform is

$$H = 170 \text{ atm}$$

Partial pressure of $CHC1_3$, $p$, is

$$p = \frac{18 \text{ mm}}{760 \text{ mm}} \times 1 \text{ atm} = 0.024 \text{ atm}$$

**TABLE 5.4 Henry's Law Constant and Temperature Correction Factors**

| Gas | Henry's constant at 20°C, atm | ΔH, $10^3$ kcal/kmol | $J$ |
|---|---|---|---|
| Ammonia | 0.76 | 3.75 | 6.31 |
| Benzene | 240 | 3.68 | 8.68 |
| Bromoform | 35 | — | — |
| Carbon dioxide | $1.51 \times 10^2$ | 2.07 | 6.73 |
| Carbon tetrachloride | $1.29 \times 10^3$ | 4.05 | 10.06 |
| Chlorine | 585 | 1.74 | 5.75 |
| Chlorine dioxide | 54 | 2.93 | 6.76 |
| Chloroform | 170 | 4.00 | 9.10 |
| Hydrogen sulflde | 515 | 1.85 | 5.88 |
| Methane | $3.8 \times 10^4$ | 1.54 | 7.22 |
| Nitrogen | $8.6 \times 10^4$ | 1.12 | 6.85 |
| Oxygen | $4.3 \times 10^4$ | 1.45 | 7.11 |
| Ozone | $5.0 \times 10^3$ | 2.52 | 8.05 |
| Sulfur dioxide | 38 | 2.40 | 5.68 |
| Trichloroethylene | 550 | 3.41 | 8.59 |
| Vinyl chloride | $1.21 \times 10^3$ | — | — |

Using Henry's law

$$p = Hx$$

$$x = p/H = 0.024 \text{ atm}/170\text{atm}$$

$$= 1.41 \times 10^{-4}$$

Step 2. Convert mole fraction to mass concentration

In 1 L of water, there are 1000/18 mol/L = 55.6 mol/L. Thus

$$x = \frac{n}{n + n_w}$$

where $n$ = number of mole for $CHC1_3$

$n_w$ = number of mole for water

Then $nx + n_w x = n$

$$n = \frac{n_w x}{1 - x} = \frac{55.6 \times 1.41 \times 10^{-4}}{1 - 1.41 \times 10^{-4}}$$

$$= 7.84 \times 10^{-3} \text{ mol/L}$$

Molecular weight of $CHC1_3 = 12 + 1 + 3 \times 35.45 = 119.4$
Concentration ($C$) of $CHC1_3$ is

$$C = 119.4 \text{ g/mol} \times 7.84 \times 10^{-3} \text{ mol/L}$$

$$= 0.94 \text{ g/L}$$

$$= 940 \text{ mg/L}$$

*Note*: In practice, the treated water is open to the atmosphere. Therefore the partial pressure of chloroform in the air is negligible. Therefore the concentration of chloroform in water is near zero or very low.

**Example 2:** The finished water in an enclosed clear well has dissolved oxygen of 6.0 mg/L. Assume it is in equilibrium. Determine the concentration of oxygen in the air space of the clear well at 17°C.

**solution:**

Step 1. Calculate Henry's constant at 17°C (290 K)

From Table 5.4 and Eq. (5.18)

$$\log H = \frac{-\Delta H}{RT} + J$$

$$= \frac{-1.45 \times 10^3 \text{ kcal/kmol}}{(1.987 \text{ kcal/kmol K})(290 \text{ K})} + 7.11$$

$$= 4.59$$

$$H = 38{,}905 \text{ atm} \times \frac{103.3 \text{ kPa}}{1 \text{ atm}}$$

$$= 4.02 \times 10^6 \text{ kPa}$$

Step 2. Compute the mole fraction of oxygen in the gas phase
Oxygen (M.W. = 32) as gas $i$ in water

$$C_i = 6 \text{ mg/L} \times 10^{-3} \text{ g/mg} \times 10^3 \text{ L/m}^3 \div 32 \text{ g/mol}$$
$$= 0.1875 \text{ mol/m}^3$$

At 1 atm for water, the molar density of water is

$$C_w = 55.6 \text{ kmol/m}^3$$

$$x_i = \frac{C_i}{C_w} = \frac{0.1875 \text{ mol/m}^3}{55.6 \times 10^3 \text{ mol/m}^3}$$

$$= 3.37 \times 10^{-6}$$

$$P_t = 1 \text{ atm}$$

From Eq. (5.17)

$$y_i = \frac{H_i xi}{P_t} = \frac{38{,}905 \text{ atm} \times 3.37 \times 10^{-6}}{1 \text{ atm}}$$

$$= 0.131$$

Step 3. Calculate partial pressure $P_i$

$$P_i = y_i P_t = 0.131 \times 1$$

$$= 0.131 \text{ atm}$$

Step 4. Calculate oxygen concentration in gas phase $C_g = n_i/V$

Using the ideal gas formula

$$P_iV = n_iRT$$

$$\frac{n_i}{V} = \frac{P_i}{RT} = \frac{0.131 \text{ atm}}{0.08206 \text{ atm L/mol K} \times (290 \text{ K})}$$

$$= 5.50 \times 10^{-3} \text{ mol/L}$$

$$= 5.50 \times 10^{-3} \text{ mol/L} \times 32 \text{ g/mol}$$

$$= 0.176 \text{ g/L}$$

$$= 176 \text{ mg/L}$$

## 6.1 Gas transfer models

Three gas-liquid mass transfer models are generally used. They are two-film theory, penetration theory, and surface renewal theory. The first theory is discussed here. The other two can be found elsewhere (Schroeder, 1977).

The two-film theory is the oldest and simplest. The concept of the two-film model is illustrated in Fig. 5.1. Flux is a term used as the mass transfer per time through a specified area. It is a function of the driving force for diffusion. The driving force in air is the difference between the bulk concentration and the interface concentration. The flux gas

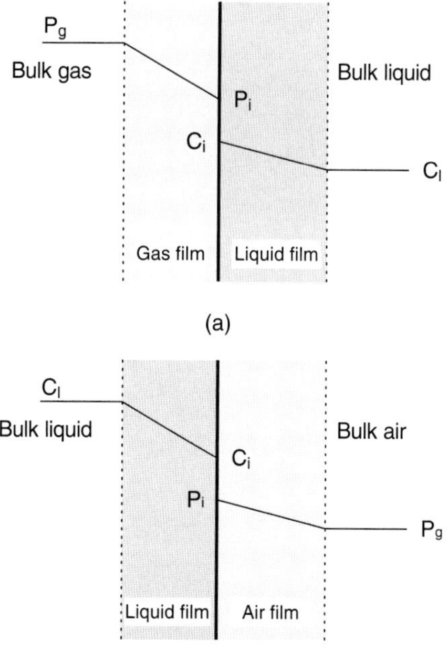


**Figure 5.1** Schematic diagram of two-film theory for (a) gas absorption; (b) gas stripping.

through the gas film must be the same as the flux through the liquid film. The flux relationship for each phase can be expressed as:

$$F = \frac{dW}{dtA} = k_g(P_g - P_i) = k_l(C_i - C_l) \tag{5.19}$$

where $F$ = flux
$W$ = mass transfer
$A$ = a given area gas-liquid transferal
$t$ = time
$k_g$ = local interface transfer coefficient for gas
$P_g$ = concentration of gas in the bulk of the air phase
$P_i$ = concentration of gas in the interface
$k_l$ = interface transfer coefficient for liquid
$C_i$ = local interface concentration at equilibrium
$C_l$ = liquid-phase concentration in bulk liquid

Since $P_i$ and $C_i$ are not measurable, volumetric mass transfer coefficients are used, which combine the mass transfer coefficients with the interfacial area per unit volume of system. Applying Henry's law and introducing $P^*$ and $C^*$ which correspond to the equilibrium concentration that would be associated with the bulk-gas partial pressure $P_g$ and bulk-liquid concentration $C_1$, respectively, we can obtain

$$F = K_G(P_g - P^*) = K_L\,(C^* - C_1) \tag{5.20}$$

The concentration differences of gas in air and water is plotted in Fig. 5.2. From this figure and Eq. (5.19), we obtain

$$\begin{aligned} P_g - P^* &= (P_g - P_i) + (P_i - P^*) \\ &= P_g - P_i + s_1(C_i - C_1) \\ &= \frac{F}{k_g} + \frac{s_1 F}{k_1} \end{aligned}$$

From Eq. (5.20)

$$\frac{F}{K_G} = P_g - P^*$$

Thus

$$\frac{F}{K_G} = \frac{F}{k_g} + \frac{s_1 F}{k_1}$$

$$\frac{1}{K_G} = \frac{1}{k_g} + \frac{s_1}{k_1}$$

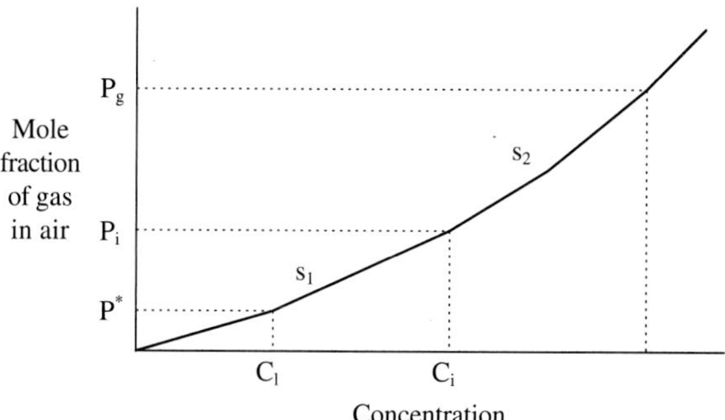


**Figure 5.2** Concentration differences of gas in air and water phases.

Similarly, we can obtain

$$\frac{1}{K_\mathrm{L}} = \frac{1}{s_2 k_\mathrm{g}} + \frac{1}{k_1}$$

In dilute conditions, Henry's law holds, the equilibrium distribution curve would be a straight line and the slope is the Henry's constant:

$$s_1 = s_2 = H$$

Finally the overall mass transfer coefficient in the air phase can be determined with

$$\frac{1}{K_\mathrm{G}} = \frac{1}{k_1} + \frac{H}{k_1} \tag{5.21}$$

For liquid phase mass transfer

$$\frac{1}{K_\mathrm{L}} = \frac{1}{k_1} + \frac{H}{Hk_\mathrm{g}} \tag{5.22}$$

where $1/k_\mathrm{l}$ and $1/k_\mathrm{g}$ are referred to as the liquid-film resistance and the gas-film resistance, respectively. For oxygen transfer, $1/k_\mathrm{l}$ is considerably larger than $1/k_\mathrm{g}$, and the liquid film usually controls the oxygen transfer rate across the interface.

The flux equation for the liquid phase using concentration of mg/L is

$$F = \frac{dc}{dt}\frac{V}{A} = K_\mathrm{L}(C_\mathrm{s} - C_t) \tag{5.23a}$$

mass transfer rate:

$$\frac{dc}{dt} = K_L \frac{A}{V}(C_\mathrm{s} - C_t) \tag{5.23b}$$

or

$$\frac{dc}{dt} = K_L a(C^* - C_\mathrm{t}) \tag{5.23c}$$

where $K_L$ = overall mass-transfer coefficient, cm/h
$A$ = interfacial area of transfer, cm$^2$
$V$ = volume containing the interfacial area, cm$^3$
$C_t$ = concentration in bulk liquid at time $t$, mg/L
$C^*$ = equilibrium concentration with gas at time $t$ as $P_t = HC_s$ mg/L
$a$ = the specific interfacial area per unit system volume, cm$^{-1}$
$K_La$ = overall volumetric mass transfer coefficient in liquid $h^{-1}$, or [g · mole/(h · cm$^3$ · atm)]

the volumetric rate of mass transfer $M$ is

$$M = \frac{A}{V}N = K_La(C^* - C) \tag{5.24}$$

$$= K_Ga(P - P^*-) \tag{5.25}$$

where $N$ = rate at which solute gas is transferred between phases, g · mol/h

**Example 1:** An aeration tank has the volume of 200 m$^3$ (7060 ft$^3$) with 4 m (13 ft) depth. The experiment is conducted at 20°C for tap water. The oxygen transfer rate is 18.7 kg/h (41.2 lb/h). Determine the overall volumetric mass transfer coefficient in the liquid phase.

**solution:**

Step 1. Determine the oxygen solubility at 20°C and 1 atm

Since oxygen in air is 21%, according to Dalton's law of partial pressure, $P$ is

$$P = 0.21 \times 1 \text{ atm} = 0.21 \text{ atm}$$

From Table 5.3, we find

$$H = 4.3 \times 10^4 \text{ atm}$$

The mole fraction $x$ from Eq. (5.14) is

$$x = \frac{P}{H} = \frac{0.21}{4.3 \times 10^4} = 4.88 \times 10^{-6}$$

In 1L of solution, the number of moles of water $n_w$ is

$$n_w = \frac{1000}{18} = 55.6 \text{g} \cdot \text{mol/L}$$

Let $n$ = moles of gas in solution

$$x = \frac{n}{n + n_w}$$

Since $n$ is so small

$$n \cong n_{w}x$$
$$= 55.6 \text{ g} \cdot \text{mol/L} \times 4.88 \times 10^{-6}$$
$$= 2.71 \times 10^{-4} \text{ g} \cdot \text{mol/L}$$

Step 2. Determine the mean hydrostatic pressure on the rising air bubbles

Pressure at the bottom is $p_{\mathrm{b}}$

1 atm = 10.345 m of water head

$$p_{\mathrm{b}} = 1 \text{atm} + 4\text{m} \times \frac{1 \text{ atm}}{10.345 \text{ m}} = (1 + 0.386)\text{atm}$$
$$= 1.386 \text{atm}$$

Pressure at the surface of aeration tank is $p_{\mathrm{s}} = 1$ atm

Mean pressure $p$ is

$$p = \frac{p_{\mathrm{s}} + p_{\mathrm{b}}}{2} = \frac{1 \text{ atm} + 1.386 \text{ atm}}{2} = 1.193 \text{ atm}$$

Step 3. Compute $C^* - C$

$$C^* = 2.71 \times 10^{-4} \text{ g} \cdot \text{mol/L} \cdot \text{atm} \times 1.193 \text{ atm}$$
$$= 3.233 \times 10^{-4} \text{ g} \cdot \text{mol/L}$$

The percentage of oxygen in the air bubbles decrease, when the bubbles rise up through the solution, assuming that the air that leaves the water contains only 19% of oxygen.

Thus

$$C^* - C = 3.223 \times 10^{-4} \text{ g} \cdot \text{mol/L} \times \frac{19}{21}$$
$$= 2.925 \times 10^{-4} \text{ g} \cdot \text{mol/L}$$

Step 4. Compute $K_{\mathrm{L}}a$

$$N = 18{,}700 \text{ g/h} \div 32/\text{mole} = 584 \text{ g} \cdot \text{mol/h}$$

Using Eq. (5.24) $$K_{L}a = \frac{N}{V(C^* - C)}$$

$$= \frac{584 \text{ g} \cdot \text{mol/h}}{200 \text{ m}^3 \times 1000 \text{ L/m}^3 \times 2.925 \times 10^{-4} \text{ g} \cdot \text{mol/L}}$$

$$= 9.98 \text{ h}^{-1}$$

Equation (5.23c) is essentially the same as Fick's first law:

$$\frac{dC}{dt} = K_{L}a(C_{\mathrm{s}} - C) \tag{5.23d}$$

where $\dfrac{dC}{dt}$ = rate of change in concentration of the gas in solution, mg/L · s

$K_{L}a$ = overall mass transfer coefficient, $s^{-1}$

$C_s$ = saturation concentration of gas in solution, mg/L

$C$ = concentration of solute gas in solution, mg/L

The value of $C_s$ can be calculated with Henry's law. The term $C_s - C$ is a concentration gradient. Rearrange Eq. (5.23d) and integrating the differential form between time from 0 to $t$, and concentration of gas from $C_0$ to $C_t$ in mg/L, we obtain

$$\int_{C_0}^{C_t} \frac{dC}{C_s - C} = K_{L}a \int_0^t dt$$

$$\ln \frac{C_s - C_t}{C_s - C} = -K_{L}at \tag{5.24a}$$

or

$$\frac{C_s - C_t}{C_s - C} = e^{-K_{L}at} \tag{5.24b}$$

When gases are removed or stripped from the solution, Eq. (5.24b) becomes

$$\frac{C_0 - C_s}{C_t - C_s} = e^{-(K_{L}at)}$$

Similarly, Eq. (5.25) can be integrated to yield

$$\ln \frac{p_t - p^*}{p_0 - p^*} = K_{G}aRTt \tag{5.25a}$$

$K_{L}a$ values are usually determined in full-scale facilities or scaled up from pilot-scale facilities. Temperature and chemical constituents in wastewater affect oxygen transfer. The temperature effects on overall mass transfer coefficient $K_{L}a$ are treated in the same manner as they were treated in the BOD rate coefficient (section 8.3 of Chapter 1.2). It can be written as

$$K_{L}a_{(T)} = K_{L}a_{(20)}\theta^{T-20} \tag{5.26}$$

where $K_{L}a_{(T)}$ = overall mass transfer coefficient at temperature $T$°C, $s^{-1}$

$K_{L}a_{(20)}$ = overall mass transfer coefficient at temperature 20°C, $s^{-1}$

Values of $\theta$ range from 1.015 to 1.040, with 1.024 commonly used.

Mass transfer coefficient are influenced by total dissolved solids in the liquid. Therefore, a correction factor $\alpha$ is applied for wastewater (Tchobanoglous and Schroeder, 1985):

$$\alpha = \frac{K_L a \text{ (wastewater)}}{K_L a \text{ (tap water)}} \tag{5.27}$$

Value of $\alpha$ range from 0.3 to 1.2. Typical values for diffused and mechanical aeration equipment are in the range of 0.4 to 0.8 and 0.6 to 1.2.

The third correction factor ($\beta$) for oxygen solubility is due to particulate, salt, and surface active substances in water (Doyle and Boyle, 1986):

$$\beta = \frac{C^*\text{(wastewater)}}{C^*\text{(tap water)}} \tag{5.28}$$

Values of $\beta$ range from 0.7 to 0.98, with 0.95 commonly used for wastewater.

Combining all three correction factors, we obtain (Tchobanoglous and Schroeder 1985)

$$\text{AOTR} = \text{SOTR}(\alpha)(\beta)(\theta^{T-20})\left(\frac{C_s - C_w}{C_{s20}}\right) \tag{5.29}$$

where AOTR = actual oxygen transfer rate under field operating conditions in a respiring system, kg · $O_2$/kW · h

SOTR = standard oxygen transfer rate under test conditions at 20°C and zero dissolved oxygen, kg · $O_2$/kW · h

$\alpha, \beta, \theta$ = defined previously

$C_s$ = oxygen saturation concentration for tap water at field operating conditions, g/m$^3$

$C_w$ = operating oxygen concentration in wastewater, g/m$^3$

$C_{s20}$ = oxygen saturation concentration for tap water at 20°C, g/m$^3$

**Example 2:** Aeration tests are conducted with tap water and wastewater at 16°C in the same container. The results of the tests are listed below. Assume the saturation DO concentrations ($C_s$) for tap water and wastewater are the same. Determine the values of $K_L a$ for tap water and wastewater and $\alpha$ values at 20°C.

Assume $\theta = 1.024$

**solution:**

Step 1. Find DO saturation concentration at 16°C

From Table 1.2

$$C_s = 9.82 \text{ mg/L}$$

| Contact time, min | DO concentration, mg/L | | $-\ln\frac{C_s - C_t}{C_s - C_0}$ | |
|---|---|---|---|---|
| | Tap water | Wastewater | Tap water | Wastewater |
| 0 | 0.0 | 0.0 | 0 | 0 |
| 20 | 3.0 | 2.1 | 0.36* | 0.24 |
| 40 | 4.7 | 3.5 | 0.65 | 0.44 |
| 60 | 6.4 | 4.7 | 1.05 | 0.65 |
| 80 | 7.2 | 5.6 | 1.32 | 0.84 |
| 100 | 7.9 | 6.4 | 1.69 | 1.05 |
| 120 | 8.5 | 7.1 | 2.01 | 1.28 |

*$\ln (C_s - C_t)/(C_s - C_0) = \ln(9.82 - 3.0)/(9.82 - 0) = 0.36$

Step 2. Calculate $-\ln(C_s - C_t)/(C_s - C_0)$, $C_0 = 0$

The values of $-\ln(C_s - C_t)/(C_s - C_0)$ calculated are listed with the raw test data

Step 3. Plot the calculated results in Step 2 in Fig. 5.3.

Step 4. Find the $K_La$ at the test temperature of 16°C, ($T$), from the above table

For tap water

$$K_La = \frac{2.01}{120 \text{ min}} \times \frac{60 \text{ min}}{1 \text{ h}}$$

$$= 1.00 \text{ h}^{-1}$$

For wastewater

$$K_La = \frac{1.28}{120 \text{ min}} \times \frac{60 \text{ min}}{1 \text{ h}}$$

$$= 0.64 \text{ h}^{-1}$$

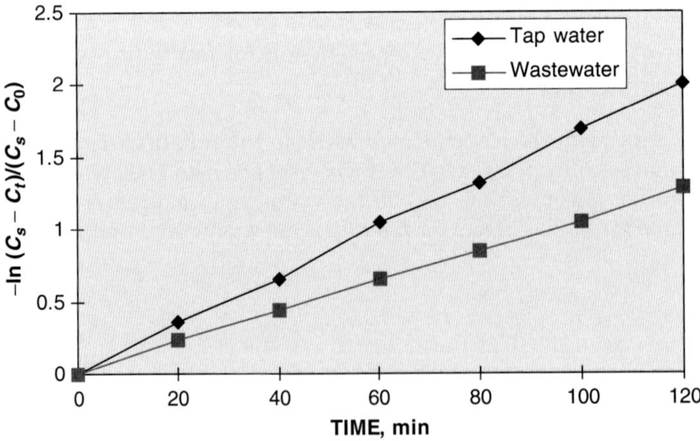


**Figure 5.3** Functional plot of the data from aeration test.

Step 5. Convert the values of $K_La$ at 20°C using $\theta = 1.024$

Using Eq. (5.26)
For tap water

$$K_La_{(T)} = K_La_{(20°C)}\theta^{20-T}$$

or

$$K_La_{(20)} = K_La_{(T)}\theta^{20-T}$$
$$= 1.00\ h^{-1}(1.024)^{20-16}$$
$$= 1.10\ h^{-1}$$

For wastewater

$$K_La_{(20)} = 0.64\ h^{-1}(1.024)^{20-16}$$
$$= 0.70\ h^{-1}$$

Step 6. Compute the $\alpha$ value using Eq. (5.27)

$$\alpha = \frac{K_La \text{ for wastewater}}{K_La \text{ for tap water}}$$
$$= \frac{0.70\ h^{-1}}{0.10\ h^{-1}}$$
$$= 0.64$$

## 6.2 Diffused aeration

Diffused aeration systems distribute the gas uniformly through the water or wastewater, in such processes as ozonation, absorption, activated-sludge process, THM removal, and river or lake reaeration, etc. It is more costly using diffused aeration for VOC removal than the air stripping column.

The two-resistance layer theory is also applied to diffused aeration. The model proposed by Mattee-Müller *et al.* (1981) is based on mass transfer flux derived from the assumption of diffused bubbles rising in a completely mixed container. The mass transfer rate for diffused aeration is

$$F = Q_G H_u C_e\left(1 - \exp\frac{K_LaV}{H_uQ_L}\right) \tag{5.30}$$

where $F$ = mass transfer rate
$Q_G$ = gas (air) flow rate, m³/s or ft³/s
$Q_L$ = flow rate of liquid (water), m³/s or ft³/s

$H_u$ = unitless Henry's constant (see, Section of Design of Packed Tower)
$C_e$ = effluent (exit) gas concentration, μg/L
$K_La$ = overall mass transfer coefficient, per time
$V$ = reaction volume (water), $m^3$ or $ft^3$

Assuming that the liquid volume in the reactor is completely mixed and the air rises as a steady state plug flow, the mass balance equation can be expressed as

$$\frac{C_e}{C_i} = \frac{1}{1 + H_u Q_G/Q_L\,[1 - \exp(-K_L aV/H_u Q_G]} \tag{5.31a}$$

or

$$\frac{C_e}{C_i} = \frac{1}{1 + H_u Q_G/Q_L\,[1 - \exp(-\theta)]} \tag{5.31b}$$

where $C_i$ = initial concentration, μg/L

$$\theta = \frac{K_L aV}{H_u Q_G}$$

If $\theta >> 1$, the transfer of a compound is with very low Henry's constant such as ammonia. Air bubbles exiting from the top of the liquid surface is saturated with ammonia in the stripping process. Ammonia removal could be further enhanced by increasing the air flow. Until $\theta < 4$, the exponent term becomes essentially zero. When the exponent term is zero, the air and water have reached an equilibrium condition and the driving force has decreased to zero at some point within the reactor vessel. The vessel is not fully used. Thus the air-to-water ratio could be increased to gain more removal.

On the other hand, if $\theta << 1$, the mass transfer efficiency could be improved by increasing overall mass transfer coefficient by either increasing the mixing intensity in the tank or by using a finer diffuser. In the case for oxygenation, $\theta < 0.1$, the improvements are required.

**Example:** A groundwater treatment plant has a capacity of 0.0438 $m^3$/s (1 MGD) and is aerated with diffused air to remove trichloroethylene with 90% design efficiency. The detention time of the tank is 30 min. Evaluate the diffused aeration system with the following given information.

$$T = 20°\text{C}$$

$$C_i = 131\ \mu\text{g/L},\ (\text{expected } C_e = 13.1\ \mu\text{g/L})$$

$$H_u = 0.412$$

$$K_L a = 44\ \text{h}^{-1}$$

**solution:**

Step 1. Compute volume of reactor $V$

$$V = 0.0438\ \text{m}^3/\text{s} \times 60\ \text{s/min} \times 30\ \text{min} = 79\ \text{m}^3$$

Step 2. Compute $Q_G$, let $Q_G = 30\ Q_L = 30V$

$$Q_G = 30 \times 79\ \text{m}^3 = 2370\ \text{m}^3$$

Step 3. Compute $\theta$

$$\theta = \frac{K_L a V}{H_u Q_G} = \frac{44 \times 79}{0.412 \times 2370} = 3.56$$

Step 4. Compute effluent concentration $C_e$ with Eq (5.31b)

$$C_e = \frac{C_i}{1 + H_u Q_G/Q_L[1 - \exp(-\theta)]}$$

$$= \frac{131\,\mu\text{g/L}}{1 + 0.412 \times 30[1 - \exp(-3.56)]}$$

$$= 10.0\ \mu\text{g/L}$$

$$\%\ \text{removal} = \frac{(131 - 10) \times 100}{131} = 92.4$$

This exceeds the expected 90%.

## 6.3 Packed towers

Recently, the water treatment industry used packed towers for stripping highly volatile chemicals, such as hydrogen sulfide and VOCs from water and wastewater. It consists of a cylindrical shell containing a support plate for the packing material. Although many materials can be used as the packing material, plastic products with various shapes and design are most commonly used due to less weight and lower cost. Packing material can be individually dumped randomly into the cylinder tower or fixed packing. The packed tower or columns are used for mass transfer from the liquid to gas phase.

Figure 5.4 illustrates a liquid-gas contacting system with a downward water velocity $L$ containing $c_1$ concentration of gas. The flows are counter current. The air velocity $G$ passes upward through the packed material containing influent $p_1$ and effluent $p_2$. There are a variety of mixing patterns, each with a different rate of mass transfer. Removal of an undesirable gas in the liquid phase needs a system height of $z$ and a selected gas flow rate to reduce the mole fraction of dissolved gas from $c_1$ to $c_2$. If there is no chemical reaction that takes place in the packed

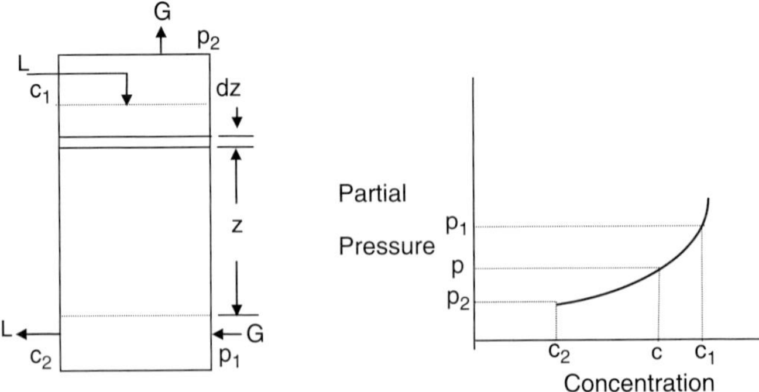


**Figure 5.4** Schematic diagram of packed column.

tower, the gas lost by water should be equal to the gas gained by air. If the gas concentration is very dilute, then

$$L\Delta c = G\Delta p \tag{5.32}$$

where $L$ = liquid velocity, m/s, $m^3/(m^2 \cdot s)$, or $mol/(m^2 \cdot s)$
$G$ = gas velocity, same as above
$\Delta c$ = change of gas concentration in water
$\Delta p$ = change in gas fraction in air

From Eq. (5.23d), for stripping ($c$ and $c_s$ will be reversed):

$$\frac{dc}{dt} = K_L a(c - c_s) \tag{5.23e}$$

then

$$dt = \frac{dc}{K_L a(c - c_s)} \tag{5.33}$$

Multiplying each side by $L$, we obtain

$$Ldt = dz = \frac{Ldc}{K_L a(c - c_s)} \tag{5.34}$$

where $dz$ is the differential of height.

The term $(c - c_s)$ is the driving force (DF) of the reaction and is constantly changing with the depth of the column due to the change of $c$ with time.

Integrating the above equation yields

$$\int dz = \frac{L}{K_L a} \int_{c_1}^{c_2} \frac{dc}{d(c - c_s)}$$

$$z = \frac{L}{K_L a} \int_{c_1}^{c_2} \int_{\mathrm{DF}} \frac{dc}{d(\mathrm{DF})} \tag{5.35}$$

The integral of DF is the same as the log mean ($DF_{lm}$) of the influent ($DF_i$) and effluent driving forces ($DF_e$). Therefore

$$z = \frac{L(c_i - c_e)}{K_L a DF_{lm}} \tag{5.36}$$

and

$$DF_{lm} = \frac{DF_e - DF_i}{\ln(DF_e/DF_i} \tag{5.37}$$

where $z$ = height of column, m
$L$ = liquid velocity, $m^3/m^2 \cdot h$
$c_i$, $c_e$ = gas concentration in water at influent and effluent, respectively, mg/L
$K_L a$ = overall mass transfer coefficient for liquid, $h^{-1}$
$DF_i$, $DF_e$ = driving force at influent and effluent, respectively, mg/L
$DF_{lm}$ = log mean of $DF_i$ and $DF_e$, mg/L

**Example 1:** Given

$T = 20°C = 293$ K
$L = 80\ m^3$ water $(m^2 \cdot h)$
$G = 2400\ m^3$ air$/(m^2 \cdot h)$
$c_i = 131$ μg/L trichloroethylene concentration in water at entrance
$c_e = 13.1$ μg/L $CCHCl_3$ concentration at exit
$K_L a = 44\ h^{-1}$

Determine the packed tower height to remove 90% of trichloroethylene by an air stripping tower.

**solution:**

Step 1. Compute the molar fraction of $CCHCl_3$ in air $p_2$

Assuming no $CCHCl_3$ present in the air the entrance, i.e.

$$p_i = 0$$

MW of $CCHCl_3$ = 131, 1 mole =131g of $CCHCl_3$ per liter.

$$c_i = 131\ \mu g/L = 131\frac{\mu g}{L} \times \frac{1\ mol/g}{131 \times 10^6\ \mu g/g} \times \frac{10^3\ L}{1\ m^3}$$

$$= 1 \times 10^{-3}\ mol/m^3$$

$$c_e = 13.1\ \mu g/L = 0.1 \times 10^{-3}\ mol/m^3 (\text{with } 90\% \text{ removal})$$

Applying Eq. (5.32)

$$L\Delta c = G\Delta p$$

$$80 \text{ m}^3 \text{ air/(m}^2 \cdot \text{h)}(1 - 0.1) \times 10^{-3} \text{ mol gas/m}^3 \text{ air} = 2400 \text{ m}^3\text{/(m}^2 \cdot \text{h)}(p_e - 0)$$

$$p_e = 3.0 \times 10^{-5} \text{ mol gas/m}^3 \text{ air}$$

Step 2. Convert $p_e$ in terms of mol gas/mol air

Let $V$ = volume of air per mole of air

$$\begin{aligned} V = \frac{nRT}{P} &= \frac{(1\,\text{mole})(0.08206 \text{ L atm/mol K}) \times (293 \text{ K})}{1\,\text{atm}} \\ &= 24.0 \text{ L} \\ &= 0.024 \text{ m}^3 \end{aligned}$$

From Step 1

$$\begin{aligned} p_e &= 3.0 \times 10^{-5} \text{mol gas/m}^3\text{air} \\ &= 3.0 \times 10^{-5} \frac{\text{mol gas}}{\text{m}^3 \text{ air}} \times \frac{0.024 \text{ m}^3 \text{ air}}{1 \text{ mol air}} \\ &= 7.2 \times 10^{-7} \text{mol gas/mol air} \end{aligned}$$

Step 3. Compute DF for gas entrance and exit
At gas influent (bottom)

$$\begin{aligned} p_i &= 0 \\ c_e &= 13.1\ \mu\text{g/L} = 0.0131 \text{ mg/L} \\ c_s &= 0 \\ \text{DF}_i &= c_e - c_s = 0.0131 \text{ mg/L} \end{aligned}$$

At gas effluent (top)

$$\begin{aligned} p_e &= 7.2 \times 10^{-7} \text{mol gas/mol air} \\ c_i &= 131\ \mu\text{g/L} = 0.131 \text{ mg/L} \\ c_s &= \text{to be determined} \end{aligned}$$

From Table 5.3, Henry's constant $H$ at 20°C

$$H = 550 \text{ atm}$$

Convert atm to atm L/mg, $H_d$

$$\begin{aligned} H_d &= \frac{H}{55{,}600 \times \text{MW}} = \frac{550\,\text{atm}}{55{,}600 \times 131\,\text{mg/L}} \\ &= 7.55 \times 10^{-5} \text{atm} \cdot \text{L/mg} \end{aligned}$$

$$c_s = \frac{p_e p_t}{H_d} = \frac{7.2 \times 10^{-7}\ \text{mol gas/mol air} \times 1\,\text{atm}}{7.55 \times 10^{-5}\,\text{atm} \cdot \text{L/mg}}$$

$$= 0.0095\ \text{mg/L}$$

$$\text{DF}_e = c_i - c_s = 0.131\ \text{mg/L} - 0.0095\ \text{mg/L}$$

$$= 0.1215\ \text{mg/L}$$

Step 4. Compute $\text{DF}_{\text{lm}}$

From Eq. (5.37)

$$\text{DF}_{\text{lm}} = \frac{\text{DF}_e - \text{DF}_i}{\ln(\text{DF}_e/\text{DF}_i)} = \frac{0.1215\ \text{mg/L} - 0.0131\ \text{mg/L}}{\ln(0.1215/0.0131)}$$
$$= 0.0487\ \text{mg/L}$$

Step 5. Compute the height of the tower $z$

From Eq. (5.36)

$$z = \frac{L(c_i - c_e)}{K_L a \text{DF}_{\text{lm}}} = \frac{80\ \text{m/h} \times (0.131 - 0.0131)\ \text{mg/L}}{44\ \text{h}^{-1} \times 0.0487\ \text{mg/L}}$$
$$= 4.4\ \text{m}$$

**Example 2** In a groundwater remediation study, volatile organic carbon removal through the vapor phase, the total hydrocarbons analyzer measured 1,1,1,-trichloroethane (Jones *et al.*, 2000). The extraction pump and air stripper combined air flow averaged 0.40m$^3$/min (14ft$^3$/min). The concentration of 1,1,1,-trichloroethane averaged 25 parts per million by volume (ppmv) of air. Determine the amount of 1,1,1,-trichloroethane removed daily. Temperature is 20°C.

**solution**

Step 1. Calculate 1,1,1,-trichloroethane concentration using the ideal gas law

$$T = 20°\text{C} = (20 + 273)\text{K} = 293\ \text{K}$$

$$n = \frac{PV}{RT} = \frac{(1\ \text{atm})(1000\ \text{L/mg}^3)}{0.082(\text{L} \cdot \text{atm/mol-K}) \times 293\ \text{K}}$$

$$= 41.6\ \text{mol/m}^3$$

This means 41.6 mole of total gases in lm$^3$ of the air.
The MW of 1,1,1,-trichloroethane ($CH_3CC1_3$) = 133
One ppmv of 1,1,1,-trichloroethane $= 41.6 \times 10^{-6}\ \text{mol/m}^3 \times 133\text{g/mol} \times 1000\ \text{mg/g}$
$= 5.53\ \text{mg/m}^3$
25 ppmv of 1,1,1,-trichloroethane $= 5.53\ \text{mg/m}^3 \times 25$
$= 138\ \text{mg/m}^3$

Step 2. Calculate the daily removal

$$\text{Daily removal} = 138 \text{ mg/m}^3 \times 0.40 \text{ m}^3\text{/min} \times 1440 \text{ min/d}$$
$$= 79500 \text{ mg/d}$$
$$= 79.5 \text{ g/d}$$

**Design of packed tower.** Process design of packed towers or columns is based on two quantities: the height of the packed column, $z$, to achieve the designed removal of solute is the product of the height of a transfer unit (HTU) and the number of transfer unit (NTU). It can be expressed as (Treybal, 1968):

$$z = (\text{HTU})(\text{NTU}) \tag{5.38}$$

The HTU refers the rate of mass transfer for the particular packing materials used. The NTU is a measure of the mass transfer driving force and is determined by the difference between actual and equilibrium phase concentrations. The height of a transfer unit is the constant portion of Eq. (5.35):

$$\text{HTU} = \frac{L}{K_{\text{L}}a} \tag{5.39}$$

The number of transfer units is the integral portion of Eq. (5.35). For diluted solutions, Henry's law holds. Substituting the integral expression for NTU with $p_1 = 0$, the NTU is

$$\text{NTU} = \frac{R}{R-1} \ln \frac{(c_1/c_2)(R-1)+1}{R} \tag{5.40}$$

where
$$R = \frac{H_{\text{u}}G}{L} \tag{5.41}$$

$=$ stripping factor, unitless when $H_{\text{u}}$ is unitless

$c_1, c_2$ = mole fraction for gas entrance and exit, respectively

Convert Henry's constant from terms of atm to unitless:

$$H_{\text{u}} = \left[H \frac{\text{atm(mol gas/mol air)}}{\text{mol gas/mol water}}\right]\left(\frac{1 \text{ mol air}}{0.082T \text{ atm L of air}}\right)\left(\frac{1 \text{ L of water}}{55.6 \text{ mole}}\right) \tag{5.42}$$

$$= H/4.56T$$

When $T = 20°C = 293$ K

$$H_u = H/4.56 \times 293 = 7.49 \times 10^{-4}H, \text{ unitless} \tag{5.42a}$$

$$G = \text{superficial molar air flow rate } (k \text{ mol/s} \cdot \text{m}^2)$$

$$L = \text{superficial molar water flow rate } (k \text{ mol/s} \cdot \text{m}^2)$$

The NTU depends upon the designed gas removal efficiency, the air-water velocity ratio, and Henry's constant. Treybal (1968) plotted the integral part (NTU) of Eq. (5.35) in Fig. 5.5. By knowing the desired removal efficiency, the stripping factor, and Henry's constant, the NTU in a packed column can be determined for any given stripping factor of the air to water flow rate ratios. It can be seen from Fig. 5.5 that when the stripping factor, $R$, is greater than 3, little improvement for the NTU occurs.

**Example:** Using the graph of Fig. 5.5 to solve Example 1 above. Given:

$$T = 20°C = 293 \text{ K}$$

$$L = 80 \text{ m}^3 \text{ water/(m}^2 \text{ column cross section h)}$$

$$G = 2400 \text{ m}^3 \text{ air/(m}^2 \text{ column cross section h)}$$

$$c_1 = 131\ \mu\text{g/L of CCHCl}_3$$

$$c_2 = 13.1\ \mu\text{g/L of CCHCl}_3$$

$$K_La = 44 \text{ h}^{-1}$$

**solution:**

Step 1. Compute HTU with Eq. (5.39)

$$HTU = \frac{L}{K_La} = \frac{80 \text{ m/h}}{44 \text{ h}^{-1}} = 1.82 \text{ m}$$

Step 2. Compute $H_u$, the unitless Henry's constant at 20°C

From Table 5.3

$$H = 550 \text{ atm}$$

Using Eq. (5.42a)

$$H_u = 7.49 \times 10^{-4}H = 7.49 \times 10^{-4} \times 550$$
$$= 0.412$$

Step 3. Compute the stripping factor $R$ using Eq. (5.41)

$$R = \frac{H_uG}{L} = \frac{0.412 \times 2400 \text{ m/h}}{80 \text{ m/h}}$$
$$= 12.36$$

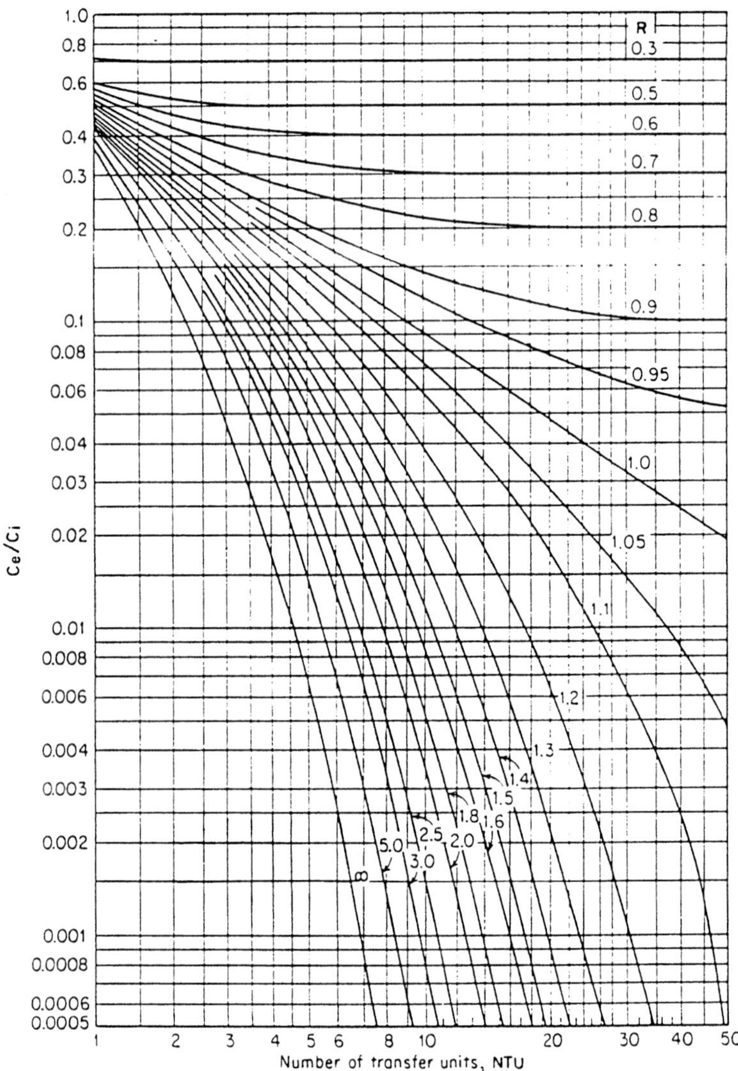


**Figure 5.5** Number of transfer units for absorbers or strippers with constant absorption or stripping factors (*D. A. Cornwell, Air Stripping and aeration. In: AWWA, Water Quality and Treatment. Copyright 1990, McGraw-Hill, New York, reprinted with permission of McGraw-Hill*).

Step 4. Find NTU from Fig. 5.5

Since
$$\frac{c_2}{c_1} = \frac{13.1\ \mu\text{g/L}}{131\ \mu\text{g/L}} = 0.1$$

Using $R = 12.36$ and $c_2/c_1 = 0.1$, from the graph in Fig. 5.5, we obtain

$$\text{NTU} = 2.42$$

Step 5. Compute the height of tower $z$ by Eq. (5.38)

$$z = (\text{HTU})\,(\text{NTU}) = 1.82 \text{ m} \times 2.42$$
$$= 4.4 \text{ m}$$

## 6.4 Nozzles

There are numerous commercially available spray nozzles. Nozzles and tray aerators are air-water contact devices. They are used for iron and manganese oxidation, removal of carbon dioxide and hydrogen sulfide from water, and removal of taste and odor causing materials.

Manufacturers may occasionally have mass transfer data such as $K_L$ and $a$ values. For an open atmosphere spray fountain, the specific interfacial area $a$ is (Calderbrook and Moo-Young, 1961):

$$a = \frac{6}{d} \tag{5.43}$$

where $d$ is the droplet diameter which ranges from 2 to 10,000 μm. Under open atmosphere conditions, $C_s$ remains constant because there is an infinite air-to-water ratio and the Lewis and Whitman equation can express the mass transfer for nozzle spray (Fair *et al.*, 1968; Cornwell, 1990)

$$C_e - C_i = (C_s - C_i)[1 - \exp(-K_L at)] \tag{5.44}$$

where $c_i$, $c_e$ = solute concentrations in bulk water and in droplets, respectively, μg/L

$t$ = time of contact between water droplets and air

= twice the rise time, $t_r$

$$= \frac{2V\sin\phi}{g} \tag{5.45a}$$

or $$t_r = V\,\text{Sin}\phi/g \tag{5.45b}$$

where $\phi$ = angle of spray measured from horizontal

$V$ = velocity of the droplet from the nozzle

$$V = C_v\sqrt{2gh} \tag{5.46}$$

where $C_v$ = velocity coefficient, 0.40 to 0.95, obtainable from the manufacturer

$g$ = gravitational acceleration, 9.81 m/s$^2$

Combine Eq. (5.45b) and Eq. (5.46)

$$gt_r = C_v\sqrt{2gh}\ \sin\phi$$

The driving head $h$ is

$$h = gt_r^2/(2C_v^2\ sin^2\phi) \tag{5.47}$$

Neglecting wind effect, the radius of the spray circle is

$$\begin{aligned} r &= Vt_r \cos\phi \\ &= V(V\sin\phi/g)\cos\phi \\ &= (V^2/g)\sin\phi\cos\phi \\ &= C_v^2 \cdot 2gh/g \cdot \sin 2\phi \\ &= 2C_v^2 h \sin 2\phi \end{aligned} \tag{5.48}$$

and the vertical rise of the spray is

$$h_r = \frac{1}{2}gt_r^2$$

From Eq. (5.47) $$h_r = \frac{1}{2}h(2C_v^2\sin^2\phi) = C_v^2 h\ \sin^2\phi \tag{5.49}$$

**Example 1:** Determine the removal percentage of trichloroethylene from a nozzle under an operating pressure of 33 psi (22.8 kPa).

The following data is given:

$$\begin{aligned} d &= 0.05\text{ cm} \\ C_v &= 0.50 \\ \phi &= 30° \\ K_L &= 0.005\text{ cm/s} \\ C_i &= 131\ \mu\text{g/L} \end{aligned}$$

**solution:**

Step 1. Determine the volumetric interfacial area $a$ by Eq. (5.43):

$$a = \frac{6}{d} = \frac{6}{0.05\text{ cm}} = 120\text{ cm}^{-1}$$

Step 2. Compute the velocity of the droplet $V$ by Eq. (5.46)

The pressure head

$$\begin{aligned} h &= 33\text{ psi} \times \frac{70.3\text{ cm of waterhead}}{1\text{ psi}} \\ &= 2320\,\text{cm} \end{aligned}$$

Using Eq. (5.46)

$$V = C_v\sqrt{2\,gh} = 0.5\sqrt{2 \times 981 \times 2320}$$
$$= 1067 \text{ cm/s}$$

Step 3. Compute the time of contact by Eq. (5.45a)

$$t = \frac{2V\sin\phi}{g} = \frac{2 \times 1067 \text{ cm/s} \sin 30}{981 \text{ cm/s}^2}$$
$$= \frac{2 \times 1067 \times 0.5}{981} \text{ s}$$
$$= 1.09 \text{ s}$$

Step 4. Compute the mass transfer $C_e - C_i$ by Eq. (5.44)

$$C_e - C_i = (C_s - C_i)[1 - \exp(-K_L at)]$$
$$C_e - 131 = (0 - 131)[1 - \exp(-0.005 \times 120 \times 1.09)]$$
$$C_e = 131 - 131 \times 0.48$$
$$= 68.1\ (\mu\text{g/L}) = 52\% \text{ remained}$$
$$= 48\% \text{ removal}$$

**Example 2:** Calculate the driving head, radius of the spray circle, and vertical rise of spray for (a) a vertical jet and (b) a jet at 45° angle. Given the expose time for water droplet is 2.2 s and $C_v$ is 0.90.

**solution:**

Step 1. For question (a), using Eq. (5.45b)

$$t_r = \frac{1}{2} \times 2.2 \text{ s} = 1.1 \text{ s}$$
$$\phi = 90^\circ$$
$$\sin\phi = \sin 90 = 1$$
$$\sin 2\phi = \sin 180 = 0$$

Using Eq. (5.47)

$$h = gt_r^2/(2C_v^2 \sin^2\phi)$$
$$= 9.81 \text{ m/s}^2\ (1.1\text{s})^2/(2 \times 0.9^2 \times 1^2)$$
$$= 7.33 \text{ m}$$

Using Eq. (5.48)

$$r = 2C_v^2 h \sin 2\phi$$
$$= 0$$

Since $\sin 2\phi = 0$, and using Eq. (5.49)

$$h_r = \frac{1}{2} g t_r^2 = \frac{1}{2} \times 9.81 \text{ m/s}^2 \times (1.2 \text{ s})^2$$
$$= 5.93 \text{ m}$$

Step 2. For question (b)

$$t_r = 1.1 \text{ s}$$
$$\phi = 45^\circ$$
$$\sin\phi = \sin 45 = 1/\sqrt{2} = 0.707$$
$$\sin 2\phi = \sin 90 = 1$$

Using Eq. (5.47)

$$h = g t_r^2 / (2 C_v^2 \sin\phi)$$
$$= 9.81 \text{ m/s}^2 \times (1.1 \text{ s})^2 / (2 \times 0.9^2 \times 0.707^2)$$
$$= 14.66 \text{ m}$$

Using Eq. (5.48)

$$r = 2 C_v^2 h \sin 2\phi$$
$$= 2(0.9)^2 (14.66 \text{ m})(1)$$
$$= 23.7 \text{ m}$$

and using Eq. (5.49)

$$h_r = C_v^2 h \sin^2\phi$$
$$= (0.9)^2 \times 14.66 \text{ m} \times (0.707)^2$$
$$= 5.93 \text{ m}$$

## 7 Solubility Equilibrium

Chemical precipitation is one of the most commonly employed methods for drinking water treatment. Coagulation with alum, ferric sulfate, or ferrous sulfate, and lime softening involve chemical precipitation. Most chemical reactions are reversible to some degree. A general chemical reaction which has reached equilibrium is commonly written by the law of mass action as

$$a\text{A} + b\text{B} \leftrightarrow c\text{C} + d\text{D} \tag{5.50a}$$

where A, B = reactants
C, D = products
$a, b, c, d$ = stoichiometric coefficients for A, B, C, D, respectively

The *equilibrium constant* $K_{eq}$ for the above reaction is defined as

$$K_{eq} = \frac{[C]^c[D]^d}{[A]^a[B]^b} \tag{5.50b}$$

where $K_{eq}$ is a true constant, called the equilibrium constant, and the square brackets signify the molar concentration of the species within the brackets. For a given chemical reaction, the value of equilibrium constant will change with temperature and the ionic strength of the solution.

For an equilibrium to exist between a solid substance and its solution, the solution must be saturated and in contact with undissolved solids. For example, at pH greater than 10, solid calcium carbonate in water reaches equilibrium with the calcium and carbonate ions in solution: consider a saturated solution of $CaCO_3$ that is in contact with solid $CaCO_3$. The chemical equation for the relevant equilibrium can be expressed as

$$CaCO_3\,(s) \leftrightarrow Ca^{2+}\,(aq) + CO_3^{2-}\,(aq) \tag{5.50c}$$

The equilibrium constant expression for the dissolution of $CaCO_3$ can be written as

$$K_{eq} = \frac{[Ca^{2+}][CO_3^{2-}]}{[CaCo_3]} \tag{5.50d}$$

Concentration of a solid substance is treated as a constant called $K_s$ in mass-action equilibrium, thus $[CaCO_3]$ is equal to $k_s$. Then

$$K_{eq}K_s = [Ca^{2+}][Co_3^{2-}] = K_{sp} \tag{5.50e}$$

The constant $K_{sp}$ is called the *solubility product constant*. The rules for writing the solubility product expression are the same as those for the writing of any equilibrium constant expression. The solubility product is equal to the product of the concentrations of the ions involved in the equilibrium, each raised to the power of its coefficient in the equilibrium equation. For the dissolution of a slightly soluble compound when the (brackets) concentration is denoted in moles, the equilibrium constant is called the solubility product constant. The general solubility product expression can be derived from the general dissolution reaction

$$A_xB_y(s) \leftrightarrow xA^{y+} + yB^{x-} \tag{5.50f}$$

and is expressed as

$$K_{sp} = [A^{y+}]^x [B^{x-}]^y \tag{5.50g}$$

Solubility product constants for various solutions at or near room temperature are presented in Appendix C.

If the product of the ionic molar concentration $[A^{y+}]^x [B^{x-}]^y$ is less than the $K_{sp}$ value, the solution is unsaturated, and no precipitation will occur. In contrast, if the product of the concentration of ions in solution is greater than $K_{sp}$ value, precipitation will occur under supersaturated condition.

**Example 1:** Calculate the concentration of $OH^-$ in a 0.20 mole (M) solution of $NH_3$ at 25°C. $K_{eq} = 1.8 \times 10^{-5}$

**solution:**

Step 1. Write the equilibrium expression

$$NH_3(aq) + H_2O(1) \leftrightarrow NH_4^+(aq) + OH^-(aq)$$

Step 2. Find the equilibrium constant

$$K_{eq} = \frac{[NH_4^+][OH^-]}{[NH_3]} = 1.8 \times 10^{-5}$$

Step 3. Tabulate the equilibrium concentrations involved in the equilibrium:

Let $x$ = concentration (M) of $OH^-$

$$NH_3\ (aq) + H_2O\ (1) \leftrightarrow NH_4^+(aq) + OH^-(aq)$$

| | $NH_3$ | $NH_4^+$ | $OH^-$ |
|---|---|---|---|
| Initial: | 0.2M | 0 M | 0 M |
| Equilibrium: | $(0.20 - x)$ M | $x$M | $x$M |

Step 4. Inserting these quantities into Step 2 to solve for $x$

$$K_{eq} = \frac{[NH_4^+][OH^-]}{[NH_3]}$$

$$= \frac{(x)(x)}{(0.20 - x)} = 1.8 \times 10^{-5}$$

or

$$x^2 = (0.20 - x)(1.8 \times 10^{-5})$$

Neglecting $x$ term, then approximately

$$x^2 \cong 0.2 \times 1.8 \times 10^{-5}$$

$$x = 1.9 \times 10^{-3}\,M$$

**Example 2:** Write the expression for the solubility product constant for (a) $A1(OH)_3$ and (b) $Ca_3(PO_4)_2$.

**solution:**

Step 1. Write the equation for the solubility equilibrium

(a) $A1(OH)_3\ (s) \leftrightarrow Al^{3+}(aq) + 3OH^-(aq)$

(b) $Ca_3\ (PO_4)_2\ (s) \leftrightarrow 3Ca^{2+}\ (aq) + 2PO_4^{3-}\ (aq)$

Step 2. Write $K_{sp}$ by using Eq. (5.50g) and from Appendix C

(a) $K_{sp} = [Al^{3+}]\ [OH^-]^3 = 2 \times 10^{-32}$

(b) $K_{sp} = [Ca^{2+}]^3\ [PO_4{}^{3-}]^2 = 2.0 \times 10^{-29}$

**Example 3:** The $K_{sp}$ for $CaCO_3$ is $8.7 \times 10^{-9}$ (Appendix C). What is the solubility of $CaCO_3$ in water in g/L?

**solution:**

Step 1. Write the equilibrium equation

$$CaCO_3\ (s) \leftrightarrow Ca^{2+}(aq) + CO_3^{2-}(aq)$$

Step 2. Set molar concentrations

For each mole of $CaCO_3$ that dissolves, 1 mole of $Ca^{2+}$ and 1 mole of $CO_3^{2-}$ enter the solution.

Let $x$ be the solubility of $CaCO_3$ in mol/L. The molar concentrations of $Ca^{2+}$ and $CO_3^{2-}$ are

$$[Ca^{2+}] = x \quad \text{and} \quad [CO_3^{2-}] = x$$

Step 3. Solve solubility $x$ in M/L

$$\begin{aligned} K_{sp} &= [Ca^{2+}][CO_3^{2-}] = 8.7 \times 10^{-9} \\ (x)(x) &= 8.7 \times 10^{-9} \\ x &= 9.3 \times 10^{-5}\ \text{M/L} \end{aligned}$$

1 mole of $CaCO_3$ = 100 g/L of $CaCO_3$

then

$$\begin{aligned} x &= 9.3 \times 10^{-5}(100\ \text{g/L}) \\ &= 0.0093\ \text{g/L} \\ &= 9.3\ \text{mg/L} \end{aligned}$$

## 8 Coagulation

Coagulation is a chemical process to remove turbidity and color producing material that is mostly colloidal particles (1 to 200 millimicrons, m$\mu$) such as algae, bacteria, organic and inorganic substances, and clay particles. Most colloidal solids in water and wastewater are negatively charged. The mechanisms of chemical coagulation involve the zeta potential derived from double-layer compression, neutralization by opposite charge, interparticle bridging, and precipitation. Destabilization of colloid particles is influenced by the *Van der Waals force* of attraction and Brownian movement. Detailed discussion of the theory of coagulation can be found elsewhere (American Society of Civil Engineers and American Water Works Association, 1990). Coagulation and flocculation processes were discussed in detail by Amirtharajah and O'Melia (1990).

Coagulation of water and wastewater generally add either aluminum, or iron salt, with and without polymers and coagulant aids. The process is complex and involves dissolution, hydrolysis, and polymerization. pH values play an important role in chemical coagulation depending on alkalinity. The chemical coagulation can be simplified as the following reaction equations:

Aluminum sulfate (alum):

$$Al_2\,(SO_4)\cdot 18H_2O + 3Ca(HCO_3)_2 \rightarrow 2A1(OH)_3\downarrow + 3CaSO_4 + 6CO_2 + 18H_2O \quad (5.51a)$$

When water does not have sufficient total alkalinity to react with alum, lime, or soda ash is usually also dosed to provide the required alkalinity. The coagulation equations can be written as below:

$$Al_2\,(SO_4)_3\cdot 18H_2O + 3Ca(OH)_2 \rightarrow 2A1(OH)_3\downarrow + 3CaSO_4 + 18H_2O \quad (5.51b)$$

$$A1_2(SO_4)_3\cdot 18H_2O + 3Na_2CO_3 + 3H_2O \rightarrow 2A1(OH)_3\downarrow + 3Na_2SO_4 + 3CO_2 + 18H_2O \quad (5.51c)$$

Ferric chloride:

$$2FeCl_3 + 3Ca(HCO_3)_2 \rightarrow 2Fe(OH)_3\downarrow + 3CaCl_2 + 6CO_2 \quad (5.51d)$$

Ferric sulfate:

$$Fe(SO_4)_3 + 3Ca(HCO_3)_2 \rightarrow 2Fe(OH)_3\downarrow + 3CaSO_4 + 6CO_2 \quad (5.51e)$$

Ferrous sulfate and lime:

$$FeSO_4\cdot 7H_2O + Ca(OH)_2 \rightarrow Fe(OH)_2 + CaSO_4 + 7H_2O \quad (5.51f)$$

followed by, in the presence of dissolved oxygen

$$4Fe(OH)_2 + O_2 + 2H_2O \rightarrow 4Fe(OH)_3\downarrow \quad (5.51g)$$

Chlorinated copperas:

$$3FeSO_4 \cdot 7H_2O + 1.5\ Cl_2 \rightarrow Fe_2(SO_4)_3 + FeCl_3 + 21H_2O \quad (5.51h)$$

followed by reacting with alkalinity as above

$$Fe_2(SO_4)_3 + 3Ca(HCO_3)_2 \rightarrow 2Fe(OH)_3 \downarrow +3CaSO_4 + 6CO_2 \quad (5.51i)$$

and

$$2FeCl_3 + 3Ca(HCO_3)_2 \rightarrow 2Fe(OH)_3 \downarrow +3CaCl_2 + 6CO_2 \quad (5.51j)$$

Each of the above reaction has an optimum pH range.

**Example 1:** What is the amount of natural alkalinity required for coagulation of raw water with dosage of 15.0 mg/L of ferric chloride?

**solution:**

Step 1. Write the reactions equation (Eq. 5.51j) and calculate MW

$$2FeCl_3 + 3Ca(HCO_3)_2 \rightarrow 2Fe(OH)_3 + 3CaCl_2 + 6CO_2$$

$$2(55.85 + 3 \times 35.45) \quad 3[40.08 + 2(1 + 12 + 48)]$$
$$= 324.4 \qquad\qquad = 486.3$$

The above equation suggests that 2 moles of ferric chloride react with 3 moles of $Ca(HCO_3)_2$.

Step 2. Determine the alkalinity needed for $X$

$$\frac{\text{mg/L } Ca(HCO_3)_2}{\text{mg/L } FeCl_2} = \frac{486.2}{324.4} = 1.50$$

$$X = 1.50 \times \text{mg/L } FeCl_2 = 1.50 \times 15.0 \text{ mg/L}$$

$$= 22.5 \text{ mg/L as } Ca(HCO_3)_2$$

Assume $Ca(HCO_3)_2$ represents the total alkalinity of the natural water. However, alkalinity concentration is usually expressed in terms of mg/L as $CaCO_3$. We need to convert $X$ to a concentration of mg/L as $CaCO_3$.

$$\text{MW of } CaCO_3 = 40.08 + 12 + 48 = 100.1$$

$$\text{MW of } Ca(HCO_3)_2 = 162.1$$

Then

$$X = 22.5 \text{ mg/L} \times \frac{100.1}{162.1} = 13.9 \text{ mg/L as } CaCO_3$$

**Example 2:** A water with low alkalinity of 12 mg/L as $CaCO_3$ will be treated with the alum-lime coagulation. Alum dosage is 55 mg/L. Determine the lime dosage needed to react with alum.

**solution:**

Step 1. Determine the amount of alum needed to react with the natural alkalinity

From Example 1

$$\text{Alkalinity} = 12 \text{ mg/L as } CaCO_3 \frac{162.1 \text{ as } Ca(HCO_3)_2}{100.1 \text{ as } CaCO_3}$$

$$= 19.4 \text{ mg/L as } Ca(HCO_3)_2$$

$$\text{MW of } A1_2(SO_4)_3 \cdot 18H_2O = 27 \times 2 + 3(32 + 16 \times 4) + 18(2 + 16)$$

$$= 666$$

$$\text{MW of } Ca(HCO_3)_2 = 162.1$$

Equation (5.51a) suggests that 1 mole of alum reacts with 3 moles of $Ca(HCO_3)_2$.

Therefore, the quantity of alum to react with natural alkalinity is $Y$:

$$Y = 19.4 \text{ mg/L as } Ca(HCO_3)_2 \times \frac{666 \text{ as alum}}{3 \times 162.1 \text{ as } Ca(HCO_3)_2}$$

$$= 26.6 \text{ mg/L as alum}$$

Step 2. Calculate the lime required

The amount of alum remaining to react with lime is (dosage $- Y$)

$$55 \text{ mg/L} - 26.6 \text{ mg/L} = 28.4 \text{ mg/L}$$

$$\text{MW of } Ca(OH)_2 = 40.1 + 2(16 + 1) = 74.1$$

$$\text{MW of CaO} = 40.1 + 16 = 56.1$$

Equation (5.51b) indicates that 1 mole of alum reacts with 3 moles of $Ca(OH)_2$. $Ca(OH)_2$ required:

$$28.4 \text{ mg/L alum} \frac{3 \times 74.1 \text{ as } Ca(OH)_2}{666 \text{ alum}}$$

$$= 9.48 \text{ mg/L as } Ca(OH)_2$$

Let dosage of lime required be $Z$

$$Z = 9.48 \text{ mg/L as } Ca(OH)_2 \times \frac{56.1 \text{ as CaO}}{74.1 \text{ as } Ca(OH)_2}$$

$$= 7.2 \text{ mg/L as CaO}$$

## 8.1 Jar test

For the jar test, chemical (coagulant) is added to raw water sample for mixing in the laboratory to simulate treatment-plant mixing conditions. Jar tests may provide overall process effectiveness, particularly to mixing intensity and duration as it affects floc size and density. It can also be used for evaluating chemical feed sequence, feed intervals, and chemical dilution ratios. A basic description of the jar test procedure and calculations is presented elsewhere (APHA 1995).

It is common to use six 2 L Gator jars with various dosages of chemical (alum, lime, etc); and one jar as the control without coagulant. Appropriate coagulant dosages are added to the 2 L samples before the rapid mixing at 100 revolution per minute (rpm) for 2 min. Then the samples and the control were flocculated at 20 rpm for 20 or more minutes and allow to settle. Water temperature, floc size, settling characteristics (velocity, etc.), color of supernatant, pH, etc. should be recorded.

The following examples illustrate calculations involved in the jar tests; prepare stock solution, jar test solution mixture.

**Example 1:** Given that liquid alum is used as a coagulant. Specific gravity of alum is 1.33. One gallon of alum weighs 11.09 pounds (5.03 kg) and contains 5.34 pounds (5.42 kg) of dry alum. Determine: (a) the alum concentration, (b) mL of liquid alum required to prepare a 100 mL solution of 20,000 mg/L alum concentration, (c) the dosage concentration of 1 mL of stock solution in a 2000 mL Gator jar sample.

**solution:**

Step 1. Determine alum concentration in mg/mL

$$\text{Alum (mg/L)} = \frac{(5.34 \text{ lb})\ (453{,}600 \text{ mg/lb})}{(1 \text{ gal})(3785 \text{ mL/gal})}$$

$$= 640 \text{ mg/mL}$$

Step 2. Prepare 100 mL stock solution having a 20,000 mg/L alum concentration

Let $x$ = mg of alum required to prepare 100 mL stock solution

$$\frac{x}{100 \text{ mL}} = \frac{20{,}000 \text{ mg}}{1000 \text{ mL}}$$

$$x = 2000 \text{ mg}$$

Step 3. Calculate mL ($y$) of liquid alum to give 2000 mg

$$\frac{y \text{ mL}}{1 \text{ mL}} = \frac{2000 \text{ mg}}{640 \text{ mg}}$$

$$y = 3.125 \text{ mL}$$

*Note*: Or add 6.25 mL of liquid alum into the 200 mL stock solution, since 3.125 mL is difficult to accurately measure.

Step 4. Find 1 mL of alum concentration ($z$) in 2000 mL sample (jar)

$$(z \text{ mg/L})\ (2000 \text{ mL}) = (20{,}000 \text{ mg/L})(1 \text{ mL})$$

$$z = \frac{20{,}000}{2000}\ (\text{mg/L})$$

$$= 10 \text{ mg/L}$$

*Note*: Actual final volume is 2001 mL; using 2000 mL is still reasonable.

**Example 2:** Assuming that 1 mL of potassium permanganate stock solution provides 1 mg/L dosage in a 2 L jar, what is the weight of potassium permanganate required to prepare 100 mL of stock solution? Assume 100% purity.

**solution:**

Step 1. Find concentration ($x$ mg/L) of stock solution needed

$$(x \text{ mg/L})\ (1 \text{ mL}) = (1 \text{ mg/L})\ (\text{as } 2000 \text{ mL})$$

$$x = 2000 \text{ mg/L}$$

Step 2. Calculate weight ($y$ mg) to prepare 100 mL (as 2000 mg/L)

$$\frac{y \text{ mg}}{100 \text{ mL}} = \frac{2000 \text{ mg}}{1000 \text{ mL}}$$

$$y = 200 \text{ mg}$$

Add 200 mg of potassium permanganate into 100 mL of deionized water to give 100 mL stock solution of 2000 mg/L concentration.

**Example 3:** Liquid polymer is used in a 2 L Gator jar test. It weighs 8.35 lb/gal. One mL of polymer stock solution provides 1 mg/L dosage. Compute how many mL of polymer are required to prepare 100 mL of stock solution.

**solution:**

Step 1. Calculate ($x$) mg of polymer in 1 mL

$$x = \frac{(8.35 \text{ lb})(453{,}600 \text{ mg/lb})}{(1 \text{ gal})(3785 \text{ mL/gal})}$$

$$= 1000 \text{ mg/mL}$$

Step 2. As in Example 2, find the weight (y mg) to prepare 100 mL of 2000 mg/L stock solution

$$\frac{y \text{ mg}}{100 \text{ mL}} = \frac{2000 \text{ mg}}{1000 \text{ mL}}$$

$$y = 200 \text{ mg (of polymer in 100 mL distilled water)}$$

Step 3. Compute volume ($z$ mL) of liquid polymer to provide 200 mg

$$z = \frac{200 \text{ mg}}{x} = \frac{200 \text{ mg} \times 1 \text{ mL}}{1000 \text{ mg}}$$

$$= 0.20 \text{ mL}$$

*Note*: Use 0.20 mL of polymer in 100 mL of distilled water. This stock solution has 2000 mg/L polymer concentration. One mL stock solution added to a 2 L jar gives 1 mg/L dosage.

## 8.2 Mixing

Mixing is an important operation for the coagulation process. In practice, rapid mixing provides complete and uniform dispersion of a chemical added to the water. Then follows a slow mixing for flocculation (particle aggregation). The time required for rapid mixing is usually 10 to 20 s. However, recent studies indicate the optimum time of rapid mixing is a few minutes.

Types of mixing include propeller, turbine, paddle, pneumatic, and hydraulic mixers. For a water treatment plant, mixing is used for coagulation and flocculation, and chlorine disinfection. Mixing is also used for biological treatment processes for wastewater. Rapid mixing for coagulant in raw water and activated-sludge process in wastewater treatment are complete mixing. Flocculation basins after rapid mixing are designed based on an ideal plug flow using first-order kinetics. It is very difficult to achieve an ideal plug flow. In practice, baffles are installed to reduce short-circuiting. The time of contact or detention time in the basin can be determined by:

For complete mixing

$$t = \frac{V}{Q} = \frac{1}{K}\left(\frac{C_i - C_e}{C_e}\right) \tag{5.52a}$$

For plug flow

$$t = \frac{V}{Q} = \frac{L}{v} = \frac{1}{K}\left(\ln\frac{C_i}{C_e}\right) \tag{5.52b}$$

where $t$ = detention time of the basin, min
$V$ = volume of basin, $m^3$ or $ft^3$
$Q$ = flow rate, $m^3/s$ or cfs
$K$ = rate constant
$C_i$ = influent reactant concentration, mg/L
$C_e$ = effluent reactant concentration, mg/L
$L$ = length of rectangular basin, m or ft
$v$ = horizontal velocity of flow, m/s or ft/s

**Example 1:** Alum dosage is 50 mg/L, 379 = 90 per day based on laboratory tests. Compute the detention times for complete mixing and plug flow reactor for 90% reduction.

**solution:**

Step 1. Find $C_e$

$$C_e = (1 - 0.9)C_i = 0.1 \times C_i = 0.1 \times 50 \text{ mg/L}$$
$$= 5 \text{ mg/L}$$

Step 2. Calculate $t$ for complete mixing

Using Eq. (5.52a)

$$t = \frac{1}{K}\left(\frac{C_i - C_e}{C_e}\right) = \frac{1}{90/\text{day}}\left(\frac{50 \text{ mg/L} - 5 \text{ mg/L}}{5 \text{ mg/L}}\right)$$
$$= \frac{1\text{day}}{90} \times \frac{1440 \text{ min}}{1 \text{ day}} \times 9$$
$$= 144 \text{ min}$$

Step 3. Calculate $t$ for plug flow

Using Eq. (5.52b)

$$t = \frac{1}{K}\left(\ln\frac{C_i}{C_e}\right) = \frac{1440 \text{ min}}{90}\left(\ln\frac{50}{5}\right)$$
$$= 36.8 \text{ min}$$

**Power requirements.** Power required for turbulent mixing is traditionally based on the velocity gradient or $G$ values proposed by Camp and Stein (1943). The mean velocity gradient $G$ for mechanical mixing is

$$G = \left(\frac{P}{\mu V}\right)^{1/2} \quad (5.53)$$

where $G$ = mean velocity gradient; velocity (ft/s)/distance (ft) is equal to per second
$P$ = power dissipated, ft · lb/s or N · m/s (W)
$\mu$ = absolute viscosity, lb · s/ft$^2$ or N · s/m$^2$
$V$ = volume of basin, ft$^3$ or m$^3$

The equation is used to calculate the mechanical power required to facilitate rapid mixing. If a chemical is injected through orifices with mixing times of approximately 1.0 s, the $G$ value is in the range of 700 to 1000/s. In practice, $G$ values of 3000 to 5000/s are preferable for rapid mixing (ASCE and AWWA, 1990).

Eq. (5.53) can be expressed in terms of horsepower (hp) as

$$G = \left(\frac{550 \text{ hp}}{\mu V}\right)^{1/2} \tag{5.54a}$$

For SI units the velocity gradient per second is:

$$G = \left(\frac{\text{kW} \times 10}{\mu V}\right)^{1/2} \tag{5.54b}$$

where kW = energy input, kW
$V$ = effective volume, m$^3$
$\mu$ = absolute viscosity, centipoise, cp
= 1 cp at 20°C (see Table 4.1a, 1cp = 0.001 N · s/m$^2$)

The equation is the standard design guideline used to calculate the mechanical power required to facilitate rapid mixing. Camp (1968) claimed that rapid mixing at $G$ values of 500 to 1000/s for 1 to 2 min produced essentially complete flocculation and no further benefit for prolonged rapid mixing. For rapid mixing the product of $Gt$ should be 30,000 to 60,000 with $t$ (time) generally 60 to 120 s.

**Example 2:** A rapid mixing tank is 1 m × 1 m × 1.2 m. The power input is 746 W (1 hp). Find the $G$ value at a temperature of 15°C.

**solution:** At 10°C, $\mu$ = 0.00113 N · s/m$^2$ (from Table 4.1a)

$$V = 1\text{m} \times 1\text{m} \times 1.2 \text{ m} = 1.2 \text{ m}^3$$

$$P = 746\ W = 746 \text{ N} \cdot \text{m/s}$$

Using Eq. (5.53):

$$G = (P/\mu V)^{0.5}$$
$$= \left(\frac{746 \text{ N} \cdot \text{m/s}}{0.00113 \text{ N} \cdot \text{s/m}^2 \times 1.2 \text{ m}^3}\right)^{0.5}$$
$$= 742 \text{ s}^{-1}$$

## 9 Flocculation

After rapid mixing, the water is passed through the flocculation basin. It is intended to mix the water to permit agglomeration of turbidity settled particles (solid capture) into larger floes which would have a mean velocity gradient ranging 20 to 70 $s^{-1}$ for a contact time of 20 to 30 min taking place in the flocculation basin. A basin is usually designed in four compartments (ASCE and AWWA 1990).

The conduits between the rapid mixing tank and the flocculation basin should maintain $G$ values of 100 to 150 $s^{-1}$ before entering the basin.

For baffled basin, the $G$ value is

$$G = \left(\frac{Q\gamma H}{\mu V}\right)^{0.5} = \left(\frac{62.4H}{\mu t}\right)^{0.5} \tag{5.55}$$

where $G$ = mean velocity gradient, $s^{-1}$
$Q$ = flow rate, $ft^3/s$
$\gamma$ = specific weight of water, 62.4 $lb/ft^3$
$H$ = head loss due to friction, ft
$\mu$ = absolute viscosity, $lb \cdot s/ft^2$
$V$ = volume of flocculator, $ft^3$
$t$ = detention time, s

For paddle flocculators, the useful power input of an impeller is directly related to the drag force of the paddles ($F$). The drag force is the product of the coefficient of drag ($C_d$) and the impeller force ($F_i$). The drag force can be expressed as (Fair *et al.*, 1968):

$$F = C_d F_i \tag{5.56a}$$

$$F_i = \rho A \frac{v^2}{2} \tag{5.57}$$

then

$$F = 0.5\, C_d\, \rho A v^2 \tag{5.56b}$$

where $F$ = drag force, lb
$C_d$ = dimensionless coeffiicient of drag
$\rho$ = mass density, lb · s$^2$/ft$^4$
$A$ = area of the paddles, ft$^2$
$v$ = velocity difference between paddles and water, fps

The velocity of paddle blades ($v_p$) can be determined by

$$v_p = \frac{2\pi rn}{60} \tag{5.58}$$

where $v_p$ = velocity of paddles, fps
$n$ = number of revolutions per minute, rpm
$r$ = distance from shaft to center line of the paddle, ft

The useful power input is computed as the product of drag force and velocity difference as below:

$$P = Fv = 0.5\,C_d\,\rho\,Av^3 \tag{5.59}$$

**Example 1:** In a baffled basin with detention time of 25 min. Estimate head loss if $G$ is 30/s, $\mu = 2.359 \times 10^{-5}$ lb · s/ft$^2$ at $T$ = 60°F (Table 4.1b).

**solution:** Using Eq. (5.55)

$$G = \left(\frac{62.4H}{\mu t}\right)^{1/2}$$

$$H = \frac{G^2\mu t}{62.4} = \frac{(30\text{s}^{-1})^2(2.359 \times 10^{-5}\ \text{lb}\cdot\text{s/ft}^2) \times (25 \times 60\ \text{s})}{62.4\ \text{lb/ft}^3}$$

$$= 0.51\ \text{ft}$$

**Example 2:** A baffled flocculation basin is divided into 16 channels by 15 around-the-end baffles. The velocities at the channels and at the slots are 0.6 and 2.0 fps (0.18 and 0.6 m/s), respectively. The flow rate is 12.0 cfs (0.34 m$^3$/s). Find (a) the total head loss neglecting channel friction; (b) the power dissipated; (c) the mean velocity gradient at 60°F (15.6°C); the basin size is 16 × 15 × 80 ft$^3$; (d) the $Gt$ value, if the detention (displacement) time is 20 min, and (e) loading rate in gpd/ft$^3$.

**solution:**

Step 1. Estimate loss of head $H$

$$\text{Loss of head in a channel} = \frac{v_1^2}{2g} = \frac{0.6^2}{2 \times 32.2} = 0.00559\ \text{(ft)}$$

$$\text{Loss of head in a slot} = \frac{2^2}{64.4} = 0.0621\ \text{(ft)}$$

(a)
$$H = 16 \times 0.00559 + 15 \times 0.0621$$
$$= 1.02 \text{ (ft)}$$

Step 2. Compute power input $P$

(b)
$$P = Q_{\gamma}H = 12 \text{ ft}^3/\text{s} \times 62.37 \text{ lb/ft}^3 \times 1.02 \text{ ft}$$
$$= 763 \text{ ft} \cdot \text{lb/s}$$

*Note*: $\gamma = 62.37 \text{ lb/ft}^3$ at 60°F (Table 5.1b)

Step 3. Compute $G$

From Table 4.1b

At 60°F,
$$\mu = 2.359 \times 10^{-5} \text{lb} \cdot \text{s/ft}^2$$
$$V = 16 \text{ ft} \times 15 \text{ ft} \times 80 \text{ ft} = 19{,}200 \text{ ft}^3$$

Using Eq. (5.53)

(c)
$$G = \left(\frac{P}{\mu V}\right)^{1/2} = \left(\frac{763 \text{ ft} \cdot \text{lb/s}}{2.359 \times 10^{-5} \text{lb} \cdot \text{s/ft}^2 \times 19{,}200 \text{ ft}^3}\right)^{1/2}$$
$$= 41.0 \text{ s}^{-1}$$

Step 4. Compute $Gt$

(d)
$$Gt = 41 \text{ s}^{-1} \times 20 \text{ min} \times 60 \text{ s/min}$$
$$= 49.200$$

Step 5. Compute loading rate

$$Q = 12 \text{ cfs} = 12 \text{ cfs} \times 0.646 \text{ MGD/cfs}$$
$$= 7.75 \text{ MGD}$$
$$= 7.75 \times 10^6 \text{ gpd}$$

(e)
$$\text{Loading rate} = Q/V = 7.76 \times 10^6 \text{ gpd}/19{,}200 \text{ ft}^3$$
$$= 404 \text{ gpd/ft}^3$$

**Example 3:** A flocculator is 16 ft (4.88 m) deep, 40 ft (12.2 m) wide, and 80 ft (24.4 m) long. The flow of the water plant is 13 MGD (20 cfs, 0.57 $\text{m}^3/\text{s}$). Rotating paddles are supported parallel to four horizontal shafts. The rotating speed is 2.0 rpm. The center line of the paddles is 5.5 ft (1.68 m) from the shaft (mid-depth of the basin). Each shaft equipped with six paddles. Each paddle blade is 10 in (25 cm) wide and 38 ft (11.6 m) long. Assume the mean velocity of the water is 28% of the velocity of the paddles and their drag coefficient is 1.9. Estimate:

(a) the difference in velocity between the paddles and water
(b) the useful power input
(c) the energy consumption per million gallons (Mgal)
(d) the detention time
(e) the value of $G$ and $Gt$ at 60°F
(f) the loading rate of the flocculator

**solution:**

Step 1. Find velocity differential $v$

Using Eq. (5.58)

$$v_p = \frac{2\pi rn}{60} = \frac{2 \times 3.14 \times 5.5\text{ft} \times 2}{60 \text{ s}}$$
$$= 1.15 \text{ fps}$$

(a)

$$v = v_p(1 - 0.28) = 1.15 \text{ fps} \times 0.72$$
$$= 0.83 \text{ fps}$$

Step 2. Find $P$

$A$ = paddle area = 4 shaft × 6 pads/shaft ×38 ft × 10/12 ft/pad = 760 ft$^2$

$$\rho = \frac{\gamma}{g} = \frac{62.4 \text{ lb/ft}^3}{32.2 \text{ ft/s}^2} = 1.938 \text{ lb}\cdot\text{s}^2/\text{ft}^4$$

Using Eq. (5.59):

(b)

$$P = 0.5C_d \rho A v^3$$
$$= 0.5 \times 1.9 \times 1.938 \text{ lb} \cdot \text{s}^2/\text{ft}^4 \times 760 \text{ ft}^2 \times (0.83 \text{ fps})^3$$
$$= 800 \text{ ft} \cdot \text{lb/s}$$

or

$$= (800/550) \text{ hp}$$
$$= 1.45 \text{ hp}$$

or

$$= 1.45 \times 0.746 \text{ kW}$$
$$= 1.08 \text{ kW}$$

Step 3. Determine energy consumption $E$

(c)

$$E = \frac{1.45\text{hp}}{13 \text{ Mgal/d}} \times \frac{24 \text{ h}}{\text{d}}$$
$$= 2.68 \text{ hp}\cdot\text{h/Mgal}$$

or

$$E = \frac{1.08 \text{ kW}}{13 \text{ Mgal/d}} \times \frac{24 \text{ h}}{\text{d}}$$
$$= 1.99 \text{ kW}\cdot\text{h/Mgal}$$

Step 4. Determine detention time $t$

$$\text{Basin volume, } V = 16 \text{ ft} \times 40 \text{ ft} \times 80 \text{ ft} = 5.12 \times 10^4 \text{ ft}^3$$
$$= 5.12 \times 10^4 \text{ ft}^3 \times 7.48 \text{ gal/ft}^3$$
$$= 3.83 \times 10^5 \text{ gal}$$

(d)

$$t = \frac{V}{Q} = \frac{3.83 \times 10^5 \text{ gal}}{13 \times 10^6 \text{ gal/d}} = 0.0295 \text{ day}$$
$$= 0.0295 \text{ day} \times 1440 \text{ min/d}$$
$$= 42.5 \text{ min}$$

Step 5. Compute $G$ and $Gt$

Using Eq. (5.53):

(e)

$$G = (P/\mu\text{V})^{0.5}$$
$$= (800/2.359 \times 10^{-5} \times 5.12 \times 10^4)^{0.5}$$
$$= 25.7 \text{ (fps/ft, or s}^{-1})$$
$$Gt = (25.7\text{s}^{-1})\ (42.5 \times 60 \text{ s})$$
$$= 65{,}530$$

Step 6. Compute the loading rate

(f)

$$\text{Loading rate} = \frac{Q}{V} = \frac{13 \times 10^6 \text{ gpd}}{5.12 \times 10^4 \text{ ft}^3}$$
$$= 254 \text{ gpd/ft}^3$$

## 10 Sedimentation

Sedimentation is one of the most basic processes of water treatment. Plain sedimentation, such as the use of a presedimentation basin (grit chamber) and sedimentation tank (or basin) following coagulation—flocculatiaon, is the most commonly used in water treatment facilities. The grit chamber is generally installed upstream of a raw water pumping station to remove larger particles or objects. It is usually a rectangular horizontal-flow tank with a contracted inlet and the bottom should have at minimum a 1:100 longitudinal slope for basin draining and cleaning purposes. A trash screen (about 2-cm opening) is usually installed at the inlet of the grit chamber.

Sedimentation is a solid—liquid separation by gravitational settling. There are four types of sedimentation: discrete particle settling (type 1), flocculant settling (type 2), hindered settling (type 3), and compression settling (type 4). Sedimentation theories for the four types are discussed in Chapter 6 and elsewhere (Gregory and Zabel, 1990).

The terminal settling velocity of a single discrete particle is derived from the forces (gravitational force, buoyant force, and drag force) that act on the particle. The classical discrete particle settling theories have been based on spherical particles. The equation is expressed as

$$u = \left(\frac{4g(\rho_p - \rho)d}{3C_{D\rho}}\right)^{1/2} \tag{5.60}$$

where $u$ = settling velocity of particles, m/s or ft/s
$g$ = gravitational acceleration, $m/s^2$ or $ft/s^2$
$\rho_p$ = density of particles, $kg/m^3$ or $lb/ft^3$
$\rho$ = density of water, $kg/m^3$ or $lb/ft^3$
$d$ = diameter of particles, m or ft
$C_D$ = coefficient of drag

The values of drag coefficient depend on the density of water ($\rho$), relative velocity ($\mu$), particle diameter ($d$), and viscosity of water ($\mu$), which gives the Reynolds number **R** as:

$$\mathbf{R} = \frac{\rho u d}{\mu} \tag{5.61}$$

The value of $C_D$ decreases as the Reynolds number increases. For **R** less than 2 (or 1), $C_D$ is related to **R** by the linear expression as follows:

$$C_D = \frac{24}{\mathbf{R}} \tag{5.62}$$

At low values of **R,** substituting Eq. (5.61) and (5.62) into Eq. (5.60) gives

$$u = \frac{g(\rho_p - \rho)d^2}{18\,\mu} \tag{5.63}$$

This expression is known as the Stokes' equation for laminar flow conditions.

In the region of higher Reynolds numbers ($2 < \mathbf{R} < 500-1000$), $C_D$ becomes (Fair *et al.*, 1968):

$$C_D = \frac{24}{\mathbf{R}} + \frac{3}{\sqrt{\mathbf{R}}} + 0.34 \tag{5.64}$$

In the region of turbulent flow ($500-1000 < \mathbf{R} < 200{,}000$,) the $C_D$ remains approximately constant at 0.44. The velocity of settling particles results in Newton's equation (ASCE and AWWA 1990):

$$u = 1.74\left[\frac{(\rho_p - \rho)gd}{\rho}\right]^{1/2} \tag{5.65}$$

When the Reynolds number is greater than 200,000, the drag force decreases substantially and $C_D$ becomes 0.10. No settling occurs at this condition.

**Example:** Estimate the terminal settling velocity in water at a temperature of 15°C of spherical silicon particles with specific gravity 2.40 and average diameter of (a) 0.05 mm and (b) 1.0 mm.

**solution:**

Step 1. Using the Stokes' equation (Eq. (5.63)) for (a)

From Table 4.1a, at $T = 15°\text{C}$

$$\rho = 999 \text{ kg/m}^3, \text{ and } \mu = 0.00113 \text{ N} \cdot \text{s/m}^2$$
$$d = 0.05 \text{ mm} = 5 \times 10^{-5} \text{ m}$$

(a)

$$u = \frac{g(\rho_p - \rho)d^2}{18\mu}$$

$$= \frac{9.81 \text{ m/s}^2(2400 - 999) \text{ kg/m}^3(5 \times 10^{-5}\text{m})^2}{18 \times 0.00113 \text{ N} \cdot \text{s/m}^2}$$

$$= 0.00169 \text{ m/s}$$

Step 2. Check with the Reynolds number (Eq. (5.61))

$$\mathbf{R} = \frac{\rho u d}{\mu} = \frac{999 \times 0.00169 \times 5 \times 10^{-5}}{0.00113}$$

$$= 0.075$$

(a) The Stokes' law applies, since $\mathbf{R} < 2$.

Step 3. Using the Stokes' law for (b), $d = 1$ mm $= 0.001$ m

$$u = \frac{9.81(2400 - 999)(0.001)^2}{18 \times 0.00113}$$
$$= 0.676 \text{ (m/s)}$$

Step 4. Check the Reynold number

Assume the irregularities of the particles $\phi = 0.85$

$$\mathbf{R} = \frac{\phi \rho u d}{\mu} = \frac{0.85 \times 999 \times 0.676 \times 0.001}{0.00113}$$
$$= 508$$

Since $\mathbf{R} > 2$, the Stokes' law does not apply. Use Eq. (5.60) to calculate $u$.

Step 5. Using Eqs. (5.64) and (5.60)

$$C_\mathrm{D} = \frac{24}{\mathbf{R}} + \frac{3}{\sqrt{\mathbf{R}}} + 0.34 = \frac{24}{508} + \frac{3}{\sqrt{508}} + 0.34$$
$$= 0.52$$

$$u^2 = \frac{4g(\rho_\mathrm{p} - \rho)d}{3C_\mathrm{D}\rho}$$

$$u^2 = \frac{4 \times 9.81 \times (2400 - 999) \times 0.001}{3 \times 0.52 \times 999}$$

$$u = 0.188 \text{ (m/s)}$$

Step 6. Recheck $\mathbf{R}$

$$\mathbf{R} = \frac{\phi \rho u d}{\mu} = \frac{0.85 \times 999 \times 0.188 \times 0.001}{0.00113}$$
$$= 141$$

Step 7. Repeat Step 5 with new $\mathbf{R}$

$$C_\mathrm{D} = \frac{24}{141} + \frac{3}{\sqrt{141}} + 0.34$$
$$= 0.76$$

$$u^2 = \frac{4 \times 9.81 \times 1401 \times 0.001}{3 \times 0.76 \times 999}$$

$$u = 0.155 \text{ (m/s)}$$

Step 8. Recheck **R**

$$\mathbf{R} = \frac{0.85 \times 999 \times 0.155 \times 0.001}{0.00113}$$

$$= 116$$

Step 9. Repeat Step 7

$$C_D = \frac{24}{116} + \frac{3}{\sqrt{116}} + 0.34$$

$$= 0.83$$

$$u^2 = \frac{4 \times 9.81 \times 1401 \times 0.001}{3 \times 0.83 \times 999}$$

$$u = 0.149 \text{ (m/s)}$$

(b) The estimated velocity is around 0.15 m/s, based on Steps 7 and 9.

### 10.1 Overflow rate

For sizing the sedimentation basin, the traditional criteria used are based on overflow rate, detention time, weir loading rate, and horizontal velocity. The theoretical detention time is computed from the volume of the basin divided by average daily flow (plug flow theory):

$$t = \frac{24V}{Q} \tag{5.66}$$

where $t$ = detention time, h
24 = 24 h/d
$V$ = volume of basin, $m^3$ or million gallon (Mgal)
$Q$ = average daily flow, $m^3$/d or Mgal/d (MGD)

The overflow rate is a standard design parameter which can be determined from discrete particle settling analysis. The overflow rate or surface loading rate is calculated by dividing the average daily flow by the total area of the sedimentation basin as follows:

$$u = \frac{Q}{A} = \frac{Q}{lw} \tag{5.67}$$

where $u$ = overflow rate, $m^3/(m^2 \cdot d)$ or gpd/ft$^2$
$Q$ = average daily flow, $m^3$/d or gpd
$A$ = total surface area of basin, $m^2$ or ft$^2$
$l$ and $w$ = length and width of basin, respectively m or ft

For alum coagulation, $u$ is usually in the range of 40 to 60 $m^3/(m^2 \cdot d)$ (or m/d) (980 to 1470 gpd/ft$^2$) for turbidity and color removal. For lime softening, the overflow rate ranges 50 to 110 m/d (1230 to 2700 gpm/ft$^2$). The overflow rate in wastewater treatment is lower, ranging from 10 to 60 m/d (245 to 1470 gpm/ft$^2$). All particles having a settling velocity greater than the overflow rate will settle and be removed.

It should be noted that rapid particle density changes due to temperature, solid concentration, or salinity can induce density current which can cause severe short-circuiting in horizontal tanks (Hudson, 1972).

**Example:** A water treatment plant has four clarifiers treating 4.0 MGD (0.175 $m^3$/s) of water. Each clarifier is 16 ft (4.88 m) wide, 80 ft (24.4 m) long, and 15 ft (4.57 m) deep. Determine: (a) the detention time, (b) overflow rate, (c) horizontal velocity, and (d) weir loading rate assuming the weir length is 2.5 times the basin width.

**solution:**

Step 1. Compute detention time $t$ for each clarifier, using Eq. (5.66)

$$Q = \frac{4\text{ MGD}}{4} = \frac{1{,}000{,}000\text{ gal}}{\text{d}} \times \frac{1\text{ ft}^3}{7.48\text{ gal}} \times \frac{1\text{ day}}{24\text{ h}}$$

$$= 5570\text{ ft}^3/\text{h}$$

$$= 92.83\text{ ft}^3/\text{min}$$

(a)

$$t = \frac{V}{Q} = \frac{16\text{ft} \times 80\text{ft} \times 15\text{ft}}{5570\text{ ft}^3/\text{h}}$$

$$= 3.447\text{ h}$$

Step 2. Compute overflow rate $u$, using Eq. (5.67)

(b)

$$u = \frac{Q}{lw} = \frac{1{,}000{,}000\text{ gpd}}{16\text{ ft} \times 80\text{ ft}}$$

$$= 781\text{ gpd/ft}^2 = 31.83\text{ m}^3/\text{m}^2 \cdot \text{d}$$

Step 3. Compute horizontal velocity $v$

(c)

$$v = \frac{Q}{wd} = \frac{92.83\text{ ft}^3/\text{min}}{16\text{ft} \times 15\text{ft}}$$

$$= 0.387\text{ ft/min} = 11.8\text{ cm/min}$$

Step 4. Compute weir loading rate $u_w$

(d)
$$u_w = \frac{Q}{2.5w} = \frac{1{,}000{,}000\ \text{gpd}}{2.5 \times 16\,\text{ft}}$$
$$= 25{,}000\ \text{gpd/ft} = 310\ \text{m}^3/\text{m}\cdot\text{d}$$

### 10.2 Inclined settlers

Inclined (tube and plate) settlers are sedimentation units that have been used for more than two decades. A large number of smaller diameter (20 to 50 mm) tubes are nested together to act as a single unit and inclined with various angles (7° to 60°). The typical separation distance between inclined plates for unhindered settling is 2 in (5 cm) with inclines of 3 to 6 ft (1 to 2 m) height. The solids or flocs settle by gravitational force. It is not necessary to use tubes and can take various forms or plates also. The materials are lightweight, generally PVC or ABC plastic (1 m × 3 m in size).

Tube settlers have proved as effective units. However, there is a tendency of clogging.

Inclined settling systems can be designed as cocurrent, countercurrent, and cross-flow. Comprehensive theoretical analyses of various flow geometries have been discussed by Yao (1976). The flow velocity of the settler module ($v$) and the surface loading rate for the inclined settler ($u$) (Fig. 5.6) are calculated as (James M. Montgomery Consulting Engineering, 1985):

$$v = \frac{Q}{A \sin\theta} \tag{5.68}$$

$$u = \frac{Qw}{A(H\cos\theta + w\cos^2\theta)} \tag{5.69}$$

where $v$ = velocity of the water in settlers, m/s or ft/s
$Q$ = flow rate, $\text{m}^3$/s or MGD
$A$ = surface area of basin, $\text{m}^2$ or $\text{ft}^2$
$\theta$ = inclined angle of the settlers
$u$ = settling velocity, m/s or ft/s
$w$ = width of settler, m or ft
$H$ = vertical height, m or ft.

**Example:** Two flocculators treat 1.0 $\text{m}^3$/s (22.8 MGD) and remove flocs large than 0.02 mm. The settling velocity of the 0.02 mm flocs is measured in the laboratory as 0.22 mm/s (0.67 in/min) at 15°C. Tube settlers of 50.8 mm (2 in) square honeycombs are inclined at a 50° angle, and its vertical height

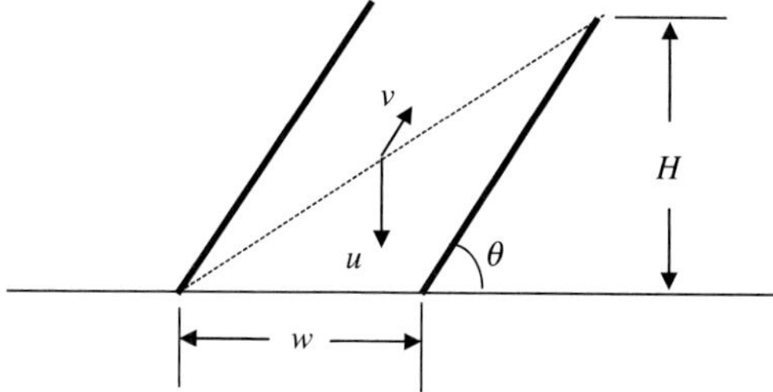


**Figure 5.6** Schematic diagram of settling section.

is 1.22 m (4 ft). Determine the basin area required for the settler module and the size of each flocculator at 15°C.

**solution:**

Step 1. Determine the area needed for the settler modules

$$Q = (1\ \text{m}^3/\text{s})/2 = 0.5\ \text{m}^3/\text{s} = 30\ \text{m}^3/\text{min}$$

$$w = 50.8\ \text{mm} = 0.0508\ \text{m}$$

$$H = 1.22\ \text{m}$$

$$\theta = 50°$$

Using Eq. (5.69)

$$u = \frac{Qw}{A(H\cos\theta + w\cos^2\theta)}$$

$$= \frac{0.5(0.0508)}{A(1.22 \times 0.643 + 0.0508 \times 0.643^2)}$$

$$= \frac{0.0315}{A}$$

Step 2. Determine $A$

In practice, the actual conditions in the settlers are not as good as under controlled laboratory ideal conditions.

A safety factor of 0.6 may be applied to determine the designed settling velocity. Thus

$$u = 0.6 \times 0.00022\ \text{m/s} = \frac{0.0315}{A}$$

$$A = 238.6\,\text{m}^2$$

$$(\text{use } 240\ \text{m}^2)$$

Step 3. Find surface loading rate $Q/A$

$$Q/A = (0.5 \times 24 \times 60 \times 60 \text{ m}^3/\text{d})/240 \text{ m}^2$$
$$= 180 \text{ m}^3/(\text{m}^2 \cdot \text{d})$$
$$= 3.07 \text{ gpm/ft}^2 \text{ (Note: } 1\text{m}^3/(\text{m}^2 \cdot \text{d}) = 0.017 \text{ gpm/ft}^2$$

Step 4. Compute flow velocity in the settlers, using Eq. (5.68)

$$v = Q/A \sin\theta = 180/0.766$$
$$= 235 \text{ (m/d)}$$
$$= 0.163 \text{ m/min}$$
$$= 0.0027 \text{ m/s}$$

Step 5. Determine size of the basin

Two identical settling basins are designed. Generally, the water depth of the basin is 4 m (13.1 ft). The width of the basin is chosen as 8.0 m (26.2 ft). The calculated length of the basin covered by the settler is

$$l = 240 \text{ m}^2/8 \text{ m} = 30 \text{ m} = 98 \text{ ft}$$

In practice, one-fourth of the basin length is left as a reserved volume for future expansion. The total length of the basin should be

$$30 \text{ m} \times \frac{4}{3} = 40 \text{ m} (131 \text{ ft})$$

Step 6. Check horizontal velocity

$$Q/A = (30 \text{ m}^3/\text{min})/(4 \text{ m} \times 8 \text{ m})$$
$$= 0.938 \text{ m/min} = 3 \text{ ft/min}$$

Step 7. Check Reynolds number (**R**) in the settler module

$$\text{Hydraulic radius } R = \frac{0.0508^2}{4 \times 0.0508} = 0.0127 \text{ m}$$

$$\mathbf{R} = \frac{vR}{\mu} = \frac{(0.0027 \text{ m/s})(0.0127 \text{ m})}{0.000001131 \text{ m}^2/\text{s}}$$

$$= 30 < 2000, \text{ thus it is in a lamella flow}$$

## 11 Filtration

The conventional filtration process is probably the most important single unit operation of all the water treatment processes. It is an operation process to separate suspended matter from water by flowing it through porous filter medium or media. The filter media may be silica sand, anthracite coal, diatomaceous earth, garnet, ilmenite, or finely woven fabric.

In early times, a slow sand filter was used. It is still proved to be efficient. It is very effective for removing flocs containing microorganisms such as algae, bacteria, virus, *Giardia,* and *Cryptosporidium.* Rapid filtration has been very popular for several decades. Filtration usually follows the coagulation–flocculation–sedimentation processes. However, for some water treatment, direct filtration is used due to the high quality of raw water. Dual-media filters (sand and anthracite, activated carbon, or granite) give more benefits than single-media filters and became more popular; even triple-media filters have been used. In Russia, up-flow filters are used. All filters need to clean out the medium by backwash after a certain period (most are based on head loss) of filtration.

The filters are also classified by allowing loading rate. Loading rate is the flow rate of water applied to the unit area of the filter. It is the same value as the flow velocity approaching the filter surface and can be determined by

$$v = Q/A \tag{5.67a}$$

where $v$ = loading rate, $m^3/(m^2 \cdot d)$ or gpm/ft$^2$
$Q$ = flow rate, $m^3$/d or ft$^3$/d or gpm
$A$ = surface area of filter, $m^2$ or ft$^2$

On the basis of loading rate, the filters are classified as slow sand filters, rapid sand filters, and high-rate sand filters. With each type of filter medium or media, there are typical design criteria for the range of loading rate, effective size, uniform coefficient, minimum depth requirements, and backwash rate. The typical loading rate for rapid sand filters is 120 $m^3/(m^2 \cdot d)$ [83 $L/(m^2 \cdot min)$ or 2 gpm/ft$^2$]. For high-rate filters, the loading rate may be four to five times this rate.

**Example:** A city is to install rapid sand filters downstream of the clarifiers. The design loading rate is selected to be 160 $m^3/(m^2 \cdot d)$ (2.7 gpm/ft$^2$). The design capacity of the waterworks is 0.35 $m^3/s$ (8 MGD). The maximum surface per filter is limited to 50 $m^2$. Design the number and size of filters and calculate the normal filtration rate.

**solution:**

Step 1. Determine the total surface area required

$$A = \frac{Q}{v} = \frac{0.35\ m^3/s\,(86{,}400\ s/d)}{160\ m^3/m^2 \cdot d}$$
$$= 189\ m^2$$

Step 2. Determine the number ($n$) of filters

$$n = \frac{189 \text{ m}^2}{50 \text{ m}} = 3.78$$

Select four filters.

The surface area ($a$) for each filter is

$$a = 189 \text{ m}^2/4 = 47.25 \text{ m}^2$$

We can use 7 m × 7 m or 6 m × 8 m, or 5.9 m × 8 m (exact)

Step 3. If a 7 m × 7 m filter is installed, the normal filtration rate is

$$v = \frac{Q}{A} = \frac{0.35 \text{ m}^3/\text{s} \times 86{,}400 \text{ s/d}}{4 \times 7 \text{ m} \times 7 \text{ m}}$$
$$= 154.3 \text{ m}^3/(\text{m}^2 \cdot \text{d})$$

### 11.1 Filter medium size

Before a filter medium is selected, a grain size distribution analysis should be performed. The sieve size and percentage passing by weight relationships are plotted on logarithmic-probability paper. A straight line can be drawn. Determine the geometric mean size ($\mu_g$) and geometric standard deviation size ($\sigma_g$). The most common parameters used in the United States to characterize the filter medium are effective size (ES) and uniformity coefficient (UC) of medium size distribution. The ES is that the grain size for which 10% of the grain ($d_{10}$) are smaller by weight. The UC is the ratio of the 60-percentile ($d_{60}$) to the 10-percentile. They can be written as (Fair *et al.*, 1968; Cleasby, 1990):

$$\text{ES} = d_{10} = \mu_g/\sigma_g^{1.282} \tag{5.70}$$

$$\text{UC} = d_{60}/d_{10} = \sigma_g^{1.535} \tag{5.71}$$

The 90-percentile, $d_{90}$, is the size for which 90% of the grains are smaller by weight. It is interrelated to $d_{10}$ as (Cleasby 1990)

$$d_{90} = d_{10}\,(10^{1.67 \log \text{UC}}) \tag{5.72}$$

The $d_{90}$ size is used for computing the required filter backwash rate for a filter medium.

**Example:** A sieve analysis curve of a typical filter sand gives $d_{10} = 0.54$ mm and $d_{60} = 0.74$ mm. What are its uniformity coefficient and $d_{90}$?

**solution:**

Step 1. $UC = d_{60}/d_{10} = 0.74\text{ mm}/0.54\text{ mm}$
$= 1.37$

Step 2. Find $d_{90}$ using Eq. (5.72)

$$d_{90} = d_{10}\,(10^{1.67 \log UC})$$
$$= 0.54\text{ mm}(10^{1.67 \log 1.37})$$
$$= 0.54\text{ mm}(10^{0.228})$$
$$= 0.91\text{ mm}$$

### 11.2 Mixed media

Mixed media are popular for filtration units. For the improvement process performance, activated carbon or anthracite is added on the top of the sand bed. The approximate specific gravity ($s$) of ilmenite sand, silica sand, anthracite, and water are 4.2, 2.6, 1.5, and 1.0, respectively. For equal settling velocities, the particle sizes for media of different specific gravity can be computed by

$$\frac{d_1}{d_2} = \left(\frac{s_2 - s}{s_1 - s}\right)^{2/3} \tag{5.73}$$

where $d_1$, $d_2$ = diameter of particles 1 and 2, respectively
$s_1$; $s_2$, $s$ = specific gravity of particles 1, 2, and water, respectively

**Example:** Estimate the particle sizes of ilmenite (specific gravity = 4.2) and anthracite (specific gravity = 1.5) which have same settling velocity of silica sand 0.60 mm in diameter (specific gravity = 2.6)

**solution:**

Step 1. Find the diameter of anthracite by Eq. (5.73)

$$d = (0.6\text{ mm})\left(\frac{2.6 - 1}{1.5 - 1}\right)^{2/3}$$
$$= 1.30\text{ mm}$$

Step 2. Determine diameter of ilmenite sand

$$d = (0.6\text{ mm})\left(\frac{2.6 - 1}{4.2 - 1}\right)^{2/3}$$
$$= 0.38\text{ mm}$$

### 11.3 Hydraulics of filter

**Head loss for fixed bed flow.** The conventional fixed-bed filters use a granular medium of 0.5 to 1.0 mm size with a loading rate or filtration velocity of 4.9 to 12.2 m/h (2 to 5 gpm/ft$^2$). When the clean water flows through a clean granular (sand) filter, the loss of head (pressure drop) can be estimated by the Kozeny equation (Fair *et al.*, 1968):

$$\frac{h}{L} = \frac{k\mu(1 - \varepsilon)^2}{g\rho\varepsilon^3}\left(\frac{A}{V}\right)v \tag{5.74}$$

where $h$ = head loss in filter depth L, m, or ft
$k$ = dimensionless Kozeny constant, 5 for sieve openings, 6 for size of separation
$g$ = acceleration of gravity, 9.81 m/s or 32.2 ft/s
$\mu$ = absolute viscosity of water, N · s/m$^2$ or lb · s/ft$^2$
$\rho$ = density of water, kg/m$^3$ or lb/ft$^3$
$\varepsilon$ = porosity, dimensionless
$A/V$ = grain surface area per unit volume of grain
= specific surface $S$ (or shape factor = 6.0 to 7.7)
= $6/d$ for spheres
= $6/\psi d_{eq}$ for irregular grains
$\psi$ = grain sphericity or shape factor
$d_{eq}$ = grain diameter of spheres of equal volume
$v$ = filtration (superficial) velocity, m/s or fps

The Kozeny (or Carmen–Kozeny) equation is derived from the fundamental Darcy–Waeisback equation for head loss in circular pipes (Eq. (4.17)). The Rose equation is also used to determine the head loss resulting from the water passing through the filter medium.

The Rose equation for estimating the head loss through filter medium was developed experimentally by Rose in 1949 (Rose, 1951). It is applicable to rapid sand filters with a uniform near spherical or spherical medium. The Rose equation is

$$h = \frac{1.067\, C_D L v^2}{\varphi g d \varepsilon^4} \tag{5.75}$$

where $h$ = head loss, m or ft
$\varphi$ = shape factor (Ottawa sand 0.95, round sand 0.82, angular sand 0.73, pulverized coal 0.73)
$C_D$ = coefficient of drag (Eq. (5.64))

Other variables are defined previously in Eq. (5.74)

Applying the medium diameter to the area to volume ratio for homogeneous mixed beds the equation is

$$h = \frac{1.067 C_D L v^2}{\varphi g \varepsilon^4} \sum \frac{x}{d_g} \tag{5.76}$$

For a stratified filter bed, the equation is

$$h = \frac{1.067 L v^2}{\varphi g \varepsilon^4} \sum \frac{C_D x}{d_g} \tag{5.77}$$

where $x$ = percent of particles within adjacent sizes
$d_g$ = geometric mean diameter of adjacent sizes

**Example 1:** A dual medium filter is composed of 0.3 m (1 ft) anthracite (mean size of 2.0 mm) that is placed over a 0.6 m (2 ft) layer of sand (mean size 0.7 mm) with a filtration rate of 9.78 m/h (4.0 gpm/ft$^2$). Assume the grain sphericity is $\psi = 0.75$ and a porosity for both is 0.40. Estimate the head loss of the filter at 15°C.

**solution:**

Step 1. Determine head loss through anthracite layer

Using the Kozeny equation (Eq. (5.74))

$$\frac{h}{L} = \frac{k\mu(1-\varepsilon)^2}{g\rho\varepsilon^3}\left(\frac{A}{V}\right)^2 v$$

where $k = 6$
$g = 9.81 \text{ m/s}^2$
$\mu/\rho = \nu = 1.131 \times 10^{-6} \text{ m}^2 \cdot \text{s}$ (Table 4.1a) at 15°C
$\omega = 0.40$
$A/V = 6/0.75d = 8/\text{d} = 8/0.002$
$v = 9.78 \text{ m/h} = 0.00272 \text{ m/s}$
$L = 0.3$ m

then

$$h = 6 \times \frac{1.131 \times 10^{-6}}{9.81} \times \frac{(1-0.4)^2}{0.4^3} \times \left(\frac{8}{0.002}\right)^2 (0.00272)(0.3)$$
$$= 0.0508 \text{ (m)}$$

Step 2. Compute the head loss passing through the sand

Most input data are the same as in Step 1, except

$$k = 5$$
$$d = 0.0007 \text{ m}$$
$$L = 0.6 \text{ m}$$

then

$$h = 5 \times \frac{1.131 \times 10^{-6}}{9.81} \times \frac{0.6^2}{0.4^3}\left(\frac{8}{d}\right)^2 (0.00272)(0.6)$$
$$= 0.3387 \times 10^{-6}/d^2$$
$$= 0.3387 \times 10^{-6}/(0.0007)^2$$
$$= 0.6918 \text{ (m)}$$

Step 3. Compute total head loss

$$h = 0.0508 \text{ m} + 0.6918 \text{ m}$$
$$= 0.743 \text{ m}$$

**Example 2:** Using the same data given in Example 1, except the average size of sand is not given. From sieve analysis, $d_{10}$ (10-percentile of size diameter) = 0.53 mm, $d_{30}$ = 0.67 mm, $d_{50}$ = 0.73 mm, $d_{70}$ = 0.80 mm, and $d_{90}$ = 0.86 mm, estimate the head of a 0.6 m sand filter at 15°C.

**solution:**

Step 1. Calculate head loss for each size of sand in the same manner as

Step 2 of Example 1

$$h_{10} = 0.3387 \times 10^{-6}/d^2 = 0.3387 \times 10^{-6}/(0.00053)^2$$
$$= 1.206 \text{ (m)}$$
$$h_{30} = 0.3387 \times 10^{-6}/(0.00067)^2$$
$$= 0.755 \text{ (m)}$$

Similarly $h_{50} = 0.635$ m

$$h_{70} = 0.529 \text{ m}$$
$$h_{90} = 0.458 \text{ m}$$

Step 2. Taking the average of the head losses given above

$$h = (1.206 + 0.755 + 0.635 + 0.529 + 0.458)/5$$
$$= 0.717 \text{ (m)}$$

*Note*: This way of estimation gives slightly higher values than Example 1.

**Head loss for a fluidized bed.** When a filter is subject to back washing, the upward flow of water travels through the granular bed at a sufficient velocity to suspend the filter medium in the water. This is fluidization. During normal filter operation, the uniform particles of sand occupy the depth $L$. During backwashing the bed expands to a depth of $L_e$. When the critical velocity ($v_c$) is reached, the pressure drops ($\Delta p = h\rho g = \gamma c$) and is equal to the buoyant force of the grain. It can be expressed as

$$h\gamma = (\gamma_s - \gamma)(1 - \varepsilon_e) L_e$$

or

$$h = L_e(1 - \varepsilon_e)(\gamma_s - \gamma)/\gamma \tag{5.78}$$

The porosity of the expanded bed can be determined by using (Fair *et al.*, 1963)

$$\varepsilon_e = (u/v_s)^{0.22} \tag{5.79}$$

where $u$ = upflow (face) velocity of the water
$v_s$ = terminal settling velocity of the particles

A uniform bed of particles will expand when

$$u = v_s \varepsilon_e^{4.5} \tag{5.80}$$

The depth relationship of the unexpanded and expanded bed is

$$L_e = L\left(\frac{1 - \varepsilon}{1 - \varepsilon_e}\right) \tag{5.81}$$

The minimum fluidizing velocity ($u_{mf}$) is the superficial fluid velocity needed to start fluidization. The minimum fluidization velocity is important in determining the required minimum backwashing flow rate. Wen and Yu (1966) proposed the $U_{mf}$ equation excluding shape factor and porosity of fluidization:

$$U_{mf} = \frac{\mu}{\rho d_{eq}}(1135.69 + 0.0408G_n)^{0.5} - \frac{33.7\mu}{\rho d_{eq}} \tag{5.82}$$

where

$$G_n = \text{Galileo number}$$

$$= d_{eq}^3 \rho(\rho_s - \rho)g/\mu^2 \quad (5.83)$$

Other variables used are expressed in Eq. (5.74).

In practice, the grain diameter of spheres of equal volume $d_{eq}$ is not available. Thus the $d_{90}$ sieve size is used instead of $d_{eg}$. A safety factor of 1.3 is used to ensure adequate movement of the grains (Cleasby and Fan, 1981).

**Example:** Estimate the minimum fluidization velocity and backwash rate for the sand filter at 15°C. The $d_{90}$ size of sand is 0.88 mm. The density of sand is 2.65 g/cm$^3$.

**solution:**

Step 1. Compute the Galileo number

From Table 4.1a, at 15°C

$$\rho = 0.999 \text{ g/cm}^3$$

$$\mu = 0.00113 \text{ N} \cdot \text{s/m}^2 = 0.00113 \text{ kg/m} \cdot \text{s} = 0.0113 \text{ g/cm} \cdot \text{s}$$

$$\mu/\rho = 0.0113 \text{ cm}^2\text{/s}$$

$$g = 981 \text{ cm/s}^2$$

$$d = 0.088 \text{ cm}$$

$$\rho_s = 2.65 \text{ g/cm}^3$$

Using Eq. (5.83)

$$G_n = d_{eq}^3 \rho(\rho_s - \rho)g/\mu^2$$

$$= (0.088)^3(0.999)(2.65 - 0.999)(981)/(0.0113)^2$$

$$= 8635$$

Step 2. Compute $U_{mf}$ by Eq. (5.82)

$$U_{mf} = \frac{0.0113}{0.999 \times 0.088}(1135.69 + 0.0408 \times 8635)^{0.5} - \frac{33.7 \times 0.0113}{0.999 \times 0.088}$$

$$= 0.627 \text{ (cm/s)}$$

Step 3. Compute backwash rate

Apply a safety factor of 1.3 to $U_{mf}$ as backwash rate

$$\begin{aligned}\text{Backwash rate} &= 1.3 \times 0.627 \text{ cm/s} = 0.815 \text{ cm/s} \\ &= 0.00815 \text{ m/s} \\ &= 0.00815 \text{ m/s} \times 86{,}400 \text{ s/d} \\ &= 704.16 \text{ m/d or } (\text{m}^3/\text{m}^2 \cdot \text{d}) \\ &= 704 \text{ m/d}/(0.01705 \text{ gpm/ft}^2) \times 1 \text{ (d/m)} \\ &= 12.0 \text{ gpm/ft}^2\end{aligned}$$

## 11.4 Washwater troughs

In the United States, in practice washwater troughs are installed at even spaced intervals (5 to 7 ft apart) above the gravity filters. The washwater troughs are employed to collect spent washwater. The total rate of discharge in a rectangular trough with free flow can be calculated by (Fair *et al.*, 1968; ASCE and AWWA, 1990)

$$Q = Cwh^{1.5} \tag{5.84a}$$

where $Q$ = flow rate, cfs
$C$ = constant (2.49)
$w$ = trough width, ft
$h$ = maximum water depth in trough, ft

For rectangular horizontal troughs of such a short length friction losses are negligible. Thus the theoretical value of the constant $C$ is 2.49. The value of $C$ may be as low as 1.72 (ASCE and AWWA, 1990).

In European practice, backwash water is generally discharged to the side and no troughs are installed above the filter. For SI units, the relationship is

$$Q = 0.808\, wh^{1.5} \tag{5.84b}$$

where $Q$ = flow rate, $m^3/(m^2 \cdot s)$
$w$ = trough width, m
$h$ = water depth in trough, m

**Example 1:** Troughs are 20 ft (6.1 m) long, 18-in (0.46 m) wide, and 8 ft (2.44 m) to the center with a horizontal flat bottom. The backwash rate is 24 in/min (0.61 m/min). Estimate: (1) the water depth of the troughs with free flow into the gullet, and (2) the distance between the top of the troughs and the 30-in sand bed. Assuming 40% expansion and 6 in of freeboard in the troughs and 8 in of thickness.

**solution:**

Step 1. Estimate the maximum water depth ($h$) in trough

$$v = 24 \text{ (in/min)} = 2 \text{ ft/60 s} = 1/30 \text{ fps}$$

$$A = 20 \text{ ft} \times 8 \text{ ft} = 160 \text{ ft}^2$$

$$Q = VA = 160/30 \text{ cfs}$$

$$= 5.33 \text{ cfs}$$

Using Eq. (5.84a)

$$Q = 2.49wh^{1.5}, \; w = 1.5 \text{ ft}$$

$$h = (Q/2.49w)^{2/3}$$

$$= [5.33/(2.49 \times 1.5)]^{2/3}$$

$$= 1.27 \text{ (ft)}$$

Say $h = 16$ in $= 1.33$ ft

Step 2. Determine the distance ($y$) between the sand bed surface and the top troughs

$$\text{Depth of sand bed} = 30 \text{ in} = 2.5 \text{ ft}$$

$$\text{Height of expansion} = 2.5 \text{ ft} \times 0.4$$

$$\text{Freeboard} = 6 \text{ in} = 0.5 \text{ ft}$$

$$\text{Thickness} = 8 \text{ in} = 0.67 \text{ ft (the bottom of the trough)}$$

$$y = 2.5 \text{ ft} \times 0.4 + 1.33 \text{ ft} + 0.5 \text{ ft} + 0.67 \text{ ft}$$

$$= 3.5 \text{ ft}$$

**Example 2:** A filter unit has surface area of 16 ft (4.88 m) wide and 30 ft (9.144 m) long. After filtering 2.88 Mgal (10,900 $m^3$) for 50 h, the filter is backwashed at a rate of 16 gpm/ft$^2$ (0.65 m/min) for 15 min. Find: (a) the average filtration rate, (b) the quantity of washwater, (c) percent of washwater to treated water, and the flow rate to each of the four troughs.

**solution:**

Step 1. Determine flow rate $Q$

$$Q = \frac{2{,}880{,}000 \text{ gal}}{50 \text{ h} \times 60 \text{ min/h} \times 16 \text{ ft} \times 30 \text{ ft}}$$

$$= 2.0 \text{ gpm/ft}^2$$

$$= \frac{2 \text{ gal} \times 0.0037854 \text{ m}^3 \times 1440 \text{ min} \times 10.764 \text{ ft}^2}{\text{min} \times \text{ft}^2 \times 1 \text{ gal} \times 1 \text{ day} \times 1 \text{ m}^2}$$

$$= 117.3 \text{ m/d}$$

say $$= 120 \text{ m/d}$$

Step 2. Determine quantity of washwater $q$

$$q = (16 \text{ gal/min ft}^2) \times 15 \text{ min} \times 16 \text{ ft} \times 30 \text{ ft}$$
$$= 115{,}200 \text{ gal}$$

Step 3. Determine percentage (%) of washwater to treated water

$$\% = 115{,}200 \text{ gal} \times 100\%/2{,}880{,}000 \text{ gal}$$
$$= 4.0\%$$

Step 4. Compute flow rate ($r$) in each trough

$$r = 115{,}200 \text{ gal}/(15 \text{ min} \times 4 \text{ troughs})$$
$$= 1920 \text{ gpm}$$

## 11.5 Filter efficiency

The filter efficiency is defined as the effective filter rate divided by the operation filtration rate as follows (AWWA and ASCE, 1998):

$$E = \frac{R_e}{R_o} = \frac{UFRV - UBWV}{UFRV} \qquad (5.84c)$$

where $E$ = filter efficiency, %
$R_e$ = effective filtration rate, gpm/ft$^2$ or m$^3$/m$^2$/h (m/h)
$R_o$ = operating filtration rate, gpm/ft$^2$ or m$^3$/m$^2$/h (m/h)
$UFRV$ = unit filter run volume, gal/ft$^2$ or L/m$^2$
$UBWV$ = unit backwash volume, gal/ft$^2$ or L/m$^2$

**Example:** A rapid sand filter operating at 3.7gpm/ft$^2$ (9.0 m/h) for 46 h. After this filter run, 295 gal/ft$^2$ (12,000 L/m$^2$) of backwash water is used. Find the filter efficiency.

**solution:**

Step 1. Calculate operating filtration rate, $R_o$

$$R_o = 3.7 \text{ gal/min/ft}^2 \times 60 \text{ min/h} \times 46 \text{ h}$$
$$= 10{,}212 \text{ gal/ft}^2$$

Step 2. Calculate effective filtration rate, $R_e$

$$R_e = (10212 - 295) \text{ gal/ft}^2$$
$$= 9917 \text{ gal/ft}^2$$

Step 3. Calculate filter efficiency, $E$, using Eq (5.84c)

$$E = 9917/10212 = 0.97$$
$$= 97\%$$

## 12 Water Softening

Hardness in water is mainly caused by the ions of calcium and magnesium. It may also be caused by the presence of metallic cations of iron, sodium, manganese, and strontium. These cations are present with anions such as $HCO_3^-$, $SO_4^{2-}$, $Cl^-$, $NO_3^-$, and $SiO_4^{2-}$. The carbonates and bicarbonates of calcium, magnesium, and sodium are called carbonate hardness or temporary hardness since it can be removed and settled by boiling of water. Noncarbonate hardness is caused by the chloride and sulfate slots of divalent cations. The total hardness is the sum of carbonate and noncarbonate hardness.

The classification of hardness in water supply is shown in Table 5.5. Although hard water has no health effects, using hard water would increase the amount of soap needed and would produce scale on bath fixtures, cooking utensils. Hardness also causes scale and corrosion in hot-water heaters, boilers, and pipelines. Moderate hard water with 60 to 120 mg/L as $CaCO_3$ is generally publicly acceptable (Clark *et al.*, 1977).

### 12.1 Lime-soda softening

The hardness in water can be removed by precipitation with lime, $Ca(OH)_2$ and soda ash, $Na_2CO_3$ and by an ion exchange process. Ion exchange is discussed in another section. The reactions of lime-soda ash precipitation for hardness removal in water and recarbonation are shown in the following equations:

Removal of free carbon dioxide with lime

$$\begin{bmatrix} CO_2 \\ H_2CO_3 \end{bmatrix} + Ca(OH)_2 \rightarrow CaCO_3\downarrow + H_2O \qquad (5.85)$$

Removal of carbonate hardness with lime

$$Ca^{2+} + 2HCO_3^- + Ca(OH)_2 \rightarrow 2CaCO_3\downarrow + 2H_2O \qquad (5.86)$$

**TABLE 5.5 Classification of Hard water**

| | mg/L as $CaCO_3$ | |
|---|---|---|
| Hardness classification | US | International |
| Soft | 0–60 | 0–50 |
| Moderate soft | | 51–100 |
| Slightly hard | | 101–150 |
| Moderate hard | 61–120 | 151–200 |
| Hard | 121–180 | 201–300 |
| Very hard | >180 | >300 |

$$Mg^{2+} + 2HCO_3^- + 2Ca(OH)_2 \rightarrow 2CaCO_3\downarrow + Mg(OH)_2\downarrow + 2H_2O \qquad (5.87)$$

Removal of noncarbonate hardness with soda ash and lime

$$Ca^{2+} + \begin{bmatrix} 2Cl^- \\ SO_4^{2-} \end{bmatrix} + Na_2CO_3 \rightarrow CaCO_3\downarrow + 2Na^+ + \begin{bmatrix} 2Cl^- \\ SO_4^{2-} \end{bmatrix} \qquad (5.88)$$

$$Mg^{2+} + \begin{bmatrix} 2Cl^- \\ SO_4^{2-} \end{bmatrix} + Ca(OH)_2 \rightarrow Mg(OH)_2\downarrow + Ca^{2+} + \begin{bmatrix} 2Cl^- \\ SO_4^{2-} \end{bmatrix} \qquad (5.89)$$

Recarbonation for pH control (pH $\cong$ 8.5)

$$CO_3^{2-} + CO_2 + H_2O \rightarrow 2HCO_3^- \qquad (5.90)$$

Recarbonation for removal of excess lime and pH control (pH $\cong$ 9.5)

$$Ca(OH)_2 + CO_2 \rightarrow CaCO_3\downarrow + H_2O \qquad (5.91)$$

$$Mg(OH)_2 + CO_2 \rightarrow MgCO_3 + H_2O \qquad (5.92)$$

Based on the above equations, the stoichiometric requirement for lime and soda ash expressed in equivalents per unit volume are as follows (Tchobanoglous and Schroeder, 1985):

$$\text{Lime required (eq/m}^3) = CO_2 + HCO_3^- + Mg^{2+} + \text{excess} \qquad (5.93)$$

$$\text{Soda ash required (eq/m}^3) = Ca^{2+} + Mg^{2+} - \text{alkalinity} \qquad (5.94)$$

Approximately 1 eq/m$^3$ of lime in excess of the stoichiometric requirement must be added to bring the pH to above 11 to ensure $Mg(OH)_2$ complete precipitation. After the removal of precipitates, recarbonation is needed to bring pH down to a range of 9.2 to 9.7.

The treatment processes for water softening may be different depending on the degree of hardness and the types and amount of chemical added. They may be single stage lime, excess lime, single stage lime-soda ash, and excess lime-soda ash processes.

Coldwell—Lawrence diagrams are based on equilibrium principles for solving water softening. The use of diagrams is an alternative to the stoichiometric method. It solves simultaneous equilibria equations and estimates the chemical dosages of lime-soda ash softening. The interested reader is referred to the American Water Works Association (1978) publication: Corrosion Control by Deposition of $CaCO_3$ Films, and to Benefield and Morgan (1990).

**Example 1:** Water has the following composition: calcium = 82 mg/L, magnesium = 33 mg/L, sodium = 14 mg/L, bicarbonate = 280 mg/L, sulfate = 82 mg/L, and chloride = 36 mg/L. Determine carbonate hardness, noncarbonate hardness, and total hardness, all in terms of mg/L of $CaCO_3$.

**solution:**

Step 1. Convert all concentration to mg/L of $CaCO_3$ using the following formula:

The species concentration in milliequivalents per liter (meq/L) is computed by the following equation

$$\text{meq/L} = \frac{\text{mg/L}}{\text{equivalent weight}} = \frac{100}{2} = 50 \text{ for } CaCO_3$$

The species concentration expressed as mg/L of $CaCO_3$ is calculated from

$$\text{mg/L as } CaCO_3 = \text{meq/L of species} \times 50$$
$$= \frac{(\text{mg/L of species}) \times 50}{\text{equivalent weight of species}}$$

Step 2. Construct a table for ions in mg/L as $CaCO_3$

| Ion species | Molecular weight | Equivalent weight | Concentration | | |
|---|---|---|---|---|---|
| | | | mg/L | meq/L | mg/L as $CaCO_3$ |
| $Ca^{2+}$ | 40 | 20.0 | 82 | 4.0 | 200 |
| $Mg^{2+}$ | 24.3 | 12.2 | 33 | 2.7 | 135 |
| $Na^+$ | 23 | 23.0 | 14 | 0.6 | 30 |
| | | | | Total: 7.3 | 365 |
| $HCO_3^-$ | 61 | 61 | 280 | 4.6 | 230 |
| $Cl^-$ | 35.5 | 35.5 | 36 | 1.0 | 50 |
| $SO_4^{2-}$ | 96.1 | 48 | 82 | 1.7 | 85 |
| | | | | Total: 7.3 | 365 |

Step 3. Construct an equivalent bar diagram for the cationic and anionic species of the water

The diagram shows the relative proportions of the chemical species important to the water softening process. Cations are placed above anions on the graph. The calcium equivalent should be placed first on the cationic scale and be followed by magnesium and other divalent species and then by the monovalent species sodium equivalent. The bicarbonate equivalent should be placed first on the anionic scale and immediately be followed by the chloride equivalent and then by the sulfate equivalent.

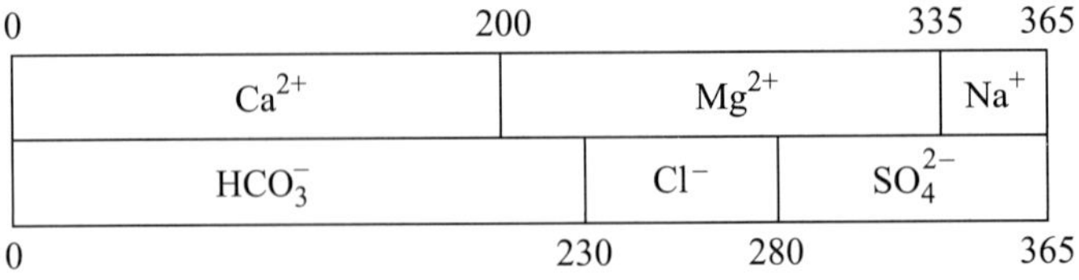

Step 4. Compute the hardness distribution

$$\text{Total hardness} = 200 + 135 = 335 \text{ (mg/L as CaCO}_3\text{)}$$

$$\text{Alkalinity (bicarbonate)} = 230 \text{ mg/L as CaCO}_3$$

$$\text{Carbonate hardness} = \text{alkalinity}$$

$$= 230 \text{ mg/L as CaCO}_3$$

$$\text{Noncarbonate hardness} = 365 - 230 = 165 \text{ (mg/L as CaCO}_3\text{)}$$

$$\text{or} = Na^+ + Cl^- + SO_4^{2-}$$

**Example 2:** (single-stage lime softening): Raw water has the following composition: alkalinity = 248 mg/L as $CaCO_3$, pH = 7.0, $\alpha_1 = 0.77$ (at $T = 10°C$), calcium = 88 mg/L, magnesium = 4 mg/L. Determine the necessary amount of lime to soften the water, if the final hardness desired is 40 mg/L as $CaCO_3$. Also estimate the hardness of the treated water.

**solution:**

Step 1. Calculate bicarbonate concentration

$$\text{Alkalinity} = [HCO_3^-] + [CO_3^{2-}] + [OH^-] + [H^-]$$

At pH = 7.0, assuming all alkalinity is in the carbonate form

$$HCO_3^- = 248\frac{\text{mg}}{\text{L}} \times \frac{1 \text{ g}}{1000 \text{ mg}} \times \frac{1 \text{ mole}}{61 \text{ g}} \times \frac{61 \text{ eq wt of HCO}_3}{50 \text{ eq wt of alkalinity}}$$
$$= 4.96 \times 10^{-3} \text{ mol/L}$$

Step 2. Compute total carbonate species concentration $C_T$ (Snoeyink and Jenkins, 1980)

$$C_T = [HCO_3^-]/\alpha_1$$
$$= 4.96 \times 10^{-3}/0.77$$
$$= 6.44 \times 10^{-3} \text{ mol/L}$$

Step 3. Estimate the carbonate acid concentration

$$C_T = [H_2CO_3] + [HCO_3^-] + [CO_3^{2-}]$$

Rearranging, and $[CO_3^{2-}] = 0$

$$[H_2CO_3] = (6.44 - 4.96) \times 10^{-3} \text{ mol/L}$$

$$= 1.48 \times 10^{-3} \text{ mol/L}$$

$$= 1.48 \times 10^{-3} \text{ mol/L} \times 1000 \times 62 \text{ mg/mol}$$

$$= 92 \text{ mg/L}$$

or

$$= 148 \text{ mg/L as CaCO}_3 \ (92 \times 50/32)$$

Step 4. Construct a bar diagram for the raw water converted concentration of Ca and Mg as $CaCO_3$

$$Ca = 88 \times 100/40 = 220 \text{ mg/L as } CaCO_3$$
$$Mg = 4 \times 100/24.3 = 16 \text{ mg/L as } CaCO_3$$

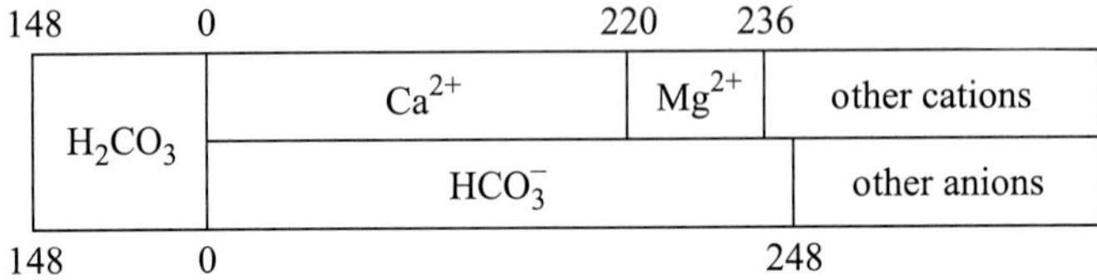


Step 5. Find hardness distribution

$$\text{Calcium carbonate hardness} = 220 \text{ mg/L}$$
$$\text{Magnesium carbonate hardness} = 16 \text{ mg/L}$$
$$\text{Total hardness} = 220 + 16 = 236 \text{ (mg/L)}$$

Step 6. Estimate the lime dose needed

Since the final hardness desired is 40 mg/L as $CaCO_3$, magnesium hardness removal would not be required. The amount of lime needed ($x$) would be equal to carbonic acid concentration plus calcium carbonate hardness:

$$x = 148 + 220$$
$$= 368 \text{ mg/L as } CaCO_3$$

or

$$= 368 \text{ mg/L as } CaCO_3 \times \frac{74 \text{ as } Ca(OH)_2}{100 \text{ as } CaCO_3}$$
$$= 272 \text{ mg/L as } Ca(OH)_2$$

or

$$= 368 \times \frac{56}{100} \text{ mg/L as CaO}$$
$$= 206 \text{ mg/L as CaO}$$

The purity of lime is 70%. The total amount of lime needed is

$$206/0.70 = 294 \text{ mg/L as CaO}$$

Step 7. Estimate the hardness of treated water

The magnesium hardness level of 16 mg/L as $CaCO_3$ remains in the softened water theoretically. The limit of calcium achievable is 30 to 50 mg/L of $CaCO_3$ and would remain in the water unless using 5% to 10% in excess of lime. Hardness levels less than 50 mg/L as $CaCO_3$ are seldom achieved in plant operation.

**Example 3:** (excess lime softening): Similar to Example 2, except for concentrations of calcium and magnesium: pH = 7.0, $\alpha_1 = 0.77$ (at $T = 10°C$). Alkalinity, calcium, and magnesium concentrations are 248, 158, and 56 mg/ L as $CaCO_3$, respectively. Determine the amount of lime needed to soften the water.

**solution:**

Step 1. Estimate $H_2CO_3$ as Example 2

$$H_2CO_3 = 148 \text{ mg/L as } CaCO_3$$

Step 2. Construct a bar diagram for the raw water

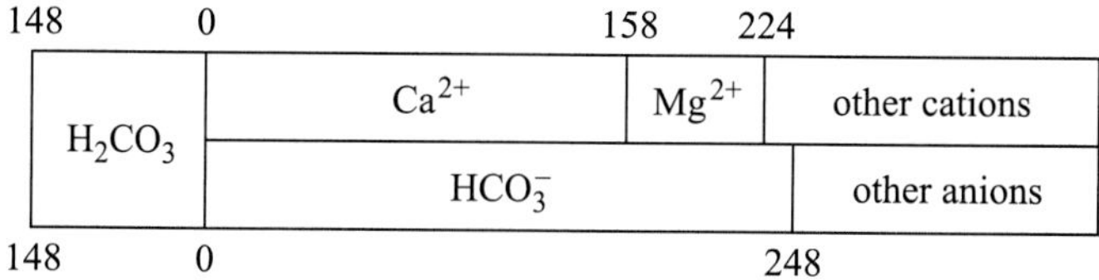


Step 3. Find hardness distribution

$$\text{Calcium carbonate hardness} = 158 \text{ mg/L}$$
$$\text{Magnesium carbonate hardness} = 66 \text{ mg/L}$$
$$\text{Total carbonate hardness} = 158 + 66 = 224 \text{ (mg/L)}$$

Comment: Sufficient lime (in excess) must be dosed to convert all bicarbonate alkalinity to carbonate alkalinity and to precipitate magnesium as magnesium hydroxide (needs 1 eq/L excess lime).

Step 4. Estimate lime required ($x$) is the sum for carbonic acid, total alkalinity, magnesium hardness, and excess lime (say 60 mg/L as $CaCO_3$) to raise pH = 11.0

$$x = (148 + 248 + 66 + 60) \text{ mg/L as } CaCO_3$$
$$= 522 \text{ mg/L as } CaCO_3$$
$$= 522 \times 74/100 \text{ mg/L as } Ca(OH)_2$$
$$= 386 \text{ mg/L as } Ca(OH)_2$$

or

$$= 522 \times 0.56 \text{ mg/L as CaO}$$
$$= 292 \text{ mg/L as CaO}$$

**Example 4:** (single stage lime-soda ash softening): A raw water has the following analysis: pH = 7.01, $T$ = 10°C, alkalinity = 248 mg/L, Ca = 288 mg/L, Mg = 12 mg/L, all as $CaCO_3$. Determine amounts of lime and soda ash required to soften the water.

**solution:**

Step 1. As in previous example, construct a bar diagram

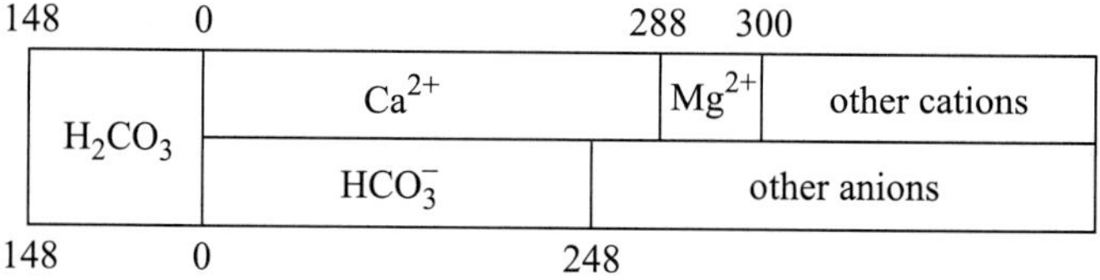

Step 2. Find the hardness distribution

$$\text{Total hardness} = 288 + 12 = 300 \text{ (mg/L as } CaCO_3)$$
$$\text{Calcium carbonate hardness, CCH} = 248 \text{ mg/L } CaCO_3$$
$$\text{Magnesium carbonate hardness} = 0$$
$$\text{Calcium noncarbonate hardness, CNH} = 288 - 248 = 40 \text{ (mg/L as } CaCO_3)$$
$$\text{Magnesium noncarbonate hardness, MNH} = 12 \text{ mg/L}$$

Step 3. Determine lime and soda ash requirements for the straight lime-soda ash process, refer to Eq. (5.85) and Eq. (5.86)

$$\text{Lime dosage} = H_2CO_3 + \text{CCH} = (148 + 248) \text{ mg/L as } CaCO_3$$
$$= 396 \text{ mg/L as } CaCO_3$$

or

$$= 396 \times 0.74 \text{ mg/L as } Ca(OH)_2$$
$$= 293 \text{ mg/L as } Ca(OH)_2$$

$$\text{and soda ash dosage} = \text{CNH (Refer to Eq. (6.88))}$$
$$= 40 \text{ mg/L as } CaCO_3$$
$$= (40 \times 106/100) \text{ mg/L as } Na_2CO_3$$
$$= 42.4 \text{ mg/L as } Na_2CO_3$$

**Example 5:** (excess lime-soda ash process): Raw water has the following analysis: pH = 7.0, $T$ = 10°C, alkalinity, calcium, and magnesium are 248, 288, and 77 mg/L as $CaCO_3$, respectively. Estimate the lime and soda ash dosage required to soften the water.

**solution:**

Step 1. Construct a bar diagram for the raw water. Some characteristics are the same as the previous example.

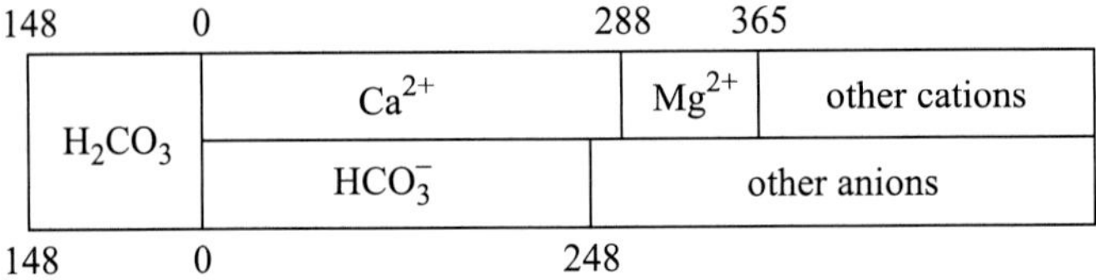


Step 2. Define hardness distribution

$$\text{Total hardness} = 288 + 77 = 365 \text{ (mg/L as } CaCO_3)$$
$$\text{Calcium carbonate hardness, CCH} = 248 \text{ mg/L as } CaCO_3$$
$$\text{Magnesium carbonate hardness, MCH} = 0$$
$$\text{Calcium noncarbonate hardness, CNH} = 288 - 248 = 40 \text{ (mg/L)}$$
$$\text{Magnesium noncarbonate hardness, MNH} = 77 \text{ mg/L}$$

Step 3. Estimate the lime and soda ash requirements by the excess lime-soda ash process

$$\text{Lime dosage} = \text{carb. acid} + \text{CCH} + 2(\text{MCH}) + \text{MNH} + \text{excess lime}$$
$$= (148 + 248 + 2(0) + 77 + 60) \text{ mg/L as } CaCO_3$$
$$= 533 \text{ mg/L as } CaCO_3$$

or

$$= 533 \times 74/100$$
$$= 394 \text{ mg/L as } Ca(OH)_2$$

and soda ash dosage for excess lime-soda ash ($Y$) is

$$Y = \text{CNH} + \text{MNH}$$
$$= (40 + 77) \text{ mg/L as } CaCO_3$$
$$= 117 \text{ mg/L as } CaCO_3$$

or

$$= 87 \text{ mg/L as } Ca(OH)_2$$

## 12.2 Pellet softening

Pellet softening has been used for softening in the Netherlands for many years. Pellet softening reactors have been installed at a number of locations in North America. The world's largest pellet softening plant, 450,000 $m^3$/d (119 MGD) of the design capacity, was installed in 2004 at the Cheng-Ching Lake Advanced Water Purification Plant in Kaohsiung, Taiwan. The pellet softening process reactors consist of square (or rectangular) or inverted conical tanks where calcium crystallizes on a suspended bed of fine sand.

Advantages of the pellet softening reactor are its small size and low installation cost. Residuals consist of small pellets that dewater readily, minimizing residuals volume. The pellet can be used for steel manufacture. However, pellet reactors should not be considered for systems high in magnesium content because magnesium hydroxide is not affected and may foul in the reactor.

Pellet softening is not proved and is not widely accepted as a treatment system in the United States. Pilot testing is advisable. Design should be carefully coordinated with the equipment manufacturer. Pellet softening processes should be designed cautiously.

**Design and operation considerations.** The pellet softening process is used for removal of only calcium carbonate ($CaCO_3$) hardness. The process is based on the crystallization of calcium carbonate on seeded sand (0.2 to 0.63 mm in size, fixed-bed height = 1.2 to 1.5 m) media. The chemical reaction is described as below:

$$Ca^{2+} + HCO_3^- + Na(OH) \rightarrow Na^+ + CaCO_3 \downarrow + H_2O \quad (5.94a)$$

Flocculated-settled water or raw water and caustic soda (NaOH) are injected (pumped) into the pressure chamber of pellet reactor through a nozzle system. Water muzzles should be used to ensure a proper distribution of the water across the surface of the reactor.

The reactor is filled with a bed of sand material (automatically seeded) that is fluidized by the water flow. The precipitated $CaCO_3$ will form a fixed layer first on the surface of the seeded sand grain and later on the created pellet surface itself. The pH value and the magnesium hardness will not influence the process.

The NaOH is mixed intensively with the water to avoid locally high supersaturation, which would lead to spontaneous nucleation of $CaCO_3$ instead of crystal growth on the sand. The chemical feed installation (NaOH dosing pumps) is controlled by the digital control system according to the calculated NaOH dosage, which is a function of the amount of calcium hardness to be removed.

Due to the growth of $CaCO_3$ on the sand, the size of the individual pellets will grow. This causes the increase in the fluidized bed height (at least height = 10 m at 100 m/h flow). To balance the height of the fluidized bed, pellets must be discharged from the reactor; also, a specific amount of new sand must be supplied to the pellet bed. The sizes of the pellets can be controlled by the equilibrium of pellet extraction and sand supply.

A small amount of supersaturated $CaCO_3$ may be carried in the outflow of the reactor. Sulfuric acid is dosed at the outflow of the reactor for the pH (7.5 to 7.9) adjustment. After softening in the pellet reactors, the softened water and the bypass raw water may be blended and then flow by gravity to the rapid sand filters.

A pellet-reactor design should comply with the following criteria:

- The water and NaOH in the pellet reactor should be properly distributed to avoid short-circuiting and to obtain plug-flow conditions in the pellet reactor.
- The water and NaOH should be mixed intensively in the presence of seed grains with a high specific surface area to achieve immediate crystallization.
- The turbulence in the reactor should be sufficiently high to prevent scaling of inlet nozzles and the reactor wall, and low enough to reduce pellet erosion.

**Sizing for reactors.** The design overflow rate is between 80 and 100 m/h ($m^3/m^2 \cdot h$). The height of the reactor is generally about 8 m. Thus, the total area of the reactors can be easily determined with the softening water flow divided by the overflow rate.

**Example:** Determine the size of pellet reactor for a plant flow of 45,400 $m^3/d$ or 12 MGD.

**solution:**

Step 1. Convert the flow unit

$$Q = 45{,}400\ m^3/d = 45400\ m^3/d \div 24\ h/d$$
$$= 1892\ m^3/h$$

Step 2. Determine the total area, $A$, for the reactors, using 80 $m^3/m^2 \cdot h$

$$A = Q/\text{overflow rate}$$
$$= 1892\ m^3/h \div 80\ m^3/m^2 \cdot h$$
$$= 23.66\ m^2$$

Using 2 square softening reactors, for each reactor would have an area of 11.83 $m^2$

The length, $L$, is for a side of the square reactor, then

$$L^2 = 11.83\ m^2$$
$$L = 3.44\ m$$

Step 3. The height of a reactor is usually 8 m; thus

The size of each of the 2 reactors is 3.44 m × 3.44 m × 8.00 m.

**Caustic soda demand.** For calculation purposes, the purity of caustic soda is used as 100%. The estimation of NaOH demand for the total raw water flow rate is calculated as follows:

$$\text{NaOH demand, mg/L} = (C_i - C_e)\ 40/100 + \text{excess}$$
$$= 0.4\ (C_i - C_e) + 20 \qquad (5.94b)$$

where $C_i$ = influent hardness, mg/L
$C_e$ = effluent hardness, mg/L
40 = molecular weight of NaOH as g/mol
100 = molecular weight of $CaCO_3$ as g/mol
20 = 0.5 mol/ NaOH, is the correction factor ($f$) for $CO_2$ neutralization and softening efficiency.
40 mg/L of NaOH are required for the neutralization of 44 mg/L.

Softening efficiency is greatly reduced by too large pellet diameter, insufficient sand seeding, and high operation velocity in the reactor.

The efficiency ($E$) of the NaOH dosage (demand) without considering the NaOH demand for $CO_2$ can be computed as:

$$E = 100\% - (\text{NaOH dosage} - 0.4(C_i - C_e))\% \qquad (5.94c)$$

**Example 1:** Estimate the NaOH demand for pellet softening of the entire raw water flow rate and softening efficiency without considering the NaOH demand for $CO_2$. The calcium hardness in influent and effluent are 250 and 80 mg/L as $CaCO_3$, respectively.

**solution:**

Step 1. Estimation of NaOH demand by using Eq. (5.94b)

$$\begin{aligned}\text{NaOH demand, mg/L} &= 0.4(C_i - C_e) + 20\\ &= 0.4(250 - 80) + 20\\ &= 88\end{aligned}$$

Step 2. Calculate softening efficiency, E, using Eq. (5.94c)

$$\begin{aligned}E &= 100\% - (\text{NaOH dosage} - 0.4(C_i - C_e))\%\\ &= 100\% - (88 - 0.4\,(250 - 80))\%\\ &= 100\% - 20\%\ (= 100\% - \text{correction factor, } f\text{, } 20\%)\\ &= 80\%\end{aligned}$$

If the softened water is blended with flocculated /settled water at the softening reactor bypass and is blended additionally at postozonation with flocculated and filtered water, both blended waters shall increase the total hardness. Therefore, the water at softening must be softened further to allow final treated water hardness within the desired concentration. The NaOH dosage for the blended water can be calculated as:

$$\text{Dosage} = \text{NaOH demand} \times Q_i/(Q_i - Q_b - Q_f) \qquad (5.94d)$$

where Dosage = NaOH dosage for the blended water, mg/L
NaOH demand = calculated by using Eq. (5.94b), mg/L
$Q_i$ = flow rate to the softening reactor
$Q_b$ = flow rate softening bypass
$Q_f$ = flow rate from filtration

**Example 2:** Estimate the NaOH dosage for the blended water. Given: $Q_i$, $Q_b$, and $Q_f$ as 1890, 105, and 95 $m^3/h$, respectively. The NaOH demand is 88 mg/L.

**solution:** Using Eq. (5.94d),

$$\begin{aligned}\text{Dosage} &= \text{NaOH demand} \times Q_i\,/\,(Q_i - Q_b - Q_f)\\ &= 88\text{ mg/L} \times 1890\,/\,(1890 - 105 - 95)\\ &= 98\text{ mg/L}\end{aligned}$$

**Estimation of the expected softener hardness.** Since 40 mg/L NaOH reduces 100 mg/L of the total hardness, the expected total hardness of at the softener, $TH_s$, can be estimated as:

$$TH_s = TH_r - (NaOH_{cal} - f) \times 100/40 \qquad (5.94e)$$

where
$TH_r$ = raw water total hardness, mg/L as $CaCO_3$
$NaOH_{cal}$ = calculated average NaOH dose, mg/L
$f$ = correction factor, 20 mg/L NaOH
40 = molecular weight of NaOH, g/mol
100 = molecular weight of $CaCO_3$, g/mol

**Example:** Given: $TH_r$ = 250 mg/L as $CaCO_3$; $NaOH_{cal}$ = 98 mg/L. Determine the total hardness at the softener, $TH_s$.

**solution:** Using Eq. (5.94e)

$$\begin{aligned} TH_s &= TH_r - (NaOH_{cal} - f) \times 100/40 \\ &= 250 \text{ mg/L} - (98 - 20) \text{ mg/L} \times 2.5 \\ &= 55 \text{ mg/L} \end{aligned}$$

**Estimation of the expected treated water hardness.** The expected treated water hardness, $TH_w$, can be estimated by the following formula:

$$TH_w = TH_r - (NaOH_{avg} - f) \times 100/40 \qquad (5.94f)$$

where
$NaOH_{avg}$ = calculated average NaOH dose for entire water flow including bypass and filtration flow rate, mg/L
Others = as previously mentioned

**Example:** Given: $TH_r$ = 250 mg/L as $CaCO_3$; $NaOH_{avg}$ = 88 mg/L. Determine the expected total hardness in the treated water, $TH_w$.

**solution:** Using Eq. (5.94f)

$$\begin{aligned} TH_w &= TH_r - (NaOH_{avg} - f) \times 100/40 \\ &= 250 \text{ mg/L} - (88 - 20) \text{ mg/L} \times 2.5 \\ &= 80 \text{ mg/L} \end{aligned}$$

**NaOH dosing pumps.** The actual delivery rate of a NaOH dosing pump is estimated as follows:

$$\begin{aligned} \text{Pump rate} &= \text{maximum pump capacity} \times \text{efficiency} \\ &\quad \times \text{\% stroke adjustment} \end{aligned} \qquad (5.94g)$$

**Example:** Given: maximum pump capacity = 720 L/h, speed efficiency = 0.85, and stroke adjustment = 50%. Estimate the actual delivery rate of the NaOH dosing pump.

**solution:** Using Eq. (5.94g)

$$\text{Actual pumping rate} = 720 \text{ L/h} \times 0.85 \times 0.50$$
$$= 306 \text{ L/h}$$

**Sulfuric acid dose rate.** Sulfuric acid is dosed into the treated water to neutralize the excess of NaOH. The dose rate is calculated as:

$$H_2SO_4 \text{ dose rate} = \text{acid flow rate} \times \text{concentration} / \text{treated water flow rate} \quad (5.94h)$$

**Example:** Given: required $H_2SO_4$ flow rate = 60 L/h; $H_2SO_4$ (50%) concentration = 690 g/L; and the treated water flow rate = 1900 $m^3$/h. Compute the dose rate in mg/L of $H_2SO_4$.

**solution:** Using Eq. (5.94h)

$$1900 \text{ m}^3\text{/h} = 1{,}900{,}000 \text{ L/h}$$
$$H_2SO_4 \text{ dose rate} = 60 \text{ L/h} \times 690 \text{ g/L} / 1{,}900{,}000 \text{ L/h}$$
$$= 0.0218 \text{ g/L}$$
$$= 21.8 \text{ mg/L}$$

**Seeding sand demand.** A lack of sand supply for crystallization has a negative impact on the efficiency of the softening (then more NaOH is needed to get the desired total hardness). This excess dosage of NaOH to be neutralized triggers an additional sulfuric acid demand.

The seeding sand is added according to the demanded ratio between effective sand size and pellet size. If the effective sand size or pellet size is altered, the sand dosing rate also must be altered.

It is essential that very fine, fine, and smaller sand particles should be removed from sand prior feeding to the reactors. If not, the fine particles would be flushed out from the reactor due to up to 100 m/h operation velocity there and enter the rapid filtration units, causing additional permanent head losses. Those fine particles will not be removed during the rapid filter backwash sequence from the filter, as the filter backwash velocity is only 40 to 50 m/h. The separation of the very fine, fine, and smaller sand particles from large particles is effective by washing the sand with a velocity of 120 m/h for a set time.

Sand can be fed (must be dry from a silo with free flow) to a reactor batchwise at a time. The expected sand dosing rates range from 3 to 7 mg of sand per liter of water treated. Using the pellet-to-sand weight ratio, the demanded dosage of seeding sand can be calculated as:

$$\text{Required sand dosage, mg/L} = (\text{TH}_r - \text{TH}_w)/\text{pellet-to-sand ratio} \quad (5.94i)$$

where $\text{TH}_r$ and $\text{TH}_w$ are stated previously.

**Example:** Determine the required sand dosage and sand butch per volume under the following conditions: The total hardness of the raw water and treated water are respective 250 and 88 mg/L as $CaCO_3$; the pellet-to-sand ratio is 25 to 1. The surface area of the sand washer is $A = 0.2\ \text{m}^2$. Specific density of sand $\gamma = 1.55$ g/L.

**solution:** Using Eq. (5.94i)

Step 1. Determine the required sand dosage, using Eq. (5.94i)

$$\begin{aligned}\text{Required sand dosage} &= (\text{TH}_r - \text{TH}_w)/\text{pellet-to-sand ratio}\\ &= (250\ \text{mg/L} - 88\ \text{mg/L})/25\\ &= 6.48\ \text{mg/L}\end{aligned}$$

Step 2. Determine sand butch per volume

Sand dosing is effected batchwise with fixed amount of sand per batch. The height of sand after washing in the sand washer should be measured frequently.

$$\text{Washer (nozzle bottom) depth measured from the top} = 235\ \text{cm}$$

$$\text{Sand level after washing measured from the top} = 173\ \text{cm}$$

$$\text{Difference, } D = 62\ \text{cm} = 0.62\ \text{m}$$

$$\begin{aligned}\text{Volume of washed sand, } V = A \times D &= 0.2\ \text{m}^2 \times 0.62\ \text{m} = 0.124\ \text{m}^3\\ &= 124\ \text{L}\end{aligned}$$

$$\begin{aligned}\text{Weight of sand, } W &= V \times \gamma = 124\ \text{L} \times 1.55\ \text{kg/L} = 192.2\ \text{kg}\\ &= 192.2 \times 10^6\ \text{mg}\end{aligned}$$

Step 3. Calculate the water can be treated with one batch of sand

$$\begin{aligned}\text{Effected volume} &= W/\text{sand dosage} = 192.2 \times 10^6\ \text{mg}/6.48\ \text{mg/L}\\ &= 29.66 \times 10^6\ \text{L}\\ &= 29{,}660\ \text{m}^3\end{aligned}$$

## 13 Ion Exchange

Ion exchange is a reversible process. Ions of a given species are displaced from an insoluble solid substance (exchange medium) by ions of another species dissolved in water. In practice, water is passed through the exchange medium until the exchange capacity is exhausted and then it is regenerated. The process can be used to remove color, hardness (calcium and magnesium), iron and manganese, nitrate and other inorganics, heavy metals, and organics.

Exchange media which exchange cations are called cationic or acid exchangers, while materials which exchange anions are called anionic or base exchangers.

Common cation exchangers used in water softening are zeolite, greensand, and polystyrene resins. However, most ion exchange media are currently in use as synthetic materials.

Synthetic ion exchange resins include four general types used in water treatment. They are strong- and weak-acid cation exchangers and strong- and weak-base anion exchangers. Examples of exchange reactions as shown below (Schroeder, 1977):

Strong acidic

$$2\text{R—SO}_3\text{H} + \text{Ca}^{2+} \leftrightarrow (\text{R—SO}_3)_2\text{Ca} + 2\text{H}^+ \tag{5.95}$$

$$2\text{R—SO}_3\text{Na} + \text{Ca}^{2+} \leftrightarrow (\text{R—SO}_3)_2\text{Ca} + 2\text{Na}^+ \tag{5.96}$$

Weak acidic

$$2\text{R—COOH} + \text{Ca}^{2+} \leftrightarrow (\text{R—COO})_2\text{Ca} + 2\text{H}^+ \tag{5.97}$$

$$2\text{R—COONa} + \text{Ca}^{2+} \leftrightarrow (\text{R—COO})_2\text{Ca} + 2\text{Na}^+ \tag{5.98}$$

Strong-basic

$$2\text{R—X}_3\text{NOH} + \text{SO}_4^{2-} \leftrightarrow (\text{R—X}_3\text{N})_2\text{SO}_4 + 2\text{OH}^- \tag{5.99}$$

$$2\text{R—X}_3\text{NCl} + \text{SO}_4^{2-} \leftrightarrow (\text{R—X}_3\text{N})_2\text{SO}_4 + 2\text{Cl}^- \tag{5.100}$$

Weak-basic

$$2\text{R—NH}_3\text{OH} + \text{SO}_4^{2-} \leftrightarrow (\text{R—NH}_3)_2\text{SO}_4 + 2\text{OH}^- \tag{5.101}$$

$$2\text{R—NH}_3\text{Cl} + \text{SO}_4^{2-} \leftrightarrow (\text{R—NH}_3)_2\text{SO}_4 + 2\text{Cl}^- \tag{5.102}$$

where in each reaction, R is a hydrocarbon polymer and X is a specific group, such as $CH_2$.

The exchange reaction for natural zeolites ($Z$) can be written as

$$\text{Na}_2\text{Z} + \begin{Bmatrix} \text{Ca}^{2+} \\ \text{Mg}^{2+} \\ \text{Fe}^{2+} \end{Bmatrix} \leftrightarrow \begin{Bmatrix} \text{Ca}^{2+} \\ \text{Mg}^{2+} \\ \text{Fe}^{2+} \end{Bmatrix} \text{Z} + 2\text{Na}^+ \tag{5.103}$$

In the cation-exchange water softening process, the hardness-causing elements of calcium and magnesium are removed and replaced with sodium by a strong-acid cation resion. Ion-exchange reactions for softening may be expressed as

$$\mathrm{Na_2R} + \left.\begin{matrix}\mathrm{Ca}\\ \mathrm{Mg}\end{matrix}\right\}\begin{matrix}\mathrm{(HCO_3)_2}\\ \mathrm{SO_4}\\ \mathrm{Cl_2}\end{matrix} \rightarrow \left.\begin{matrix}\mathrm{Ca}\\ \mathrm{Mg}\end{matrix}\right\}\mathrm{R} + \left\{\begin{matrix}\mathrm{2\ NaHCO_3}\\ \mathrm{Na_2SO_4}\\ \mathrm{2\ NaCl}\end{matrix}\right. \tag{5.104}$$

where R represents the exchange resin. They indicate that when a hard water containing calcium and magnesium is passed through an ion exchanger, these metals are taken up by the resin, which simultaneously gives up sodium in exchange. Sodium is dissolved in water. The normal rate is 6 to 8 gpm/ft$^2$ (350 to 470 m/d) of medium.

After a period of operation, the exchanging capacity would be exhausted. The unit is stopped from operation and regenerated by backwashing with sodium chloride solution and rinsed. The void volume for backwash is usually 35% to 45% of the total bed volume.

The exchange capacity of typical resins are in the range of 2 to 10 eq/kg. Zeolite cation exchangers have the exchange capacity of 0.05 to 0.1 eq/kg (Tchobanoglous and Schroeder, 1985). During regeneration, the reaction can be expressed as:

$$\left.\begin{matrix}\mathrm{Ca}\\ \mathrm{Mg}\end{matrix}\right\}\mathrm{R} + \mathrm{2NaCl} \rightarrow \mathrm{Na_2R} + \left.\begin{matrix}\mathrm{Ca}\\ \mathrm{Mg}\end{matrix}\right\}\mathrm{Cl_2} \tag{5.105}$$

The spherical diameter in commercially available ion exchange resins is of the order of 0.04 to 1.0 mm. The most common size ranges used in large treatment plant are 20 to 50 mesh (0.85 to 0.3 mm) and 50 to 100 mesh (0.3 to 0.15 mm) (James M. Montgomery Consulting Engineering, 1985). Details on the particle size and size range, effective size, and uniform coefficient are generally provided by the manufacturers.

The affinity of exchanges is related to charge and size. The higher the valence, the greater affinity and the smaller the effective size the greater the affinity. For a given sense of similar ions, there is a general order of affinity for the exchanger. For synthetic resin exchangers, relative affinities of common ions increase as shown in Table 5.6.

The design of ion exchange units is based upon ion exchange equilibria. The generalized reaction equation for the exchange of ions A and on a cation exchange resin can be expressed as

$$n\mathrm{R^-A^+} + \mathrm{B}^{n+} \leftrightarrow \mathrm{R}n^-\mathrm{B}^{n+}n\mathrm{A}^+ \tag{5.106}$$

where $\mathrm{R}^-$ = an anionic group attached to exchange resin
$\mathrm{A}^+, \mathrm{B}^{n+}$ = ions in solution

**TABLE 5.6 Selectivity Scale for Cations on Eight Percent Cross-Linked Strong-Acid Resin and for the Anions on Strong-Base Resins**

| Cation | Selectivity | Anion | Selectivity |
|---|---|---|---|
| $Li^+$ | 1.0 | $HPO_4^{2-}$ | 0.01 |
| $H^+$ | 1.3 | $CO_3^{2-}$ | 0.03 |
| $Na^+$ | 2.0 | $OH^-$ (type I) | 0.06 |
| $UO^{2+}$ | 2.5 | $F^-$ | 0.1 |
| $NH_4^+$ | 2.6 | $SO_4^{2-}$ | 0.15 |
| $K^+$ | 2.9 | $CH_3COO^-$ | 0.2 |
| $Rb^+$ | 3.2 | $HCO_3^-$ | 0.4 |
| $Cs^+$ | 3.3 | $OH^-$ (type II) | 0.65 |
| $Mg^{2+}$ | 3.3 | $BrO_3^-$ | 1.0 |
| $Zn^{2+}$ | 3.5 | $Cl^-$ | 1.0 |
| $Co^{2+}$ | 3.7 | $CN^-$ | 1.3 |
| $Cu^{2+}$ | 3.8 | $NO_2^-$ | 1.3 |
| $Cd^{2+}$ | 3.9 | $HSO_4^-$ | 1.6 |
| $Ni^{2+}$ | 3.9 | $Br^-$ | 3 |
| $Be^{2+}$ | 4.0 | $NO_3^-$ | 4 |
| $Mn^{2+}$ | 4.1 | $I^-$ | 8 |
| $Pb^{2+}$ | 5.0 | $SO_4^{2-}$ | 9.1 |
| $Ca^{2+}$ | 5.2 | $SeO_4^{2-}$ | 17 |
| $Sr^{2+}$ | 6.5 | $CrO_4^{2-}$ | 100 |
| $Ag^+$ | 8.5 | | |
| $Pb^{2+}$ | 9.9 | | |
| $Ba^{2+}$ | 11.5 | | |
| $Ra^{2+}$ | 13.0 | | |

SOURCES: James M. Montgomery Consulting Engineering (1985), Clifford (1990)

The equilibrium expression for this reaction is

$$\mathrm{K}_{\mathrm{A}\to\mathrm{B}} = \frac{[\mathrm{R}_n^- \mathrm{B}^{n+}][\mathrm{A}^+]^n}{[\mathrm{R}^- \mathrm{A}^+]^n[\mathrm{B}^{n+}]} = \frac{q_{\mathrm{B}} C_{\mathrm{A}}}{q_{\mathrm{A}} C_{\mathrm{B}}} \qquad (5.107)$$

where $K_{\mathrm{A}\to\mathrm{B}} = K_{\mathrm{A}}^{\mathrm{B}}$ = selectivity coefficient, a function of ionic strength and is not a true constant

$[\mathrm{R}^- \mathrm{A}^+]$, $[\mathrm{R}^- \mathrm{B}^{n+}]$ = mole fraction of $A^+$ and $B^+$ exchange resin, respectively, overbars represent the resin phase, or expressed as $[\overline{A}]$, and $[\overline{B}]$

$[\mathrm{A}^+]$, $[\mathrm{B}^{n+}]$ = concentration of $A^+$ and $B^+$ in solution, respectively, mol/L

$q_{\mathrm{A}}$, $q_{\mathrm{B}}$ = concentration of A and B on resin site, respectively, eg/L

$C_{\mathrm{A}}$, $C_{\mathrm{B}}$ = concentration of A and B in solution, respectively, mg/L

The selectivity constant depends upon the valence, nature, and concentration of the ion in solution. It is generally determined in laboratory of specific conditions measured.

For monovalent/monovalent ion exchange process such as

$$H^+R^- + Na^+ \leftrightarrow Na^+R^- + H^+$$

The equilibrium expression is (by Eq. (5.107))

$$K_{HR \to NaR} = \frac{[Na^+R^-][H^+]}{[H^+R^-][Na^+]} = \frac{q_{Na}C_H}{q_H C_{Na}} \tag{5.108}$$

and for divalent/divalent ion exchange processes such as

$$2(Na^+R^-) + Ca^{2+} \leftrightarrow Ca^{2+}R^{2-} + 2Na^+$$

then using Eq. (5.107)

$$K_{NaR \to CaR} = \frac{[Ca^{2+}R][Na^+]^2}{[Na^+R]^2[Ca^{2+}]} = \frac{q_{Ca}C_{Na}^2}{q_{Na}^2 C_{Ca}} \tag{5.109}$$

Anderson (1975) rearranged Eq. (5.107) using a monovalent/monovalent exchange reaction with concentration units to equivalent fraction as follows:

1. In the solution phase: Let $C =$ total ionic concentration of the solution, eq/L. The equivalent ionic fraction of ions $A^+$ and $B^+$ in solution will be

$$X_{A^+} = [A^+]/C \text{ or } [A^+] = CX_{A^+} \tag{5.110}$$

$$X_{B^+} = [B^+]/C \text{ or } [B^+] = CX_{B^+} \tag{5.111}$$

then

$$[A^+] + [B^+] = 1$$

or

$$[A^+] = 1 - [B^+] \tag{5.112}$$

2. In the resin phase: Let $\overline{C}$ = total exchange capacity of the resin per unit volume, eq/L. Then we get

$$\overline{X}_{A^+} = [R^-A^+]/\overline{C} \text{ or } [R^-A^+] = \overline{C}\,\overline{X}_{A^+} \tag{5.113}$$

$$= \text{equivalent fraction of the } A^+ \text{ ion in the resin}$$

and

$$\overline{X}_{B^+} = [R^-B^+]/\overline{C} \text{ or } [R^-B^+] = \overline{C}\,\overline{X}_{B^+} \tag{5.114}$$

also

$$\overline{X}_{A^+} + \overline{X}_{B^+} = 1$$

or

$$\overline{X}_{A^+} = 1 - \overline{X}_{B^+} \tag{5.115}$$

Substitute Eqs. (5.110), (5.111), (5.113), and (5.114) into Eq. (5.107) which yields

$$K_A^B = \frac{[\overline{CX}_{B^+}][CX_{A^+}]}{[\overline{CX}_{A^+}][CX_{B^+}]}$$

or

$$K_A^B = \frac{(\overline{CX}_{B^+})(CX_{A^+})}{(\overline{CX}_{A^+})(CX_{B^+})}$$

$$\frac{\overline{X}_{B^+}}{\overline{X}_{A^+}} = K_A^B \frac{X_{B^+}}{X_{A^+}} \tag{5.116}$$

Substituting for Eqs. (5.112) and (5.115) in Eq. (5.116) gives

$$\frac{\overline{X}_{B^+}}{1 - \overline{X}_{B^+}} = K_A^B \frac{X_{B^+}}{1 - X_{B^+}} \tag{5.117}$$

If the valence is $n$, Eq. (5.117) will become

$$\frac{\overline{X}_{B^{n+}}}{(1 - \overline{X}_{B^{n+}})^n} = K_A^B \left(\frac{\overline{C}}{C}\right)^{n-1} \frac{X_{B^{n+}}}{(1 - X_{B^{n+}})^n} \tag{5.118}$$

**Example 1:** Determine the meq/L of $Ca^{2+}$ if $Ca^{2+}$ concentration is 88 mg/L in water.

**solution:**

Step 1. Determine equivalent weight (EW)

$$\begin{aligned} \text{EW} &= \text{molecular weight/electrical charge} \\ &= 40/2 \\ &= 20 \text{ g/equivalent weight (or mg/meq)} \end{aligned}$$

Step 2. Compute meq/L

$$\begin{aligned} \text{meq/L} &= \text{(mg/L)/EW} \\ &= \text{(88 mg/L)/(20 mg/meq)} \\ &= 4.4 \end{aligned}$$

**Example 2:** A strong-base anion exchange resin is used to remove nitrate ions from well water which contain high chloride concentration. Normally bicarbonate and sulfate is presented in water (assume they are negligible). The total resin capacity is 1.5 eq/L. Find the maximum volume of water that can be treated per liter of resin. The water has the following composition in meq/L:

| | |
|---|---|
| $Ca^{2+} = 1.4$ | $CI^- = 3.0$ |
| $Mg^{2+} = 0.8$ | $SO_4^{2-} = 0.0$ |
| $Na^+ = 2.6$ | $NO_3^- = 1.8$ |
| Total cations = 4.8 | Total anions = 4.8 |

**solution:**

Step 1. Determine the equivalent fraction of nitrate in solution

$$X_{NO_3^-} = 1.8/4.8 = 0.38$$

Step 2. Determine selective coefficient for sodium over chloride from Table 5.5

$$K_{Cl}^{NO_3} = 4/1 = 4$$

Step 3. Compute the theoretical resin available for nitrate ion by Eq. (5.117)

$$\frac{\overline{X}_{NO_3^-}}{1 - \overline{X}_{NO_3^-}} = 4 \times \frac{0.38}{1 - 0.38} = 2.45$$

$$\overline{X}_{NO^-_3} = 0.71$$

It means that 71% of resin sites will be used

Step 4. Compute the maximum useful capacity $Y$

$$Y = 1.5 \text{ eq/L} \times 0.71 = 1.065 \text{ eq/L} = 1065 \text{ meq/L}$$

Step 5. Compute the volume of water ($V$) that can be treated per cycle

$$V = \frac{1065 \text{ meq/L of resin}}{1.8 \text{ meq/L of water}}$$

$$= 592 \text{ L of water/L of resin}$$

$$= 592\frac{\text{L}}{\text{L}} \times \frac{1 \text{ gal}}{3.785 \text{ L}} \times \frac{28.32 \text{ L}}{1 \text{ ft}^3}$$

$$= 4429 \text{ gal of water/ft}^3 \text{ of resin}$$

**Example 3:** A strong-acid cation exchanger is employed to remove calcium hardness from water. Its wet-volume capacity is 2.0 eq/L in the sodium form. If calcium concentrations in the influent and effluent are 44 mg/L (2.2 meq/L) and 0.44 MG/l, respectively, find the equivalent weight (meq/L) of the component in the water if given the following:

| Cations | | Anions | |
|---|---|---|---|
| $Ca^{2+}$ | 2.2 | $HCO_3^-$ | 2.9 |
| $Mg^{2+}$ | 1.0 | $Cl^-$ | 3.1 |
| $Na^+$ | 3.0 | $SO_4^=$ | 0.2 |
| Total cations | 6.2 | Total anions | 6.2 |

**solution:**

Step 1. Determine the equivalent fraction of $Ca^{2+}$

$$X_{Ca^{2+}} = 2.2/6.2 = 0.35$$

Step 2. Find the selectivity coefficient $K$ for calcium over sodium from Table 5.5

$$K_{Na}^{Ca} = 5.2/2.0 = 2.6$$

Step 3. Compute the theoretical resin composition with respect to the calcium ion

$$\overline{C}/C = 6.2/6.2 = 1.0$$

Using Eq. (5.118), $n = 2$

$$\frac{\overline{X}_{Ca^{2+}}}{(1 - \overline{X}_{Ca^{2+}})^2} = 2.6 \frac{0.35}{(1 - 0.35)^2} = 2.15$$

$$\overline{X}_{Ca^{2+}} = 0.51$$

This means that a maximum of 51% of the resin sites can be used with calcium ions from the given water. At this point, the water and resin are at equilibrium with each other.

Step 4. Compute the limiting useful capacity of the resin ($Y$)

$$Y = 2.0 \text{ eq/L} \times 0.51$$

$$= 1.02 \text{ eq/L (or 1020 meq/L)}$$

Step 5. Compute the maximum volume ($V$) of water that can be treated per cycle

$$V = \frac{1020 \text{ meq/L of resin (for Ca)}}{2.2 \text{ meq/L of calcium in water}}$$

$$= 464 \text{ L of water/L of resin}$$

$$= 3472 \text{ gal of water/ft}^3 \text{of resin}$$

### 13.1 Leakage

Leakage is defined as the appearance of a low concentration of the undesirable ions in the column effluent during the beginning of the exhaustion. It comes generally from the residual ions in the bottom resins due to incomplete regeneration. Leakage will occur for softening process. A water softening column is usually not fully regenerated due to inefficient use of salt to completely regenerate the resin on the sodium form. The leakage depends upon the ionic composition of the bed bottom and the composition of the influent water.

A detailed example of step-by-step design method for a fix bed ion exchange column for water softening is given by Benefield *et al.* (1982).

**Example:** The bottom of the water softener is 77% in the calcium form after regeneration. The strong-acid cation resin has a total capacity of 2.0 eq/L and selective coefficient for calcium over sodium is 2.6. Determine the initial calcium leakage the water composition has as follows (in meq/L):

$$Ca^{2+} = 0.4 \qquad Cl^- = 0.4$$

$$Mg^{2+} = 0.2 \qquad SO_4^{2-} = 0.4$$

$$Na^+ = 1.2 \qquad HCO_3^- = 1.0$$

$$\text{Total cations} = 1.8 \qquad \text{Total anions} = 1.8$$

**solution:**

Step 1. Determine the equivalent fraction of calcium ion from the given:

$$\overline{X}_{Ca^{2+}} = 0.77$$

$$\overline{C} = 2.0 \text{ eq/L}$$

$$C = 2 \text{ meq/L} = 0.002 \text{ eq/L}$$

$$K_{Na}^{Ca} = 2.6$$

Using Eq. (5.118)

$$\frac{0.77}{(1 - 0.77)^2} = 2.6\left(\frac{2}{0.002}\right)\left[\frac{X_{Ca}}{(1 - X_{Ca})^2}\right]$$

$$\frac{X_{Ca}}{(1 - X_{Ca})^2} = 0.0056$$

Step 2. Calculate the initial calcium leakage $y$

$$y = 0.0056\, c = 0.0056 \times 1.8 \text{ meq/L}$$

$$= 0.010 \text{ meq/L}$$

$$= 0.010 \frac{\text{meq}}{\text{L}} \times \frac{20 \text{ mg as Ca}}{\text{meq}}$$

$$= 0.20 \text{ mg/L as Ca}$$

$$= 0.20 \text{ mg/L as Ca} \frac{100 \text{ as CaCO}_3}{40 \text{ as Ca}}$$

$$= 0.50 \text{ mg/L as CaCO}_3$$

### 13.2 Nitrate removal

Nitrate, $NO_3^-$, is a nitrogen–oxygen ion that occurs frequently in nature as the result of interaction between nitrogen in the atmosphere and living things on earth. It is a portion of the nitrogen cycle. When plant and animal proteins are broken down, ammonia and nitrogen gas are released. Ammonia is subsequently oxidized to nitrite ($NO_2^-$) and nitrate by bacterial action (Fig. 1.1).

High concentration of nitrate in drinking water may cause methemoglobinemia, especially for infants. Nitrates interfere with the body's ability to take oxygen from the air and distribute it to body cells. The United States EPA's standard for nitrate in drinking water is 10 mg/L. In Illinois, the water purveyor must furnish special bottle water (low nitrate) to infants if the nitrate level of the finished water exceeds the standard.

The sources of nitrates to water are due to human activities. Agricultural activities, such as fertilizer application and animal feedlots, are major contributors of nitrate and ammonia to be washed off (runoff) and percolate to soil with precipitation. The nitrates that are polluted can subsequently flow into surface waters (streams and lakes) and into groundwater. Municipal sewage effluents improperly treat domestic wastewaters, and draining of septic tanks may cause nitrate contamination in the raw water source of a waterworks. Identification of pollution sources and protective alternatives should be taken.

If the nitrate concentration in the water supply is excessive, there are two approaches for solving the problem, i.e. treatment for nitrate removal and nontreatment alternatives. Nontreatment alternatives may include (1) use new water source, (2) blend with low nitrate waters, (3) connect to other supplier(s), and (4) organize in a regional system.

Nitrate is not removed by the conventional water treatment processes, such as coagulation–flocculation, sedimentation, filtration, activated carbon adsorption, and disinfection (chlorination and ozonation). Some treatment technologies, such as ion exchange, reverse osmosis, electrodialysis, microbial denitrification, and chemical reduction, can remove nitrates from drinking water. An ion exchange process design is given as the following examples.

Two types (fixed bed and continuous) of ion exchangers are used. The fixed bed exchange is used mostly for home and industry uses and is controlled by a flow totalizer with an automatic regeneration cycle at approximately 75% to 80% of the theoretical bed capacity. Continuous ion exchangers are employed by larger installations, such as waterworks, that provide continuous product water and require minimum bed volume. A portion of the resin bed is withdrawn and regenerated outside of the main exchange vessel.

In the design of an ion exchange system for nitrate removal, raw water quality analyses and pilot testing are generally required. The type of resin and resin capacity, bed dimensions, and regenerant requirement must be determined. Basic data, such as the design flow rate through the exchanger, influent water quality, total anions, and suggested operating conditions for the resin selected (from the manufacturer) must be known. Analysis of water quality may include nitrate, sulfate, chloride, bicarbonate, calcium carbonate, iron, total suspended solids, and total organic carbon. For Duolite A-104, for example, the suggested design parameters are listed below (Diamond Shamrock Co., 1978):

| Parameters | Recommended values |
|---|---|
| Minimum bed depth | 30 in |
| Backwash flow rate | 2–3 gpm/ft$^2$ |
| Regenerant dosage | 15–18 lb sodium chloride (NaCl) per ft$^3$ resin |
| Regenerant concentration | 10–12% NaCl by weight |
| Regenerant temperature | Up to 120°F or 49°C |
| Regenerant flow rate | 0.5 gpm/ft$^3$ |
| Rinse flow rate | 2 gpm/ft$^3$ |
| Rinse volume | 50–70 gal/ft$^3$ |
| Service flow rate | Up to 5 gpm/ft$^3$ |
| Operating temperature | Salt form, up to 180°F or 85°C |
| pH limitation | None |

In design processes, resin capacity, bed dimensions, and regenerant requirements must be computed. Resin capacity determines the quantity of resin needed in the ion exchanger and is computed from a pilot study. Data is provided by the manufacturer. For example, the operating capacity of Duloite A-104 resin for nitrate removal is quite dependent upon the nitrate, sulfate, and total anion concentrations to calculate the corrected resin capacity. The raw or uncorrected resin capacity is determined by using the manufacturer's graph (Fig. 5.7). This capacity must be adjusted downward to reflect the presence of sulfate in the water (Fig. 5.8), because sulfate anions will be exchanged before nitrate (US EPA, 1983).

Once the adjusted resin capacity is determined for the water to be treated, the required bed volume of the ion exchange resin can be

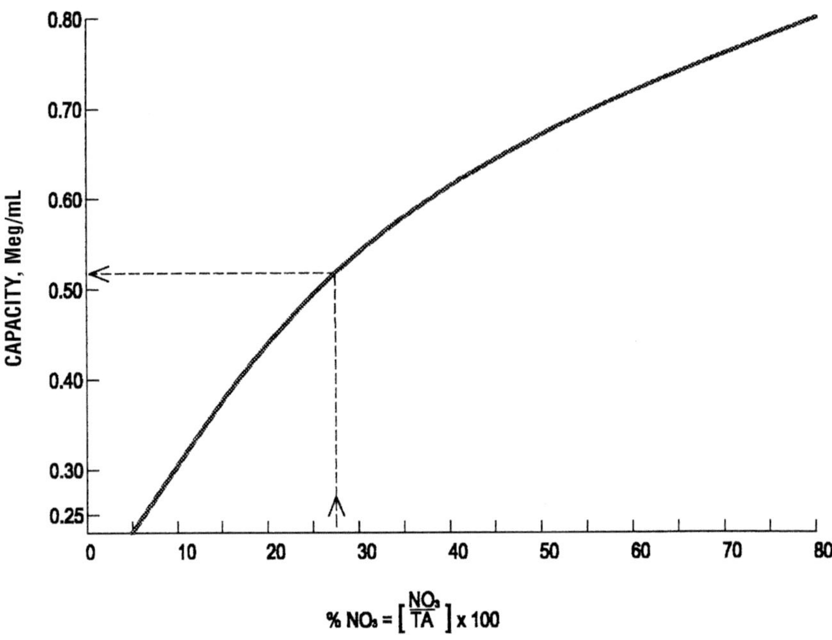


**Figure 5.7** Relationship of $NO_3$/TA and unadjusted resin capacity for A-104 resin (*US EPA, 1983*).

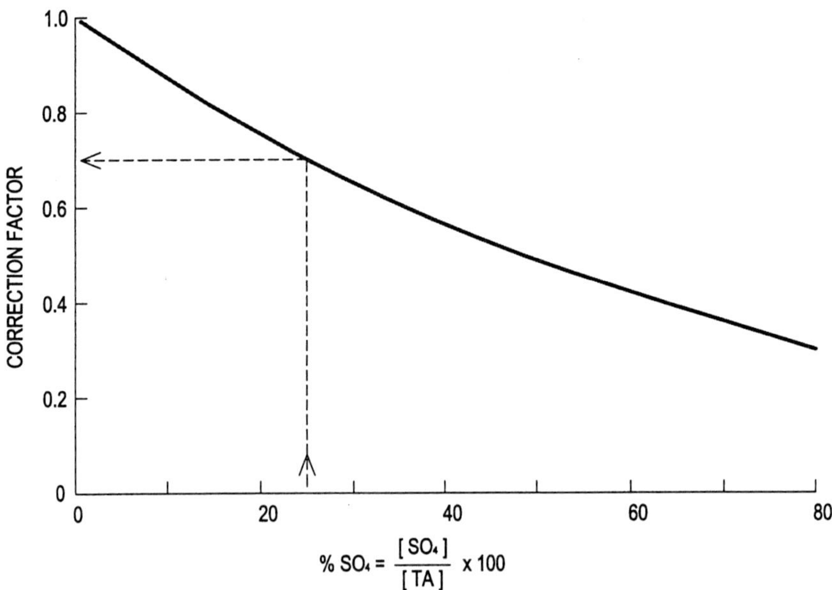


**Figure 5.8** Correction factor for A-104 resin (*US EPA, 1983*).

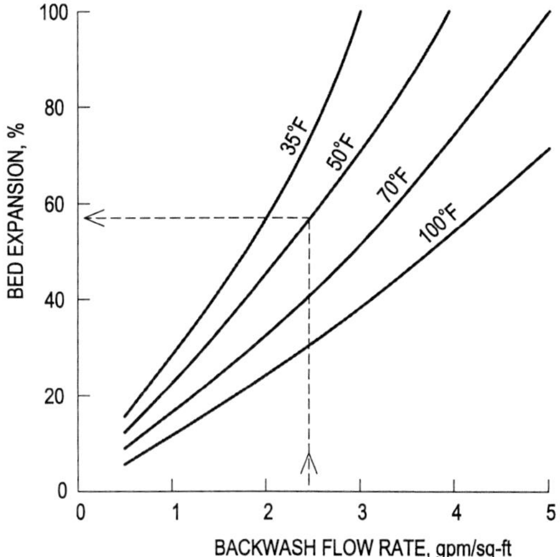


**Figure 5.9** Bed expansion curve for A-104 resin (modified from *US EPA, 1983*).

calculated. The bed volume is the amount of nitrate that must be removed in each cycle divided by the adjusted resin capacity. Using this bed volume and minimum depth requirement, the surface area of resin can be calculated. A standard size containment vessel can be selected with the closest surface area. Using this standard size, the depth of the resin can be recalculated. The final height of the vessel should be added to a bed expansion factor during backwashing with a design temperature selected (Fig. 5.9).

Once the bed volume and dimensions are computed, the regeneration system can be designed. The regeneration system must determine: (1) the salt required per generation cycle, (2) the volume of brine produced per cycle, (3) the total volume of the brine storage tank, and (4) the time needed for the regeneration process. The design of an ion exchanger for nitrate removal is illustrated in the following three examples.

**Example 1:** A small public water system has the maximum daily flow rate of 100,000 gal/d, maximum weekly flow of 500,000 gal/week, and maximum nitrate concentration of 16 mg/L. The plant treats 100,000 gal of water and operates only 7 h/d and 5d/week (assume sufficient storage capacity for weekend demand). Finished water after ion exchange process with 0.6 mg/L of nitrate–nitrogen ($NO_3$-N) will be blended with untreated water 16 mg/L of $NO_3$-N to produce a finished water of 8 mg/L or less of $NO_3$-N. The $NO_3$-N standard is 10 mg/L. Determine the flow rate of ion exchanger in gpm and the blending rate.

**solution:**

Step 1. Compute the quantity of water that passes the ion exchanger, $q_1$

Let untreated flow = $q_2$ and $c$, $c_1$, and $c_2$ = $NO_3$-N concentrations in finished, treated (ion exchanger) and untreated waters, respectively.

Then

$$c = 8 \text{ mg/L}$$
$$c_1 = 0.6 \text{ mg/L}$$
$$c_2 = 16 \text{ mg/L}$$

by mass balance

$$c_1 q_1 + c_2 q_2 = 100{,}000c$$
$$0.6q_1 + 16q_2 = 100{,}000 \times 8 \qquad \text{(a)}$$

since

$$q_2 = 100{,}000 - q_1 \qquad \text{(b)}$$

Substituting (b) into (a)

$$06q_1 + 16(100{,}000 - q_1) = 800{,}000$$
$$15.4q_1 = 800{,}000$$
$$q_1 = 51{,}950 \text{ (gpd)}$$

Step 2. Compute flow rate in gpm to the exchanger (only operates 7 h/d)

$$q_1 = 51{,}950 \frac{\text{gal}}{\text{d}} \times \frac{24 \text{ h}}{7 \text{ h}} \times \frac{1 \text{ day}}{1440 \text{ min}}$$
$$= 124 \text{ gpm}$$

Step 3. Compute blending ratio, br

$$\text{br} = \frac{51{,}950 \text{ gpd}}{100{,}000 \text{ gpd}}$$
$$\cong 0.52$$

**Example 2:** The given conditions are the same as in Example 1. The anion concentration of the influent of the ion exchanger are: $NO_3$-N = 16 mg/L, $SO_4$ = 48 mg/L, Cl = 28 mg/L, and $HCO_3$ = 72 mg/L. Determine the total quantity of nitrate to be removed daily by the ion exchanger and the ratios of anions.

**solution:**

Step 1. Compute total anions and ratios

| Anions | Concentration, mg/L | Milliequivalent weight, mg/meq | Concentration, meq/L | Percentage |
|---|---|---|---|---|
| $NO_3$-N | 16 | 14.0 | 1.14 | 27.7 |
| $SO_4$ | 48 | 48.0 | 1.00 | 24.4 |
| Cl | 28 | 35.5 | 0.79 | 19.2 |
| $HCO_3$ | 72 | 61.0 | 1.18 | 28.7 |
| Total | | | 4.11 | |

In the last column above for $NO_3$-N

$$\% = 1.14 \times 100/4.11 = 27.7$$

Step 2. Determine nitrate to be removed daily

The total amount of nitrates to be removed depends on the concentrations of influent and effluent, $c_i$ and $c_e$ and can be determined by

$$\text{nitrates removed (meq/d)} = [(c_i - c_e)\ \text{meq/L}] \times q_1(\text{L/d})$$

Since

$$c_i = 1.14\ \text{meq/L}$$

$$c_e = 0.6\ \text{mg/L} \div 14\ \text{mg/meq}$$

$$= 0.043\ \text{meq/L}$$

$$\text{Nitrates removed} = (1.14 - 0.043)\ \text{meq/L} \times 51{,}950\ \text{gpd} \times 3.785\ \text{L/gal}$$

$$= 215{,}700\ \text{meq/d}$$

If the ion exchanger is going to operate with only one cycle per day, the quantity of nitrate to be removed per cycle is 215,700 meq.

**Example 3:** Design a 100,000 gpd ion exchanger for nitrate removal using the data given in Examples 1 and 2. Find resin capacity, resin bed dimension, and regenerant requirements (salt used, brine production, the total volume of the brine storage tank, and regeneration cycle operating time) for the ion exchanger.

**solution:**

Step 1. Determine the uncorrected (unadjusted) volume of resin needed

(a) In Example 2, the nitrate to total anion ratio is 27.7%. Using 27.7% in Fig. 5.7, we obtain the uncorrected resin capacity of 0.52 meq/($NO_3$ mL) resin.

(b) In Example 2, the sulfate to total anion ratio is 24.4%. Using 25% in Fig. 5.8, we obtain the correction factor of resin capacity due to sulfate is 0.70.

(c) The adjusted resin capacity is downward to

$$0.52 \text{ meq/mL} \times 0.70 = 0.364 \text{ meq/mL}$$

(d) Convert meq/mL to meq/ft$^3$

$$0.364 \text{ meq/mL} \times 3785 \text{ mL/gal} \times 7.48 \text{ gal/ft}^3$$
$$= 10{,}305 \text{ meq/ft}^3$$

Step 2. Determine the bed volume, BV

In Example 2, nitrate to be removed in each cycle is 215,700 meq

$$\text{BV} = 215{,}700 \text{ meq} \div 10{,}305 \text{ meq/ft}^3$$
$$= 20.93 \text{ ft}^3$$

Step 3. Check the service flow rate, SFR

In Example 1, the ion exchange unit flow rate =124 gpm

$$\text{SFR} = 124 \text{ gpm/BV} = 124 \text{ gpm}/20.93 \text{ ft}^3$$
$$= 5.9 \text{ gpm/ft}^3$$

This value exceeds the manufacturer's maximum SFR of 5 gpm/ft$^3$. In order to reduce the SFR, we can modify it either by: (a) increasing the exchanger operating time per cycle, or (b) increasing the bed volume.

(a) Calculate adjusted service time, AST

$$\text{AST} = \frac{7 \text{ h}}{\text{cycle}} \times \frac{5.9 \text{ gpm/ft}^3}{5.0 \text{ gpm/ft}^3}$$
$$= 8.3 \text{ h/cycle}$$

Calculate the adjusted flow rate for the new cycle

In Example 1, $q_1 = 51{,}950$ gpd or 51,950 gal/cycle

$$\text{then, flow rate} = q_1/\text{AST}$$
$$= 51{,}950 \text{ gal/cycle}/(8.3 \text{ h/cycle} \times 60 \text{ min/h})$$
$$= 104 \text{ gpm}$$

Alternatively

(b) Increase the bed volume and retain the initial unit flow rate

$$\text{Adjusted BV} = 20.93 \text{ ft}^3 \times (5.9 \text{ gpm/ft}^3/5.0 \text{ gpm/ft}^3)$$
$$= 24.7 \text{ ft}^3$$
$$\text{say} \quad = 25 \text{ ft}^3$$

In comparison between alternatives (a) and (b), it is more desirable to increase BV using 7 h/d operation. Thus BV = 25 $ft^3$ is chosen.

Step 4. Determine bed dimensions.

(a) Select the vessel size

Based on the suggested design specification, the minimum bed depth ($z$) is 30 in or 2.5 feet. The area, $A$, would be

$$A = \text{BV}/z = 25\ \text{ft}^3/2.5\ \text{ft}$$
$$= 10.0\ \text{ft}^2$$

For a circular vessel, find the diameter

$$A = \pi D^2/4$$
$$D = \sqrt{4A/\pi} = \sqrt{4 \times 10.0\ \text{ft}^2/3.14}$$
$$= 3.57\ \text{ft}$$

The closest premanufactured size is 3.5 ft in diameter with a correspon-ding cross-sectional area of 9.62 $ft^2$. The resin height, $h$, would then be recalculated as

$$h = 25\ \text{ft}^3/9.62\ \text{ft}^2$$
$$= 2.6\ \text{ft}$$

(b) Adjusting for expansion during backwash

Since allowable backwash flow rate is 2 to 3 gpm/$ft^2$. If the backwash rate is selected as 2.5 gpm/$ft^2$ with the backwash water temperature of 50°F, the percent bed expansion would be 58% as read from the curve in Fig. 5.9. Thus, the design height ($H$) of the vessel should be at least

$$H = 2.6\ \text{ft} \times (1 + 0.58)$$
$$= 4.11\ \text{ft}$$

*Note*: We use 3.5 ft diameter and 4.11 ft (at least) height vessel.

Step 5. Determine regenerant requirements

(a) Choice salt (NaCl) required per regeneration cycle since regenerant dosage is 15 to 18 lb of salt per cubic foot of resin. Assume 16 lb/$ft^3$ is selected.

The amount of salt required is

$$\text{Salt} = 16\ \text{lb/ft}^3 \times 25\ \text{ft}^3$$
$$= 400\ \text{lb}$$

We need 400 lb of salt per cycle.

(b) Calculate volume of brine used per cycle

On the basis of the specification, the regenerant concentration is 10% to 12% NaCl by weight. Assume it is 10%, then

$$\text{Salt concentration, \%} = \frac{\text{wt. of salt} \times 100}{\text{total wt. of brine}}$$

$$\text{total wt. of brine} = 400 \text{ lb} \times 100/10$$
$$= 4000 \text{ lb}$$

(c) Calculate total volume of brine

$$\text{Weight of water} = (4000 - 400) \text{ lb/cycle}$$
$$= 3600 \text{ lb}$$

$$\text{Volume of water} = \frac{\text{wt. of water}}{\text{density of water}} = \frac{3600 \text{ lb}}{62.4 \text{ lb/ft}^3}$$
$$= 57.69 \text{ ft}^3$$

The specific weight of salt is 2.165

$$\text{Volume of salt} = 400 \text{ lb}/(62.4 \text{ lb/ft}^3 \times 2.165)$$
$$= 2.96 \text{ ft}^3$$
$$\text{Total volume of the brine} = 57.69 \text{ ft}^3 + 2.96 \text{ ft}^3$$
$$= 60.65 \text{ ft}^3$$
$$\text{Total volume in gallons} = 60.65 \text{ ft}^3 \times 7.48 \text{ gal/ft}^3$$
$$= 454 \text{ gal}$$

*Note*: The total volume of brine generated is 454 gal per cycle. The total volume of the brine storage tank should contain sufficient capacity for 3 to 4 generations. Assume 3 cycles of brine produced should be stored, then the total brine tank volume ($V$) would be

$$V = 454 \text{ gal/cycle} \times 3 \text{ cycle}$$
$$= 1362 \text{ gal}$$

(d) Calculate time required for regeneration cycle

The regeneration flow rate ($Q$) is 0.5 gpm/ft$^3$resin specified by the manufacturer. Then

$$Q = 0.5 \text{ gpm/ft}^3 \times 25 \text{ ft}^3$$
$$= 12.5 \text{ gpm}$$

The regeneration time ($t$) is the volume of the brine per cycle divided by the flow rate $Q$, thus

$$t = 454 \text{ gal}/12.5 \text{ gal/min}$$
$$= 36.3 \text{ min}$$

## 14 Iron and Manganese Removal

Iron (Fe) and manganese (Mn) are abundant elements in the earth's crust. They are mostly in the oxidized state (ferric, $Fe^{3+}$, and $Mn^{4+}$) and are insoluble in natural waters. However, under reducing conditions (i.e. where dissolved oxygen is lacking and carbon dioxide content is high), appreciable amounts of iron and manganese may occur in surface water, groundwater, and in water from the anaerobic hypolimnion of stratified lakes and reservoirs. The reduced forms are soluble divalent ferrous ($Fe^{2+}$) and manganous ($Mn^{2+}$) ions that are chemically bound with organic matter. Iron and manganese get into natural water from dissolution of rocks and soil, from acid mine drainage, and from corrosion of metals. Typical iron concentrations in groundwater are 1.0 to 10 mg/L, and typical concentrations in oxygenated surface waters are 0.05 to 0.2 mg/L. Manganese exists less frequently than iron and in smaller amounts. Typical manganese values in natural water range from 0.1 to 1.0 mg/L (James M. Montgomery Consulting Engineers, 1985). Voelker (1984) reported that iron and manganese levels in groundwaters in the American Bottoms area of southwestern Illinois ranged from <0.01 to 82.0 mg/L and <0.01 to 4.70 mg/L, respectively, with mean concentrations of 8.4 mg/L and 0.56 mg/L, respectively.

Generally, iron and manganese in water are not a health risk. However, in public water supplies they may discolor water, and cause taste and odors. Elevated iron and manganese levels also can cause stains in plumbing, laundry, and cooking utencils. Iron and manganese may also cause problems in water distribution systems because metal depositions may result in pipe encrustation and may promote the growth of iron bacteria which may in turn cause tastes and odors. Iron and manganese may also cause difficulties in household ion exchange units by clogging and coating the exchange medium.

To eliminate the problems caused by iron and manganese, the US Environmental Protection Agency (1987) has established secondary drinking water standards for iron at 0.3 mg/L and for manganese at 0.05 mg/L. The Illinois Pollution Control Board (IPCB, 1990) has set effluent standards of 2.0 mg/L for total iron and 1.0 mg/L for total manganese.

It is considered that iron and manganese are essential elements for plants and animals. The United Nation Food and Agriculture Organization recommended maximum levels for iron and manganese in irrigation waters are 0.5 and 0.2 mg/L, respectively.

The techniques for removing iron and manganese from water are based on the oxidation of relatively soluble Fe(II) and Mn(II) to the insoluble Fe(III) and Mn(III,IV) and the oxidation of any organic-complex compounds. This is followed by filtration to remove the Fe(III) and processes for iron and manganese removal are discussed elsewhere (Rehfeldt *et al.*, 1992).

Four major techniques for iron and manganese removal from water are: (1) oxidation–precipitation–filtration, (2) manganese zeolite process, (3) lime softening–settling–filtration, and (4) ion exchange. Aeration–filtration, chlorination–filtration, and manganese zeolite process are commonly used for public water supply.

### 14.1 Oxidation

Oxidation of soluble (reduced) iron and manganese can be achieved by aeration, chlorine, chlorine dioxide, ozone, potassium permanganate ($KMnO_4$) and hydrogen peroxide ($H_2O_2$). The chemical reactions can be expressed as follows:

For aeration (Jobin and Ghosh, 1972; Dean, 1979)

$$Fe^{2+} + \tfrac{1}{4}O_2 + 2OH^- + \tfrac{1}{2}H_2O \rightarrow Fe(OH)_3 \tag{5.119}$$

$$Mn^{2+} + O_2 \rightarrow MnO_2 + 2e \tag{5.120}$$

In theory, 0.14 mg/L of oxygen is needed to oxidize 1 mg/L of iron and 0.29 mg/L of oxygen for each mg/L of manganese.

For chlorination (White, 1972):

$$2Fe(HCO_3)_2 + Cl_2 + Ca(HCO_3)_2 \rightarrow 2Fe(OH)_3 + CaCl_2 + 6CO_2 \tag{5.121}$$

$$MnSO_4 + Cl_2 + 4NaOH \rightarrow MnO_2 + 2NaCl + Na_2SO_4 + 2H_2O \tag{5.122}$$

The stoichiometric amounts of chlorine needed to oxidize each mg/L of iron and manganese are 0.62 and 1.3 mg/L, respectively.

For potassium permanganate oxidation (Ficek, 1980):

$$3Fe^{2+} + MnO_4 + 4H^+ \rightarrow MnO_2 + 2Fe^{3+} + 2H_2O \tag{5.123}$$

$$3Mn^{2+} + 2MnO_4 + 2H_2O \rightarrow 5MnO_2 + 4H^+ \tag{5.124}$$

In theory, 0.94 mg/L of $KMnO_4$ is required for each mg/L of soluble iron and 1.92 mg/L for 1 mg/L of soluble manganese. In practice, a 1 to 4% solution of $KMnO_4$ is fed in at the low-lift pump station or at the rapid-mix point. It should be totally consumed prior to filtration. However, the required $KMnO_4$ dosage is generally less than the theoretical values (Humphrey, and Eikleberry, 1962; Wong, 1984). Secondary oxidation reactions occur as

$$Fe^{2+} + MnO \rightarrow Fe^{3+} + Mn_2O_3 \tag{5.125}$$

$$Mn^{2+} + MnO_2 \cdot 2H_2O \rightarrow Mn_2O_3 \cdot X(H_2O) \tag{5.126}$$

and

$$2Mn^{2+} + MnO_2 \rightarrow Mn_3O_4 \cdot X(H_2O) \tag{5.127a}$$

or

$$MnO \cdot Mn_2O_3 \cdot X(H_2O) \tag{5.127b}$$

For hydrogen peroxide oxidation (Kreuz, 1962):

1. Direct oxidation

$$2Fe^{2+} + H_2O_2 + 2H^+ \rightarrow 2Fe^{3+} + 2H_2O \quad (5.128)$$

2. Decomposition to oxygen

$$H_2O_2 \rightarrow \tfrac{1}{2}O_2 + H_2O \quad (5.129)$$

followed by oxidation

$$2Fe^{2+} + \tfrac{1}{2}O_2 + H_2O \rightarrow 2Fe^{3+} + 2OH^- \quad (5.130)$$

In either case, 2 moles of ferrous iron are oxidized per mole of hydrogen peroxide, or 0.61 mg/L of 50% $H_2O_2$ oxidizes 1 mg/L of ferrous iron.

For oxidation of manganese by hydrogen peroxide, the following reaction applies (H.M. Castrantas, FMC Corporation, Princeton, NJ, personal communication).

$$Mn^{2+} + H_2O_2 + 2H^+ + 2e \rightarrow Mn^{3+} + e + H_2O \quad (5.131)$$

To oxidize 1 mg/L of soluble manganese, 1.24 mg/L of 50% $H_2O_2$ is required.

Studies on groundwater by Rehfeldt *et al.* (1992) reported that oxidant dosage required to meet the Illinois recommended total iron standard follow the order of Na, O, Cl as $Cl_2 > KMnO_4 > H_2O_2$. The amount of residues generated by oxidants at the critical dosages follow the order of $KMnO_4 > NaOCl > H_2O_2$, with $KMnO_4$ producing the largest, strongest, and densest flocs.

For ozonation: It is one of the strongest oxidants used in the water industry for disinfection purposes. The oxidation potential of common oxidants relative to chlorine is as follows (Peroxidation Systems, 1990):

| | |
|---|---|
| Fluoride | 2.32 |
| Hydroxyl radical | 2.06 |
| Ozone | 1.52 |
| Hydrogen peroxide | 1.31 |
| Potassium permanganate | 1.24 |
| Chlorine | 1.00 |

Ozone can be very effective for iron and manganese removal. Because of its relatively high capital costs and operation and maintenance costs, the ozonation process is rarely employed for the primary purpose of oxidizing iron and manganese. Since ozone is effective in oxidizing trace toxic organic matter in water, preozonation instead of prechlorination is becoming popular. In addition, many water utilities are using ozone for disinfection purposes. Ozonation can be used for two purposes: disinfection and metal removal.

**Manganese zeolite process.** In the manganese zeolite process, iron and manganese are oxidized to the insoluble form and filtered out, all in one unit, by a combination of sorption and oxidation. The filter medium can be manganese greensand, which is a purple-black granular material processed from glauconitic greens, and/or a synthetic formulated product. Both of these compositions are sodium compounds treated with a manganous solution to exchange manganese for sodium and then oxidized by $KMnO_4$ to produce an active manganese dioxide. The greensand grains in the filter become coated with the oxidation products. The oxidized form of greensand then adsorbs soluble iron and manganese, which are subsequently oxidized with $KMnO_4$. One advantage is that the greensand will adsorb the excess KMnO4 and any discoloration of the water.

**Regenerative-batch process.** The regenerative-batch process uses manganese-treated greensand as both the oxidant source and the filter medium. The manganese zeolite is made from $KMnO_4$-treated greensand zeolite. The chemical reactions can be expressed as follows (Humphrey and Eikleberry, 1962; Wilmarth, 1968):

Exchange:

$$\mathrm{NaZ} + \mathrm{Mn}^{2+} \rightarrow \mathrm{MnZ} + 2\mathrm{Na}^{+} \tag{5.132}$$

Generation:

$$\mathrm{MnZ} + \mathrm{KMnO_4} \rightarrow \mathrm{Z \cdot MnO_2} + \mathrm{K}^{+} \tag{5.133}$$

Degeneration:

$$\mathrm{Z \cdot MnO_2} + \begin{matrix} \mathrm{Fe}^{2+} \\ \mathrm{Mn}^{2+} \end{matrix} \rightarrow \mathrm{Z \cdot MnO_3} + \begin{matrix} \mathrm{Fe}^{3+} \\ \mathrm{Mn}^{3+} \\ \mathrm{Mn}^{4+} \end{matrix} \tag{5.134}$$

Regeneration:

$$\mathrm{Z \cdot MnO_3} + \mathrm{KMnO_4} \rightarrow \mathrm{Z \cdot MnO_2} \tag{5.135}$$

where NaZ is greensand zeolite and $Z \cdot MnO_2$ is manganese zeolite. As the water passes through the mineral bed, the soluble iron and manganese are oxidized (degeneration). Regeneration is required after the manganese zeolite is exhausted.

One of the serious problems with the regenerative-batch process is the possibility of soluble manganese leakage. In addition, excess amounts of $KMnO_4$ are wasted, and the process is not economical for water high in metal content. Manganese zeolite has an exchange capacity of 0.09 lb of iron or manganese per cubic foot of material, and the flow rate to the exchanger is usually 3.0 gpm/ft$^3$. Regeneration needs approximately 0.18 lb $KMnO_4$ /ft$^3$ of zeolite.

**Continuous process.** For the continuous process, 1% to 4% $KMnO_4$ solution is continuously fed ahead of a filter containing anthracite be (6- to 9-in thick), manganese-treated greensand (24 to 30 in), and gravel. The system takes full advantage of the higher oxidation potential of $KMnO_4$ as compared to manganese dioxide. In addition, the greensand can act as a buffer. The $KMnO_4$ oxidizes iron, manganese, and hydrogen sulfide to the insoluble state before the water reaches the manganese zeolite bed.

Greensand grain has a smaller effective size than silica sand used in filters and can result in comparatively higher head loss. Therefore, a layer of anthracite is placed above the greensand to prolong filter runs by filtering out the precipitate. The upper layer of anthracite operates basically as a filter medium. When iron and manganese deposits build up, the system is backwashed like an ordinary sand filter. The manganese zeolite not only serves as a filter medium but also as a buffer to oxidize any residual soluble iron and manganese and to remove any excess unreacted $KMnO_4$. Thus a $KMnO_4$ demand test should be performed. The continuous system is recommended for waters where iron predominates, and the intermittent regeneration system is recommended for groundwater where manganese predominates (Inversand Co., 1987).

**Example 1:** Theoretically, how many mg/L each of ferrous iron and soluble manganese can be oxidized by 1 mg/L of potassium permanganate?

**solution:**

Step 1. For $Fe^{2+}$, using Eq. (5.123)

$$3Fe^{2+} + 4H^+ + KMnO_4 \rightarrow MnO_2 + 2Fe^{3+} + K^+ + 2H_2O$$

MW $3 \times 55.85$ $\quad$ $39.1 + 54.94 + 4 \times 16$

$= 167.55$ $\quad$ $= 158.14$

$X$ mg/L $\quad$ 1 mg/L

By proportion

$$\frac{X}{1} = \frac{167.55}{158.14}$$

$$X = 1.06 \text{ mg/L}$$

Step 2. For $Mn^{2+}$, using Eq. (5.124)

$$3Mn^{2+} + 2H_2O + 2KMnO_4 \rightarrow 5MnO_2 + 4H^+ + 2K^+$$

| | | |
|---|---|---|
| $M_1W_1$ | $3 \times 54.94$ | $2 \times 158.14$ |
| | $= 164.82$ | $= 316.28$ |
| | $y$ | 1 mg/L |

then

$$y = \frac{164.82}{316.28} = 0.52 \text{ mg/L}$$

**Example 2:** A groundwater contains 3.6 mg/L of soluble iron and 0.78 mg/L of manganese. Find the dosage of potassium permanganate required to oxidize the soluble iron and manganese.

**solution:** From Example 1 the theoretical potassium permanganate dosage are 1.0 mg/L per 1.06 mg/L of ferrous iron and 1.0 mg/L per 0.52 mg/L of manganese. Thus

$$\begin{aligned} KMnO_4 \text{ dosage} &= \frac{1.0 \times 3.6 \text{ mg/L}}{1.06} + \frac{1.0 \times 0.78 \text{ mg/L}}{0.52} \\ &= 4.9 \text{ mg/L} \end{aligned}$$

## 15 Activated Carbon Adsorption

### 15.1 Adsorption isotherm equations

The Freundlich isotherm equation is an empirical equation, which gives an accurate description of adsorption of organic adsorption in water. The equation under constant temperature equilibrium is (US EPA, 1976; Brown and LeMay, 1981; Lide, 1996)

$$q_e = KC_e^{1/n} \qquad (5.136)$$

or

$$\log q_e = \log K + (1/n)\log C_e \qquad (5.136a)$$

where $q_e$ = quantity of absorbate per unit of absorbent, mg/g
$C_e$ = equilibrium concentration of adsorbate in solution, mg/L
$K$ = Freundlich absorption coefficient, (mg/g) $(L/mg)^{1/n}$
$n$ = empirical coefficient

The constant $K$ is related to the capacity of the absorbent for the absorbate. $1/n$ is a function of the strength of adsorption. The molecule that accumulates, or adsorbs, at the surface is called an *adsorbate*; and the solids on which adsorption occurs is called *adsorbent*. Snoeyink (1990) compiled the values of $K$ and $1/n$ for various organic compounds from the literature which is listed in Appendix D.

From the adsorption isotherm, it can be seen that, for fixed values of $K$ and $C_e$, the smaller the value of $1/n$, the stronger the adsorption capacity. When $1/n$ becomes very small, the adsorption tends to be independent to $C_e$. For fixed values of $C_e$ and $1/n$, the larger the $K$ value, the greater the adsorption capacity $q_e$.

Another adsorption isotherm developed by Langmuir assumed that the adsorption surface is saturated when a monolayer has been absorbed. The Langmuir adsorption model is

$$q_e = \frac{abC_e}{1 + bC_e} \tag{5.137}$$

or

$$\frac{C_e}{q_e} = \frac{1}{ab} + \frac{C_e}{a} \tag{5.138}$$

where $a$ = empirical coefficient
$b$ = saturation coefficient, $m^3/g$

Other terms are the same as defined in the Freundlich model. The coefficient $a$ and $b$ can be obtained by plotting $C_e/q_e$ versus $C_e$ on arithmetic paper from the results of a batch adsorption test with Eq. (5.138).

Adsorption isotherm can be used to roughly estimate the granular activated carbon (GAC) loading rate and its GAC bed life. The bed life $Z$ can be computed as

$$Z = \frac{(q_e)_0 \times \rho}{(C_0 - C_1)} \tag{5.139}$$

where $Z$ = bed life, L $H_2O$/L GAC
$(q_e)_0$ = mass absorbed when $C_e = C_0$, mg/g of GAC
$\rho$ = apparent density of GAC, g/L
$C_0$ = influent concentration, mg/L
$C_1$ = average effluent concentration for entire column run, mg/L

$C_1$ would be zero for a strongly absorbed compound that has a sharp breakthrough curve, and is the concentration of the nonadsorbable compound presented.

The rate at which GAC is used or the carbon usage rate (CUR) is calculated as follows:

$$\text{CUR(g/L)} = \frac{C_0 - C_1}{(q_e)_0} \tag{5.140}$$

**Contact time.** The contact time of GAC and water may be the most important parameter for the design of a GAC absorber. It is commonly used as the empty-bed contact time (EBCT) which is related to the flow rate, depth of the bed and area of the GAC column. The EBCT is calculated by taking the volume ($V$) occupied by the GAC divided by the flow rate ($Q$):

$$\text{EBCT} = \frac{V}{Q} \tag{5.141}$$

or

$$\text{EBCT} = \frac{H}{Q/A} = \frac{\text{The depth of GAC bed}}{\text{Loading rate}} \tag{5.142}$$

where EBCT = empty-bed contact time, s
$V$ = HA = GAC volume, ft$^3$ or m$^3$
$Q$ = flow rate, cfs or m$^3$/s
$Q/A$ = surface loading rate, cfs/ft$^2$ or m$^3$/(m$^2 \cdot$ s)

The critical depth of a bed ($H_{cr}$) is the depth which creates the immediate appearance of an effluent concentration equal to the breakthrough concentration, $C_b$. The $C_b$ for a GAC column is designated as the maximum acceptable effluent concentration or the minimum datable concentration. The GAC should be replaced or regenerated when the effluent quality reaches $C_b$. Under these conditions, the minimum EBCT (EBCT$_{min}$) will be:

$$\text{EBCT}_{\min} = \frac{H_{cr}}{Q/A} \tag{5.143}$$

The actual contact time is the product of the EBCT and the interparticle porosity (about 0.4 to 0.5).

Powdered activated carbon (PAC) has been used for taste and order removal. It is reported that PAC is not as effective as GAC for organic compounds removal. The PAC minimum dose requirement is

$$D = \frac{C_i - C_e}{q_e} \tag{5.144}$$

where $D$ = dosage, g/L
$C_i$ = influent concentration, mg/L
$C_e$ = effluent concentration, mg/L
$q_e$ = absorbent capacity = $KC_e^{1/n}$, mg/g

**Example 1:** In a dual media filtration unit, the surface loading rate is 3.74 gpm/ft$^2$. The regulatory requirement of EBCT$_{min}$ is 5.5 min. What is the critical depth of GAC?

**solution:**

$$Q = 3.74 \text{ gpm} = 3.74 \frac{\text{gal}}{\text{min}} \times \frac{1 \text{ ft}^3}{7.48 \text{ gal}}$$
$$= 0.5 \text{ ft}^3/\text{min}$$
$$Q/A = 0.5 \text{ ft}^3/\text{min}/1 \text{ ft}^2 = 0.5 \text{ ft/min}$$

From Eq. (5.143), the minimum EBCT is

$$\text{EBCT}_{\text{min}} = \frac{H_{\text{cr}}}{Q/A}$$

and rearranging

$$H_{\text{cr}} = \text{EBCT}_{\text{min}}(Q/A)$$
$$= 5.5 \text{ min} \times 0.5 \text{ ft/min}$$
$$= 2.75 \text{ ft}$$

**Example 2:** A granular activated carbon absorber is designed to reduce 12 μg/L of chlorobenzene to 2 μg/L. The following conditions are given: $K = 100$ (mg/g) (L/mg)$^{1/n}$, $1/n = 0.35$ (Appendix D), and $\rho_{\text{GAC}} = 480$ g/L. Determine the GAC bed life and carbon usage rate.

**solution:**

Step 1. Compute equilibrium adsorption capacity, $(q_e)_0$

$$(q_e)_0 = KC^{1/n}$$
$$= 100 \text{ (mg/g) (L/mg)}^{1/n} \times (0.002 \text{ mg/L})^{0.35}$$
$$= 11.3 \text{ mg/g}$$

Step 2. Compute bed life, $Z$

Assume:

$$C_T = 0 \text{ mg/L}$$
$$Z = \frac{(q_e)_0 \times \rho}{(C_0 - C_1)}$$
$$= \frac{11.3 \text{ mg/g} \times 480 \text{ g/L of GAC}}{(0.012 - 0.002) \text{ mg/L of water}}$$
$$= 542{,}400 \text{ L of water/L of GAC}$$

Step 3. Compute carbon usage rate, CUR

$$\text{CUR} = \frac{(C_0 - C_1)}{(q_e)_0}$$
$$= \frac{(0.012 - 0.002) \text{ mg/L}}{11.3 \text{ mg/g}}$$
$$= 0.00088 \text{ g GAC/L water}$$

**Example 3:** The maximum contaminant level (MCL) for trichloroethylene is 0.005 mg/L. $K = 28$ (mg/g) (L/mg)$^{1/n}$, and $1/n = 0.62$. Compute the PAC minimum dosage needed to reduce trichloroethylene concentration from 33 μg/L to MCL.

**solution:**

Step 1. Compute $q_e$ assuming PAC will equilibrate with 0.005 mg/L ($C_e$) trichloroethylene concentration

$$\begin{aligned} q_e &= KC_e^{1/n} \\ &= 28\ (\text{mg/g})\ (\text{L/mg})^{0.62} \times (0.005\ \text{mg/L})^{0.62} \\ &= 1.05\ \text{mg/g} \end{aligned}$$

Step 2. Compute required minimum dose, $D$

$$\begin{aligned} D &= \frac{(C_0 - C_e)}{q_e} \\ &= \frac{0.033\ \text{mg/L} - 0.005\ \text{mg/L}}{1.05\ \text{mg/g}} \\ &= 0.027\ \text{g/L} \end{aligned}$$

or

$$= 27\ \text{mg/L}$$

## 16 Membrane Processes

In the water industry, the membrane systems have generally been utilized in water and for higher quality water production. Membranes also are being used for municipal and industrial wastewater applications, for effluent filtration to achieve high-quality effluent for sensitive receiving waters, for various water reuse applications, as a space-saving technology, and for on-site treatment. In the last decade, tremendous improvements of equipment developments, more stringent effluent limit, and market economics have led to membranes becoming a mainstream treatment technology. There has been significant growth and increase in application of membrane technologies, and it is expected that membrane processes will be used even more in the future as more stringent drinking water quality standards will likely become enforced.

The use of semipermeable membranes having pore size as small as 3 Å (1 angstrom $= 10^{-8}$ cm) can remove dissolved impurities in water or wastewater. Desalting of seawater is an example. The process that allows water to pass through the membranes is called "osmosis" or hyperfiltration. On the other hand, the process in which ions and solute that pass through a membrane is called "dialysis" which is also used in the medical field.

### 16.1 Process description and operations

The membrane processes were originally developed based on the theory of reverse osmosis. Osmosis is the natural passage of water through a semipermeable membrane from a weaker solution to a stronger solution, to equalize the concentration of solutes in both sides of the membrane. Osmotic pressure is the driving force for osmosis to occur. In reverse osmosis, an external pressure greater than the osmotic pressure is applied to the solution. This high pressure causes water to flow against the natural direction through the membrane, thus producing high-quality demineralized water while rejecting the passage of dissolved solids.

The driving force for a membrane process can be pressure, electrical voltage, concentration gradient, temperature, or combinations of the above. The first two driving forces are employed for water and wastewater membrane filtrations. Membrane processes driven by pressure are microfiltration (MF), utrafiltration (UF), nanofiltration (NF), and reverse osmosis (RO); while processes driven by electric current are electrodialysis (ED) and electrodialysis reversal (EDR). ED is originally used for medical purposes. EDR is generally used for water and wastewater treatments. The vacuum-driven process typically applies to MF and UF only.

**Microfiltration.** Microfiltration uses microporous membranes that have effective pore sizes in the range of 0.07 to1.3 μm and typically have actual pore size of 0.45 μm (Bergman, 2005). The particle-removal range is between 0.05 and 1 μm. Flow through a microporous membrane can occur without the application of pressure on the feed side of the membrane, but in most water and wastewater applicants, a small pressure difference across the membrane produces significant increases in flux, which is required for water production. MF membranes are capable of removing particles with sizes down to 0.1 to 0.2 μm. As granular filtration, MF system filters out turbidity, algae, bacteria, *Giardia* cysts, *Cryptosporidium* oocysts, and all particulate matters in water treatment (Bergman, 2005; Movahed, 2006a, 2006b). It is also most often used to separate suspended and colloidal solids from wastewater.

**Ultrafiltration.** Ultrafiltration uses membranes that have effective pore sizes of 0.005 to 0.25 μm (Bergman, 2005). UF units are capable of separating some large-molecular-weight dissolved organics, colloids, macromolecules, asbestos, and some viruses from water and wastewater by pressure (Qasim *et al.*, 2000; Bergman, 2005, Movahed, 2006b). The UF process is designed to remove colloidalized particles, in the range from 0.005 to 0.1 μm (Movahed, 2006a). It is not effective for demineralization purposes; however, most turbidity-causing particles, viruses, and most organic substances such as large molecular NOMs (natural organic matters, with coagulation) and proteins are within the removal range of UF.

In the case of food processing, UF units can be used to separate proteins and carbohydrates from the wastewater. The proteins and carbohydrates may then be reused in the process or sold as a by-product. Another use of UF units is the separation of emulsified fats, oils, and grease from wastewater (Alley, 2000). The UF process retains nonionic material and generally passes most ionic matter depending on molecular weight cutoff of the membrane.

Both MF and UF membrane systems generally use hollow fibers that can be operated in the outside-in or inside-out direction of flow. Low pressure (5 to 35 psi or 34 to 240 kPa) or vacuum (–1 to –12 psi or –7 to –83 kPa for outside-in membranes only) can be used as the driving force across the membrane (Movahed, 2006a; Bergman, 2005). MF and UF membranes are most commonly made from various organic polymers such as cellulose derivatives, polysulfones, polypropylene, and polyvinylidene fluoride (PVDF). Physical configurations of MF and UF membranes include hollow fiber, spiral wound, cartridge, and tubular constructions. Membrane technologies are also widely used for wastewater treatment (Movahed, 2006a).

Most MF and UF facilities operate with high recoveries of 90% to 98%. Full-scale systems have demonstrated the efficient performance of both MF and UF as feasible treatment alternatives to conventional granular media filtrations. Both MF and UF units can also be used as a pretreatment for NF or RO or to replace the conventional media filtration.

**Nanofiltration.** "Nano" means one-thousandth of a million ($10^{-9}$). One nanometer (1 nm) = $10^{-9}$ m = $10^{-3}$ μm. Nanofiltration uses membranes that have a larger effective pore size (0.0009 to 0.005 μm or about 1 to 5 nm) than that of RO membranes (Qasim *et al.*, 2000; Bergman, 2005). NF and RO separation properties are typically expressed in molecular weight cutoff (MWCO) as the unit of mass in daltons. The typical masses used for NF and RO are 200 to 1000 daltons and <100 daltons, respectively (Movahed, 2006c).

The typical feed pressures for NF system are 50 to 150 psi (340 to 1030 kPa) (Bergman, 2005). The osmotic pressure equation still applies to NF, but the osmotic pressure difference across the membrane will be lower. It could be 10 to 20 psi (Movahed, 2006c). The osmotic pressure differential across the membrane is taken into account in the equation for water transport through the membrane.

The major equipment for an NF process is similar to that for an RO system, and includes a membrane module with support system, a high-pressure pump and piping, and a concentrate-handling and disposal system. The NF is only periodically backwashed by using either water or air as with a granular filtration system. Membranes are usually cleaned automatically as required.

**New development.** NF processes reject molecules in the range from 0.001 to 0.01 μm. Such a process can be used for softening and for reduction of total dissolved solids (TDS) and TOC. A lower-pressure RO technology called nanofiltration, also known as "membrane softening," has also been widely used for treatment of hard, high-color, and high-organic-content feedwater (Movahed, 2006a). NF also removes NOM, disinfection by-products (DBPs), and radionuclides (Movahed, 2006a; Crittenden *et al.*, 2005).

Cregorski (2006) reported that in early 2006, researchers at the Lawrence Livermore National Laboratory created a membrane made of carbon nanotubes and silicon which may offer an even less-expensive desalination technology. The nanotubes, special molecules made of carbon atoms in a unique arrangement, are hollow and more than 50,000 times thinner than a human hair. Billions of these tubes act as pores in the membrane allowing liquids and gases to rapidly pass through while the tiny pore size blocks larger molecules.

The research team was able to measure flows of liquids by making a membrane on a silicon chip with carbon nanotube pores, ideal for desalination. The membrane was created by filling with a ceramic matrix material into the gaps between aligned carbon nanotubes. The pores are so small that only six water molecules could fit across their diameter.

According to Olgica Bakajin, lead scientist for the National Laboratory nanotube membrane research project, "The gas and water flows that we measured are 100 to 10,000 times faster than what classical models predict. In regards to desalination, the process is usually performed using an RO membrane process, which does require a large amount of pressure and energy. According to the research team, the more permeable nanotube membranes could reduce energy costs of desalination by up to 75% compared to current RO technology. We hope that this technology will be ready for the market soon".

**Reverse osmosis.** Reverse osmosis is a membrane separation process, which through the application of pressure, reverses the natural phenomenon of osmosis. Osmosis is the flow of water from an area of low ionic concentration to an area of high ionic concentration. Under an applied pressure, water is forced through a semipermeable membrane from an area of high concentration to one of low concentration. The semipermeable membrane rejects the solutes in the water while allowing the clean water to pass through.

The effective pore sizes of RO membranes range from 0.1 to 1.2 nm (0.0001 to 0.0012 μm) (Qasim *et al.*, 2000; Bergman, 2005). This pore size allows the passage of only water. In order for RO to take place, the applied pressure must be greater than the naturally occurring osmotic pressure. Osmotic pressure is a characteristic of the solution and is a function of the molar concentration of the solute, the number of ions formed when the solute dissociates, and the solution temperature.

Reverse osmosis modules have been developed in four different types of designs: plate and frame, large tube, spiral wound, and hollow fiber. Many types of membranes for RO have been developed, but cellulose acetate and polyamide are currently the most widely used membrane materials in RO systems. Typical membranes are approximately 100 μm-thick having a surface skin of about 0.2-μm thickness that serves as the rejecting surface. The remaining layer is porous and spongy, and serves as backing material. The hollow fine fibers have outer and inner diameters of 50 and 25 μm, respectively (Qasim *et al.*, 2000).

Reverse osmosis processes are mostly used for demineralization in industrial water supply or for desalinization of seawater and brackish water in drinking water supply. The RO cartridges have been widely installed for tap water at homes and offices in Asian countries, such as Taiwan and Japan. Even coin-operated RO water supplies are popular on streets.

Reverse osmosis membrane technology has been successfully used since the 1970s for brackish and seawater desalination. In comparison with other membranes, the RO membrane rejects most solute ions and molecules, while allowing water of vary low mineral content to pass through (Movahed, 2006b). Of course, this process also works as a barrier for cysts, bacteria, viruses, NOMs, DBPs, etc.

In the RO process, the feedwater is forced by hydrostatic pressure through membranes while impurities remain behind. The purified water (permeable or product water) emerges at near atmospheric pressure. The waste (as concentrate) remains at its original pressure and is collected in a concentrated mineral stream. The pressure difference is called osmotic pressure that is a function of the solute concentration, characteristics, and temperature. The operating pressure ranges from 800 to 1200 psi (5520 to 8270 kPa or 54.4 to 81.6 atm) for desalting seawater and from 300 to 600 psi (2400 to 4100 kPa or 20.4 to 40.8 atm) for brackish water desaltation (Montgomery Consulting Engineers, 1985; Bergman, 2005).

The RO process produces a concentrated reject stream in addition to the clean permeate product. By-product water or the "concentrate" may range from 10% to 60% of the raw water pumped to the RO unit. For most brackish water and ionic contaminant removal applications, the concentrate is in the 10% to 25% range, while for seawater, it could be as high as 50%. Typical RO/NF elements are in spiral wound element configuration, while EDR is in stacks containing membrane sheets.

**Electrodialysis reversal.** Electrodialysis is an electrochemical separation process in which ions are transferred through ion exchange membrane by means of an electrical driving force. When a DC voltage is applied across a pair of electrodes, positively charged ions move toward the cathode, while negatively charged ions move toward the anode. Membranes are placed between the electrodes to form several compartments. Flow spacers are placed between the membranes to support the membranes and to create turbulent flow. Water flows along the spacers' flow paths across the surface of the membranes, rather than through the membranes as in RO (von Gottberg, 1999). Impurities are electrically removed from the water. EDR is an ED process, except by reversing polarity two to four times per hour; the system provides constant, automatic self-cleaning that enables improving treatment efficiency and less downtime for periodic cleaning.

The configuration of EDR is in stacks containing membrane sheets. The nonpressure EDR has also been widely used for removal of dissolved substances and contaminants. In the United States, RO and EDR first have been used for water utilities ranging from 0.5 to 12 MGD (1.94 to 45.4 $m^3$/d); most are located in the East Coast and West Coast (Ionic, 1993).

## 16.2 Design considerations

Membrane filtration technologies have emerged as viable options for addressing current and future drinking water regulations related to the treatment of surface water, groundwater under the influence, and water reuse applications for microbial, turbidity, NOM, and DBP removal. Membrane systems are now available in several different forms and sizes, each uniquely fitting a particular need and application.

The basic theory, process fundamentals, membrane fouling, and process design for membrane filtration are given elsewhere (Crittenden *et al.*, 2005; Bergman, 2005). The design engineer must understand the fundamentals of membrane material, modules, fouling, scaling, and performance to evaluate new technologies with the objective of the project. Design assistance is often available from the equipment manufacturers.

Membrane processes that satisfy treatment objectives at the lowest possible life-cycle cost should be selected. The following procedures should be considered in membrane system selection (Bergman, 2005; Crittenden *et al.*, 2005; Qasim *et al.*, 2000):

- Review the historical water quality data.
- Identify overall project goals and define the correct treatment.
- Evaluate source water quality characteristics and select pretreatment processes.
- Assess the need for bench or pilot testing.
- Select membrane system after consultation with the equipment manufacturers and pilot testing.
- Select feedwater pretreatment or disinfection requirements (methods to control fouling).
- Select the basic performance criteria: plant capacity, salinity of feedwater, flux rate, recovery rate, rejection, applied pressure, system life, performance level, feedwater temperature, and permeate solute concentration based on the best and worst conditions.
- Determine product water quality and quantity requirements and blending options (for example, split-flow treatment).
- Establish process design criteria and develop specifications for major system components.
- Calculate system size.
- Select the type of membrane and determine the array configuration.
- Determine ancillary and support facilities, such as transfer piping, pumping facilities, chemical storage facilities, laboratory space, buildings, energy recovery system, and electrical systems.
- Determine power requirement.
- Evaluate hydraulic grade line and waste wash water disposal options.
- Select permeate posttreatment requirements.
- Estimate system economics, including amortization of capital and operation and maintenance cost, supplies and chemicals, and concentrate disposal.
- Select equipment and procedures for process monitoring.

### 16.3 Membrane performance

The osmotic pressure theoretically varies in the same manner as the pressure of an ideal gas (James Montgomery Consulting Engineers, 1985; Conlon, 1990):

$$\pi = \frac{nRT}{\nu} \qquad (5.145a)$$

or (Applegate, 1984):

$$\pi = 1.12\, T \Sigma\, m \tag{5.145b}$$

where $\pi$ = osmotic pressure, psi
$n$ = number of moles of solute
$R$ = universal gas constant
$T$ = absolute temperature, °C + 273
$\nu$ = molar volume of water
$\Sigma m$ = sum of molarities of all ionic and nonionic constituent in solution

At equilibrium, the pressure difference between the two sides of the RO membranes equals the osmotic pressure difference. In low solute concentration, the osmotic pressure ($\pi$) of a solution is given by the following equation (US EPA, 1996):

$$\pi = C_s RT \tag{5.145c}$$

where $\pi$ = osmotic pressure, psi
$C_s$ = concentration of solutes in solution, mol/cm$^3$ or mol/ft$^3$
$R$ = ideal gas constant, ft · lb/mol K
$T$ = absolute temperature, K = °C + 273

When dilute and concentrated solutions are separated by a membrane, the liquid tends to flow through the membrane from the dilute to the concentrated side until equilibrium reaches sides of the membrane. The liquid and water passages (flux of water) through the membrane are a function of pressure gradient (Allied Signal, 1970):

$$F_w = W(\Delta P - \Delta\pi) \tag{5.146}$$

and for salt and solute flow through a membrane:

$$F_S = S(C_1 - C_2) \tag{5.147}$$

where $F_w$, $F_S$ = liquid and salt fluxes across the membrane, respectively, g/cm$^2$/s or lb/ft$^2$/s; cm$^3$/cm$^2$/s or gal/ft$^2$/d
$W$ = flux rate coefficients, empirical, depend on membrane characteristics, solute type, salt type, and temperature, s/cm or s/ft
$\Delta P$ = drop in total water pressure across a membrane, atm or psi
$\Delta\pi$ = osmotic pressure gradient, atm or psi
$\Delta P - \Delta\pi$ = net driving force for liquid pass through membrane

$C_1, C_2$ = salt concentrations on both sides of membrane, g/cm$^3$ or lb/ft$^3$
$C_1 - C_2$ = concentration gradient, g/cm$^3$ or lb/ft$^3$

The liquid or solvent flow depends upon the pressure gradient. The salt or solute flow depends upon the concentration gradient. Also, both are influenced by the membrane types and characteristics.

The terms "water flux" and "salt flux" are the quantities of water and salt that can pass through the membrane per unit area per unit time. The percent of feedwater recovered terms water recovery factor, $R$:

$$\text{Recovery, } R = \frac{Q_\text{p}}{Q_\text{f}} \times 100\% \tag{5.148}$$

where $Q_\text{p}$ = product water flow, m$^3$/d or gpm
$Q_\text{f}$ = feedwater flow, m$^3$/d or gpm

Water recovery for most RO plants is designed in the range of 75% to 90% for brackish water and of 30% to 50% for seawater (Movahed, 2006c).

The percent of salt rejection (Rej) is:

$$\begin{aligned}\text{Rej} &= 100 - \text{salt passage (SP)} \\ &= \left(1 - \frac{\text{product concentration}}{\text{feedwater concentration}}\right) \times 100\% \end{aligned} \tag{5.149}$$

$$\text{SP} = \frac{C_\text{p}}{C_\text{f}} \times 100\% \tag{5.149a}$$

$$C_\text{p} = \frac{C_f(1 - \text{Rej})}{R} \text{ (Qasim \textit{et al.}, 2000)} \tag{5.149b}$$

where $C_\text{p}$ and $C_\text{f}$ = TDS concentrations in permeate and feedwater, respectively.

Salt rejection is commonly measured by the total dissolved solids (TDS). Measurements of conductivity are sometimes used in lieu of TDS. In RO system, salt rejection can be achieved at 9% or more for seawaters; while NF could be as low as 60% to 70% (Movahed, 2006c). Solute rejection varies with feedwater concentration, membrane types, water recovery rate, chemical valance of the ions in the solute, and other factors.

**Example 1:** A city's water demand is 26,500 m$^3$/d (7 MGD). What is the source water (feedwater) flow required for a brackish water RO, if the plant recovery rate is 98%?

**solution:**

$$\text{Recovery} = \frac{Q_p}{Q_f} \times 100\%$$

rearranging:

$$Q_f = (26{,}500\ \text{m}^3/\text{d}) \times 100/98$$

$$= 27{,}000\ \text{m}^3/\text{d}$$

or

$$= 7.13\ \text{MGD}$$

**Example 2:** For a brackish water RO treatment plant, the feedwater applied is 53,000 $m^3/d$ (14 MGD) to the membrane and the product water yields 42,400 $m^3/d$ (11.2 MGD). What is the percentage of concentrate rate?

**solution:**

$$\text{Rate of concentrate produced} = \text{feedwater rate} - \text{product rate}$$

$$= Q_f - Q_p$$

$$= (53{,}000 - 42{,}400)\ \text{m}^3/\text{d}$$

$$= 10{,}600\ \text{m}^3/\text{d}$$

$$\text{Percentage of concentrate rate} = \frac{10{,}600\ \text{m}^3/\text{d}}{53{,}000\ \text{m}^3/\text{d}} \times 100\%$$

$$= 20\%$$

**Example 3:** At a brackish water RO treatment plant, the total dissolved solids concentrations for the pretreated feedwater and the product water are 2860 and 89 mg/L, respectively. Determine the percentages of salt rejection and salt passage.

**solution:**

Step 1: Compute salt rejection using Eq. (5.149)

$$\text{Salt rejection} = \left(1 - \frac{\text{product conc.}}{\text{feedwater conc.}}\right) \times 100\%$$

$$= \left(1 - \frac{89}{2860}\right) \times 100$$

$$= 96.9\%$$

Step 2. Compute salt passage

$$\text{Salt passage} = 100\% - \text{salt rejection}$$

$$= (100 - 96.9)\%$$

$$= 3.1\ \%$$

**Example 4:** A pretreated feedwater to a brackish water RO process contains 2600 mg/L of TDS. The flow is 0.25 $m^3/s$ (5.7 MGD). The designed TDS concentration of the product water is no more than 450 mg/L. The net pressure is 40 atm. The membrane manufacturer provides that the membrane has a water flux rate coefficient of $1.8 \times 10^{-6}$ s/m and a solute mass transfer rate of $1.2 \times 10^{-6}$ m/s. Determine the membrane area required.

**solution:**

Step 1. Compute flux of water

$$\text{Given: } \Delta P - \Delta\pi = 40 \text{ atm} = 40 \text{ atm} \times 101.325 \text{ kPa/atm}$$
$$= 4053 \text{ kPa}$$
$$= 4053 \text{ kg/m} \cdot \text{s}^2$$
$$W = 1.8 \times 10^{-6} \text{ s/m}$$

Using Eq (6.146)

$$F_w = W(\Delta P - \Delta\pi) = 1.8 \times 10^{-6} \text{ s/m } (4053 \text{ kg/m} \cdot \text{s}^2)$$
$$= 7.3 \times 10^{-3} \text{ kg/m}^2 \cdot \text{s}$$

Step 2. Estimate membrane area

$$Q = F_w A$$

rearranging:

$$A = Q/Fw = (0.25 \text{ m}^3\text{/s}) \times (1000 \text{ kg/m}^3)/(7.3 \times 10^{-3} \text{ kg/m}^2 \cdot \text{s})$$
$$= 34{,}250 \text{ m}^2$$

Step 3. Determine permeate TDS ($C_p$) with above area

$$C_f = 2600 \text{ mg/L} = 2600 \text{ g/m}^3 = 2.6 \text{ kg/m}^3$$
$$F_s = S(C_f - C_p)$$
$$QC_p = F_s A = S(C_f - C_p)A$$
$$C_p = \frac{SC_f A}{Q + SA} = \frac{1.2 \times 10^{-6} \text{ m/s} \times 2.6 \text{ kg/m}^3 \times 34{,}250 \text{ m}^2}{0.25 \text{ m}^3\text{/s} + 1.2 \times 10^{-6} \text{ m/s} \times 34{,}250 \text{ m}^2}$$
$$= 0.367 \text{ kg/m}^3$$
$$= 367 \text{ g/m}^3$$
$$= 367 \text{ mg/L}$$

The TDS concentration of the product water is lower than the limit desired. For the purpose of economy, blending some feedwater to the product water can reduce membrane area.

Step 4. Estimate blended flows

Using $C_p = 0.367\ \text{kg/m}^3$, estimate TDS concentration of the product water. The TDS in the blended water is limited to $0.45\ \text{kg/m}^3$.

Mass balance equation would be

$$(Q_f + Q_p)\ (0.45\ \text{kg/m}^3) = Q_f\ (2.6\ \text{kg/m}^3) + Q_p\ (0.367\ \text{kg/m}^3)$$

$$2.15Q_f - 0.083Q_p = 0 \tag{1}$$

and

$$Q_f + Q_p = 0.25\ \text{m}^3/\text{s}$$

$$Q_p = 0.25 - Q_f \tag{2}$$

Substituting (2) into (1)

$$2.15Q_f - 0.083\ (0.25 - Q_f) = 0$$

$$2.067Q_f = 0.02075$$

$$Q_f = 0.010\ (\text{m}^3/\text{s})$$

then $$Q_p = (0.25 - 0.01)\ \text{m}^3/\text{s} = 0.24\ \text{m}^3/\text{s}$$

The water supply will use blending $0.24\ \text{m}^3/\text{s}$ of RO product water and $0.01\ \text{m}^3/\text{s}$ of feedwater (4%).

Step 5. Compute the required membrane area

$$A = \frac{Q_p}{f_w}$$

$$= \frac{0.24\ \text{m}^3/\text{s} \times 1000\ \text{kg/m}^3}{7.3 \times 10^{-3}\ \text{kg/m}^2 \cdot \text{s}}$$

$$= 32{,}880\ \text{m}^2$$

**Example 5:** Determine the sizes of a 4 MGD (15,140 $\text{m}^3$/d) RO system to produce a final blended water TDS of <300 mg/L. The design average TDS in feedwater is 1100 mg/L. The other design criteria (Crittenden *et al.*, 2005) and conditions are given below:

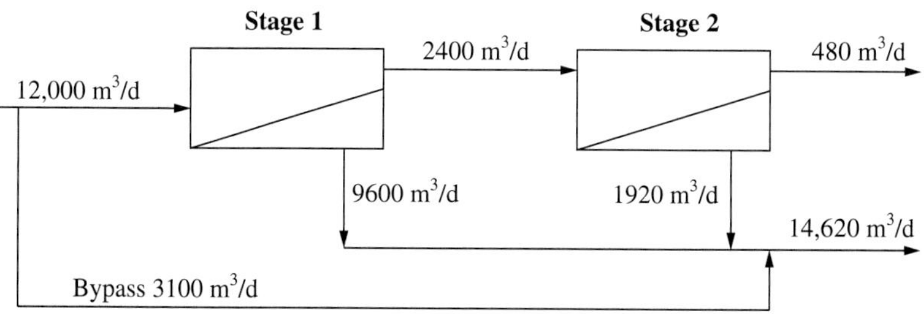

| | |
|---|---|
| Plant average flow rate, $Q$ | 15,100 $m^3/d$ |
| Feedwater pressure | 1035 $kN/m^2$ (150 psi, 10.3 bar) |
| Feedwater temperature | 20°C |
| System recovery factor, $R$ | 80% for both stages 1and 2 |
| Salt rejection, Rej | 96% |
| System design criteria | |
| Number of RO skids | 3 |
| Capacity per skid | 5040 $m^3/d$ |
| Membrane area per element | 32.5 $m^2$ |
| Element per pressure vessel | 7 maximum |
| Number of stages per skid | 2 |
| Average permeate flux for stage 1 | 22 $L/m^2 \cdot h$ |
| Average permeate flux for stage 2 | 17 $L/m^2 \cdot h$ |

**solution:**

Step1. Calculate expected permeate TDS concentration in stage 1, $C_{p1}$, using Eq. (5.194b)

$$C_{p1} = C_f(1–\text{Rej})/R = 1100 \text{ mg/L} (1–0.96)/0.80$$
$$= 55 \text{ mg/L}$$

Step 2. For stage 1 per skid: Calculate permeate flow rate ($Q_{p1}$), concentrate flow rate $Q_c$, and TDS in concentrate $C_c$ (by mass balance)

$$Q_{p1} = QR = 5040 \text{ m}^3/\text{d} \times 0.80 = 4030 \text{ m}^3/\text{d}$$
$$Q_{c1} = 5040 \text{ m}^3/\text{d} - 4030 \text{ m}^3/\text{d} = 1010 \text{ m}^3/\text{d}$$

then
$$1010 \text{ m}^3/\text{d} \times C_c + 4030 \text{ m}^3/\text{d} \times 55 \text{ mg/L} = 5040 \text{ m}^3/\text{d} \times 1100 \text{ mg/L}$$
$$C_c = (5040 \times 1100 - 4030 \times 55)/1010$$
$$= 5270 \text{ (mg/L)}$$

Step 3. For stage 2 per skid: Same as Steps 1 and 2, but using feedwater flow rate of 1010 $m^3/d$ and TDS of 5270 mg/L (which is actual feed to second stage), we calculate TDS concentration and flow rate, for both permeate ($C_{p2}$ and $Q_{p2}$) and concentrate ($C_{c2}$ and $Q_{c2}$) for the stage 2

$$C_{p2} = 5270 \text{ mg/L} \times 0.04/0.8 = 264 \text{ mg/L}$$
$$Q_{p2} = 1010 \text{ m}^3/\text{d} \times 0.8 = 808 \text{ m}^3/\text{d} \quad \text{(say 810 m}^3/\text{d)}$$
$$Q_{c2} = 1010 \text{ m}^3/\text{d} - 810 \text{ m}^3/\text{d} = 200 \text{ m}^3/\text{d}$$

and
$$200 \text{ m}^3/\text{d} \times C_{c2} + 810 \text{ m}^3/\text{d} \times 264 \text{ mg/L} = 1010 \text{ m}^3/\text{d} \times 5270 \text{ mg/L}$$
$$C_{c2} = 25{,}540 \text{ mg/L}$$

Step 4. Compute combined permeate flow rate $Q_p$, and its TDS concentration $C_p$

$$Q_p = Q_{p1} + Q_{p2} = 4030 \text{ m}^3/\text{d} + 810 \text{ m}^3/\text{d} = 4840 \text{ m}^3/\text{d}$$

and $$Q_pC_p = Q_{p1}C_{p1} + Q_{p2}C_{p2}$$

$$4840\ \text{m}^3/\text{d} \times C_p = 4030\ \text{m}^3/\text{d} \times 55\ \text{mg/L} + 810\ \text{m}^3/\text{d} \times 264\ \text{mg/L}$$

$$C_p = 90\ \text{mg/L}$$

Step 5. Determine the fraction ($F$) of the plant flow rate for bypass flow $Q_b$ from feedwater to blend with the final permeate and to meet finished water TDS <300 mg/L requirement (same $Q_{\text{final}}$)

$$F \times 1100\ \text{mg/L} + (1 - F) \times 90\ \text{mg/L} = 1 \times 300\ \text{mg/L}$$

$$F = 0.208$$

$$Q_b = 0.208 \times 5040\ \text{m}^3/\text{d} = 1050\ \text{m}^3/\text{d}$$

Step 6. Compute feedwater flow rate to be treated ($Q_t$) by RO system to meet the TDS <300 mg/L requirement

$$Q_t = 5040\ \text{m}^3/\text{d} - 1050\ \text{m}^3/\text{d} = 3990\ \text{m}^3/\text{d} \quad (\text{say } 4000\ \text{m}^3/\text{d})$$

Step 7. Determine RO system per skid

Feedwater flow rate = 4000 $\text{m}^3/\text{d}$

Permeate flow rate = 3200 $\text{m}^3/\text{d}$

Concentrate flow rate = 800 $\text{m}^3/\text{d}$

System recovery = 80%

Number of RO skids = 3

Permeate capacity per skid = 3200 $\text{m}^3/\text{d}$

Step 8. Sizing RO for stage 1 per skid

$$\text{Capacity per skid} = 4000\ \text{m}^3/\text{d} = 4000\ \text{m}^3/\text{d} \times 1000\ \text{L/m}^3 \div 24\ \text{h/d}$$
$$= 167{,}000\ \text{L/h}$$

$$\text{Permeate flow for each element} = 32.5\ \text{m}^2 \times 22\ \text{L/m}^2 \cdot \text{h} = 715\ \text{L/h}$$

$$\text{Number of elements} = (167{,}000\ \text{L/h}) / (715\ \text{L/h}) = 237$$

Provide 240 elements. Using 6 elements per pressure vessel, 40 pressure vessels are needed.

Step 9. Sizing RO for stage 2 per skid

$$\text{Capacity per array} = 4000\ \text{m}^3/\text{d} \times 0.2 = 800\ \text{m}^3/\text{d} \times 1000\ \text{L/m}^3 \div 24\ \text{h/d}$$
$$= 33{,}300\ \text{L/h}$$

$$\text{Flux rate for each element} = 32.5\ \text{m}^2 \times 17\ \text{L/m}^2 \cdot \text{h} = 552.5\ \text{L/h}$$

$$\text{Number of elements} = (33{,}300\ \text{L/h})/(552.5\ \text{L/h}) = 60$$

Provide 60 elements. There will be 10 pressure vessels that use 6 elements per pressure vessel.

### 16.4 Silt density index

Silt density index (SDI) test involves water filtering in the dead end mode with no cross-flow at a constant pressure of 207 kPa (30 psig). It uses disc membranes rated at 0.45 μm for a 15-min test with a 500-mL water sample. The SDI is calculated as (Chellam *et al.*, 1997):

$$\mathrm{SDI} = \left(1 - \frac{t_\mathrm{i}}{t_\mathrm{f}}\right)\frac{100}{15} \tag{6.150}$$

where SDI = silt density index
$t_\mathrm{i}$ = time initially needed to filter 500 mL of sample
$t_\mathrm{f}$ = time needed to filter 500 mL at the end of the 15-min test period

The SDI has been widely used as a rough estimation of the potential for colloidal fouling of NF and RO systems. Generally, the feedwaters for NF and RO should have SDI of 4 or less. (Most manufacturers have a 4 as warranty limit.)

**Example:** The time initially required to filter 500 mL of dual-media filter effluent is 14.5 s. The time required to filter 500 mL of the same water sample at the end of the 15-min test period is 35 s. Calculate the SDI.

**solution:**

$$\mathrm{SDI} = \left(1 - \frac{t_\mathrm{i}}{t_\mathrm{f}}\right)\frac{100}{15} = \left(1 - \frac{14.5\ \mathrm{s}}{35\ \mathrm{s}}\right) \times \frac{100}{15}$$

$$= 3.9$$

## 17 Residual from Water Plant

A water treatment plant not only produces drinking water, but is also a solids generator. The residual (solids or wastes) comes principally from clarifier basins and filter backwashes. These residuals contain solids which are derived from suspended and dissolved solids in the raw water with the addition of chemicals and chemical reactions.

Depending on the treatment process employed, waste from water treatment plants can be classified as alum, iron, or polymer sludge from coagulation and sedimentation; lime sludge or brine waste from water softening; backwash wastewater and spent granular activated carbon from filtration; and wastes from iron and manganese removal process, ion exchange process, diatomaceous earth filters, microstrainers, and membranes.

The residual characteristics and management of water plant sludge and environmental impacts are discussed in detail elsewhere (Lin and Green, 1987).

Prior to the 1970s, residuals from a water treatment plant were disposed of in a convenient place, mostly in surface water. Under the 1972 Water Pollution Control Act, the discharge of a water plant residuals requires a National Pollution Discharge Elimination System (NPDES) permit. Residuals directly discharged into surface water is prohibited by law in most states. Proper residual (solids) handling, disposal, and/or recovery is necessary for each waterwork nowadays. Engineers must make their best estimate of the quantity and quality of residuals generated from the treatment units.

**Example 1:** A conventional water treatment plant treats an average flow of 0.22 $m^3/s$ (5.0 MGD). The total suspended solids (TSS) concentration in raw (river) water averages 88 mg/L. The TSS removal through sedimentation and filtration processes is 97%. Alum is used for coagulation/sedimentation purpose. The average dosage of alum is 26 mg/L. Assume the aluminum ion is completely converted to aluminum hydroxide, (a) Compute the average production of alum sludge. (MW: Al = 27, S = 32, O = 16, and H = 1). (b) Estimate the volume of residual produced daily, assuming the specific gravity of 4.0% solids is 1.02.

**solution:**

Step 1. Determine quantity ($q_1$) settled for TSS

$$\begin{aligned} \text{TSS} &= 88 \text{ mg/L} = 88 \text{ g/m}^3 \\ q_1 &= \text{flow} \times \text{TSS} \times \%\text{ removal} \\ &= 0.22 \text{ m}^3\text{/s} \times 88 \text{ g/m}^3 \times 0.97 \\ &= 18.8 \text{ g/s} \end{aligned}$$

Step 2. Determine $A1(OH)_3$ generated, $q_2$

The reaction formula for alum and natural alkalinity in water using Eg. (5.51a):

$$A1_2(SO_4)_3 \cdot 18H_2O + 3Ca(HCO_3)_2 \rightarrow 2A1(OH)_3 + 3CaSO_4 + 6CO_2 + 18H_2O$$

$$27 \times 2 + 3(32 + 16 \times 4) + 18(18) \qquad 2(27 + 17 \times 3)$$

$$= 666 \qquad\qquad = 156$$

The above reaction formula suggests that each mole of dissolved $Al^{3+}$ dosed will generate 2 moles of $A1(OH)_3$

$$q_2 = \text{alum dosage} \times \text{flow} \times \left[\frac{156 \text{ g Al(OH)}_3}{666 \text{ g alum}}\right]$$

$$= 26 \text{ g/m}^3 \times 0.22 \text{ m}^3\text{/s} \times \frac{156}{666}$$

$$= 1.34 \text{ g/s}$$

Step 3. Compute total residual generated $q$ for (a)

$$q = q_1 + q_2 = 18.8 \text{ g/s} + 1.34 \text{ g/s}$$

$$= 20.14 \text{ g/s}$$

or

$$= 20.14 \frac{\text{g}}{\text{s}} \times \frac{1 \text{ kg}}{1000 \text{ g}} \times \frac{86{,}400 \text{ s}}{1 \text{ day}}$$

$$= 1740 \text{ kg/d}$$

or

$$= 1740 \frac{\text{kg}}{\text{d}} \times \frac{2.194 \text{ lb}}{1 \text{ kg}}$$

$$= 3817 \text{ lb/d}$$

Step 4. Compute the volume generated daily for (b)

$$\text{Volume} = 1740 \text{ kg/d} \div (1000 \text{ kg/m}^3 \times 1.02 \times 0.04)$$

$$= 42.6 \text{ m}^3\text{/d}$$

or

$$\text{Volume} = 3818 \text{ lb/d} \div (62.4 \text{ lb/ft}^3 \times 1.02 \times 0.04)$$

$$= 1500 \text{ ft}^3\text{/d}$$

**Example 2:** A river raw water is coagulated with a dosage of 24 mg/L of ferrous sulfate and an equivalent dosage of 24 mg/L of ferrous sulfate and an equivalent dosage of lime. Determine the sludge generated per $m^3$ of water treated.

**solution:** From Eqs. (5.51 f) and (5.51 g), 1 mole of ferrous sulfate reacts with 1 mole of hydrated lime and this produces 1 mole of $Fe(OH)_3$ precipitate.

$$\text{MW of FeSO}_4 \cdot 7\text{H}_2\text{O} = 55.85 + 32 + 4 \times 16 + 7 \times 18$$

$$= 277.85$$

$$\text{MW of Fe(OH)}_3 = 55.85 + 3 \times 17$$

$$= 106.85$$

By proportion

$$\frac{\text{sludge}}{\text{F.S. dosage}} = \frac{106.85}{277.85} = 0.3846$$

$$\begin{aligned}\text{Sludge} &= \text{F.S. dosage} \times 0.3846 \\ &= 24 \text{ mg/L} \times 0.3846 \\ &= 9.23 \text{ mg/L} \\ &= 9.23 \frac{\text{mg}}{\text{L}} \times \frac{1\text{g}}{1000 \text{ mg}} \times \frac{1000 \text{ L}}{1\text{m}^3} \\ &= 9.23 \text{ g/m}^3\end{aligned}$$

or

$$= 9.23 \frac{\text{g}}{\text{m}^3} \times \frac{0.0022 \text{ lb}}{1 \text{ g}} \times \frac{1 \text{ m}^3}{2.642 \times 10^{-4} \text{ Mgal}}$$

$$= 77.0 \text{ lb/Mgal}$$

*Note*: Mgal = million gallons.

**Example 3:** A 2-MGD (0.088 $m^3$/s) water system generates 220 lb/Mgal (26.4 mg/L) solids in dry weight basis. The residual solids concentration by weight is 0.48%. Estimate the rate of residual solids production and pumping rate if the settled residual is withdrawn every 5 min/h.

**solution:**

Step 1. Calculate residual production

$$\frac{(220 \text{ lb/Mgal})(2 \text{ MGD})}{(8.34 \text{ lb/gal})(0.0048)} = 10{,}990 \text{ gpd} = 7.63 \text{ gpm}$$

Step 2. Calculate pump capacity required

$$7.63 \text{ gpm} \times 60/5 = 92 \text{ gpm}$$

### 17.1 Residual production and density

The quantity of residuals (called sludge) generated from clarifiers are related to coagulants used, raw water quality, and process design. The residual production for whole water plant as the weight per unit volume treated (g/$m^3$ or lb/Mgal) and its solids concentration should be determined. These values with the average flow rate will be used for the estimation of daily residual production. Then the residual production rate can be employed for the design of residual (sludge) treatment units, commonly intermitted sand drying bed. The density of residuals (pounds per gallon) can be estimated using the following equation:

$$\text{w} = \frac{8.34}{C_s/\rho_s + c/\rho} \quad \text{(US customary units)} \qquad (5.151\text{a})$$

$$w = \frac{1}{c_s/\rho_s + c/\rho} \quad \text{(SI units)} \tag{5.151b}$$

where $w$ = density of residuals, lb/gal or kg/L
$c_s$ = weight fractional percent of solids
$c$ = weight fractional percent of water
$\rho_s$ = specific gravity of solid
$\rho$ = specific gravity of water (approximately 1.0)

**Example 1:** Calculate the sludge density (lb/gal) for a 6% sludge composed of solids with a specific gravity of 2.48.

**solution:**

Step 1. Find specific gravity of 6% residual (sludge) $\rho_r$

Weight fraction of sludge, solids, and water are 1.0, 0.06, and 0.94, respectively. Using mass balance, we obtain

$$\frac{1}{\rho_r} = \frac{0.06}{\rho_s} + \frac{0.94}{\rho} = \frac{0.06}{2.48} + \frac{0.94}{1.0}$$

$$\rho_r = 1.037$$

Step 2. Determine the density of the residue

$$\text{Density} = \rho_r \times 8.34 \text{ lb/gal} = 1.037 \times 8.34 \text{ lb/gal}$$

$$= 8.65 \text{ lb/gal}$$

or

$$= 1037 \text{ g/L}$$

*Note*: This solution is a kind of proof of Eq. (5.151).

**Example 2:** A water plant treats 4.0 MGD (0.175m$^3$/s) and generates 240 lb of residuals per million gallons (Mgal) treated (28.8 g/m$^3$). The solid concentration of the residuals is 0.5%. Estimate: (a) the daily production of residuals, (b) the pump capacity, if the pump operates for 12 min each hour, and (c) the density of the residuals, assuming the specific density of the residual is 2.0 with 4% solids.

**solution:**

Step 1. Determine residual production rate $q$

$$q = \frac{(4 \text{ MGD})(240 \text{ lb/Mgal})}{(0.005)(8.34 \text{ lb/gal})}$$

(a)

$$= 23{,}020 \text{ gpd}$$
$$= 16.0 \text{ gpm}$$

Step 2. Find pump capacity $q_p$

$$q_p = 16 \text{ gpm} \times \frac{60 \text{ min}}{12 \text{ min}}$$

(b) $$= 80 \text{ gpm}$$

Step 3. Estimate residual density $w$, using Eq. (5.151)

$$w = \frac{8.34}{0.04/2.0 + 0.96/1.0}$$

(c) $$= 8.51 \text{ lb/gal}$$

**Example 3:** In a 1 MGD (0.0438 $m^3/s$) waterworks, 1500 lb of dry solids are generated per million gallons of water treated. The sludge is concentrated to 3% and then applied to an intermittent sand drying bed at a 18-in depth, with 24 beds used per year. Determine the surface area needed for the sand drying bed.

**solution:**

Step 1. Calculate yearly 3% sludge volume, $v$

$$\begin{aligned} v &= 1500 \text{ lb/Mgal} \times 1 \text{ Mgal/d}/0.03 \\ &= 50{,}000 \text{ lb/d} \\ &= 50{,}000 \text{ lb/d} \times 365 \text{ d}/(62.41\text{b/ft}^3) \\ &= 292{,}500 \text{ ft}^3 = 8{,}284 \text{ m}^3 \end{aligned}$$

Step 2. Calculate area $A$ ($D$ = depth, $n$ = number of applications/year)

$$v = ADn$$

or $$A = v/Dn = 292{,}500 \text{ ft}^3/(1.5 \text{ ft} \times 24)$$

$$= 8125 \text{ ft}^2$$

$$= 755 \text{ m}^2$$

## 18 Disinfection

Disinfection is a process to destroy disease-causing organisms, or pathogens. Disinfection of water can be done by boiling the water, ultraviolet radiation, and chemical inactivation of the pathogen. In the water treatment processes, pathogens and other organisms can be partly physically eliminated through coagulation, flocculation, sedimentation, and filtration, in addition to the natural die-off. After filtration, to ensure pathogen-free water, the chemical addition of chlorine (so called chlorination), rightly or wrongly, is most widely used for disinfection of drinking water. This less expensive and powerful disinfection of drinking water provides more benefits than its short coming due to disinfection by-products (DBPs). DBPs have to be controlled. The use of ozone and ultraviolet for disinfection of water and wastewater is increasing in the United States.

Chlorination serves not only for disinfection, but as an oxidant for other substances (iron, manganese, cyanide, etc.) and for taste and odor control in water and wastewater. Other chemical disinfectants include chlorine dioxide, ozone, bromine, and iodine. The last two chemicals are generally used for personal application, not for the public water supply.

### 18.1 Chemistry of chlorination

**Free available chlorine.** Effective chlorine disinfection depends upon its chemical form in water. The influencing factors are temperature, pH, and organic content in the water. When chlorine gas is dissolved in water, it rapidly hydrolyzes to hydrochloric acid (HCl) and hypochlorous acid (HOCl)

$$\mathrm{Cl_2 + H_2O \leftrightarrow H^+ + Cl^- + HOCl} \tag{5.152}$$

The equilibrium constant is

$$K_{\mathrm{H}} = \frac{[\mathrm{H^+}][\mathrm{Cl^-}][\mathrm{HOCl}]}{[\mathrm{Cl_{2(aq)}}]}$$

$$= 4.48 \times 10^{-4} \text{ at } 25°\text{C (White, 1972)} \tag{5.153}$$

The dissolution of gaseous chlorine, $\mathrm{Cl_{2(g)}}$, to form dissolved molecular chlorine, $\mathrm{Cl_{2(aq)}}$ follows Henry's law and can be expressed as (Downs and Adams, 1973)

$$\mathrm{Cl_{2(g)}} = \frac{\mathrm{Cl_{2(aq)}}}{\mathrm{H(mol/L\,atm)}} = \frac{[\mathrm{Cl_{2(aq)}}]}{P_{\mathrm{Cl_2}}} \tag{5.154}$$

where $[\mathrm{Cl_{2(aq)}}]$ = molar concentration of $\mathrm{Cl_2}$
$P_{\mathrm{Cl_2}}$ = partial pressure of chlorine in atmosphere

The distribution of free chlorine between HOCl and $\mathrm{OCl^-}$ is presented in Fig. 5.10. The disinfection capabilities of HOCl is generally higher than that of $\mathrm{OCl^-}$ (Water, 1978).

$$H = \text{Henry's law constant, mol/L atm}$$

$$= 4.805 \times 10^{-6} \exp\left(\frac{2818.48}{T}\right) \tag{5.155}$$

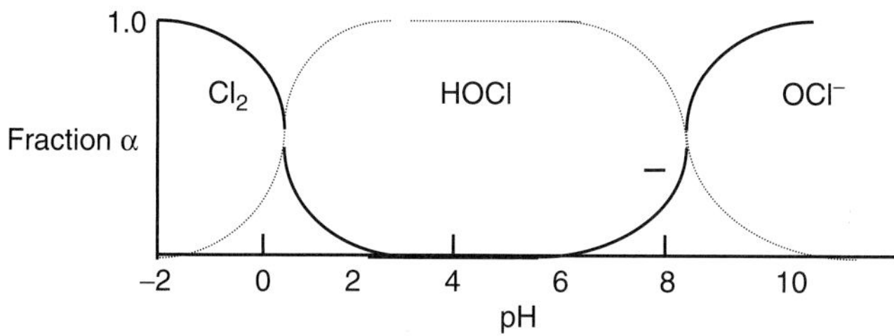


**Figure 5.10** Distribution of chlorine species (*Snoeyink and Jenkins, 1980*).

Hypochlorous acid is a weak acid and subject to further dissociation to hypochlorite ions ($OCl^-$) and hydrogen ions:

$$HOCl \leftrightarrow OCl^- + H^+ \tag{5.156}$$

and its acid dissociation constant $K_a$ is

$$K_a = \frac{[OCl^-][H^+]}{[HOCl]} \tag{5.157}$$

$$= 3.7 \times 10^{-8} \text{ at } 25°C$$

$$= 2.61 \times 10^{-8} \text{ at } 20°C$$

The values of $K_a$ for hypochlorous acid is a function of temperature in kelvins as follows (Morris, 1966):

$$\ln K_a = 23.184 - 0.058T - 6908/T \tag{5.158}$$

**Example 1:** The dissolved chlorine in the gas chlorinator is 3900 mg/L at pH 4.0. Determine the equilibrium vapor pressure of chlorine solution at 25°C (use $K_H = 4.5 \times 10^{-4}$).

**solution:**

Step 1.

Since pH = 4.0 < 5, from Fig. 5.10, the dissociation of HOCl to $OCl^-$ is not occurring. The available chlorine is $[Cl_2] + [HOCl]$:

$$[Cl_2] + [HOCl] = 3900 \text{ mg/L} = \frac{3.9 \text{ g/L}}{70.9 \text{ g/mol}}$$

$$= 0.055 \text{ mol/L}$$

$$[HOCl] = 0.055 - [Cl_2]$$

Step 2. Compute $[H^+]$

$$pH = 4.0 = \log \frac{1}{[H^+]}$$

$$[H^+] = 10^{-4}$$

Step 3. Apply $K_H$ at 25°C

$$K_H = 4.5 \times 10^{-4} = \frac{[H^+][Cl^-][HOCl]}{[Cl_2]}$$

Substitute $[H^+]$ from Step 2

$$4.5 \times 10^{-8} = \frac{[Cl^-][HOCl]}{[Cl_2]}$$

From Eq. (5.152), it can be expected that, for each mole of HOCl produced, one mole of $Cl^-$ will also be produced. Assuming there is no chloride in the feedwater, then

$$[Cl^-] = [HOCl] = 0.055 - [Cl_2]$$

Step 4. Solve the above two equation

$$4.5 \times 10^8 = \frac{(0.055 - [Cl_2])^2}{[Cl_2]}$$

$$0.0002/2[Cl_2]^{1/2} = 0.055 - [Cl_2]$$

$$[Cl_2] + 0.0002/2[Cl_2]^{1/2} - 0.055 = 0$$

$$[Cl_2]^{1/2} = \frac{-0.0002/2 \pm \sqrt{0.000212^2 + 0.22}}{2}$$

$$= 0.2345 \times 10^{-6}$$

$$[Cl_2] = 0.055 \times 10^{-11} (\text{mol/L})$$

Step 5. Compute $H$ using Eq. (5.155)

$$H = 4.805 \times 10^{-6} \exp\left(\frac{2818.48}{25 + 273}\right)$$

$$= 0.062 \text{ mol/L} \cdot \text{atm}$$

*Note*: [HOCl] in negligible

Step 6. Compute the partial pressure of chlorine gas

$$P_{Cl_2} = [Cl_2]/H = (0.055 \times 10^{-11}\ \text{mol/L})\ (0.062\ \text{mol/(L} \cdot \text{atm))}$$

$$= 0.89 \times 10^{-11}\ \text{atm}$$

or

$$= 8.9 \times 10^{-12}\ \text{atm} \times 101{,}325\ \text{Pa/atm}$$

$$= 9.0 \times 10^{-7}\ \text{Pa}$$

**Example 2:** A water treatment plant uses 110 lb/d (49.9 kg/d) of chlorine to treat 10.0 MGD (37,850 $m^3$/d or 0.438 $m^3$/s) of water. The residual chlorine after 30 min contact time is 0.55 mg/L. Determine the chlorine dosage and chlorine demand of the water.

**solution:**

Step 1. Calculate chlorine demand

Since 1 mg/L = 8.341b/Mgal

$$\text{Chlorine dosage} = \frac{110\ \text{lb/d}}{10.0\ \text{Mgal/d}} \times \frac{1\ \text{mg/L}}{8.34\ \text{lb/Mgal}}$$

$$= 1.32\ \text{mg/L}$$

or solve by SI units:

Since 1 mg/L = 1 g/$m^3$

$$\text{Chlorine dosage} = (49.9\ \text{kg/d}) \times (1000\ \text{g/kg})/(37{,}850\ \text{m}^3\text{/d})$$

$$= 1.32\ \text{g/m}^3$$

$$= 1.32\ \text{mg/L}$$

Step 2. Calculate chlorine demand

$$\text{Chlorine demand} = \text{chlorine dosage} - \text{chlorine residual}$$

$$= 1.32\ \text{mg/L} - 0.55\ \text{mg/L}$$

$$= 0.77\ \text{mg/L}$$

**Example 3:** Calcium hypochlorite (commercial high-test calcium hypochlorite, HTH) containing 70% available chlorine is used for disinfection of a new main. (a) Calculate the weight (kilograms) of dry hypochlorite powder needed to prepare a 2.0% hypochlorite solution in a 190-L (50 gal) container. (b) The volume of the new main is 60,600 L (16,000 gal). How much of the 2 percent hypochlorite solution will be used with a feed rate of 55 mg/L of chlorine?

**solution:**

Step 1. Compute kg of dry calcium hypochlorite powder needed

$$\text{Weight} = \frac{190\ \text{L} \times 1.0\ \text{kg/L} \times 0.02}{0.70}$$

$$= 5.42\ \text{kg}$$

Step 2. Compute volume of 2% hypochlorite solution needed

2% solution = 20,000 mg/L =20 g/L
Let $X$ = 2% hypochlorite solution needed
Then (20,000 mg/L) ($X$ L) = (55 mg/L) (60,600 L)
$X$ = 167 L

**Combined available chlorine.** Chlorine reacts with certain dissolved constituents in water, such as ammonia and amino nitrogen compounds to produce chloramines. These are referred to as combined chlorine. In the presence of ammonium ions, free chlorine reacts in a stepwise manner to form three species: monochloramine, $NH_2Cl$; dichloramine, $NHCl_2$; and trichloramine or nitrogen trichloride, $NCl_3$.

$$NH_4^+ + HOCl \leftrightarrow NH_2Cl + H_2O + H^+ \tag{5.159}$$

or

$$NH_{3(aq)} + HOCl \leftrightarrow NH_2C1 + H_2O \tag{5.160}$$

$$NH_2Cl + HOCl \leftrightarrow NHCl_2 + H_2O \tag{5.161}$$

$$NHCl_2 + HOCl \leftrightarrow NCl_3 + H_2O \tag{5.162}$$

At high pH, the reaction for forming dichloramine from monochloramine is not favored. At low pH, other reactions occur as follows:

$$NH_2Cl + H^+ \leftrightarrow NH_3Cl^+ \tag{5.163}$$

$$NH_3Cl^+ + NH_2Cl \leftrightarrow NHCl_2 + NH_4^+ \tag{5.164}$$

Free residual chlorine is the sum of [HOCl] and [$OCl^-$]. In practice, free residual chlorine in water is 0.5 to 1.0 mg/L. The term "total available chlorine" is the sum of "free chlorine" and "combined chlorine." Chloramines are also disinfectants, but they are more slower to act than that of hypochlorite.

Each chloramine molecule reacts with two chlorine atoms. Therefore, each mole of mono-, di-, and trichloramine contains 71, 142, and 213 g of available chlorine, respectively. The molecular weight of these three chloramines are respectively 51.5, 85.9, and 120.4. Therefore, the chloramines contain 1.38, 1.65, and 1.85 g of available chlorine per gram of chloramine, respectively.

The pH of the water is the most important factor on the formation of chloramine species. In general, monochloramine is formed at pH above 7. The optimum pH for producing monochloramine is approximately 8.4.

**Breakpoint chlorination.** When the molar ratio of chlorine to ammonia is greater than 1.0, there is a reduction of chlorine and oxidation of ammonia. A substantially complete oxidation—reduction process occurs in ideal conditions by a 2:1 ratio and results in the disappearance of all ammonium ions with excess free chlorine residual. This is called the breakpoint phenomenon. As shown in Fig. 5.11, chlorine reacts with easily oxidized constituents, such as iron, hydrogen sulfide, and some organic matter. It then continues to oxidize ammonia to form chloramines and chloroorganic compound below a $Cl_2$:$NH_4^+$ ratio of 5.0 (which is around the peak). The destruction of chloramines and chloroorganic compounds are between the ratio of 5.0 and 7.6. The ratio at 7.6 is the breakpoint. All chloramines and other compounds are virtually oxidized. Further addition of chlorine becomes free available chlorine, HOCl, and $OCl^-$. At this region, it is called breakpoint chlorination.

The breakpoint chlorination can be used as a means of ammonia nitrogen removal from waters and wastewaters. The reaction is:

$$2NH_3 + 3HOCl \leftrightarrow N_2\uparrow + 3H^+ + 3Cl^- + 3H_2O \quad (5.165)$$

or

$$2NH_3 + 3Cl_2 \leftrightarrow N_2\uparrow + 6HCl \quad (5.166)$$

$$NH_3 + 4Cl_2 + 3H_2O \leftrightarrow NO_3^- + 8Cl^- + 9H^+ \quad (5.167)$$

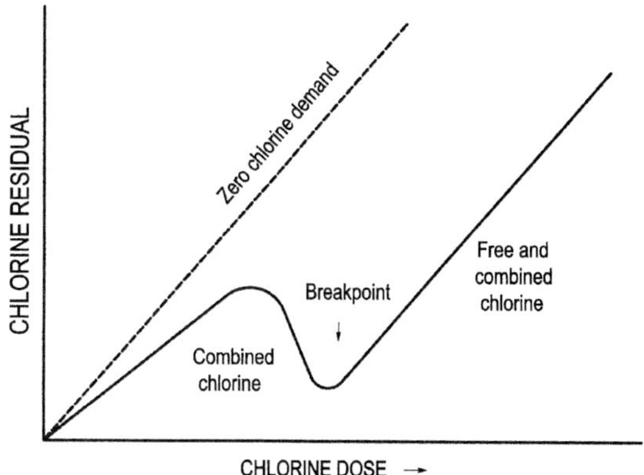


**Figure 5.11** Theoretical drawing of breakpoint chlorination curve.

In practice, the breakpoint does not occur at a $Cl_2$: $NH_4^-$ mass ratio of around 10:1, and mass dose ratio of 15:1 or 20:1 is applied (White, 1972; Hass, 1990).

McKee *et al.* (1960) developed the relationship of available chlorine in the form of dichloramine to available chlorine in the form of monochloramine as follows:

$$A = \frac{BM}{1 - [1 - BM(2 - M)]^{1/2}} - 1 \tag{5.168}$$

where $A$ = ratio of available chlorine in di- to monochloramine

$M$ = molar ratio of chlorine (as $Cl_2$) added to ammonia-N present

$B = 1 - 4K_{eq}[H^+]$ (5.169)

where the equilibrium constant $k_{eq}$ is from

$$H^+ + 2NH_2Cl \leftrightarrow NH_4^+ + NHCl_2 \tag{5.170}$$

$$K_{eq} = \frac{[NH_4^+][NHCl_2]}{[H^+][NH_2Cl]^2} \tag{5.171}$$

$$= 6.7 \times 10^5 \text{ L/mol at } 25°C$$

The relationship (Eq. (5.168)) is pH dependent. When pH decreases and the Cl:N dose ratio increases, the relative amount of dichloramine also increases.

**Example:** The treated water has pH of 7.4, a temperature of 25°C, and a free chlorine residual of 1.2 mg/L.

Chloramine is planned to be used in the distribution system. How much ammonia is required to keep the ratio of dichloramine to monochloramine of 0.15, assuming all residuals are not dissipated yet.

**solution:**

Step 1. Determine factor $B$ by Eq. (5.171)

$$K_{eq} = 6.7 \times 10^5 \text{ L/mol at } 25°C$$
$$pH = 7.4, [H^+] = 10^{-7.4}$$

From Eq. (5.169)

$$B = 1 - 4K_{eq}[H^+] = 1 - 4(6.7 \times 10^5)(10^{-7.4})$$
$$= 0.893$$

Step 2. Compute molar ratio $M$

$$A = 0.15$$

Using Eq. (5.168)

$$A = \frac{BM}{1 - [1 - BM\,(2 - M)]^{1/2}} - 1$$

$$0.15 = \frac{0.893M}{1 - [1 - 0.893M\,(2 - M)]^{-1/2}} - 1$$

$$1 - [1 - 0893M(2 - M)]^{1/2} = 0.777M$$

$$(1 - 0.777M)^2 = 1 - 0.893M(2 - M)$$

$$0.29M^2 - 0.232M = 0$$

$$M = 0.80$$

Step 3. Determine the amount of ammonia nitrogen (N) to be added

$$\text{Mole wt. of } Cl_2 = 70.9 \text{ g}$$

$$1.2 \text{ mg/L of } Cl_2 = \frac{1.2 \text{ mg/L}}{70.9 \text{ g/mol}} \times 0.001 \text{ g/mg}$$

$$= 1.7 \times 10^{-5} \text{ mol/L}$$

$$M = \frac{1.7 \times 10^{-5} \text{ mol/L}}{NH_3}$$

$$NH_3 = 1.7 \times 10^{-5}(\text{mol/L})/0.80$$

$$= 2.12 \times 10^{-5} \text{ mol/L}$$

$$= 2.12 \times 10^{-5} \text{ mol/L} \times 14 \text{ g of N/mol}$$

$$= 0.30 \text{ mg/L as N}$$

**Chlorine dioxide.** Chlorine dioxide ($ClO_2$) is an effective disinfectant (1.4 times that of the oxidation power of chlorine). It is an alternative disinfectant of potable water, since it does not produce significant amounts of trihalomethanes (THMs) which chlorine does. Chlorine dioxide has not been widely used for water and wastewater disinfection in the United States. In western Europe, the use of chlorine dioxide is increasing.

Chlorine dioxide is a neutral compound of chlorine in the +IV oxidation state, and each mole of chlorine dioxide yields five redox equivalents (electrons) on being reduced to chlorine ions. The chemistry of chlorine dioxide in water is relatively complex. Under acid conditions, it is reduced to chloride:

$$ClO_2 + 5e + 4H^+ \leftrightarrow Cl^- + 2H_2O \qquad (5.172)$$

Under reactively neutral pH found in most natural water it is reduced to chlorite:

$$ClO_2 + e \leftrightarrow ClO_2^- \tag{5.173}$$

In alkaline solution, chlorine dioxide is disproportioned into chlorite, $ClO_2^-$, and chlorate, $ClO_3$:

$$2ClO_2 + 2OH^- \leftrightarrow ClO_2^- + ClO_3^- + H_2O \tag{5.174}$$

Chlorine dioxide is generally generated at the water treatment plant prior to application. It is explosive at elevated temperatures and on exposure of light or in the presence of organic substances. Chlorine dioxide may be generated by either the oxygeneration of lower-valence compounds (such as chlorites or chlorous acid) or the reduction of a more oxidized compound of chlorine. Chlorine dioxide may be generated by oxidation of sodium chlorite with chlorine gas by the following reaction.

$$2NaClO_2 + Cl_2 \leftrightarrow 2NaCl + 2ClO_2 \tag{5.175}$$

The above equation suggests that 2 moles of sodium chlorite react with 1 mole of chlorine to produce 2 moles of chlorine dioxide.

Chlorine dioxide is only 60% to 70% pure (James M. Montgomery Consulting Engineering, 1985). Recent technology has improved it to in excess of 95%. It is a patented system that reacts gaseous chlorine with a concentrated sodium chlorite solution under a vacuum. The chlorine dioxide produced is then removed from the reaction chamber by a gas ejector, which is very similar to the chlorine gas vacuum feed system. The chlorine dioxide solution concentration is 200 to 1000 mg/L and contains less than 5% excess chlorine (ASCE and AWWA, 1990).

**Example:** A water treatment plant has a chlorination capacity of 500 kg/d (1100 lb/d) and is considering the switch to using chlorine dioxide for disinfection. The existing chlorinator would be employed to generate chlorine dioxide. Determine the theoretical amount of chlorine dioxide that can be generated and the daily requirement of sodium chlorite.

**solution:**

Step 1. Find $ClO_2$ generated by Eq. (5.175)

$$
\begin{array}{lcccc}
 & 2NaClO_2 & + \; Cl_2 & \leftrightarrow 2NaCl \; + & 2ClO_2 \\
\text{MW} & 180.9 & 70.9 & & 134.9 \\
 & y & 500\ \text{kg/d} & & x
\end{array}
$$

$$x = \frac{500 \text{ kg/d}}{70.9 \text{ g}} \times 134.9 \text{ g}$$
$$= 951 \text{ kg/d}$$
$$= 2100 \text{ lb/d}$$

Step 2. Determine $NaClO_2$ needed

$$y = \frac{500 \text{ kg/d}}{70.9 \text{ g}} \times 18.9 \text{ g}$$
$$= 1276 \text{ kg/d}$$
$$= 2810 \text{ lb/d}$$

## 18.2 Ozonation

Ozone ($O_3$) is a blue (or colorless) unstable gas at the temperature and pressure encountered in water and wastewater treatment processes. It has a pungent odor. Ozone is one of the most powerful oxidizing agents. It has been used for the purposes of disinfection of water and wastewater and chemical oxidation and preparation for biological processing, such as nitrification.

Ozonation of water for disinfection has been used in France, Germany, and Canada. Interest in ozonation in water treatment is increasing in the US due to THMs problems of chlorination. Ozone is an allotrope of oxygen and is generated by an electrical discharge to split the stable oxygen–oxygen covalent bond as follows:

$$3O_2 + \text{energy} \leftrightarrow 2O_3 \tag{5.176}$$

The solubility of ozone in water follows Henry's law. The solubility constant $H$ is (ASCE and AWWA, 1990):

$$H = (1.29 \times 10^6/T) - 3721 \tag{5.177}$$

where $T$ is temperature in Kelvin.

The maximum gaseous ozone concentration that can be generated is approximately 50 mg/L (or 50 g/m$^3$). In practice, the maximum solution of ozone in water is only 40mg/L (US EPA, 1986). Ozone production in ozone generators from air is from 1% to 2 % by weight, and from pure oxygen ranges from 2% to 10% by weight. One percent by weight in air is 12.9 g/m$^3$ ($\cong$ 6000 ppm) and 1% by weight in oxygen is 14.3 g/m$^3$ ($\cong$ 6700 ppm). Residual ozone in water is generally

0.1 to 2 g/m$^3$ (mg/L) of water (Rice, 1986). The ozone residual present in water decays rapidly

### 18.3 Disinfection kinetics

Although much research has been done on modeling disinfection kinetics, the classical Chick's law or Chick–Watson model is still commonly used. Basic models describing the rate of microorganism destruction or natural die-off is Chick's law, a first-order chemical reaction:

$$\ln(N/N_0) = -kt \tag{5.178a}$$

or

$$-dN/dt = kt \tag{5.178b}$$

where $N$ = microbial density present at time $t$
$N_0$ = microbial density present at time 0 (initial)
$k$ = rate constant
$t$ = time
$-dN/dt$ = rate of decrease in microbial density

Watson refined Chick's equation and develop an empirical equation including changes in the concentration of disinfectant as the following relationship:

$$k = k'C^n \tag{5.179}$$

where $k'$ = corrected die-off rate, presumably independent of $C$ and $N$
$C$ = disinfection concentration
$n$ = coefficient of dilution

The combination of Eqs. (5.178a) and (5.179) yields

$$\ln(N/N_0) = -k'C^n t \tag{5.180a}$$

or

$$-dN/dt = k'tC^n \tag{5.180b}$$

The disinfection kinetics is influenced by temperature, and the Arrhenius model is used for temperature correction as (US EPA, 1986):

$$k'_T = k'_{20}\theta^{(T-20)} \tag{5.181}$$

where $k'_T$ = rate constant at temperature $T$, °C
$k'_{20}$ = rate constant at 20°C
$\theta$ = empirical constant

The Chick–Watson relationship can plot the disinfection data of the survival rate ($N/N_0$) on a logarithmic scale against time, and a straight line can be drawn. However, in practice, most cases do not yield a straight line relationship. Hom (1972) refined the Chick–Watson model and developed a flexible, but highly empirical kinetic formation as follows:

$$-dN/dt = k'Nt^m C^n \tag{5.182}$$

where $m$ = empirical constant
= 2 for *E. coli*

Other variables are as previous stated

For changing concentrations of the disinfectant, the disinfection efficiency is of the following form (Fair *et al.,* 1968):

$$C^n t_p = \text{constant} \tag{5.183}$$

where $t_\text{p}$ is the time required to produce a constant percent of kill or die-off. This approach has evolved into a "CT" (concentration × contact time) regulation to ensure a certain percentage kills of *Crypotosporidium, Giardia,* and viruses. The percentage kills are expressed in terms of log removals.

### 18.4 *CT* values

On June 29, 1989, the US Environmental Protection Agency (1989b) published the final Surface Water Treatment Regulations (SWTR), which specifies that water treatment plants using surface water as a source or groundwater under the influence of surface water, and without filtration, must calculate *CT* values daily. The *CT* value refers to the value of disinfectant content or disinfectant residual, *C* (mg/L), multiplied by the contact time, *T* (minutes, or min). Disinfectants include chlorine, chloramines, ozone, and chlorine dioxide.

**Regulations.** The US EPA reaffirms its commitment to the current Safe Drinking Water Act, which includes regulations related to disinfection and pathogenic organism control for water supplies. Public water supply utilities must continue to comply with current rules while new rules for microorganisms and disinfectants/disinfection by-products are being developed.

The Surface Water Treatment Rule requires treatment for *Giardia* and viruses of all surface water and groundwater under the direct influence of surface water. Public water systems are required to comply with a new operating parameter, the so-called *CT* value, an indicator of the effectiveness of the disinfection process. This parameter depends on pH and temperature to remove or inactivate *Giardia lamblia* (a protozoan) and viruses that could pass through water treatment unit processes.

Requirements (US EPA, 1999) for treatment (using the best available technology) such as disinfection and filtration are established in place of a maximum contaminant level (MCL) for *Giardia, Legionella,* heterotrophic plate count (HPC), and turbidity. Total treatment of the system must achieve at least 99.9% (3-log) removal or inactivation of *G. lamblia* cysts and 99.99% (4-log) removal of inactivation of viruses. To avoid nitration, a system must demonstrate that its disinfection conditions can achieve the above efficiencies every day of operation except during any one day each month. Filtered water turbidity must at no time exceed 5 NTU (nephelo-metric turbidity units), and 95 percent of the turbidity measurements must meet 0.5 NTU for conventional and direct filtration and 1.0 NTU for slow sand filtration and diatomaceous earth filtration. Turbidity must be measured every 4 h by grab sampling or continuous monitoring. *Cryptosporidium* was not regulated by the SWTR in 1989 because sufficient data about the organism were lacking at the time.

The residual disinfectant in finished water entering the distribution system should not be less than 0.2 mg/L for at least 4 h. The residual disinfectant at any distribution point should be detectable in 95% of the samples collected in a month for any two consecutive months. A water plant may use HPC in lieu of residual disinfectant. If HPC is less than 500 per 100 mL, the site is considered to have a detectable residual for compliance purposes (US EPA, 1989a).

There are some criteria for avoiding filtration. The water plant must calculate the *CT* value daily. The water plant using filtration may or may not need to calculate *CT* values, depending on state primary requirements. The primary agency credits the slow sand filter, which produces water with turbidity of 0.6 to 0.8 NTU, with 2-log *Giardia* and virus removal. Disinfection must achieve an additional 1-log *Giardia* and 2-log virus removal/inactivation to meet the overall treatment objective.

The Interim Enhanced Surface Water Treatment Rule (US EPA, 1999a) defines a disinfection profile as a compilation of daily *Giardia* and/or virus log inactivations over a period of a year or more (may be 3 years). Inactivation of pathogens is typically reported in orders of magnitude inactivation of organisms on a logarithmic scale.

The disinfection profile is a graphical representation of the magnitude of daily *Giardia* or virus log inactivation that is developed, in part, based on daily measurements of operational data (disinfectant residual concentration, contact time, pH, and temperature). A plot of daily log inactivation values versus time provides a visual representation of the log inactivation the treatment system achieved over time. Changes in log inactivation due to disinfectant residual concentrations, temperature, flow rate, or other changes can be seen from this plot (US EPA, 1999b).

**Determination of *CT* values.** Calculation of *CT* values is a complicated procedure depending on the configuration of the whole treatment system, type and number of point applications, and residual concentration of disinfectant (Pontius, 1993).

A water utility may determine the inactivation efficiency based on one point of disinfectant residual measurement prior to the first customer, or on a profile of the residual concentration between the point of disinfectant application and the first customer. The first customer is the point at which finished water is first consumed. To determine compliance with the inactivation (*Giardia* and viruses) requirements, a water system must calculate the *CT* value(s) for its disinfection conditions during peak hourly flow once each day that it is delivering water to its first customers. To calculate *CT* value, *T* is the time (in minutes) that the water moves, during the peak hourly flow, between the point of disinfectant application and the point at which *C*, residual disinfectant concentration, is measured prior to the first customer. The residual disinfectant concentration, *C*, pH, and temperature should be measured each day during peak hourly flow at the effluents of treatment units (each segment or section) and at the first customer's tap. The segments may include the inflow piping system, rapid mixer, flocculators, clarifiers, filters, clearwells, and distribution systems.

The contact time, *T*, is between the application point and the point of residual measured or effective detention time (correction factor from hydraulic retention time, HRT). The contact time in pipelines can be assumed equivalent to the hydraulic retention time and is calculated by dividing the internal volume of the pipeline by the peak hourly flow rate through the pipeline. Due to short circuiting, the contact time in mixing tanks, flocculators, settling basins, clearwells, and other tankages should be determined from tracer studies or an equivalent demonstration. The time determined from tracer studies to be used for calculating *CT* is $T_{10}$. In the calculation of *CT* values, the contact time, designated by $T_{10}$, is the time needed for 10% of the water to pass through the basin or reservoir. In other words, $T_{10}$ is the time (in minutes) that 90% of the water (and microorganisms in the water) will be exposed to disinfectant in the disinfection contact chamber.

The contact time $T_{10}$ is used to calculate the *CT* value for each section. The $CT_{cal}$ is calculated for each point of residual measurement. Then the inactivation ratio, $CT_{cal}/CT_{99.9}$ or $CT_{cal}/CT_{99}$ should be calculated for each section. $CT_{99.9}$ and $CT_{99}$ are the *CT* values to achieve 99.9% and 99% inactivation, respectively. Total inactivation ratio is the sum of the inactivation ratios for each section (sum $CT_{cal}/CT_{99.9}$). If the total inactivation ratio is equal to or greater than 1.0, the system provides more than 99.9% inactivation of *Giardia* cysts and meets the disinfection performance requirement. In fact, many water plants meet the $CT$ requirement for 3-log *Giardia* inactivation because the plant configurations allow for adequate contact time.

Calculation of $CT$ values is a complicated procedure and depends on the configuration of the entire treatment system, type and number of point applications, and residual concentration of disinfectant (US EPA, 1999). The water temperature, pH, type and concentration of disinfectant, and effective retention time (ERT) determine the $CT$ values. Flow patterns and tank configurations of flocculators, clarifiers, and clearwells (location of influent and effluent pipe, and the baffling condition) affect the ERT. The ERT is different from the HRT in most instances.

The theoretical contact time, $T_{\text{theo}}$, can be determined as (assuming completely mixed tank):

$$T_{\text{theo}} = V/Q \tag{5.184}$$

where $T_{\text{theo}}$ = time (= HRT), minutes (min)
V = volume of tank or pipe, gallons or $m^3$
Q = flow rate, gallons per minute, gpm or $m^3/s$

The flow rate can be determined by a theoretical calculation or by using tracer studies. An empirical curve can be developed for $T_{10}$ against $Q$. For the distribution pipelines, all water passing through the pipe is assumed to have detention time equal to the theoretical (Eq. (5.184)) or mean residence time at a particular flow rate, i.e., the contact time is 100% of the time the water remains in the pipe.

In mixing basins, storage reservoirs, and other treatment plant process units, the water utility is required to determine the contact time for the calculation of the $CT$ value through tracer studies or other methods approved by the primacy agency, such as the state EPA.

The calculation of contact time $T$ for treatment units is not a straightforward process of dividing the volume of tank or basin by the flow through rate (Eq. 5.184). Only a partial credit is given to take into account short circuiting that occurs in tanks without baffles. The instructions on calculating the contact time (out of the scope of this manual) are given in Appendix C of the Guidance Manual (US EPA, 1989a).

Baffling is used to maximize utilization of basin volume, increase the plug flow zone in the basin, and to minimize short circuiting. Some form of baffling is installed at the inlet and outlet of the basin to evenly distribute flow across the basin. Additional baffling may be installed within the interior of the basin. Baffling conditions generally are classified as poor, average, or superior and are based on results of tracer studies for use in determining $T_{10}$ from the theoretical detention time of a tank. The $T_{10}/T$ fractions associated with various degrees of baffling conditions are

**TABLE 5.7 Classification of Baffling and $T_{10}/T$ Values**

| Condition of baffling | Description of baffling | $T_{10}/T$ |
|---|---|---|
| Unbaffled (mixed flow) | No baffle, agitated basin, very low length-to-width ratio, high inlet and outlet flow velocities | 0.1 |
| Poor | Single or multiple unbaffled inlets and outlet, no intrabasin baffles | 0.3 |
| Average | Only baffled inlet or outlet, with some intrabasin baffles | 0.5 |
| Superior | Perforated inlet baffle, serpentine or perforated intrabasin baffles, outlet weir, or perforated launders | 0.7 |
| Perfect (plug flow) | Very high length-to-width ratio (pipeline flow), perforated inlet, outlet, and intrabasin baffles | 1.0 |

SOURCE: Table C-5 of the US EPA Guidance Manual (US EPA, 1989a)

shown in Table 5.7. In practice, theoretical $T_{10}/T$ values of 1.0 for plug flows and 0.1 for mixed flow are seldom achieved due to the effect of dead space. Superior baffling conditions consist of at least a baffled inlet and outlet, and possibly some intrabasin baffling to redistribute the flow throughout the basin's cross section. Average baffling conditions include intrabasin baffling and either a baffled inlet or outlet. Poor baffling conditions refer to basins without intrabasin baffling and unbaffled inlets and outlets.

The procedure to determine the inactivation capability of a water plant is summarized as follows (US EPA, 1989a, 1990, 1999; IEPA, 1992):

1. Determine the peak hourly flow rate ($Q$ in gpm) for each day from monitoring records.
2. Determine hydraulic detention time: HRT $= T = V/Q$.
3. Calculate the contact time ($T_{10}$) for each disinfection segment based on baffling factors or tracer studies.
4. Find correction factor, $T_{10}/T$, look at the actual baffling conditions to obtain $T_{10}$ from Table 5.6.
5. Compute effective retention time (ERT) $=$ HRT $\times$ $(T_{10}/T)$.
6. Measure the disinfectant residual: $C$ in mg/L (with pH, and temperature in °C) at any number of points within the treatment train during peak hourly flow (in gpm).
7. Calculate $CT$ value ($CT_{\text{cal}}$) for each point of residual measurement (using ERT for $T$) based on actual system data.
8. Find $CT_{99.9}$ or $CT_{99.99}$ value from Tables 5.7 or 5.9 (from Appendix E of the Manual, US EPA, 1989a) based on water temperature, pH, residual disinfectant concentration, and $\log_{10}$ removal $=$ 3 or 4.

9. Compute the inactivation ratio, $CT_{\mathrm{cal}}/CT_{99.9}$ and $CT_{\mathrm{cal}}/CT_{99.99}$ for *Giardia* and viruses, respectively.
10. Calculate the estimated log inactivation by multiplying the ratio of step 9 by 3 for *Giardia* and by 4 for viruses, because $CT_{99.9}$ and $CT_{99.99}$ are equivalent to 3-log and 4-log inactivations, respectively.
11. Sum the segment log inactivation of *Giardia* and viruses (such as rapid mixing tanks, flocculators, clarifiers, filters, clearwell, and pipelines) to determine the plant total log inactivations due to disinfection.
12. Determine whether the inactivations achieved are adequate. If the sum of the inactivation ratios is greater than or equal to one, the required 3-log inactivation of *Giardia* cysts and 4-log viruses inactivation have been achieved.
13. The total percent of inactivation can be determined as:

$$y = 100 - 100/10^x \tag{5.185}$$

where $y$ = % inactivation
$x$ = log inactivation

Tables 5.8 and 5.9 present the $CT$ values for achieving 99.9% and 99% inactivation of *G. lamblia*. Table 5.10 presents $CT$ values for achieving 2-, 3-, and 4-log inactivation of viruses at pH 6 through 9 (US EPA, 1989b). The SWTR Guidance Manual did not include $CT$ values at pH above 9 due to the limited research results available at the time of rule promulgation. In November 1997, a new set of proposed rules was developed for the higher pH values, up to pH of 11.5 (Federal Register, 1997).

**Example 1:** What are the percentages of inactivation for 2- and 3.4-log removal of *Giardia lamblia*?

**solution:** Using Eq. (5.185)

$$y = 100 - 100/10^x$$

as $x = 2$

$$y = 100 - 100/10^2 = 100 - 1 = 99(\%)$$

as $x = 3.4$

$$y = 100 - 100/10^{3.4} = 100 - 0.04 = 99.96(\%)$$

**Example 2:** A water system of 100,000 gpd (0.0044 $m^3/s$) using a slow sand filtration system serves a small town of 1000 persons. The filter effluent turbidity

**TABLE 5.8** ***CT*** **Values [(mg/l)min] for Achieving 99.9% (3 log) Inactivation of *Giardia lamblia***

| Disinfectant, mg/L | pH | Temperature, °C 0.5 or <1 | 5 | 10 | 15 | 20 | 25 |
|---|---|---|---|---|---|---|---|
| Free chlorine | | | | | | | |
| ≤0.4 | 6 | 137 | 97 | 73 | 49 | 36 | 24 |
| | 7 | 195 | 139 | 104 | 70 | 52 | 35 |
| | 8 | 277 | 198 | 149 | 99 | 74 | 50 |
| | 9 | 390 | 279 | 209 | 140 | 105 | 70 |
| 1.0 | 6 | 148 | 105 | 79 | 53 | 39 | 26 |
| | 7 | 210 | 149 | 112 | 75 | 56 | 37 |
| | 8 | 306 | 216 | 162 | 108 | 81 | 56 |
| | 9 | 437 | 312 | 236 | 156 | 117 | 78 |
| 1.6 | 6 | 157 | 109 | 83 | 56 | 42 | 28 |
| | 7 | 226 | 155 | 119 | 79 | 59 | 40 |
| | 8 | 321 | 227 | 170 | 116 | 87 | 58 |
| | 9 | 466 | 329 | 236 | 169 | 126 | 82 |
| 2.0 | 6 | 165 | 116 | 87 | 58 | 44 | 29 |
| | 7 | 236 | 165 | 126 | 83 | 62 | 41 |
| | 8 | 346 | 263 | 182 | 122 | 91 | 61 |
| | 9 | 500 | 353 | 265 | 177 | 132 | 88 |
| 3.0 | 6 | 181 | 126 | 95 | 63 | 47 | 32 |
| | 7 | 261 | 182 | 137 | 91 | 68 | 46 |
| | 8 | 382 | 268 | 201 | 136 | 101 | 67 |
| | 9 | 552 | 389 | 292 | 195 | 146 | 97 |
| $ClO_2$ | 6–9 | 63 | 26 | 23 | 19 | 15 | 11 |
| Ozone | 6–9 | 2.9 | 1.9 | 1.43 | 0.95 | 0.72 | 0.48 |
| Chloramine | 6–9 | 3800 | 2200 | 1850 | 1500 | 1100 | 750 |

SOURCE: Abstracted from Tables E-1 to E-6, E-8, E-10, and E-12 of the US EPA Guidance Manual (US EPA, 1989a)

values are 0.4 to 0.6 NTU and pH is about 7.5. Chlorine is dosed after filtration and prior to the clearwell. The 4-in (10 cm) transmission pipeline to the first customer is 1640 ft (500 m) in distance. The residual chlorine concentrations in the clearwell and the distribution main are 1.6 and 1.0 mg/L, respectively. The volume of the clearwell is 70,000 gal (265 $m^3$). Determine *Giardia* inactivation at a water temperature of 10°C at the peak hour flow of 100 gpm.

**solution:**

Step 1.
An overall inactivation of 3 logs for *Giardia* and 4 logs for viruses is required. The Primacy Agency can credit the slow sand filter process, which produces water with turbidity ranging from 0.6 to 0.8 NTU, with a 2-log *Giardia* and virus inactivation. For this example, the water system meets the turbidity standards. Thus, disinfection must achieve an additional

**TABLE 5.9** ***CT* Values [(mg/L) min] for Achieving 90% (1 log) Inactivation of *Giardia lamblia***

| Disinfectant, mg/L | pH | Temperature, °C 0.5 or <1 | 5 | 10 | 15 | 20 | 25 |
|---|---|---|---|---|---|---|---|
| Free chlorine | | | | | | | |
| ≤0.4 | 6 | 46 | 32 | 24 | 16 | 12 | 8 |
| | 7 | 65 | 46 | 35 | 23 | 17 | 12 |
| | 8 | 92 | 66 | 50 | 33 | 25 | 17 |
| | 9 | 130 | 93 | 70 | 47 | 35 | 23 |
| 1.0 | 6 | 49 | 35 | 26 | 18 | 13 | 9 |
| | 7 | 70 | 50 | 37 | 25 | 19 | 12 |
| | 8 | 101 | 72 | 54 | 36 | 27 | 18 |
| | 9 | 146 | 104 | 78 | 52 | 39 | 26 |
| 1.6 | 6 | 52 | 37 | 28 | 19 | 14 | 9 |
| | 7 | 75 | 52 | 40 | 26 | 20 | 13 |
| | 8 | 110 | 77 | 58 | 39 | 29 | 19 |
| | 9 | 159 | 112 | 84 | 56 | 42 | 28 |
| 2.0 | 6 | 55 | 39 | 29 | 19 | 15 | 10 |
| | 7 | 79 | 55 | 41 | 28 | 21 | 14 |
| | 8 | 115 | 81 | 61 | 41 | 30 | 20 |
| | 9 | 167 | 118 | 88 | 59 | 46 | 29 |
| 3.0 | 6 | 60 | 42 | 32 | 21 | 16 | 11 |
| | 7 | 87 | 61 | 46 | 30 | 23 | 15 |
| | 8 | 127 | 89 | 67 | 45 | 36 | 22 |
| | 9 | 184 | 130 | 97 | 65 | 49 | 32 |
| Chlorine dioxide | 6–9 | 21 | 8.7 | 7.7 | 6.3 | 5.0 | 3.7 |
| Ozone | 6–9 | 0.97 | 0.63 | 0.48 | 0.32 | 0.24 | 0.16 |
| Chloramine | 6–9 | 1270 | 735 | 615 | 500 | 370 | 250 |

SOURCE: Abstracted from Tables E-l to E-6, E-8, E-10, and E-12 of the US EPA Guidance Manual (US EPA, 1989a)

1-log *Giardia* and 2-log virus removal/inactivation to meet the overall treatment efficiency.

Step 2. Calculate $T_{10}$ at the clearwell (one-half of volume used, see Step 7(a) of Example 5)

$T_{10}$ can be determined by the trace study at the peak hour flow or by calculation

$$\begin{aligned} T_{10} &= V_{10}/Q = 0.1 \times (70{,}000\text{gal}/2)/(100\text{gal/min}) \\ &= 35 \text{ min} \end{aligned}$$

Step 3. Calculate $CT_{\text{cal}}$ in the clearwell

$$CT_{\text{cal}} = 1.6 \text{ mg/L} \times 35 \text{ min} = 56 \text{ (mg/L) min}$$

**TABLE 5.10** ***CT* Values [(mg/L) min] for Achieving Inactivation of Viruses at pH 6 through 9**

| Disinfectant mg/L | Log inactivation | Temperature, °C | | | | | |
|---|---|---|---|---|---|---|---|
| | | ≤1 | 5 | 10 | 15 | 20 | 25 |
| Free chlorine | 2 | 6 | 4 | 3 | 2 | 1 | 1 |
| | 3 | 9 | 6 | 4 | 3 | 2 | 1 |
| | 4 | 12 | 8 | 6 | 4 | 3 | 2 |
| Chlorine dioxide | 2 | 8.4 | 5.6 | 4.2 | 2.8 | 2.1 | 1.4 |
| | 3 | 25.6 | 17.1 | 12.8 | 8.6 | 6.4 | 4.3 |
| | 4 | 50.1 | 33.4 | 25.1 | 16.7 | 12.5 | 8.4 |
| Ozone | 2 | 0.9 | 0.6 | 0.5 | 0.3 | 0.25 | 0.15 |
| | 3 | 1.4 | 0.9 | 0.8 | 0.5 | 0.4 | 0.25 |
| | 4 | 1.8 | 1.2 | 1.0 | 0.6 | 0.5 | 0.3 |
| Chloramine | 2 | 1243 | 857 | 643 | 428 | 321 | 214 |
| | 3 | 2063 | 1423 | 1067 | 712 | 534 | 365 |
| | 4 | 2883 | 1988 | 1491 | 994 | 746 | 497 |

SOURCE: Modified from Tables E-7, E-9, E-ll, and E-13 of the US EPA Guidance Manual (US EPA, 1989a)

Step 4. Calculate $CT_{\text{cal}}/CT_{99.9}$

From Table 5.8, for 3-log removal for 1.6 mg/L chlorine residual at 10°C and pH 7.5

$$CT_{99.9} = 145 \text{ (mg/L) min (by linear proportion between pH 7 and 8)}$$

then

$$CT_{\text{cal}}/CT_{99.9} = 56/145 = 0.38$$

Step 5. Calculate contact time at transmission main

$$Q = 100{,}000 \text{ gal/d} \times 1 \text{ ft}^3/7.48 \text{ gal} \times 1 \text{ day}/440 \text{ min}$$
$$= 9.28 \text{ ft}^3/\text{min}$$

$$A = 3.14(4 \text{ in}/2/12 \text{ in/ft})^2 = 00872 \text{ ft}^2$$

$$v = \frac{Q}{A} = \frac{9.28 \text{ ft}^3/\text{min}}{0.0872 \text{ ft}^2}$$
$$= 106 \text{ ft/min}$$

$$T = \text{length}/v = 1640 \text{ ft}/106 \text{ ft/min}$$
$$= 15.5 \text{ min}$$

where $A$ is the cross-sectional area of the 4-in pipe, and $v$ is the flow velocity.

Step 6. Calculate $CT_{\text{cal}}$ and $CT_{\text{cal}}/CT_{99.9}$ for the pipeline

$$\begin{aligned} CT_{\text{cal}} &= 1.0 \text{ mg/L} \times 15.5 \text{ min} \\ &= 15.5 \text{ (mg/L) min} \end{aligned}$$

From Table 5.8, for 3-log removal of 1.0 mg/L chlorine residual at 10°C and pH 7.5

$$CT_{99.9} = 137 \text{ (mg/L) min}$$

then $$CT_{\text{cal}}/CT_{99.9} = 15.5/137 = 0.11 \text{ (log)}$$

Step 7. Sum of $CT_{\text{cal}}/CT_{99.9}$ from Steps 4 and 6

$$\text{Total } CT_{\text{cal}}/CT_{99.9} = 0.38 + 0.11 = 0.49 \text{ (log)}$$

Because this calculation is based on a 3-log removal, the ratio needs to be multiplied by 3. The equivalent *Giardia* inactivation then is

$$3 \times \text{total } CT_{\text{cal}}/\text{CT}_{99.9} = 3 \times 0.49 \text{ log} = 1.47 \text{ log}$$

The 1.47-log *Giardia* inactivation by chlorine disinfection in this system exceeds the 1-log additional inactivation needed to meet the overall treatment objectives. If a system uses chloramine and is able to achieve *CT* values for 99.9% inactivation of *Giardia* cysts, it is not always appropriate to assume that 99.99% or greater inactivation of viruses also is achieved.

**Example 3:** For a 2000-ft long, 12-in transmission main with a flow rate of 600 gpm, what is the credit of inactivation of *Giardia* and viruses using chlorine dioxide ($ClO_2$) residual = 0.6 mg/L, at 5°C and pH 8.5.

**solution:**

Step 1. Calculate $CT_{\text{cal}}/CT_{99.9}$ for transmission pipe

$$\begin{aligned} T &= \pi r^2 L/Q \\ &= 3.14 \text{ (0.5 ft)}^2 \text{ (2000 ft) (7.48 gal/ft}^3\text{)/(600 gal/min)} \\ &= 19.6 \text{ min} \end{aligned}$$

where *r* is the radius of the pipe and *L* is the length of the pipe.

$$\begin{aligned} CT_{\text{cal}} &= 0.6 \text{mg/L} \times 19.6 \text{ min} \\ &= 11.7 \text{ (mg/L) min} \end{aligned}$$

At 5°C and pH of 8.5, using $ClO_2$, from Table 5.7

$$\begin{aligned} CT_{99.9} &= 26 \text{ (mg/L) min} \\ \text{Cl}_{\text{cal}}/\text{CT}_{99.9} &= 11.7/26 = 0.45 \text{ (log)} \end{aligned}$$

Step 2. Calculate log inactivation for *Giardia* ($X$) and viruses ($Y$)

$$X = 3(CT_{cal}/CT_{99.9}) = 3 \times 0.45 \log$$
$$= 1.35 \log$$
$$Y = 4\,(CT_{cal}/CT_{99 \cdot 9}) = 4 \times 0.45 \log$$
$$= 1.8 \log$$

**Example 4:** A water system pumps its raw water from a remote lake. No filtration is required due to good water quality. Chlorine is dosed at the pumping station near the lake. The peak pumping rate is 320 gpm (0.02 $m^3$/s). The distance from the pumps to the storage reservoir (tank) is 3300 ft (1000 m); the transmission pipe is 10 in in diameter. The chlorine residual at the outlet of the tank is 1.0 mg/L ($C$ for the tank). The $T_{10}$ for the tank is 88 min at the peak flow rate determined by a tracer study. Assuming the inactivation for the service connection to the first customer is negligible, determine the minimum chlorine residue required at the inlet of the tank ($C$ at the pipe) to meet *Giardia* 3-log removal at 5°C and pH 7.0.

**solution:**

Step 1. Calculate $CT_{cal}$ for the pipe

$$Q = 320 \text{ gpm} = 320 \text{ gal/min} \times 0.1337 \text{ ft}^3\text{/gal}$$
$$= 42.8 \text{ ft}^3\text{/min}$$
$$T = \pi r^2 L/Q = 3.14\,(0.417 \text{ ft})^2 \times 3300 \text{ ft}/(42.8 \text{ ft}^3\text{/min})$$
$$= 42 \text{ min}$$
$$CT_{cal} \text{ (pipe)} = 42\, C_p \text{(mg/L)min}$$

where $C_p$ is chlorine residual at the end of the pipe.

Step 2. Calculate $CT_{cal}$ for the tank, $CT_{cal}$ (tank)

$$CT_{cal} \text{ (tank)} = 1.0 \text{ mg/L} \times 88 \text{ min} = 88 \text{ (mg/L) min}$$

Step 3. Find $CT_{99.9}$ for *Giardia* removal

At water temperature of 5°C, pH 7.0, and residual chlorine of 1.0 mg/L, from Table 5.7:

$$CT_{99.9} = 149 \text{ (mg/L) min}$$

Step 4. Calculate chlorine residue required at the end of the pipe (tank inlet), $C_p$

$$CT_{cal} \text{ (pipe)} + CT_{cal} \text{ (tank)} = CT_{99.9}$$

From Steps 1 to 3, we obtain

$$42 \text{ min} \times C_p\text{(mg/L)} + 88\text{(mg/L)min} = 149\text{(mg/L)min}$$
$$C_p = 1.45 \text{ mg/L (minimum required)}$$

**Example 5:** In a 1-MGD water plant (1 MGD = 694 gpm = 0.0438 $m^3$/s), water is pumped from a lake and prechlorinated with chlorine dioxide at the lake site. The source water is pretreated with chlorine at the intake near the plant. The peak flow is 600 gpm. The worst conditions to be evaluated are at 5°C and pH 8.0. Gaseous chlorine is added at the plant after filtration (clearwell inlet). The service connection to the first customer is located immediately next to the clearwell outlet. The following conditions also are given:

| Treatment unit | Volume, gal | Outlet residual $Cl_2$, mg/L | Baffling conditions |
|---|---|---|---|
| Rapid mixer | 240 | 0.4 | Baffled at inlet and outlet |
| Flocculators | 18,000 | 0.3 | Baffled at inlet and outlet and with horizontal paddles |
| Clarifiers | 150,000 | 0.2 | Only baffled at outlet |
| Filters | 6600 | 0.1 | |
| Clearwell | 480,000 | 0.3 (free)<br>1.6 (combined) | |

Chlorine and ammonia are added at the clearwell inlet. Residual free and combined chlorine measured at the outlet of the clearwell are 0.3 and 1.6 mg/L, respectively. Estimate the inactivation level of *Giardia* and viruses at each treatment unit.

**soution:**

Step 1. Calculate $CT_{\text{cal}}$ for the rapid mixer

$$\text{HRT} = \frac{V}{Q} = \frac{240\ \text{gal}}{600\ \text{gal/min}}$$

$$= 0.4\ \text{min}$$

Assuming average baffling condition and using Table 5.6, we obtain the correction factor $(T_{10}/T) = 0.5$ for average baffling.

Thus

$$\begin{aligned}\text{ERT} &= \text{HRT}\ (T_{10}/T) = 0.4\ \text{min} \times 0.5 \\ &= 0.2\ \text{min}\end{aligned}$$

Use this ERT as $T$ for $CT$ calculation

$$\begin{aligned}CT_{\text{cal}} &= 0.4\ \text{mg/L} \times 0.2\ \text{min} \\ &= 0.08\ (\text{mg/L})\ \text{min}\end{aligned}$$

Step 2. Determine $CT_{\text{cal}}/CT_{99\cdot 9}$ for the rapid mixer

Under the following conditions:

$$\begin{aligned}\text{pH} &= 8.0 \\ T &= 5°\text{C}\end{aligned}$$

$$\text{residual chlorine} = 0.4 \text{ mg/L}$$
$$\text{log inactivation} = 3$$

From Table 5.7, we obtain:

$$CT_{99.9} = 198 \text{ (mg/L) min}$$
$$CT_{\text{cal}}/CT_{99\cdot 9} = (0.08 \text{ (mg/L) min})/(198 \text{ (mg/L) min})$$
$$= 0.0004$$

Step 3. Calculate the log inactivation of *Giardia* for the rapid mixer

$$\text{log removal} = 3\ (CT_{\text{cal}}/CT_{99.9})$$
$$= 0.0012$$

Step 4. Similar to Steps 1 to 3, determine log inactivation of *Giardia* for the flocculators

$$\text{HRT} = \frac{V}{Q} = \frac{18{,}000 \text{ gal}}{600 \text{ gpm}}$$
$$= 30 \text{ min}$$

with a superior baffling condition from Table 5.6, the correction $T_{10}/T = 0.7$

$$\text{ERT} = \text{HRT}\ (T_{10}/T) = 30 \text{ min} \times 0.7$$
$$= 21 \text{ min}$$

then

$$CT_{\text{cal}} = 0.3 \text{ mg/L} \times 21 \text{ min}$$
$$= 6.3 \text{ (mg/L) min}$$
$$\text{log removal} = 3\ [6.3 \text{ (mg/L) min}]/[198 \text{ (mg/L) min}]$$
$$= 0.095$$

Step 5. Similarly, for the clarifiers

$$\text{HRT} = \frac{150{,}000 \text{ gal}}{600 \text{ gpm}} = 250 \text{ min}$$

Baffling condition is considered as poor, thus

$$T_{10}/T = 0.3$$
$$\text{ERT} = 250 \text{ min} \times 0.3 = 75 \text{ min}$$
$$CT_{\text{cal}} = 0.2 \text{ mg/L} \times 75 \text{ min} = 15 \text{ (mg/L) min}$$

$$\text{log removal} = 3 \times 15/198 = 0.227$$

Step 6. Determine *Giardia* removal in the filters

$$\text{HRT} = \frac{6600 \text{ gal}}{600 \text{ gpm}} = 11 \text{ min}$$

$$T_{10}/T = 0.5$$

(Teefy and Singer, 1990. The volume of the filter media is subtracted from the volume of filters with good-to-superior baffling factors.)

$$\text{ERT} = 11 \text{ min} \times 0.5 = 5.5 \text{ min}$$

$$CT_{\text{cal}} = 0.1 \text{ mg/L} \times 5.5 \text{ min} = 0.55 \text{ (mg/L) min}$$

$$\text{log removal} = 3 \times 0.55/198 = 0.008$$

Step 7. Determine *Giardia* removal in the clearwell

There are two types of disinfectant; therefore, similar calculations should be performed for each disinfectant.

(a) Free chlorine inactivation.
In practice, the minimum volume available in the clearwell during the peak demand period is approximately one-half of the working volume (Illinois EPA, 1992). Thus, one-half of the clearwell volume (240,000 gal) will be used to calculate the HRT:

$$\text{HRT} = 240{,}000 \text{ gal/600 gpm} = 400 \text{ min}$$

The clearwell is considered a poor baffling condition, because it has no inlet, interior, or outlet baffling.

$$T_{10}/T = 0.3 \text{ (Table 5.6), calculate ERT}$$

$$\text{ERT} = \text{HRT} \times 0.3 = 400 \text{ min} \times 0.3 = 120 \text{ min}$$

The level of inactivation associated with free chlorine is

$$CT_{\text{cal}} = 0.3 \text{ mg/L} \times 120 \text{ min} = 36 \text{ (mg/L) min}$$

From Table 5.7, $CT_{99.9} = 198$ (mg/L) min

The log inactivation is

$$\text{log removal} = 3 \times CT_{\text{cal}}/CT_{99.9} = 3 \times 36/198$$
$$= 0.545$$

(b) Chloramines inactivation

$$CT_{\text{cal}} = 1.6 \text{ mg/L} \times 120 \text{ min}$$
$$= 192 \text{ (mg/L) min}$$

For chloramines, at pH $= 8$, $T = 5°\text{C}$, and 3-log inactivation

$$CT_{99.9} = 2200 \text{ (mg/L) min (from Table 5.7)}$$

Log inactivation associated with chloramines is

$$\text{log removal} = 3 \times 192/2200$$
$$= 0.262$$

The log removal at the clearwell is

$$\text{log removal} = 0.545 + 0.262$$
$$= 0.807$$

Step 8. Summarize the log *Giardia* inactivation of each unit in the water treatment plant

| Unit | Log inactivation | |
|---|---|---|
| Rapid mixing | 0.0012 | (Step 3) |
| Flocculators | 0.095 | (Step 4) |
| Clarifiers | 0.227 | (Step 5) |
| Filters | 0.008 | (Step 6) |
| Clearwell | 0.807 | (Step 7) |
| Total | 1.138 | |

The sum of log removal is equal to 1.138, which is greater than 1. Therefore, the water system meets the requirements of providing a 3-log inactivation of *Giardia* cysts. The viruses inactivation is estimated as follows.

Step 9. Estimate viruses inactivation in the rapid mixer

Using $CT_{\text{cal}}$ obtained in Step 1

$$CT_{\text{cal}} = 0.08 \text{ (mg/L) min}$$

Referring to Table 5.9 ($T = 5°\text{C}$, 4 log),

$$CT_{99.99} = 8 \text{ (mg/L) min}$$

The level of viruses inaction in the rapid mixer is

$$\begin{aligned}\text{log removal} &= 4 \times CT_{\text{cal}}/CT_{99.99} = 4 \times 0.08/8 \\ &= 0.04\end{aligned}$$

Step 10. Estimate viruses inactivation in the fiocculators

$$\begin{aligned}CT_{\text{cal}} &= 6.3 \text{ (mg/L) min (Step 4)} \\ \text{log removal} &= 4 \times 6.3/8 \\ &= 3.15\end{aligned}$$

Step 11. Estimate viruses inactivation in the clarifiers

$$\begin{aligned}CT_{\text{cal}} &= 15 \text{ (mg/L) min (Step 5)} \\ \text{log removal} &= 4 \times 15/8 \\ &= 7.5\end{aligned}$$

Step 12. Estimate viruses inactivation in the filters

$$\begin{aligned}CT_{\text{cal}} &= 0.55 \text{ (mg/L) min (Step 6)} \\ \text{log removal} &= 4 \times 0.55/8 \\ &= 0.275\end{aligned}$$

Step 13. Estimate viruses inactivation in the clearwell

For free chlorine

$$\begin{aligned}CT_{\text{cal}} &= 36 \text{ (mg/L) min (Step 6(a))} \\ \text{log removal} &= 4 \times 36/8 \\ &= 18\end{aligned}$$

For chloramines

$$\begin{aligned}CT_{\text{cal}} &= 192 \text{ (mg/L) min (Step 6(b))} \\ CT_{99.99} &= 1988 \text{ (mg/L) min (Table 5.9)} \\ \text{log removal} &= 4 \times 192/1988 \\ &= 0.39 \\ \text{Total log removal} &= 18 + 0.39 = 18.39\end{aligned}$$

Step 14. Sum of log inactivation for viruses in the plant is

$$\begin{aligned}\text{Plant log removal} &= \text{sum of Steps 9 to 13} \\ &= 0.04 + 3.15 + 7.5 + 0.275 + 18.39 \\ &= 29.35\end{aligned}$$

*Note*: Total viruses inactivation is well above 1.

The water system meets the viruses inactivation requirement.

In summary, a brief description of regulations in the United States and determination of $CT$ values are presented. Under the SWTR, all surface water supplies and groundwater supplies that are under the influence of surface water must calculate $CT$ values daily during peak hourly flow. A minimum of 3-log *Giardia lamblia* and 4-log virus removal and/or inactivation performance must be achieved at all times to comply with the existing SWTR. The $CT$ values are used to evaluate the achievement of disinfection and to determine compliance with the SWTR. They also are used to compute the log inactivation of *Giardia* and viruses during water treatment and to construct a disinfection profile.

To calculate the $CT$ value, operation data, e.g. disinfectant concentration, water pH, and water temperature, should be measured daily during the peak hourly flow. Contact time ($T_{10}$), actual calculated $CT$ value ($CT_{\mathrm{cal}}$), 3-log *Giardia* inactivation ($CT_{99.9}$ from Table 5.7), and/or 4-log virus inactivation (from Table 5.9), $CT_{\mathrm{cal}}/CT_{99.9}$, and $CT_{\mathrm{cal}}/CT_{99.99}$ should be calculated for the estimated segment log inactivations. The total plant log inactivation is the sum of all segment log inactivations. The daily log inactivation values can be used to develop a disinfection profile for the treatment system.

## 18.5 Disinfection by-products

**Water disinfection and its by-products.** Chlorine has been widely used as a disinfectant in water treatment process to kill or inactivate waterborne pathogens. Chlorine disinfection of public drinking water had dramatically reduced outbreaks of illness. In 1974, chloroform (trichloromethane) was discovered as a disinfection by-product (DBP) resulting from the interaction of chlorine with natural organic matter in water. This finding raised a serious dilemma that water chlorination clearly reduced the risk of infectious diseases and might also result in the formation of potentially harmful DBP with this exposure in drinking water. DBP problem is currently most concerned by water supply professionals.

Chlorine, chlorine dioxide, chloramines, ozone, and potassium permanganate are used in US water treatment plants as a disinfectant. Table 5.11 lists some of the microorganisms targeted by disinfection practice and some of more appropriate disinfectants for each microorganism. Based on the types of disinfectants used, the numbers of water supplies are roughly (keep changing) 22,000 (91.6%), 360 (1.5%), 140 (0.6%), 300 (1.3%), and 1200 (5.0%), respectively. The disinfectants are often so powerful that they nonselectively react with other substances in the water. There are actually thousands of DBPs. When natural waters are disinfected, more than 100 potentially toxic halogenated compounds can be created (Gray *et al.*, 2001). The compounds of concern are halogenated methanes, haloacetic acids, and nitrosamines.

**TABLE 5.11 Effects of Disinfectants on Waterborne Pathogens**

| Disinfectants | Bacteria, such as coliform (*E. coli*), *Legionella* | *Giardia lamblia* cysts | *Cryptosporidium parvum* oocysts | Viruses |
|---|---|---|---|---|
| (Health effects) | Legionnaire's disease, GI,* death | GI, death | GI, death | GI, death |
| Chlorine | X | X | | X |
| Chlorine dioxide | X | X | X | X |
| Ozone | X | X | X | X |
| Chloramine | X | | | |

*astroenteric disease

Two classes of DBPs that dominate the identifiable organic matter; trihalomethanes (THMs) and haloacetic acids (haloacetates, HAAs) are of regulatory interest. THMs are a group of compounds with three halogen atoms. Only chlorinated and brominated ones are routinely found in potable water. Chloroform ($CHCl_3$) is one of a class of compounds of THMs. In the United States, since 1974, additional DBPs have been identified in potable water and concerns have intensified about health risks resulting from exposures to them. Along with chloroform, three species that are of most concern are bromodichloromethane ($CHBrCl_2$), dibromochloromethane ($CHBr_2Cl$), and tribromomethane (bromoform, $CHBr_3$). The above four species are called trihalomethanes. The sum of these four species is expressed as total trihalomethanes (TTHMs). THMs are regulated under the Stage 2 DBP rule (Table 5.11).

HAAs (Table 5.12) are also formed during chlorination of water. Like THMs, HAAs are also linked with increased incidence of cancer in laboratory animals (Herren-Freund *et al.*, 1987; Xu *et al.*, 1995). Unlike THMs, HAAs are capable of dissociating in water. HAAs are >99% ionized (deprotonated) to the haloacetate anions under drinking water conditions. However, they are regulated and usually reported in terms of the parent acids rather than the carboxylate anions (US EPA, 2001). HAAs account for about 13% of the halogenated organic matter after disinfection (Weinberg, 1999).

Bromate is formed from the ozonation of source waters that contain bromide. In ozonated water supplies, a variety of aldehydes and ketones abound as well as some carboxylic acids. In addition to these organic products, inorganic species are also found. These include oxyanions of halogens, such as chlorite, chlorate, and bromate, which can be formed by a variety of oxidizing disinfectants. Bromate is of particular interest since it is suspected of posing one of the highest cancer risks of any DBP (US EPA, 2001).

**TABLE 5.12 Haloacetic Acids Found in Potable Water**

| Haloacetic acids | Chemical formula | Grouping* |
|---|---|---|
| Chloroacetic acid | $ClCH_2COOH$ | HAA5, 6, 9 |
| Dichloroacetic acid | $Cl_2CHCOOH$ | HAA5, 6, 9 |
| Trichloroacetic acid | $Cl_3CCOOH$ | HAA5, 6, 9 |
| Bromoacetic acid | $BrCH_2COOH$ | HAA5, 6, 9 |
| Dibromoacetic acid | $Br_2CHCOOH$ | HAA5, 6, 9 |
| Tribromoacetic acid | $Br_3CCOOH$ | HAA9 |
| Bromochloroacetic acid | BrClCHCOOH | HAA6, 9 |
| Bromodichloroacetic acid | $BrCl_2CCOOH$ | HAA9 |
| Dibromochloroacetic acid | BrClCCOOH | HAA9 |

NOTES: *HAA5 is the sum of the concentrations of mono-, di-, and trichloroacetic acids and mono- and dibromoacetic acids. HAA5 concentrations (as the sum) are regulated under the Stage 2 DBP Rule. HAA6 data must be obtained and reported under the Information Collection Rule (ICR). HAA9 data are encouraged to obtain and report under the ICR, but not required.

Due to concern about these DBPs for more than 3 decades, some DBPs have been regulated and/or subjected to monitoring rules aimed to meet the simultaneous goal of disinfecting water and controlling DBPs. Table 5.13 presents regulated compounds along with their important information. It should be noted that only a very small subset of the much larger list of substances have been identified as DBPs.

The Stage 2 DBP rule issued in January 2006, had (1) lowered the TTHMs MCL 0.10 to 0.08 mg/L; (2) established an MCL of 0.06 mg/L for five haloacetic acids (HAA5), including, monochloroacetic acid, dichloroacetic acid, trichloroacetic acid, monobromoacetic acid, and dibromoacetic acid; (3) added an MCLG of 0.07 mg/L for (mono)chloroacetic acid; (4) established an MCL of 0.010 mg/L for bromate; and (5) established an MCL of 1.0 mg/L for chlorite.

**Control of DBPs and pathogens.** US EPA's current drinking water research program (in-house/contracted) is more sophisticated than it was in the 1980s. For example, when the treatment technology manual was published in 1981, it reported primarily on treatment-oriented research. Twenty-five years later, the technology research program includes source water protection, treatment technology, and distribution system studies. The research also reflects a concern over balancing the risks of potential carcinogenic exposure against the risks from microbial infection. Techniques for controlling DBPs and waterborne pathogens below have mostly followed the US EPA (2001) report. The controlling techniques include source protection, treatment process modification (enhanced coagulation), alternative disinfectants, activated carbon, other granular filtration, membrane filtration process, control in drinking water distribution system, etc.

**TABLE 5.13 National Primary Drinking Water Regulations Regulated Levels for DBP and Residual Disinfectants**

| Compound | MCLG, mg/L | MCL, mg/L | By-products of | Potential health effect |
|---|---|---|---|---|
| Total trihalomethanes (TTHMs) | | 0.080 LRAA | Chlorination and chloramination | Cancer and other effects |
| Chloroform | 0.07 | | Chlorination and chloramination | Cancer, liver, kidney, and reproductive effects |
| Bromodichloromethane | 0 | | Chlorination and chloramination | Cancer, liver, kidney, and reproductive effects |
| Bromoform | 0 | | Ozonation, chlorination, and chloramination | Cancer, nervous system, liver, and kidney effects |
| Dibromochloromethane | 0.06 | | Chlorination and chloramination | Nervous system, kidney, liver, and reproductive effects |
| Haloacetic acids (HAA5) | | 0.060 LRAA | Chlorination, and chloramination | Cancer and other effects |
| (Mono)chloroacetic acid | 0.07 | | Chlorination and chloramination | Cancer and other effects |
| Dichloroacetic acid | 0 | | Chlorination and chloramination | Cancer and other effects |
| Trichloroacetic acid | 0.02 | | Chlorination and chloramination | Possible cancer and other effects |
| Bromate | 0 | 0.010 | Ozonation | Cancer |
| Chlorite | 0.8 | 1.0 | Chlorine dioxide | Hemolytic anemia |
| Residual disinfectant | MRDLG, mg/L | MRDL, mg/L | | |
| Chlorine | 4.0 | 4 | | |
| Chloramines | 4.0 | 4 | | |
| Chlorine dioxide | 0.8 | 0.8 | | |

NOTE: MCLG = maximum contaminant level goal, MCL = maximum contaminant level,
TTHMs = the sum of the concentrations of chloroform, bromodichloromethane, dibromochloromethane, and bromoform, LRAA = a locational running annual average,
HAA5 = the sum of the concentrations of mono-, di-, and trichloroacetic acids and mono- and dibromoacetic acids,
MRDLG = maximum residual disinfectant level goal, MRDL = maximum residual disinfectant level.

SOURCE: US EPS, 2001; www.epa.gov/OGWDW/regs.html, 2007

**Source water protection.** Many of the drinking water utilities in the United States had invested a great deal of time, energy, and capital into source water protection (SWP). In addition, efforts were made to develop mechanisms for protection against the impact of sudden changes in influent water quality. Some of these mechanisms include investment in excess capacity and development of emergency procedures.

Source water protection measures are the important step to prevent contamination and reduce the need for treatment of drinking water supplies. SWP includes managing potential contamination sources and developing contingency plans that identify alternate drinking water sources. A community may decide to develop an SWP program based on the results of a source water assessment, which includes delineation of the area to be protected and an inventory of the potential contaminants within that area. SWP from quality degradation by microbial contaminants (i.e. bacteria, protozoa, viruses, helminthes, and fungi) is any activity undertaken to minimize the frequency, magnitude, and duration of occurrence of pathogens or indicators (e.g. indicator microorganisms, such as TC, FC, and FS (Lin, 2002); or turbidity) in source waters. SWP may also, by reducing the concentration of natural organic matter (NOM), a DBP precursor, reduce the formation of DBP.

SWP strategies comprise the first stage in the multiple-barrier approach to protecting the quality of drinking water. Other major drinking-water-quality protection barriers include water quality monitoring and selective source withdrawal, water treatment processes for removal or inactivation of pathogens and control of DBP formation, water distribution practices for preventing intrusion or regrowth of pathogens, and point-of-use treatment where required.

SWP strategies are a specific subset of a large watershed protection strategy applied when the protected receiving water is used as a water supply. Conceptually, watershed protection is heavily linked to pollution prevention, contaminant source identification, and risk management. Although watershed management does not have a universally accepted definition and connotes alternate approaches, each interpretation has an underpinning of holistic approaches to prevent or mitigate threats to the receiving water over a geographic region defined by a common hydrology.

Many groundwater supplies have proven to be vulnerable as well, resulting in various states implementing wellhead protection programs. The 1986 amendments included provisions for Protection of Ground Water Sources of Water. Two programs were set up under this requirement: the "Sole Source Aquifer Demonstration program," to establish demonstration programs to protect critical aquifer areas from degradation; and the

"Wellhead Protection Program," which required states to develop programs for protecting areas around public water supply wells to prevent contamination from residential, industrial, and farming-use activities.

In the 1996 amendments to the SDWA, protection of source waters was given greater emphasis to strengthen protection against microbial contaminants, particularly *Cryptosporidium,* while reducing potential health risks due to DBPs. Two major threats to source water quality with respect to DBP control and microbial protection are natural organic matter and microbial pathogens. From a waterborne outbreak and public health viewpoint, both *Giardia* and *Cryptosporidium* are currently of primary concern.

Managing microbial contaminant risks in watersheds requires identification and quantification of organisms. Because of difficulties in association with assaying for specific pathogens, monitoring programs have been tested for indicator organisms, including total coliform (TC), fecal coliform (FC), and fecal streptococcus (FS), to identify possible fecal contamination in water. The analytical methods for indicators are easier, faster, and more cost effective than methods for specific pathogenic organisms. The limitations of relying on indicator organisms for determining the presence of pathogens include the occurrence of false positives and the fact that indicators measure bacteria that live not only in human enteric tracts, but also in the enteric tracts of other animals (Toranzos and McFeters, 1997; Lin, 2002).

The potential sources of microbial pathogens in source water are many and varied, including nonpoint runoff, discharges from treated and untreated wastewaters, storm water runoff, and combined sewer overflows. Animal feed lot is the main source of *Cryptosporidium.* Many factors affect the types of organisms found and the concentrations at which they are detected. These include watershed contributions, treatment plant efficiency, and length of antecedent dry weather period. These treatment technologies can be both sources of contamination as well as protection of source water quality. In addition to the installation of wastewater treatment and combined overflow systems, there are passive pollution prevention and mitigation techniques called BMPs. The BMP techniques can be found elsewhere (US EPA, 1990, 1992) and vary dramatically in application, ranging from social practices to engineering applications.

In summary, long-term and effective management for pathogens and DBP is source water protection to control NOM, the DBP precursor. SWP includes effective watershed protection and water resources management strategies. It involves controlling algal growth and the production of organic carbon precursors in our raw water sources (lakes and rivers) by limiting nutrient discharges into these bodies of water.

Once NOM is controlled, there would be benefits to the water treatment system, such as less pretreatment and easier down-stream treatment, more effective unit process, and cheaper water treatment costs. Coagulant dosage requirements will be reduced and then less sludge will be produced if there is less dissolved organic carbon (DOC) in the raw water, because, for most waters, the concentration of DOC controls coagulant dosage requirements. Chlorine dioxide doses will be lowered if there is less DOC in the raw water to consume chlorine dioxide and generate chlorite. Powdered and/or granular activated carbon dosage rates for the control of taste- and odor-causing organic compounds will be reduced if there is less DOC in the raw water to compete with these target tracing organic pollutants for adsorption sites on the activated carbon. Membrane fouling will be lowered and ultraviolet irradiation for microbial inactivation will be more effective if there are less DOC and ultraviolet-adsorbing organics in the raw water. Less chlorine will be required for disinfection, and lower concentrations of all halogenated organic by-products—not just the THMs and HAAs—will be formed (Singer, 2006).

**Use of alternative disinfectants.** As discussed previously, chlorination of drinking water results in the formation of numerous DBPs, several of which are regulated. Water systems seeking to meet MCLs for regulated DBPs might consider various approaches to limiting DBPs such as removing precursor compounds before the disinfectant is applied, using less chlorine, using alternative disinfectants to chlorine, or removing DBPs after their formation. Combinations of these approaches might also be considered. As mentioned previously, removing DBPs after their formation is a method that is generally not considered, but no matter which approach is selected, the effectiveness of the disinfection process must not be jeopardized.

Studies have been conducted by or funded by the US EPA's Office of Research and Development in Cincinnati to examine the use of three alternative oxidants: chloramines, chlorine dioxide ($ClO_2$), and ozone.

*Chloramines.* Based on the researches cited (Steven *et al.,* 1989; Miltner, 1990), the formation of halogen-containing DBPs by chloramines is significantly lower than by free chlorine. An exception is the formation of cyanogens chloride with chloramination. Formation of nonhalogenated DBPs such as aldehydes and assimilable organic carbon (AOC) is found to be minimal with chloramination.

Chloramines are the second most commonly used final disinfectant in drinking water treatment after free chlorine. Although generally not as effective a disinfectant as free chlorine, an advantage of chloramination is minimization of the formation of DBPs.

*Chlorine dioxide.* Chlorine dioxide is another widely used disinfectant in drinking water treatment. It has long been used for taste and odor control and for iron and manganese control, and has gained in acceptance as an effective disinfectant.

An advantage of $ClO_2$ treatment is minimization of the formation of DBPs; it does this by oxidation of DBP precursors and by relatively minimal formation of DBPs themselves. The formation of DBPs by $ClO_2$ is also significantly lower than with free chlorine and with chloramines (Stevens *et al.*, 1989). $ClO_2$ oxidizes DBP precursors in the treated water to the extent that lower concentrations of DBPs are formed with subsequent chlorination. Using $ClO_2$ results in the formation of nonhalogenated DBPs such as aldehydes, ketones, and AOC.

The use of $ClO_2$ can result in the formation of chlorite and chlorate. Chlorite can be controlled by GAC and through the use of reducing agents. Sulfite and metabisulfite can reduce chlorite, but may form chlorate. Thiosulfate can reduce chlorite without forming chlorate. Ferrous ion can reduce chlorate, but pH adjustment is required to minimize chlorate formation. The use of a reducing agent like thiosulfate or ferrous ion can complicate the application of postdisinfectants.

*Ozone.* Ozone is a less commonly used disinfectant in drinking water treatment in the United States. Among the many benefits of ozonation of drinking water are effective inactivation of microbes, taste and odor control, iron and manganese control, oxidation of DBP precursors, and the enhancement of biological oxidation in filters. However, ozone results in the formation of bromate and of biodegradable organic matter (BOM). Bromate is regulated under the D/DBP Rule. BOM includes ozonation by-products (OBPs) like aldehydes, keto acids, carboxylic acids, AOC, and biodegradable dissolved organic carbon (BDOC). These OBPs may be responsible for regrowth of bacteria in distribution systems and can be controlled in-plant if biological oxidation is allowed to occur in downstream filters. (Refer to the section of DBP control through biological filtration.)

The formation of halogen-containing DBPs by ozone is significantly lower than that by free chlorine. Ozone can form bromo-DBPs like bromoform ($CHBr_3$), bromoacetic acid, and dibromoacetic acid, but at relatively low concentrations. Ozone oxidizes DBP precursors such that lower concentrations of TTHM, HAA6, HAN4 (haloacetonitrile 4), and TOX (total organic halide) are formed with subsequent chlorination. However, preozonation alters the nature of the precursors to the extent that higher concentrations of chloral hydrate, chloropicrin, and 1,1,1-TCP are formed with subsequent chlorination (Miltner, 1990).

Ozone converts portions of the humic fraction to nonhumic compounds and converts portions of the higher-molecular-weight fraction to lower-molecular-weight compounds. Examples of lower-molecular-weight materials formed by ozone are aldehydes, keto acids, AOC, and BDOC. Concentrations of these may be appreciable and necessitate control to ensure distribution system biostability. Generally, much of the formation of these OBPs occurs at lower ozone doses. Ozone staging (when it is applied in the treatment plant) can influence OBP formation. Ozonation of raw water results in higher OBP formation than ozonation of downstream waters in which some ozone demand has been removed (Miltner, 1993).

NOM and bromide are the two principal precursors of DBPs in raw water supplies. Bromate can result from the use of ozone. Bromate concentration increases with increasing dissolved ozone residual and with increasing bromide. Bromide control strategies may involve more energy-intensive membrane processes, such as RO. Singer (2006) pointed out that source water protection involves creative solutions to minimizing saltwater intrusion into our raw (surface and ground) water supplies because such saltwater intrusion is the primary source of bromide.

*Ultraviolet (UV) irradiation.* Carlson et al. (1985) reported that cysts of *Giardia muris* were significantly more resistant to UV treatment than *E. coli* or *Yersinia* spp. Results suggested that hydraulic short circuiting and entrapped air in UV reactors may decrease the efficiency of inactivation. *G. lamblia* cysts were found to be resistant to high doses of germicidal UV irradiation (Rice and Hoff, 1981). There has been renewed interest in the use of UV light for inactivating waterborne protozoan parasites. A study suggested that UV irradiation is an effective treatment option for inactivating oocysts of *Crytosporidium* (Clancy et al., 1998). UV irradiation was not considered for controlling DBP. Currently, due to the improvement of technology, UV has been commercially available to be used to control DBP.

**Enhanced coagulation.** Coagulation has historically been used for the control of particulates in drinking water, while simultaneously controlling organic carbon. With the D/DBP Rule, many water systems will move from conventional to enhanced coagulation and expand their coagulation objectives from removing turbidity to removing total organic carbon (TOC) as well. It is anticipated that many systems will be able to meet the requirements of enhanced coagulation for TOC removal with only moderate changes in conventional coagulation. Making the change from conventional to TOC-optimized coagulation generally results in improved removal of heterotrophic plate count (HPC) bacteria, TC, FC, FS, viruses, *Cryptosporidium parvum* oocysts (3.1 to 7.0 μm), *Cryptosporidium* oocyst-sized particles, *Giardia* cyst-sized particles (8.2 to 13.2 μm), total plate count (TPC) organisms, and bacterial endospores.

The removal of TOC results in improved removal of precursors for TTHM, HAA6, CH, HAN4, CP, the surrogate TOX, and chlorine demand (Miltner, 1994; Dryfuse *et al.,* 1995). When coagulated, raw waters higher in TOC and in specific ultraviolet absorbance tend to have higher percent removals of DBP precursors. As a treatment technique, enhanced coagulation to remove TOC can result in the removal of precursors for these DBPs. As a BAT, enhanced coagulation can result in the control of TTHM and HAA5. Many systems should be able to meet the requirements of enhanced coagulation for TOC removal with moderate changes to conventional coagulation.

Most DOC and DBP precursors are in the larger-molecular-size range and are in the humic fraction. Conventional coagulation removes a greater percentage of the >3K MS (molecular size) fraction than the smaller-sized fractions. Enhancing coagulation brings about small improvements in the >3K MS range and the <0.5K MS range; the greatest improvement with enhanced coagulation is in the 0.5K to 3K MS range (Dryfuse *et al.,* 1995). Conventional coagulation removes a greater percentage of the humic fraction than the nonhumic fraction. Enhancing coagulation brings about similar improvement in the removal of both fractions.

A general concern associated with coagulation is that, although it lowers the concentrations of DBP precursor, it shifts the distribution of the DBPs formed by chlorination toward the more brominated species. This shift becomes even greater when enhanced or optimized coagulations practiced.

Water systems switching from conventional to enhanced coagulation may achieve longer filter run times (FRTs), but the trade-off will be greater amounts of sludge production. Systems practicing enhanced coagulation should also consider pH adjustment ahead of the filter to achieve longer FRTs. The disadvantage with enhanced coagulation is that when alum is the coagulant, higher levels of dissolved aluminum will enter the distribution system.

**Filtration.** The US EPA researches have been conducted with the use of bench-, pilot-, and field-scale studies to investigate various aspects of surface water filtration for microbiological removal with various granular media. Studies found that two of these technologies, slow sand filtration (SSF) and diatomaceous earth (DE) filtration, are especially applicable to small systems. In addition, the application of granular filtration, which is utilized by medium to large systems, has been investigated for removal of *Giardia* and *Cryptosporidium.* These studies have considered the various operational conditions that enhance removal efficiency as well as the conditions under which removal efficiency deteriorates.

In the period between 1980 and 1990, *Giardia* cysts and *Cryptosporidium* oocysts were the primary organisms being considered. The effect of various water quality conditions and particle and pathogen loadings were evaluated. It was shown that, in combination with effective chemical addition and coagulation, these processes are capable of efficiently removing efficiency in SSF and conventional filtration, but had little effect n DE performance. Filtration processes are capable of removing high levels of those organisms, but that removal is dependent on the operation of those processes. Poor chemical addition and coagulation reduced the treatment efficiency for removing microorganism and other particles. Temperature was another factor that affected removal. The filtration studies conducted by US EPA concluded that good turbidity reduction and good particle removal paralleled good microorganism removal.

One measure of filtration efficiency developed by Water Supply and Water Resources Division (WSWRD), and now widely used, was that of measuring aerobic endospore removal in a water treatment process. Although not a surrogate for removal of a specific organism, endospore removal was seen as a measure of the overall filtration performance. Studies conducted by the WSWRD showed that endospore removal tracked particle removal, turbidity reduction, and pathogen removal. In studies, if good removal of endospores was achieved, good removal of particle and pathogens was observed. In some cases, good removal of pathogens occurred when poor removal of endospores was demonstrated, but no studies showed poor removal of pathogens when good removal of spores was achieved. Thus, removal of spores may be considered a conservative measure of particle and pathogen removal.

**Biological filtration for DBP control.** DBP control through biological filtration or biofiltration is defined as the removal of DBP precursor material (PM) by bacteria attached to the filter media. Dissolved organic matter (DOM), which is a part of the PM, is utilized by the filter bacteria as a substrate for cell maintenance, growth, and replication. This effect makes the PM utilized by bacteria unavailable to react with chlorine to form DBPs. The prerequisite for maximizing bacterial substrate utilization in filters is the absence of chlorine in the filter influent or backwash water.

Sand, anthracite, or GAC can be colonized by bacteria. Since anthracite and sand are considered inert because neither interacts chemically with PM, removal of PM is due solely to biological activity. GAC will initially remove DOM through adsorption and biological substrate utilization until its adsorptive capacity has been exhausted. After that point, PM removal is achieved only through substrate utilization, and the GAC is defined as biological activated carbon (BAC). All drinking water filters will become biologically active in the absence of applied disinfectant residuals. The process of biological colonization and substrate utilization is enhanced by ozonating filter influent water.

Data collected, to date, indicate that biologically active filters remove significant amounts of PM and that preozonated biofilters remove more PM than do nonozonated filters. The resulting reductions in trihalomethane formation potential (THMFP) and haloacetic acid formation potential (HAAFP) should help many drinking water utilities meet the 80 (THM) and 60 (HAA) μg/L limits mandated under the Stage II D/DBP Rule (US EPA, 2001).

**Use of GAC.** GAC can be used as part of a multimedia filter to remove particulates (filter adsorber) or a postfilter to remove specific contaminants (postfilter adsorber). When used in a filter adsorber mode, the filters are backwashed periodically to alleviate head loss, but the carbon itself is regenerated infrequently, it at all. When used in a postfilter mode, the carbon bed is rarely backwashed and is regenerated as often as needed to control the contaminant(s) of interest.

Activated carbon in a filter adsorber application removes pathogens by the same mechanisms as any other filter media. It does not remove particulates/pathogens to any greater degree than other filter media types; so, it is never recommended for particulate/pathogen removal alone.

Activated carbon is an effective process for removing DBP precursors. It is not designed for pathogen removal. The effectiveness for precursor removal is dependent on a number of design issues such as carbon type, filter location, filter depth, filter flow rate, and blending choices. It is also dependent on a number of water quality issues such as initial precursor concentration, precursor adsorbility, precursor adsorption kinetics, temperature, dissolved oxygen levels, pH, and bromide concentration. Although much progress has been made on the modeling of GAC system performance, the applicability of this technology to any drinking water utility will need to be determined by pilot testing (US EPA, 2001).

**Use of membranes.** Certain type of membranes can be very effective for controlling DBPs, while others are specifically designed to remove particulates/pathogens. For example, reverse osmosis (RO) membranes are very tight and are typically used to remove salts from seawater and brackish waters. Due to their tight membrane structure, they require high pressures to operate effectively.

Nonofiltration (NF) membranes are not as tight as RO membranes, but have been found to remove a large percentage of DBP precursors. Because they are not as tight, they can be operated at lower pressures (typically 5 to 9 bars) than RO membranes while achieving the same, or greater, flux. Ultrafiltration (UF) and microfiltration (MF) membranes are typically used for particulate/pathogen removal only. Details of MF, UF, NF, and RO are discussed in the membrane filtration section.

Speth (1998) concludes that activated carbon is an effective process for removing DBP precursors, but is not effective for pathogen removal. RO and NF are effective processes for removing DBP precursors, and UF and MF membranes are excellent for removing pathogens and particulates and, under some conditions, could be considered as a replacement for conventional filtration.

Membrane technologies are effective processes for removing DBP precursors. As stated above, MF and UF membranes are excellent for removing pathogens. NF and RO membranes are excellent for removing DBP precursors, but are not counted on as a pathogen barrier due to blending and glue-line-failure issues. These problems have been overcome by new membrane technologies. Reducing the effect of membrane foulants is a main consideration in the design of NF and RO plants, especially for plants treating surface waters. Although much progress has been made on modeling and scaling up small-membrane results, the applicability of this technology will need to be determined by pilot testing (US EPA, 2001).

**For small systems.** Small water supply systems have many problems that make compliance with drinking water standards more difficult than for medium and large systems. The pilot- and full-scale research efforts have been made to address some of these needs. Because small systems often lack of the financial, technical, and managerial capabilities of larger systems and are responsible for the majority of the SDWA violations, they have been targeted in several federal rules and regulations.

Filtration and disinfection of water supplies are highly effective public health practices. EPA's in-house research has focused primarily on filtration and disinfection technologies that are considered to be viable alternatives to conventional package plants (flocculation, coagulation, media filtration, postchlorination). Conventional package plants are site-specific and require a high level of operator skill to properly maintain appropriate chemical dosage and flow rates, especially when used to treat surface water. These difficulties, in conjunction with the other small system problems, have results. Field demonstration projects have been used to characterize some of the problems that can occur for even the best technology when conditions are not optimal. The remote monitoring and control research efforts result from the fact that many rural systems are located in topographically difficult areas or separated by large distances from other systems, thus precluding any consolidation or regionalization efforts. The software and sensing systems developed as a product of this research will allow individual treatment units to be monitored and operated from a certain location. This approach is called the *electric circuit rider concept.* One technique that has promise for improving the effectiveness of systems in the field is the use of supervisory control and data acquisition system (US EPA, 2001).

Results from the EPA research indicate that MF, UF, and RO systems are effective technologies for the removal of pathogens while still being affordable for small systems. New disinfection technologies appear to provide improvements over current systems in handling chemicals and consistency of performance. This is an area undergoing rapid change. Many organisms are readily removed and inactivated in the laboratory, but under field conditions, the same effectiveness cannot be taken for granted.

**Control in distribution system.** Virtually for any surface contact with the water in a distribution system, one can find bioflilms. Biofilms are formed in distribution system pipelines when microbial cells attach to pipe surface and multiply to form a film or slime layer on the pipe. Probably within seconds of entering the distribution system, large particles, including microorganisms, adsorb to the clean pipe surface. Some microorganisms can adhere directly to the pipe surface via appendages that extend from the cell membrane; other bacteria form a capsular material of extracellular polysaccharides (EPS), sometimes called a glycocalyx, that anchors the bacteria to the pipe surface (Geldreich, 1988). The organisms take advantage of the macromolecules attached to the pipe surface for protection and nourishment. The water flowing past carries nutrients (carbon-containing molecules, as well as other elements) that are essential for the organisms' survival and growth (US EPA, 1992).

Biofilms are complex and dynamic microenvironments, encompassing processes such as metabolism, growth, and product formation, and finally detachment, erosion, or "soughing" of the biofilm from the surface. The rate of biofilm formation and its release into a distribution system can be affected by many factors, including surface layers begin to slough off into the water (Geldreich and Rice, 1987). The pieces of biofilm released into the water may continue to provide protection for the organisms until they can colonize a new section of the distribution system.

The most common organisms found in biofilms are nonpathogenic heterotrophic bacteria. Some bacteria that live in biofilms may cause esthetic problems with water quality, including off-tastes, odors, and colored water problems. Few organisms living in distribution system biofilms pose a threat to the average consumer. Bacteria, viruses, fungi, protozoa, and other invertebrates have been isolated from drinking water biofilms (US EPA, 1992). The fact that such organisms are present within distribution system biofilms shows that, although water treatment is intended to remove all pathogenic bacteria, treatment does not produce a sterile water. Consequently, if opportunistic pathogens survive in biofilms, they could potentially cause disease in individuals with low-immunity or compromised immune systems. Factors related to increased survival of bacteria in chlorinated water include attachment to surfaces, encapsulation, aggregation, low-nutrient growth conditions, and strain variation.

Research has been conducted using the simulators to determine the effect of water quality changes on biofilms in the distribution systems. Results have indicated that biofilms are resistant to disinfection regardless of the agent used. Other work has demonstrated that lowering system pH can reduce biofilm densities; however, such an effect is transient if system pHs are returned to normal operating ranges (US EPA, 2001). It is reported that, recently, the Denver City Water developed an uncommon procedure—ozone disinfection of water mains for common field operation.

**Cost of control.** Chapter 14 (US EPA, 2001) discusses the conditions and causes of the formation of DBPs, and compares the various treatment techniques and associated costs for both controlling DBPs and ensuring microbial safety. Making direct comparisons among the various alternatives is difficult. For example, moving the point-of-disinfection (chlorination) would seem to be the lowest cost option. NF, although the most expensive technology for precursor removal, has the advantage of removing other contaminants such as total dissolved solids and various inorganics. Therefore, it might be used for achieving other treatment goals in addition to removing DBP precursors. For example, NF effectively removed microorganisms, thus serving as an alternative for chemical disinfection. Although the cost of enhanced coagulation was not evaluated, it could be very effective if a utility is only slightly out of compliance. However, in addition to increased coagulation costs, an additional cost may be associated with sludge handling. Clearly, modifying the disinfection process is the lowest cost option for controlling DBPs, but as noted, there are by-products and problems associated with the use of some of the alternate disinfectants. For example, chloramination is not as good a disinfectant as chlorine, and ozone may enhance regrowth of some organisms.

Retrofitting may be fairly easy with chloramination. For example, to switch from chlorine to chloramines may only require the addition of ammonia feed equipment. However, the use of ozone will require construction of expensive ozone contactors. The use of chlorine dioxide will probably require the use of a reducing agent such as ferrous chloride, which was not included in this costing analysis. The technologies discussed would normally be applied incrementally to a utility's existing treatment. Therefore, the base cost associated with an assumed conventional treatment system and the incremental costs associated with various DBP control alternatives have been summarized in Table 5.14 (US EPA, 2001).

**TABLE 5.14 Incremental Costs of Disinfection By-Product Control in c/m³ (c/1000 gal)**

| | Design flow in m³/d (average flow) [design flow in MGD (average flow)] | | | |
|---|---|---|---|---|
| | 387.5 (189) | 3785 (1890) | 37,850 (26,495) | 378,500 (264,950) |
| Treatment process | [0.10 (0.05)] | [1.0 (0.50)] | [10,0 (7.0)] | [100 (70)] |
| Conventional treatment | 142 (539) | 47.5 (180) | 12.6 (48) | 8.51 (32) |
| Conventional treatment and nanofiltration | 205 (778) | 88.7 (336) | 40.4 (882) | 32.1 (122) |
| Conventional treatment plus GAC ($C_e$ = 100 mg/L)* | 203 (772)– 216 (818) | 70.8 (268)– 82.1 (311) | 28.3 (101)– 32.7 (124) | 15.7 (150)– 19.2 (73) |
| Conventional treatment plus GAC ($C_e$ = 50 mg/L) | 206 (781)– 226 (856) | 73.1 (276)– 91.2 (345) | 29.4 (111)– 36.0 (136) | 16.2 (61)– 22.5 (85) |
| Conventional treatment plus | | | | |
| -chlorine/chloramines | 147 (556) | 47.8 (181) | 12.6 (48) | 8.5 (32) |
| -ozone/chlorine | 163 (619) | 49.7 (188) | 14.0 (53) | 9.3 (35) |
| -ozone/chloramines | 168 (636) | 50.0 (189) | 14.0 (53) | 9.3 (35) |
| -chlorine dioxide/chlorine | 167 (633) | 48.9 (185) | 12.6 (48) | 8.5 (32) |
| -chlorine dioxide/chloramine | 172 (651) | 49.1 (186) | 12.9 (49) | 8.5 (32) |

*GAC: granular activated carbon, $C_e$: effluent concentration
SOURCE: US EPA, 2001

Regulations to control contaminants in drinking water in the United States are expected to become more and more stringent. Forthcoming DBP regulations will affect virtually every community water system in the United States. There are various ways to control DBPs, one of which is to use an alternative to chlorine to control halogenated by-products. However, when this is done, one has to consider the consequences. The unit processes considered are those that are effective for precursor removal or for the use of alternate disinfectants. More efficient treatment will be required to meet future regulations. Also, water treatment managers will have to become more knowledgeable about various treatment options that are cost-effective in order for them to meet present and future regulations. Although essentially exempt in the past, small water systems will be required to comply with future regulations. Singer (2006) claimed that the ultimate solution to DBPs will be more expensive and will necessitate the use of more aggressive management and control strategies.

## 19 Water Fluoridation

Fluoride occurs naturally in water at low concentrations. Water fluoridation is the intentional addition of fluoride to drinking water. During the past 5 decades, water fluoridation has been proven to be both wise and a most cost-effective method to prevent dental decay. The ratio of benefits (reductions in dental bills) to cost of water fluoridation is 50:1 (US Public Health Service, 1984). In order to prevent decay, the optimum fluoride concentration has been established at 1mg/L (American Dental Association, 1980). The benefits of fluoridation last a lifetime. While it is true that children reap the greatest benefits from fluoridation, adults benefit also. Although there are benefits, the opponents of water fluoridation exist. The charges and the facts are discussed in the manual (US PHS, 1984).

Fluoride in drinking water is regulated under section 1412 of the Safe Drinking Water Act (SDWA). In 1986, US EPA promulgated an enforceable standard, a Maximum Contamination Level (MCL) of 4 mg/L for fluoride. This level is considered as protective of crippling skeletal fluorosis, an adverse health effect. A review of human data led the EPA (1990) to conclude that there was no evidence that fluoride in water presented a cancer risk in humans. A nonenforceable Secondary Maximum Contamination Level (SMCL) of 2 mg/L was set to protect against objectionable dental fluorisis, a cosmetic effect.

## 19.1 Fluoride chemicals

Fluoride is a pale yellow noxious gaseous halogen. It is the thirteenth most abundant element in the earth's crust and is not found in a free state in nature. The three most commonly used fluoride compounds in the water system are hydrofluosilicic acid ($H_2SiF_6$), sodium fluoride (NaF), and sodium silicofluoride ($Na_2SiF_6$). Fluoride chemicals, like chlorine, caustic soda, and many other chemicals used in water treatment can cause a safety hazard. The operators should observe the safe handling of the chemical.

The three commonly used fluoride chemicals have virtually 100 % dissociation (Reeves, 1986, 1990):

$$NaF \leftrightarrow Na^+ + F^- \tag{5.186}$$

$$Na_2SiF_6 \leftrightarrow 2Na^+ + SiF_6^{2-} \tag{5.187}$$

The $SiF_6$ radical will be dissociated in two ways: hydrolysis mostly dissociates very slowly

$$SiF_6^{2-} + 2H_2O \leftrightarrow 4H^+ + 6F^- + SiO_2\downarrow \tag{5.188}$$

and/or

$$SiF_6^{2-} \leftrightarrow 2F^- + SiF_4\uparrow \tag{5.189}$$

Silicon tetrafluoride ($SiO_4$) is a gas which will easily volatilize out of water when present in high concentrations. It also reacts quickly with water to form silicic acid and silica ($SiO_2$):

$$SiF_4 + 3H_2O \leftrightarrow 4HF + H_2SiO_3 \tag{5.190}$$

and

$$SiF_4 + 2H_2O \leftrightarrow 4HF + SiO_2 \downarrow \tag{5.191}$$

then

$$HF \leftrightarrow H^+ + F^- \tag{5.192}$$

Hydrofluosilicic acid has a dissociation very similar to sodium silicofluoride:

$$H_2SiF_6 \leftrightarrow 2HF + SiF_4\uparrow \tag{5.193}$$

then it follows Eqs. (5.190), (5.191), and (5.192).

Hydrofluoric acid (HF) is very volatile and will attack glass and electrical parts and will tend to evaporate in high concentrations.

### 19.2 Optimal fluoride concentrations

The manual (Reeves, 1986) presents the recommended optimal fluoride levels for fluoridated water supply systems based on the average and the maximum daily air temperature in the area of the involved school and community. For community water system in the United States the recommended fluoride levels are from 0.7 mg/L in the south to 1.2 mg/L to the north of the country. The recommended fluoride concentrations for schools are from four to five times greater than that for communities.

The optimal fluoride concentration can be calculated by the following equation:

$$F\,(\text{mg/L}) = 0.34/E \tag{5.194}$$

where $E$ is the estimated average daily water consumption for children through age 10, in ounces of water per pound of body weight. The value of $E$ can be computed from

$$E = 0.038 + 0.0062\,T \tag{5.195}$$

where $T$ is the annual average of maximum daily air temperature in degrees Fahrenheit.

**Example 1:** The annual average of maximum daily air temperature of a city is 60.5°F (15.8°C). What should be the recommended optimal fluoride concentration?

**solution:**

Step 1. Determine $E$ by Eq. (5.195)

$$E = 0.038 + 0.0062T = 0.038 + 0.0062 \times 60.5 = 0.413$$

Step 2. Calculate recommended $F$ by Eq. (5.194)

$$F = 0.34/E = 0.34/0.413 = 0.82 \text{ (mg/L)}$$

**Example 2:** Calculate the dosage of fluoride to the water plant in Example 1, if the naturally occurring fluoride concentration is 0.03 mg/L.

**solution:** The dosage can be obtained by subtracting the natural fluoride level in water from the desired concentration:

$$\text{Dosage} = 0.82 \text{ mg/L} - 0.03 \text{ mg/L} = 0.79 \text{ mg/L}$$

As for other chemicals used in water treatment, the fluoride chemicals are not 100% pure. The purity of a chemical is available from the manufacturers. The information on the molecular weight, purity ($p$), and the available fluoride ion (AFI) concentration for the three commonly used fluoride chemicals are listed in Table 5.15. The AFI is determined by the weight of fluoride portion divided by the molecular weight.

**Example 3:** What is the percent available fluoride in the commercial hydrofluosilicic acid (purity, $p = 23\%$)?

**solution:** From Table 5.15

$$\begin{aligned} \%\text{ available } F &= p \times \text{AFI} = 23 \times 0.792 \\ &= 18.2 \end{aligned} \tag{5.196}$$

### 19.3 Fluoride feed rate (dry)

Fluoride can be fed into treated water by (1) dry feeders with a mixing tank (for $Na_2SiF_6$), (2) direct solution feeder for $H_2SiF_6$, (3) saturated

**TABLE 5.15 Purity and Available Fluoride Ion (AFI) Concentrations for the Three Commonly Used Chemicals for Water Fluoridation**

| Chemical | Molecular weight | Purity, % | AFI |
|---|---|---|---|
| Sodium fluoride, NaF | 42 | 98 | 0.452 |
| Sodium silicofluoride, $Na_2SiF_6$ | 188 | 98.5 | 0.606 |
| Hydrofluosilicic acid, $H_2SiF_6$ | 144 | 23 | 0.792 |

solution of NF, and (4) unsaturated solution of NF or $Na_2SiF_6$. The feed rate, FR, can be calculated as follows:

$$\text{FR (lb/d)} = \frac{\text{dosage (mg/L)} \times \text{plant flow (MGD)} \times 8.34\ \text{lb/gal}}{p \times \text{AFI}} \tag{5.197a}$$

or

$$\text{FR (lb/min)} = \frac{\text{(mg/L)} \times \text{gpm} \times 8.34\ \text{lb/gal}}{1{,}000{,}000 \times p \times \text{AFI}} \tag{5.197b}$$

**Example 1:** Calculate the fluoride feed rate in lb/d and g/min, if 1.0 mg/L of fluoride is required using sodium silicofluoride in a 10 MGD (6944 gpm, 0.438 $m^3/s$) water plant assuming zero natural fluoride content.

**solution:** Using Eq. (5.197a) and data from Table 5.15, feed rate FR is

$$\begin{aligned} \text{FR} &= \frac{1.0\ \text{mg/L} \times 10\text{MGD} \times 8.34\ \text{lb/gal}}{0.985 \times 0.606} \\ &= 139.7\ \text{lb/d} \\ \text{or} \quad &= 139.7\ \text{lb/d} \times 453.6\ \text{g/lb} \times (1\ \text{d}/1440\ \text{min}) \\ &= 44.0\ \text{g/min} \end{aligned}$$

**Example 2:** A water plant has a flow of 1600 gpm (0.1$m^3/s$). The fluoride concentration of the finished water requires 0.9 mg/L using sodium fluoride in a dry feeder. Determine the feed rate if 0.1 mg/L natural fluoride is in the water.

**solution:** Using Eq. (5.197), and data from Table 5.15

$$\begin{aligned} \text{F} &= \frac{(0.9 - 0.1)\text{mg/L} \times 1600\text{gpm} \times 8.34\ \text{lb/gal}}{1{,}000{,}000 \times 0.98 \times 0.452} \\ &= 0.024\ \text{lb/min} \\ \text{or} \quad &= 10.9\ \text{g/min} \end{aligned}$$

**Example 3:** If a water plant treats 3500 gpm (5 MGD, 0.219 $m^2/s$) of water with a natural fluoride level of 0.1 mg/L. Find the hydrofluosilicic acid feed rate (mL/min) with the desired fluoride level of 1.1 mg/L.

**solution:**

Step 1. Calculate FR in gal/min

$$\begin{aligned} \text{FR} &= \frac{(1.1 - 0.1)\ \text{mg/L} \times 3500\ \text{gpm}}{1{,}000{,}000 \times 0.23 \times 0.792} \\ &= 0.0192\ \text{gpm} \end{aligned}$$

Step 2. Convert gpm into mL/min

$$\begin{aligned}\text{FR} &= 0.0192 \text{ gpm}\\ &= 0.0192 \text{ gal/min} \times 3785 \text{ mL/gal}\\ &= 72.7 \text{ mL/min}\end{aligned}$$

### 19.4 Fluoride feed rate for saturator

In practice 40 g of sodium fluoride will dissolve in 1 L of water. It gives 40,000 mg/L (4% solution) of NF. The AFI is 0.45. Thus the concentration of fluoride in the saturator is 18,000 mg/L (40,000 mg/L × 0.45). The sodium fluoride feed rate can be calculated as

$$\text{FR(gpm or L/min)} = \frac{\text{dosage(mg/L)} \times \text{plant flow(gpm or L/min)}}{18{,}000\,\text{mg/L}} \quad (5.198)$$

**Example 1:** In a 1 MGD plant with 0.2 mg/L natural fluoride, what would the fluoride (NF saturator) feed rate be to maintain 1.2 mg/L of fluoride in the water?

**solution:** By Eq. (5.198)

$$\begin{aligned}\text{FR} &= \frac{(1.2 - 0.2)\text{ mg/L} \times 1{,}000{,}000\text{ gpd}}{18{,}000\text{ mg/L}}\\ &= 55.6\text{ gpd}\end{aligned}$$

*Note*: Approximately 56 gal of saturated NF solution to treat 1 Mgal of water at 1.0 mg/L dosage.

**Example 2:** Convert the NF feed rate in Example 1 to mL/min.

**solution:**

$$\begin{aligned}\text{FR} &= 55.6\text{ gal/d} \times 3785\text{ mL/gal} \times (1\text{d}/1440\text{ min})\\ &= 146\text{ mL/min}\end{aligned}$$

or

$$= 2.5\text{ mL/s}$$

*Note*: Approximately 2.5 mL/s (or 150 mL/min) of saturated sodium fluoride solution to 1 MGD flow to obtain 1.0 mg/L of fluoride concentration.

### 19.5 Fluoride dosage

Fluoride dosage can be calculated from Eqs. (5.196) and (5.198) by rearranging those equations as follows:

$$\text{Dosage (mg/L)} = \frac{\text{FR(lb/d)} \times p \times \text{AFI}}{\text{plant flow(MGD)} \times 8.34\text{ lb/(Mgal} \cdot \text{mg/L)}} \quad (5.199)$$

and for the saturator:

$$\text{Dosage(mg/L)} = \frac{\text{FR(gpm)} \times 18{,}000 \text{ mg/L}}{\text{plant flow(gpm)}} \tag{5.200}$$

For a solution concentration, $C$, of unsaturated NF solution

$$C = \frac{18{,}000 \text{ mg/L} \times \text{ solution strength(\%)}}{4\%} \tag{5.201}$$

**Example 1:** A water plant (1 MGD) feeds 22 lb/d of sodium fluoride into the water. Calculate the fluoride dosage.

**solution:** Using Eq. (5.199) with Table 5.15:

$$\begin{aligned}\text{Dosage} &= \frac{22\,\text{lb/d} \times 0.98 \times 0.452}{1{,}000{,}000\,\text{gal/d} \times 8.34\,\text{lb/(Mgal}\cdot\text{mg/L)}}\\ &= 1.17\,\text{mg/L}\end{aligned}$$

**Example 2:** A water plant adds 5 gal of sodium fluoride from its saturator to treat 100,000 gal of water. Find the dosage of the solution.

**solution:** Using Eq. (5.200)

$$\begin{aligned}\text{Dosage} &= \frac{5 \text{ gal} \times 18{,}000 \text{ mg/L}}{100{,}000 \text{ gal}}\\ &= 0.9 \text{ mg/L}\end{aligned}$$

**Example 3:** A water system uses 400 gpd of a 2.8% solution of sodium fluoride in treating 500,000 gpd of water. Calculate the fluoride dosage.

**solution:**

Step 1. Find fluoride concentration in NF solution using Eq. (5.201)

$$\begin{aligned}C &= \frac{18{,}000 \text{ mg/L} \times 2.8\%}{4\%}\\ &= 12{,}600 \text{ mg/L}\end{aligned}$$

Step 2. Calculate fluoride dosage by Eq. (5.200)

$$\begin{aligned}\text{Dosage} &= \frac{400 \text{ gpd} \times 12{,}600 \text{ mg/L}}{500{,}000 \text{ gpd}}\\ &= 10.08 \text{ mg/L}\end{aligned}$$

**Example 4:** A water system adds 6.5 lb/d per day of sodium silicofluoride to fluoridate 0.5 MGD of water. What is the fluoride dosage?

**solution:** Using Eq. (5.199), $p = 0.985$, and AFI = 0.606

$$\text{Dosage} = \frac{6.5 \text{ lb/d} \times 0.985 \times 0.606}{0.5 \text{ MGD} \times 8.34 \text{ lb/(Mgal}\cdot\text{mg/L)}}$$
$$= 0.93 \text{ mg/L}$$

**Example 5:** A water plant uses 2.0 lb of sodium silicofluoride to fluoridate 160,000 gal water. What is the fluoride dosage?

**solution:**

$$\text{Dosage} = \frac{2.0 \text{lb} \times 0.985 \times 0.606 \times 0.606 \times 10^6 \text{ mg/L}}{0.16 \text{ Mgal/d} \times 8.34 \text{ lb/(Mgal}\cdot\text{mg/L)}}$$
$$= 0.9 \text{ mg/L}$$

**Example 6:** A water plant feeds 100 lb of hydrofluosilicic acid of 23% purity during 5 days to fluoridate 0.437 MGD of water. Calculate the fluoride dosage.

**solution:**

$$\text{FR} = 100\text{lb}/5 \text{ d} = 20 \text{ lb/d}$$

$$\text{Dosage} = \frac{\text{FR} \times p \times \text{AFI}}{(\text{MGD}) \times 8.34 \text{ lb/(Mgal}\cdot\text{mg/L)}}$$
$$= \frac{20 \text{ lb/d} \times 0.23 \times 0.792}{0.437 \text{ MGD} \times 8.34 \text{ lb/(Mgal}\cdot\text{mg/L)}} \quad \text{(obtain AFI} = 0.792 \text{ from Table 5.15)}$$
$$= 1.0 \text{ mg/L}$$

**Example 7:** A water system uses 1200 lb of 25% hydrofluosilicic acid in treating 26 Mgal of water. The natural fluoride level in the water is 0.1 mg/L. What is the final fluoride concentration in the finished water.

**solution:**

$$\text{Plant dosage} = \frac{1200 \text{ lb} \times 0.25 \times 0.792}{26 \times 10^6 \text{gal} \times 8.34 \text{ lb/(Mgal}\cdot\text{mg/L)}}$$
$$= 1.1 \text{ mg/L}$$
$$\text{Final floride} = (1.1 + 0.1) \text{ mg/L}$$
$$= 1.2 \text{ mg/L}$$

## 20 Health Risks

Human health risk is the probability that a given exposure or a series of exposures may have or will damage the health of individuals exposed. This section discusses the possible risk of drinking water, and assessing and managing risks by the public water suppliers through the rules and guidelines of the US EPA. This section specifically covers assessing and managing risks, radionuclides, and the value of health advisories concerning drinking water.

## 20.1 Risk

Risk is the potential for realization of unwanted adverse consequences or events. In general terms, human health risk is the probability of injury, disease, or death under a given chemical or biological exposure or under series of exposures. Risk may be expressed in quantitative term (zero to one). In many cases, it can only be described as, high, low, or trivial.

We do not live in a risk-free world, but in a chemical world. There are more than 65,000 chemicals produced, and they are increasing in number every year. Through use and abuse, many of those chemical products will end up in our environment—water, air, and land. These chemicals include organics and inorganics that are used in industries (including water treatment plants), pharmaceuticals, agricultures (insecticides), home, personal cosmic purposes, etc. Even terrorism may threaten our drinking water with chemical contamination.

Certain areas on the earth's crest and rocks contain high levels of some naturally occurring chemical elements, such as lead, mercury, fluoride, and sulfur. In addition, radionuclides such as natural radium, $Ra^{226}$, $Ra^{228}$; radon, Rn; uranium, U, etc.; and man-made radioactive substances occur throughout the world. Many contaminants may end up in source waters. Trace amounts of these contaminants might be present in the drinking water.

All human activities involve some degree of risk. Table 5.16 lists the risks of some common activities. In addition, pathogen contamination in food and drinking liquids poses risks. Waterborne disease outbreaks *(Giardia, Cryptosporidium,* acute gastroenteritis, *E. Coli, etc.)* have occurred in the past and can threaten at anytime (Hass, 2002; Lin, 2002).

Sand filtration, disinfection, and application of drinking water standards reduce waterborne diseases to protect public health. It was discovered that DBPs result from the reaction of chlorine with natural organic matter (NOM) in source water. The risk of carcinogenic DBPs is a dilemma for water utilities. Alternative treatment measures and microbial control have to be properly managed (see Section 19).

The term "safe," in its common usage, means "without risk." In technical terms, however, this common usage is misleading because science cannot ascertain the conditions under which a given chemical exposure is likely to be absolutely without a risk of any type.

Human health risk is the likelihood (or probability) that a given chemical exposure or series of exposures may damage the health of exposed individuals. Chemical risk assessment involves the complex analysis of

**TABLE 5.16 Annual Risk of Death from Selected Common Human Activities**

| Activities | Individual risk, per year | Lifetime risk* |
|---|---|---|
| Automobile accident | 1/4500 | 1/65 |
| Lightening | 1/2000,000 | |
| Coal mining | | |
| Accident | 1/770 | 1/17 |
| Black lung disease | 1/125 | 1/3 |
| Truck driving accident | 1/10,000 | 1/222 |
| Falls | 1/13,000 | 1/186 |
| Home accidents | 1/83,000 | 1/130 |
| Others | 1/1,000,000 | |
| Smoking | | |
| Flying | | |
| Drinking diet soda | | |
| Living near nuclear power plant | | |
| Living in stone or brick building | | |
| Use of microwaves | | |
| Eating 100 charcoal broiled steaks | | |

*Calculated based on 70-year lifetime and 45-year work exposure
SOURCE: US EPA, 1986; R. Begole, State Farm Insurance, personal communication

exposures that have taken place in the past, the adverse health effects of which may or may not have already occurred. It also involves prediction of the likely consequences of exposures that have not yet occurred.

## 20.2 Risk assessment

Risk assessment is a quantitative evaluation process of health or/and environmental risks determining the potential risks associated with exposure to a type of human hazard—physical, chemical, or biological. There are four components to every (complete) risk assessment (US EPA, 1986):

1. Hazard identification:
   - Review and analyze toxic data (animal and human studies, and negative epidemiological studies).
   - Weigh the evidence that a substance causes various toxic effects.
   - Evaluate whether toxic effects in one setting will occur in other settings.

2. Dose-response evaluation:
   - Perform an estimate of the quantitative relationship between the amount of exposure to a hazardous identified substance and the extent of toxic injury or disease.
   - Extrapolate from high dose to low dose.
   - Extrapolate test animals to humans.

3. Human exposure evaluation:
   - Investigate the inhalation, ingestion, and skin contact of a toxic chemical.
   - Conduct case studies: how many people are exposed and through which routes; what is the magnitude, duration, and time of exposure.
   - Perform epidemiological studies within different geographic areas.
4. Risk characterization:
   - Integrate the data and analyses of the above three evaluations to estimate potential carcinogenic risk and noncarcinogenic health effects.

**Example:** In most public health evaluations, it is assumed that an individual adult or child consumes 2 or 1 L of water, respectively, each day through all uses. The average body weights for a men, women, and children are assumed to be 70 kg (154 lb), 50 kg (110 lb), and 10 kg (22 lb), respectively. A toxic substance is present at 0.7 mg/L in water. Determine the daily dose of the three groups for toxicological purposes.

**solution:**

Step 1. Calculate daily intake (DI) of toxic substance

For an adult

$$\begin{aligned} \text{DI} &= 0.7\ \text{mg/L} \times 2\ \text{L/d} \\ &= 1.4\ \text{mg/d} \end{aligned}$$

For a child

$$\begin{aligned} \text{DI} &= 0.7\ \text{mg/L} \times 1\ \text{L/d} \\ &= 0.7\ \text{mg/d} \end{aligned}$$

Step 2. Compute daily dose (DD) per unit weight

For a man

$$\begin{aligned} \text{DD} &= (1.4\ \text{mg/d})/70\ \text{kg} \\ &= 0.02\ \text{mg/kg} \cdot \text{d} \end{aligned}$$

For a woman

$$\begin{aligned} \text{DD} &= (1.4\ \text{mg/d})/50\ \text{kg} \\ &= 0.028\ \text{mg/kg} \cdot \text{d} \end{aligned}$$

For a child

$$\begin{aligned} \text{DD} &= (0.7\ \text{mg/d})/10\ \text{kg} \\ &= 0.07\ \text{mg/kg} \cdot \text{d} \end{aligned}$$

*Note*: A child gets the highest dose.

**Dose-response models.** The dose-response models for cancer death estimation have been summarized (IUPAC, 1933; Klaassen, 2001). The most frequently used ones are one-hit, Weibull, and multistage models shown as follows:

**One-hit (one-stage) model.** The one-hit mechanistic model is based on the somatic mutation theory in which only one hit of some minimum critical amount of a carcinogen at a cellular target for critical cellular interaction is required for a cell to be altered (become cancerous). The probability statement for this model is

$$P(d) = 1 - \exp(-bd) \tag{5.202}$$

where $P(d)$ = the probability of cancer death from a continuous dose rate, unitless
$b$ = constant
$d$ = dose rate, μg/L
$bd$ = the number of hits occurring during a minute

**Weibull model.** The Weibull model is expressed as

$$P(d) = 1 - \exp(-\,bd^k) \tag{5.203}$$

where $k$ = critical number of hits for the toxic cellular response
Others are the same as the above.

**Multistage model.** Armitage and Doll (1957) developed a multistage model for carcinogenesis that was based on these equations and on the hypothesis that a series of ordered stages was required before a cell could undergo mutation, initiation, transformation, and progression to form a tumor. This relationship was generalized by Crump (1980) by maximizing the likelihood function over polynomials so that the probability statement is

$$P(d) = 1 - \exp(-\,(b_0 + b_1 d^1 + b_2 d^2 + \cdots\cdots\cdots + b_n d^k)) \tag{5.204}$$

where $P(d)$ = the probability of cancer death from a continuous dose rate, unitless
$b$'s = constants, $(\text{mg/kg} \cdot \text{d})^{-1}$
$k$ = the number of dose groups (biological stages)
$d$ = dose rate, μg/L

If the true value of $b_1$ is replaced with $b_1{}^*$ (the upper confidence limit of $b_1$), then a linearized multistage model can be derived where the expression is dominated by $(bd^*)d$ at low doses. The slope on this confidence interval, $q_1$*, is used by US EPA for quantitative cancer assessment. To obtain an upper 95% confidence interval on risk, the $q_1$* value (risk/Δ dose in $(\text{mg/kg} \cdot \text{d})^{-1}$) is multiplied by the amount of exposure (mg/kg · d). Thus, the upper-bound estimate on risk $R$ is calculated as

$$R = q_1^* \ (\text{mg/kg} \cdot \text{d})^{-1} \times \text{exposure} \ (\text{mg/kg} \cdot \text{d}) \qquad (5.205)$$

This relationship has been used to calculate a "virtually safe dose" (VSD), which represents the lower 95% confidence limit on a dose that gives an "acceptable level" of risk (e.g. upper confidence limit for $10^{-6}$ excess risk). The integrated risk information system (IRIS) developed by the US EPA gives $q$* values for many environmental carcinogens (US EPA, 2000a). Because both the $q_1$* and VSD values are calculated using 95% confidence intervals, the values are believed to represent conservative, protective estimates.

The US EPA has utilized the linearized multistage model to calculate "unit risk estimates" in which the upper confidence limit on increased individual lifetime risk of cancer for 70-kg human, breathing 1 μg/m$^3$ of contaminated air or drinking 2 L/d of water containing 1 ppm (1 mg/L), is estimated over a 70-year life span.

The human equivalent dosages are derived by multiplying the corresponding animal dosages by $(W_h/W_a)^{1/3}$. Where $W_h$ and $W_a$ are average body weight of men and test animals, respectively.

When the response and human equivalent dose data are fit to the linearized multistage model, the 95% upper limit on the largest linear term is $q_1$*. Then,

$$q_1^* = q_{1a}^* \ (W_h/W_a)^{1/3}$$

To estimate the 95% lower level of dose concentration, $d$, corresponding to a 95% confidence level of risk $R$,

$$R = 1 - \exp(-\, dq_1^*) \qquad (5.206)$$

$$\exp\,(-\, dq_1^*) = 1 - R$$

then

$$d \ (\text{in mg/kg} \cdot \text{d}) = (1/q_1^*) \times \ln(1 - R) \qquad (5.207)$$

To solve for $d$ in μg/L with a 70-kg man drinking 2 L/d of water, a factor is applied.

$$1 \text{ mg/kg} \cdot \text{d} = 1 \text{ mg/kg} \cdot \text{d} \times 1000 \ \mu\text{g/mg} \times 70 \text{ kg} \div 2 \text{ L/d} = 35{,}000 \ \mu\text{g/L}$$

then

$$d \ (\text{in } \mu\text{g/L}) = 35{,}000 \ \mu\text{g}(\text{mg/kg} \cdot \text{d})^{-1} \div q_1^*(\text{mg/kg} \cdot \text{d})^{-1} \times \ln(1 - R) \qquad (5.208)$$

**Example:** The highest value of $q_1$* for a dose-response study for vinyl chloride in the diet is 2.3 (mg/kg · d)$^{-1}$. Determine the corresponding lowest concentration, $d$, when $R$ is set as $10^{-4}$, $10^{-5}$, and $10^{-6}$.

**solution:** Using Eq. (5.208)

Step 1. Setting $R = 10^{-4}$

$$\begin{aligned} d \text{ (in } \mu\text{g/L)} &= (-35{,}000/2.3) \times \ln(1 - 10^{-4}) \\ &= 6.09 \end{aligned}$$

Step 2. Setting $R = 10^{-5}$

$$\begin{aligned} d \text{ (in } \mu\text{g/L)} &= (-35{,}000/2.3) \times \ln(1 - 10^{-5}) \\ &= 0.61 \end{aligned}$$

Step 3. Setting $R = 10^{-6}$

$$\begin{aligned} d \text{ (in } \mu\text{g/L)} &= (-35{,}000/2.3) \times \ln(1 - 10^{-6}) \\ &= 0.061 \end{aligned}$$

**Radionuclides in drinking water.** Radionuclides may occur naturally and may be man-made (about 200% in number more than natural ones). Natural radionuclides in drinking water supplies include radium (Ra-226 and Ra-228), radon (Rn-222), uranium (U-238), lead (Pb-210), polonium (Po-210), and thorium (Th-230 and Th-232). They omit one or more of the three types of nuclear radiation, i.e. alpha, beta, and gamma rays in air, food, and drinking water. The radiation attacks bone, red bone marrow, gonads, breast, kidney, lung, skin (burn), etc. Most radionuclides have long decay half-lives. For example, the half-lives for U-238 and Ra-228 are respectively of $4.5 \times 10^9$ and 6.7 years. The radioactivity is expressed in curies (Ci). One curie is the number of particles per second from one gram of radium. Dose is the energy of a particle. One rad of dose is 100 ergs per gram of energy deposited.

The MCLGs for radionuclides are set as zero. The MCLs for gross alpha particle activity, combined with Ra-226 and Ra-228, and man-made (approximately 200) radionuclides are 15 pCi/L, 5 pCi/L, and 4 millirem/yr. respectively. One pCi (pico Curie) is $10^{-12}$ Ci.

The drinking water equivalent level (DWEL, in pCi/L) for radionuclides is defined as (US EPA, 1986)

$$\text{DWEL} = \frac{(\text{NOAEL})\,(\text{animalfactor}, f_1)(\text{BW})}{(\text{UF})(\text{WC})(\text{humanfactor}, f_2)} \tag{5.209}$$

where NOAEL = no observed adverse effect level
BW = body weight, 70 kg for adult male
UF = uncertainty factor, 10, 100, or 100
WC = daily water consumption, 2L/d for adult

**Example:** Determine DWEL of natural uranium for a man. Given: NOAEL = 1 mg/kg · d, $f_1 = 0.01$, $f_2 = 0.05$, and UF = 100.

**solution:** Usually, BW = 70 kg and WC = 2 L/d are used for assessing risks.

Using Eq. (5.209)

$$\text{DWEL} = \frac{(\text{NOAEL})\,(\text{animalfactor}, f_1)(\text{BW})}{(\text{UF})(\text{WC})(\text{humanfactor}, f_2)}$$

$$= \frac{(1\ \text{mg/kg}\cdot\text{d})(0.01)(70\ \text{kg})}{(100)(2\ \text{L/d})(0.05)}$$

$$= 0.07\ \text{mg/L} \times 1000\ \mu\text{g/mg}$$

$$= 70\ \mu\text{g/L}$$

### 20.3 Risk management

Risk management is defined as decisions about whether an assessed risk is sufficiently high to present a public health concern and about the appropriate methods for control of a risk judged to be significant. In other words, it is the process of deciding what to do about the problems.

Risk management involves the following:

- find and decide the problems from risk assessment (type and concentration of contaminants),
- assume knowledge of health risks,
- factor in the hydrology feasibility study and capital and operating (site-specific) cost analyses,
- decide and evaluate the measures (alternative control strategies) for solving the problems,
- reexamine exposure issues previously dealt with in risk assessment, and
- evaluate and monitor the results for the corrective measures for meeting drinking water standards control strategies.

The basic categories of alternative control strategies may include source control, treatment, combination of the above two, and short-term strategies. The short-term control strategies are the use of bottled water, point-of-use treatment (RO or ion exchange), or issuance of a boil water order. The disadvantages of point-of-use units (in-home and workplace water treatment) are maintenance upkeep problems and higher cost.

Some bottled waters have questionable water quality (although most meet MCLs) and are not labeled as to sources and treatment processes. In foreign countries and even in the United States, some are bottled from the tap water (public drinking water systems).

Several best management practices for performing in the watershed may be used for controlling raw water source to reduce or eliminate contaminants. Water supply utilities may locate new sources of supply. If so, the new sources can be blended with or replace the existing water source. Interceptor well(s) upstream of the source water may be used to protect water supply wells. The wastewaters from the interceptor wells need proper disposal or treatment.

If the contaminant source is a leaking storage tank, fix or remove the tank; pump the well until contaminant levels drop, and then treat the contaminated soil.

**Treatment strategies.** Treatment involves best available treatment techniques to reduce contaminant to meet MCLs. Conventional treatment processes combined with additional techniques can reduce contaminant levels (Table 5.16; see US EPA, 1986):

a. Processes for inorganic removal

- Conventional treatment
- Lime softening
- Iron exchange
- Reverse osmosis

**TABLE 5.16 Treatment Processes for Removal of Harmful Contaminants**

| Process | Removes |
|---|---|
| Conventional | As (V) at pH $<7.5$; Cd at pH $> 8.5$; Cr (III); Pb; Ag at pH $<8.0$, pathogens, turbidity |
| Lime softening | As (V) at pH $= 10 - 10.8$; Ba at pH $= 9.5 - 10.8$ ; Cd; Cr(III) at pH $>10.5$ ; Pb; Ag; F; V |
| Reverse osmosis | All inorganics; commonly used for As (III), As(V); Ba, Cd, Cr(III)' Cr(VI); F, Pb, Hg, $NO_3$, Se (IV), Se(VI); Ag, Ra, U |
| Ion exchange, cation | Ba, Cd, Cr(III), Ag, Ra |
| Ion exchange, anion | As(V), Cr(VI); $NO_3$, Se(IV), Se(VI), U |
| Activated carbon | Volatile organics (benzene, vinyl chloride, carbon tetrachloride, TCE, PCE, etc.), chlorinated aromatics (PCB, dichlorobenzene), pesticides (aldicarb, chlordane, DBCP), DBP |
| Activated alumina | As(V), F, Se(IV) |
| Aeration | Volatile organics, Rn |
| Air stripping | Volatile organics, Rn |
| Boiling | Pathogens, bacterias |
| Membrane (nano-) | Pathogens, bacterias, DBP |

- Activated alumina
- Electrodialysis

b. Processes for organic removal

- Conventional treatment
- Aeration (diffused air, packed column, slate-tray)
- Adsorption (granular or powder activated carbon, resins)
- Oxidation (disinfection)
- Reverse osmosis
- Biodegradation
- Boiling

The specific processes can be added to the existing treatment processes. An excellent overview and case studies, including capital and operating costs of the above processes used for risk management and health advisories, are presented in the US EPA (1986).

**Example 1:** Based on lack of significant decrease in cholinesterase activity in rats, the NOAEL for aldicarb sulfoxide in rats is 0.125 mg/kg · d. For HA calculation purpose, EPA assumes:

$$\begin{aligned} \text{Body weight, BW} &= 70 \text{ kg for a man} \\ \text{BW} &= 10 \text{ kg for a child} \\ \text{Daily water consumed, WC} &= 2 \text{ L/d for man} \\ \text{WC} &= 1 \text{ L/d for 10-kg child} \\ \text{Uncertainty factor, UF} &= 100 \text{ for use with animal NOAEL} \end{aligned}$$

This example illustrates how the US EPA determines the HA numbers for pesticide aldicarb.

**solution:**

Step 1. Calculate 1-day health advisory (HA)

For the 10-kg child: Using Eq. (5.13)

$$\begin{aligned} \text{One-day HA} &= \frac{(\text{NOAEL})\,(\text{BW})}{(\text{UF})\,(\text{WC})} \\ &= \frac{(0.125 \text{ mg/kg}\cdot\text{d})\,(10 \text{ kg})}{(100)(1 \text{ L/d})} \\ &= 0.012 \text{ mg/L or } 12\ \mu\text{g/L} \end{aligned}$$

Step 2. Determine 10-day health advisory

Since aldicarb is metabolized and excreted rapidly (>90% in urine alone in a 24-h period following a single dose), the 1- and 10-day HA values would not be expected to differ to any extent. Therefore, the 10-day HA will be the same as the 1-day HA (12 μg/L).

Step 3. Longer-term health advisory

For the 10-kg child:

$$\text{Longer-term HA} = \frac{(0.125\ \text{mg/kg}\cdot\text{d})\ (10\ \text{kg})}{(100)\ (1\ \text{L/d})}$$

$$= 0.012\ \text{mg/L}\ (12\ \mu\text{g/L})$$

Step 4. Determine lifetime health advisory

a. Calculate RfD using Eq. (5.13a)

$$\text{RfD*} = \frac{\text{NOAEL}}{\text{UF}} = \frac{0.125\ \text{mg/kg}\cdot\text{d}}{100}$$

$$= 0.00125\ \text{mg/kg}\cdot\text{d}$$

*RRfD = risk reference dose: estimate of daily exposure to the human population that appears to be without appreciable risk of deleterious noncarcinogenic effects over a lifetime of exposure.

The lifetime health advisory proposed above reflects the assumption that 100 % of the exposure to aldicarb residues is via drinking water. Since aldicarb is used on food crops, the potential exists for dietary exposure also, lacking compound-specific data on actual relative source contribution; it may be assumed that drinking water contributes 20% of an adult's daily exposure to aldicarb. The lifetime health advisory for the 70-kg adult would be 9 μg/L, taking this relative source contribution into account.

**Example 2:** There are no suitable data available to estimate 1-day, 10-day, and long-term health advisories. The study of the subacute exposure to trichloroethylene (TTC) via inhalation 5 days a week for 14 weeks by adult rats identified a LOAEL 55 ppm (300 mg/m$^3$). Derive the DWEL for TTC.

**Solution:**

Step 1: Determine total absorbed dose (TAD)

$$\text{TAD} = \frac{(300\ \text{mg/m}^3)\ (8\ \text{m}^3\text{/d})\ (5/7)\ (0.3)}{(70\ \text{kg})} = 7.35\ \text{mg/kg}\cdot\text{d}$$

where 300 mg/m$^3$ = LOAEL
8 m$^3$/d = volume of air inhaled during the exposure period

5/7 = conversion factor for adjusting from 5 d/week exposure to a daily dose
0.3 = ratio of the dose absorbed.
70 kg = assumed weight of adult

Step 2: Determine RfD

$$\text{Rfd} = \frac{(7.35 \text{ mg/kg}\cdot\text{d})}{(100)(10)} = 0.00735 \text{ mg/kg}\cdot\text{d}$$

where 7.35 mg/kg · d = TAD
100 = uncertainty factor appropriate for use with data from an animal study
10 = uncertainty factor appropriate for use in conversion of LOAFL to NOAEL

Step 3: Determine DWEL

$$\text{DWEL} = \frac{(0.00735 \text{ mg/kg}\cdot\text{d})(70 \text{ kg})}{\text{L/d}} = 0.26 \text{ mg/L} \quad (260\ \mu\text{g/L})$$

where 0.00735 mg/kg · d = RRfD
70 kg = assumed weight of protected individual
2 L/d = assumed volume of water ingested by 70-kg adult

The estimated excess cancer risk associated with lifetime exposure to drinking water containing trichloroethylene at 260 μg/L is approximately $1 \times 10^{-4}$. This estimate represents the upper 90% confidence limit from extrapolations prepared by EPA's Carcinogen Assessment Group using the linearized multistage model. The actual risk is unlikely to exceed this value, but there is considerable uncertainty as to the accuracy of risk calculated by using this methodology.

**Example 3:** A subchronic toxicity study of vinyl chloride monomer (VCM) dissolved in soybean oil was administrated 6 days a week for 13 weeks, by average to male and female wister rats. The NOAEL in the study was identified as 30 mg/kg · d.

Estimate the values of HA for VCM.

**solution:**

Step 1. Estimate 1-day HA

There are insufficient data for estimation of a 1-day HA. The 10-day HA is proposed as a conservative estimate for a 1-day HA.

Step 2. Determine 10-day HA (as well as 1-day HA)

For a 10-kg child:

$$\text{Ten-day HA} = \frac{(\text{NOAEL})\,(6/7)\,(\text{BW})}{(\text{UF})\,(\text{WC})}$$

$$= \frac{(30\ \text{mg/kg}\cdot\text{d})(6/7)(10\ \text{kg})}{(100)\,(1\ \text{L/d})}$$

$$= 2.6\ \text{mg/L or } 2600\ \mu\text{g/L}$$

where 6/7 = expansion of 6 d/week treatment in the study to 7 d/week to represent daily exposure

Others = as previous examples

*Note*: This HA is equivalent to 2.6 mg/d or 0.26 mg/kg · d.

Step 3. Determine the longer-term HA

Using Eq. (5.13),

for a child

$$\text{Longer-term HA} = \frac{(0.13\ \text{mg/kg}\cdot\text{d})(10\ \text{kg})}{(100)\,(1\text{L/d})}$$

$$= 0.013\ \text{mg/L or } 13\ \mu\text{g/kg}\cdot\text{d}$$

for an adult

$$\text{Longer-term HA} = \frac{(0.13\ \text{mg/kg}\cdot\text{d})(70\ \text{kg})}{(100)\,(2\ \text{L/d})}$$

$$= 0.046\ \text{mg/L or } 46\ \mu\text{g/L}$$

*Note*: This HA is equivalent to 92 μg/d or 1.3 μg/kg · d.

Step 4. Determine the lifetime HA

Because vinyl chloride is classified as a human carcinogen (Group A of US EPA, Group 1 of International Agency for Research on Cancer), a lifetime HA is not recommended.

## References

Alley, E. R. 2000. *Water Quality Control Handbook*. New York: McGraw-Hill.

Allied Signal. 1970. *Fluid systems, Reverse osmosis principals and applications*. Allied Signal: San Diego, California.

American Dental Association. 1980. *Fluoridation facts*. American Dental Association: G21 Chicago, Illinois.

American Public Health Association, American Water Works Association (AWWA), Water Environment Federation. 1998. *Standard methods for the examination of water and wastewater.* 20th edn.

American Society of Civil Engineers (ASCE) and American Water Works Association. 1990. *Water treatment plant design.* 2nd edn. New York: McGraw-Hill.

American Water Works Association (AWWA). 1978. *Corrosion control by deposition of $CaCO_3$ films.* Denver, Colorado: AWWA.

American Water Works Association and American Society of Civil Engineers. 1998. *Water treatment plant design,* 3rd edn. New York: McGraw-Hill.

Amirtharajah, A. and O'Melia, C. R. 1990. Coagulation processes: Destabilization, mixing, and flocculation. In: AWWA, *Water quality and treatment.* New York: McGraw-Hill.

Anderson, R. E. 1975. Estimation of ion exchange process limits by selectivity calculations. *AICHE Symposium Series* 71: 152, 236.

Applegate, L. 1984. Membrane separation processes. *Chem. Eng.* June 11, 1984, pp. 64–69.

Benefield, L. D. and Morgan, J. S. 1990. Chemical precipitation. In: AWWA, *Water quality and treatment.* New York: McGraw-Hill.

Benefield, L. D., Judkins, J. F. and Weand, B. L. 1982. *Process chemistry for water and wastewater treatment.* Englewood Cliffs, New Jersey: Prentice-Hall.

Bergman, R. A. 2005. Membrane processes. In: *Water Treatment Plant Design,* 4th edn. New York: McGraw-Hill.

Brown, T. L. and LeMay, H. E. Jr. 1981. *Chemistry: the central science,* 2nd edn., Englewood Cliffs, New Jersey: Prentice-Hall.

Calderbrook, P. H. and Moo Young, M. B. 1961. The continuous phase heat and mass transfer properties of dispersions. *Chem. Eng. Sci.* **16:** 39.

Camp, T. R. 1968. Floe volume concentration. *J. Amer. Water Works Assoc.* **60**(6): 656–673.

Camp, T. R. and Stein, P. C. 1943. Velocity gradients and internal work in fluid motion. *J. Boston Soc. Civil Engineers.* **30:** 219.

Carlson, D. A. et al., 1985. *Ultraviolet disinfection of water for small water supplies.* EPA/600/285/092, Cincinnati, Ohio: US EPA.

Chellam, S., Jacongelo, J. G., Bonacquisti, T. P. and Schauer, B. A. 1997. Effect of pretreatment on surface water nanofiltration. *J. Amer. Water Works Assoc.* **89**(10): 77–89.

Clancy, J. L. et al., 1998. UV light inactivation of *Crytosporidium* oocysts. *J. AWWA* **90**(9): 92–102.

Clark, J. W. and Viessman, W. Jr. 1966. *Water supply and pollution control.* Scranton, Pennsylvania: International Textbook Co.

Clark, J. W., Viessman, W. Jr. and Hammer, M. J. 1977. *Water supply and pollution control.* New York: IEP-A Dun-Donnelley.

Cleasby, J. L. 1990. Filtration. In: AWWA, *Water quality and treatment.* New York: McGraw-Hill.

Cleasby, J. L. and Fan, K. S. 1981. Predicting fluidization and expansion of filter media. *J. Environ. Eng. Div. ASCE* **107** (EE3): 355–471.

Clifford, D. A. 1990. Ion exchange and inorganic adsorption: In: AWWA, *Water quality and treatment,* New York: McGraw-Hill.

Conlon, W. J. 1990. Membrane processes. In: AWWA, *Water quality and treatment,* New York: McGraw-Hill.

Cornwell, D. A. 1990. Airstripping and aeration. In: AWWA, *Water quality and treatment,* New York: McGraw-Hill.

Cotruvo, J. A. and Vogt, C. D. 1990. Rationale for water quality standards and goals. In: AWWA, *Water quality and treatment.* New York: McGraw-Hill.

Crittenden, J. C., Trussell, R. R., Hand, D. W., Howe, K. J. and Tchobanoglous, G. 2005. *Water Treatment Principles and Design,* 5th edn. Hoboken, NJ: John Wiley.

Dean, J. A. 1979. *Lange's handbook of chemistry.* 12th edn. New York: McGraw-Hill.

Diamond Shamrock Company. 1978. Duolite A104-Data leaflet. Functional Polymers Division, DSC, Cleveland, Ohio.

Downs, A. J. and Adams, C. J. 1973. *The chemistry of chlorine, bromine, iodine and astatine.* Oxford: Pergamon.

Doyle, M. L. and Boyle, W. C. 1986. Translation of clean to dirty water oxygen transfer rates. In: Boyle, W. C. (ed.) *Aeration systems, design, testing, operation, and control.* Park Ridge, New Jersey: Noyes.

Dryfuse, M. J. 1995. An evaluation of conventional and optimized coagulation for TOC removal and DBP control in buls and fractional waters. M.S. Thesis, Department of Civil and Environmental Engineering. Cincinnati, Ohio: University of Cincinnati.

Fair, G. M., Geyer, J. C. and Morris, J. C. 1963. *Water supply and waste-water disposal.* New York: John Wiley.

Fair, G. M., Geyer, J. C. and Okun, D. A. 1966. *Water and wastewater engineering,* Vol. 1: Water supply and wastewater removal. New York: John Wiley.

Fair, G. M., Geyer, J. C. and Okun, D. A. 1968. *Water and wastewater engineering,* Vol. 2: Water purification and wastewater treament and disposal. New York: John Wiley.

Federal Register. 1991. Environmental Protection Agency 40 CFR Part 141, 142, and 143. *National Primary Drinking Water Regulations; Final Rule* **56**(20): 3531–3533.

Federal Register. 1997. Environmental Protection Agency 40 CFR, Part 141 and 142 National Primary Drinking Water Regulation: Interim Enhanced Surface Water Treatment Rule Notice of Data Availability; Proposed Rule 62 (212): 59521–59540, Nov. 3, 1997.

Ficek, K. J. 1980. Potassium permanganate for iron and manganese removal. In: Sanks, R. L. (ed.), *Water treatment plant design.* Ann Arbor, Michigan: Ann Arbor Science.

Geldreich, E. E. and Rice, E. W. 1987. Occurrence, significance and detection of Klebsiellain water systems. *J. AWWA* **79**(5):74.

Geldreich, E. E. 1988. Coliform noncompliance nightmares in water supply distribution system. In: *Water Quality: A realistic perspective,* Chapter 3. Lansing, Michigan: University of Michigan.

Gregory, R. and Zabel, T. F. 1990. Sedimentation and flotation. In AWWA: *Water quality and treatment.* New York: McGraw-Hill.

Gregorski, T. 2006. A Cost-Effective Solution. In: The Global Impact of Membrane Filtration Treatment Systems. *Water & Wastes Digest*, Fall 2006, p. 5.

Haas, C. N. 1990. Disinfection. In: AWWA, *Water quality and treatment.* New York: McGraw-Hill.

Haas, C. N. 2002. Risks posed by pathogens in drinking water. In: *ASCE, Control of microorganisms in drinking water,* S. Lingireddy (ed.). Reston, VA: ASCE.

Hammer, M. J. 1975. *Water and waste-water technology.* New York: John Wiley.

Herren-Freund, S. L., Pereira, M. A., Khoury, M. D. and Olson, G. 1987. The carcinogenicity of trichloroethylene and its metabolites, trichloracetic acid and dichloracetic acid, in mouse liver. *Toxicology and Applied Pharmacology* **90**: 183–189.

Hom, L. W. 1972. Kinetics of chlorine disinfection in an eco-system. *J. Sanitary Eng. Div., Pro. Amer. Soc. Civil Eng.* **98**(1): 183.

Hudson, H. E., Jr. 1972. Density considerations in sedimentation. *J. Amer. Water Works Assoc.* **64**(6): 382–386.

Humphrey, S. D. and Eikleberry, M. A. 1962. Iron and manganese removal using $KMnO_4$. *Water and Sewage Works* **109** (R. N.): R142–R144.

Illinois Environmental Protection Agency (IEPA). 1992. *An example analysis of CT value.* Springfield, Illinois: IEPA.

Illinois Pollution Control Board. 1990. Rules and regulations. Title 35: Environmental Protection, Subtitle C: Water Pollution, Chapter I, Springfield, Illinois: IEPA.

International Fire Service Training Association. 1993. *Water supplies for fire protection,* 4th edn. Stillwater, Oklahoma: Oklahoma State University.

Inversand Company. 1987. *Manganese greensand.* Clayton, New Jersey: Inversand Company.

Ionic, Inc. 1993. *Ionic public water supply systems.* Bulletin no. 141. Watertown, Massachusetts.

James M. Montgomery Consulting Engineering, Inc. 1985. *Water treatment principles and design.* New York: John Wiley.

Jobin, R. and Ghosh, M. M. 1972. Effect of buffer intensity and organic matter on oxygenation of ferrous iron. *J. Amer. Water Works Assoc.* **64** (9): 590–595.

Jones, B. H., Mersereau-Kempf, J. and Shreve, D. 2000. Multiphase multimethod remediation. *WEF Industrial Wastewater* **8**(4): 21–25.

Kreuz, D. F. 1962. Iron oxidation with $H_2O_2$ in secondary oil recovery produce plant, FMC Corp. Internal report, 4079-R, Philadelphia, PA.

Lide, D. R. 1996. *CRC Handbook of chemistry and physics,* 77th edn. Boca Raton, Florida: CRC Press.

Lin, S. D. 2002. The indicator concept and its application in water supply. In: *ASCE, Control of microorganisms in drinking water,* S. Lingireddy (ed.). Reston, VA: ASCE.

Lin, S. D. and Green, C. D. 1987. *Waste from water treatment plants: Literature plants: Literature review, results of an Illinois survey and effects of alum sludge application to cropland.* ILENR/RE-WR-87/18. Springfield, Illinois: Illinois Department of Energy and Natural Resources.

Mattee-Müller, C, Gujer, W. and Giger, W. 1981. Transfer of volatile substances from water to the atmosphere. *Water Res.* **5**(15): 1271–1279.

McKee, J. E., Brokaw, C. J. and McLaughlin, R. T. 1960. Chemical and colicidal effects of halogens in sewage. *J. Water Pollut. Control Fed.* **32**(8): 795–819.

Miltner, R. J., Rice, E. W. and Stevens, A. A. 1990. Pilot-scale investigation of the formation and control of disinfection by-products. *Proceeding of American Water Works Association Annual Conference,* July 17–21, Cincinnati, Ohio.

Miltner, R. J. 1993. Transformation of NOM during water treatment. *Proceeding of American Water Works Association Research Foundation Workshop on NOM,* September. Chamonix, France.

Moorman, J. 2006. Drinking water regulatory update - Understanding the LT2 rule and Stage 2 DBP rule. *Water & Wastes Digest* **46**(9): 14–15.

Morris, J. C. 1966. The acid ionization constant of HOCl from 5°C to 35°C. *J. Phys. Chem.* **70**(12): 3789.

Movahed, B. 2006a. What can membrane technologies do for you? U.S. Water News, July 2006, p.12.

Movahed, B. 2006b. Advocating membrane technology. *Water & Wastes Digest.* Fall 2006, p. 18.

Movahed, B. 2006c. Personal communication.

National Board Fire Underwriters (NBFU). 1974. *Standard schedule for grading cities and towns of the United States with reference to their fire defenses and physical conditions.* New York: NBFU, American Insurance Association.

Peroxidation System, Inc. 1990. *Perox-pure-process description.* Tucson, Arizona: Peroxidation System Inc.

Pontius, F. W. 1993. Configuration, operation of system affects C × T value. *Opflow* **19**(8): 7–8.

Qasim, S. R., Motley, E. M. and Zhu, G. 2000. *Water Works Engineering.* Upper saddle River, New Jersey: Prentice-Hall PTR.

Reeves, T. G. 1986. *Water fluoridation—A manual for engineers and technicians.* US Public Health Service, 00-4789. Atlanta, Georgia: US PHS.

Reeves, T. G. 1990. Water fluoridation. In: *Water quality and treatment.* AWWA. New York: McGraw-Hill.

Rehfeldt, K. R., Raman, R. K., Lin, S. D. and Broms, R. E. 1992. *Asessment of the proposed discharge of ground water to surface waters of the American Bottoms area of Southwestern Illinois.* Contract Report 539. Champaign, Illinois. Illinois State Water Survey.

Rice, E. W. and Hoff, J. C. 1981. Inactivation of *Giardia lamblia* cysts by ultraviolet irradiation. *Applied and Environmental Microbiology* **43**(1): 250–251.

Rice, R. G. 1986. Instruments for analysis of ozone in air and water. In: Rice, R. G., Bolly, L. J. and Lacy, W. J. (ed.) *Analytic aspects of ozone treatment of water and wastewater.* Chelsea, Michigan: Lewis.

Rose, H. E., 1951. On the resistance coefficient—Reynolds number relationship for fluid flow through a bed of granular material. *Proc. Inst. Mech. Engineers (London),* pp. 154–160.

Schroeder, E. D. 1977. *Water and wastewater treatment.* New York: McGraw-Hill.

Singer, P. G. 2006. DBPs in drinking water: Additional scientific and policy considerations for public health protection. *J. AWWA* **98**(10): 74–80.

Snoeyink, V. L. 1990. Adsorption of organic compounds. In: AWWA, *Water quality and treatment.* New York: McGraw-Hill.

Snoeyink, V. L. and Jenkins, D. 1980. *Water chemistry.* New York: John Wiley.

Speth, T. F. 1989. Evaluation of nanofiltration foulants from treated surface waters. Ph.D. dissertation. Cincinnati, Ohio: University of Cincinnati.

Steel, E. W. and McGhee, T. J. 1979. *Water supply and sewerage,* 5th edn. New York: McGraw-Hill.

Stevens, A. A., Moore, L. A. and Miltner, R. J. 1989. Formation and control of non-THM disinfection by-products. *J. AWWA* **81**(8): 54.

Sullivan, P. J., Agardy, F.J. and Clark, J. J. J. 2005. *The Environmental Science of Drinking Water.* Burlington, Massachusetts, Elsevier Butterworth-Heinemann.

Tchobanoglous, G. and Schroeder, E. D. 1985. *Water quality.* Reading, Massachusetts: Addison-Wesley.

Teefy, S. M. and Singer, P. C. 1990. Performance of a disinfection scheme with the SWTR. *J. Amer. Water Works Assoc.* **82**(12): 88–98.

Toranzos, G. A. and McFeters, G. A. 1997. Detection of indicator microorganisms in environmental freshwaters and drinking waters. In: *Manual of Environmental Microbiology.* Hurst, C. J. *et al.* (eds.). Washington, DC: American Society for Microbiology Press.

Treybal, R. D. 1968. *Mass-transfer operations.* New York: McGraw-Hill.

US Environmental Protection Agency. 1987. *Workshops on assessment and management of drinking water contamination.* EPA/600/M-86/026. Cincinnati, Ohio: US EPA.

US Environmental Protection Agency. 1990. *The lake and reservoir restoration guidance manual,* 2nd edn. EPA-440/4-90-006. Washington, DC: US EPA.

US Environmental Protection Agency. 1991. *The watershed protection approach–an overview.* EPA/503/9-92/002. Washington, DC: US EPA.

US Environmental Protection Agency. 1992. *Control of biofilm growth in drinking water distribution systems.* EPA/625/R-92/001. Washington, DC: US EPA.

US Environmental Protection Agency. 2001. *Controlling disinfection by-products and microbial contaminants in drinking water.* EPA/600/R-01/110. Washington, DC: US EPA.

US Environmental Protection Agency. 2001. *Trihalomethanes in drinking water: sampling, analysis, monitoring and compliance.* EPA/570/9-83-002. Washington, DC: US EPA.

US Environmental Protection Agency (US EPA). 1976. Translation of reports on special problems of water technology, vol. 9—Adsorption. EPA-600/9-76-030, Cincinnati, Ohio: US EPA.

US Environmental Protection Agency (US EPA). 1983. Nitrate removal for small public water systems. EPA 570/9-83-009. Washington, DC: US EPA.

US Environmental Protection Agency (US EPA). 1986. *Design manual: Municipal wastewater disinfection.* EPA/625/1-86/021. Cincinnati, Ohio: US EPA.

US Environmental Protection Agency (US EPA). 1987. The safe drinking water act—Program summary. Region 5, Washington DC: US EPA.

US Environmental Protection Agency (US EPA). 1989a. Guidance manual for compliance with the filtration and disinfection requirements for public water systems using surface water sources. USEPA, Oct. 1989 ed. Washington, DC: US EPA.

US Environmental Protection Agency (US EPA). 1989b. National primary drinking water regulations: Filtration and disinfection; turbidity, *Giardia lamblia,* viruses *Legionella,* and heterotrophic bacteria: Proposed Rules, Fed. Reg., 54 FR:124:27486, Washington DC: US EPA.

US Environmental Protection Agency (US EPA). 1990. Environmental pollution control alternatives: Drinking water treatment for small communities. EPA/625/5-90/025, Cincinnati, Ohio: US EPA.

US Environmental Protection Agency (US EPA). 1996. Capsule Report—Reverse osmosis process. EPA/625/R-96/009, Washington DC: US EPA.

US Environmental Protection Agency (US EPA). 1999a. National primary drinking water regulations: Interim Enhanced Surface Water Treatment Rule: Notice of Data Availability; Proposed Rule. EPA 62 FR 212:59485, Nov. 3, 1999, Washington DC: US EPA.

US Environmental Protection Agency (US EPA). 1999b. Disinfection profiling and benchmarking guidance manual. EPA 815-R-99-013, Washington DC: US EPA.

US Environmental Protection Agency (US EPA). 2000. Current drinking water standards. http://www.epa.gov/safewater/mcl.html, updated 07/24/2000. Washington, DC: US EPA.

US Public Health Service (US PHS). 1984. *Fluoridation engineering manual.* Atlanta, Georgia: US Department of Health and Human Services, Washington DC: US PHS.

US Public Health Service. Homestudy course 30/7-G, Water fluoridation. Atlanta, G.A. Center of Disease Control, US PHS.

Viessman, W. Jr. and Hammer, M. J. 1993. *Water supply and pollution control,* 5th edn. New York: Harper Collins.

Voelker, D. C. 1984. Quality of water in the alluvial aquifer, American Bottoms, East St Louis, Illinois. Water Resources Investigation Report 84-4180. Chicago: US Geological Survey.

Von Gottberg, A. 1999. Advances in electrodialysis reversal systems. In: *Proceedings of 1999 AWWA Membrane Technology Conference*, February 29–March 3, 1999. Long Beach, CA: AWWA.

Weinberg, H. S. 1999. Disinfection byproducts in drinking water, the analytical challenge. *Analytical Chemistry* 71: 801A–808A.

Wen, C. Y. and Yu, Y. H. 1966. Mechanics of fluidization. *Chem. Eng. Prog. Symp. Series* 62: 100–111.

Water, G. C. 1978. *Disinfection of wastewater and water for reuse.* New York: Van Nostrand Reinhold.

White, G. C. 1972. *Handbook of chlorination.* New York: Litton Educational.

Wilmarth, W. A. 1968. Removal of iron, manganese, and sulfides. *Water and Wast Eng.* **5** (8): 52–61.

Wong, J. M. 1984. Chlorination-filtration for iron and manganese removal. *J. Amer. Water Works Assoc.* **76**(1): 76–79.

Yao, K. M. 1976. Theoretical study of high-rate sedimentation. *J. Water Pollut. Control Fed.* **42**(2): 218–228.

Chapter

# 6

# Wastewater Engineering

## 1 What Is Wastewater?

"Wastewater," also known as "sewage," originates from household wastes, human and animal wastes, industrial wastewaters, storm runoff, and groundwater infiltration. Wastewater, basically, is the flow of used water from a community. It is 99.94% water by weight (Water Pollution Control Federation, 1980). The remaining 0.06% is material dissolved or suspended in the water. It is largely the water supply of a community after it has been fouled by various uses.

## 2 Characteristics of Wastewater

An understanding of physical, chemical, and biological characteristics of wastewater is very important in design, operation, and management of collection, treatment, and disposal of wastewater. The nature of wastewater includes physical, chemical, and biological characteristics which depend on the water usage in the community, the industrial and commercial contributions, weather, and infiltration/inflow.

### 2.1 Physical properties of wastewater

When fresh, wastewater is gray in color and has a musty and not unpleasant odor. The color gradually changes with time from gray to black. Foul and unpleasant odors may then develop as a result of septic sewage. The most important physical characteristics of wastewater are its temperature and its solids concentration.

Temperature and solids content in wastewater are very important factors for wastewater treatment processes. Temperature affects chemical reaction and biological activities. Solids, such as total suspended solids (TSS), volatile suspended solids (VSS), and settleable solids, affect the operation and sizing of treatment units.

**Solids.** Solids comprise matter suspended or dissolved in water and wastewater. Solids are divided into several different fractions and their concentrations provide useful information for characterization of wastewater and control of treatment processes.

**Total solids.** Total solids (TS) is the sum of total suspended solids and total dissolved solids (TDS). Each of these groups can be further divided into volatile and fixed fractions. Total solids is the material left in the evaporation dish after it has dried for at least 1h or overnight (preferably) in an oven at 103 to 105°C and is calculated according to *Standard Methods* (APHA *et al.*, 1995)

$$\text{mg TS/L} = \frac{(A - B) \times 1000}{\text{sample volume, mL}} \tag{6.1}$$

where $A$ = weight of dried residue plus dish, mg
$B$ = weight of dish, mg
1000 = conversion of 1000 mL/L

*Note*: Samples can be natural waters, wastewaters, even treated water.

**Total suspended solids.** Total suspended solids (TSS) are referred to as nonfilterable residue. The TSS is a very important quality parameter for water and wastewater and is a wastewater treatment effluent standard. The TSS standards for primary and secondary effluents are usually set at 30 and 12 mg/L, respectively. TSS is determined by filtering a well-mixed sample through a 0.2 $\mu$m pore size, 24 mm diameter membrane; the membrane filter is placed in a Gooch crucible, and the residue retained on the filter is dried in an oven for at least 1h at a constant weight at 103 to 105°C. It is calculated as

$$\text{mg TSS/L} = \frac{(C - D) \times 1000}{\text{sample volume, mL}} \tag{6.2}$$

where $C$ = weight of filter and crucible plus dried residue, mg
$D$ = weight of filter and crucible, mg

**Total dissolved solids.** Dissolved solids are also called filterable residues. Total dissolved solids in raw wastewater are in the range of 250 to 850 mg/L.

TDS is determined as follows. A well-mixed sample is filtered through a standard glass fiber filter of 2.0 $\mu$m normal pore size, and the filtrate is evaporated for at least 1h in an oven at 180 $\pm$ 2°C. The increase in dish weight represents the total dissolved solids, which is calculated as

$$\text{mg TDS/L} = \frac{(E - F) \times 1000}{\text{sample volume, mL}} \tag{6.3}$$

where $E$ = weight of dried residue plus dish, mg
$F$ = weight of dish, mg

**Fixed and volatile solids.** The residue from TS, TSS, or TDS tests is ignited to constant weight at 550°C. The weight lost on ignition is called volatile solids, whereas the remaining solids represent the fixed total, suspended, or dissolved solids. The portions of volatile and fixed solids are computed by

$$\text{mg volatile solids/L} = \frac{(G - H) \times 1000}{\text{sample volume, mL}} \qquad (6.4)$$

$$\text{mg fixed solids/L} = \frac{(H - I) \times 1000}{\text{sample volume, mL}} \qquad (6.5)$$

where $G$ = weight of residue plus crucible before ignition, mg
$H$ = weight of residue plus crucible or filter after ignition, mg
$I$ = weight of dish or filter, mg

The determination of the volatile portion of solids is useful in controlling wastewater treatment plant operations because it gives a rough estimation of the amount of organic matter present in the solid fraction of wastewater, activated sludge, and in industrial waste.

Determination of volatile and fixed solids does not distinguish precisely between organic and inorganic matter. Because the loss on ignition is not confined only to organic matter, it includes losses due to decomposition or volatilization of some mineral salts. The determination of organic matter can be made by tests for biochemical oxygen demand (BOD), chemical oxygen demand (COD), and total organic carbon (TOC).

**Settleable solids.** Settleable solids is the term applied to material settling out of suspension within a defined time. It may include floating material, depending on the technique. Settled solids may be expressed on either a volume (mL/L) or a weight (mg/L) basis.

The volumetric method for determining settleable solids is as follows. Fill an Imhoff cone to the 1-L mark with a well-mixed sample. Settle for 45 min, gently agitate the sample near the sides of the Imhoff cone with a rod or by spinning, then continue to settle for an additional 15 min and record the volume of settleable solids in the cones as mL/L.

Another test to determine settleable solids is the gravimetric method. First, determine total suspended solids as stated above. Second, determine nonsettleable suspended solids from the supernatant of the same sample which has settled for 1 h. Then determine TSS (mg/L) of this supernatant liquor; this gives the nonsettleable solids. The settleable solids can be calculated as

$$\text{mg settleable solids/L} = \text{mg TSS/L} - \text{mg nonsettleable solids/L} \qquad (6.6)$$

**Example:** A well-mixed 25 mL of raw wastewater is used for TS analyses. A well-mixed 50 mL of raw wastewater is used for suspended solids analyses. Weights (wt.) of evaporating dish with and without the sample either dried, evaporated, or ignited were determined to constant weight according to *Standard Methods* (APHA *et al.,* 1998). The laboratory results are

Tare wt. of evaporating dish = 42.4723 g
Wt. of dish plus residue after evaporation at 105°C = 42.4986 g
Wt. of dish plus residue after ignition at 550°C = 42.4863 g
Tare wt. of filter plus Gooch crucible = 21.5308 g
Wt. of residue and filter plus crucible after drying at 105°C = 21.5447 g
Wt. of residue and filter plus crucible after ignition at 550°C = 21.5349 g

Compute the concentrations of total solids, volatile solids, fixed solids, total suspended solids, volatile suspended solids, and fixed suspended solids.

**solution:**

Step 1. Determine total solids by Eq. (6.1)

$$\text{Sample size} = 25 \text{ mL}$$

$$A = 42{,}498.6 \text{ mg}$$

$$B = 42{,}472.3 \text{ mg}$$

$$\text{TS} = \frac{(A - B) \text{ mg} \times 1000 \text{ mL/L}}{25 \text{ mL}}$$

$$= (42{,}498.6 - 42{,}472.3) \times 40 \text{ mg/L}$$

$$= 1052 \text{ mg/L}$$

Step 2. Determine total volatile solids by Eq. (6.4)

$$G = 42{,}498.6 \text{ mg}$$

$$H = 42{,}486.3 \text{ mg}$$

$$\text{VS} = (42{,}498.6 - 42{,}486.3) \times 1000/25$$

$$= 492 \text{ mg/L}$$

Step 3. Determine total fixed solids

$$\text{FS} = \text{TS} - \text{VS} = (1052 - 492) \text{ mg/L} = 560 \text{ mg/L}$$

Step 4. Determine total suspended solids by Eq. (6.2)

$$C = 21{,}544.7 \text{ mg}$$

$$D = 21{,}530.8 \text{ mg}$$

$$\text{Sample size} = 50 \text{ mL}$$

$$\text{TSS} = (\text{C} - D) \times 1000/50$$

$$= (21{,}544.7 - 21{,}530.8) \times 20$$

$$= 278 \text{ mg/L}$$

Step 5. Determine volatile suspended solids by Eq. (6.4)

$$G = 21{,}544.7 \text{ mg}$$
$$H = 21{,}534.9 \text{ mg}$$
$$\begin{aligned}\text{VSS} &= (G - H) \times 1000/50 \\ &= (21{,}544.7 - 21{,}534.9) \times 20 \\ &= 196 \text{ mg/L}\end{aligned}$$

Step 6. Determine fixed suspended solids by Eq. (6.5)

$$H = 21{,}534.9 \text{ mg}$$
$$I = 21{,}530.8 \text{ mg}$$
$$\begin{aligned}\text{FSS} &= (H - I) \times 1000/50 \\ &= (21{,}534.9 - 21{,}530.8) \times 20 \\ &= 82 \text{ mg/L}\end{aligned}$$

or

$$\begin{aligned}\text{FSS} &= \text{TSS} - \text{VSS} = (278 - 196) \text{ mg/L} \\ &= 82 \text{ mg/L}\end{aligned}$$

### 2.2 Chemical constituents of wastewater

The dissolved and suspended solids in wastewater contain organic and inorganic material. Organic matter may include carbohydrates, fats, oils, grease, surfactants, proteins, pesticides and other agricultural chemicals, volatile organic compounds, and other toxic chemicals (household and industrial). Inorganics may include heavy metals, nutrients (nitrogen and phosphorus), pH, alkalinity, chlorides, sulfur, and other inorganic pollutants. Gases such as carbon dioxide, nitrogen, oxygen, hydrogen sulfide, and methane may be present in a wastewater.

Normal ranges of nitrogen levels in domestic raw wastewater are 25 to 85 mg/L for total nitrogen (the sum of ammonia, nitrate, nitrite, and organic nitrogen); 12 to 50 mg/L ammonia nitrogen; and 8 to 35 mg/L organic nitrogen (WEF 1996a). The organic nitrogen concentration is determined by a total kjeldahl nitrogen (TKN) analysis, which measures the sum of organic and ammonia nitrogen. Organic nitrogen is then calculated by subtracting ammonia nitrogen from the TKN measurement.

Typical total phosphorus concentrations of raw wastewater range from 2 to 20 mg/L, which includes 1 to 5 mg/L of organic phosphorus and 1 to 15 mg/L of inorganic phosphorus (WEF 1996a). Both nitrogen and phosphorus in wastewater serve as essential elements for biological growth and reproduction during wastewater treatment processes and in the natural water.

The strength (organic content) of a wastewater is usually measured as 5-days biochemical oxygen demand ($BOD_5$), chemical oxygen demand, and total organic carbon. The $BOD_5$ test measures the amount of oxygen required to oxidize the organic matter in the sample during 5 days of biological stabilization at 20°C. This is usually referred to as the first stage of carbonaceous BOD (CBOD), not nitrification (second phase). Secondary wastewater treatment plants are typically designed to remove CBOD, not for nitrogenous BOD (except for advanced treatment). The $BOD_5$ of raw domestic wastewater in the United States is normally between 100 and 250 mg/L. It is higher in other countries. In this chapter, the term $BOD_5$ represents 5-days BOD, unless stated otherwise.

The ratio of carbon, nitrogen, and phosphorus in wastewater is very important for biological treatment processes, where there is normally a surplus of nutrients. The commonly accepted BOD/N/P weight ratio for biological treatment is 100/5/1; i.e. 100 mg/L BOD to 5 mg/L nitrogen to 1 mg/L phosphorus. The ratios for raw sanitary wastewater and settled (primary) effluent are 100/17/5 and 100/23/7, respectively.

**Example 1:** Calculate the pounds of $BOD_5$ and TSS produced per capita per day. Assume average domestic wastewater flow is 100 gallons (378 L) per capita per day (gpcpd), containing $BOD_5$ and TSS concentrations in wastewater of 200 and 240 mg/L, respectively.

**solution:**

Step 1. For BOD

$$\begin{aligned} \text{BOD} &= 200 \text{ mg/L} \times 1 \text{ L}/10^6 \text{ mg} \times 100 \text{ gal/(c} \cdot \text{d)} \times 8.34 \text{ lb/gal} \\ &= 0.17 \text{ lb/(c} \cdot \text{d)} \\ &= 77 \text{ g/(c} \cdot \text{d)} \end{aligned}$$

Step 2. For TSS

$$\begin{aligned} \text{TSS} &= 240 \text{ mg/L} \times 10^{-6} \text{ L/mg} \times 100 \text{ gal/(c} \cdot \text{d)} \times 8.34 \text{ lb/gal} \\ &= 0.20 \text{ lb/(c} \cdot \text{d)} \\ &= 90 \text{ g/(c} \cdot \text{d)} \end{aligned}$$

*Note*: 100 gal/(c · d) of flow and 0.17 lb $BOD_5$ per capita per day are commonly used for calculation of the population equivalent of hydraulic and BOD loadings for other wastewaters.

In newly developed communities, generally designed wastewater flow is 120 gpcpd. With the same concentrations, the loading rate for design purpose would be

$$\begin{aligned} \text{BOD} &= 0.20 \text{ lb/(c} \cdot \text{d)} \\ &= 90 \text{ g/(c} \cdot \text{d)} \\ \text{TSS} &= 0.24 \text{ lb/(c} \cdot \text{d)} \\ &= 110 \text{ g/(c} \cdot \text{d)} \end{aligned}$$

**Example 2:** An example of industrial waste has an average flow of 1,230,000 gpd (4656 $m^3$/d) with $BOD_5$ of 9850 lb/d (4468 kg/d). Calculate BOD concentration and the equivalent populations of hydraulic and BOD loadings.

**solution:**

Step 1. Determine BOD concentration

$$Q = 1.23 \text{ MGD}$$

$$\text{BOD} = \frac{9850 \text{ lb/d}}{1.23 \text{ Mgal/d} \times 8.34 \text{ lb/(Mgal} \cdot \text{mg/L)}}$$
$$= 960 \text{ mg/L}$$

Note: MGD = million gallons per day
Mgal = million gallons

Step 2. Calculate equivalent populations, E. P.

$$\text{BOD E.P.} = \frac{9850 \text{ lb/d}}{0.17 \text{ lb/c/d}}$$
$$= 57{,}940 \text{ persons (capitas)}$$

$$\text{Hydraulic E.P.} = \frac{1{,}230{,}000 \text{ gpd}}{100 \text{ g/c/d}}$$
$$= 12{,}300 \text{ persons}$$

For the SI units calculations:

Step 3. Compute BOD concentration

$$\text{BOD} = \frac{4468 \text{ kg/d} \times 10^6 \text{ mg/kg}}{4656 \text{ m}^3\text{/d} \times 10^3 \text{ L/m}^3}$$
$$= 960 \text{ mg/L}$$

Step 4. Compute E.P., refer to Example 1

$$\text{BOD E.P.} = \frac{4468 \text{ kg/d} \times 1000 \text{ g/kg}}{77 \text{ g/c/d}}$$
$$\cong 58{,}000 \text{ capitas(persons)}$$

$$\text{Hydraulic E.P.} = \frac{4656 \text{ m}^3\text{/d} \times 1000 \text{ L/m}^3}{378.5\text{/c/d}}$$
$$= 12{,}300 \text{ persons}$$

**Chemical oxygen demand.** Concept and measurement of BOD are fully covered in Chapter 1. Since the $BOD_5$ test is time consuming, chemical oxygen demand is routinely performed for wastewater treatment operations after the relationship between $BOD_5$ and COD has been developed for a wastewater treatment plant. Many regulatory agencies accept the COD test as a tool of the wastewater treatment operation. The total organic carbon test is also a means for defining the organic content in wastewater.

The chemical oxygen demand is a measurement of the oxygen equivalent of the organic matter content of a sample that is susceptible to oxidation by a strong chemical oxidant, such as potassium dichromate. For a sample, COD can be related empirically to BOD, organic carbon, or organic matter.

After the correlation of $BOD_5$ and COD has been established, the COD test is useful for controlling and monitoring wastewater treatment processes. The COD test takes 3 to 4 h rather than 5 days for BOD data. The COD results are typically higher than the BOD values. The correlation between COD and BOD varies from plant to plant. The BOD:COD ratio also varies across the plant from influent to process units to effluent. The ratio is typically 0.5:1 for raw wastewater and may drop to as low as 0.1:1 for well-stabilized secondary effluent. The normal COD range for raw wastewater is 200 to 600 mg/L (WEF 1996a).

**COD test.** Most types of organic matter in water or wastewater are oxidized by a boiling mixture of sulfuric and chromic acids with excess of potassium dichromate ($K_2Cr_2O_7$). After 2-h refluxing, the remaining unreduced $K_2Cr_2O_7$ is titrated with ferrous ammonium sulfate to measure the amount of $K_2Cr_2O_7$ consumed and the oxidizable organic matter is calculated in terms of oxygen equivalent. The COD test can be performed by either the open reflux or the closed reflux method. The open reflux method is suitable for a wide range of wastewaters, whereas the closed reflux method is more economical but requires homogenization of samples containing suspended solids to obtain reproducible results.

The COD test procedures (open reflux) are (APHA *et al.*, 1998):

- Place appropriate size of sample in a 500-mL refluxing flask use
  50 mL if COD $<$ 900 mg/L
  use less sample and dilute to 50 mL if COD $>$900 mg/L
  use large size of sample if COD $<$50 mL
- Add 1 g of $HgSO_4$ and several glass beads
- Slowly add 5 mL sulfuric acid reagent and mix to dissolve $HgSO_4$
- Cool the mixture
- Add 25.0 mL of 0.0417 mole $K_2Cr_2O_7$ solution and mix
- Attach the refluxing flask to condenser and turn on cooling water
- Add 70 mL sulfuric acid reagent through open end of condenser, then swirl and mix
- Cover open end of condenser with a small beaker
- Reflux for 2 h (usually 1 blank with several samples)

- Cool and wash down condenser with distilled water to about twice volume
- Cool to room temperature
- Add 2–3 drops ferroin indicator
- Titrate excess $K_2Cr_2O_7$ with standard ferrous ammonium sulfate (FAS)
- Take as the end point of the titration of the first sharp color change from blue-green to reddish brown
- Calculate COD result as (APHA *et al.*, 1998)

$$\text{COD as mg } O_2/\text{L} = \frac{(A - B) \times M \times 8000}{\text{mL sample}} \tag{6.7}$$

where $A$ = mL FAS used for blank
$B$ = mL FAS used for sample
$M$ = molarity of FAS to be determined daily against standard $K_2Cr_2O_7$ solution ≅ 0.25 mole

**Example:** The results of a COD test for raw wastewater (50 mL used) are given. Volumes of FAS used for blank and the sample are 24.53 and 12.88 mL, respectively. The molarity of FAS is 0.242. Calculate the COD concentration for the sample.

**solution:** Using Eq. (6.7)

$$\text{COD as mg } O_2/\text{L} = \frac{(A - B) \times M \times 8000}{\text{mL sample}}$$

$$= (24.53 - 12.88) \times 0.242 \times 8000/50$$

$$= 451$$

### 2.3 Biological characteristics of wastewater

The principal groups of microorganisms found in wastewater are bacteria, fungi, protozoa, microscopic plants and animals, and viruses. Most microorganisms (bacteria, protozoa) are responsible and are beneficial for biological treatment processes of wastewater. However, some pathogenic bacteria, fungi, protozoa, and viruses found in wastewater are of public concern.

**Indicator bacteria.** Pathogenic organisms are usually excreted by humans from the gastrointestinal tract and discharge to wastewater. Water-borne diseases include cholera, typhoid, paratyphoid fever, diarrhea, and dysentery. The number of pathogenic organisms in wastewaters is generally low in density and they are difficult to isolate and identify. Therefore, indicator bacteria such as total coliform (TC), fecal coliform (FC), and fecal streptococcus (FS) are used as indicator organisms. The concept of indicator bacteria and enumeration of bacterial density are discussed in Chapter 1.

Tests for enumeration of TC, FC, and FS can be performed by multiple-tube fermentation (most probable number, MPN) or membrane filter methods; the test media used are different for these three groups of indicators. Most regulatory agencies have adopted fecal coliform density as an effluent standard because FC is mostly from fecal material.

## 3 Sewer Systems

Sewers are underground conduits to convey wastewater and stormwater to a treatment plant or to carry stormwater to the point of disposal. Sewers can be classified into three categories: sanitary, storm, and combined. Community sewer systems, according to their discharging types, can be divided into separated and combined sewer systems.

### 3.1 Separated sewer system

Separated sewers consist of sanitary sewers and stormwater sewer networks separately. Sanitary sewers carry a mixture of household and commercial sewage, industrial wastewater, water from groundwater infiltration/inflow, basement and foundation drainage connections, and cross-connections between sanitary sewers and stormwater drainage. Separated sanitary sewers should be free of stormwater, but they seldom are.

Storm sewers are commonly buried pipes that convey storm drainage. They may include open channel elements and culverts, particularly when drainage areas are large.

Storm sewer networks convey mainly surface storm runoff from roofs, streets, parking lots, and other surfaces toward the nearest receiving water body. An urban drainage system with separated sewers is more expensive than a combined sewer system because it uses two parallel conduits. Sanitary and storm sewers are usually designed to operate under gravity flow conditions. Pressure or vacuum sewers are rare.

Storm sewers are dry much of the time. When rainfalls are gentle, the surface runoffs are usually clear and low flows present no serious problem. However, flooding runoffs wash and erode unprotected areas and create siltation.

Illicit connections from roofs, yards, and foundations drain flow to sanitary sewers through manhole covers that are not tight. The flow rates vary and are as high as 70 gal/(c · d) and average 30 gal/(c · d). A rainfall of 1 in (2.54 cm)/ h on 1200 $ft^2$ (111 $m^2$) of roof produces a flow of 12.5 gal/min (0.788 L/s) or 18,000 gal/d. Spreading over an acre (0.4047 ha), 1 in/h equals 1.008 $ft^3$/s (28.5 L/s) of runoff. Leakage through manhole covers can add 20 to 70 gal/min (1.26 to 4.42 L/s) to the sewer when there is as much as 1 in of water on the streets (Fair *et al.*, 1966).

**Example:** A house has a roof of 30 ft by 40 ft (9.14 m × 12.19 m) and is occupied by four people. What is the percentage of rainfall of 1 in/h, using a rate of leakage through manholes of 60 gal/(c · d)?

**solution:**

Step 1. Compute stormwater runoff

$$\text{Runoff} = 30\text{ ft} \times 40\text{ ft} \times 1\text{ in/h} \times (1\text{ ft}/12\text{ in}) \times (24\text{ h/d})$$
$$\times\, 7.48\text{ gal/ft}^3 \div 4\text{ c} = 4488\text{ gal/(c}\cdot\text{d)}$$
$$\text{Say} = 4500\text{ gal/(c}\cdot\text{d)}$$

Step 2. Determine percent of leakage

$$\% = 60\text{ gal/(c}\cdot\text{d)} \times 100/4500\text{ gal/(c}\cdot\text{d)}$$
$$= 1.33$$

### 3.2 Combined sewers

Combined sewers are designed for collection and conveyance of both sanitary sewage, including industrial wastewaters, and storm surface runoff in one conduit. Combined sewer systems are common in old US and European urban communities. In the United States, there has been a trend to replace combined sewers by separate sewer systems. The dry-weather flow (sanitary and industrial wastewaters, plus street drainage from washing operations and lawn sprinkling and infiltration) is intercepted to the treatment facility. During storm events, when the combined wastewater and stormwater exceed the capacity of the treatment plant, the overflow is bypassed directly to the receiving water body without treatment or is stored for future treatment. The overflow is the so-called combined sewer overflow (CSO).

The average $BOD_5$ in stormwater is approximately 30 mg/L. The average $BOD_5$ in combined sewer overflows is between 60 and 120 mg/L (Novotny and Chesters, 1981). Since the CSO bypass may result in significant pollution, even during high river flows, it will create a hazardous threat to downstream water users. Thus, proper management of CSO is required (e.g. storage and later treatment). One obvious solution is to replace combined sewers by separate sewers; however, costs and legalities of shifting are major concerns.

**Example:** The following information is given:

Separate sanitary sewer flow = 250 L/(c · d)

Population density = 120 persons/ha

$BOD_5$ of raw wastewater = 188 mg/L

$BOD_5$ of plant effluent or stormwater = 30 mg/L

A storm intensity = 25 mm/d (1 in/d)

Impervious urban watershed = 72%

Compare flow and $BOD_5$ pollution potential produced by wastewater after treatment and by the stormwater.

**solution:**

Step 1. Compute flow and $BOD_5$ loads for wastewater

$$\begin{aligned}\text{Flow} &= 250\ \text{L/(c}\cdot\text{d)} \times 120\ \text{c/ha}\\ &= 30{,}000\ \text{L/(d}\cdot\text{ha)}\\ &= 30\ \text{m}^3\text{/(d}\cdot\text{ha)}\\ \text{Raw wastewater BOD}_5 &= 30{,}000\ \text{L/(d}\cdot\text{ha)} \times 188\ \text{mg/L}\\ &= 5{,}640{,}000\ \text{mg/(d}\cdot\text{ha)}\\ &= 5640\ \text{g/(d}\cdot\text{ha)}\\ \text{Effluent BOD}_5 &= 30{,}000\ \text{L/(d}\cdot\text{ha)} \times 30\ \text{mg/L}\\ &= 900\ \text{g/(d}\cdot\text{ha)}\\ \text{BOD}_5\ \text{removed} &= (5640 - 900)\ \text{g/(d}\cdot\text{ha)}\\ &= 4740\ \text{g/(d}\cdot\text{ha)}\ (84\%\ \text{removal})\end{aligned}$$

Step 2. Compute flow and $BOD_5$ load for stormwater

$$\begin{aligned}\text{Flow} &= 25\ \text{mm/d} \times 10{,}000\ \text{m}^2\text{/ha} \times 0.72 \times 0.001\ \text{m/mm} \times 1000\ \text{L/m}^3\\ &= 180{,}000\ \text{L/(d}\cdot\text{ha)}\\ &= 180\ \text{m}^3\text{/(d}\cdot\text{ha)}\\ \text{BOD}_5\ \text{load} &= 180{,}000\ \text{L/(d}\cdot\text{ha)} \times 30\ \text{mg/L}\\ &= 5{,}400{,}000\ \text{mg/(d}\cdot\text{ha)}\\ &= 5{,}400\ \text{g/(d}\cdot\text{ha)}\end{aligned}$$

Step 3. Compare $BOD_5$ load

$$\begin{aligned}\text{Storm water BOD}_5\text{: effluent BOD}_5 &= 5400 : 900\\ &= 6 : 1\end{aligned}$$

## 4 Quantity of Wastewater

The quantity of wastewater produced varies in different communities and countries, depending on a number of factors such as water uses, climate, lifestyle, economics, etc. Metcalf and Eddy, Inc. (1991) lists typical municipal water uses and wastewater flow rates in the United States including domestic, commercial, institutional, and recreational facilities and various institutions.

A typical wastewater flow rate from a residential home in the United States might average 70 gal (265 L) per capita per day (gal/(c · d) or gpcpd). Approximately 60% to 85% of the per capita consumption of water becomes wastewater. Commonly used quantities of wastewater flow rates that form miscellaneous types of facilities are listed in Table 6.1 (Illinois EPA, 1997).

Municipal wastewater is derived largely from the water supply. Some water uses, such as for street washing, lawn watering, fire fighting, and leakage from water supply pipelines, most likely do not reach the sewers. The volume of wastewater is added to by infiltration and inflows.

Wastewater flow rates for commercial developments normally range from 800 to 1500 gal/(acre · d) (7.5 to 14 $m^3$/(ha · d)), while those for industries are 1000 to 1500 gal/(acre · d) (9.4 to 14 $m^3$/(ha · d)) for light industrial developments and 1500 to 3000 gal/(acre · d) (14 to 28 $m^3$/(ha · d)) for medium industrial developments.

Water entering a sewer system from ground through defective connections, pipes, pipe joints, or manhole wells is called infiltration. The amount of flow into a sewer from groundwater, infiltration, may range from 100 to 10,000 gal/(d · in · mile) (0.093 to 9.26 $m^3$/(d · cm · km)) or more (Metcalf and Eddy, Inc. 1991). Construction specifications commonly permit a maximum infiltration rate of 500 gal/(d · in · mile) (0.463 $m^3$/(d · cm · km)) of pipe diameter. The quantity of infiltration may be equal to 3 to 5% of the peak hourly domestic wastewater flow, or approximately 10% of the average daily flow. With better pipe joint material and tight control of construction methods, infiltration allowance can be as low as 200 gal/(d · mile · in) of pipe diameter (Hammer, 1986).

Inflow to a sewer includes steady inflow and direct inflow. Steady inflow is water drained from springs and swampy areas, discharged from foundation, cellar drains, and cooling facilities, etc. Direct inflow is from stormwater runoff direct to the sanitary sewer.

**Example 1:** Convert to the SI units for the construction allowable infiltration rate of 500 gal/(d · mile) per in of pipe diameter.

**solution:**

$$500 \text{ gal/(d} \cdot \text{mile} \cdot \text{in)} = \frac{500 \text{ gal/d} \times 0.003785 \text{ m}^3\text{/gal}}{1 \text{ mile} \times 1.609 \text{ km/mile} \times 1 \text{ in} \times 2.54 \text{ cm/in}}$$

$$= 0.463 \text{ m}^3\text{/(d} \cdot \text{km) per cm of pipe diameter}$$

**TABLE 6.1 Typical Wastewater Flow Rates for Miscellaneous Facilities**

| Type of establishment | Gallons per person per day (unless otherwise noted) |
|---|---|
| Airports (per passenger) | 5 |
| Bathhouses and swimming pools | 10 |
| Camps: | |
| Campground with central comfort station | 35 |
| With flush toilets, no showers | 25 |
| Construction camps (semipermanent) | 50 |
| Day camps (no meals served) | 15 |
| Resort camps (night and day) with limited plumbing | 50 |
| Luxury camps | 100 |
| Cottages and small dwellings with seasonal occupancy | 75 |
| Country clubs (per resident member) | 100 |
| Country clubs (per nonresident member present) | 25 |
| Dwellings: | |
| Boarding houses | 50 |
| (additional for nonresident boarders) | 10 |
| Rooming houses | 40 |
| Factories (gallons per person, per shift, exclusive of industrial wastes) | 35 |
| Hospitals (per bed space) | 250 |
| Hotels with laundry (two persons per room) per room | 150 |
| Institutions other than hospitals including nursing homes (per bed space) | 125 |
| Laundries—self service (gallons per wash) | 30 |
| Motels (per bed) with laundry | 50 |
| Picnic parks (toilet wastes only per park user) | 5 |
| Picnic parks with bathhouses, showers and flush toilets (per park user) | 10 |
| Restaurants (toilet and kitchen wastes per patron) | 10 |
| Restaurants (kitchen wastes per meal served) | 3 |
| Restaurants (additional for bars and cocktail lounges) | 2 |
| Schools: | |
| Boarding | 100 |
| Day, without gyms, cafeterias, or showers | 15 |
| Day, with gyms, cafeterias, and showers | 25 |
| Day, with cafeterias, but without gyms or showers | 20 |
| Service stations (per vehicle served) | 5 |
| Swimming pools and bathhouses | 10 |
| Theaters: | |
| Movie (per auditorium set) | 5 |
| Drive-in (per car space) | 10 |
| Travel trailer parks without individual water and sewer hook-ups (per space) | 50 |
| Travel trailer parks with individual water and sewer hook-ups (per space) | 100 |
| Workers: | |
| Offices, schools and business establishments (per shift) | 15 |

SOURCE: Illinois EPA (1997)

**Example 2:** The following data are given:

Sewered population = 55,000
Average domestic wastewater flow = 100 gal/(c · d)
Assumed infiltration flow rate = 500 gal/(d · mile) per in of pipe diameter
Sanitary sewer systems for the city:
4-in house sewers = 66.6 miles
6-in building sewers = 13.2 miles
8-in street laterals = 35.2 miles
12-in submains = 9.8 miles
18-in mains = 7.4 miles

Estimate the infiltration flow rate and its percentage of the average daily and peak hourly domestic wastewater flows.

**solution:**

Step 1. Calculate the average daily flow ($Q$) and peak hourly flow ($Q_p$)

Assuming $Q_p = 3Q$

$$\begin{aligned} Q &= 100\ \text{gal/(c} \cdot \text{d)} \times 55{,}000\ \text{persons} \\ &= 5500{,}000\ \text{gal/d} \\ Q_p &= 5500{,}000\ \text{gal/d} \times 3 \\ &= 16{,}500{,}000\ \text{gal/d} \end{aligned}$$

Step 2. Compute total infiltration flow, $I$

$$\begin{aligned} I &= \text{infiltration rate} \times \text{length} \times \text{diameter} \\ &= 500\ \text{gal/(d} \cdot \text{mile} \cdot \text{in)} \\ &\quad \times (66.6 \times 4 + 13.2 \times 6 + 35.2 \times 8 + 9.8 \times 12 + 7.4 \times 18)\ \text{mile} \cdot \text{in} \\ &= 439{,}000\ \text{gal/d} \end{aligned}$$

Step 3. Compute percentages of infiltration to daily average and peak hourly flows

$$\begin{aligned} I/Q &= (439{,}000\ \text{gal/d})/(5{,}500{,}000\ \text{gal/d}) \times 100\% \\ &= 8.0\% \\ I/Q_p &= (439{,}000\ \text{gal/d})/(16{,}500{,}000\ \text{gal/d}) \times 100\% \\ &= 2.66\% \end{aligned}$$

### 4.1 Design flow rates

The average daily flow (volume per unit time), maximum daily flow, peak hourly flow, minimum hourly and daily flows, and design peak flow are generally used as the basis of design for sewers, lift stations, sewage (wastewater) treatment plants, treatment units, and other wastewater handling facilities. Definitions and purposes of flow are given as follows.

The design average flow is the average of the daily volumes to be received for a continuous 12-month period of the design year. The average flow may be used to estimate pumping and chemical costs, sludge generation, and organic-loading rates.

The maximum daily flow is the largest volume of flow to be received during a continuous 24-hour period. It is employed in the calculation of retention time for equalization basin and chlorine contact time.

The peak hourly flow is the largest volume received during a 1h period, based on annual data. It is used for the design of collection and interceptor sewers, wet wells, wastewater pumping stations, wastewater flow measurements, grit chambers, settling basins, chlorine contact tanks, and pipings. The design peak flow is the instantaneous maximum flow rate to be received. The peak hourly flow is commonly assumed as three times the average daily flow.

The minimum daily flow is the smallest volume of flow received during a 24-h period. The minimum daily flow is important in the sizing of conduits where solids might be deposited at low flow rates.

The minimum hourly flow is the smallest hourly flow rate occurring over a 24-h period, based on annual data. It is important to the sizing of wastewater flowmeters, chemical-feed systems, and pumping systems.

**Example:** Estimate the average and maximum hourly flow for a community of 10,000 persons.

Step 1. Estimate wastewater daily flow rate

Assume average water consumption = 200 L/(c · d)

Assume 80% of water consumption goes to the sewer

$$\text{Average wastewater flow} = 200 \text{ L/(c} \cdot \text{d)} \times 0.80 \times 10{,}000 \text{ persons} \times 0.001 \text{ m}^3\text{/L}$$
$$= 1600 \text{ m}^3\text{/d}$$

Step 2. Compute average hourly flow rate

$$\text{Average hourly flow rate} = 1600 \text{ m}^3\text{/d} \times 1 \text{ d/24 h}$$
$$= 66.67 \text{ m}^3\text{/h}$$

Step 3. Estimate the maximum hourly flow rate

Assume the maximum hourly flow rate is three times the average hourly flow rate, thus

$$\text{Maximum hourly flow rate} = 66.67 \text{ m}^3\text{/h} \times 3$$
$$= 200 \text{ m}^3\text{/h}$$

## 5 Urban Stormwater Management

In the 1980s, stormwater detention or retention basin became one of the most popular and widely used best management practices (BMPs) for quality enhancement of stormwater. In the United States, Congress mandated local governments to research ways to reduce the impact of separate storm sewer systems and CSO discharges on all receiving water bodies. In Europe, most communities use combined sewer systems, with some separate storm sewer systems in newer suburban communities. The CSO problem has received considerable attention in Europe.

Both quality and quantity of stormwater should be considered when protecting water quality and reducing property damage and traffic delay by urban flooding. Stormwater detention is an important measure for both quality and quantity control. Temporarily storing or detaining stormwater is a very effective method. Infiltration practices are the most effective in removing stormwater pollutants (Livingston, 1995). When stormwater is retained long enough, its quality will be enhanced.

Several publications (US EPA 1974c, 1983a, (NIPC) 1992, WEF and ASCE 1992, Wanielista and Yousef, 1993, Urbonas and Stahre, 1993, Pitt and Voorhees, 1995, Shoemaker *et al.,* 1995, Terstriep and Lee, 1995, Truong and Phua, 1995) describe stormwater management plans and design guidelines for control (e.g. detention or storage facilities) in detail. Storage facilities include local disposal, inlet control at source (rooftops, parking lots, commercial or industrial yards, and other surfaces), on-site detention (swales or ditches, dry basins or ponds, wet ponds, concrete basins, underground pipe packages or clusters), in-line detention (concrete basins, excess volume in the sewer system, pipe packages, tunnels, underground caverns, surface ponds), off-line storage (direct to storage), storage at treatment plant, and constructed wetland.

Two of the largest projects for urban stormwater management are given. Since the early 1950s, Toronto, Canada, has expanded with urban development. The city is located at the lower end of a watershed. In 1954, Hurricane Hazel brought a heavy storm (6 in. in 1 hour). Subsequently, overflow from the Don River flooded the city. Huge damage and losses occurred. Thereafter, an urban management commission was formed and two large flood control reservoirs were constructed. Natural techniques were applied to river basin management.

After Hurricane Hazel in 1954, the Government of Ontario created "Conservation Authorities" to manage water quality in major drainage basins in the province. Their initial focus was on flood control. It gradually expanded to what we now call integrated watershed management. In the late 1960s, the Government of Canada initiated federal-provincial comprehensive river basin planning exercises in various basins across the country. The results of these comprehensive river basin, or "aquatic ecosystem" planning exercises were mixed, but much was learned about technical matters, public participation, and combining technical, economic, environmental, and social issues.

In Chicago, USA, before the 1930s combined sewers were built, as in other older municipalities. About 100 storms per year caused a combination of raw wastewater and stormwater to discharge into Chicagoland waterways. Approximately 500,000 homes had chronic flooding problems. Based on the need for pollution and flood controls, the mass $3.66 billion Tunnel and Reservoir Plan (TARP) was formed. Also known as Chicago's Deep Tunnel, this project of intergovernmental efforts has undergone more than 25 years of planning and construction. The two-phased plan was approved by the water district commissions in 1972. Phase 1 aimed to control pollution and included 109 miles (175.4 km) of tunnels and three dewatering pumping stations. Phase 2 was designed for flood control and resulted in construction of three reservoirs with a total capacity of 16 billion gal (60.6 billion L) (Carder, 1997).

The entire plan consisted of four separate systems—Mainstream, Calumet, Des Plaines, and O'Hare. TARP has a total of 130 miles (209 km) of tunnels, 243 vertical drop shafts, 460 near-surface collecting structures, three pumping stations, and 126,000 acre · ft (155,400,000 $m^3$) of storage in three reservoirs. The reservoir project began in the middle of the 1990s (F. W. Dodge Profile, 1994; Kirk, 1997).

The tunnels were excavated through the dolomite rock strata, using tunnel boring machines. The rough, excavated diameter of the tunnel is about 33 ft (10 m). Tunnel depths range from 150 to 360 ft (45.7 to 110 m) below the ground surface and their diameter ranges from 9 to 33 ft (2.7 to 10 m). The construction of the Mainstream tunnel started in 1976 and it went into service in 1985.

The TARP's mainstream pumping station in Hodgkins, Illinois, is the largest pumping station in the United States. It boasts six pumps: four with a combined capacity of 710 Mgal/d (31.1 $m^3$/s) and two with a combined capacity of 316 Mgal/d (13.8 $m^3$/s). The stored wastewater is pumped to wastewater treatment plants. The TARP system received the 1986 Outstanding Civil Engineering Achievement award from the American Society of Civil Engineers (Robison, 1986) and the 1989 Outstanding Achievement in Water Pollution Control award from the Water Pollution Control Federation (current name, Water Environment Federation). It is now one of the important tour sites in Chicago.

### 5.1 Urban drainage systems

Urban drainage systems or storm drainage systems consist of flood runoff paths, called the major system. Major (total) drainage systems are physical facilities that collect, store, convey, and treat runoff that exceeds the minor systems. It is composed of paths for runoff to flow to a receiving stream. These facilities normally include detention and retention facilities, street storm sewers, open channels, and special structures such as inlets, manholes, and energy dissipators (Metcalf and Eddy, Inc. 1970; WEF and ASCE, 1992).

The minor or primary system is the portion of the total drainage system that collects, stores, and conveys frequently occurring runoff, and provides relief from nuisance and inconvenience. The minor system includes streets, sewers, and open channels, either natural or constructed. It has traditionally been carefully planned and constructed, and usually represents the major portion of the urban drainage infrastructure investment. The major drainage system is usually less controlled than the minor system and will function regardless of whether or not it has been deliberately designed and/or protected against encroachment, including when the minor system is blocked or otherwise inoperable (WEF and ASCE, 1992). The minor system is traditionally planned and designed to safely convey runoff from storms with a specific recurrence interval, usually 5 to 10 years (Novotny *et al.*, 1989). In the United States, flood drainage systems are included in flood insurance studies required by the Office of Insurance and Hazard Mitigation of the Federal Emergency Management Agency.

## 6 Design of Storm Drainage Systems

Urban drainage control facilities have progressed from crude drain ditches to the present complex systems. The systems start from surface runoff management control facilities, such as vegetation covers, detention or storage facilities, sedimentation basin, filtration, to curbs, gutters, inlets, manholes, and underground conduits.

Managing surface runoff in urban areas is a complex and costly task. Hydrology, climate, and physical characteristics of the drainage area should be considered for design purposes. Hydrologic conditions of precipitate–runoff relationship (such as magnitude, frequency, and duration of the various runoffs, maximum events) and local conditions (soil types and moisture, evapotranspiration, size, shape, slope, land-use, etc.) of the drainage (watershed) should be determined.

Clark and Viessman (1966) presented an example of hydraulic design of urban storm drainage systems to carry 10-year storm runoffs from eight inlets. A step-by-step solution of the problem was given.

## 7 Precipitation and Runoff

Precipitation includes rainfall, snow, hail, and sleet. It is one part of the hydrologic cycle. Precipitation is the primary source of water in springs, streams and rivers, lakes and reservoirs, groundwater, and well waters.

The US National Weather Service maintains observation stations throughout the United States. The US Geological Survey operates and maintains a national network of stream-flow gage stations throughout the country's major streams. State agencies, such as the Illinois State Water Survey and the Illinois Geological Survey, also have some stream flow data similar to the national records. In other countries, similar governmental agencies also maintain long-term precipitation, watershed runoff, stream flow data, and groundwater information.

Upon a catchment area (or watershed), some of the precipitation runs off immediately to streams, lakes or lower land areas, while some evaporates from land to air or penetrates the ground (infiltration). Snow remains where it falls, with some evaporation. After snow melts, the water may run off as surface and groundwater.

Rainwater or melting snow that travels as overland flow across the ground surface to the nearest channel is called surface runoff. Some water may infiltrate into the soil and flow laterally in the surface soil to a lower land water body as interflow. A third portion of the water may percolate downward through the soil until it reaches the ground-water. Overland flows and interflow are usually grouped together as direct runoff.

## 7.1 Rainfall intensity

The rainfall intensity is dependent on the recurrence interval and the time of concentration. The intensity can be determined from the cumulative rainfall diagram. From records of rainfall one can format the relationship between intensity, and frequency as well as duration. Their relationship is expressed as (Fair *et al.*, 1966, Wanielista, 1990)

$$i = \frac{at^m}{(b + d)^n} \tag{6.8}$$

where

$i$ = rainfall intensity, in/h or cm/h
$t$ = frequency of occurrences, yr
$d$ = duration of storm, min
$a, b, m, n$ = constants varying from place to place

By fixing a frequency of occurrence, the intensity–duration of $n$-year rainstorm events can be drawn. For many year storm events, the intensity–duration–frequency curves (Fig. 6.1) can be formatted for future use. The storm design requires some relationship between expected rainfall intensity and duration.

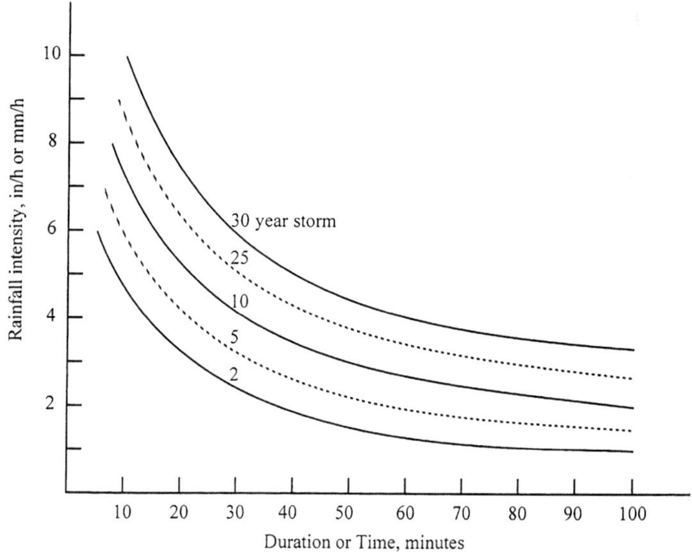


**Figure 6.1** Intensity–duration–frequency curves.

Let $A = at^m$ and $n = 1$. Eliminating $t$ in Eq. (6.8) the intensity–duration relationship (Talbot parabola formula) is

$$i = \frac{A}{B + d} \tag{6.9}$$

where $i$ = rainfall intensity, mm/h or in/h

$d$ = the duration, min or h

$A, B$ = constants

Rearrange the Talbot equation for a straight line

$$d = \frac{A}{i} - B$$

Use the following two equations to solve constants $A$ and $B$:

$$\begin{cases} \sum d = A\sum\frac{1}{i} - nB \\ \sum\frac{d}{i} = A\sum\frac{1}{i^2} - nB\sum\frac{1}{i} \end{cases}$$

From the data of rainfall intensity and duration, calculate $\Sigma d$, $\Sigma(1/i)$, $\Sigma(1/i^2)$, $\Sigma(d/i;)$ then $A$ and $B$ will be solved. For example, for a city of 5-year storm, $i$(mm/h) = 5680/[42 + $d$ (min)]. Similarly, a curve such as that in Fig. 6.1 can be determined (42 + $d$ (min)).

### 7.2 Time of concentration

The time for rainwater to flow overland from the most remote area of the drainage area to an inlet is called the inlet time, $t_i$. The time for water to flow from this inlet through a branch sewer to the storm sewer inlet is called the time of sewer flow, $t_s$. The time of concentration for a particular inlet to a storm sewer, $t$, is the sum of the inlet time and time of sewer flow, i.e.

$$t = t_i + t_s \tag{6.10}$$

The inlet time can be estimated by

$$t_i = C(L/S\, i^2)^{1/3} \tag{6.11}$$

where $t_i$ = time of overland flow, min
$L$ = distance of overland flow, ft
$S$ = slope of land, ft/ft
$i$ = rainfall intensity, in/h
$C$ = coefficient
= 0.5 for paved areas
= 1.0 for bare earth
= 2.5 for turf

The time of concentration is difficult to estimate since the different types of roof, size of housing, population density, land coverage, topography, and slope of land are very complicated. In general, the time of concentration may be between 5 and 10 min for both the mains and the submains.

**Example:** The subdrain sewer line to a storm sewer is 1.620 km (1 mile). The average flow rate is 1 m/s. Inlet time from the surface inlet through the overland is 7 min. Find the time of concentration to the storm sewer.

**solution:**

Step 1. Determine the time of sewer flow, $t_s$

$$t_s = \frac{L}{v} = \frac{1620 \text{ m}}{1 \text{ m/s}} = 1620 \text{ s} = 27 \text{ min}$$

Step 2. Find $t$

$$\begin{aligned} t &= t_i + t_s = 7 \text{ min} + 27 \text{ min} \\ &= 34 \text{ min} \end{aligned}$$

### 7.3 Estimation of runoff

The quantity of stormwater runoff can be estimated by the rational method or by the empirical formula. Each method has its advantages. The rational formula for the determination of the quantity of stormwater runoff is

$$Q = CIA \tag{6.12a}$$

where $Q$ = peak runoff from rainfall, $\mathrm{ft^3/s}$
$C$ = runoff coefficient, dimensionless
$I$ = rainfall intensity, in rain/h
$A$ = drainage area, acres

The formula in SI units is

$$Q = 0.278\ CIA \tag{6.12b}$$

where $Q$ = peak runoff, $\mathrm{m^3/s}$
$C$ = runoff coefficient, see Table 6.2
$I$ = rainfall intensity, mm/h
$A$ = drainage area, $\mathrm{km^2}$

The coefficient of runoff for a specific area depends upon the character and the slope of the surface, the type and extent of vegetation, and other factors. The approximate values of the runoff coefficient are shown in Table 6.2. There are empirical formulas for the runoff coefficient. However, the values of $C$ in Table 6.2 are most commonly used.

**Example:** In a suburban residential area of 1000 acres (4.047 $\mathrm{km^2}$), the rainfall intensity–duration of a 20-min 25-year storm is 6.1 in/h (155 mm/h). Find the maximum rate of runoff.

**TABLE 6.2 Coefficient of Runoff for Various Surfaces**

| Type of surface | Flat slope <2% | Rolling slope 2–10% | Hilly slope <10% |
|---|---|---|---|
| Pavements, roofs | 0.90 | 0.90 | 0.90 |
| City business areas | 0.80 | 0.85 | 0.85 |
| Dense residential areas | 0.60 | 0.65 | 0.70 |
| Suburban residential areas | 0.45 | 0.50 | 0.55 |
| Earth areas | 0.60 | 0.65 | 0.70 |
| Grassed areas | 0.25 | 0.30 | 0.30 |
| Cultivated land | | | |
| clay, loam | 0.50 | 0.55 | 0.60 |
| sand | 0.25 | 0.30 | 0.35 |
| Meadows and pasture lands | 0.25 | 0.30 | 0.35 |
| Forests and wooded areas | 0.10 | 0.15 | 0.20 |

SOURCE: Perry (1967)

**solution:** From Table 6.2, $C = 0.50$ (for rolling suburban)
Using Eq. (6.12a)

$$Q = CIA$$

$$= 0.5 \times 6.1 \text{ in/h} \times 1000 \text{ acre}$$

$$= 3050 \frac{\text{acre} \cdot \text{in}}{\text{h}} \times \frac{1 \text{ h}}{3600 \text{ s}} \times \frac{1 \text{ ft}}{12 \text{ in}} \times \frac{43{,}560 \text{ ft}^2}{1 \text{ acre}} = 3075 \text{ ft}^3/\text{s}$$

*Note*: In practice, a factor of 1.0083 conversion is not necessary. The answer could be just 3050 $\text{ft}^3/\text{s}$.

## 8 Stormwater Quality

The quality and quantity of stormwaters depend on several factors—intensity, duration, and area extent of storms. The time interval between successive storms also has significant effects on both the quantity and quality of stormwater runoff. Land contours, urban location permeability, land uses and developments, population densities, incidence and nature of industries, size and layout of sewer systems, and other factors are also influential.

Since the 1950s, many studies on stormwater quality indicate that runoff quality differs widely in pattern, background conditions, and from location to location. Wanielista and Yousef (1993) summarized runoff quality on city street, lawn surface, rural road, highway, and by land-use categories. Rainwater quality and pollutant loading rates are also presented.

### 8.1 National urban runoff program

In 1981 and 1982, the US EPA's National Urban Runoff Program (NURP) collected urban stormwater runoff data from 81 sites located in 22 cities throughout the United States. The data covered more than 2300 separate storm events. Data was evaluated for solids, oxygen demand, nutrients, metals, toxic chemicals, and bacteria. The median event mean concentrations (EMC) and coefficients of variance for ten standard parameters for four different land use categories are listed in Table 6.3 (US EPA 1983a). It was found that lead, copper, and zinc are the most significant heavy metals found in urban runoff and showed the highest concentrations. The EMC for a storm is determined by flow-weighted calculation.

### 8.2 Event mean concentration

Individual pollutants are measured on the flow-weighted composite to determine the event mean concentrations. The event mean concentration

**TABLE 6.3 Mean EMCS and Coefficient of Variance for all NURP Sites by Land Use Category**

| | Residential | | Mixed | | Commercial | | Open/Nonurban | |
|---|---|---|---|---|---|---|---|---|
| Pollutant | Median | CV | Median | CV | Median | CV | Median | CV |
| BOD (mg/L | 10.0 | 0.41 | 7.8 | 0.52 | 9.3 | 0.31 | – | – |
| COD (mg/L) | 73 | 0.55 | 65 | 0.58 | 57 | 0.39 | 40 | 0.78 |
| TSS (mg/L) | 101 | 0.96 | 67 | 1.14 | 69 | 0.35 | 70 | 2.92 |
| Total lead (μg/L) | 144 | 0.75 | 114 | 1.35 | 104 | 0.68 | 30 | 1.52 |
| Total copper (μ/L) | 33 | 0.99 | 27 | 1.32 | 29 | 0.81 | – | – |
| Total zinc (μg/L) | 135 | 0.84 | 154 | 0.78 | 226 | 1.07 | 195 | 0.66 |
| Total kjeldahl nitrogen (μg/L) | 1900 | 0.73 | 1288 | 0.50 | 1179 | 0.43 | 965 | 1.00 |
| $NO_2$–N+$NO_3$–N (μg/L) | 736 | 0.83 | 558 | 0.67 | 572 | 0.48 | 543 | 0.91 |
| Total P (μg/L) | 383 | 0.69 | 263 | 0.75 | 201 | 0.67 | 121 | 1.66 |
| Soluble P (μg/L) | 143 | 0.46 | 56 | 0.75 | 80 | 0.71 | 26 | 2.11 |

*Notes*: EMC = event mean concentration
NURP = National urban runoff program
CV = coefficient of variance
SOURCE: US EPA (1983a)

is defined as the event loading for a specific constituent divided by the event stromwater volume. It is expressed as

$$C = \frac{W}{V} \tag{6.13}$$

where $C$ = event mean concentration, mg/L
$W$ = total loading per event, mg
$V$ = volume per event, L

The total loading for a storm event is determined by the sum of the loadings during each sampling period, the loading being the flow rate (volume) multiplied by the concentration. The loading per event is

$$W = \sum_{i=1}^{n} V_i C_i \tag{6.14}$$

where $W$ = loading for a storm event, mg
$n$ = total number of samples taken during a storm event
$V_i$ = volume proportional to flow rate at time $i$, L
$C_i$ = concentration at time $i$, mg/L

**Example:** An automatic sampler was installed at the confluence of the main tributary of a lake. The following sampling time and tributary flow rate were measured in the field and the laboratory. The results are shown in Table 6.4.

**TABLE 6.4 Data Collected from Stormwater Sampling**

| Time of collection (1) | Creek flow, L/s (2) | Total phosphorus, mg/L (3) |
|---|---|---|
| 19:30 | 0 | 0 |
| 19:36 | 44 | 52 |
| 19:46 | 188 | 129 |
| 19:56 | 215 | 179 |
| 20:06 | 367 | 238 |
| 20:16 | 626 | 288 |
| 20:26 | 752 | 302 |
| 20:36 | 643 | 265 |
| 20:46 | 303 | 189 |
| 20:56 | 105 | 89 |
| 21:06 | 50 | 44 |

The watershed covers mainly agricultural lands. The flow rate and total phosphorus (TP) concentrations listed are the recorded values subtracted by the dry flow rate and TP concentration during normal flow period. Estimate the EMC and TP concentration for this storm event.

**solution:**

Step1. Calculate sampling interval $\Delta t$ = time intervals of 2 successive samplings in column 1 of Table 6.5.

Step 2. Calculate mean flow $q$ (col. 2 of Table 6.5)

$q$ = average of 2 successive flows measured in column 2 of Table 6.4

Step 3. Calculate mean runoff volume, $\Delta V$

In Table 6.5, col. 3 = col. 1 × 60 × col. 2

**TABLE 6.5 Calculations for Total Phosphorus Loading**

| Sampling interval, $\Delta t$ min (1) | Mean flow, L/s (2) | Mean runoff volume $\Delta V$, L (3) | Mean TP, μg/L (4) | Loading $\Delta W$, mg (5) |
|---|---|---|---|---|
| 6 | 22 | 7,920 | 26 | 206 |
| 10 | 116 | 69,600 | 90.5 | 6,300 |
| 10 | 202 | 121,200 | 153.5 | 18,604 |
| 10 | 291 | 174,600 | 288 | 50,285 |
| 10 | 497 | 298,200 | 263 | 78,427 |
| 10 | 689 | 413,400 | 295 | 121,953 |
| 10 | 638 | 382,800 | 283.5 | 108,524 |
| 10 | 473 | 283,800 | 227 | 64,423 |
| 10 | 204 | 122,400 | 139 | 17,014 |
| 10 | 78 | 46,800 | 66.5 | 2,713 |
| | | Σ = 1,920,720 | | Σ = 468,449 |

Step 4. Calculate total volume of the storm

$$V = \Sigma\Delta V = 1{,}920{,}720 \text{ L}$$

Step 5. Calculate mean total phosphorus, TP (col. 4 of Table 6.5)

TP = average of 2 successive TP values in Table 6.4

Step 6. Calculate loading (col. 5 of Table 6.5)

$$\Delta W = (\Delta V \text{ in L})(\text{TP in } \mu\text{g/L})(1 \text{ mg}/1000\ \mu\text{g})$$

Step 7. Calculate total load by Eq. (6.14)

$$W = \Sigma\Delta W = 468{,}449 \text{ mg}$$

Step 8. Calculate EMC

$$C = W/V = 468{,}449 \text{ mg}/1{,}920{,}720 \text{ L}$$

$$= 0.244 \text{ mg/L}$$

*Note*: If the TP concentrations are not flow weighted, the mathematic mean of TP is 0.183 mg/L which is less than the EMC.

## 8.3 Street and road loading rate

Rainfall will carry pollutants from the atmosphere and remove deposits on impervious surfaces. The rainfall intensity, duration, and storm runoff affect the type and the amount of pollutants removed. According to the estimation by the US EPA (1974a), the times required for 90% particle removal from impervious surfaces are 300, 90, 60, and 30 min, respectively for 0.1, 0.33, 0.50, and 1.00 in/h storms. Average loading of certain water quality parameters and some heavy metals from city street, highway, and rural road are given in Table 6.6. Loading rates vary

**TABLE 6.6 Average Loads (kg/curb · km · d) of Water Quality Parameters and Heavy Metals on Streets and Roads**

| Parameter | Highway | City street | Rural road |
|---|---|---|---|
| $BOD_5$ | 0.900 | 0.850 | 0.140 |
| COD | 10.000 | 5.000 | 4.300 |
| $PO_4$ | 0.080 | 0.060 | 0.150 |
| $NO_3$ | 0.015 | 0.015 | 0.025 |
| N (total) | 0.200 | 0.150 | 0.055 |
| Cr | 0.067 | 0.015 | 0.019 |
| Cu | 0.015 | 0.007 | 0.003 |
| Fe | 7.620 | 1.360 | 2.020 |
| Mn | 0.134 | 0.026 | 0.076 |
| Ni | 0.038 | 0.002 | 0.009 |
| Pb | 0.178 | 0.093 | 0.006 |
| Sr | 0.018 | 0.012 | 0.004 |
| Zn | 0.070 | 0.023 | 0.006 |

SOURCE: Wanielista and Youset (1993).

with local conditions. Loads from highways are larger than those from city streets and rural roads. Loading is related to daily traffic volume.

**Example:** A storm sewer drains a section of a downtown city street 260 m (850 ft) and 15 m (49 ft) wide with curbs on both sides. Estimate the $BOD_5$ level in the stormwater following a 60-min storm of 0.5 in/h. Assume there is no other source of water contribution or water losses; also, it has been 4 days since the last street cleansing.

**solution:**

Step 1. Determine BOD loading

From Table 6.6, average BOD loading for a city street is 0.850 kg/(curb km · d)

$$= 850{,}000 \text{ mg/(curb km} \cdot \text{d)}$$

$$\text{Total street length} = 0.26 \text{ km} \times 2 \text{ curb} = 0.52 \text{ km} \cdot \text{curb}$$

$$\text{Total BOD loading} = 850{,}000 \text{ mg/(curb km} \cdot \text{d)} \times 0.52 \text{ km} \cdot \text{curb} \times 4\text{d}$$

$$= 1{,}768{,}000 \text{ mg}$$

A 0.5 in/h storm of 60 min duration will remove 90% of pollutant; thus

$$\text{Runoff BOD mass} = 1{,}768{,}000 \text{ mg} \times 0.9$$

$$= 1{,}591{,}000 \text{ mg}$$

Step 2. Determine the volume of runoff

$$\text{Area of street} = 260 \text{ m} \times 15 \text{ m} = 3900 \text{ m}^2$$

$$\text{Volume} = 0.5 \text{ in/h} \times 1 \text{ h} \times 0.0254 \text{ m/in} \times 3900 \text{ m}^2$$

$$= 49.53 \text{ m}^3$$

$$= 49{,}530 \text{ L}$$

Step 3. Calculate BOD concentration

$$\text{BOD} = 1{,}591{,}000 \text{ mg} \div 49{,}530 \text{ L}$$

$$= 32.1 \text{ mg/L}$$

### 8.4 Runoff models

Several mathematical models have been proposed to predict water quality characteristics from storm runoffs. A consortium of research constructions proposed an urban runoff model based on pollutants deposited on street surfaces and washed off by rainfall runoff. The amount of pollutants transported from a watershed ($dP$) in any time interval ($dt$) is proportional to the amount of pollutants remaining in the watershed ($P$). This assumption is a first-order reaction (discussed in Chapter 1):

$$-\frac{dP}{dt} = kP \tag{6.15}$$

Integrating Eq. (6.15) to form an equation for mass remaining

$$P = P_0 e^{-kt} \tag{6.16}$$

The amount removed (proposed model) is

$$P_0 - P = P_0(1 - e^{-kt}) \tag{6.17}$$

where $P_0$ = amount of pollutant on the surface initially present, lb
$P$ = amount of pollutant remaining after time $t$, lb
$P_0 - P$ = amount of pollutant washed away in time $t$, lb
$k$ = transport rate constant, per unit time
$t$ = time

The transport rate constant $k$ is a factor proportional to the rate of runoff. For impervious surfaces,

$$k = br$$

where $b$ = constant
$r$ = rainfall excess

To determine $b$, it was assumed that a uniform runoff of 0.5 in/h would wash 90% of the pollutant in 1 h (as stated above). We can calculate $b$ from Eq. (6.16)

$$P/P_0 = 0.1 = e^{-brt} = e^{-b(0.5\text{ in/h})\,(1\text{h})} = e^{-0.5b}$$

$$2.304 = 0.5b\ (\text{in})$$

$$b = 4.6\text{ in}$$

Substituting $b$ into Eq. 6.17 leads to the equation for impervious surface areas

$$P_0 - P = P_0(1 - e^{-4.6rt}) \tag{6.18}$$

Certain modifications to the basic model (Eq. (6.18)) for predicting total suspended solids and BOD have been proposed to refine the agreement between the observed and predicted values for these parameters. The University of Cincinnati Department of Civil Engineering (1970) developed a mathematical model for urban runoff quality based essentially on the same principles and assumptions as in Eq. (6.18). The major difference is that an integral solution was developed by this group instead of the stepwise solution suggested by the consortium. The amount of a pollutant remaining on a runoff surface at a particular time, the rate of runoff at that time, and the general characteristics of the watershed were found to be given by the following relationship

$$P = P_0 e^{-kV_t} \tag{6.19}$$

where $P_0$ = amount of pollutant initially on the surface, lb
$P$ = amount of pollutant remaining on the surface at time $t$, lb
$V_t$ = accumulated runoff water volume up to time $t$

$$= \int_0^t q\,dt$$

$q$ = runoff intensity at time $t$
$k$ = constant characterizing the drainage area

Many computer simulation models for runoff have been proposed and have been discussed by McGhee (1991).

## 9 Sewer Hydraulics

The fundamental concepts of hydraulics can be applied to both sanitary sewers and stormwater drainage systems. Conservation of mass, momentum, energy, and other hydraulic characteristics of conduits are discussed in Chapter 4 and elsewhere (Metcalf and Eddy, Inc. 1991; WEF and ASCE, 1992) WEF and ASCE, (1992) also provide the design guidelines for urban stormwater drainage systems, including system layout, sewer inlets, street and intersection, drainage ways (channels, culverts, bridges), erosion control facilities, check dams, energy dissipators, drop shaft, siphons, side-overflow weirs, flow splitters, junctions, flap gates, manholes, pumping, combined sewer systems, evaluation and mitigation of combined sewer overflows, stormwater impoundments, and stormwater management practices for water quality enhancement, etc.

## 10 Sewer Appurtenances

The major appurtenances used for wastewater collection systems include street (stormwater) inlets, catch basins, manholes, building connection, flushing devices, junctions, transitions, inverted siphons, vertical drops, energy dissipators, overflow and diversion structure, regulators, outlets, and pumping stations.

### 10.1 Street inlets

Street inlets are structures to transfer stormwater and street cleansing wastewater to the storm sewers. The catch basin is an inlet with a basin which allows debris to settle. There are four major types of inlet, and multiple inlets. Location and design of inlets should consider traffic safety (for both pedestrian and vehicle) and comfort. Gutter inlet is more efficient than curb inlet in capturing gutter flow, but clogging by debris is a problem. Combination inlets are better. Various manufactured inlets and assembled gratings are available.

Street inlets are generally placed near the corner of the street, depending on the street length. The distance between inlets depends on the amount of stormwater, the water depth of the gutter, and the depression to the gutter. The permissible depth of stormwater in most U.S. cities is limited to 6 in (15 cm) on residential streets.

Flow in the street gutter can be calculated by the Manning formula, modified for a triangular gutter cross section (McGhee, 1991):

$$Q = K(z/n)s^{1/2}y^{8/3} \tag{6.20}$$

where $Q$ = gutter flow, $m^3/s$ or $ft^3/s$
$K$ = constant = $22.61\ m^3/(min \cdot m) = 0.38\ m^3/(s \cdot m)$
or = $0.56\ ft^3/(s \cdot ft)$
$z$ = reciprocal of the cross transverse slope of the gutter
$n$ = roughness coefficient
= 0.015 for smooth concrete gutters
$s$ = slope of the gutter
$y$ = water depth in the gutter at the curb

The water depth at the curve can be calculated from the flow and the street cross section and slope. The width over which the water will spread is equal to $yz$.

**Example 1:** A street has a longitudinal slope of 0.81%, a transverse slope of 3.5%, a curb height of 15 cm (6 in), and a coefficient of surface roughness ($n$) of 0.016. The width of the street is 12 m (40 ft). Under storm design conditions, 4 m of street width should be kept clear during the storm. Determine the maximum flow that can be carried by the gutter.

**solution:**

Step 1. Calculate the street spread limit, $w$

$$w = (12\ m - 4\ m)/2 = 4\ m$$

Step 2. Calculate the curb depth ($d$) with the spread limit for a transverse slope of 3.5%

$$d = 4\ m \times 0.035 = 0.14\ m = 14\ cm$$

The street gutter flow is limited by either the curb height (15 cm) or the curb depth with the spread limit. Since

$$d = 14\ cm$$

a curb height of 14 cm is the limit factor for the gutter flow.

Step 3. Calculate the maximum gutter flow $Q$ by Eq. (6.20)

$$z = 1/0.035 = 28.57$$

$$\begin{aligned} Q &= K(z/n)s^{1/2}y^{8/3} \\ &= 0.38\ \text{m}^3/(\text{s}\cdot\text{m})(28.57/0.016)\ (0.0081)^{1/2}(0.14\ \text{m})^{8/3} \\ &= 0.323\ \text{m}^3/\text{s} \\ &= 11.4\ \text{ft}^3/\text{s} \end{aligned}$$

**Example 2:** The stormwater flow of a street gutter at an inlet is 0.40 $\text{m}^3$/s (14.1 $\text{ft}^3$/s). The longitudinal slope of the gutter is 0.01 and its cross transverse slope is 0.025. The value of roughness coefficient is 0.016. Estimate the water depth in the gutter at the curb.

**solution:**

Step 1. Calculate the value of $z$

$$z = 1/0.025 = 40$$

Step 2. Find $y$ by Eq. (6.20)

$$\begin{aligned} 0.40\ \text{m}^3/\text{s} &= 0.38\ \text{m}^3/(\text{s}\cdot\text{m})\ (40/0.016)\ (0.01)^{1/2}\ (y\ \text{m})^{8/3} \\ y^{8/3} &= 0.00421\ \text{m} \\ y &= 0.129\ \text{m} \\ &= 5.06\ \text{in} \end{aligned}$$

## 10.2 Manholes

Manholes provide an access to the sewer for inspection and maintenance operations. They also serve as ventilation, multiple pipe intersections, and pressure relief. Most manholes are cylindrical in shape.

The manhole cover must be secured so that it remains in place and avoids a blow-out during peak flooding period. Leakage from around the edges of the manhole cover should be kept to a minimum.

For small sewers, a minimum inside diameter of 1.2 m (4 ft) at the bottom tapering to a cast-iron frame that provides a clear opening usually specified as 0.6 m (2 ft) has been widely adopted (WEF and ASCE, 1992). For sewers larger than 600 mm (24 in), larger manhole bases are needed. Sometimes a platform is provided at one side, or the manhole is simply a vertical shaft over the center of the sewer. Various sizes and types of manholes are available locally.

Manholes are commonly located at the junctions of sanitary sewers, at changes in grades or alignment except in curved alignments, and at locations that provide ready access to the sewer for preventive maintenance and emergency service. Manholes are usually installed at street intersections (Parcher, 1988).

Manhole spacing varies with available sanitary sewer maintenance methods. Typical manhole spacings range from 90 to 150 m (300 to 500 ft) in straight lines. For sewers larger than 1.5 m (5 ft), spacings of 150 to 300 m (500 to 1000 ft) may be used (ASCE and WPCF, 1982).

Where the elevation difference between inflow and outflow sewers exceeds about 0.5 m (1.5 ft), sewer inflow that is dropped to the elevation of the outflow sewer by an inside or outside connection is called a drop manhole (or drop inlet). Its purpose is to protect workers from the splashing of wastewater, objectionable gases, and odors.

## 10.3 Inverted siphons (depressed sewers)

A sewer that drops below the hydraulic gradient to pass under an obstruction, such as a railroad cut, subway, highway, conduit, or stream, is often called an inverted siphon. More properly, it should be called a depressed sewer. Because a depressed sewer acts as a trap, the velocity of sewer flow should be greater than 0.9 m/s (3 ft/s) or more for domestic wastewater, and 1.25 to 1.5 m/s (4 to 5 ft/s) for stormwater, to prevent deposition of solids (Metcalf and Eddy, Inc. 1991). Thus, sometimes, two or more siphons are needed with an inlet splitter box.

In practice, minimum diameters for depressed sewers are usually the same as for ordinary sewers: 150 or 200 mm (6 or 8 in) in sanitary sewers, and about 300 mm (12 in) in storm sewers (Metcalf and Eddy, Inc. 1991).

The determination of the pipe size for depressed sewers is the same as for water and wastewater mains. The size depends upon the maximum wastewater flow and the hydraulic gradient.

Due to high velocities in depressed sewers, several pipes in parallel are commonly used. For example, it may be that a small pipe may be designed large enough to carry the minimum flow; a second pipe carries the difference between the minimum and average flow (or maximum dry-weather flow); and a third pipe carries peak flow above the average flow. Depressed sewers can be constructed of ductile iron, concrete, PVC, or tile.

**Example:** Design a depressed sewer system using the following given conditions:

- Diameter of gravity sewer to be connected by depressed sewer = 910 mm (36 in)
- Slope of incoming sewer, $S = 0.0016$ m/m (ft/ft)
- Minimum flow velocity in depressed sewer = 0.9 m/s (3 ft/s)
- Length of depressed sewer = 100 m (328 ft)
- Maximum sewer deression = 2.44 m (8 ft)

- Design flows:
  minimum flow = 0.079 $m^3$/s (2.8 $ft^3$/s)
  average flow = 0.303 $m^3$/s (10.7 $ft^3$/s) = maximum dry-weather flow
  full (maximum) flow = capacity of gravity sanitary sewer
- Design three inverted siphons from the inlet chamber
  (1) to carry minimum flow
  (2) to carry flows from minimum to average
  (3) to carry all flows above the average flow
- Available fall from invert to invert = 1.0 m (3.3 ft)
- Available head loss at inlet = 125 mm (0.5 ft)
- Available head loss for friction in depressed sewer = 1.0 m (3.3 ft)
- Available hydraulic grade line = 1 m/100 m = 0.01 m/m
- $n = 0.015$ (ductile-iron pipe)

*Note*: The above information is required to design a depressed sewer system.

**solution:**

Step 1. Design the depressed sewer

(a) Calculate velocity and flow of the 910 mm *(D)* sewer for full flow

$$\text{by Eq. (4.21)} \quad R = D/4 = 910 \text{ mm}/4 = 227.5 \text{ mm}$$
$$= 0.2275 \text{ m}$$
$$\text{by Eq. (4.22)} \quad V = (1/n)R^{2/3}S^{1/2}$$
$$= (1/0.015)\,(0.2275)^{2/3}(0.0016)^{1/2}$$
$$= 0.994 \text{ (m/s)}$$
$$\text{Flow } Q = AV = 3.14\,(0.455 \text{ m})^2\,(0.994 \text{ m/s})$$
$$= 0.646 \text{ m}^3/\text{s}$$

(b) Determine the size of the small depressed sewer pipe to carry the minimumflow ($d$ = diameter of the pipe)

$$Q = \pi(d/2)^2(1/n)(d/4)^{2/3}S^{1/2}$$
$$= (0.3115/n)\,d^{8/3}\,S^{1/2}$$
$$= 0.079 \text{ (m}^3\text{/s)}$$

and

$$S = 0.01 \text{ m/m}$$

then

$$0.079 = (0.3115/0.015)\,d^{8/3}(0.01)^{1/2}$$
$$0.079 = 2.077d^{8/3}$$

$$d^{8/3} = 0.038$$
$$d = 0.293 \text{ (m)}$$
$$\cong 300 \text{ mm}$$
$$= 12 \text{ in}$$

Using a 12-in (300 mm) pipe will just carry the 0.079 $m^3/s$ flow Check velocity

$$V = (0.397/n)\, d^{2/3} S^{1/2}$$
$$= (0.397/0.015)\,(0.304)^{2/3}(0.01)^{1/2}$$
$$= 2.647\,(0.304)^{2/3}$$
$$= 1.20 \text{ (m/s) (verified, } >0.9 \text{ m/s)}$$

*Note*: A nomograph for the Manning equation can be used without calculation.

$$0.397 = 1/4^{2/3}$$

(c) Determine the size of the second depressed sewer pipe for maximum dry-weather flow above the minimum flow

$$Q = (0.303 - 0.079)\ m^3/s$$
$$= 0.224\ m^3/s$$
$$0.224 = 2.077\, d^{8/3} \text{ (refer to step (b))}$$
$$d^{8/3} = 0.1078$$
$$d = 0.434 \text{ (m)}$$
$$= 17.1 \text{ in}$$

A standard 18-in (460 mm) pipe would be used. Check velocity of 460 mm pipe, from Step 1b

$$V = 2.647\,(0.460)^{2/3} = 1.58 \text{ (m/s)}$$

The capacity of the 460 mm pipe would be (refer to step (b))

$$Q = 2.077\,(0.46)^{8/3}$$
$$= 0.262\ (m^3/s)$$

(d) Determine the size of the third pipe to carry the peak flow, from steps (a) and (c). The third pipe must carry (0.646 − 0.079 − 0.262) $m^3/s$

$$= 0.305\ m^3/s$$

The size ($d$) required would be

$$d^{8/3} = 0.305/2.077$$
$$d^{8/3} = 0.1468$$
$$d = 0.487 \text{ (m)}$$

The size of 500-mm (20-in) diameter standard is chosen.
The capacity and velocity of a 500-mm pipe with 0.01 hydraulic slope is

$$
\begin{aligned}
Q &= 2.077\ (0.50)^{8/3} \\
&= 0.327\ (\text{m}^3/\text{s}) \\
V &= 2.647\ (0.50)^{2/3} \\
&= 1.67\ (\text{m/s})
\end{aligned}
$$

(e) Calculate total capacity of the three pipes (300, 460, and 500 mm)

$$
\begin{aligned}
Q &= (0.079 + 0.262 + 0.327)\ \text{m}^3/\text{s} \\
&= 0.668\ \text{m}^3/\text{s}\ (0.646\ \text{m}^3/\text{s is needed})
\end{aligned}
$$

Step 2. Design the inlet and outlet chambers

These depressed sewer pipes are connected from the inlet chamber and outlet chamber. Weirs (2 m in length) are installed to divide the chamber into three portions. The design detail can be found elsewhere (Metcalf and Eddy, Inc. 1981).

## 11 Pumping Stations

The pumping station must be able to adjust to the variations of wastewater flows. The smallest capacity pump should be able to pump from the wet well and discharge at a self-cleansing velocity of about 0.6 m/s (2 ft/s). It should be connected to a 100-mm (4-in) force main which would have a capacity of approximately 280 L/min or 74 gal/min. The wet well capacity should contain sufficient wastewater to permit the pump to run for at least 2 min and restart not more than once in 5 min (Steel and McGhee, 1979). The pump running time ($t_r$) and the filling time ($t_f$) in the wet well are computed as

$$t_r = \frac{V}{D - Q} \tag{6.21}$$

and

$$t_f = \frac{V}{Q} \tag{6.22}$$

where $t_r$ = pump running time, min
$V$ = storage volume of wet well, L or gal
$D$ = pump discharge, at peak flow, L/min ($\text{m}^3$/s) or gpm, L/min or gpm
$Q$ = average daily, L/min ($\text{m}^3$/s) or gpm inflow, $\text{m}^3$/s or gpd
$t_f$ = filling time with the pump off, min

and the total cycle time ($t$) is

$$t = t_{\mathrm{r}} + t_{\mathrm{f}} = \frac{V}{D - Q} + \frac{V}{Q} \tag{6.23}$$

The starting limitations on pump motors usually dictate the minimum size of a well. The wet well should be large enough to prevent pump motors from overheating due to extensive cycling, but small enough to accommodate cycling times that will reduce septicity and odor problems. Typically, submersible pumps can cycle four to ten times per hour. A maximum storage volume for a cycling time should be no more than 30 min.

If the selected pumps have a capacity equal to the peak (maximum) flow rate, the volume of a wet well is calculated as (WEF, 1993a)

$$V = TQ/4 \tag{6.24}$$

where $V$ = storage volume of wet well, gal
$T$ = pump cycle time, min
$Q$ = peak flow, gal/min

**Example 1:** A subdivision generates an average daily wastewater flow of 144,000 L/d (38,000 gal/d). The minimum hourly flow rate is 20,000 L/d (5300 gal/d) and the peak flow is 500,000 L/d (132,000 gal/d). Determine the pumping conditions and the size of a wet well.

**solution:**

Step 1. Determine pump capacity for peak flow

$$D = 500{,}000 \text{ L/day} \times 1 \text{ day/1440 min}$$
$$= 347 \text{ L/min}$$

Step 2. Calculate the minimum volume ($V_1$) for 2-min running time

$$V_1 = 347 \text{ L/min} \times 2 \text{ min}$$
$$= 694 \text{ L}$$

Step 3. Calculate volume ($V_2$) for 5-min cycle using Eq. (6.23)

Average flow $Q$ = 144,000 L/d = 100 L/min

$$t = \frac{V_2}{D - Q} + \frac{V_2}{Q}$$

$$5 \text{ min} = \frac{V_2}{(347 - 100)\text{L/min}} + \frac{V_2}{100\text{L/min}}$$

$$100V_2 + 247V_2 = 5 \times 247 \times 100$$

$$V_2 = 356 \text{ L}$$

Step 4. Determine the control factor

Since $V_1 < V_2$, therefore the pump running time is the control factor. Say $V_1$ = 700 L for design.

Step 5. Calculate the actual time of the pumping cycle

$$t = \frac{700\text{ L}}{(347 - 100)\text{L/min}} + \frac{700\text{ L}}{100\text{ L/min}}$$
$$= 9.83\text{ min}$$

Step 6. Determine size of wet well

A submergence of 0.3 m (1 ft) above the top of the suction pipe is required for an intake velocity of 0.6 m/s (2 ft/s). The depth between the well bottom and the top of submergence is 0.5 m (1.6 ft). If a 1.2-m (4-ft) diameter of wet well is chosen, surface area is 1.13 $m^2$ (12.2 $ft^2$).
For storage

$$V_2 = 700\text{ L} = 0.7\text{ m}^3$$

the depth would be

$$d = 0.7/1.13 = 0.62\text{ (m)}$$

Typically, 0.6 m (2 ft) of freeboard is required

$$\text{Thus total depth of the wet well} = (0.50 + 0.62 + 0.60)\text{ m}$$
$$= 1.72\text{ m}$$
$$= 5.6\text{ ft}$$

**Example 2:** Wastewater is collected from a subdivision of 98.8 acre (40.0 ha) area that consists of 480 residential units and 2.2 acres (0.89 ha) of commercial center. Each of the two pumps will be cycled, alternately, four times per hour. Determine the volume of wet well needed.

**solution:**

Step 1. Determine domestic sewer flow $q_1$

Assume the residential units have 3.5 persons (United States) and each produces 100 gal/d (378 L/d)

$$q_1 = 100\text{ gal/(c} \cdot \text{d)} \times 3.5\text{ person/unit} \times 480\text{ unit}$$
$$= 168{,}000\text{ gal/d}$$

Step 2. Estimate commercial area contribution $q_2$

Assume 1500 gal/d · per acre (468 $m^3$/d)

$$q_2 = 1500\text{ gal/(d} \cdot \text{a)} \times 2.2\text{ a}$$
$$= 3300\text{ gpd}$$

Step 3. Estimate infiltration/inflow (I/I) $q_3$

Assume I/I is 1000 gal/(d · a) or (9.35 m$^3$/(d · ha))

$$q_3 = 1000 \text{ gal/(d} \cdot \text{a)} \times 98.8 \text{ a}$$
$$= 98{,}800 \text{ gal/d}$$

Step 4. Determine total sewer flow $q$

This average daily flow

$$q = q_1 + q_2 + q_3$$
$$= (168{,}000 + 3300 + 98{,}800) \text{ gal/d}$$
$$= 270{,}100 \text{ gal/d}$$
$$= 188 \text{ gal/min}$$

Step 5. Estimate population equivalent (PE) and peak flow ($Q$)

$$\text{PE} = 270{,}100 \text{ gal/d/100 gal/(c} \cdot \text{d)}$$
$$= 2700 \text{ persons}$$

Take

$$Q = 3.5q \text{ for the peak flow}$$
$$Q = 188 \text{ gal/min} \times 3.5$$
$$= 658 \text{ gal/min}$$

*Note*: The selected pump should be able to deliver 658 gal/min. The pipe diameter that will carry the flow and maintain at least 2.0 ft/s (0.6 m/s) of velocity should be selected by the manufacturer's specification. For this example a 10-in (254-mm) pressure class ductile iron pipe will be used.

Step 6. Determine the volume of the wet well, $V$

The selected pumps can cycle four times per hour. Alternating each pump between starts gives 8 cycles per hour. The time between starts, $T$, is

$$T = 60 \text{ min/8} = 7.5 \text{ min}$$

It means that one pump is capable of starting every 7.5 min. Using Eq. (6.24)

$$V = TQ/4$$
$$= 7.5 \text{ min} \times 658 \text{ (gal/min)/4}$$
$$= 1230 \text{ gal } (=4662 \text{ L})$$
$$= 164 \text{ ft}^3$$

Wet wells are typically available in cylindrical sections of various sizes. In this example, a 6-ft (1.8-m) diameter gives 28.3 ft$^2$ of surface area.

The depth $D$ of the wet well is

$$D = 164/28.3$$
$$= 5.8 \text{ ft}$$

*Note*: One foot of freeboard should be added. Thus the well is 6 ft in diameter and 6.8 ft in depth.

## 12 Sewer Construction

Conduit material for sewer construction consists of two types: rigid pipe and flexible pipe. Specified rigid materials include asbestos–cement, cast iron, concrete, and vitrified clay. Flexible materials include ductile iron, fabricated steel, corrugated aluminum, thermoset plastic (reinforced plastic mortar and reinforced thermosetting resin), and thermoplastic. Thermoplastic consists of acrylonitrile–butadiene–styrene (ABS), ABS composite, polyethylene (PE), and polyvinyl chloride (PVC). Their advantages, disadvantages, and applications are discussed in detail elsewhere (ASCE and WPCF, 1982, WEF, and ASCE, 1993a).

Nonpressure sewer pipe is commercially available in the size range from 4 to 42 in (102 to 1067 mm) in diameter and 13 ft (4.0 m) in length. Half-length sections of 6.5 ft (2 m) are available for smaller size pipes.

### 12.1 Loads and buried sewers

Loads on sewer lines are affected by conditions of flow, groundwater, adjacent earth, and superimposed situation. Loads include hydraulic loads, earth loads, groundwater loads, and superimposed loads (weight and impact of vehicles or other structure). Crushing strength of the sewer material, type of bedding, and backfill load are important factors.

**Marston's equation.** Figure 6.2 illustrates common cuts used for sewer pipe installations. Marston's equation is widely used to determine the vertical load on buried conduits caused by earth forces in all of the most

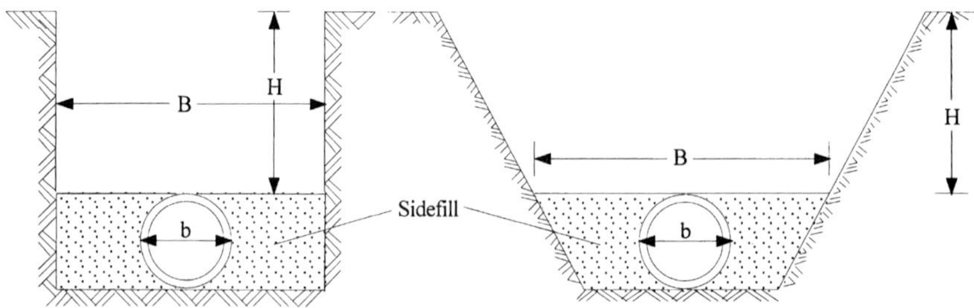


**Figure 6.2** Common trench cuts for sewer pipes.

commonly encountered construction conditions (Marston, 1930). The general form of Marston's formula is

$$W = CwB^2 \tag{6.25}$$

where $W$ = vertical load on pipe as a result of backfill, lb per linear ft
$C$ = dimensionless load coefficient based on the backfill and ratio of trench depth to width; a nomograph can be used
$w$ = unit weight of backfill, lb/ft$^3$
$B$ = width of trench at top of sewer pipe, ft (see Fig. 6.2)

The load coefficient $C$ can be calculated as

$$C = \frac{1 - e^{-2k\mu'(H/B)}}{2k\mu'} \tag{6.26}$$

where $e$ = base of natural logarithms
$k$ = Rankine's ratio of lateral pressure to vertical pressure

$$k = \frac{\sqrt{\mu^2 + 1} - \mu}{\sqrt{\mu^2 + 1} + \mu} = \frac{1 - \sin\Phi}{1 + \sin\Phi}$$

$\mu = \tan\Phi$
= coefficient of internal friction of backfill material
$\mu' = \tan\Phi'$
= coefficient of friction between backfill material and the sides of the trench $\leq \mu$
$H$ = height of backfill above pipe, ft (see Fig. 6.2)

**Load on sewer for trench condition.** The load on a sewer conduit for the trench condition is affected directly by the soil backfill. The load varies widely over different soil types, from a minimum of approximately 100 lb/ft$^3$ (1600 kg/m$^3$) to a maximum of about 135 lb/ft$^3$ (2160 kg/m$^3$) (WEF and ASCE, 1992). The unit weight (density) of backfill material is as follows (McGhee, 1991):

100 lb/ft$^3$ (1600 kg/m$^3$) for dry sand, and sand and damp topsoil;

115 lb/ft$^3$ (1840 kg/m$^3$) for saturated topsoil and ordinary sand;

120 lb/ft$^3$ (1920 kg/m$^3$) for wet sand and damp clay; and

130 lb/ft$^3$ (2080 kg/m$^3$) for saturated clay.

The average maximum unit weight of soil which will constitute the backfill over the sewer pipe may be determined by density measurements in advance of the structural design of the sewer pipe. A design value of not less than 120 or 125 lb/ft$^3$ (1920 or 2000 kg/m$^3$) is recommended (WEF and ASCE, 1992).

The load on a sewer pipe is also influenced by the coefficient of friction between the backfill and the side of the trench ($\mu'$) and by the coefficient of internal friction of the backfill soil ($\mu$). For most cases these two values are considered the same for design purposes. But, if the backfill is sharp sand and the sides of the trench are sheeted with finished lumber, $\mu$ may be substantially greater than $\mu'$. Unless specific information to the contrary is available, values of the products $k\mu$ and $k\mu'$ may be assumed to be the same and equal to 0.103. If the backfill soil is slippery clay, $k\mu'$ and $k\mu'$ are equal to 0.110 (WEF and ASCE, 1992).

The values of the product $k\mu'$ in Eq. (6.26) range from 0.10 to 0.16 for most soils; specifically, 0.110 for saturated clay, 0.130 for clay, 0.150 for saturated top soil, 0.165 for sand and gravel, and 0.192 for cohesionless granular material (McGhee, 1991). Graphical solutions of Eq. (6.26) are presented elsewhere (ASCE and WPCF, 1982).

**Example:** A 20-in (508-mm) ductile iron pipe is to be installed in an ordinary trench of 10 ft (3.05 m) depth at the top of the pipe and 4 ft (1.22 m) wide. The cut will be filled with damp clay. Determine the load on the sewer pipe.

**solution:**

Step 1. Compute load coefficient $C$ by Eq. (6.26)

$$k\mu' = 0.11$$

$$H/B = 3.05 \text{ m}/1.22 \text{ m} = 2.5$$

$$C = \frac{1 - e^{-2k\mu' H/B}}{2k\mu'} = \frac{1 - e^{-2(0.11)(2.5)}}{2(0.11)}$$

$$= 1.92$$

Step 2. Compute the load $W$ by Eq. (6.25)

$$w = 120 \text{ lb/ft}^3 = 1920 \text{ kg/m}^3$$

$$W = CwB^2 = 1.92 \times 1920 \text{ kg/m}^3 \times (1.22 \text{ m})^2$$

$$= 5487 \text{ kg/m}$$

$$= 3687 \text{ lb/ft}$$

## 13 Wastewater Treatment Systems

As discussed in Chapters 1 and 2, the natural waters in streams, rivers, lakes, and reservoirs have a natural waste assimilative capacity to remove solids, organic matter, even toxic chemicals in the wastewater. However, it is a long process.

Wastewater treatment facilities are designed to speed up the natural purification process that occurs in natural waters and to remove contaminants in wastewater that might otherwise interfere with the natural process in the receiving waters.

Wastewater contains varying quantities of suspended and floating solids, organic matter, and fragments of debris. Conventional wastewater treatment systems are combinations of physical and biological (sometimes with chemical) processes to remove its impurities.

The alternative methods for municipal wastewater treatment are simply classified into three major categories: (1) primary (physical process) treatment, (2) secondary (biological process) treatment, and (3) tertiary (combination of physical, chemical, and biological process) or advanced treatment. As can be seen in Fig. 6.3, each category should include previous treatment devices (preliminary), disinfection, and sludge management (treatment and disposal). The treatment devices shown in the preliminary treatment are not necessarily to be included, depending on the wastewater characteristics and regulatory requirements.

For over a century, environmental engineers and aquatic scientists have developed wastewater treatment technologies. Many patented treatment process and package treatment plants have been developed and applied. The goal of wastewater treatment processes is to produce clean effluents and to protect public health, natural resources, and the ambient environment.

The Ten States Recommended Standards for Sewage Works, adopted by the Great Lakes—Upper Mississippi River Board (GLUMRB), was revised five times as a model for the design of wastewater treatment plants and for the recommended standards for other regional and state agencies. The original members of the ten states were Illinois, Indiana, Iowa, Michigan, Minnesota, Missouri, New York, Ohio, Pennsylvania, and Wisconsin. Recently, Ontario, Canada was added as a new member. The new title of the standards is "Recommended Standards for Wastewater Facilities—Policies for design, review, and approval of plans and specifications for wastewater collection and treatment facilities," 1996 edition, by Great Lakes—Upper Mississippi River Board of State and Provincial Public Health and Environmental Managers.

## 13.1 Preliminary treatment systems

Preliminary systems are designed to physically remove or cut up the larger suspended and floating materials, and to remove the heavy inorganic solids and excessive amounts of oil and grease. The purpose of preliminary treatment is to protect pumping equipment and the subsequent treatment units. Preliminary systems consist of flow measurement devices and regulators (flow equalization), racks and screens, comminuting

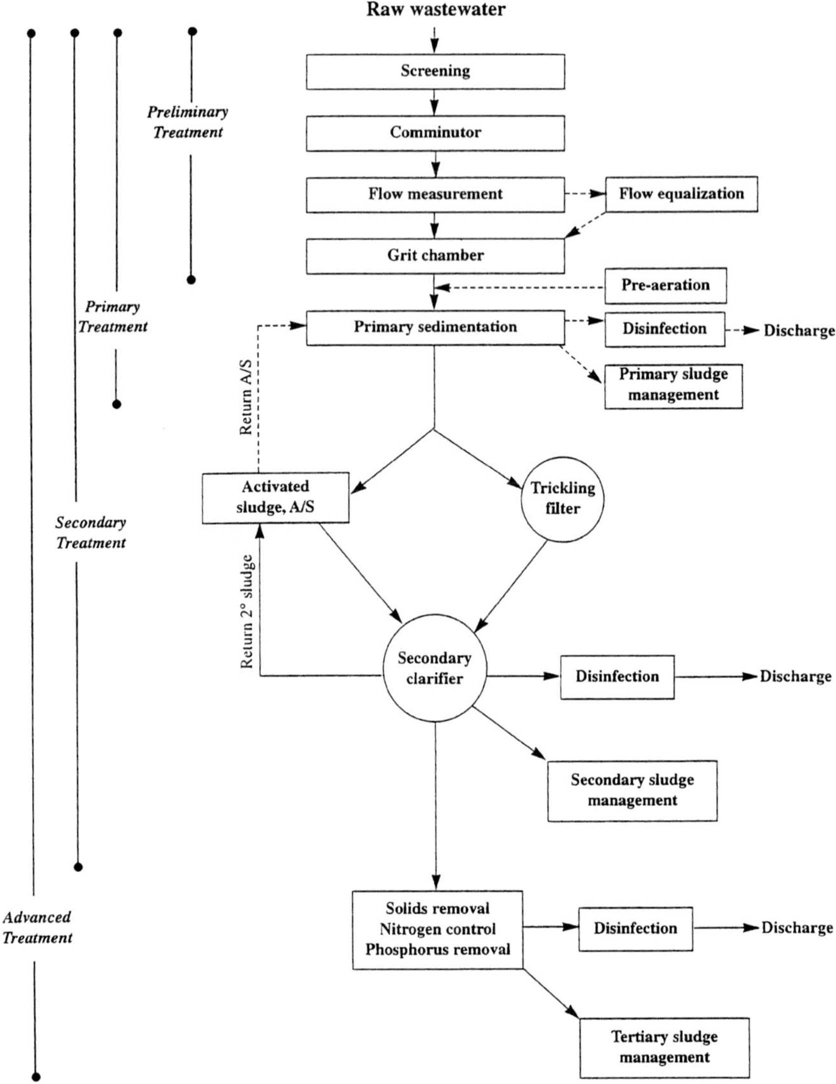


**Figure 6.3** Flow chart for wastewater treatment processes.

devices (grinders, cutters, and shredders), flow equalization, grit chambers, preaeration tanks, and (possibly) chlorination. The quality of wastewater is not substantially improved by preliminary treatment.

## 13.2 Primary treatment systems

The object of primary treatment is to reduce the flow velocity of the wastewater sufficiently to permit suspended solids to settle, i.e. to remove settleable materials. Floating materials are also removed by skimming. Thus, a primary treatment device may be called a settling tank (or basin). Due to variations in design and operation, settling tanks can be divided into four groups: plain sedimentation with mechanical sludge removal, two story tanks (Imhoff tank, and several patented units), upflow clarifiers with mechanical sludge removal, and septic tanks. When chemicals are applied, other auxiliary units are needed. Auxiliary units such as chemical feeders, mixing devices, and flocculators (New York State Department of Health, 1950) and sludge (biosolids) management (treatment and dispose of) are required if there is no further treatment.

The physical process of sedimentation in settling tanks removes approximately 50% to 70% of total suspended solids from the wastewater. The $BOD_5$ removal efficiency by primary system is 25% to 35%. When certain coagulants are applied in settling tanks, much of the colloidal as well as the settleable solids, or a total of 80% to 90% of TSS, is removed. Approximately 10% of the phosphorus corresponding insoluble is normally removed by primary settling. During the primary treatment process, biological activity in the wastewater is negligible.

Primary clarification is achieved commonly in large sedimentation basins under relatively quiescent conditions. The settled solids are then collected by mechanical scrapers into a hopper and pumped to a sludge treatment unit. Fats, oils, greases and other floating matter are skimmed off from the basin surface. The settling basin effluent is discharged over weirs into a collection conduit for further treatment, or to a discharging outfall.

In many cases, especially in developing countries, primary treatment is adequate to permit the wastewater effluent discharge, due to proper receiving water conditions or to the economic situation. Unfortunately, many wastewaters are untreated and discharged in many countries. If primary systems only are used, solids management and disinfection processes should be included.

## 13.3 Secondary treatment systems

After primary treatment the wastewater still contains organic matter in suspended, colloidal, and dissolved states. This matter should be removed before discharging to receiving waters, to avoid interfering with subsequent downstream users.

Secondary treatment is used to remove the soluble and colloidal organic matter which remains after primary treatment. Although the removal of those materials can be effected by physicochemical means providing further removal of suspended solids, secondary treatment is commonly referred to as the biological process.

Biological treatment consists of application of a controlled natural process in which a very large number of microorganisms consume soluble and colloidal organic matter from the wastewater in a relatively small container over a reasonable time. It is comparable to biological reactions that would occur in the zone of recovery during the self-purification of a stream.

Secondary treatment devices may be divided into two groups: attached and suspended growth processes. The attached (film) growth processes are trickling filters, rotating biologic contactors (RBC) and intermittent sand filters. The suspended growth processes include activated sludge and its modifications, such as contact stabilization (aeration) tanks, sequencing batch reactors, aerobic and anaerobic digestors, anaerobic filters, stabilization ponds, and aerated lagoons. Secondary treatment can also be achieved by physical–chemical or land application systems.

Secondary treatment processes may remove more than 85% of $BOD_5$ and TSS. However, they are not effective for the removal of nutrients (N and P), heavy metals, nonbiodegradable organic matter, bacteria, viruses, and other microorganisms. Disinfection is needed to reduce densities of microorganisms. In addition, a secondary clarifier is required to remove solids from the secondary processes. Sludges generated from the primary and secondary clarifiers need to undergo treatment and proper disposal.

## 13.4 Advanced treatment systems

Advanced wastewater treatment is defined as the methods and processes that remove more contaminants from wastewater than the conventional treatment. The term advanced treatment may be applied to any system that follows the secondary, or that modifies or replaces a step in the conventional process. The term tertiary treatment is often used as a synonym; however, the two are not synonymous. A tertiary system is the third treatment step that is used after primary and secondary treatment processes.

Since the early 1970s, the use of advanced wastewater treatment facilities has increased significantly in the United States. Most of their goals are to remove nitrogen, phosphorus, and suspended solids (including $BOD_5$) and to meet certain regulations for specific conditions. In some areas where water supply sources are limited, reuse of waste-water is becoming more important. Also, there are strict rules and regulations regarding the removal of suspended solids, organic matter, nutrients, specific toxic compounds and refractory organics that cannot be achieved by conventional secondary treatment systems as well as some industrial wastewater; thus, advanced wastewater treatment processes are needed.

In the U.S. federal standards for secondary effluent are BOD 30 mg/L and TSS 30 mg/L. In Illinois, the standards are more stringent: BOD 20 mg/L and TSS 25 mg/L for secondary effluent. In some areas, 10 to 12 (BOD = 10 mg/L and TSS = 12 mg/L) standards are implied. For ammonia nitrogen standards, very complicated formulas depending on the time of the year and local conditions are used.

In the European Community, the European Community Commission for Environmental Protection has drafted the minimum effluent standards for large wastewater treatment plants. The standards include: $BOD_5$ <25 mg/L, COD <125 mg/L, suspended solids <35 mg/L, total nitrogen <10 mg/L, and phosphorus <1 mg/L. Stricter standards are presented in various countries. The new regulations were expected to be ratified in 1998 (Boehnke *et al.,* 1997).

TSS concentrations less than 20 mg/L are difficult to achieve by sedimentation through the primary and secondary systems. The purpose of advanced wastewater treatment techniques is specifically to reduce TSS, TDS, BOD, organic nitrogen, ammonia nitrogen, total nitrogen, or phosphorus. Biological nutrient removal processes can eliminate nitrogen or phosphorus, and any combination.

Advanced processes use some processes for the drinking water treatment. These include chemical coagulation of wastewater, wedge-wire screens, granular media filters, diatomaceous earth filters, microscreening, and ultrafiltration and nanofiltration, which are used to remove colloidal and fine-size suspended solids.

For nitrogen control, techniques such as biological assimilation, nitrification (conversion of ammonia to nitrogen and nitrate), and denitrification, ion exchange, breakpoint chlorination, air stripping are used. Soluble phosphorus may be removed from wastewater by chemical precipitation and biological (bacteria and algae) uptake for normal cell growth in a control system. Filtration is required after chemical and biological processes. Physical processes such as reverse osmosis and ultrafiltration also help to achieve phosphorus reduction, but these are primarily for overall dissolved inorganic solids reduction. Oxidation ditch, Bardenpho process, anaerobic/oxidation (A/O) process, and other patented processes are available.

The use of lagoons, aerated lagoons, and natural and constructed wetlands is an effective method for nutrients (N and P) removal.

Removal of some species of groups of toxic compounds and refractory organics can be achieved by activated carbon adsorption, air stripping, activated sludge powder, activated-carbon processes, and chemical oxidation. Conventional coagulation–sedimentation–filtration and biological treatment (trickling filter, RBC, and activated sludge) processes are also used to remove the priority pollutants and some refractory organic compounds.

### 13.5 Compliance with standards

The National Pollutant Discharge Elimination System (NPDES) has promulgated discharge standards to protect and preserve beneficial uses of receiving water bodies based on water quality criteria, or technology-based limits, or both. The receiving water quality criteria are typically established for a 7-day, 10-year, low-flow period.

Table 6.7 shows the national minimum performance standards for secondary treatment and its equivalency for public owned treatment works

**TABLE 6.7 Minimum National Performance Standards for Public Owned Treatment Works (Secondary Treatment and Its Equivalency)**

| Parameter | 30-day average shall not exceed | 7-day average shall not exceed |
|---|---|---|
| *Conventional secondary treatment processes* | | |
| 5-day biochemical oxygen demand,* $BOD_5$ | | |
| Effluent, mg/L | 30 | 45 |
| Percent removal† | 85 | – |
| 5-day carbonaceous biochemical oxygen demand,* $CBOD_5$ | | |
| Effluent, mg/L | 25 | 40 |
| Percent removal† | 85 | – |
| Suspended solids, SS | | |
| Effluent, mg/L | 30 | 45 |
| Percent removal† | 85 | – |
| pH | 6.0 to 9.0 at all times | – |
| Whole effluent toxicity | Site specific | – |
| Fecal coliform, MPN/100 mL | 200 | 400 |
| *Stabilization ponds and other equivalent of secondary treatment* | | |
| 5-day biochemical oxygen demand,* $BOD_5$ | | |
| Effluent, mg/L | 45 | 65 |
| Percent removal† | 65 | – |
| 5-day carbonaceous biochemical oxygen demand,* $CBOD_5$ | | |
| Effluent, mg/L | 40 | 60 |
| Percent removal† | 65 | – |
| Suspended solids, SS | | |
| Effluents, mg/L | 45 | 65 |
| Percent removal† | 65 | – |
| pH | 6.0 to 9.0 at all times | – |
| Whole effluent toxicity | Site specific | – |
| Fecal coliform, MPN/100 mL | 200 | 400 |

*Notes*: *Chemical oxygen demand (COD) or total organic carbon (TOC) may be substituted for $BOD_5$ when a long-term $BOD_5$: COD or $BOD_5$:TOC correlation has been demonstrated.

†Percent removal may be waived on a case-by-case basis for combined sewer service areas and for separated sewer areas not subject to excessive inflow and infiltration (I/I) where the base flow plus infiltration is ≤ 120 gpd/capita and the base flow plus I/I is ≤ 275 gpd/capita.

SOURCE: Federal Register, 1991 (40 CFR 133; 49FR 37006, September 20, 1984; revised through July 1, 1991. http/www.access.gpo.gov/nara/cfr/index.htm/, 1999)

(POTW) (Federal Register, 1991). The secondary treatment regulation was established from 40 Code of Regulation, Part 133, 49 Federal Register 37006, on September 20, 1984; and revised through July 1, 1991.

The values in Table 6.7 are still relevant as far as national standards go; however, they really do not have much applicability in Illinois and many other states because state standards for most facilities are more stringent than the national standards. For example, in Illinois, most POTWs must meet monthly averages of 10/12 (10 mg/L of $BOD_5$ and 12 mg/L of SS) effluent standards if the dilution factor is less than 5:1. The daily maximum effluent concentrations are 20/24. For lagoon effluents, the standards are 30/30 if the dilution factor is greater than 5:1. The NPDES applies to each facility of interest because most have some water quality based effluent limits, especially for ammonia. The Illinois state effluent standards are in Title 35, Subtile C, Chapter II, Part 370 (IEPA, 1997) which can be found on the Illinois Pollution Control Board's web site at http:/www.state.il.us/title35/35conten.htm.

## 14 Screening Devices

The wastewater from the sewer system either flows by gravity or is pumped into the treatment plant. Screening is usually the first unit operation at wastewater treatment plants. The screening units include racks, coarse screens, and fine screens. The racks and screens used in preliminary treatment are to remove large objects such as rags, plastics, paper, metals, dead animals, and the like. The purpose is to protect pumps and to prevent solids from fouling subsequent treatment facilities.

### 14.1 Racks and screens

Coarse screens are classified as either bar racks (trash racks) or bar screens, depending on the spacing between the bars. Bar racks have clear spacing of 5.08 to 10.16 cm (2.0 to 4.0 in), whereas bar screens typically have clear spacing of 0.64 to 5.08 cm (0.25 to 2.0 in). Both consist of a vertical arrangement of equally spaced parallel bars designed to trap coarse debris. The debris captured on the bar screen depends on the bar spacing and the amount of debris caught on the screen (WEF, 1996a).

Clear openings for manually cleaned screens between bars should be from 25 to 44 mm (1 to $1\frac{3}{4}$ in). Manually cleaned screens should be placed on a slope of 30 to 45 degrees to the horizontal. For manually or mechanically raked bar screens, the maximum velocities during peak flow periods should not exceed 0.76 m/s (2.5 ft/s) (Ten States Standards (GLUMRB) 1996; Illinois EPA, 1998).

Hydraulic losses through *bar racks* are a function of approach (upstream) velocity, and the velocity through the bars (downstream),

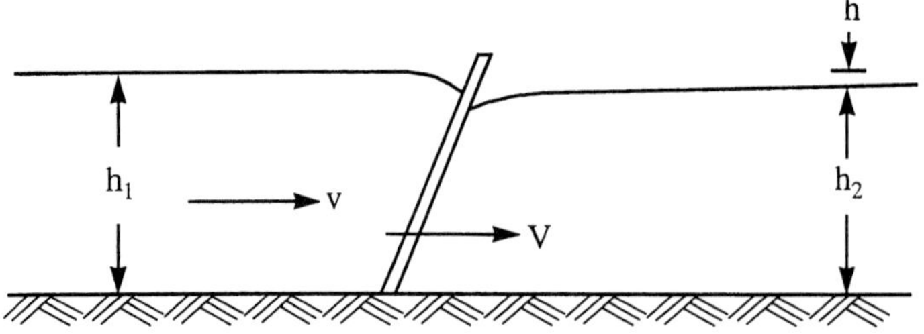


**Figure 6.4** Profile for wastewater flowing through a bar screen.

with a discharge coefficient. Referring to Fig. 6.4, Bernoulli's equation can be used to estimate the headloss through bar racks:

$$h_1 + \frac{v^2}{2g} = h_2 + \frac{V^2}{2g} + \Delta h \tag{6.28}$$

and

$$h = h_1 - h_2 = \frac{V^2 - v^2}{2gC^2} \tag{6.29}$$

where $h_1$ = upstream depth of water, m or ft
$h_2$ = downstream depth of water, m or ft
$h$ = headloss, m or ft
$V$ = flow velocity through the bar rack, m/s or ft/s
$v$ = approach velocity in upstream channel, m/s or ft/s
$g$ = acceleration due to gravity, 9.81 m/s$^2$ or 32.2 ft/s$^2$

The headloss is usually incorporated into a discharge coefficient $C$; a typical value of $C = 0.84$, thus $C^2 = 0.7$. Equation (6.29) becomes (for bar racks)

$$h = \frac{1}{0.7}\left(\frac{V^2 - v^2}{2g}\right) \tag{6.30}$$

Kirschmer (1926) proposed the following equation to describe the headloss through racks:

$$H = B\left(\frac{w}{b}\right)^{4/3} \frac{v^2}{2g} \sin\theta \tag{6.31}$$

where $H$ = headloss, m
$w$ = maximum width of the bar facing the flow, m
$b$ = minimum clear spacing of bars, m
$v$ = velocity of flow approaching the rack, m/s

$g$ = gravitational acceleration, 9.81 m/s$^2$
$\theta$ = angle of the rack to the horizontal
$B$ = bar shape factor, as follows

| Bar type | $B$ |
|---|---|
| Sharp-edged rectangular | 2.42 |
| Rectangular with semicircular face | 1.83 |
| Circular | 1.79 |
| Rectangular with semicircular upstream and downstream faces | 1.67 |
| Tear shape | 0.76 |

The maximum allowable headloss for a rack is about 0.60 to 0.70 m. Racks should be cleaned when headloss is more than the allowable values.

**Example 1:** Compute the velocity through a rack when the approach velocity is 0.60 m/s (2 ft/s) and the measured headloss is 38 mm (0.15 in)

**solution:** Using Eq. (6.30)

$$h = \frac{V^2 - v^2}{0.7(2g)}$$

$$0.038\,\text{m} = \frac{V^2 - (0.6\text{ m/s})^2}{0.7(2 \times 9.81\text{ m/s})^2}$$

$$V^2 = 0.882$$

$$V = 0.94\text{ m/s}$$

$$= 3.08\text{ ft/s}$$

**Example 2:** Design a coarse screen and calculate the headloss through the rack, using the following information:

Peak design wet weather flow = 0.631 m$^3$/s (10,000 gal/min)

Velocity through rack at peak wet weather flow = 0.90 m/s (3 ft/s)

Velocity through rack at maximum design dry weather flow = 0.6 m/s (2 ft/s)

$\theta = 60°$, with a mechanical cleaning device

Upstream depth of wastewater = 1.12 m (3.67 ft)

**solution:**

Step 1. Calculate bar spacing and dimensions

(a) Determine total clear area ($A$) through the rack

$$A = \frac{\text{peak flow}}{v}$$

$$= \frac{0.631 \text{ m}^3/\text{s}}{0.90 \text{ m/s}}$$

$$= 0.70 \text{ m}^2$$

(b) Calculate total width of the opening at the rack, $w$

$$w = A/d = 0.70 \text{ m}^2/1.12 \text{ m}$$

$$= 0.625 \text{ m}$$

(c) Choose a 25-mm clear opening

(d) Calculate number of opening, $n$

$$n = w/\text{opening} = 0.625 \text{ m}/0.025 \text{ m}$$

$$= 25$$

*Note*: Use 24 bars with 10 mm (0.01 m) width and 50 mm thick.

(e) Calculate the width ($W$) of the chamber

$$\text{width} = 0.625 \text{ m} + 0.01 \text{ m} \times 24$$

$$= 0.865 \text{ m}$$

(f) Calculate the height of the rack

$$\text{height} = 1.12 \text{ m}/\sin 60° = 1.12 \text{ m}/0.866$$

$$= 1.29 \text{ m}$$

Allowing at least 0.6 m of freeboard, a 2-m height is selected.

(g) Determine the efficiency coefficient, EC

$$\text{EC} = \frac{\text{clear opening}}{\text{width of the chamber}}$$

$$= 0.625 \text{ m}/0.865 \text{ m}$$

$$= 0.72$$

*Note*: The efficiency coefficient is available from the manufacturer.

Step 2. Determine headloss of the rack by Eq. (6.31)

Select rectangular bars with semicircular upstream face, thus

$$B = 1.83$$

$$w/b = 1$$

$$\sin\theta = \sin 60° = 0.866$$

$$H = B\left(\frac{\omega}{b}\right)^{4/3}\frac{v^2}{2g}\sin\theta$$

$$= 1.83 \times 1 \times [(0.9\ \text{m/s})^2/(2 \times 9.81\ \text{m})] \times 0.866$$

$$= 0.065\ \text{m}$$

*Note*: If we want to calculate the headloss through the rack at 50% clogging, many engineers use an approximate method. When the rack becomes half-plugged, the area of the flow is reduced to one half, and velocity through the rack is doubled. Thus the headloss will be 0.260 m (four times 0.065 m).

## 14.2 Fine screens

Fine screens are used more frequently in wastewater treatment plants for preliminary treatment or preliminary/primary treatment purposes. Fine screens typically consist of wedge-wire, perforated plate, or closely spaced bars with openings 1.5 to 6.4 mm (0.06 to 0.25 in). Fine screens used for preliminary treatment are rotary or stationary-type units (US EPA 1987a).

The clean water headloss through *fine screens* may be obtained from manufacturers' rating tables, or may be computed by means of the common orifice equation

$$h = \frac{1}{2g}\left(\frac{v}{C}\right)^2 = \frac{1}{2}\left(\frac{Q}{CA}\right)^2 \qquad (6.32)$$

where $h$ = headloss, m or ft
$v$ = approach velocity, m/s or ft/s
$C$ = coefficient of discharge for the screen
$g$ = gravitational acceleration, m/s$^2$ or ft/s$^2$
$Q$ = discharge through the screen, m$^3$/s or ft$^3$/s
$A$ = area of effective opening of submerged screen, m$^2$ or ft$^2$

Values of $C$ depend on the size and milling of slots, the diameter and weave of the wire, and the percentage of open area. They must be determined experimentally. A typical value of $C$ for a clean screen is 0.60. The headloss of clean water through a clean screen is relatively less. However, the headloss of wastewater through a fine screen during operation depends on the method and frequency of cleaning, the size and quantity of suspended solids in the wastewater, and the size of the screen opening.

The quantity of screenings generated at wastewater treatment plants varies with the bar opening, type of screen, wastewater flow, characteristics of served communities, and type of collection system. Roughly,

3.5 to 35 L (0.93–9.25 gal) of screenings is produced from 1000 $m^3$ (264, 200 gal) wastewater treated. Screenings are normally 10% to 20% dry solids, with bulk density of 640 to 1120 kg/$m^3$ (40 to 70 lb/$ft^3$) (WEF and ASCE, 1991a).

## 15 Comminutors

As an alternative to racks or screens, a comminutor or shredder cuts and grinds up the coarse solids in the wastewater to about 6 to 10 mm (1/4 to 3/8 in) so that the solids will not harm subsequent treatment equipment. The chopped or ground solids are then removed in primary sedimentation basins. A comminutor consists of a fixed screen and a moving cutter. Comminution can eliminate the messy and offensive screenings for solids handling and disposal. However, rags and large objects cause clogging problems.

Comminutors are installed directly in the wastewater flow channel and are equipped with a bypass so that the unit can be isolated for service maintenance. The sizes, installations, operation, and maintenance of the comminutors are available from the manufactures.

## 16 Grit Chamber

Grit originates from domestic wastes, stormwater runoff, industrial wastes, pumpage from excavations, and groundwater seepage. It consists of inert inorganic material such as sand, cinders, rocks, gravel, cigarette filter tips, metal fragments, etc. In addition grit includes bone chips, eggshells, coffee grounds, seeds, and large food wastes (organic particles). These substances can promote excessive wear of mechanical equipment and sludge pumps, and even clog pipes by deposition.

Composition of grit varies widely, with moisture content ranging from 13% to 63%, and volatile content ranging from 1% to 56%. The specific gravity of clean grit particles may be as high as 2.7 with inert material, and as low as 1.3 when substantial organic matter is agglomerated with inert. The bulk density of grit is about 1600 kg/$m^3$ or 100 lb/$ft^3$ (Metcalf and Eddy, Inc. 1991).

Grit chambers should be provided for all wastewater treatment plants, and are used on systems required for plants receiving sewage from combined sewers or from sewer systems receiving a substantial amount of ground garbage or grit. Grit chambers are usually installed ahead of pumps and comminuting devices.

Grit chambers for plants treating wastewater from combined sewers usually have at least two hand cleaned units, or a mechanically cleaned unit with bypass. There are three types of grit settling chamber: hand cleaned, mechanically cleaned, and aerated or vortex-type degritting units. The chambers can be square, rectangular, or circular. A velocity of 0.3 m/s (1 ft/s) is commonly used to separate grit from the organic material. Typically, 0.0005 to 0.00236 $m^3$/s (1 to 5 $ft^3$/min) of air per foot of chamber length is required for a proper aerated grit chamber; or 4.6 to 7.7 L/s per meter of length. The transverse velocity at the surface should be 0.6 to 0.8 m/s or 2 to 2.5 ft/s (WEF, 1996a).

Grit chambers are commonly constructed as fairly shallow longitudinal channels to catch high specific gravity grit (1.65). The units are designed to maintain a velocity close to 0.3 m/s (1.0 ft/s) and to provide sufficient time for the grit particle to settle to the bottom of the chamber.

**Example:** The designed hourly average flow of a municipal wastewater plant is 0.438 $m^3$/s (10 Mgal/d). Design an aerated grit chamber where the detention time of the peak flow rate is 4.0 min (generally 3 to 5 min).

**solution:**

Step 1. Determine the peak hourly flow $Q$

Using a peaking factor of 3.0

$$\begin{aligned} Q &= 0.438\,\text{m}^3/\text{s} \times 3 \\ &= 1.314\ \text{m}^3/\text{s} \\ &= 30\ \text{Mgal/d} \end{aligned}$$

Step 2. Calculate the volume of the grit chamber

Two chambers will be used; thus, for each unit

$$\begin{aligned} \text{Volume} &= 1.314\ \text{m}^3/\text{s} \times 4\ \text{min} \times 60\ \text{s/min} \div 2 \\ &= 157.7\ \text{m}^3 \\ &= 5570\ \text{ft}^3 \end{aligned}$$

Step 3. Determine the size of a rectangular chamber

Select the width of 3 m (10 ft), and use a depth-to-width ratio of 1.5:1 (typically 1.5:1 to 2.0:1)

$$\begin{aligned} \text{Depth} &= 3\ \text{m} \times 1.5 = 4.5\ \text{m} \\ &= 15\ \text{ft} \\ \text{Length} &= \text{volume}/(\text{depth} \times \text{width}) = 157.7\ \text{m}^3/(4.5\ \text{m} \times 3\text{m}) \\ &= 11.7\ \text{m} \\ &= 36\ \text{ft} \end{aligned}$$

*Note*: Each of the two chambers has a size of 3 m × 4.5 m × 11.7 m or 10 ft × 15 ft × 36 ft.

Step 4. Compute the air supply needed

Use 5 std ft$^3$/min (scfm) or (0.00236 m$^3$/s per ft (0.3 m) length.

$$\begin{aligned} \text{Air needed} &= 0.00236\ \text{m}^3/(\text{s}\cdot\text{ft}) \times 36\ \text{ft} \\ &= 0.085\ \text{m}^3/\text{s} \\ \text{or} &= 5\ \text{ft}^3/\text{min}\cdot\text{ft} \times 36\ \text{ft} \\ &= 180\ \text{ft}^3/\text{min} \end{aligned}$$

Step 5. Estimate the average volume of grit produced

Assume 52.4 mL/m$^3$ (7 ft$^3$/Mgal) of grit produced

$$\begin{aligned} \text{Volume of grit} &= 52.4\ \text{mL/m}^3 \times 0.438\ \text{m}^3/\text{s} \times 86{,}400\ \text{s/d} \\ &= 1{,}980{,}000\ \text{mL/d} \\ &= 1.98\ \text{m}^3/\text{d} \\ \text{or} &= 7\ \text{ft}^3/\text{Mgal} \times 10\ \text{Mgal/d} \\ &= 70\ \text{ft}^3/\text{d} \end{aligned}$$

## 17 Flow Equalization

The Parshall flume is commonly used in wastewater treatment plants. Methods of flow measurement are discussed in Chapter 4.

The incoming raw wastewater varies with the time of the day, the so-called diurnal variation, ranging from less than one half to more than 200% of the average flow rate. A storm event increases the flow. Flow equalization is used to reduce the sudden increase of inflow and to balance the fluctuations in the collection system or in the in-plant storage basins. This benefits the performance of the downstream treatment processes and reduces the size and cost of treatment units.

Flow equalization facilities include the temporary storage of flows in existing sewers, the use of in-line or on-line separate flow-equalization facilities or retention basins.

The volume for a flow equalization basin is determined from mass diagrams based on average diurnal flow patterns.

**Example:** Determine a flow equalization basin using the following diurnal flow record:

| Time | Flow, $m^3/s$ | Time | Flow, $m^3/s$ |
|---|---|---|---|
| Midnight | 0.0492 | Noon | 0.1033 |
| 1 | 0.0401 | 1 p.m. | 0.0975 |
| 2 | 0.0345 | 2 | 0.0889 |
| 3 | 0.0296 | 3 | 0.0810 |
| 4 | 0.0288 | 4 | 0.0777 |
| 5 | 0.0312 | 5 | 0.0755 |
| 6 | 0.0375 | 6 | 0.0740 |
| 7 | 0.0545 | 7 | 0.0700 |
| 8 | 0.0720 | 8 | 0.0688 |
| 9 | 0.0886 | 9 | 0.0644 |
| 10 | 0.0972 | 10 | 0.0542 |
| 11 | 0.1022 | 11 | 0.0513 |

**solution 1:**

Step 1. Compute the average flow rate $Q$

$$Q = \sum q/24 = 0.0655\ \text{m}^3/\text{s}$$

Step 2. Compare the observed flows and average flow from the data shown above

The first observed flow to exceed $Q$ is at 8 a.m.

Step 3. Construct a table which is arranged in order, beginning at 8 a.m.

See Table 6.8. Calculations of columns 3 to 6 are given in the following steps.

Step 4. For col. 3, convert the flows to volume for 1 h time interval

$$\begin{aligned}\text{Volume} &= 0.072\ \text{m}^3/\text{s} \times 1\ \text{h} \times 3600\ \text{s/h}\\ &= 259.2\ \text{m}^3\end{aligned}$$

Step 5. For col. 4, for each row, calculate average volume to be treated

$$\begin{aligned}\text{Volume} &= Q \times 1\ \text{h} \times 3600\ \text{s/h}\\ &= 0.0655\ \text{m}^3/\text{s} \times 1\ \text{h} \times 3600\ \text{s/h}\\ &= 235.8\ \text{m}^3\end{aligned}$$

Step 6. For col. 5, calculate the excess volume needed to be stored

$$\text{col. 5} = \text{col. 3} - \text{col. 4}$$

$$\text{Example: } 259.2\ \text{m}^3 - 235.8\ \text{m}^3 = 23.4\ \text{m}^3$$

Step 7. For col. 6, calculate the cumulative sum of the difference (col. 5)

**TABLE 6.8 Analysis of Flow Equalization**

| (1) Time | (2) Flow $m^3/s$ | (3) Volume in, $m^3$ | (4) Volume out, $m^3$ | (5) Storage $m^3$ | (6) Σ storage, $m^3$ |
|---|---|---|---|---|---|
| 8 a.m. | 0.072 | 259.2 | 235.8 | 23.4 | 23.4 |
| 9 | 0.0886 | 318.96 | 235.8 | 83.16 | 106.56 |
| 10 | 0.0972 | 349.92 | 235.8 | 114.12 | 220.68 |
| 11 | 0.1022 | 367.92 | 235.8 | 132.12 | 352.8 |
| 12 | 0.1033 | 371.88 | 235.8 | 136.08 | 488.88 |
| 1 p.m. | 0.0975 | 351 | 235.8 | 115.2 | 604.08 |
| 2 | 0.0889 | 320.04 | 235.8 | 84.24 | 688.32 |
| 3 | 0.081 | 291.6 | 235.8 | 55.8 | 744.12 |
| 4 | 0.0777 | 279.72 | 235.8 | 43.92 | 788.04 |
| 5 | 0.0755 | 271.8 | 235.8 | 36 | 824.04 |
| 6 | 0.0740 | 266.4 | 235.8 | 30.6 | 854.64 |
| 7 | 0.0700 | 252 | 235.8 | 16.2 | 870.84 |
| 8 | 0.0688 | 247.68 | 235.8 | 11.88 | 882.72 |
| 9 | 0.0644 | 231.84 | 235.8 | –3.96 | 878.76 |
| 10 | 0.0542 | 195.12 | 235.8 | –40.68 | 838.08 |
| 11 | 0.0513 | 184.68 | 235.8 | –51.12 | 786.96 |
| 12 | 0.0492 | 177.12 | 235.8 | –58.68 | 728.28 |
| 1 a.m. | 0.0401 | 144.36 | 235.8 | –91.44 | 636.84 |
| 2 | 0.0345 | 124.2 | 235.8 | –111.6 | 525.24 |
| 3 | 0.0296 | 106.56 | 235.8 | –129.24 | 396 |
| 4 | 0.0288 | 103.68 | 235.8 | –132.12 | 263.88 |
| 5 | 0.0312 | 112.32 | 235.8 | 123.48 | 140.4 |
| 6 | 0.0375 | 135 | 235.8 | –100.8 | 39.6 |
| 7 | 0.0545 | 196.2 | 235.8 | –39.6 | 0 |

Example: For the second time interval, the cumulative storage (cs) is

$$\text{cs} = 23.4 \text{ m}^3 + 83.16 \text{ m}^3 = 106.56 \text{ m}^3$$

*Note*: The last value for the cumulative storage should be zero. Theoretically, it means that the flow equalization basin is empty and ready to begin the next day's cycle.

Step 8. Find the required volume for the basin

The required volume for the flow equalization basin for this day is the maximum cumulative storage. In this case, it is 882.72 $m^3$ at 8 p.m. (col. 6, Table 6.8). However, it is common to provide 20% to 50% excess capacity for unexpected flow variations, equipment, and solids deposition. In this case, we provide 35% excess capacity. Thus the total storage volume should be

$$\text{Total basin volume} = 882.72 \text{ m}^3 \times 1.35$$
$$= 1192 \text{ m}^3$$

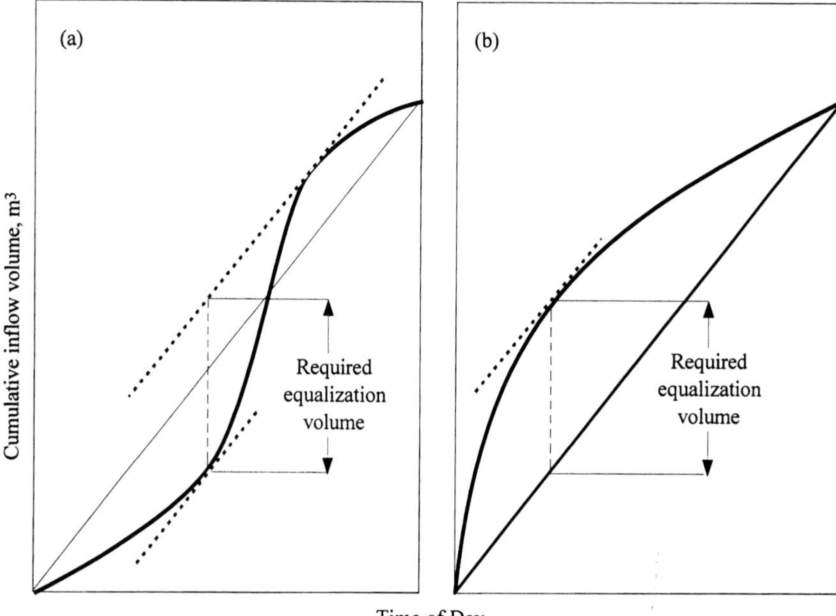


**Figure 6.5** Mass diagram for determining required equalization volume.

**solution 2:** Graphic method (Fig. 6.5).

Step 1. Calculate cumulative volumes as in solution 1

Step 2. Plot time of day at $X$-axis (starting at midnight) versus cumulative volume at $Y$-axis to produce a mass curve

Step 3. Connect the origin and the final point on the mass curve

This gives the daily average flow rate ($m^3$/d or Mgal/d).

Step 4. Draw two lines parallel to the average flow rate and tangent to the mass curve at the highest and lowest points

Step 5. Determine the required volume for the flow equalization basin

The vertical distance between two parallels drawn in Step 4 is the required basin capacity.

*Note*: In the above example, the storage starts to fill in at 8 a.m. and it is empty 24 h later. At the highest point of the cumulative volume draw a tangent line (Fig. 6.5b). The distance between this tangent line and the average flow line is the storage volume required.

## 18 Sedimentation

Sedimentation is the process of removing solid particles heavier than water by gravity settling. It is the oldest and most widely used unit operation in water and wastewater treatments. The terms sedimentation, settling, and clarification are used interchangeably. The unit sedimentation basin may also be referred to as a sedimentation tank, clarifier, settling basin, or settling tank.

In wastewater treatment, sedimentation is used to remove both inorganic and organic materials which are settleable in continuous-flow conditions. It removes grit, particulate matter in the primary settling tank, and chemical flocs from a chemical precipitation unit. Sedimentation is also used for solids concentration in sludge thickeners.

Based on the solids concentration, and the tendency of particle interaction, there are four types of settling which may occur in waste-water settling operations. The four categories are discrete, flocculant, hindered (also called zone), and compression settlings. They are also known as types 1, 2, 3, and 4 sedimentation, respectively. Some discussion of sedimentation is covered in Chapter 5. The following describes each type of settling.

### 18.1 Discrete particle sedimentation (type 1)

The plain sedimentation of a discrete spherical particle, described by Newton's law, can be applied to grit removal in grit chambers and sedimentation tanks. The terminal settling velocity is determined as (also in Chapter 5, Eq. (5.60))

$$v_s = \left[\frac{4g(\rho_s - \rho)d}{3C_D\rho}\right]^{1/2} \tag{6.33}$$

where $v_s$ = terminal settling velocity, m/s or ft/s
$\rho_s$ = mass density of particle, kg/m$^3$ or lb/ft$^3$
$\rho$ = mass density of fluid, kg/m$^3$ or lb/ft$^3$
$g$ = acceleration due to gravitation, 9.81 m/s$^2$ or 32.2 ft/s$^2$
$d$ = diameter of particle, mm or in
$C_D$ = dimensionless drag coefficient

The drag coefficient $C_D$ is not constant. It varies with the Reynolds number and the shape of the particle. The Reynolds number $\mathbf{R} = vd\rho/\mu$, where $\mu$ is the absolute viscosity of the fluid, and the other terms are defined as above.

$C_D$ varies with the effective resistance area per unit volume and shape of the particle. The relationship between $\mathbf{R}$ and $C_D$ is as follows

$$1 > \mathbf{R}: \quad C_D = \frac{24}{\mathbf{R}} = \frac{24\mu}{v\rho d} \tag{6.34}$$

$$1 < \mathbf{R} < 1000: \quad C_D = \frac{24}{\mathbf{R}} + \frac{3}{\mathbf{R}^{0.5}} + 0.34 \tag{6.35}$$

$$\text{or} = \frac{18.5}{\mathbf{R}^{0.5}} \tag{6.36}$$

$$\mathbf{R} > 1000: \quad C_D = 0.34 \text{ to } 0.40 \tag{6.37}$$

For small **R** (<1 or 2) with laminar flow. Equation (6.34) is applied. Equation (6.35) or (6.36) is applicable for **R** up to 1000, which includes all situations of water and wastewater treatment processes. For fully developed turbulent settling use $C_D = 0.34$ to 0.40 (Eq. (6.37)).

When the Reynolds number is less than 1, substitution of Eq. (6.34) for $C_D$ in Eq. (6.33) yields Stoke's law (Eq. (5.63))

$$v_s = \frac{g(\rho_s - \rho)d^2}{18\mu} \tag{6.38}$$

Discrete particle settling refers to type 1 sedimentation. Under quiescent conditions, suspended particles in water or wastewater exhibit a natural tendency to agglomerate, or the addition of coagulant chemicals promotes flocculation. The phenomenon is called flocculation–sedimentation or type 2 sedimentation. For flocculated particles the principles of settling are the same as for a discrete particle, but settling merely occurs at a faster rate.

## 18.2 Scour

The horizontal velocity in grit chambers or in sedimentation tanks must be controlled to a value less than what would carry the particles in traction along the bottom. The horizontal velocity of fluid flow just sufficient to create scour is described as (Camp, 1946)

$$V = \left[\frac{8\beta(s-1)gd}{f}\right]^{1/2} \tag{6.39}$$

where $V$ = horizontal velocity, m/s
$\beta$ = constant for the type of scoured particles
= 0.04 for unigranular material
= 0.06 for sticky interlocking material
$s$ = specific gravity of particle
$g$ = acceleration due to gravity, 9.81 m/s$^2$
$d$ = diameter of particle, m
$f$ = Darcy–Weisbach friction factor, 0.02 to 0.03

The $f$ values are a function of the Reynolds number and surface characteristics of the settled solids. The horizontal velocity in most sedimentation tanks is well below that which would cause scour. In grit chambers, scour is an important factor for design.

**Example:** Determine the surface overflow rate and horizontal velocity of a grit chamber to remove the grit without removing organic material. Assume that grit particles have a diameter of 0.2 mm (0.01 in) and a specific gravity of 2.65 (sand, silt, and clay); and organic material has the same diameter and a specific gravity of 1.20. Assume $C_D = 10$.

**solution:**

Step 1. Compute the terminal settling velocity, using Eq. (6.33)

$$C_D = 10$$

$$d = 0.2 \text{ mm} = 0.02 \text{ cm}$$

$$v_s = \left[\frac{4g(\rho_s - \rho)d}{3C_D\rho}\right]^{1/2}$$

$$= \left[\frac{4 \times 981 \times (2.65 - 1) \times 0.02}{3 \times 10 \times 1}\right]^{1/2}$$

$$= 2.08\,(\text{cm/s})$$

*Note*: This will be the surface overflow rate to settle grit, not organic matter.

Step 2. Compute the horizontal velocity ($V_1$) just sufficient to cause the grit particles to scour

Use $\beta = 0.06$ and $f = 0.03$. Using Eq. (6.39)

$$V_1 = \left[\frac{8\beta(s - 1)gd}{f}\right]^{1/2}$$

$$= \left[\frac{8 \times 0.06 \times (2.65 - 1) \times 981 \times 0.02}{0.03}\right]^{1/2}$$

$$= 22.8\,(\text{cm/s})$$

Step 3. Compute the scouring velocity $V_2$ for organic material, using Eq. (6.39)

$$V_2 = \left[\frac{8 \times 0.06(1.20 - 1) \times 981 \times 0.02}{0.03}\right]^{1/2}$$

$$= 7.9\,(\text{cm/s})$$

*Note*: The grit chamber is designed to have a surface overflow rate (settling velocity) of 2.1 cm/s and a horizontal velocity less than 22.8 cm/s but greater than 7.9 cm/s. Under these conditions, the grit will be removed and organic matter will not. If the horizontal velocity is close to the scour velocity, the grit will be reasonably clean.

## 18.3 Sedimentation tank (basin) configuration

Sedimentation tanks can be rectangular, square or circular. Imhoff tanks perform the dual function of settling and aerobic treatment with two-story chambers; however, the Imhoff tank is old technology and is no longer allowed in the developed countries.

For a continuous flow sedimentation tank, the shape can be either rectangular or circular. Camp (1953) divided the ideal sedimentation tank into four zones which affect settling, namely the inlet zone, theoretical effective settling zone, sludge zone (beneath the settling zone), and outlet zone (Fig. 6.6). The inlet and outlet condition and tank geometry influence short circuiting, which can be minimized in narrow rectangular horizontal flow basins. Short circuiting is a common problem in circular radial flow clarifiers.

Figure 6.6 illustrates an ideal rectangular continuous horizontal flow settling tank. The inlet zone uniformly distributes wastewater flows and solids over the cross-sectional area of the tank in such a manner that flow through the settling zone follows horizontal paths to prevent short circuiting. In the settling zone, a uniform concentration of particles settles at terminal settling velocity to the sludge zone at the bottom of the tank. In the real world, there is no theoretical effective settling zone. Particle settling vectors are difficult to predict. However, it is usually assumed that the flow of wastewater through the settling zone is steady

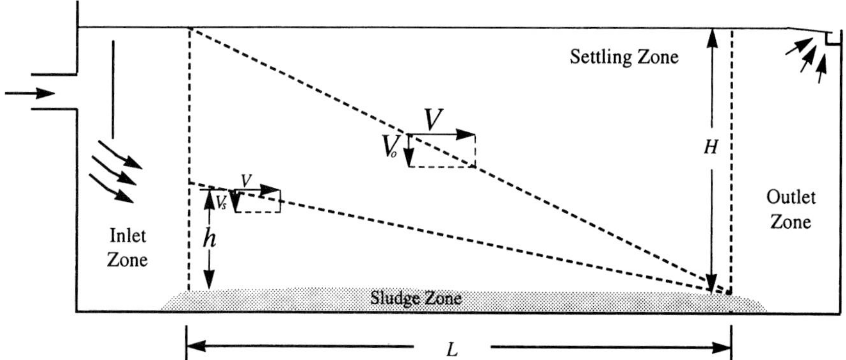


**Figure 6.6** Sketch of the discrete particle settling in an ideal settling tank.

and that the concentration of each sized particle is uniform throughout the cross section normal to the flow direction. The sludge zone is a region for storing the settled sediments below the settling zone. This zone may be neglected for practical purposes, if mechanical equipment continually removes the sediment. In the outlet zone, the supernatant (clarified effluent) is collected through an outlet weir and discharged to further treatment units or to the outfall.

In the design of clarifiers, a particle terminal velocity $V_0$ is used as a design overflow settling velocity, which is the settling velocity of the particle which will settle through the total effective depth $H$ of the tank in the theoretical detention time. All particles that have a terminal velocity ($V_s$) equal to or greater than $V_0$ will be removed. The surface overflow rate of wastewater is (Stoke's law)

$$V_0 = Q/A = Q/WL \tag{6.40}$$

$$= \frac{g(\rho_s - \rho)d^2}{18\mu} \tag{6.41}$$

where $Q$ = flow, $m^3$/d or gal/d
$A$ = surface area of the settling zone, $m^2$ or $ft^2$
$V_0$ = overflow rate or surface loading rate, $m^3/(m^2 \cdot d)$ or $gal/(ft^2 \cdot d)$
$W, L$ = width and length of the tank, m or ft

This is called type 1 settling. Flow capacity is independent of the depth of a clarifier. Basin depth $H$ is a product of the design overflow velocity and detention time $t$

$$H = V_0 t \tag{6.42}$$

The flow through velocity $V_f$ is

$$V_f = Q/HW \tag{6.43}$$

where $H$ is the depth of the settling zone. The retention time $t$ is

$$t = Volume/Q \tag{6.44}$$

The removal ratio $r$ (or fraction of removal) of particles having a settling velocity equal to $V_s$ will be $h/H$. Since depth equals the product of the settling velocity and retention time $t$ (Fig. 6.6)

$$r = \frac{h}{H} = \frac{V_s t}{V_0 t} = \frac{V_s}{V_0} \tag{6.45}$$

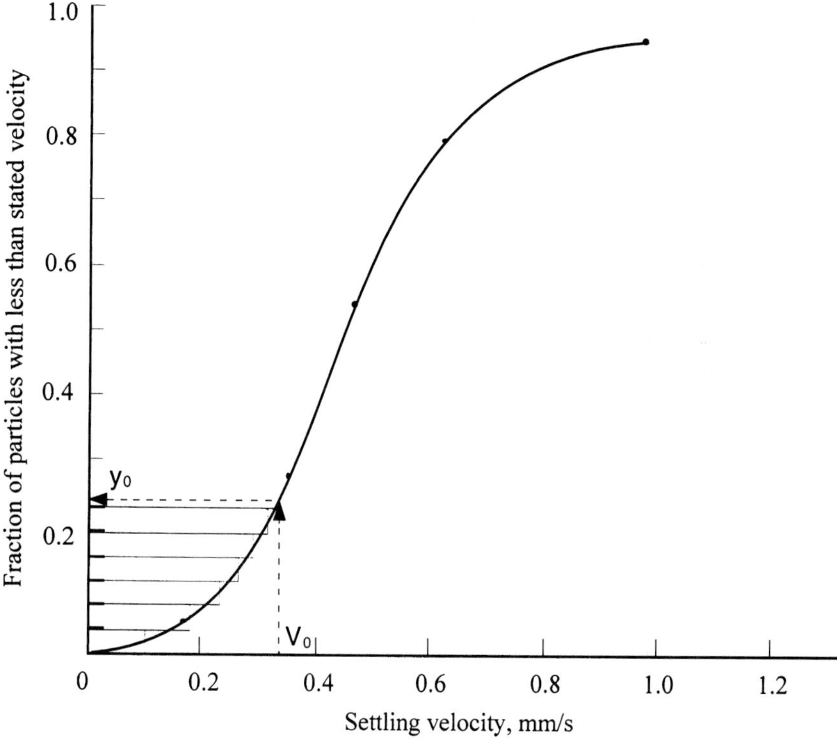


**Figure 6.7** Cumulative particles removal versus settling velocity curve.

where $r$ is the fraction of the particles with settling velocity $V_s$ that are removed. This means that in a horizontal flow particles with settling velocity $V_s$ less than $V_0$ will also be removed if they enter the settling zone at a depth less than $H$.

The settling velocity distribution for a suspension sample can be determined from a column settling test. The data obtained from the test can be used to construct a cumulative settling velocity frequency distribution curve, as shown in Fig. 6.7.

For a given clarification flow rate $Q$, only those particles having settling velocity $\geq V_0$ $(= Q/A)$ will be completely removed. Let $y_0$ represent the portion of particles with a settling velocity $<V_0$; then the percentage removed will be $1 - y_0$. Also, for each size particle with $V_s < V_0$ its proportion of removal, expressed as Eq. (6.45), is equal to $r = V_s/V_0$. When considering various particle sizes in this group, the percentage of removal is

$$\int_0^{y_0} \frac{V_s}{V_0} dy$$

The overall fraction of particles removed, $F$, would be

$$F = (1 - y_0) + \frac{1}{V_0}\int_0^{y_0} V_s dy \tag{6.46}$$

Approximation:

$$F = 1 - y_0 + \frac{V_0 + V_1}{2V_0}(y_0 - y_1) + \frac{V_1 + V_2}{2V_0}(y_1 - y_2) + \cdots$$

$$+ \frac{V_i + V_{i+1}}{2V_0}(y_1 - y_{i+1})$$

$$F = 1 - y_0 + \frac{1}{V_0}\sum V\Delta y \tag{6.47}$$

where $y_0$ = fraction of particles by weight with $V_s \geq V_0$
$i$ = $i$th particle

**Example:** (type 1): A clarifier is designed to have a surface overflow rate of 28.53 $m^3/(m^2 \cdot d)$ (700 $gal/(ft^2 \cdot d)$). Estimate the overall removal with the settling analysis data and particle size distribution in columns 1 and 2 of Table 6.9. The wastewater temperature is 15°C and the specific gravity of the particles is 1.20.

**solution:**

Step 1. Determine settling velocities of particles by Stoke's law, Eq. (6.41)

From Table 4.1a, at 15°C

$$\mu = 0.00113\ \text{N} \cdot \text{s/m}^2 = 0.00113\ \text{kg/(s} \cdot \text{m)}$$

$$\rho = 0.9990$$

$$V = \frac{g(\rho_s - \rho)d^2}{18\mu}$$

$$= \frac{9.81\ \text{m/s}^2(1200 - 999)\ \text{kg/m}^3 \times d^2}{18 \times 0.00113\ \text{kg/(s} \cdot \text{m)}}$$

$$= 96{,}942\ d^2\ \text{m/s}$$

where $d$ is in m

Step 2. Calculate $V$ for each particle size (col. 3 of Table 6.9)

**TABLE 6.9 Results of Settling Analysis Test and Estimation of Overall Solids Removal**

| Particle size mm | Weight fraction < size, % | Settling velocity $V$, mm/s |
|---|---|---|
| 0.10 | 12 | 0.968 |
| 0.08 | 18 | 0.620 |
| 0.07 | 35 | 0.475 |
| 0.06 | 72 | 0.349 |
| 0.05 | 86 | 0.242 |
| 0.04 | 94 | 0.155 |
| 0.02 | 99 | 0.039 |
| 0.01 | 100 | 0.010 |

For $d$ = 0.1 mm = 0.0001 m

$$\begin{aligned} V &= 96{,}942\,(0.0001)^2 \\ &= 0.000969\ (\text{m/s}) \\ &= 0.968\ \text{mm/s} \end{aligned}$$

Similarly, calculate the settling velocities for other particle sizes

Step 3. Construct the settling velocities versus cumulative distribution curve shown in Fig. 6.7

Step 4. Calculate designed settling velocity $V_0$

$$\begin{aligned} V_0 &= 28.53\ \text{m/d} \\ &= 28{,}530\ \text{mm/d} \times 1\ \text{d}/86{,}400\ \text{s} \\ &= 0.33\ \text{mm/s} \end{aligned}$$

*Note*: All particles with settling velocities greater than 0.33 mm/s (700 gal/(d · ft$^2$)) will be removed.

Step 5. Find the fraction $(1 - y_0)$

From Fig. 6.7 we read $y_0 = 0.25$ at $V_0 = 0.33$ mm/s

then $$1 - y_0 = 1 - 0.25 = 0.75$$

Step 6. Graphical determination of $\Sigma V \Delta y$

Referring to Fig. 6.7

| $\Delta y$ | 0.04 | 0.04 | 0.04 | 0.04 | 0.04 | 0.04 | 0.01 |
|---|---|---|---|---|---|---|---|
| $V$ | 0.09 | 0.17 | 0.23 | 0.26 | 0.28 | 0.31 | 0.33 |
| $V\Delta y$ | 0.0036 | 0.0068 | 0.0092 | 0.0104 | 0.0112 | 0.0124 | 0.0033 |

$$\sum V\Delta y = 0.0569$$

Step 7. Determine overall removal $R$

Using Eq. (6.47)

$$F = (1 - y_0) + \frac{1}{V_0}\sum V\Delta y$$

$$= 0.75 + 0.0569/0.33$$

$$= 0.92$$

$$= 92\%$$

### 18.4 Flocculant settling (type 2)

In practice, the actual settling performance cannot be adequately predicted because of unrealistic assumptions on ideal discrete particle settling. Under quiescent conditions, suspended particles in water or wastewater exhibit a natural tendency to agglomerate. Also, suspended solids in wastewater are not discrete particles and vary more than light and small particles, as they contact and agglomerate and grow in size. As coalescence of flocculation occurs, including chemical coagulation and biological flocs, the mass of the particles increases and they settle faster. This phenomenon is called flocculant or type 2 sedimentation.

The flocculation process increases removal efficiency but it cannot be adequately expressed by equations. Settling-column analysis is usually used to determine the settling characteristics of flocculated particles. A column can be of any diameter and equal in length to the proposed clarifier. Satisfactory results can be achieved with 15 cm (6 in) diameter plastic tube 3 m (10 ft) in height (Metcalf and Eddy, Inc. 1991). Sampling ports are uniformly spaced (45 to 60 cm or 1.5 to 2 ft) along the length of the column. The test suspension is placed in the settle-column and allowed to settle in a quiescent manner. The initial suspended solids concentration is measured. Samples are withdrawn from the sampling ports at various selected time intervals from different depths. Analyses of SS are performed for each sample, and the data used to calculate the percentage of removal is plotted as a number against time and depth. Between the plotted points, curves of equal percent removal are drawn. The results of settling-column analyses are presented in Fig. 6.8. Use of the curves in Fig. 6.8 is illustrated in the following example.

**Example:** Using the settling curves of Fig. 6.8, determine the overall removal of solids in a sedimentation basin (type 2 flocculant settling) with a depth equal to the test cylinder and at a detention time of 25 min. The total depth is 2.5 m.

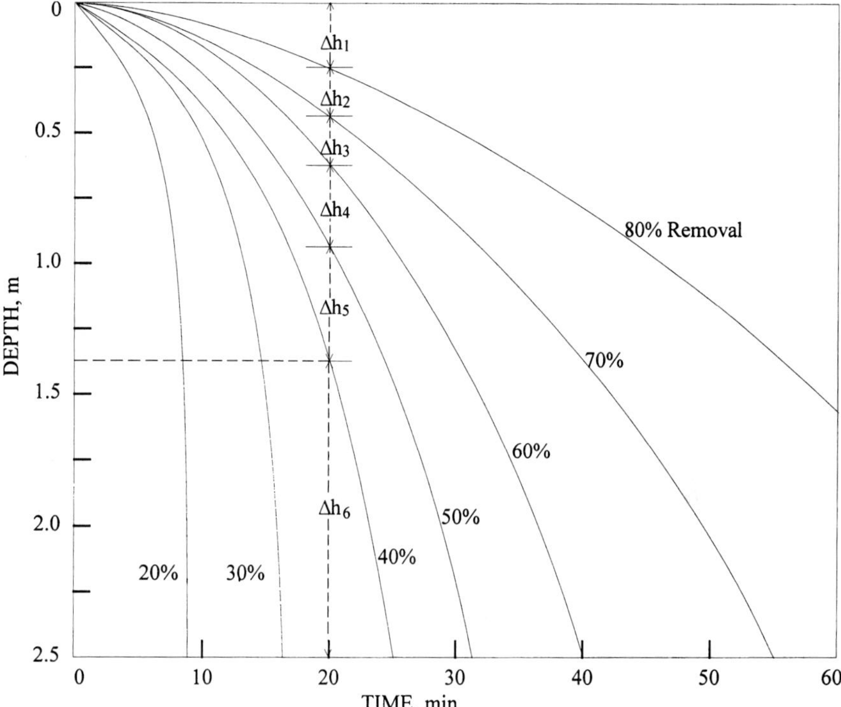


**Figure 6.8** Settling trajectory characteristics for flocculent particles.

**solution:**

Step 1. From Fig. 6.8, 40% of the particles will have a settling velocity of 0.1 m/min (2.5 m/25 min)

At $t$ = 25 min, the volume of the test cylinder within $\Delta h_6$ has 40% removal.

Step 2: Determine percent removal of each volume of the tank

In the volume of the tank corresponding to $\Delta h_5$ between 50% and 40% removal will occur. Similarly, in the tank volume corresponding to $\Delta h_4$ between 60% and 50% will be removed. In like fashion, this is applied to other tank volumes.

Step 3. Calculate the overall removal

Since $1/h = 1/2.5 = 0.4$

$$\Delta h_1 = 0.23 \text{ m}$$

$$\Delta h_2 = 0.14 \text{ m}$$

$$\Delta h_3 = 0.20 \text{ m}$$

$$\Delta h_4 = 0.32 \text{ m}$$

$$\Delta h_5 = 0.50 \text{ m}$$

$$F = 40 + \frac{\Delta h_5}{h}\left(\frac{40 + 50}{2}\right) + \frac{\Delta h_4}{h}\left(\frac{50 + 60}{2}\right) + \frac{\Delta h_3}{h}\left(\frac{60 + 70}{2}\right)$$

$$+ \frac{\Delta h_2}{h}\left(\frac{70 + 80}{2}\right) + \frac{\Delta h_1}{h}\left(\frac{80 + 100}{2}\right)$$

$$= 40 + 0.4(0.5 \times 45 + 0.32 \times 55 + 0.20 \times 65$$

$$+ 0.14 \times 75 + 0.23 \times 90)$$

$$= 73.7\% \text{ removal}$$

## 18.5 Hindered sedimentation (type 3)

In systems with high concentrations of suspended solids, the velocity fields of closely spaced particles are obstructed, causing an upward displacement of the fluid and hindered or zone settling (type 3) and compression settling (type 4). In addition, discrete (free) settling (type 1) and flocculant settling (type 2) occur. This settling phenomenon of concentrated suspensions (such as activated sludge) is illustrated in a graduated cylinder, as shown in Fig. 6.9.

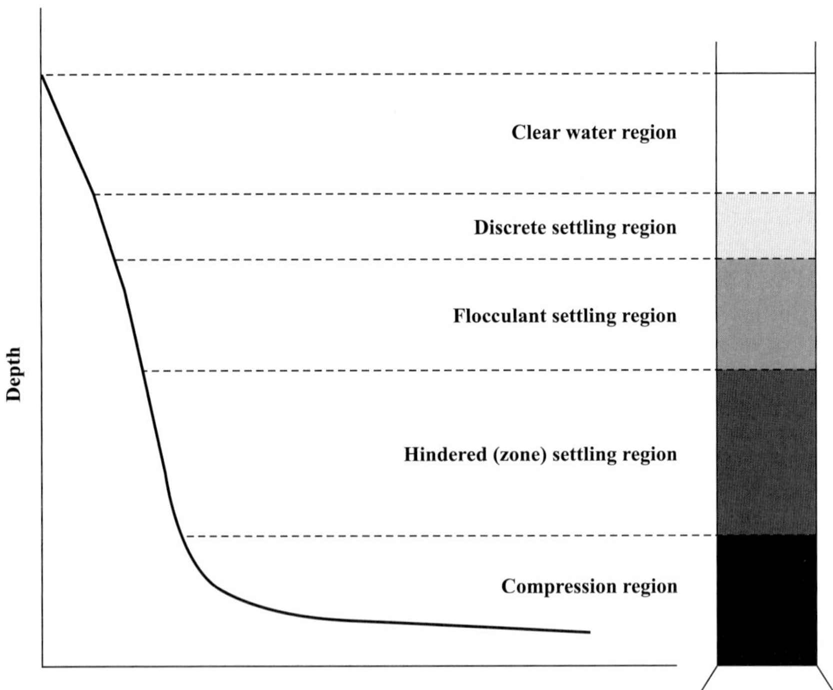


**Figure 6.9** Schematic drawing of settling regions for concentrated suspensions.

Hindered (zone) settling occurs in sludge thickeners and at the bottom of a secondary clarifier in biological treatment processes. The velocity of hindered settling is estimated by (Steel and McGhee, 1979)

$$v_h/v = (1 - C_v)^{4.65} \tag{6.48}$$

where $v_h$ = hindered settling velocity, m/s or ft/s
$v$ = free settling velocity, calculated by Eq. (6.33) or (6.38)
$C_v$ = volume of particles divided by the volume of the suspension

Equation (6.48) is valid for Reynolds numbers less than 0.2, which is generally the situation in hindered settling.

A typical curve of interface height versus time for activated sludge is shown in Fig. 6.10. From A to B, there is a hindered settling of the particles and this is called liquid interface. From B to C there is a deceleration marking the transition from hindered settling into the compression

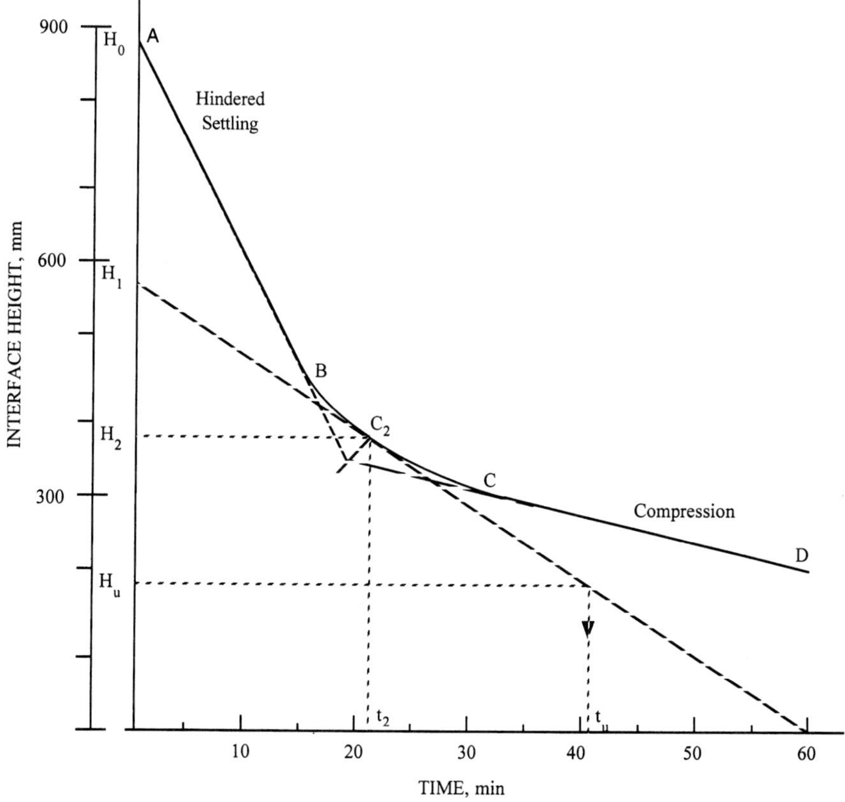


**Figure 6.10** Graphic analysis of interface.

zone. From C to D there is a compression zone where settling depends on compression of the sludge blanket.

The system design for handling concentrated suspensions for hindered settling must consider three factors: (1) the area needed for discrete settling of particles at the top of the clarifier; (2) the area needed for thickening (settling of the interface between the discrete and hindered settling zones); and (3) the rate of sludge withdrawal. The settling rate of the interface is usually the controlling factor.

Column settling tests, as previously described, can be used to determine the area needed for hindered settling. The height of the interface is plotted against time, as shown in Fig. 6.10. The area needed for clarification is

$$A = Q/v_s \qquad (6.49)$$

where $A$ = surface area of the settling zone, $m^2$ or $ft^2$
$Q$ = overflow rate, $m^3/s$ or gal/min
$v_s$ = subsidence rate in the zone of hindering settling, mm/s or in/s

A value of $v_s$ is determined from batch settling column test data by computing the slope of the hindered settling portion of the interface height versus time curve (Fig. 6.10). The area needed for thickening is obtained from the batch settling test of a thick suspension. The critical area required for adequate thickening is (Rich, 1961)

$$A = \frac{Qt_u}{H_0} \qquad (6.50)$$

where $A$ = area needed for sludge thickening, $m^2$ or $ft^2$
Q = flow into settling tank, $m^3/s$ or $ft^3/s$
$t_u$ = time to reach a desired underflow or solids concentration, s
$H_0$ = depth of the settling column (initial interface height), m or ft

From Fig. 6.10, the critical concentration ($C_2$) is determined by extending the tangent from the hindered and compression settling lines to their point of intersection and bisecting the angle formed. The bisector intersects the subsidence curve at $C_2$ which is the critical concentration. The critical concentration controls the sludge-handling capacity of the tank at a height of $H_2$.

A tangent is drawn to the subsidence curve at $C_2$ and the intersection of this tangent with depth $H_u$, required for the desired underflow (or solids concentration $C_u$), will yield the required retention time $t_u$. Since the total weight of solids in the system must remain constant, i.e.

$C_0H_0A = C_uH_uA$, the height $H_u$ of the particle–liquid interface at the underflow desired concentration $C_u$ is

$$H_u = \frac{C_0H_0}{C_u} \tag{6.51}$$

The time $t_u$ can be determined as:
Draw a horizontal line through $H_u$ and draw a tangent to the subsidence settling curve at $C_2$. Draw a vertical line from the point of intersection of the two lines drawn above to the time axis to find the value of $t_u$. With this value of $t_u$, the area needed for thickening can be calculated using Eq. (6.50). The area required for clarification is then determined. The larger of the two calculated areas is the controlling factor for design from Eqs. (6.49) and (6.50).

**Example:** The batch-settling curve shown in Fig. 6.10 is obtained for an activated sludge with an initial solids concentration $C_0$ of 3600 mg/L. The initial height of the interface in the settling column is 900 mm. This continuous inflow to the unit is 380 m$^3$/d (0.10 Mgal/d). Determine the surface area required to yield a thickened sludge of 1.8 percent by weight. Also determine solids and hydraulic loading rate.

**solution:**

Step 1. Calculate $H_u$ by Eq. (6.51)

$$C_u = 1.8\% = 18{,}000 \text{ mg/L}$$

$$H_u = \frac{C_0H_0}{C_u} = \frac{3600 \text{ mg/L} \times 900 \text{ mm}}{18{,}000 \text{ mg/L}}$$

$$= 180 \text{ mm}$$

Step 2. Determine $t_u$

Using the method described above to find the value of $t_u$

$$t_u = 41 \text{ min} = 41 \text{ min}/1440 \text{ min/d}$$

$$= 0.0285 \text{ day}$$

Step 3. Calculate the area required for the thickening, using Eq. (6.50)

$$A = \frac{Qt_u}{H_0} = \frac{380 \text{ m}^3\text{/d} \times 0.0285 \text{ day}}{0.90 \text{ m}}$$

$$= 12.02 \text{ m}^2$$

$$= 129 \text{ ft}^2$$

Step 4. Calculate the subsidence velocity $v_s$ in the hindered settling portion of the curve

In 10 min

$$v_s = \frac{(900 - 617)\text{ mm}}{10\text{ min} \times 60\text{ s/min}}$$

$$= 0.47\text{ mm/s}$$

$$= 40.6\text{ m/d}$$

Step 5. Calculate the area required for clarification

Using Eq. (6.49)

$$A = Q/v_s = 380\text{ m}^3/\text{d} \div 40.6\text{ m/d}$$

$$= 9.36\text{ m}^2$$

Step 6. Determine the controlling area

From comparison of areas calculated from Steps 3 and 5, the larger area is the controlling area. Thus the controlling area is the thickening area of 12.02 $m^2$ (129 $ft^2$)

Step 7. Calculate the solids loading

$$C_0 = 3600\text{ mg/L} = 3600\text{ g/m}^3 = 3.6\text{ kg/m}^3$$

$$\text{Solids weight} = QC_0 = 380\text{ m}^3/\text{d} \times 3.6\text{ kg/m}^3$$

$$= 1368\text{ kg/d}$$

$$= 3016\text{ lb/d}$$

$$\text{Solids loading rate} = 1368\text{ kg/d} \div 12.02\text{ m}^2$$

$$= 114\text{ kg/(m}^2\cdot\text{ d)}$$

$$= 23.3\text{ lb/(ft}^2\cdot\text{ d)}$$

Step 8. Determine the hydraulic (overflow) loading rate

$$\text{Hydraulic loading rate} = 380\text{ m}^3/\text{d} \div 12.02\text{ m}^2$$

$$= 31.6\text{ m}^3/(\text{m}^2\cdot\text{ d}) = 31.6\text{ m/d}$$

$$\text{or} = 100{,}000\text{ gal/d} \div 129\text{ ft}^2$$

$$= 776\text{ gal/(ft}^2\cdot\text{ d)}$$

### 18.6 Compression settling (type 4)

When the concentration of particles is high enough to bring the particles into physical contact with each other, compression settling will occur. Consolidation of sediment at the bottom of the clarifier is extremely slow. The rate of settlement decreases with time due to increased resistance to flow of the fluid.

The volume needed for the sludge in the compression region (thickening) can also be estimated by settling tests. The rate of consolidation in this region has been found to be approximately proportional to the difference in sludge height $H$ at time $t$ and the final sludge height $H_\infty$ obtained after a long period of time, perhaps 1 day. It is expressed as (Coulson and Richardson, 1955)

$$\frac{dH}{dt} = i(H - H_\infty) \tag{6.52}$$

where $H$ = sludge height at time $t$
$i$ = constant for a given suspension
$H_\infty$ = final sludge height

Integrating Eq. (6.52) between the limits of sludge height $H_t$ at time $t$ and $H_1$ at time $t_1$, the resulting expression is

$$H_t - H_\infty = (H_1 - H_\infty)e^{-i(t-t_1)} \tag{6.53}$$

or

$$i(t - t_1) = \ln(H_t - H_\infty) - \ln(H_1 - H_\infty) \tag{6.54}$$

A plot of ln $[(H_t - H_\infty] - \ln(H_1 - H_\infty)$ versus $(t - t_1)$ is a straight line having the slope $-i$. The final sludge height $H_\infty$ depends on the liquid surface film which adheres to the particles.

It has been found that gentle stirring serves to compact sludge in the compression region by breaking up the floc and permitting water to escape. The use of mechanical rakes with 4 to 5 revolutions per hour will serve this purpose. Dick and Ewing (1967) reported that gentle stirring also helped to improve settling in the hindered settling region.

## 19 Primary Sedimentation Tanks

Primary treatment has traditionally implied a sedimentation process to separate the readily settleable and floatable solids from the wastewater. The treatment unit used to settle raw wastewater is referred to as the primary sedimentation tank (basin), primary tank (basin), or primary clarifier. Sedimentation is the oldest and most widely used process in the effective treatment of wastewater.

After the wastewater passes the preliminary processes, it enters sedimentation tanks. The suspended solids that are too light to fall out in the grit chamber will settle in the tank over a few hours. The settled sludge is then removed by mechanical scrapers, or pumped. The floatable substances on the tank surface are removed by a surface skimmer device. The effluent flows to the secondary treatment units or is discharged off (not in the United States and some countries).

The primary sedimentation tank is where the flow velocity of the wastewater is reduced by plain sedimentation. The process commonly removes particles with a settling rate of 0.3 to 0.6 mm/s (0.7 to 1.4 in/min). In some cases, chemicals may be added. The benefits of primary sedimentation are reduced suspended solids content, equalization of sidestream flow, and BOD removal. The overflow rate of the primary sedimentation tanks ranges from 24.5 to 49 $m^3/(m^2 \cdot d)$ (600 to 1200 gal/(d · $ft^2$)). The detention time in the tank is usually 1 to 3 h (typically 2 h). Primary tanks (or primary clarifiers) should remove 90% to 95% of settleable solids, 50% to 60% of total suspended solids, and 25% to 35% of the $BOD_5$ (NY Department of Health, 1950).

Settling characteristics in the primary clarifier are generally characterized by type 2 flocculant settling. The Stokes formula for settling velocity cannot be used because the flocculated particles are continuously changing in shape, size, and specific gravity. Since no mathematical equation can describe flocculant settling satisfactorily, laboratory analyses of settling-column tests are commonly used to generate design information.

Some recommended standards for the design of primary clarifiers are as follows (GLUMRB–Ten States Standards, 1996; Illinois EPA, 1998). Multiple tanks capable of independent operation are desirable and shall be provided in all plants where design average flows exceed 380 $m^3$/d (100,000 gal/d). The minimum length of flow from inlet to outlet should be 3.0 m (10 ft) unless special provisions are made to prevent short circuiting. The side depth for primary clarifiers shall be as shallow as practicable, but not less than 3.0 m (10 ft). Hydraulic surface settling rates (overflow rates) of the clarifier shall be based on the anticipated peak hourly flow. For normal domestic wastewater, the overflow rate, with some indication of BOD removal, can be obtained from Fig. 6.11. If waste-activated sludge is returned to the primary clarifier, the design surface settling rate shall not exceed 41 $m^3/(m^2 \cdot d)$ (1000 gal/(d · $ft^2$)). The maximum surfaced settling rate for combined sewer overflow and bypass settling shall not exceed 73.3 $m^3/(m^2 \cdot d)$ (1800 gal/(d · $ft^2$)), based on peak hourly flow. Weir loading rate shall not exceed 250 $m^3$/d linear meter (20,000 gal/(d · ft)), based on design peak hourly flow for plants having a design average of 3785 $m^3$/d (1 Mgal/d) or less. Weir loading rates shall not exceed 373 $m^3/(d^2 \cdot m)$ (30,000 gal/(d · ft)), based on peak design hourly flow for plants having a design average flow greater than 3785 $m^3$/d (1.0 Mgal/d). Overflow rates, side water depths, and weir loading rates recommended by various institutions for primary settling tanks are listed elsewhere (WEF and ASCE, 1991a).

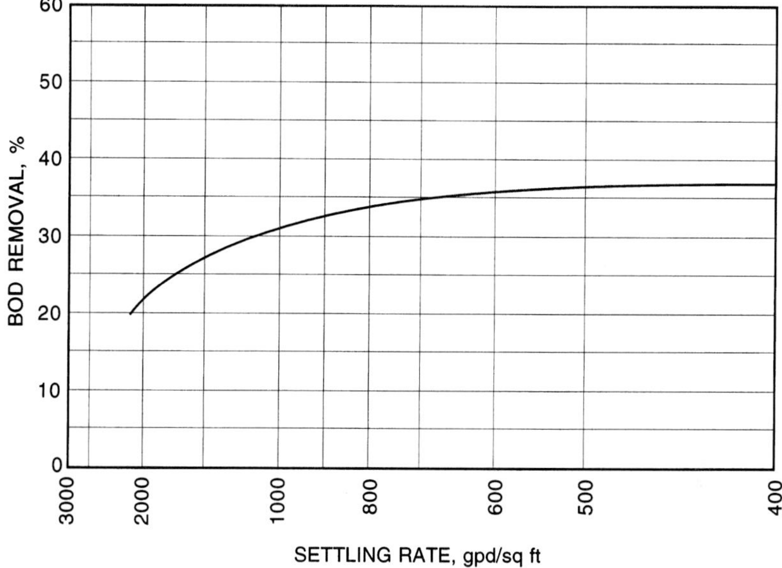


**Figure 6.11** $BOD_5$, removal in primary settling tank *(source: Illinois EPA 1998).*

In cases where a reliable loading–performance relationship is not available, the primary tank design may be based on the overflow rates and side water depths listed in Table 6.10. The design surface settling is selected on the basis of Fig. 6.11 and Table 6.10. The hydraulic

**TABLE 6.10 Typical Design Parameters for Primary Clarifiers**

| | | Surface settling rate, $m^3/(m^2 \cdot d)$ (gal/(d · ft$^2$)) | | |
|---|---|---|---|---|
| Type of treatment | Source | Average | Peak | Depth, m (ft) |
| Primary settling followed by secondary treatment | US EPA, 1975a | 33–19 (800–1200) | 81–122 (2000–3000) | 3–3.7 (10–12) |
| | GLUMEB–Ten States Standards and Illinois EPA, 1998 | 600 | Figure 6.11 | minimum 2.1 (7) |
| Primary settling with waste activated sludge return | US EPA, 1975a | 24–33 (600–800) | 49–61 (1200–1500) | 3.7–4.6 (12–15) |
| | Ten States Standards, GLUMRB, 1996 | ≤ 41 (≤ 1000) | ≤ 61 (≤ 1500) | 3.0 (10) minimum |

detention time $t$ in the primary clarifier can be calculated from Eq. (6.44). The hydraulic detention times for primary clarifier design range from 1.5 to 2.5 h, typically 2 h. Consideration should be made for low flow period to ensure that longer detention times will not cause septic conditions. Septic conditions may cause a potential odor problem, stabilization and loading to the downstream treatment processes. In the cold climatic region, a detention time multiplier should be included when wastewater temperature is below 20°C (68°F). The multiplier can be calculated by the following equation (Water Pollution Control Federation, 1985a)

$$M = 1.82e^{-0.03T} \tag{6.55}$$

where $M$ = detention time multiplier
$T$ = temperature to wastewater, °C

In practice, the linear flow-through velocity (scour velocity) is limited to 1.2 to 1.5 m/min (4 to 5 ft/min) to avoid resuspension of settled solids in the sludge zone (Theroux and Betz, 1959). Camp (1946) suggested that the critical scour velocity can be computed by Eq. (6.39). Scouring velocity may resuspend settled solids and must be avoided with clarifier design. Camp (1953) reported that horizontal velocities up to 18 ft/min (9 cm/s) may not create scouring; but design horizontal velocities should still be designed substantially below 18 ft/min. As long as scouring velocities are not approached, solids removal in the clarifier is independent of the tank depth.

**Example 1:** Determine the detention time multipliers for wastewater temperatures of 12 and 6°C.

**solution:** Using Eq. (6.55), for $T$ = 12°C

$$\begin{aligned} M &= 1.82e^{-0.03\times12} = 1.82 \times 0.70 \\ &= 1.27 \end{aligned} \tag{6.56}$$

For $T$ = 6°C

$$\begin{aligned} M &= 1.82 \times e^{-0.03\times6} = 1.82 \times 0.835 \\ &= 1.52 \end{aligned}$$

**Example 2:** Two rectangular settling tanks are each 6 m (20 ft) wide, 24 m (80 ft) long, and 2.7 m (9 ft) deep. Each is used alternately to treat 1900 $m^3$ (0.50 Mgal) in a 12-h period. Compute the surface overflow (settling) rate, detention time, horizontal velocity, and outlet weir loading rate using H-shaped weir with three times the width.

**solution:**

Step 1. Determine the design flow $Q$

$$Q = \frac{1900\ \text{m}^3}{12\ \text{h}} \times \frac{24\ \text{h}}{1\ \text{day}}$$

$$= 3800\ \text{m}^3/\text{d}$$

Step 2. Compute surface overflow rate $v_0$

$$v_0 = Q/A = 3800\ \text{m}^3/\text{d} \div (6\ \text{m} \times 24\ \text{m})$$

$$= 26.4\ \text{m}^3/(\text{m}^2 \cdot \text{d})$$

$$= 650\ \text{gal}/(\text{d} \cdot \text{ft}^2)$$

Step 3. Compute detention time $t$

$$\text{Tank volume } V = 6\ \text{m} \times 24\ \text{m} \times 2.1\text{m} \times 2$$

$$= 604.8\ \text{m}^3$$

$$t = V/Q = 604.8\ \text{m}^3/(3800\ \text{m}^3/\text{d})$$

$$= 0.159\ \text{day}$$

$$= 3.8\ \text{h}$$

Step 4. Compute horizontal velocity $v_h$

$$v_h = \frac{3800\ \text{m}^3/\text{d}}{6\ \text{m} \times 2.1\ \text{m}}$$

$$= 301\ \text{m/d}$$

$$= 0.209\ \text{m/min}$$

$$= 0.686\ \text{ft/min}$$

Step 5. Compute outlet weir loading, $wl$

$$wl = \frac{3800\ \text{m}^3/\text{d}}{6\ \text{m} \times 3}$$

$$= 211\ \text{m}^3/(\text{d} \cdot \text{m})$$

$$= 17{,}000\ \text{gal}/(\text{d} \cdot \text{ft})$$

### 19.1 Rectangular basin design

Multiple units with common walls shall be designed for independent operation. A bypass to the aeration basin shall be provided for emergency conditions. Basin dimensions are to be designed on the basis of surface overflow (settling) rate to determine the required basin surface area. The area required is the flow divided by the selected overflow rate. An overflow rate of 36 $m^3/(m^2 \cdot d)$ (884 gal/($ft^2 \cdot d$)) at average design flow is generally acceptable.

Basin surface dimensions, the length ($l$) to width ($w$) ratio ($l/w$), can be increased or decreased without changing the volume of the basin. The greater the $l/w$ ratio, the better the basin conforms to plug flow conditions. Also, for greater $l/w$ ratio, the basin has a proportionally larger effective settling zone and smaller percent inlet and outlet zones. Increased basin length allows the development of a more stable flow. Best conformance to the plug flow model has been reported by a basin with $l/w$ ratio of 3:1 or greater (Aqua-Aerobic Systems, 1976).

Basin design should be cross-checked for detention time for conformance with recommended standards by the regulatory agencies.

For bean bridge (cross the basin) design, a commercially available economical basin width can be used, such as 1.5 m (5 ft), 3.0 m (10 ft), 5.5 m (18 ft), 8.5 m (28 ft) or 11.6 m (38 ft).

The inlet structures in rectangular clarifiers are placed at one end and are designed to dissipate the inlet velocity to diffuse the flow equally across the entire cross section of the basin and to prevent short circuiting. Typical inlets consist of small pipes with upward ells, perforated baffles, multiple ports discharging against baffles, a single pipe turned back to discharge against the headwall, simple weirs, submerged weirs sloping upward to a horizontal baffle, etc. (Steel and McGhee, 1979; McGhee, 1991). The inlet structure should be designed not to trap scum or settling solids. The inlet channel should have a velocity of 0.3 m/s (1 ft/s) at one half design flow (Ten States Standards, GLUMRB, 1996).

Baffles are installed 0.6 to 0.9 m (2 to 3 ft) in front of inlets to assist in diffusing the flow and submerged 0.45 to 0.60 m (1.5 to 2 ft) with 5 cm (2 in) water depth above the baffle to permit floating material to pass. Scum baffles are placed ahead of outlet weirs to hold back floating material from the basin effluent and extend 15 to 30 cm (6 to 12 in) below the water surface (Ten States Standards, GLUMRB, 1971). Outlets in a rectangular basin consist of weirs located toward the discharge end of the basin. Weir loading rates range from 250 to 373 $m^3/(d \cdot m)$ (20,000 to 30,000 gal/($d \cdot ft$)) (Ten State Standards, GLUMRB, 1996).

Walls of settling tanks should extend at least 150 mm (6 in) above the surrounding ground surface and shall provide not less than 300 mm (12 in) freeboard (GLUMRB, 1996).

Mechanical sludge collection and withdrawal facilities are usually designed to assure rapid removal of the settled solids. The minimum slope of the side wall of the sludge hopper is 1.7 vertical to 1 horizontal. The hopper bottom dimension should not exceed 600 mm (2 ft). The sludge withdrawal line is at least 150 mm (6 in) in diameter and has a static head of 760 mm (30 in) or greater with a velocity of 0.9 m/s (3 ft/s) (Ten States Standards, GLUMRB, 1996).

**Example:** Design a primary clarification system for a design average wastewater flow of 7570 $m^3$/d (2.0 Mgal/d) with a peak hourly flow of 18,900 $m^3$/d (5.0 Mgal/d) and a minimum flow of 4540 $m^3$/d (1.2 Mgal/d). Design a multiple units system using Ten States Standards for an estimated 35% $BOD_5$ removal at the design flow.

**solution:**

Step 1. List Ten States Standards

Referring to Fig. 6.11, for 35% BOD removal:

$$\text{Surface settling rate } v = 28.5 \text{ m}^3/(\text{m}^2 \cdot \text{d}) \text{ or } (700 \text{ gal/(d} \cdot \text{ft}^2))$$

$$\text{Minimum depth} = 3.0 \text{ m (10 ft)}$$

$$\text{Maximum weir loading} = 124 \text{ m}^3/(\text{d} \cdot \text{m}) \text{ or } (10{,}000 \text{ gal/(d} \cdot \text{ft))}$$

for average daily flow

Step 2. Determine tank dimensions

1. Surface area needed

Use two settling tanks, each with design flow of 3785 $m^3$/d (1.0 Mgal/d)

$$\text{Surface area } A = Q/v = 3785 \text{ m}^3/\text{d} \div 28.5 \text{ m}^3/(\text{m}^2 \cdot \text{d})$$

$$= 132.8 \text{ m}^2$$

2. Determine length l and width $w$ using $l/w$ ratio of 4/1

$$(w)(4w) = 132.8$$

$$w^2 = 33.2$$

$$w = 5.76 \text{ (m)}$$

Standard economic widths of bean bridge are: 1.52 m (5 ft), 3.05 m (10 ft), 5.49 m (18 ft), 8.53 m (28 ft), and 11.58 m (38 ft).
Select standard 5.49 m (18 ft width); $w = 5.49$ m

$$5.49l = 132.8$$

$$l = 24.19 \text{ (m)}$$

$$= 79.34 \text{ ft}$$

Say area is 5.5 m × 24 m (18 ft × 79 ft)
Compute tank surface area $A$

$$A = 132 \text{ m}^2 = 1422 \text{ ft}^2$$

3. Compute volume $V$ with a depth of 3 m

$$V = 132\ \text{m}^2 \times 3\ \text{m}$$
$$= 396\ \text{m}^3 = 14{,}000\ \text{ft}^3$$

Step 3. Check detention time $t$

At average design flow

$$t = V/Q = 396\ \text{m}^3/(3785\ \text{m}^3/\text{d})$$
$$= 0.105\ \text{day} \times 24\ \text{h/d}$$
$$= 2.5\ \text{h}$$

At peak flow

$$t = 2.5\ \text{h} \times (2/5)$$
$$= 1.0\ \text{h}$$

Step 4. Check overflow rate at peak flow

$$v = 28.5\ \text{m}^3/(\text{m}^2 \cdot \text{d}) \times (5/2)$$
$$= 71.25\ \text{m}^3/(\text{m}^2 \cdot \text{d})$$

Step 5. Determine the length of outlet weir

$$\text{Length} = \text{flow} \div \text{weir loading rate}$$
$$= 3785\ \text{m}^3/\text{d} \div 124\ \text{m}^3/(\text{d} \cdot \text{m})$$
$$= 30.5\ \text{m}$$

It is 5.5 times the width of the tank.

### 19.2 Circular basin design

Inlets in circular or square basins are typically at the center and the flow is directed upward through a baffle that channels the wastewater (influent) toward the periphery of the tank. Inlet baffles are 10% to 20% of the basin diameter and extend 0.9 to 1.8 m (3 to 6 ft) below the wastewater surface (McGhee, 1991).

Circular basins have a higher degree of turbulence than rectangular basins. Thus circular basins are more efficient as flocculators.

A typical depth of sidewall of a circular tank is 3 m (10 ft). As shown in Fig. 6.12, the floor slope of the tank is typically 300 mm (12 in) horizontal to 25 mm (1 in) vertical (Aqua-Aerobic Systems, 1976).

Outlet weirs extend around the periphery of the tank with baffles extending 200 to 300 mm (8 to 12 in) below the wastewater surface to retain floating material (McGhee, 1991). Overflow weirs shall be located to optimum actual hydraulic detention time and minimize short circuiting.

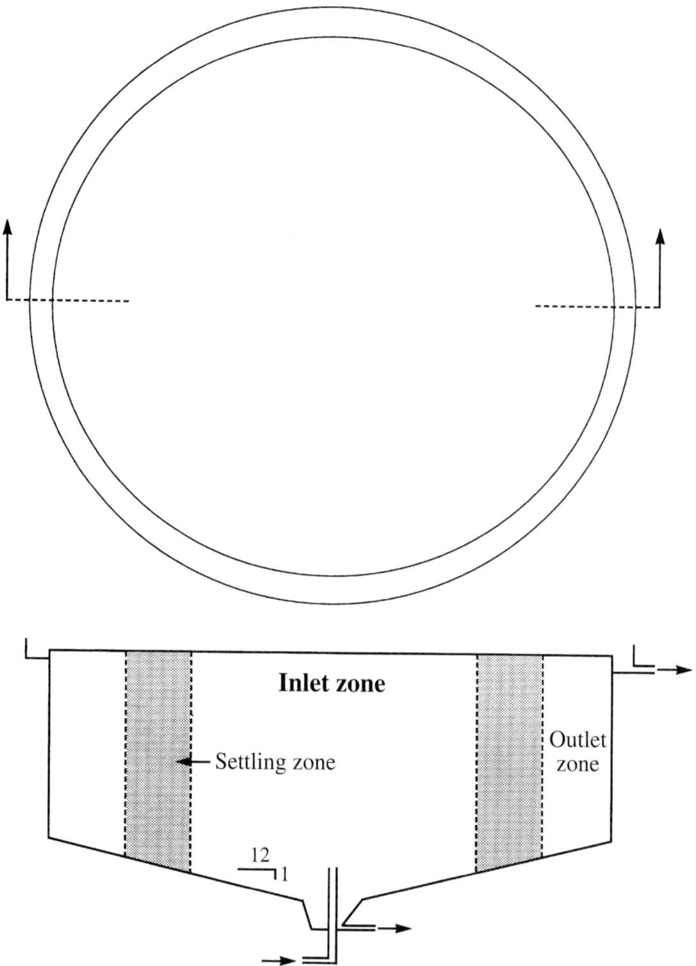


**Figure 6.12** Typical circular basin design-plan view and cross section.

Peripheral weirs shall be placed at least 300 cm (1 ft) from the well (Ten States Standards, GLUMRB, 1996).

**Example 1:** Design circular clarifiers using English system units with the same given information as in the example for rectangular clarifiers design.

**solution:**

Step 1. Calculate surface area $A$

Design 2 circular clarifiers, each treating 1.0 Mgal/d with an overflow rate of 700 gal/(d · ft$^2$).

$$A = 1{,}000{,}000 \text{ (gal/d)}/700 \text{ gal/(d} \cdot \text{ft}^2)$$
$$= 1429 \text{ ft}^2 = 132.8 \text{ m}^2$$

Step 2. Determine the tank radius $r$

$$\pi r^2 = 1429$$
$$r^2 = 1429/3.14 = 455$$
$$r = 21.3 \text{ ft}$$

Use $d$ = 44 ft (13.4 m) diameter tank

Surface area = 1520 ft$^2$ (141 m$^2$)

Step 3. Check overflow rate

$$\text{Overflow rate} = 1{,}000{,}000 \text{ (gal/d)}/1520 \text{ ft}^2$$
$$= 658 \text{ gal/(d} \cdot \text{ft}^2)$$
$$= 26.8 \text{ m}^3/(\text{m}^2 \cdot \text{d})$$

From Fig. 6.11, $BOD_5$ removal for an overflow rate of 658 gal/(d · ft$^2$) is 35%.

Step 4. Determine detention time $t$

Use a wastewater side wall depth of 10 ft and, commonly, plus 2-ft freeboard for wind protection

$$t = \frac{1520 \text{ ft}^2 \times 10 \text{ ft} \times 7.48 \text{ gal/ft}^3}{1{,}000{,}000 \text{ gal}}$$
$$= 0.114 \text{ day}$$
$$= 2.73 \text{ h}$$

Step 5. Calculate weir loading rate

Use an inboard weir with diameter of 40 ft (12.2 m)

$$\text{Length of periphery} = \pi \times 40 \text{ ft} = 125.6 \text{ ft}$$
$$\text{Weir loading} = 1{,}000{,}000 \text{ (gal/d)}/125.6 \text{ ft}$$
$$= 7960 \text{ gal/(d} \cdot \text{ft)} = 98.8 \text{ m}^3/\text{m} \cdot \text{d}$$

Step 6. Calculate the number ($n$) of V notches

Use 90° standard V notches at rate of 8 in center-to-center of the launders.

$$n = 125.6 \text{ ft} \times 12 \text{ in/ft} \div 8 \text{ in}$$
$$= 188$$

Step 7. Calculate average discharge per notch at average design flow $q$

$$q = 1{,}000{,}000 \text{ gal/d} \times (1 \text{ d}/1440 \text{ min}) \div 188$$
$$= 3.7 \text{ gal/min} = 14 \text{ L/min}$$

**Example 2:** If the surface overflow rate is 40 $m^3/(m^2 \cdot d)$ [982 $gal/(d \cdot ft^2)$] and the weir overflow rate is 360 $m^3/(d \cdot m)$ [29,000 $gal/(d \cdot ft)$], determine the maximum radius for a circular primary clarifier with a single peripheral weir.

**solution**

Step 1. Compute the area ($A$) required with a flow ($Q$)

Let $r$ = the radius of the clarifier

$$A = \pi r^2 = Q/40 \text{ m}^3/(\text{m}^2 \cdot \text{d}) \qquad \text{or} \qquad Q = 40\,\pi r^2$$

Step 2. Compute the weir length (= $2\pi r$)

$$2\pi r = Q/360 \text{ m}^3/(\text{d} \cdot \text{m})$$

or

$$Q = 720\pi r \text{ m}^3/(\text{d} \cdot \text{m})$$

Step 3. Compute $r$ by solving the equations in Steps 1 and 2

$$\pi r^2 = 720\pi r\,[\text{m}^3/(\text{d} \cdot \text{m})]/40\,[\text{m}^3/(\text{m}^2 \cdot \text{d})]$$
$$r = 18 \text{ m (59 ft)}$$

## 20 Biological (Secondary) Treatment Systems

The purpose of primary treatment is to remove suspended solids and floating material. In many situations in some countries, primary treatment with the resulting removal of approximately 40% to 60% of the suspended solids and 25% to 35% of $BOD_5$, together with removal of material from the wastewater, is adequate to meet the requirement of the receiving water body. If primary treatment is not sufficient to meet the regulatory effluent standards, secondary treatment using a biological process is mostly used for further treatment due to its greater removal efficiency and less cost than chemical coagulation. Secondary treatment processes are intended to remove the soluble and colloidal organics (BOD) which remain after primary treatment and to achieve further removal of suspended solids and, in some cases, also to remove nutrients such as phosphorus and nitrogen. Biological treatment processes provide similar biological activities to waste assimilation, which would take place in the receiving waters, but in a reasonably shorter time. Secondary treatment may remove more than 85% of $BOD_5$ and suspended matter, but is really not effective for removing nutrients, nonbiodegradable organics, heavy metals, and microorganisms.

Biological treatment systems are designed to maintain a large active mass and a variety of microorganisms, principally bacteria (and fungi, protozoa, rotifers, algae, etc.), within the confined system under favorable environmental conditions, such as dissolved oxygen, nutrient, etc. Biological treatment processes are generally classified mainly as suspended growth processes (activated sludge, Fig. 6.13), attached (film) growth processes (trickling filter and rotating biological contactor, RBC), and dual-process systems (combined). Other biological wastewater treatment processes include the stabilization pond, aerated lagoon, contaminant pond, oxidation ditch, high-purity oxygen activated sludge, biological nitrification, denitrification, and phosphorus removal units.

In the suspended biological treatment process, under continuous supply of air or oxygen, living aerobic microorganisms are mixed thoroughly with the organics in the wastewater and use the organics as food for their growth. As they grow, they clump or flocculate to form an active mass of microbes. This is so-called biologic floc or activated sludge.

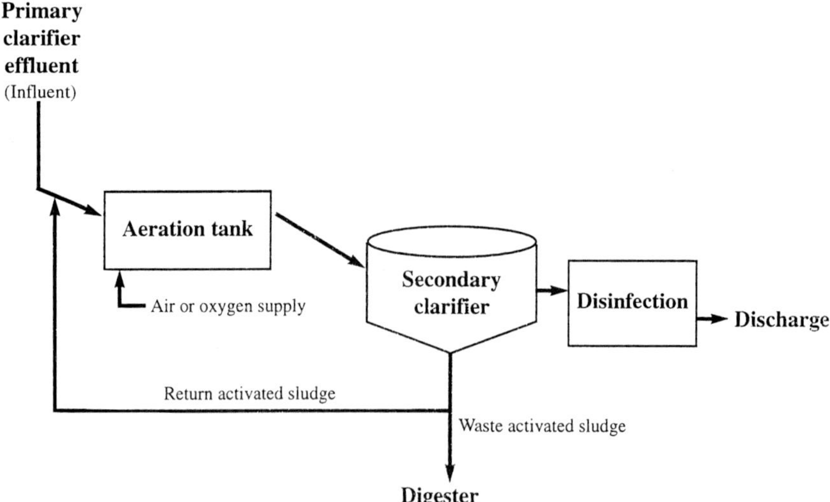


**Figure 6.13** Conventional activated sludge process.

## 20.1 Cell growth

Each unicellular bacterium grows and, after reaching a certain size, divides to produce two complete individuals by binary fission. The period of time required for a growing population to double is called the generation time. With each generation, the total number increases as the power (exponent) of 2, i.e. $2^0$, $2^1$, $2^2$, $2^3$ . . . The exponent of 2 corresponding to any given number is the logarithm of that number to the base 2 ($\log_2$). Therefore, in an exponentially growing culture, the $\log_2$ of the number of cells increasing in proportion to time, is often referred to as logarithmic growth.

The growth rate of microorganisms is affected by environmental conditions, such as DO levels, temperature, nutrient levels, pH, micro-bial community, etc. Exponential growth does not normally continue for a long period of time. A general growth pattern for fission-reproduction bacteria in a batch culture is sketched in Fig. 6.14. They are the lag phase, the exponential (logarithmic) growth phase, the maximum (stationary) phase, and the death phase.

When a small number of bacteria is inoculated into a fixed volume of vessel with culture medium, bacteria generally require time to acclimatize to their environmental condition. For this period of time, called the lag phase, the bacterial density is almost unchanged. Under excess

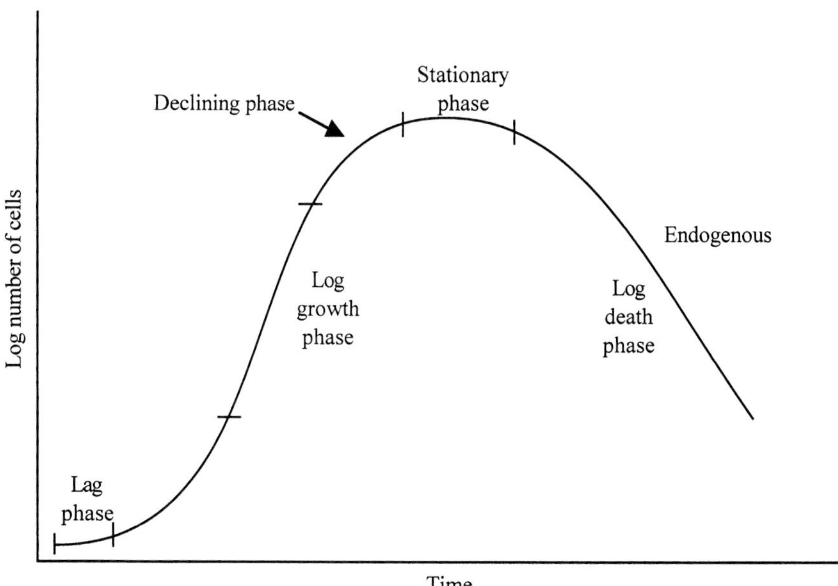


**Figure 6.14** Bacterial density with growth time.

food supply, a rapid increase in number and mass of bacteria occurs in the log growth phase. During this maximum rate of growth a maximum rate of substrate removal occurs. Either some nutrients become exhausted, or toxic metabolic products may accumulate. Subsequently, the growth rate decreases and then ceases. For this period, the cell number remains stationary at the stationary phase. As bacterial density increases and food runs short, cell growth will decline. The total mass of protoplasm exceeds the mass of viable cells, because bacteria form resistant structures such as endospores.

In the endogenous growth period, the microorganisms compete for the limiting substrate even to metabolize their own protoplasm. The rate of metabolism decreases and starvation occurs. The rate of death exceeds the rate of reproduction; cells also become old, die exponentially, and lysis. Lysis can occur in which the nutrients from the dead cells diffuse out to furnish the remaining living cells as food. The results of cell lysis decrease both the population and the mass of microorganisms.

In the activated-sludge process, the balance of food to microorganisms is very important. Metabolism of the organic matter results in an increased mass of microorganisms in the process. The excess mass of microorganisms must be removed (waste sludge) from the system to maintain a proper balance between mass of microorganisms and substrate concentration in the aeration basin.

In both batch and continuous culture systems the growth rate of bacterial cells can be expressed as

$$r_g = \mu X \tag{6.57}$$

where $r_g$ = growth rate of bacteria, mg/(L · d)
$\mu$ = specific growth rate, per day
$X$ = mass of microorganism, mg/L

since

$$\frac{dX}{dt} = r_g \tag{6.58}$$

Therefore

$$\frac{dX}{dt} = \mu X \tag{6.59}$$

**Substrate limited growth.** Under substrate limited growth conditions, or for an equilibrium system, the quantity of solids produced is equal to that lost, and specific growth rate (the quantity produced per day) can be expressed as the well-known Monod equation

$$\mu = \mu_{\mathrm{m}} \frac{S}{K_{\mathrm{s}} + S} - k_{\mathrm{d}} \tag{6.60}$$

where $\mu$ = specific growth rate, per day
$\mu_{\mathrm{m}}$ = maximum specific growth rate, per day
$S$ = concentration for substrate in solution, mg/L
$K_{\mathrm{s}}$ = half velocity constant, substrate concentration at one-half the maximum growth rate, mg/L
$k_{\mathrm{d}}$ = cell decay coefficient reflecting the endogenous burn-up of cell mass, mg/(mg · d)

Substituting the value of $\mu$ from Eq. (6.60) into Eq. (6.57), the resulting equation for the cell growth rate is

$$r_{\mathrm{g}} = \frac{\mu_{\mathrm{m}} S X}{K_{\mathrm{s}} + S} - k_{\mathrm{d}} \tag{6.61}$$

## 21 Activated-Sludge Process

The activated-sludge process was first used in Manchester, England. This process is perhaps the most widely used process for secondary treatment of wastewaters. Recently, the activated-sludge process has been applied as a nitrification and denitrification process, and modified with an anoxic and anaerobic section for phosphorus removal.

A basic suspended-growth (activated-sludge) process is depicted in Fig. 6.13. The wastewater continuously enters an aeration tank where previously developed biological flocs are brought into contact with the organic material of the wastewater. Air or oxygen-enriched air is continuously injected into the aeration tank as an oxygen source to keep the system aerobic and the activated sludge in suspension. Approximately 8 $\mathrm{m}^3$ of air is required for each $\mathrm{m}^3$ of wastewater.

The microbes in the activated sludge consist of a gelatinous matrix of filamentous and unicellular bacteria which are fed on protozoa. The predominant bacteria, such as *Pseudomonas,* utilize carbohydrate and hydrocarbon wastes, whereas *Bacillus, Flavobacterium,* and *Alcaligenes* consume protein wastes. When a new activated-sludge system is first started, it should be seeded with activated sludge from an existing operating plant. If seed sludge is not available, activated sludge can be prepared by simply continuously aerating, settling, and returning settled solids to the wastewater for a few weeks (4 to 6 weeks) (Alessi *et al,* 1978; Cheremisinoff, 1995).

The microorganisms utilize the absorbed organic matter as a carbon and energy source for cell growth and convert it to cell tissue, water, and oxidized products (mainly carbon dioxide, $CO_2$). Some bacteria attack the original complex substance to produce simple compounds as their waste products. Other bacteria then use these waste products to produce simpler compounds until the food is used up.

The mixture of wastewater and activated sludge in the aeration basis is called mixed liquor. The biological mass (biomass) in the mixed liquor is called the mixed liquor suspended solids (MLSS) or mixed liquor volatile suspended solids (MLVSS). The MLSS consists mostly of microorganisms, nonbiodegradable suspended organic matter, and other inert suspended matter. The microorganisms in MLSS are composed of 70% to 90% organic and 10% to 30% inorganic matter (Okun, 1949; WEF and ASCE, 1996a). The types of bacterial cell vary, depending on the chemical characteristics of the influent waste-water tank conditions and the specific characteristics of the microorganisms in the flocs. Microbial growth in the mixed liquor is maintained in the declining or endogenous growth phase to insure good settling properties.

After a certain reaction time (4 to 8 h), the mixed liquor is discharged from the aeration tank to a secondary sedimentation basin (settling tank, clarifier) where the suspended solids are settled out from the treated wastewater by gravity. However, in a sequencing batch reactor (SBR), mixing and aeration in the aeration tank are stopped for a time interval to allow MLSS to settle and to decant the treated wastewater; thus a secondary clarifier is not needed in an SBR system. Most concentrated biological settled sludge is recycled back to the aeration tank (so-called return activated sludge, RAS) to maintain a high population of microorganisms to achieve rapid breakdown of the organics in the wastewater. The volume of RAS is typically 20% to 30% of the wastewater flow. Usually more activated sludge is produced than return sludge.

The treated wastewater is commonly chlorinated and dechlorinated, then discharged to receiving water or to a tertiary treatment system (Fig. 6.3). The preliminary, primary, and activated-sludge (biological) processes are included in the so-called secondary treatment process.

### 21.1 Aeration periods and BOD loadings

The empirical design of activated sludge is based on BOD loading, food-to-microorganism ratio (F/M), sludge age, and aeration period. Empirical design concepts are still acceptable. The Ten States Standards states that when activated-sludge process design calculations are not submitted, the aeration tank capacities and permissible loadings for the several adoptions of the processes shown in Table 6.11 are used as simple design criteria. Those values apply to plants receiving diurnal load

**TABLE 6.11 Permissible Aeration Tank Capacities and Loadings**

| Process | Organic ($BOD_5$) loading lb/(d · 1000 $ft^3$) (kg/d · $m^3$)* | F/M ratio, lb $BOD_5$/d per lb MLVSS | MLSS, mg/L‡ |
|---|---|---|---|
| Conventional step aeration complete mix | 40(0.64) | 0.2–0.5 | 1000–3000 |
| Contact stabilization† | 50† (0.80) | 0.2–0.6 | 1000–3000 |
| Extended aeration single-stage nitrification | 15(0.24) | 0.05–0.1 | 3000–5000 |

Note: Loadings are based on the influent organic load to the aeration tank at plant design average $BOD_5$.
† Total aeration capacity includes both contact and reaeration capacities; normally the contact zone equals 30% to 35% of the total aeration capacity.
‡The values of MLSS are dependent on the surface area provided for secondary settling and the rate of sludge return as well as the aeration processes.
SOURCE: GLUMRB (Ten States Standards) (1996)

ratios of design peak hourly $BOD_5$ to design $BOD_5$ ranging from 2:1 to 4:1 (GLUMRB, 1996).

The aeration period is the retention time of the influent wastewater flow in the aeration basin and is expressed in hours. It is computed from the basin volume divided by the average daily flow excluding the return sludge flow. For normal domestic sewage, the aeration period commonly ranges from 4 to 8 h with an air supply of 0.5 to 2.0 $ft^3$/gal (3.7 to 15.0 $m^3/m^3$) of wastewater. The return-activated sludge is expressed as a percentage of the wastewater influent of the aeration tank.

The organic ($BOD_5$) loading on an aeration basin can be computed using the BOD in the influent wastewater without including the return sludge flow. BOD loadings are expressed in terms of lb BOD applied per day per 1000 $ft^3$ or [kg/(d · $m^3$)] of liquid volume in the aeration basin and in terms of lb BOD applied per day per lb of mixed liquid volatile suspended solids. The latter is called the food-to-microorganism ratio.

BOD loadings per unit volume of aeration basin vary widely from 10 to more than 100 lb/1000 $ft^3$ (0.16 to 1.6 kg/(d · $m^3$)), while the aeration periods correspondingly vary from 2.5 to 24 h (Clark *et al.*, 1977). The relationship between the two parameters is directly related to the BOD concentration in the wastewater.

### 21.2 F/M ratio

The F/M ratio is used to express BOD loadings with regard to the microbial mass in the process. The value of the F/M ratio can be calculated by the following equation:

$$\text{F/M} = \frac{\text{BOD, lb/d}}{\text{MLSS, lb}} \tag{6.62}$$

$$= \frac{Q\ (\text{Mgal/d}) \times \text{BOD (mg/L)} \times 8.34\ (\text{lb/gal})}{V\ (\text{Mgal}) \times \text{MLSS (mg/L)} \times 8.34\ (\text{lb/gal})} \tag{6.63}$$

or

$$\text{F/M} = \frac{\text{BOD, kg/d}}{\text{MLSS, kg}} \tag{6.64}$$

where F/M = food-to-microorganism ratio, kg (lb) of BOD per day per kg (lb) of MLSS
$Q$ = wastewater flow, $m^3$/d or Mgal/d
BOD = wastewater 5-day BOD, mg/L
$V$ = liquid volume of aeration tank, $m^3$ or Mgal
MLSS = mixed liquor suspended solids in the aeration tank, mg/L

In Eqs. (6.62) and (6.63), some authors use mixed liquor volatile suspended solids instead of MLSS. The MLVSS is the volatile portion of the MLSS and ranges from 0.75 to 0.85. Typically they are related, for design purposes, by MLVSS = 0.80 × MLSS. The use of MLVSS may more closely approximate the total active biological mass in the process.

The F/M ratio is also called the sludge loading ratio (SLR). The equation for the calculation of the SLR is (Cheremisinoff, 1995)

$$\text{SLR} = \frac{24\ \text{BOD}}{\text{MLVSS}\ (t)(1 + R)} \tag{6.65}$$

where SLR = sludge loading, g of BOD/d per g of MLVSS
BOD = wastewater BOD, mg/L
MLVSS = mixed liquor volatile suspended solids, mg/L
$t$ = retention time, d
$R$ = recyle ratio

**Example 1:** An activated-sludge process has a tank influent BOD concentration of 140 mg/L, influent flow of 5.0 Mgal/d (18,900 $m^3$/d) and 35,500 lb (16,100 kg) of suspended solids under aeration. Calculate the F/M ratio.

**solution:**

Step 1. Calculate BOD in lb/d

$$\begin{aligned}\text{BOD} &= Q \times \text{BOD} \times 8.34\\ &= 5.0\ \text{Mgal/d} \times 140\ \text{mg/L} \times 8.34\ \text{lb/gal}\\ &= 5838\ \text{lb/d}\end{aligned}$$

Step 2. Calculate the volatile SS under aeration

Assume VSS is 80% of TSS

$$\begin{aligned}\text{MLVSS} &= 35{,}500\ \text{lb} \times 0.8\\ &= 28{,}400\ \text{lb}\end{aligned}$$

Step 3. Calculate F/M ratio

Using Eq. (6.62)

$$\begin{aligned}\text{F/M} &= (5838\ \text{lb/d})/(28{,}400\ \text{lb})\\ &= 0.206\ \text{lb BOD/d per lb MLVSS}\end{aligned}$$

**Example 2:** Convert the BOD concentration of 160 mg/L in the primary effluent into BOD loading rate in terms of kg/m$^3$ and lb/1000 ft$^3$. If this is used for 24 h high rate aeration, what is the rate for 6 h aeration?

**solution:**

Step 1. Calculate BOD loading in kg/m$^3$

$$\begin{aligned}160\ \text{mg/L} &= \frac{160\ \text{mg} \times (1\ \text{g}/1000\ \text{mg})}{1\ \text{L} \times (1\ \text{m}^3/1000\ \text{L})}\\ &= 160\ \text{g/m}^3\\ &= 0.16\ \text{kg/m}^3\end{aligned}$$

Step 2. Calculate BOD loading in lb/1000 ft$^3$

$$\begin{aligned}0.16\ \text{kg/m}^3 &= \frac{0.16\ \text{kg} \times 2.205\ \text{lb/kg}}{1\ \text{m}^3 \times 35.3147\ \text{ft}^3/\text{m}^3}\\ &= 0.01\ \text{lb/ft}^3\\ &= \frac{0.01\ \text{lb} \times 1000}{1000\ \text{ft}^3}\\ &= 10\ \text{lb/1000 ft}^3\end{aligned}$$

Thus $1\ \text{lb/1000 ft}^3 = 0.016\ \text{kg/m}^3$

Step 3. Calculate loading for 6-h aeration

$$0.16 \text{ kg/(d} \cdot \text{m}^3) \times \frac{24 \text{ h}}{6 \text{ h}} = 0.64 \text{ kg/(d} \cdot \text{m}^3)$$

and

$$10.0 \text{ lb/d/1000 ft}^3 \times \frac{24 \text{ h}}{6 \text{ h}} = 40 \text{ lb/(d} \cdot \text{m}^3)$$

*Note*: Refer to Table 6.11 to meet Ten States Standards.

The influent BOD to the conventional activated-sludge process is limited to 160 mg/L for 6-h aeration.

## 21.3 Biochemical reactions

The mechanism of removal of biodegradable organic matter in aerobic suspended-growth systems can be expressed by the energy production or respiration equation

$$\underset{\text{(CHONS)}}{\text{Organic matter}} + \underset{\text{(heterotrophic)}}{\text{bacteria}} + O_2 \rightarrow CO_2 + H_2O + NH_4^+ + \underset{\text{(energy)}}{\text{new cells}} \tag{6.66}$$

Further nitrification process can take place by selected autotrophs with oxidation of ammonia to nitrate and protoplasm synthesis

$$NH_4^+ + O_2 + CO_2 + HCO_3 \xrightarrow[\text{Energy}]{\text{Bacteria}} NO_3^- + H_2O + H^+ \quad \underset{\text{(protoplasm)}}{\text{new cells}} \tag{6.67}$$

The oxidation of protoplasm is a metabolic reaction which breaks down the protoplasm into elemental constituents, so that cells die. This is called endogenous respiration or cell maintenance, as follows:

$$\text{Protoplasm} + O_2 \rightarrow CO_2 + NH_3 + H_2O + \text{dead cells} \tag{6.68}$$

## 21.4 Process design concepts

The activated-sludge process has been used extensively in its original basic form as well as in many modified forms. The process design considerations include hydraulic retention time (HRT) for reaction kinetics; wastewater characteristics; environmental conditions, such as temperature, pH, and alkalinity; and oxygen transfer.

Single or multiple aeration tanks are typically designed for completed mixed flow, plug flow, or intermediate patterns and sized to provide an HRT in the range of 0.5 to 24 h or more (WEF and ASCE, 1991a).

In the past, designs of activated-sludge processes were generally based on empirical parameters such as $BOD_5$ (simplified as BOD) loadings and aeration time (hydraulic retention time). In general, short HRTs were used for weak wastewaters and long HRTs for strong wastewaters. Nowadays, the basic theory and design parameters for the activated-sludge process are well developed and generally accepted. The different design approaches were proposed by researchers on the basis of the concepts of $BOD_5$, mass balance, and microbial growth kinetics (McKinney, 1962; Eckenfelder, 1966; Jenkins and Garrison, 1968; Eckenfelder and Ford, 1970; Lawrence and McCarty, 1970; Ramanathan and Gaudy, 1971; Gaudy and Kincannon, 1977; Schroeder, 1977; Bidstrup and Grady, 1988).

Solution of the theoretical sophisticated design equations and computer models requires knowledge of microbial metabolism and growth kinetics, with pilot studies to obtain design information. Alternatives to such studies are: (1) to assume certain wastewater characteristics and embark on a semiempirical design; and (2) to use an entirely empirical approach relying on regulatory recommended standards (WEF and ASCE, 1991a).

## 21.5 Process mathematical modeling

For almost half a century, numerous design criteria utilizing empirical and rational parameters based on biological kinetic equations have been developed for suspended-growth systems. A survey of major consulting firms in the United States indicates that the basic Lawrence and McCarty (1970) model is most widely used. Details of its development can be obtained in the references (Lawrence and McCarty, 1970; Grady and Lim, 1980; Qasim, 1985; Metcalf and Eddy, Inc. 1991). The basic Lawrence and McCarty design equations used for sizing suspended-growth systems are listed below.

**Complete mix with recycle.** The flow in a reactor is continuously stirred. The contents of the reactor are mixed completely. It is called the complete-mix reactor or continuous flow stirred tank reactor. Ideally, it is uniform throughout the tank. If the mass input rate into the reactor remains constant, the content of the effluent remains constant.

For a complete-mix system, the mean hydraulic retention time (HRT) $\theta$ for the aeration tank is

$$\theta = V/Q \qquad (6.69)$$

where $\theta$ = hydraulic retention time, day
$V$ = volume of aeration tank, $m^3$
$Q$ = influent wastewater flow, $m^3/d$

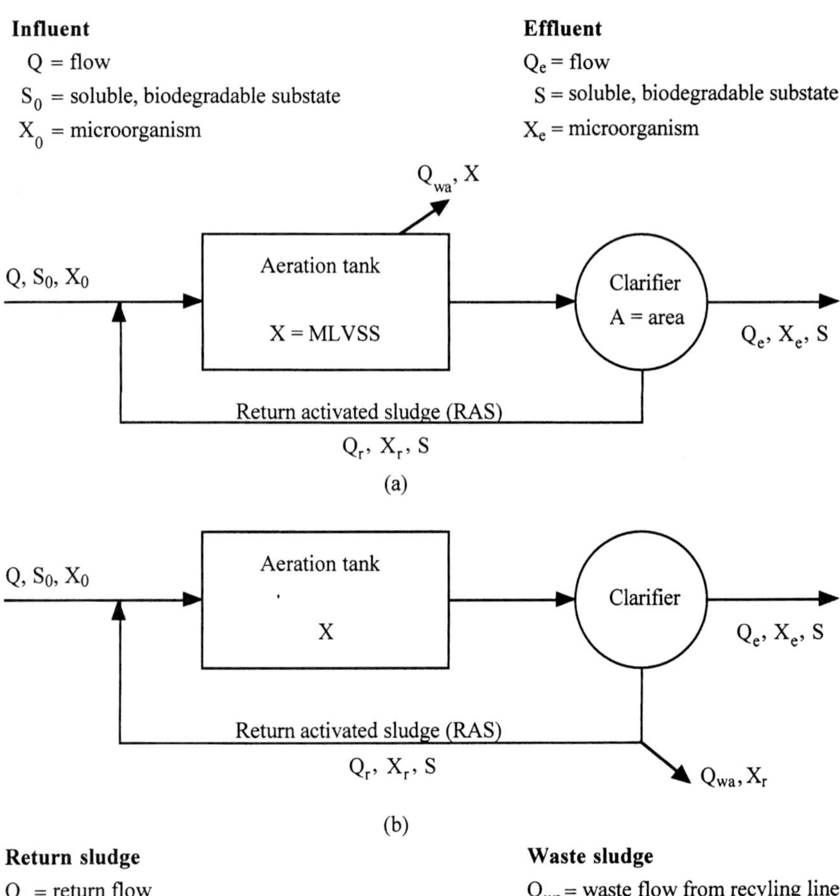


**Figure 6.15** Schematic chart of complete-mix activated sludge reactor: (a) sludging wasting from the aeration tank; (b) sludge wasting from return sludge line.

Referring to Fig. 6.15a, the mean cell residence time $\theta_c$ (or sludge age or SRT) in the system is defined as the mass of organisms in the aeration tank divided by the mass of organisms removed from the system per day, and is expressed as

$$\theta_c = \frac{X}{(\Delta X/\Delta t)} \tag{6.70}$$

$$\theta_c = \frac{VX}{Q_{wa}X + Q_eX_e} = \frac{\text{total mass SS in reactor}}{\text{SS wasting rate}} \tag{6.71}$$

where $\theta_c$ = mean cell residence time based on solids in the tank, day
$X$ = concentration of MLVSS maintained in the tank, mg/L
$\Delta X/\Delta t$ = growth of biological sludge over time period $\Delta t$, mg/(L · d)
$Q_{wa}$ = flow of waste sludge removed from the aeration tank, $m^3$/d
$Q_e$ = flow of treated effluent, $m^3$/d
$X_e$ = microorganism concentration (VSS) in effluent, mg/L

For system-drawn waste sludge from the return sludge line (Fig. 6.15b), the mean cell residence time would be

$$\theta_c = \frac{VX}{Q_{wr}X_r + Q_eX_e} \qquad (6.72)$$

where $Q_{wr}$ = flow of waste sludge from return sludge line, $m^3$/d
$X_r$ = microorganism concentration in return sludge line, mg/L

**Microorganism and substrate mass balance.** Because the term $V \times$ MLSS in Eq. (6.63) is a function of SRT or $\theta_c$ and not HRT or return sludge ratio, the F/M ratio is also a function only of SRT. Therefore, operation of an activated-sludge plant at constant SRT will result in operation at a constant F/M ratio.

The mass balance for the microorganisms in the entire activated sludge system is expressed as the rate of accumulation of the micro-organisms in the inflow plus net growth, minus that in the outflow. Mathematically, it is expressed as (Metcalf and Eddy, Inc. 1991)

$$V\frac{dX}{dt} = QX_0 + V(r'_g) - (Q_{wa}X + Q_eX_e) \qquad (6.73)$$

where $V$ = volume of aeration tank, $m^3$
$dX/dt$ = rate of change of microorganisms concentration (VSS), mg/(L · $m^3$ · d)
$Q$ = influent flow, $m^3$/d
$X_0$ = microorganisms concentration (VSS) in influent, mg/L
$X$ = microorganisms concentration in tank, mg/L
$r'_g$ = net rate of microorganism growth (VSS), mg/(L · d)
Other terms are as in the above equations.

The net rate of bacterial growth is expressed as

$$r'_g = Yr_{su} - k_dX \qquad (6.74)$$

where $Y$ = maximum yield coefficient over finite period of log growth, mg/mg
$r_{su}$ = substrate utilization rate, mg/(m$^3$ · d)
$k_d$ = endogenous decay coefficient, per day

Substituting Eq. (6.74) into Eq. (6.73), and assuming the cell concentration in the influent is zero and steady-state conditions, this yields

$$\frac{Q_{wa}X + Q_e X_e}{VX} = -Y\frac{r_{su}}{X} - k_d \tag{6.75}$$

The left-hand side of Eq. (6.75) is the inverse of the mean cell residence time $\theta_c$ as defined in Eq. (6.71); thus

$$\frac{1}{\theta_c} = -Y\frac{r_{su}}{X} - k_d \tag{6.76}$$

The term $1/\theta_c$ is the net specific growth rate.

The term $r_{su}$ can be computed from the following equation

$$r_{su} = \frac{Q}{V}(S_0 - S) = \frac{S_0 - S}{\theta} \tag{6.77}$$

where $S_0 - S$ = mass concentration of substrate utilized, mg/L
$S_0$ = substrate concentration in influent, mg/L
$S$ = substrate concentration in effluent, mg/L
$\theta$ = hydraulic retention time (Eq. (6.69)), day

**Effluent microorganism and substrate concentrations.** The mass concentration of microorganisms $X$ in the aeration tank can be derived by substituting Eq. (6.77) into Eq. (6.76)

$$X = \frac{\theta_c Y(S_0 - S)}{\theta(1 + k_d\theta_c)} = \frac{\mu_m(S_0 - S)}{k(1 - k_d\theta_c)} \tag{6.78}$$

where $\mu_m$ = maximum specific growth rate, per day
$k$ = maximum rate of substrate utilization per unit mass of microorganism, per day

Substituting for $\theta$ from Eq. (6.69) for (6.78) and solving for the reactor (aeration tank) volume yields

$$V = \frac{\theta_c QY(S_0 - S)}{X(1 + k_d\theta_c)} \tag{6.79}$$

**TABLE 6.12 Ranges and Typical Biological Kinetic Coefficients for the Activated-Sludge Process for Domestic Wastewater**

| Coefficient | Range | Typical value |
|---|---|---|
| $k$, per day | 11–20 | 5 |
| $k_d$, per day | 0.025–0.075 | 0.06 |
| $K_s$, mg/L $BOD_5$ | 25–100 | 60 |
| mg/L COD | 15–70 | 40 |
| $Y$, mg VSS/mg $BOD_5$ | 0.4–0.8 | 0.6 |

SOURCE: Metcalf and Eddy, Inc. (1991), Techobanoglous and Schroeder (1985)

The substrate concentration in effluent $S$ can also be determined from the substrate mass balance by the following equation

$$S = \frac{K_s(1 + \theta_c k_d)}{\theta_c(Yk - k_d) - 1} \tag{6.80}$$

where $S$ = effluent substrate (soluble $BOD_5$) concentration, mg/L
$K_s$ = half-velocity constant, substrate concentration at one half the maximum growth rate, mg/L
Other terms are as mentioned in previous equations.

The ranges of typical biological kinetic coefficients for activated-sludge systems for domestic wastewater are given in Table 6.12. When the kinetic coefficients are available, Eqs. (6.78) and (6.80) can be used to predict densities of effluent microorganisms and substrate (soluble $BOD_5$) concentrations, respectively. They do not take into account influent suspended solids concentrations (primary effluent). They can be used to evaluate the effectiveness of various treatment system changes.

Substituting the value of $X$ given by Eq. (6.78) for $r'_g$ in Eq. (6.74), and dividing by the term $S_0 - S$ which corresponds to the value of $r_{su}$ expressed as concentration value, the observed yield in the system with recycle is

$$Y_{obs} = \frac{Y}{1 + K_d\theta_c \text{ or } \theta_{ct}} \tag{6.81}$$

where $Y_{obs}$ = observed yield in the system with recycle, mg/mg
$\theta_{ct}$ = mean of all residence times based on solids in the aeration tank and in the secondary clarifier, day
Other terms are defined previously.

**Process design and control relationships.** To predict the effluent biomass and $BOD_5$ concentration, the use of Eqs. (6.78) and (6.80) is difficult because many coefficients have to be known. In practice, the relationship between specific substrate utilization rate, mean cell residence time, and the food to microorganism (F/M) ratio is commonly used for activated-sludge process design and process control.

In Eq. (6.76), the term $(-r_{su}/X)$ is called the specific substrate utilization rate (or food to microorganisms ratio), $U$. Applying $r_{su}$ in Eq. (6.77), the specific substrate utilization rate can be computed by

$$U = -\frac{r_{su}}{X} \tag{6.82}$$

$$U = \frac{Q(S_0 - S)}{VX} = \frac{S_0 - S}{\theta X} \tag{6.83}$$

The term $U$ is substituted for the term $(-r_{su}/X)$ in Eq. (6.76). The resulting equation becomes

$$\frac{1}{\theta_c} = YU - k_d \tag{6.84}$$

The term $1/\theta_c$ is the net specific growth rate and is directly related to $U$, the specific substrate utilization rate.

In order to determine the specific substrate utilization rate, the substrate utilized and the biomass effective in the utilization must be given. The substrate utilized can be computed from the difference between the influent and the effluent $BOD_5$ or COD.

In the complete-mix activated-sludge process with recycle, waste sludge (cells) can be withdrawn from the tank or from the recycling line. If waste sludge is withdrawn from the tank and the VSS in the effluent $X_e$ is negligible $(Q_eX_e \approx 0)$, Eq. (6.71) will (if $X_e$ is very small) be approximately rewritten as

$$\theta_c \approx \frac{VX}{Q_{wa}X} \tag{6.85}$$

or

$$Q_{wa} \approx \frac{V}{\theta_c} \tag{6.86}$$

The flow rate of waste sludge from the sludge return line will be approximately

$$Q_{wr} = \frac{VX}{\theta_c X_r} \tag{6.87}$$

where $X_r$ is the concentration (in mg/L) of sludge in the sludge return line.

In practice, the food-to-microorganism (F/M) ratio is widely used and is closely related to the specific substrate utilization rate $U$. The F/M (in per day) ratio is defined as the influent soluble $BOD_5$ concentration ($S_0$) divided by the product of hydraulic retention time $\theta$ and MLVSS concentration $X$ in the aeration tank

$$\text{F/M} = \frac{S_0}{\theta X} = \frac{QS_0}{VX} = \frac{\text{mg BOD}_5\text{/d}}{\text{mg MLVSS}} \tag{6.88}$$

F/M and $U$ are related by the efficiency $E$ of the activated-sludge process as follows:

$$U = \frac{(\text{F/M})E}{100} \tag{6.89}$$

The value of $E$ is determined by

$$E = \frac{S_0 - S}{S_0} \times 100 \tag{6.90}$$

where $E$ = process efficiency, %
$S_0$ = influent substrate concentration, mg/L
$S$ = effluent substrate concentration, mg/L

**Sludge production.** The amount of sludge generated (increased) per day affects the design of the sludge treatment and disposal facilities. It can be calculated by

$$P_x = Y_{obs}Q(S_0 - S) \div (1000\,\text{g/kg}) \qquad \text{(SI units)} \tag{6.91}$$

$$P_x = Y_{obs}Q(S_0 - S)(8.34) \qquad \text{(British system)} \tag{6.92}$$

where $P_x$ = net waste activated sludge (VSS), kg/d or lb/d
$Y_{obs}$ = observed yield (Eq. (6.81)), g/g or lb/lb
$Q$ = influent wastewater flow, $m^3$/d or Mgal/d
$S_0$ = influent soluble $BOD_5$ concentration, mg/L
$S$ = effluent soluble $BOD_5$ concentration, mg/L
8.34 = conversion factor, lb/(Mgal) · (mg/L)

**Oxygen requirements in the process.** The theoretical oxygen requirement in the activated sludge is determined from $BOD_5$ of the wastewater and the amount of microorganisms wasted per day from the process. The biochemical reaction can be expressed as below

$$\begin{array}{lcl} C_5H_7NO_2 & + \quad 5O_2 & \rightarrow 5CO_2 + 2H_2O + NH_3 + \text{energy} \\ 113 & 5 \times 32 = 160 & \\ \text{organism cells} & & \\ 1 & 1.42 & \end{array} \tag{6.93}$$

Equation (6.93) suggests that the $BOD_u$ (ultimate BOD) for one mole of cells requires 1.42 (160/113) moles of oxygen. Thus the theoretical oxygen requirement to remove the carbonaceous organic matter in wastewater for an activated-sludge process is expressed as (Metcalf and Eddy, Inc. 1991).

$$\text{Mass of } O_2/\text{d} = \text{total mass of } BOD_u \text{ used} - 1.42 \text{ (mass of organisms wasted, } p_x)$$

$$\text{kg } O_2/\text{d} = \frac{Q(S_0 - S)}{(1000\text{ g/kg})\, f} - 1.42P_x \quad \text{(SI units)} \tag{6.94a}$$

$$\text{kg } O_2/\text{d} = \frac{Q(S_0 - S)}{1000\text{ g/kg}}\left(\frac{1}{f} - 1.42Y_{\text{obs}}\right) \tag{6.94b}$$

$$\text{lb } O_2/\text{d} = Q(S_0 - S) \times 8.34\left(\frac{1}{f} - 1.42Y_{\text{obs}}\right) \quad \text{(US customary units)} \tag{6.95}$$

where $Q$ = influent flow, $m^3$/d or Mgal/d
$S_0$ = influent soluble $BOD_5$ concentration, mg/L
$S$ = effluent soluble $BOD_5$ concentration, mg/L
$f$ = conversion factor for converting $BOD_5$ to $BOD_u$
$Y_{obs}$ = observed yield, g/g or lb/lb
8.34 = conversion factor, lb/(Mgal · (mg/L))

When nitrification is considered, the total oxygen requirement is the mass of oxygen per day for removal of carbonaceous matter and for nitrification. It can be calculated as

$$\text{kg } O_2/\text{d} = \frac{Q(S_0 - S)}{1000\text{ g/kg}}\left(\frac{1}{f} - 1.42\,Y_{\text{obs}}\right) + \frac{Q(N_0 - N)}{1000\text{ g/kg}} \quad \text{(SI units)} \tag{6.96}$$

$$\text{lb } O_2/\text{d} = 8.34[Q(S_0 - S)(1/f - 1.42Y_{\text{obs}}) + 4.75(N_0 - N)] \quad \text{(US customary units)} \tag{6.97}$$

where $N_0$ = influent total kjeldahl nitrogen concentration, mg/L
$N$ = effluent total kjeldahl nitrogen concentration, mg/L
4.75 = conversion factor for oxygen requirement for complete oxidation of TKN

Oxygen requirements generally depend on the design peak hourly $BOD_5$, MLSS, and degree of treatment. Aeration equipment must be able to maintain a minimum of 2.0 mg/L of dissolved oxygen concentration in the mixed liquor at all times and provide adequate mixing. The normal air requirements for all activated-sludge systems, except extended aeration, are 1.1 kg of oxygen (93.5 $m^3$ of air) per kg $BOD_5$ or 1.1 lb of oxygen (1500 $ft^3$ of air) per lb $BOD_5$, for design peak aeration tank loading. That is 94 $m^3$ of air per kg of $BOD_5$ (1500 $ft^3$/lb $BOD_5$) at standard conditions of temperature, pressure, and humidity. For the extended aeration process, normal air requirements are 128 $m^3$/kg $BOD_5$ or 2050 $ft^3$/lb $BOD_5$ (GLUMRB, 1996).

For F/M ratios greater than 0.3 $d^{-1}$, the air requirements for conventional activated-sludge systems amount to 31 to 56 $m^3$/kg (500 to 900 $ft^3$/lb) of $BOD_5$ removed for coarse bubble (nonporous) diffusers and 25 to 37 $m^3$/kg (400 to 600 $ft^3$/lb) $BOD_5$ removal for fine bubble (porous) diffusers. For lower F/M ratios, endogenous respiration, nitrification, and prolonged aeration increase air use to 75 to 112 $m^3$/kg (1200 to 1800 $ft^3$/lb) of $BOD_5$ removal. In practice, air requirements range from 3.75 to 15.0 $m^3$ air/$m^3$ water (0.5 to 2 $ft^3$/gal) with a typical value of 7.5 $m^3/m^3$ or 1.0 $ft^3$/gal (Metcalf and Eddy, Inc. 1991).

**Example 1a:** Design a complete-mix activated-sludge system.

Given:

Average design flow = 0.32 $m^3$/s (6.30 Mgal/d)
Peak design flow = 0.80 $m^3$/s (18.25 Mgal/d)
Raw wastewater $BOD_5$ = 240 mg/L
Raw wastewater TSS = 280 mg/L
Effluent $BOD_5$ $\leq$ 20 mg/L
Effluent TSS $\leq$ 24 mg/L
Wastewater temperature = 20°C

Operational parameters and biological kinetic coefficients:

Design mean cell residence time $\theta_c$ = 10 d
MLVSS = 2400 mg/L (can be 3600 mg/L)
VSS/TSS = 0.8
TSS concentration in RAS = 9300 mg/L
$Y$ = 0.5 mg VSS/mg$BOD_5$
$k_d$ = 0.06/d
$BOD_5$/ultimate $BOD_u$ = 0.67

Assume:

1. BOD (i.e. $BOD_5$) and TSS removal in the primary clarifiers are 33% and 67%, respectively.
2. Specific gravity of the primary sludge is 1.05 and the sludge has 4.4% of solids content.
3. Oxygen consumption is 1.42 mg per mg of cell oxidized.

**solution:**

Step 1. Calculate BOD and TSS loading to the plant

$$\text{Design flow } Q = 0.32 \text{ m}^3/\text{s} \times 86{,}400 \text{ s/d}$$
$$= 27{,}648 \text{ m}^3/\text{d}$$
$$\text{Since } 1 \text{ mg/L} = 1 \text{ g/m}^3 = 0.001 \text{ kg/m}^3$$
$$\text{BOD loading} = 0.24 \text{ kg/m}^3 \times 27{,}648 \text{ m}^3/\text{d}$$
$$= 6636 \text{ kg/d}$$
$$\text{TSS loading} = 0.28 \text{ kg/m}^3 \times 27{,}648 \text{ m}^3/\text{d}$$
$$= 7741 \text{ kg/d}$$

Step 2. Calculate characteristics of primary sludge

$$\text{BOD removed} = 6636 \text{ kg/d} \times 0.33 = 2190 \text{ kg/d}$$
$$\text{TSS removed} = 7741 \text{ kg/d} \times 0.67 = 5186 \text{ kg/d}$$
$$\text{Specific gravity of sludge} = 1.05$$
$$\text{Solids concentration} = 4.4\% = 0.044 \text{ kg/kg}$$
$$\text{Sludge flow rate} = \frac{5186 \text{ kg/d}}{1.05 \times 1000 \text{ kg/m}^3} \div 0.044 \text{ kg/kg}$$
$$= 112 \text{ m}^3/\text{d}$$

Step 3. Calculate flow, BOD, and TSS in primary effluent (secondary influent)

$$\text{Flow} = \text{design flow, } 27{,}648 \text{ m}^3/\text{d} - 112 \text{ m}^3/\text{d}$$
$$= 27{,}536 \text{ m}^3/\text{d} = Q \text{ for Step 6}$$
$$\text{BOD} = 6636 \text{ kg/d} - 2190 \text{ kg/d}$$
$$= 4446 \text{ kg/d}$$
$$= \frac{4446 \text{ kg/d} \times 1000 \text{ g/kg}}{27{,}536 \text{ m}^3/\text{d}}$$

$$= 161.5 \text{ g/m}^3$$

$$= 161.5 \text{ mg/L} = S_0$$

$$\text{TSS} = 7741 \text{ kg/d} - 5186 \text{ kg/d}$$

$$= 2555 \text{ kg/d}$$

$$= \frac{2555 \text{ kg/d} \times 1000 \text{ g/kg}}{27{,}536 \text{ m}^3\text{/d}}$$

$$= 92.8 \text{ g/m}^3$$

$$= 92.8 \text{ mg/L}$$

Step 4. Estimate the soluble $BOD_5$ escaping treatment, $S$, in the effluent

Use the following relationship

$$\text{Effluent BOD} = \text{influent soluble BOD escaping treatment, } S + \text{BOD of effluent suspended solids}$$

(a) Determine the $BOD_5$ of the effluent SS (assuming 63% biodegradable)

$$\text{Biodegradable effluent solids} = 24 \text{ mg/L} \times 0.63 = 15.1 \text{ mg/L}$$

$$\text{Ultimate BOD}_u \text{ of the biodegradable effluent solids} = 15.1 \text{ mg/L} \times 1.42 \text{ mg O}_2\text{/mg cell}$$

$$= 21.4 \text{ mg/L}$$

$$\text{BOD}_5 = 0.67 \text{ BOD}_u = 0.67 \times 21.4 \text{ mg/L}$$

$$= 14.3 \text{ mg/L}$$

(b) Solve for influent soluble $BOD_5$ escaping treatment

$$20 \text{ mg/L} = S + 14.3 \text{ mg/L}$$

$$S = 5.7 \text{ mg/L}$$

Step 5. Calculate the treatment efficiency $E$ using Eq. (6.90)

(a) The efficiency of biological treatment based on soluble BOD is

$$E = \frac{S_0 - S}{S_0} \times 100 = \frac{(161.5 - 5.7 \text{ mg/L}) \times 100\%}{161.5 \text{ mg/L}}$$

$$= 96.5\%$$

(b) The overall plant efficiency including primary treatment is

$$E = \frac{(240 - 20)\text{ mg/L} \times 100\%}{240\text{ mg/L}}$$

$$= 91.7\%$$

Step 6. Calculate the reactor volume using Eq. (6.79)

$$V = \frac{\theta_c QY(S_0 - S)}{X(1 + k_d\theta_c)}$$

$$\theta_c = 10\text{ d}$$

$$Q = 27{,}536\text{ m}^3/\text{d (from Step 3)}$$

$$Y = 0.5\text{ mg/mg}$$

$$S_0 = 161.5\text{ mg/L (from Step 3)}$$

$$S = 5.7\text{ mg/L (from Step 4b)}$$

$$X = 2400\text{ mg/L}$$

$$k_d = 0.06\text{ d}^{-1}$$

$$V = \frac{(10\text{ days})(27{,}536\text{ m}^3/\text{d})(0.5)(161.5 - 5.7)\text{ mg/L}}{(2400\text{ mg/L})(1 + 0.06\text{ day}^{-1} \times 10\text{ days})}$$

$$= 5586\text{ m}^3$$

$$= 1.48\text{ Mgal}$$

Step 7. Determine the dimensions of the aeration tank

Provide 4 rectangular tanks with common walls. Use width-to-length ratio of 1:2 and water depth of 4.4 m with 0.6 m freeboard

$$w \times 2w \times (4.4\text{ m}) \times 4 = 5586\text{ m}^3$$

$$w = 12.6\text{ m}$$

$$\text{width} = 12.6\text{ m}$$

$$\text{length} = 25.2\text{ m}$$

$$\text{water depth} = 4.4\text{ m (total tank depth} = 5.0\text{ m)}$$

*Note*: In the Ten States Standards, liquid depth should be 3 to 9 m (10 to 30 ft). The tank size would be smaller if a higher design value of MLVSS were used.

Step 8. Calculate the sludge wasting flow rate from the aeration tank

Using Eq. (6.71), also $V = 5586\ \text{m}^3$ and VSS = 0.8 SS

$$\theta_c = \frac{VX}{Q_{wa}X + Q_eX_e}$$

$$10\ \text{days} = \frac{(5586\ \text{m}^3)(2400\ \text{mg/L})}{Q_{wa}(3000\ \text{mg/L}) + (27{,}536\ \text{m}^3/\text{d})(24\ \text{mg/L} \times 0.8)}$$

$$Q_{wa} = 270\ \text{m}^3/\text{d}$$

$$= 0.0715\ \text{Mgal/d}$$

Step 9. Estimate the quantity of sludge to be wasted daily

(a) Calculate observed yield
Using Eq. (6.81)

$$Y_{obs} = \frac{Y}{1 + K_d\theta_c} = \frac{0.5}{1 + 0.06 \times 10}$$

$$= 0.3125$$

(b) Calculate the increase in the mass of MLVSS for Eq. (6.91)

$$p_x = Y_{obs}Q(S_0 - S) \times (1\ \text{kg}/1000\ \text{g})$$

$$= 0.3125 \times 27{,}536\ \text{m}^3/\text{d} \times (161.5 - 5.7)\ \text{g/m}^3 \times 0.001\ \text{kg/g}$$

$$= 1341\ \text{kg/d}$$

*Note*: A factor of 8.34 lb/Mgal is used if $Q$ is in Mgal/d.

(c) Calculate the increase in MLSS (or TSS), $p_{ss}$

$$p_{ss} = 1341\ \text{kg/d} \div 0.8$$

$$= 1676\ \text{kg/d}$$

(d) Calculate TSS lost in the effluent, $p_e$

$$p_e = (27{,}536 - 270)\ \text{m}^3/\text{d} \times 24\ \text{g/m}^3 \div 1000\ \text{g/kg}$$

$$= 654\ \text{kg/d}$$

*Note*: Flow is less sludge wasting rate from Step 8.

(e) Calculate the amount of sludge that must be wasted

$$\text{Wastewater sludge} = p_{ss} - p_e$$

$$= 1676\ \text{kg/d} - 654\ \text{kg/d}$$

$$= 1022\ \text{kg/d}$$

Step 10. Estimate return activated sludge rate

Using a mass balance of VSS, $Q$ and $Q_r$ are the influent and RAS flow rates, respectively.

$$\text{VSS in aerator} = 2400 \text{ mg/L}$$

$$\text{VSS in RAS} = 9300 \text{ mg/L} \times 0.8 = 7440 \text{ mg/L}$$

$$2400\,(Q + Q_r) = 7440\,Q_r$$

$$Q_r/Q = 0.4762$$

$$Q_r = 0.4762 \times 27{,}536 \text{ m}^3\text{/d}$$

$$= 13{,}110 \text{ m}^3\text{/d}$$

$$= 0.152 \text{ m}^3\text{/s}$$

Step 11. Check hydraulic retention time (HRT = $\theta$)

$$\theta = V/Q = 5586 \text{ m}^3/(27{,}536 \text{ m}^3\text{/d})$$

$$= 0.203 \text{ d} \times 24 \text{ h/d}$$

$$= 4.87 \text{ h}$$

*Note*: The preferred range of HRT is 5–15 h.

Step 12. Check F/M ratio using $U$ in Eq. (6.83)

$$U = \frac{S_0 - S}{\theta X} = \frac{161.5 \text{ mg/L} - 5.7 \text{ mg/L}}{(0.203 \text{ day})(2400 \text{ mg/L})}$$

$$= 0.32 \text{ day}^{-1}$$

Step 13. Check organic loading rate and mass of ultimate $\text{BOD}_u$ utilized

$$\text{Loading} = \frac{QS_0}{V} = \frac{27{,}536 \text{ m}^3\text{/d} \times 161.5 \text{ g/m}^3}{5586 \text{ m}^3 \times 1000 \text{ g/kg}} = 0.80 \text{ kg BOD}_5/(\text{m}^3 \cdot \text{d})$$

$$\text{BOD}_5 = 0.67\ \text{BOD}_u \quad \text{(given)}$$

$$\text{BOD}_u \text{ used} = Q(S_0 - S)/0.67$$

$$= \frac{27{,}536 \text{ m}^3\text{/d} \times (161.5 - 5.7) \text{ g/m}^3}{0.67 \times 1000 \text{ g/kg}}$$

$$= 6403 \text{ kg/d}$$

Step 14. Compute theoretical oxygen requirements

The theoretical oxygen required is calculated from Eq. (6.94a)

$$O_2 = \frac{Q(S_0 - S)}{(1000 \text{ g/kg}) f} - 1.42 p_x$$

$$= 6403 \text{ kg/d (from Step 13)} - 1.42 \times 1341 \text{ kg/d (from Step 9b)}$$

$$= 4499 \text{ kg/d}$$

Step 15. Compute the volume of air required

Assume that air weighs 1.202 kg/m$^3$ (0.075 lb/ft$^3$) and contains 23.2% oxygen by weight; the oxygen transfer efficiency for the aeration equipment is 8%; and a safety factor of 2 is used to determine the actual volume for sizing the blowers.

(a) The theoretical air required is

$$\text{Air} = \frac{4499 \text{ kg/d}}{1.202 \text{ kg/m}^3 \times 0.232 \text{ g } O_2\text{/g air}}$$

$$= 16{,}200 \text{ m}^3\text{/d}$$

(b) The actual air required at an 8% oxygen transfer efficiency

$$\text{Air} = 16{,}200 \text{ m}^3\text{/d} \div 0.08$$

$$= 202{,}000 \text{ m}^3\text{/d}$$

$$= 140 \text{ m}^3\text{/min}$$

$$= 4950 \text{ ft}^3\text{/min}$$

(c) The design air required (with a factor of safety 2) is

$$\text{Air} = 140 \text{ m}^3\text{/min} \times 2$$

$$= 280 \text{ m}^3\text{/min}$$

$$= 9900 \text{ ft}^3\text{/min}$$

$$= 165 \text{ ft}^3\text{/s or (cfs)}$$

Step 16. Check the volume of air required per unit mass $BOD_5$ removed, and per unit volume of wastewater and aeration tank, using the actual value obtained in Step 15b

(a) Air supplied per kg of $BOD_5$ removed

$$\text{Air} = \frac{202{,}000 \text{ m}^3\text{/d} \times 1000 \text{ g/kg}}{27{,}536 \text{ m}^3\text{/d} \times (161.5 - 5.7) \text{ g/m}^3}$$

$$= 47.1 \text{ m}^3 \text{ of air/kg } BOD_5$$

$$= 754 \text{ ft}^3\text{/lb}$$

(b) Air supplied per m$^3$ of wastewater treated

$$\text{Air} = \frac{202{,}000\ \text{m}^3/\text{d}}{27{,}536\ \text{m}^3}$$

$$= 7.34\ \text{m}^3\ \text{air/m}^3\ \text{wastewater}$$

$$= 0.98\ \text{ft}^3\ \text{air/gal wastewater}$$

(c) Air supplied per m$^3$ of aeration tank

$$\text{Air} = \frac{202{,}000\ \text{m}^3/\text{d}}{5586\ \text{m}^3}$$

$$= 36.2\ \text{m}^3/(\text{m}^3 \cdot \text{d})$$

$$= 36.2\ \text{ft}^3/(\text{ft}^3 \cdot \text{d})$$

**Example 1b:** Design secondary clarifiers using the data in Example la and the MLSS settling test results. The MLSS setting data is derived from a pilot plant study and is shown below:

| MLSS, mg/L | 1200 | 1800 | 2400 | 3300 | 4000 | 5500 | 6800 | 8100 |
|---|---|---|---|---|---|---|---|---|
| Initial settling velocity, m/h | 4.1 | 3.1 | 2.1 | 1.2 | 0.77 | 0.26 | 0.13 | 0.06 |

**solution:**

Step 1. Plot the MLSS settling curve (Fig. 6.16) on log-log paper from the observed data

Step 2. Construct the gravity solid-flux curve from Fig. 6.17

Data in columns 1 and 2 of the following table is adopted from Fig. 6.16. Values in col. 3 are determined by col. 1 × col. 2. Plot the solids–flux curve using MLSS concentration and calculate solids flux as shown in Fig. 6.17.

| (1) MLSS concentration $X$, mg/L or g/m$^3$ | (2) Initial settling velocity $V_1$, m/h | (3) Solids flux $XV_1$, kg/(m$^2$ · h) |
|---|---|---|
| 1000 | 4.2 | 4.20 |
| 1500 | 3.7 | 5.55 |
| 2000 | 2.8 | 5.60 |
| 2500 | 2.0 | 5.00 |
| 3000 | 1.5 | 4.50 |
| 4000 | 0.76 | 3.04 |
| 5000 | 0.41 | 2.05 |
| 6000 | 0.22 | 1.32 |
| 7000 | 0.105 | 0.74 |
| 8000 | 0.062 | 0.50 |
| 9000 | 0.033 | 0.30 |

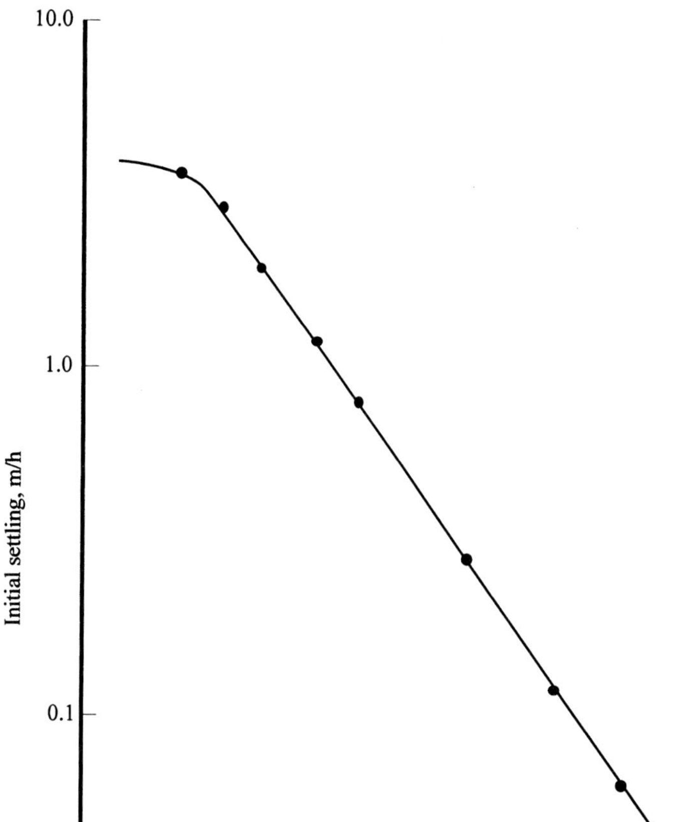


**Figure 6.16** Solids settling rate: experimental results for design of secondary clarifier.

Step 3. Determine the limiting solids flux value

From Fig. 6.17, determine the limiting solids flux (SF) for an underflow concentration of 9300 mg/L. This is achieved by drawing a tangent to the solids flux curve from 9300 mg/L (the desired underflow) solids concentration. The limiting solids flux value is

$$1.3\ \text{kg/(m}^2 \cdot \text{h)} \equiv 31.2\ \text{kg/(m}^2 \cdot \text{d)} \equiv 6.4\ \text{lb/(ft}^2 \cdot \text{d)}$$

Step 4. Calculate design flow to the secondary clarifiers, $Q$

From Steps 8 and 10 of Example 1a

$$\begin{aligned} Q &= \text{average design flow} + \text{return sludge flow} - \text{MLSS wasted} \\ &= (27{,}563 + 12{,}942 - 270)\ \text{m}^3/\text{d} \\ &= 40{,}235\ \text{m}^3/\text{d} \\ &= 0.466\ \text{m}^3/\text{s} \end{aligned}$$

Use two clarifiers, each one with flow of 20,200 m$^3$/d.

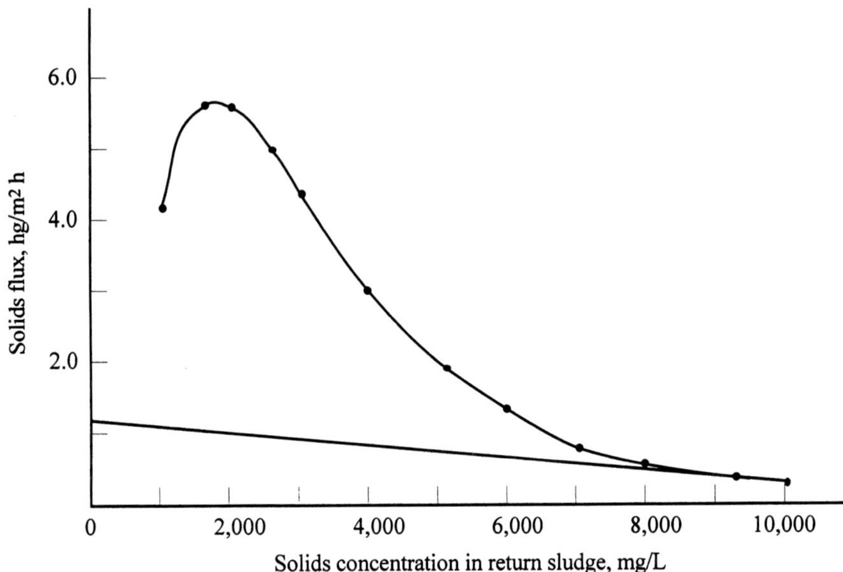


**Figure 6.17** Solids flux curve.

Step 5. Compute the area $A$ and diameter $d$ of the clarifier

$$A = \frac{QX}{\text{SF}} \tag{6.98}$$

where $A$ = area of the secondary clarifier, $m^2$ or $ft^2$
$Q$ = influent flow of the clarifier, $m^3/h$ or gal/h
$X$ = MLSS concentration, $kg/m^3$ or $lb/ft^3$
SF = limiting solids flux, $kg/(m^2 \cdot h)$ or $lb/(ft^2 \cdot h)$

From Eq. (6.98), for each clarifier

$$Q = 20{,}200 \text{ m}^3/\text{d} = 841.7 \text{ m}^3/\text{h}$$

$$\text{MLSS} = (2400/0.8) \text{ mg/L} = 3.0 \text{ kg/m}^3$$

$$\text{SF} = 1.3 \text{ kg/(m}^2 \cdot \text{h) (from Step 3)}$$

Therefore

$$A = \frac{QX}{\text{SF}} = \frac{841.7 \text{ m}^3/\text{h} \times 3.0 \text{ kg/m}^3}{1.3 \text{ kg/(m}^2 \cdot \text{h)}}$$

$$= 1942 \text{ m}^2$$

$$A = \pi(d/2)^2 = 1942 \text{ m}^2$$

$$d = \sqrt{1942 \text{ m}^2 \times 4 \div 3.14}$$
$$= 49.7 \text{ m} \approx 50 \text{ m}$$
$$= 164 \text{ ft}$$

Step 6. Check the surface overflow rate at average design flow

$$\text{Overflow rate} = \frac{Q}{A} = \frac{20{,}200 \text{ m}^3/\text{d}}{1942 \text{ m}^2}$$
$$= 10.4 \text{ m}^3/(\text{m}^2 \cdot \text{d})$$
$$= 255 \text{ gal}/(\text{d} \cdot \text{ft}^2)$$

*Note*: This is less than the design criteria of 15 $\text{m}^3/(\text{m}^2 \cdot \text{d})$

Step 7. Check the clarifier's area for clarification requirements

From Step 6, the calculated surface overflow rate

$$Q/A = 10.4 \text{ m}^3/(\text{m}^2 \cdot \text{d}) = 0.433 \text{ m/h}$$

From Fig. 6.16, the MLSS concentration corresponding to a 0.433 m/h of settling rate is 4700 mg/L. The design MLSS is only 2400 mg/L. The area is sufficient.

Step 8. Check the surface overflow rate at peak design flow

$$\text{Peak flow} = 0.80 \text{ m}^3/\text{s} \times 86{,}400 \text{ s/d} + 13{,}112 \text{ m}^3/\text{d}$$
$$= 82{,}232 \text{ m}^3/\text{d}$$

$$\text{Overflow rate} = \frac{82{,}062 \text{ m}^3/\text{d}}{1942 \text{ m}^2 \times 2}$$
$$= 21.2 \text{ m}^3/(\text{m}^2 \cdot \text{d})$$

Step 9. Determine recycle ratio required to maintain MLSS concentration at 3000 mg/L

$$(Q + Q_r) \times 3000 = QX + Q_r X_u$$

where $Q$ = influent flow, $\text{m}^3$/d or Mgal/d
$Q_r$ = recycle flow, $\text{m}^3$/d or Mgal/d
$X$ = influent SS concentration, mg/L
$X_u$ = underflow SS concentration, mg/L

$$Q(3000 - X) = Q_r(X_u - 3000)$$

$$\frac{Q_r}{Q} = \frac{3000 - X}{X_u - 3000} = \alpha = \text{recycle ratio}$$

when $X_u$ = 9300 mg/L (from Step 3) and

$$X = 93 \text{ mg/L (from Step 3 of Example 1a, 92.8 mg/L)}$$

$$\alpha = \frac{3000 \text{ mg/L} - 93 \text{ mg/L}}{9300 \text{ mg/L} - 3000 \text{ mg/L}}$$

$$= 0.46$$

Step 10. Estimate the depth required for the thickening zone

The total depth of the secondary clarifier is the sum of the required depths of the clear water zone, the solids thickening zone, and the sludge storage zone. In order to estimate the depth of the thickening zone, it is assumed that, under normal conditions, the mass of solids retained in the secondary clarifier is 30% of the mass of solids in the aeration basin, and that the average concentration of solids in the sludge zone is 7000 mg/L (Metcalf and Eddy, Inc. 1991). The system has 4 aeration tanks and 2 clarifiers.

(a) Compute total mass of solids in each aeration basin

$$\text{MLSS} = 3000 \text{ mg/L} = 3.0 \text{ kg/m}^3$$

Refer to Step 7 of Example 1a

$$\text{Solids in each aeration tank} = 3.0 \text{ kg/m}^3 \times 4.4 \text{ m} \times 12.6 \text{ m} \times 25.2 \text{ m}$$
$$= 4191 \text{ kg}$$

(b) Compute the mass of solids in a clarifier for 2 days (see Step 11)

$$\text{Solids in each clarifier} = 4191 \text{ kg} \times 0.3 \times 2$$
$$= 2515 \text{ kg}$$

(c) Compute depth of sludge zone

$$\text{Depth} = \frac{\text{mass}}{\text{area} \times \text{concentration}} = \frac{2515 \text{ kg} \times 1000 \text{ g/kg}}{1942 \text{ m}^2 \times 7000 \text{ g/m}^3}$$

$$= 0.19 \text{ m}$$

$$\approx 0.2 \text{ m}$$

Step 11. Estimate the depth of the sludge storage zone

This zone is provided to store excess solids at peak flow conditions or at a period over which the sludge-processing facilities are unable to handle the sludge quantity. Assume storage capacity for 2 days' sustained peak flow (of 2.5 average flow) and for 7 days' sustained peak BOD loading (of 1.5 BOD average).

(a) Calculate total volatile solids generated under sustained BOD loading. Using Eq. (6.91)

$$P_x = Y_{obs}Q(S_0 - S) \div (1000 \text{ g/kg})$$

$$Y_{obs} = 0.3125 \text{ (from Step 9(a) of Example 1a)}$$

$$Q = 2.5\ (0.32 \text{ m}^3\text{/s}) = 0.8 \text{ m}^3\text{/s} = 69{,}120 \text{ m}^3\text{/d}$$

$$S_0 = 161.5 \text{ mg/L} \times 1.5 = 242 \text{ mg/L}$$

$$S = 5.7 \text{ mg/L} \times 1.5 = 9 \text{ mg/L}$$

$$P_x = 0.3125 \times 69{,}120 \text{ m}^3\text{/d} \times (242 - 9) \text{ g/m}^3 \div (1000 \text{ g/kg})$$

$$= 5033 \text{ kg/d}$$

(b) Calculate mass of solid for 2-day storage

$$\text{Total solids stored} = 5033 \text{ kg/d} \times 2 \text{ d} \div 0.8$$

$$= 12{,}580 \text{ kg}$$

(c) Calculate stored solids per clarifier

$$\text{Solids to be stored} = 12{,}580 \text{ kg} \div 2$$

$$= 6290 \text{ kg}$$

(d) Calculate total solids in each secondary clarifier

$$\text{Total} = 6290 \text{ kg} + 2515 \text{ kg (from Step 10b)}$$

$$= 8805 \text{ kg}$$

(e) Calculate the required depth for sludge storage in the clarifier

$$\text{Depth} = \frac{8505 \text{ kg} \times 1000 \text{ g/kg}}{7000 \text{ g/m}^3 \times 1942 \text{ m}^2}$$

$$= 0.63 \text{ m}$$

Step 12. Calculate total required depth of a clarifier

The depth of clear water and settling zone is commonly 1.5 to 2 m. Provide 2 m for this example.

$$\text{Total required depth of a clarifier} = 2 \text{ m} + 0.2 \text{ m} + 0.63 \text{ m}$$

$$= 2.83 \text{ m}$$

$$\approx 3 \text{ m}$$

With addition of 0.65 m freeboard,

$$\text{the total water depth of the clarifier} = 3.65 \text{ m}$$

$$= 12 \text{ ft}$$

Step 13. Check hydraulic retention time of the secondary clarifier

$$\text{Diameter} = 50 \text{ m (from Step 5)}$$
$$\text{Volume} = 3.14\ (50 \text{ m}/2)^2 \times 3.0 \text{ m}$$
$$= 5888 \text{ m}^3$$

Under average design flow plus recirculation, from Step 4

$$\text{HRT} = \frac{5888 \text{ m}^3 \times 24 \text{ h/d}}{40{,}400 \text{ m}^3\text{/d (from Step 4)}}$$
$$= 3.5 \text{ h}$$

At peak design flow with recirculation, from Step 8

$$\text{HRT} = \frac{5888 \text{ m}^3 \times 24 \text{ h/d}}{82{,}062 \text{ m}^3\text{/d (from Step 8)}}$$
$$= 1.72 \text{ h}$$

**Plug-flow with recycle.** Conventional and many modified activated-sludge processes are designed approximately like the plug-flow model (Fig. 6.18). In a time plug-flow reactor, all fluid particles entering the reactor stay in the reactor for an equal amount of time and pass through the reactor in sequence. Those that enter first leave first. It is assumed that there is no mixing in the lateral direction. In practice, a time plug-flow reactor is difficult to obtain because of longitudinal dispersion. In a real plug-flow, some particles may make more passes through the reactor as a result of recycling.

The mathematical modeling for the plug-flow process is difficult. The most accepted useful kinetic model of the plug-flow reactor for the activated-sludge process was developed by Lawrence and McCarty (1969). They

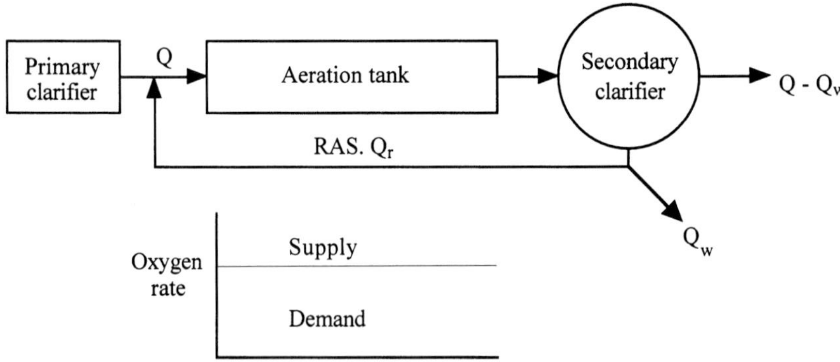


**Figure 6.18** Conventional activated-sludge process and DO in aeration tank.

made two assumptions: (1) The concentration of microorganisms in the influent of the reactor is approximately the same as that in the effluent of the reactor. This assumption is valid only when $\theta_c/\theta > 5$. (2) The rate of substrate utilization $r_{su}$ as wastewater flows through the aeration tank is expressed as

$$r_{su} = \frac{kSX}{K_s + S} \tag{6.99}$$

The reverse of the mean cell residence time is

$$\frac{1}{\theta_c} = \frac{Yk(S_0 - S)}{(S_0 - S) + (1 + \alpha)K_s \ln(S_i/S)} - k_d \tag{6.100}$$

and

$$S_i = \frac{S_0 + \alpha S}{1 + \alpha} \tag{6.101}$$

where
$r_{su}$ = substrate utilization rate, mg/L
$S$ = effluent concentration of substrate, mg/L
$S_0$ = influent concentration of substrate, mg/L
$X$ = average concentration of microorganisms in the reactor, mg/L
$k$, $K_s$, $Y$ = kinetic coefficients as defined previously
$\theta_c$ = mean cell residence time, day
$\alpha$ = recycle ratio
$S_i$ = influent concentration in reactor after dilution with recycle flow

## 21.6 Operation and control of activated-sludge processes

The aeration basin is the heart of the activated-sludge process. Process operation and system controls usually need well trained operators, with operational parameters such as air or oxygen supply, F/M ratio, and mass balance to return activated sludge. However, some nontheoretical indexes such as sludge volume index, sludge density index, sludge settleability, and sludge age are widely used and valuable tools for the daily process of operation and control.

**Sludge volume index.** The sludge volume index (SVI) is the volume of mL occupied by 1 g of suspension after 30 min of settling. Although SVI is not theoretically supported, experience has shown it is useful in routine process control. SVI typically is used to monitor the settling characteristics of activated sludge and can impact on return sludge rate and MLSS.

SVI is calculated from the laboratory test results of the suspended solids concentration of a well mixed sample of the suspension and the 30-min settled sludge volume. The sludge volume is measured by filling a 1-liter graduated cylinder to the 1.0-liter mark, allowing settling for 30 min, and then reading the volume of settled solids. The mixed liquid suspended solids is determined by filtering, drying, and weighing a sample of the mixed liquor as stated in the previous section in this chapter. The value of SVI can be calculated by the following formula (Standard Methods—APHA *et al.*, 1998).

$$\text{SVI} = \frac{\text{SV} \times 1000 \text{ mg/g}}{\text{MLSS}} \tag{6.102}$$

or

$$\text{SVI} = \frac{\text{Wet settled sludge, mL/L}}{\text{Dry sludge solid, mg/L}} \tag{6.102a}$$

where SVI = sludge volume index, mL/g
SV = settled sludge volume, mL/L
MLSS = mixed liquor suspended solids, mg/L
1000 = milligrams per gram, mg/g

Typical values of SVI for domestic activated-sludge plants operating with an MLSS concentration of 2000 to 3500 mg/L range from 80 to 150 mL/g (Davis and Cornwell, 1991).

The SVI is an important factor in process design. It limits the tank MLSS concentration and return sludge rate.

**Sludge density index.** Sludge density index is used in a way similar to the sludge volume index to indicate the settleability of a sludge in a secondary clarifier or effluent. The weight in grams of 1 mL of sludge, after settling for 30 min, is calculated as

$$\text{SDI} = 100/\text{SVI} \tag{6.103}$$

where SDI = sludge density index, g/mL
SVI = sludge volume index, mL/g

A sludge with good settling characteristics has an SDI of between 1.0 and 2.0, whereas an SDI of 0.5 indicates a bulky or nonsettleable sludge (Cheremisinoff, 1995).

**Return activated sludge.** Return activated sludge is the settled activated sludge that is collected in the secondary clarifier and returned to the aeration tank to mix with the influent wastewater.

**TABLE 6.13 Guidelines for Return Activated Sludge Flow Rate**

| Type of process | Percent of design average flow | |
|---|---|---|
| | Minimum | Maximum |
| Conventional | 15 | 100 |
| Carbonaceous stage of separate-stage nitrification | 15 | 100 |
| Step-feed aeration | 15 | 100 |
| Complete-mix | 15 | 100 |
| Contact stabilization | 50 | 150 |
| Extended aeration | 50 | 150 |
| Nitrification stage of separate stage nitrification | 50 | 200 |

SOURCE: GLUMRB (Ten States Standards) (1996)

The efficiency of the activated-sludge process is measured by BOD removal, which is directly related to the volatile activated-sludge solids in the aeration basin. The purpose of sludge return is to maintain a sufficient concentration of activated sludge in the aeration tank. The RAS makes it possible for the microorganisms to be in the treatment system longer than the flowing wastewater. The RAS flow for a conventional activated-sludge system is usually 20% to 40% of the tank influent flow rate. Table 6.13 lists typical ranges of RAS flow rates for some of the activated-sludge processes. Sludge volume index (Eq. (6.102)) is empirical and is used to control the rate of return sludge. The minimum percentage of return activated sludge is related to SVI and the solids concentration in the mixed liquor and is expressed as (Clark and Viessman, 1966)

$$\% \text{ of return sludge} = \frac{100}{100/(\text{SVI})P - 1} \tag{6.104}$$

where SVI = sludge volume index, mL/g
$P$ = percentage of solids in the mixed liquor

**Example 1:** Determine the aeration basin dimensions for a town of 20,000 population. Assume the mixed liquor suspended solids is 2600 mg/L; BOD loading rate is 0.48 kg/(d · m$^3$) or 0.03 lb/(d · ft$^3$); SVI = 110 mL/g; and $P$ = 2600 mg/L.

**solution:**

Step 1. Calculate total BOD loading ($L$) on aeration basin

Using average BOD contribution of 0.091 kg/(person · d) or 0.20 lb/(c · d)

$$\begin{aligned}\text{Total BOD load to the wastewater}\\ \text{treatment plant} &= 20{,}000 \text{ person} \times 0.091 \text{ kg/(person} \cdot \text{d)}\\ &= 1820 \text{ kg/d}\end{aligned}$$

Assume 30% BOD removal in the primary clarifiers.

The total daily BOD load on aeration basin would be

$$L = 1820 \text{ kg/d} \times (1.0 - 0.3) = 1274 \text{ kg/d}$$

Step 2. Calculate percentage return activated sludge

Using Eq. (6.104)

$$\text{SVI} = 110 \text{ mL/g}$$

$$P = 2600 \text{ mg/L} = 0.26\%$$

$$\% \text{ return} = \frac{100}{100/(\text{SVI})P - 1} = \frac{100}{100/(100 \times 0.26) - 1}$$

$$= 40$$

Step 3. Determine the total BOD loading on the basin

Assume the BOD concentration for both the RAS and the influent are the same.

Allowing an additional 40% of return sludge, the total BOD loading would be

$$1274 \text{ kg/d} \times 1.40 = 1784 \text{ kg/d}$$

Step 4. Determine the volume $V$ required for the aeration tank

$$V = \frac{\text{Total BOD loading}}{\text{Allowed loading per m}^3}$$

$$= \frac{1784 \text{ kg/d}}{0.48 \text{ kg/(d} \cdot \text{m}^3)}$$

$$= 3717 \text{ m}^3$$

Step 5. Determine the dimensions of the aeration tank

Select eight tanks, the water depth of 4.4 m, and add 0.6 m for the freeboard, and width of 7 m. The length of a tank is

$$\text{length} = \frac{3717 \text{ m}^3}{8 \times 4.4 \text{ m} \times 7 \text{ m}}$$

$$= 15 \text{ m}$$

*Note:* Each tank measures 5 m × 7 m × 15 m.

**Example 2:** Determine the aeration tank based on 6 h of aeration period, using the data given in Example 1. Assume 0.53 $\text{m}^3$/(person · d) or 140 gal/(c · d) of waste flow.

**solution:**

Step 1. Calculate plant flow rate $Q$

$$Q = 0.53\ \text{m}^3/(\text{person} \cdot \text{d}) \times 20{,}000\ \text{persons}$$
$$= 10{,}600\ \text{m}^3/\text{d}$$

Step 2. Plus 40% of $Q_r$

$$Q + Q_r = 1.4 \times 10{,}600\ \text{m}^3/\text{d}$$
$$= 14{,}840\ \text{m}^3/\text{d}$$

Step 3. Calculate the required volume $V$ of a tank

$$V = 14{,}840\ \text{m}^3/\text{d} \times (1\ \text{day}/24\ \text{h}) \times 6\ \text{h}$$
$$= 3710\ \text{m}^3$$

*Note:* For these two examples, the tank volume determined is based on either organic loading or hydraulic loading; these give almost identical results.

Step 4. Determine tank dimensions

Use eight tanks with depth of 5 m, width of 7 m, and length of 15 m, as in Example 1.

**Return activated-sludge flow rate.** The RAS, at a constant flow rate, is independent of the tank influent flow rate and changes the MLSS in the tank continuously. The MLSS concentration will be at a maximum during low influent flows and at a minimum during peak influent flows. The secondary clarifier must constantly change the depth of the sludge blanket. To maintain a good F/M ratio, a programmed control device should be installed to maintain a constant percent of RAS flow to the tank influent flow. However, any change in the activated-sludge quality will require a different RAS flow rate because of the changing settling characteristics of the sludge.

The relationship between maximum RAS concentration ($X_r$) and SVI can be derived from Eq. (6.69) using settled sludge volume SV = 1000 mL/L. The suspended solids concentration in RAS can be calculated by

$$X_r = \left(\frac{\text{SV} \times 1000\ \text{mg/g}}{\text{SVI}}\right)$$

$$= \frac{1000\ \text{mL/L} \times 1000\ \text{mg/g}}{\text{SVI mL/g}} \tag{6.105}$$

$$X_r = \frac{10^6}{\text{SVI}}\ \text{mg/L}$$

**Sludge settleability.** Another method of calculating RAS flow rate is based on the settleability approach. The settleability is defined as the percentage of volume occupied by the sludge after settling for 30 min. The RAS flow rate is related to settled sludge volume as follows:

$$1000Q_r = (SV)(Q + Q_r) \tag{6.106}$$

or

$$Q_r = \frac{(SV)Q}{1000 - SV} \tag{6.107}$$

where 1000 = factor mg/L
$Q_r$ = flow of return activated sludge, $m^3$/s or Mgal/d
$Q$ = flow of tank influent, $m^3$/s or Mgal/d
SV = settled sludge volume (30-min settling), mL/L

**Example:** In practice, the operator checks the VSS concentration in the return activated sludge at least once every shift and makes the appropriate RAS flow rate adjustment. The previous operator recorded that RAS flow was 44 gal/min (240 $m^3$/d) with the VSS in RAS of 5800 mg/L. The on-duty operator determines the VSS in RAS as 5500 mg/L. What should the RAS flow rate be adjusted to?

**solution:** Since the VSS in the RAS is decreasing, the sludge wasting flow should be increased proportionally to waste the amount of VSS.

$$Q_{adj} \times 5550 \text{ mg/L} = 44 \text{ gal/min} \times 5800 \text{ mg/L}$$

$$\begin{aligned} Q_{adj} &= 44 \times 5800/5550 \text{ gal/min} \\ &= 46 \text{ gal/min} \\ &= 250 \text{ m}^3\text{/d} \end{aligned}$$

**Aeration tank mass balance.** In practice, the maximum RAS pumping capacity is commonly designed as 100% of design average flow (use 150% for oxidation ditch). The required RAS pumpage can be determined by aeration tank mass balance. Assume new cell growth is negligible and there is no accumulation in the aeration tank. Also, the solids concentration in the tank influent is negligible compared to the MLSS, $X$, in the tank. Thus mass of inflow is equal to that of outflow. This can be expressed as

$$X_r Q_r = X(Q + Q_r) \tag{6.108}$$

or

$$Q_r = \frac{X}{X_r - X} Q \tag{6.109}$$

$$X_r = \frac{X(Q + Q_r)}{Q_r} = \frac{\text{MLSS}(Q + Q_r)}{Q_r} \tag{6.110}$$

where $X_r$ = return activated-sludge suspended solids, mg/L
$Q_r$ = flow of RAS, $m^3$/s or Mgal/d
$X$ = mixed liquor suspended solids (MLSS), mg/L
$Q$ = flow of secondary influent (or plant flow), $m^3$/s or Mgal/d

The values of $X$ and $X_r$ include both the volatile and nonvolatile (inert) fractions. Equations (6.106) and (6.110) assume no loss of suspended solids in the basin. Equation (6.109) is essentially the same as Eq. (6.107).

**Example 1:** Determine the return activated-sludge flow as a percentage of the influent flow 10.0 Mgal/d (37,850 $m^3$/d). The sludge settling volume in 30 min is 255 mL.

**solution:**

Step 1. Compute RAS flow $Q_r$ in %

Using Eq. (6.107)

$$Q_r, \% = \frac{\text{SV}}{1000 - \text{SV}} = \frac{255 \text{ mL/L} \times 100\%}{1000 \text{ mL/L} - 255 \text{ mL/L}}$$

$$= 34\%$$

Step 2. Compute RAS flow rate in Mgal/d

$$Q_r = 0.34\ Q = 0.34 \times 10 \text{ Mgal/d}$$

$$= 3.40 \text{ Mgal/d}$$

$$= 12{,}900 \text{ m}^3\text{/d}$$

**Example 2:** The MLSS concentration in the aeration tank is 2800 mg/L. The sludge settleability test showed that the sludge volume, settled for 30 min in a 1-L graduated cylinder, is 285 mL. Calculate the sludge volume index and estimate the SS concentration in the RAS and the required return sludge ratio.

**solution:**

Step 1. Calculate SVI

Using Eq. (6.102)

$$\text{SVI} = \frac{\text{SV} \times 1000 \text{ mg/g}}{\text{MLSS}} = \frac{285 \text{ mL/L} \times 1000 \text{ mg/g}}{2800 \text{ mg/L}}$$

$$= 102 \text{ mL/g}$$

This is in the typical range of 80 to 150 mL/g

Step 2. Calculate SS in RAS

Using Eq. (6.105)

$$X_r = \frac{1{,}000{,}000}{\text{SVI}} = \frac{1{,}000{,}000\ (\text{mL/L})(\text{mg/g})}{102\ \text{mL/g}}$$

$$= 9804\ \text{mg/L}$$

$$= 0.98\%$$

Step 3. Calculate $Q_r/Q$

Using Eq. (6.107)

$$\frac{Q_r}{Q} = \frac{\text{SV}}{1000 - \text{SV}} = \frac{285\ \text{mg/L}}{(1000 - 285)\ \text{mg/L}}$$

$$= 0.40$$

$$= 40\%\ \text{return}$$

**Example 3:** Compute the return activated-sludge flow rate in m$^3$/d and as a percentage of the influent flow of 37,850 m$^3$/d (10 Mgal/d). The laboratory results show that the SVI is 110 mg/L and the MLSS is 2500 mg/L.

**solution:**

Step 1. Compute the suspended solids in RAS based on the SVI

Using Eq. (6.105)

$$X_r \text{ in RAS} = \frac{1{,}000{,}000\ \text{mg/L}}{\text{SVI}} = \frac{1{,}000{,}000\ \text{mg/L}}{110}$$

$$= 9090\ \text{mg/L}$$

Step 2. Compute RAS flow rate $Q_r$, based on SVI

Using Eq. (6.109)

$$Q_r = \frac{X}{X_r - X} Q = \frac{2500\ \text{mg/L} \times 37{,}850\ \text{m}^3/\text{d}}{9090\ \text{mg/L} - 2500\ \text{mg/L}}$$

$$= 14{,}360\ \text{m}^3/\text{d}$$

$$= 3.79\ \text{Mgal/d}$$

Step 3. Compute RAS flow as percentage of influent flow

$$\text{RAS flow, \%} = \frac{Q_r \times 100\%}{Q} = \frac{3.79\ \text{Mgal/d} \times 100\%}{10.0\ \text{Mgal/d}}$$

$$= 37.9\%$$

$$\text{Say} = 38\%\ \text{of influent flow}$$

**Waste activated sludge.** The excess of activated sludge generated from the secondary clarifier must be wasted, usually from the return sludge line. The waste activated sludge (WAS) is discharged to either the primary clarifiers or to sludge thickeners. Withdrawing mixed liquor directly from the aeration basin or from the basin effluent line is an alternative method of sludge wasting. The amount of WAS removed affects mixed liquor settleability, oxygen consumption, growth rate of the microorganisms, nutrient quantities, input occurrence of foaming and sludge bulking effluent quality, and possibility of nitrification.

The purpose of WAS is to maintain a given food-to-microorganisms ratio or mean cell residence time of the system. It should remove just the amount of microorganisms that grow in excess of the microorganism death rate.

Sludge wasting can be accomplished on an intermittent or continuous basis. The parameters used for control guidelines are F/M ratio, mean cell residence time, MLVSS, and sludge age.

**Secondary clarifier mass balance.** Although it assumes the sludge blanket is level in the secondary clarifier, the secondary clarifier mass balance approach is a useful tool to determine the RAS flow rate. The calculations are based on the secondary clarifier (Fig. 6.15). Assuming the sludge blanket in the clarifier is not changed and the effluent suspended solids are negligible, the SS entering the clarifier is equal to the SS leaving. The relationship can be written as

$$(Q + Q_{wr})\,(\text{MLSS}) = Q_{wa}(\text{WAS}) + Q_{wr}(\text{RAS})$$

$$Q(\text{MLSS}) - Q_{wa}(\text{WAS}) = Q_{wr}(\text{RAS} - \text{MLSS})$$

$$Q_{wr} = \frac{Q(\text{MLSS}) - Q_{wa}(\text{WAS})}{\text{RAS} - \text{MLSS}} \tag{6.111}$$

where $Q$ = flow of tank influent, $m^3/d$
$Q_{wr}$ = return activated-sludge flow, $m^3/d$
$Q_{wa}$ = waste activated-sludge flow rate, $m^3/d$
MLSS = mixed liquor suspended solids, mg/L
WAS = SS of waste activated sludge, mg/L
RAS = SS of return activated sludge, mg/L

*Note*: WAS is equal to RAS (Fig. 6.15a).

**Sludge age.** Sludge age is a measure of the length of time a particle of suspended solids has been retained in the activated-sludge process. It is also referred to as mean cell residence time and is an operational parameter related to the F/M ratio. Sludge age is defined as that of the suspended solids under aeration (kg or lb) divided by the suspended solids added (kg/d or lb/d) and is calculated by the following equation which relates MLSS in the system to the influent suspended solids

$$\text{Sludge age} = \frac{\text{SS under aeration, kg or lb}}{\text{SS added, kg/d or lb/d}}$$

$$\text{Sludge age} = \frac{\text{MLSS} \times V}{\text{SS}_\text{w} \times Q_\text{w} + \text{SS}_\text{e} \times Q_\text{e}} \tag{6.112}$$

where Sludge age = mean cell residence time, day
MLSS = mixed liquor suspended solids, mg/L
$V$ = volume of aeration basin, $m^3$ or Mgal
$SS_w$ = suspended solids in waste sludge, mg/L
$Q_w$ = flow of waste sludge, $m^3$/d or Mgal/d
$SS_e$ = suspended solids in wastewater effluent, mg/L
$Q_e$ = flow of wastewater effluent, $m^3$/d or Mgal/d

*Note*: Equations (6.112) and (6.72) are the same.

Sludge age can also be expressed in terms of the volatile portion of suspended solids, which is more representative of biological mass. Solids retention time in an activated-sludge system is measured in days, whereas the liquid aeration period is in hours. In most activated-sludge plants, sludge age ranges from 3 to 8 days (US EPA, 1979). The liquid aeration periods vary from 3 to 30 h (Hammer, 1986). Wastewater passes through the aeration tank only once and rather quickly, whereas the resultant biological growths and extracted waste organics are repeatedly recycled from the secondary clarifier back to the aeration basin.

**Example 1:** An activated-sludge system has an influent flow of 22,700 $m^3$/d (6.0 Mgal/d) with suspended solids of 96 mg/L. Three aeration tanks hold 1500 $m^3$ (53,000 $ft^3$) each with MLSS of 2600 mg/L. Calculate the sludge age for the system.

**solution:**

Step 1. Calculate the SS under aeration, MLSS × $V$

$$\text{MLSS} = 2600 \text{ mg/L} = 2600 \text{ g/m}^3 = 2.6 \text{ kg/m}^3$$

$$\begin{aligned} \text{MLSS} \times V &= 2.6 \text{ kg/m}^3 \times 1500 \text{ m}^3 \times 3 \\ &= 11{,}700 \text{ kg} \\ &= 25{,}800 \text{ lb} \end{aligned}$$

Step 2. Calculate the SS added

$$\begin{aligned} \text{SS added} &= \text{Influent flow} \times \text{SS conc.} \\ &= 22{,}700\ \text{m}^3/\text{d} \times 0.096\ \text{kg/m}^3 \\ &= 2180\ \text{kg/d} \\ &= 4800\ \text{lb/d} \end{aligned}$$

Step 3. Calculate sludge age

Using Eq. (6.112)

$$\text{Sludge age} = \frac{\text{SS under aeration}}{\text{SS added}} = \frac{11{,}700\ \text{kg}}{2180\ \text{kg/d}}$$

$$= 5.4\ \text{days}$$

*Note:* It is in the typical range of 3 to 8 days.

**Example 2:** In an activated-sludge system, the solids under aeration are 13,000 kg (28,700 lb); the solids added are 2200 kg/d (4850 lb/d); the return activated-sludge SS concentration is 6600 mg/L; the desired sludge age is 5.5 days; and the current waste activated sludge (WAS) is 2100 kg/d (4630 lb/d). Calculate the WAS flow rate using the sludge age control technique.

**solution:**

Step 1. Calculate the desired SS under aeration for the desired sludge age of 5.5 days

$$\begin{aligned} \text{SS} &= \text{solid added} \times \text{sludge age} \\ &= 2200\ \text{kg/d} \times 5.5\ \text{days} \\ &= 12{,}100\ \text{kg} \\ &= 26{,}800\ \text{lb} \end{aligned}$$

Step 2. Calculate the additional suspended solids removed per day

$$\text{SS aerated} - \text{SS desired} = 13{,}000\ \text{kg} - 12{,}100\ \text{kg} = 900\ \text{kg}$$

Step 3. Calculate the additional WAS flow ($q$) to maintain the desired sludge age

$$\begin{aligned} q\ (\text{SS in RAS}) &= 900\ \text{kg/d} \\ q &= (900\ \text{kg/d})/(6.6\ \text{kg/m}^3) \\ &= 136.4\ \text{m}^3/\text{d} = 0.095\ \text{m}^3/\text{min} \\ &= 25.0\ \text{gal/min} \end{aligned}$$

Step 4. Calculate total WAS flow

$$\begin{aligned} \text{Current flow} &= (2100\ \text{kg/d})/(6.6\ \text{kg/m}^3) \\ &= 318.2\ \text{m}^3/\text{d} = 0.221\ \text{m}^3/\text{min} \\ &= 58.4\ \text{gal/min} \end{aligned}$$

$$\text{Total WAS flow} = (0.095 + 0.221)\ \text{m}^3/\text{min}$$
$$= 0.316\ \text{m}^3/\text{min}$$
$$= 83.4\ \text{gal/min}$$

**Example 3:** The aeration basin volume is 6600 m$^3$ (1.74 Mgal). The influent flow to the basin is 37,850 m$^3$/d (10.0 Mgal/d) with BOD of 140 mg/L. The MLSS is 3200 mg/L with 80% volatile portion. The SS concentration in the return activated sludge is 6600 mg/L. The current waste activated-sludge flow rate is 340 m$^3$/d (0.090 Mgal/d). Determine the desired WAS flow rate using the F/M ratio control technique with a desired F/M ratio of 0.32.

**solution:**

Step 1. Calculate BOD loading

$$\text{BOD} = 140\ \text{mg/L} = 140\ \text{g/m}^3 = 0.14\ \text{kg/m}^3$$
$$\text{Loading} = \text{BOD} \times \text{flow}$$
$$= 0.14\ \text{kg/m}^3 \times 37{,}850\ \text{m}^3/\text{d}$$
$$= 5299\ \text{kg/d}$$
$$= 11{,}680\ \text{lb/d}$$

Step 2. Calculate the desired MLVSS with the desired F/M = 0.32

$$\text{Desired MLVSS} = \text{BOD loading/(F/M) ratio}$$
$$= (5299\ \text{kg/d})/(0.32\ \text{kg/(d} \cdot \text{kg))}$$
$$= 16{,}560\ \text{kg}$$

Step 3. Calculate the desired MLSS

$$\text{Desired MLSS} = \text{MLVSS}/0.80 = 16{,}560\ \text{kg}/0.80$$
$$= 20{,}700\ \text{kg}$$

Step 4. Calculate actual MLSS under aeration

$$\text{Actual MLSS} = \text{concentration} \times \text{basin volume}$$
$$= 3.2\ \text{kg/m}^3 \times 6600\ \text{m}^3$$
$$= 21{,}120\ \text{kg}$$

Step 5. Calculate the additional solids to be removed daily

$$(\text{actual} - \text{desired})\ \text{MLSS} = 21{,}120\ \text{kg/d} - 20{,}700\ \text{kg/d}$$
$$= 420\ \text{kg/d}$$

Step 6. Calculate the additional WAS flow $q$ required

$$q = \text{additional SS removed/SS in RAS}$$
$$= (420\ \text{kg/d})/(6.6\ \text{kg/m}^3)$$
$$= 63.6\ \text{m}^3/\text{d}$$

Step 7. Calculate the total WAS flow $Q$

$$Q = (340 + 63.6)\ \text{m}^3/\text{d}$$
$$= 403.6\ \text{m}^3/\text{d} = 0.28\ \text{m}^3/\text{min}$$
$$= 74\ \text{gal/min}$$

**Example 4:** Calculate the waste activated-sludge flow rate using the mean cell residence time (MCRT) method. The given conditions are the same as those in Example 3. In addition, the desired MCRT is 7.5 days and the SS level in the effluent is 12 mg/L.

**solution:** (use English system)

Step 1. Calculate SS concentration in the aeration tank

$$\text{MLSS} = 3200\ \text{mg/L}$$
$$V = 1.74\ \text{Mgal}$$
$$\text{SS} = 1.74\ \text{Mgal} \times 3200\ \text{mg/L} \times 8.34\ \text{lb/(Mgal} \cdot \text{mg/L)}$$
$$= 46{,}440\ \text{lb} = 21{,}070\ \text{kg}$$

Step 2. Determine SS lost in the effluent

$$\text{SS}_\text{e} = 12\ \text{mg/L}$$
$$Q_\text{e} = 10\ \text{Mgal/d}$$
$$\text{SS}_\text{e} \times Q_\text{e} = 10\ \text{Mgal/d} \times 12\ \text{mg/L} \times 8.34\ \text{lb/(Mgal} \cdot \text{mg/L)}$$
$$= 1000\ \text{lb/d} = 453\ \text{kg/d}$$

Step 3. Calculate the desired SS in waste sludge

Using Eq. (6.113)

$$\text{MCRT, day} = \frac{\text{SS in aerator, lb}}{\text{SS wasted, lb/d} + \text{SS in effluent, lb/d}}$$

$$7.5\ \text{days} = \frac{46{,}440\ \text{lb}}{\text{SS wasted, lb/d} + 1000\ \text{lb/d}}$$

$$\text{SS wasted} = (46{,}440\ \text{lb} - 7500\ \text{lb})/7.5\ \text{days}$$
$$= 5192\ \text{lb/d} = 2355\ \text{kg/d}$$

Step 4. Calculate the WAS flow rate $Q_\text{w}$

$$\text{SS}_\text{w} = 6600\ \text{mg/L}$$
$$\text{SS wasted} = Q_\text{w} \times \text{SS}_\text{w} \times 8.34$$

$$Q_\text{w} = \frac{\text{SS wasted}}{\text{SS}_\text{w} \times 8.34} = \frac{5192\ \text{lb/d}}{6600\ \text{mg/L} \times 8.34\ \text{lb/(Mgal} \cdot \text{mg/L)}}$$
$$= 0.0943\ \text{Mgal/d}$$

$$= 0.0943 \text{ Mgal/d} \times 694\text{(gal/min)/(Mgal/d)}$$
$$= 65.4 \text{ gal/min}$$
$$= 356 \text{ m}^3\text{/d}$$

**Example 5:** The given operating conditions are the same as in Example 3. The desired MLVSS is 16,500 kg (36,400 lb). The SS concentration in RAS is 6600 mg/L. Using MLVSS as a control technique, calculate the desired WAS flow rate. The current WAS flow rate is 420 $m^3$/d (0.11 Mgal/d)

**solution:**

Step 1. Calculate actual MLVSS

$$\text{Tank volume} = 6600 \text{ m}^3$$
$$\text{VSS/TSS} = 0.80$$
$$\text{MLSS} = 3200 \text{ mg/L} = 3.2 \text{ kg/m}^3$$
$$\text{Actual MLVSS} = \text{tank volume} \times \text{volatiles} \times \text{MLSS}$$
$$= 6600 \text{ m}^3 \times 0.8 \times 3.2 \text{ kg/m}^3$$
$$= 16{,}900 \text{ kg}$$

Step 2. Calculate additional VSS to be wasted daily

$$\text{Wasted} = \text{actual} - \text{desired (Step 3 of Example 3)}$$
$$= (16{,}900 - 16{,}500) \text{ kg/d}$$
$$= 400 \text{ kg/d}$$

Step 3. Calculate additional WAS flow, $q$

$$\text{VSS in RAS} = 6600 \text{ mg/L} \times 0.8 = 5280 \text{ mg/L}$$
$$= 5.28 \text{ kg/m}^3$$
$$q = (400 \text{ kg/d})/(5.28 \text{ kg/m}^3)$$
$$= 75.8 \text{ m}^3\text{/d} = 13.9 \text{ gal/min}$$

Step 4. Calculate the desired WAS flow rate

$$Q = \text{current} + \text{additional flow}$$
$$= (420 + 75.8) \text{ m}^3\text{/d}$$
$$= 495.8 \text{ m}^3\text{/d}$$
$$= 91 \text{ gal/min}$$

**Sludge bulking.** A desirable activated sludge is one which settles rapidly leaving a clear, odorless, and stable supernatant. The efficiency of treatment achieved in an activated-sludge process depends directly on settleability of the sludge in the secondary clarifier. At times, poorly settling sludge and foam formation cause the most common operational problems of the activated-sludge process. Sludge with poorly flocculated (pin) particles or buoyant filamentous growths increases its volume and does not settle well in the clarifier. Light sludge in the clarifier then spills over the weirs and is carried away in the effluent. The concentrations of BOD and suspended solids increase in the effluent. This phenomenon is called sludge bulking and it frequently occurs unexpectedly.

Sludge bulking is caused by: (1) the growth of filamentous organisms, or organisms that can grow in filamentous form under adverse conditions, and (2) adverse environmental conditions such as excessive flow, insufficient aeration, short circulating of aeration tanks, lack of nutrients, septic influent, presence of toxic substances, or overloading. *Nocardia amarae, Microthrix parvicella, N. amarae-like* organisms, *N. pinensis*-like organisms, and type 0092 were most commonly found in foam and bulk sludge. *Nocardia* growth is supported by high sludge age, low F/M ratio, and higher rather than lower wastewater temperature (Droste, 1997). Fungi are mostly filamentous microorganisms. Some bacteria such as *Beggiatoa, Thiotrix,* and *Leucothrix* grow in filamentous sheaths.

There are no certain rules for prevention and control of sludge bulking. If bulking develops, the solution is to determine the cause and then either eliminate or correct it or take compensatory steps in operation control. Some remedial steps may be taken, e.g. changing parameters such as wastewater characteristics, BOD loading, dissolved oxygen concentration in the aeration basin, return sludge pumping rate, microscopic examination of organisms (check for protozoa, rotifers, filamentous bacteria, and nematodes) in clarifiers' and other operating units. Chlorination of RAS in the range of 2 to 3 mg/L of chlorine per 1000 mg/L of MLVSS is suggested (Metcalf and Eddy, Inc. 1991). Hydroperoxide also can be used as an oxidant. Reducing the suspended solids in the aeration basin, increasing air supply rate, or increasing the BOD loading (which may depress filamentous growth), and addition of lime to the mixed liquor for pH adjustment are the remedial methods which have been used. The F/M ratio should be maintained at 0.5 to 0.2.

## 21.7 Modified activated-sludge processes

Numerous modifications (variations) of the conventional activated-sludge plug flow process have been developed and proved effective for removal of BOD and/or nitrification. The modified processes include tapered aeration, step-feed aeration, complete-mix extended aeration, modified aeration, high-rate aeration, contact stabilization, Hatfield process, Kraus process, sequencing batch reactor, high-purity oxygen, oxidation ditch, deep shaft reactor, single-stage nitrification, and separate stage nitrification. Except for a few processes, most modified processes basically differ on the range of the F/M ratio maintained and in the introduction locations for air supply and wastewater. The design parameters and operational characteristics for various activated-sludge processes are presented elsewhere (Metcalf and Eddy, Inc. 1991, WEF and ASCE, 1991a). Standard and modified processes are discussed below.

**Conventional process.** The earliest activated-sludge process was contained in long narrow tanks with air for oxygen supply and mixing through diffusers at the bottom of the aeration tank. A conventional plug-flow process consists of a primary clarifier, an aeration tank with air diffusers for mixing the wastewater and the activated sludge in the presence of dissolved oxygen, a secondary settling basin for solids removal, and a sludge-return line from the clarifier bottom to the influent of the aeration tank (Fig. 6.18). The return activated sludge is mixed with the incoming wastewater and passes through the aeration tank in a plug-flow fashion. High organics concentration and high microbial solids at the head end of the aeration tank lead to a high oxygen demand. The conventional process is more susceptible to upset from shock loads and toxic materials, and is used for low-strength domestic wastewaters.

The conventional process is designed to treat 0.3 to 0.6 kg $BOD_5$ applied/($m^3 \cdot d$) (20 to 40 lb/(1000 cu · ft · d)) in the United States with MLSS of 1500 to 3000 mg/L. The aeration period ranges between 4 and 8 h with return activated-sludge flow ratios of 0.27 to 0.75. The cell residence time ($\theta_c$) is 5 to 15 days and the F/M ratio is 0.2 to 0.4 per day. The process generally removed 85% to 95% of $BOD_5$ and produces highly nitrified effluent (WEF and ASCE, 1991a).

**Example:** The operational records at a conventional activated-sludge plant are shown as average values as below.

| | |
|---|---|
| Wastewater flow $Q$ | 7570 $m^3$/d (2.0 Mgal/d) |
| Wastewater temperature | 20°C |
| Volume of aeration tanks | 2260 $m^3$ (79,800 $ft^3$) |
| Influent $BOD_5$ (say BOD) | 143 mg/L |
| Influent total suspended solids | 125 mg/L |
| Influent total solids | 513 mg/L |
| Effluent total solids | 418 mg/L |
| Effluent TSS | 24 mg/L |
| Effluent BOD | 20 mg/L |
| Return sludge flow $Q_r$ | 3180 $m^3$/d (0.84 Mgal/d) |
| MLSS | 2600 mg/L |
| TSS in waste sludge | 8900 mg/L |
| Volume of waste sludge | 200 $m^3$/d (0.053 Mgal/d) |

Calculate the following operational parameters: (1) volumetric BOD loading rate; (2) F/M ratio; (3) hydraulic retention time $\theta$; (4) cell residence time $\theta_c$; (5) return activated-sludge ratio; and (6) removal efficiencies for BOD, TSS, and total solids.

**solution:**

Step 1. Calculate volumetric BOD loading rate

$$\text{Influent BOD} = 143 \text{ mg/L} = 143 \text{ g/m}^3$$
$$= 0.143 \text{ kg/m}^3$$

$$\text{BOD load} = \frac{\text{amount}}{\text{volume}} = \frac{7570 \text{ m}^3\text{/d} \times 0.143 \text{ kg/m}^3}{2260 \text{ m}^3}$$
$$= 0.48 \text{ kg/(m}^3 \cdot \text{d)}$$

*Note*: It is in the range of 0.3 to 0.6 kg/($m^3 \cdot$ d).

Step 2. Calculate F/M ratio using Eq. (6.64)

$$\text{Assume MLVSS} = 0.8 \text{ MLSS} = 0.8 \times 2600 \text{ mg/L}$$
$$= 2080 \text{ mg/L}$$

$$\text{F/M} = \frac{7570 \text{ m}^3\text{/d} \times 143 \text{ mg/L BOD applied}}{2260 \text{ m}^3 \times 2080 \text{ mg/L MLVSS}}$$
$$= 0.23 \text{ kg BOD applied/(kg MLVSS} \cdot \text{d)}$$
$$= 0.23 \text{ lb BOD applied/(lb MLVSS} \cdot \text{d)}$$

*Note*: $0.2 < \text{F/M} < 0.4$.

Step 3. Calculate aeration time HRT

$$\text{HRT} = V/Q = 2260 \text{ m}^3 \div 7570 \text{ m}^3\text{/d}$$
$$= 0.30 \text{ day}$$
$$= 7.17 \text{ h}$$

*Note*: 4 h < HRT < 8 h.

Step 4. Calculate cell residence time $\theta_c$

$$\text{TSS in the aeration tank} = 7570 \text{ m}^3\text{/d} \times 2.6 \text{ kg/m}^3$$
$$= 19{,}682 \text{ kg/d}$$

$$\text{TSS in sludging wasting} = 200 \text{ m}^3\text{/d} \times 8.9 \text{ kg/m}^3$$
$$= 1780 \text{ kg/d}$$

$$\text{TSS in effluent} = 7570 \text{ m}^3\text{/d} \times 0.024 \text{ kg/m}^3$$
$$= 182 \text{ kg/d}$$

$$\theta_c = \frac{19{,}682 \text{ kg/d}}{1780 \text{ kg/d} + 182 \text{ kg/d}}$$
$$= 10.0 \text{ days}$$

*Note*: 5 days $< \theta_c <$ 15 days.

Step 5. Calculate RAS ratio

$$Q_r/Q = 3180 \text{ m}^3/\text{d} \div 7570 \text{ m}^3/\text{d}$$
$$= 0.42$$
$$= 42\%$$

*Note*: $0.25 < Q_r/Q < 0.75$.

Step 6. Calculate removal efficiencies

$$\text{BOD} = \frac{(143 - 20) \text{ mg/L} \times 100\%}{143 \text{ mg/L}}$$
$$= 86\%$$

$$\text{TSS} = \frac{(125 - 22)\text{mg/L} \times 100\%}{125 \text{ mg/L}}$$
$$= 82\%$$

$$\text{TS} = \frac{(513 - 418) \text{ mg/L} \times 100\%}{513 \text{ mg/L}}$$
$$= 19\%$$

**Tapered aeration.** Since oxygen demands decrease along the length of the plug-flow reactor, the tapered aeration process attempts to match the oxygen supply to demand by adding more air at the influent end of the aeration tank than at the effluent end (Fig. 6.19). This is obtained by varying the diffuser spacing. The best results can be achieved by supplying 55% to 75% of the total air supply to the first half of the tank (Al-Layla *et al.*, 1980). Advantages of tapered aeration are those of

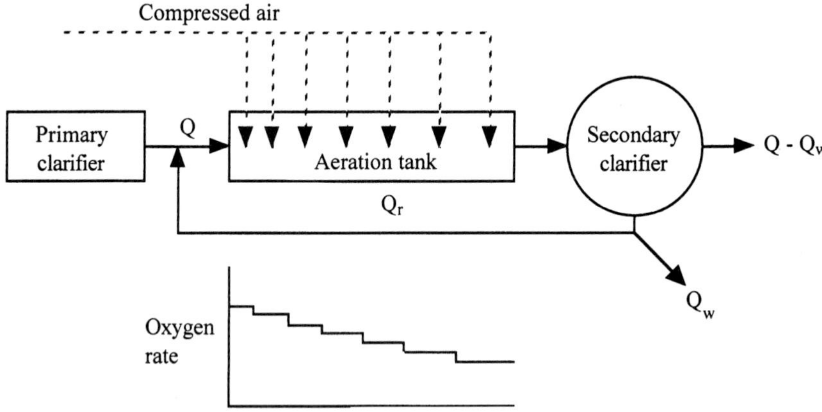


**Figure 6.19** Conventional process with tapered aeration.

reducing below-capacity and operational costs, providing better operational control, and inhibiting nitrification if desired.

**Step aeration.** In step (step-feed) aeration activated-sludge systems, the incoming wastewater is distributed to the aeration tank at a number (3 to 4) of points along the plug-flow tank, and the return activated sludge is introduced at the head of the aeration tank. The organic load is distributed over the length of the tank (Fig. 6.20), thus avoiding the

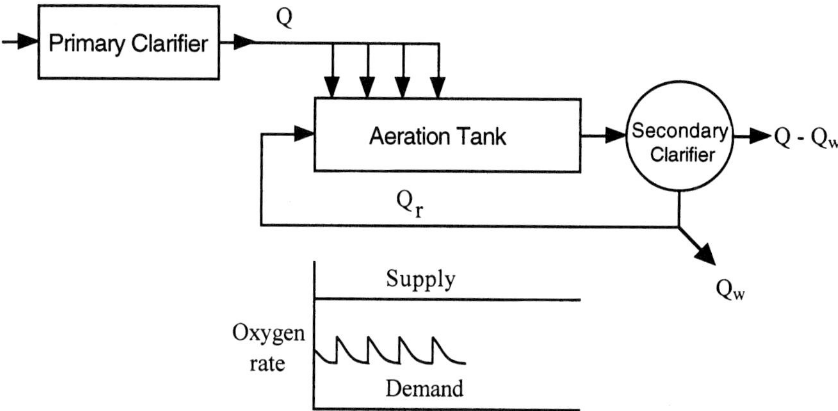


OR

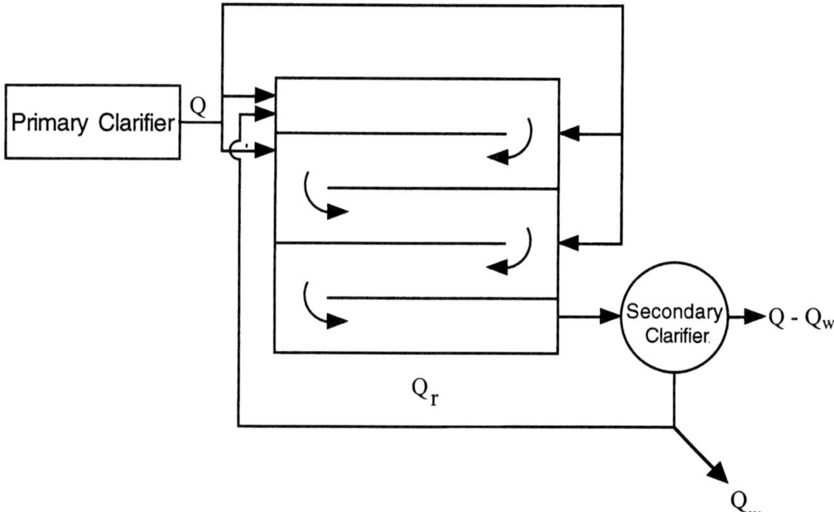


**Figure 6.20** Step aeration activated-sludge process.

locally lengthy oxygen demand encountered in conventional and tapered aeration. This process leads to shorter retention time and lower activated-sludge concentrations in the mixed liquor. The process is used for general application.

The $BOD_5$ loading rates are higher, 0.6 to 1.0 kg/($m^3$ · d) (40 to 60 lb/(1000 $ft^3$ · d)) and MLSS are also higher, 2000 to 3500 mg/L. The hydraulic retention time is shorter by 3 to 5 h. The following parameters for the step aeration process are the same as for a conventional plant: $\theta_c = 5$ to 15 days; F/M = 0.2 to 0.4 per day; $Q_r/Q = 0.25$ to 0.75; and BOD removal = 85% to 95%.

This process, with treatment efficiency practically equivalent to that of the conventional activated-sludge process, can be carried out in about one-half of the aeration time while maintaining the sludge age within proper limits of 3 to 4 days. The construction cost and the area required for step aeration are less than for the conventional process. Operational costs are about the same for both the conventional and step aeration processes.

**Complete-mix process.** Complete-mix processes disperse the influent wastewater and return activated sludge uniformly throughout the aeration tank. The shape of the reactor is not important. With careful selection of aeration and mixing equipment, the process should provide practically complete mixing. The oxygen demand is also uniform throughout the tank (Fig. 6.21).

Complete-mix processes protect against hydraulic and organic shock loadings commonly encountered in the process. Toxic materials are usually diluted below their threshold concentration. The treatment efficiency of the complete-mix process is comparable to that of the conventional process (85% to 95% BOD removal). The complete-mix process has increased in popularity for the treatment of industrial wastewaters. The process is used for general application, but is susceptible to filamentous growths.

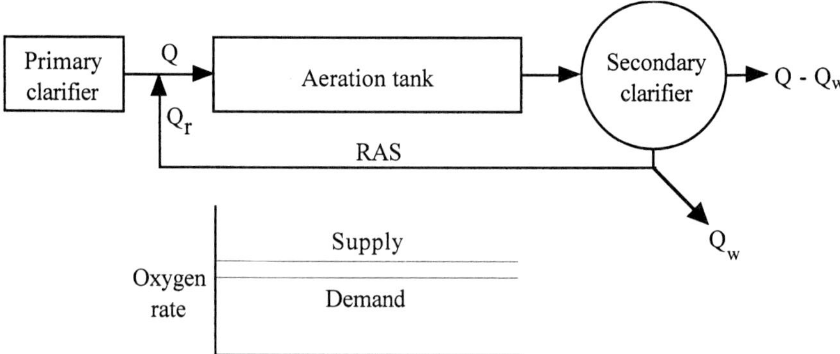


**Figure 6.21** Complete-mix activated-sludge process.

The design criteria for the complete-mix process are: 0.45 to 2.0 kg $BOD_5/(m^3 \cdot d)$ (28 to 125 lb/(1000 $ft^3 \cdot d$)), F/M = 0.2 to 0.6 per day, $\theta_c = 5$ to 15 days, MLSS = 2500 to 6500 mg/L, HRT = 3 to 5 h, $Q_r/Q = 0.25$ to 1.0 (Metcalf and Eddy, Inc. 1991). A design example for the complete-mix activated-sludge process is given in the section on mathematical modeling.

**Extended aeration.** The extended aeration process is a complete-mix activated-sludge process operated at a long HRT ($\theta$ = 16 to 24 or 36 h) and a high cell residence time (sludge age $\theta_c$ = 20 to 30 days). The process may be characterized as having a long aeration time, high MLSS concentration, high RAS pumping rate, and low sludge wastage. Extended aeration is typically used in small plants for plant flows of 3780 $m^3$/d (1 Mgal/d) or less, such as schools, villages, subdivisions, trailer parks, etc. Many extended aeration plants are prefabricated units or so-called package plants (Guo *et al.*, 1981). The process is flexible and is also used where nitrification is required.

The influent wastewater may be only screened and degritted without primary sedimentation. The extended aeration system allows the plant to operate effectively over widely varying flows and organic loadings without upset. Organic loading rates are designed as 0.1 to 0.4 kg $BOD_5/(m^3 \cdot d)$ (6 to 25 lb/(1000 $ft^3 \cdot d$)). The other design parameters are: MLSS = 1500 to 5000 mg/L, and $Q_r/Q = 0.5$ to 1.50 (WEF and ASCE, 1992). The extended aeration process has a food to microorganisms ratio (F/M) from 0.05 to 0.20 kg $BOD_5$ per kg MLSS-day. It is developed to minimize waste activated sludge production by providing a large endogenous decay of the sludge mass. The process is so designed that the mass of cells synthesized per day equals the mass of cells endogenous decayed per day. Therefore there is theoretically no cell mass production.

The extended aeration process has a food to microorganism ratio from 0.05 to 0.20 kg $BOD_5$ per kg MLSS pre day. It is developed to minimize waste activated sludge production by providing a large endogenous decay of the sludge mass. The process is so designed that the mass of cells synthesized per day equals the mass of cells endogenous decay per day. Therefore there is theoretically no cell mass products.

Secondary clarifiers must be designed to handle the variations in hydraulic loadings and high MLSS concentration associated with the system. The overflow rates range from 8 to 24 $m^3/(m^2 \cdot d)$ (200 to 600 gal/($d \cdot ft^2$)) and with long retention time. Sludge may be returned to the aeration tank through a slot opening or by an air-lift pump (Hammer, 1986). Floating materials and used sludge on the surface of the clarifier can be removed by a skimming device.

Since the process maintains a high concentration of microorganisms in the aeration tank for a long time, endogenous respiration plays a major role in activated sludge quality. More dissolved oxygen (DO) is required. Nitrification may occur in the system. The volatile portion of the sludge remaining is not degraded at the same rate as normal activated sludge and thereby results in a lower BOD exertion rate. The effluent of extended aeration often meets BOD standards (75% to 95% removal) but does not meet TSS standards due to continuous loss of pinpoint flocs (rising sludge). To overcome TSS loss, periodic sludge wasting is required. It has been suggested that the average MLSS concentration should not fall below 2000 mg/L (Guo *et al.*, 1981). In cold climates, heat lost in the extended aeration system must be controlled. However, Guo *et al.* (1981) pointed out that insufficient staff, inadequate training, or both are the most common causes of poor performance of the extended aeration process, especially in small package plants.

**Example:** An extended aeration activated-sludge plant without sludge wasting has a $BOD_5$ loading rate of 0.29 kg/(m$^3$ · d) (18 lb/(1000 ft$^3$ · d)) with aeration for 24 h. The daily MLSS buildup in the aeration basin is measured as 55 mg/L. Determine the percentage of the influent $BOD_5$ that is converted to MLSS and retained in the basin. Estimate the time required to increase the MLSS from 1500 to 5500 mg/L (this range is in the design criteria).

**solution:**

Step 1. Calculate influent BOD concentration in mg/L

$$\begin{aligned}\text{BOD} &= 0.29\ \text{kg/(m}^3\cdot\text{d)} \times 1\ \text{day} \\ &= 290\ \text{g/m}^3 \\ &= 290{,}000\ \text{mg/1000 L} \\ &= 290\ \text{mg/L}\end{aligned}$$

Step 2. Determine the percentage

$$\frac{\text{MLSS buildup}}{\text{influent BOD}} = \frac{55\ \text{mg/L} \times 100\%}{290\ \text{mg/L}}$$

$$= 19\%$$

Thus 81% of the influent BOD retained in the basin

Step 3. Estimate buildup time required, $t$

$$t = \frac{5500\ \text{mg/L} - 1500\ \text{mg/L}}{55\ \text{mg/(L}\cdot\text{d)}}$$

$$= 72.7\ \text{days}$$

**Short-term aeration.** Short-term aeration or modified aeration is a plug-flow pretreatment process. The process has extremely high loading rates of 1.2 to 2.4 kg $BOD_5/(m^3 \cdot d)$ (75 to 150 $lb/(1000\ ft^3 \cdot d)$) with an F/M ratio of 1.5 to 5.0 kg BOD/d per kg MLVSS. The volumetric loadings are low, ranging from 200 to 1000 mg/L of MLSS. The HRT ($\theta$) are 1.5 to 3 h; sludge age $\theta_c = 0.2$ to 0.5 days; and $Q_r/Q = 0.05$ to 0.25. Short retention time and low sludge age lead to a poor effluent quality and relatively high solids production. This process can be used as the first stage of a two-stage nitrification process and is used for an intermediate degree of treatment.

The short-term aeration process offers cost savings of construction by increasing BOD loading rates and reducing the required volume of aeration tank. The process produces a relatively large amount of MLVSS which may cause disposal problems. If the sludge is allowed to remain in the system, the $BOD_5$ removal efficiency ranges from 50% to 75% (Chermisinoff, 1995).

**High rate aeration.** For some cases it may not be necessary to treat the wastewater to the high degree of effluent quality achieved by conventional or other improved processes. High rate aeration is an application of the complete-mix activated-sludge process with a short HRT ($\theta = 0.5$ to 2.0 h), a short sludge age ($\theta_c = 5$ to 10 days), a high sludge recycle ratio ($Q_r/Q = 1.0$ to 5.0), and an organic loading rate of 1.6 to 16 kg $BOD_5/m^3 \cdot d$ (100 to 1000 $lb/(1000\ ft^3 \cdot d)$). The MLSS concentration in the aeration tanks ranges from 4000 to 10,000 mg/L, and the F/M ratios are 0.4 to 1.5 per day which are higher than those for the conventional process (WEF and ASCE, 1991a). The process reduces the cost of construction.

Complete-mix aeration and the hydraulic thickening action of a rapid sludge return are mandatory to offset decreased sludge settleability of the biological flocs (Bruce and Merkens, 1973). It is designed to maintain the biomass in the growth phase. Poor performance of a high-rate aeration system is usually due to insufficient aeration capacity to supply adequate dissolved oxygen during peak loading periods. Subsequently, suspended flocs are carried over to the secondary clarifier effluent. Also, inadequate RAS flow rates and high solids flux may cause sludge to be washed out through the clarifier.

Well-operated high-rate aeration processes can produce effluent quality comparable to that of a conventional plant. A $BOD_5$ removal efficiency of 75% to 90% can be achieved. A high-rate single-stage aeration system can be used for general application and to partially remove carbonaceous $BOD_5$ at the first stage of a two-stage nitrification system.

**Contact stabilization.** The contact stabilization process or biosorption was developed to take advantage of the adsorptive properties of activated sludge. A schematic flow diagram of the contact stabilization system is

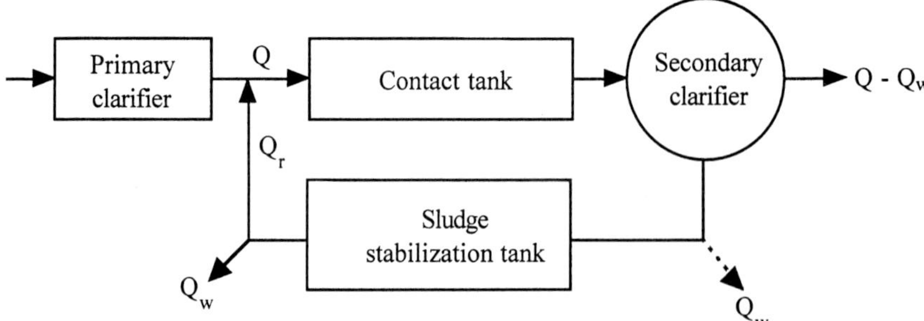


**Figure 6.22** Schematic flow diagram of contact stabilization process.

shown in Fig. 6.22. Returned sludge which has been aerated (3 to 6 h) in the stabilization basin (44,000 to 10,000 mg/L of MLSS) for stabilization of previously adsorbed organic matter is mixed with influent wastewater for a brief period (20 to 40 min). BOD removal from wastewater takes place on immediate contact with a high concentration (1000 to 4000 mg/L of MLSS in contact tank) of stabilized activated sludge. This adsorbs suspended and colloidal solids quickly, but not dissolved organic matter. Following the contact period, the activated sludge is separated from the mixed liquor in a clarifier. A small portion of the sludge is wasted while the remainder flows to the stabilization tank. During the stabilization period, the stored organic matter is utilized for cell growth and respiration; as a result, it becomes stabilized or activated again, then is recycled.

Contact stabilization was initially used to provide a partial treatment (60% to 75% $BOD_5$ removal) at larger coastal plants. It has been designed for package systems for industrial waste application and is used for expansion of existing systems and package plants.

The process is a plug-flow system and is characterized by relatively short retention times (0.5 to 3 h) with very high organic loading of 1.0 to 1.2 kg $BOD_5/(m^3 \cdot d)$ (62 to 75 lb/(1000 $ft^3 \cdot d$)). Compared to the conventional activated-sludge process, oxygen requirements are lower but waste sludge quantities are higher. Designs may either include or omit primary treatment. In general, the process has a $BOD_5$ removal efficiency of 80% to 90% (WEF and ASCE, 1992). This process can be effectively used as the first stage of a multiple system.

The disadvantages of the process are:

1. Sensitive to variations in organic and hydraulic loadings due to its short HRT and low MLSS concentration;
2. Neither as economical nor as efficient in BOD removal.

These shortcomings may result in noncompliance with effluent quality standards.

**Hatfield process.** The Hatfield process (developed by William Hatfield of the Decatur Sanitary District, Illinois) differs from the contact stabilization process. The process aerates anaerobic digester supernatant or sludge (rich in nitrogen), with (all) return sludge from the secondary clarifier then fed back to the aeration tank (Fig. 6.23). The process is used to treat wastewater with low nitrogen levels and high carbonaceous material levels, such as some industrial wastewaters. Supplying an aerobic digester effluent to the aeration tank fortifies the MLSS with amino acids and other nitrogenous substances.

The Hatfield process has the advantage, as in the contact stabilization system, of being able to maintain a large amount of microorganisms under aeration in a relatively small aeration tank. Heavier types of solids are produced in the aeration tank, which can prevent bulking problems. However, the Hatfield process is not widely used in the United States.

**Kraus process.** The conventional activated-sludge plant at Peoria, Illinois, was found to be operated improperly due to sludge bulking because of a heavy load of carbohydrates from breweries, packing houses, and paper mills. Kraus (1955) improved the process performance by aerating the mixture of anaerobic digester supernatant and a small portion of the return activated sludge in a separate aeration basin (Fig. 6.23). Some portion of the return activated sludge bypasses the sludge aeration basin and is introduced directly to the mixed liquor aeration tank. This is a modification of the Hatfield process and lies between the conventional system and the Hatfield process.

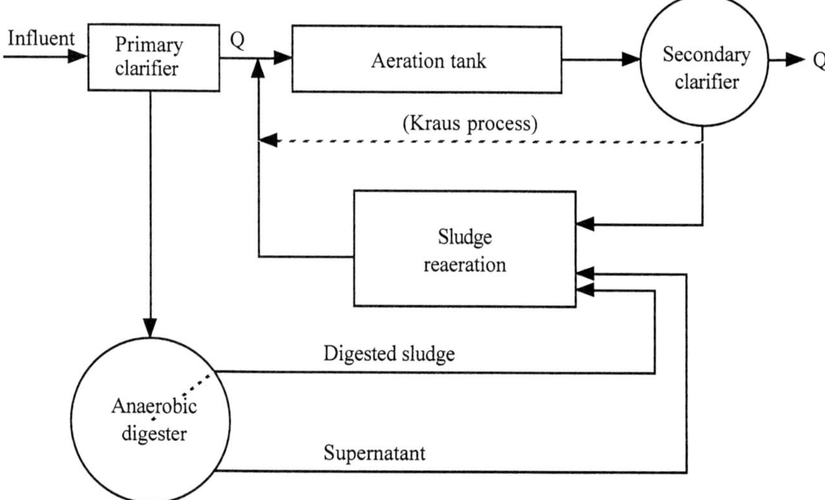


**Figure 6.23** Schematic flow diagram of Hatfield activated-sludge process and Kraus process.

The MLSS become highly nitrified materials which are brought into contact with the influent wastewater in the aeration tank. The $BOD_5$ loading is 1.8 kg/(m$^3$ · d) (112 lb/(1000 ft$^3$ · d)) with 90% removal.

The specific gravity of the digester solids is greater than that of the activated sludge, and the settling characteristics are thus improved.

Since high dissolved oxygen is required for the Kraus process, in order to cope with high oxygen demand, Kraus applied a dual system in the sludge reaeration basin in which there was a combination of coarse-bubble aeration at the top of the basin and fine-bubble aeration at the bottom of the basin. The process is used for high-strength wastewaters with low nitrogen levels.

The design criteria for the Kraus process are: $BOD_5$ loading = 0.6 to 1.6 kg/(m$^3$ · d) (40 to 100 lb/(1000 ft$^3$ · d)), F/M = 0.3 to 0.8 per day, MLSS = 2000 to 3000 mg/L, HRT = 4 to 8 days, $Q_r/Q$ = 0.5 to 1.0.

**Sequencing batch reactor.** A sequencing batch reactor (SBR) is a periodically operated, fill-and-draw activated-sludge system. The unit process used in the SBR and the conventional activated-sludge system are essentially the same. Both have aeration and sedimentation. In the activated-sludge plant, the processes are taking place simultaneously in separate basins, whereas in the SBR the aeration and clarification processes are carried out sequentially in the same basin. The system is used for small communities with limited land.

Each reactor in an SBR system has five discrete periods (steps) in each cycle: fill, react (aeration), settle (sedimentation/clarification), draw (decant), and idle (Herzbrun *et al.,* 1985). Biological activities are initiated as the influent wastewater fills the basin. During the fill and react period, the wastewater is aerated in the same manner as in a conventional activated-sludge system. After the reaction step, the mixed liquor suspended solids are allowed to settle in the same basin. The treated supernatant is withdrawn during the draw period. The idle period, the time between the draw and fill, may be zero or some certain period (days). The process is flexible and can remove nitrogen and phosphorus. Since the development of SBR technology in the early 1960s, a number of process modifications have been made to achieve specific treatment objectives.

In an SBR operation, sludge wasting is an important task relating to system performance. It usually takes place during the settle and idle phases. There is no return activated-sludge system in the SBR system.

**High-purity oxygen system.** The pure oxygen activated-sludge process was first studied by Okun in 1947 (Okun, 1949). The process achieved commercial status in the 1970s. Currently a large number of high-purity oxygen activated-sludge plants have been put into operation.

The process has been developed in an attempt to match oxygen supply and oxygen demand and to use a high-rate process through maintenance of high concentrations of biomass. The major components of the process are an oxygen generator, using high-purity oxygen in lieu of air, a specially compartmented aeration basin, a clarifier, pumps for recirculating activated sludge, and sludge disposal facilities. Oxygen is generated by manufacturing liquid oxygen cryogenically on-site for large plants and by pressure swing adsorption for small plants. Standard cryogenic air separation involves liquefaction of air followed by fractional distillation to separate the major components of oxygen and nitrogen.

The aeration basin is divided into compartments by baffles and is covered with a gastight cover. An agitator is included in each compartment (Fig. 6.24). High-purity oxygen is fed concurrently with wastewater flow. Influent wastewater, return activated sludge, and oxygen gas under slight pressure are introduced to the first compartment. A dissolved oxygen concentration of 4 to 10 mg/L is normally maintained in the mixed liquor. Successive aeration compartments are connected to each other through submerged ports. Exhausted waste gas is a mixture of nitrogen, carbon dioxide, and approximately 10% of oxygen supplied and vented from the last stage of the system. Effluent mixed liquor is then settled in a secondary clarifier. Settled activated sludge is recirculated to the aeration basin, with some wasted. The process is used for general application with high-strength wastes with limited space.

Design parameters of high-purity oxygen systems are: $BOD_5$ loading = 1.6 to 3.2 kg/(m$^3$ · d) (100 to 200 lb/(1000 ft$^3$ · d)), MLSS = 3000 to 8000 mg/L, F/M = 0.25 to 1.0 per day, HRT = 1 to 3 days, $\theta_c$ = 8 to 20 days, and $Q_r/Q$ = 0.25 to 0.5. The $BOD_5$ removal efficiency for the process is 85% to 95% (WEF and ASCE, 1992).

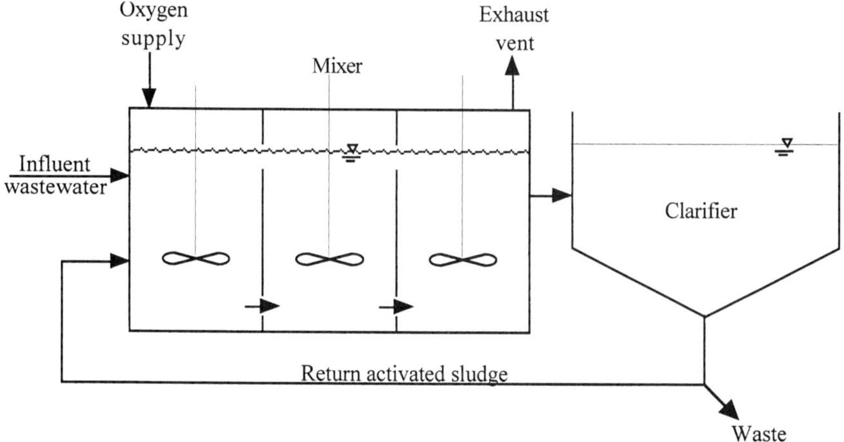


**Figure 6.24** Schematic flow diagram of a high-purity activated-sludge system.

Advantages of the high-purity oxygen activated-sludge process include higher oxygen transfer gradient and rates, smaller reaction chambers, improved biokinetics, the ability to treat high-strength soluble wastewater, greater tolerance for peak organic loadings, reduced energy requirements, reduction in bulking problems, a better settling sludge (1% to 2% solids), and effective odor control.

In a comparison of air and pure oxygen systems, researchers reported little or no significant difference in carbonaceous $BOD_5$ removal kinetics with DO concentration about 1 to 2 mg/L. Lower pH values (6.0 to 6.5) occur with high-purity oxygen and loss of alkalinity, with nitrate production. Covered aeration tanks have a potential explosion possibility.

**Example:** The high-purity oxygen activated-sludge process is selected for treating a high-strength industrial wastewater due to limited space. Determine the volume of the aeration tank and check with design parameters, using the following data.

| | |
|---|---|
| Design average flow | 1 Mgal/d (3785 $m^3$/d) |
| Influent BOD | 280 mg/L |
| Influent TSS | 250 mg/L |
| F/M | 0.70 lb BOD applied/(lb MLVSS · d) |
| MLSS | 5500 mg/L |
| VSS/TSS | 0.8 |
| Maximum volumetric BOD loading | 200 lb/(1000 $ft^3$ · d) = 3.2 kg/($m^3$ · d) |
| Minimum aeration time | 2 h |
| Minimum mean cell residence time | 3 days |

**solution:**

Step 1. Determine the tank volume

Using Eq. (6.88)

$$\text{F/M} = \frac{S_0}{\theta X} = \frac{QS_0}{VX}$$

$$V = \frac{QS_0}{(\text{F/M})X}$$

$$\text{F/M} = 0.7/\text{day}$$

where $X = 5500 \text{ mg/L} \times 0.8 = 4400 \text{ mg/L}$
8.34 = conversion factor, lb/(Mgal · (mg/L))

$$\text{Therefore } V = \frac{1 \text{ Mgal/d} \times 280 \text{ mg/L} \times 8.34}{(0.7/\text{day}) \times 4400 \text{ mg/L} \times 8.34}$$

$$= 0.0909 \text{ Mgal}$$
$$= 344 \text{ m}^3$$
$$= 12{,}150 \text{ ft}^3$$

Step 2. Check the BOD loading

$$\text{BOD loading} = \frac{1.0 \text{ Mgal/d} \times 280 \text{ mg/L} \times 8.34 \text{ lb/(Mgal} \cdot \text{(mg/L))}}{12.15 \times 1000 \text{ ft}^3}$$
$$= 192 \text{ lb/(1000 ft}^3 \cdot \text{d)}$$
$$< 200 \text{ lb/(1000 ft}^3 \cdot \text{d) (OK)}$$
$$= 3.08 \text{ kg/(m}^3 \cdot \text{d)}$$

Step 3. Check the aeration time

$$\text{HRT} = \frac{V}{Q} = \frac{0.0909 \text{ day} \times 24 \text{ h/d}}{1}$$
$$= 0.0909 \text{ day} \times 24 \text{ h/d}$$
$$= 2.2 \text{ h}$$
$$> 2.0 \text{ h (OK)}$$

Step 4. Check mean cell residence time (sludge age)

$$\text{Influent VSS } X_{\text{i}} = 0.8 \times 250 \text{ mg/L}$$
$$= 200 \text{ mg/L}$$
$$\theta_c = \frac{VX}{QX_{\text{i}}} = \frac{0.0909 \text{ Mgal} \times 4400 \text{ mg/L}}{1 \text{ Mgal/d} \times 200 \text{ mg/L}}$$
$$= 2.0 \text{ days}$$
$$< 3.0 \text{ days (OK)}$$

**Oxidation ditch.** The oxidation ditch process was developed by Pasveer (1960) in Holland and is a modification of the conventional activated-sludge plug-flow process. The system is especially applicable to small communities needing low-cost treatment. The oxidation ditch is typically an extended aeration mode. It consists typically of a single or closed-loop elongated oval channel with a liquid depth of 1.2 to 1.8 m (4 to 6 ft) and 45-degree sloping sidewalls (Fig. 6.25).

Wastewater is given preliminary treatment such as screening, comminution, or grit removal (usually without primary treatment). The wastewater is introduced into the ditch and is aerated using mechanical aerator(s) (Kessener brush) which are mounted across the channel for an extended period of time with long HRT of 8 to 36 h (typically 24 h) and SRT (solids residence time) of 20 to 30 days (WEF and ASCE, 1992).

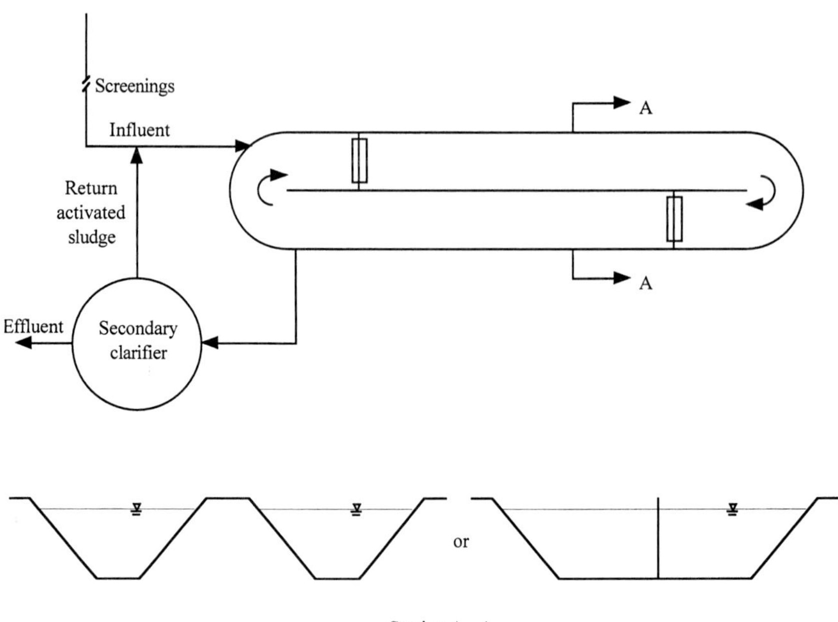


**Figure 6.25** Schematic diagram of oxidation ditch activated-sludge process.

The aerators provide mixing and circulation in the ditch, and oxygen transfer. The mechanical brush operates in the 60 to 110 rev/min range and keeps the liquid in motion at velocities from 0.24 to 0.37 m/s (0.8 to 1.2 ft/s), sufficient to prevent deposition of solids. A large fraction of the required oxygen is supplied by transfer through the free surface of the wastewater rather than by aeration. As the wastewater passes the aerator, the DO concentration rises and then falls while the wastewater traverses the circuit.

The $BOD_5$ loading in the oxidation ditch is usually as low as 0.08 to 0.48 kg/($m^3 \cdot$ d) (5 to 30 lb/(1000 $ft^3 \cdot$ d)). The F/M ratios are 0.05 to 0.30 per day. The MLSS are in the range of 1500 to 5000 mg/L. The activated sludge in the oxidation ditch removes the organic matter and converts it to cell protoplasm. This biomass will be degraded for long solids residence time.

Separate or interchannel clarifiers are used for separating and returning MLSS to the ditch. Recommendations for unit size and power requirements are available from equipment manufacturers.

The oxidation ditch is popular because of its high $BOD_5$ removal efficiency (85% to 95%) and easy operation. Nitrification occurs also in the system. The oxidation ditch is a flexible process and is used for small communities where large land area is available. However, the surface aerators may become iced during the winter months. Electrical heating must be installed at the critical area. During cold weather the operation must prevent loss of efficiency, mechanical damage, and danger to the operators.

**Example 6:** Estimate the effluent $BOD_5$ to be expected of an oxidation ditch treating a municipal wastewater with an influent BOD of 220 mg/L. Use the following typical kinetic coefficients: $k = 5/\text{d}$; $K_s = 60$ mg/L of BOD; $Y = 0.6$ mg VSS/mg BOD; $k_d = 0.06/\text{d}$; and $\theta_c = 20$ d.

**solution:**

Step 1. Estimate effluent substrate (soluble $BOD_5$) concentration From Eq. (6.80)

$$S = \frac{K_s(1 + \theta_c k_d)}{\theta_c(Yk - k_d) - 1}$$

$$= \frac{60\ \text{mg/L}(1 + 20\ \text{days} \times 0.06/\text{day})}{20\ \text{days}(0.6 \times 5/\text{days} - 0.06/\text{day}) - 1}$$

$$= 2.3\ \text{mg/L}$$

Step 2. Estimate TSS in the effluent

Using Eq. (6.78)

$$X = \frac{\theta_c Y(S_0 - S)}{\theta(1 + k_d \theta_c)}$$

$$= \frac{20\ \text{days} \times 0.6(220 - 2.3)\ \text{mg/L}}{20\ \text{days}(1 + 0.06/\text{day} \times 20\ \text{days})}$$

$$= 59.4\ \text{mg/L}$$

Step 3. Estimate the effluent BOD

It is commonly used that 0.63 mg BOD will be exerted (used) for each mg of TSS. Then, the effluent total BOD is

$$\text{BOD} = 2.3\ \text{mg/L} + 0.63 \times 59.4\ \text{mg/L}$$

$$= 39.7\ \text{mg/L}$$

**Deep shaft reactor.** A new variation of the activated-sludge process, developed in England, the deep shaft reactor is used where costs of land are high. It is used for general application with high-strength wastes. There are a number of installations in Japan. It has been marketed in the United States and Canada.

The deep shaft reactor consists of a vertical shaft about 120 to 150 m (400 to 500 ft) deep, utilizing a U-tube aeration system (Fig. 6.26). The

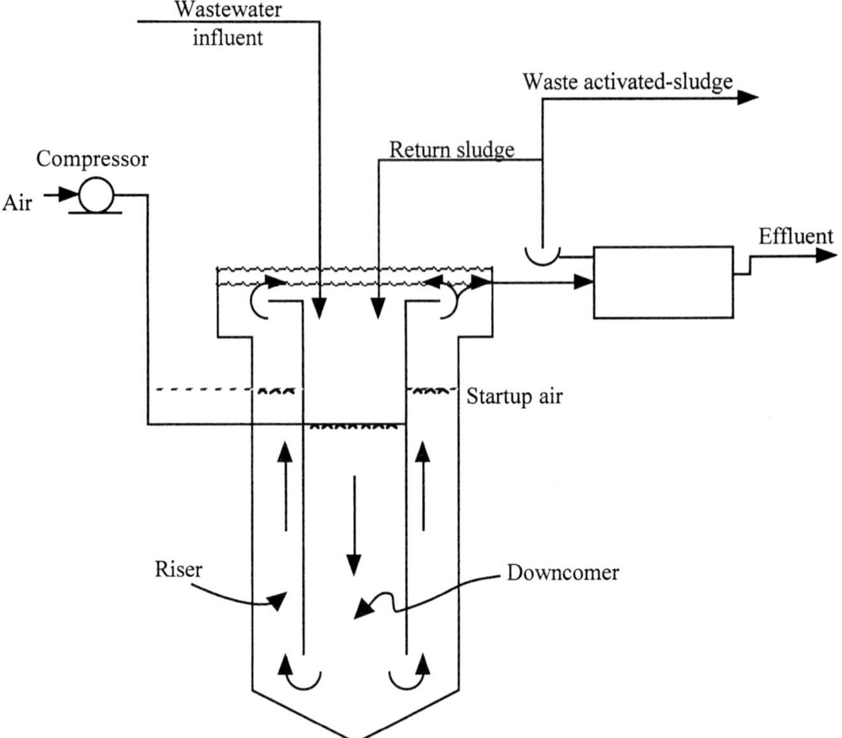


**Figure 6.26** Schematic diagram of deep shaft activated-sludge system.

shaft replaces the primary clarifiers and aeration tanks. The deep shaft is lined with a steel shell and fitted with a concentric pipe to form an annular reactor. Wastewater, return activated sludge, and air is forced down the center of the shaft and allowed to rise upward through the annulus (riser). The higher pressures obtained in the deep shaft improve the oxygen transfer rate and more oxygen can be dissolved, with beneficial effects on substrate utilization. DO concentration can range from 25 to 60 mg/L (Droste, 1997). Microbial reaction rates are not affected by the high pressures. Temperature in the system is fairly constant throughout the year. The flow in the shaft is a plug-flow mode. The F/M ratios are 0.5 to 5.0 per day and the HRT ranges from 0.5 to 5 h.

A flotation tank serves for the separation of the biomass (solids) from the supernatant. Most of the solids are recirculated to the deep shaft reactor, and some are wasted to an aerobic digester. The effluent of the reactor is supersaturated with air when exposed to atmospheric pressure. The release of dissolved air from the mixed liquor forms bubbles that float suspended solids to the surface of the flotation tank.

The advantages of the deep shaft reactor include lower construction and operational costs, less land requirement, capability to handle big organic loads, and suitability for all climatic conditions.

**Biological nitrification.** Conversion of ammonia to nitrite and nitrate can be achieved by biological nitrification. Biological nitrification is performed either in single-stage activated-sludge carbon oxidation and ammonia reduction or in separate continuous-flow stir-tank reactors or plug-flow.

The single-stage nitrification process is used for general application for nitrogen control where inhibitory industrial wastes are not present. The process also removes 85% to 95% of BOD.

The design criteria for single-stage activated-sludge nitrification are: volumetric BOD loading = 0.08 to 0.32 kg/($m^3 \cdot$ d) (5 to 20 lb/(1000 $ft^3 \cdot$ d)); MLSS = 1500 to 3500 mg/L; $\theta$ = 6 to 15 h; $\theta_c$ = 8 to 12 days; F/M = 0.10 to 0.25 per day; and $Q_r/Q$ = 0.50 to 1.50 (Metcalf and Eddy, Inc. 1991; WEF and ASCE, 1991a).

The advantage of the separate-stage suspended-growth nitrification system is its flexibility to be optimized to conform to the nitrification needs. The separate-stage process is used to update existing systems where nitrogen standards are stringent or where inhibitory industrial wastes are present and can not be removed in the previous stages.

The design criteria for the separate-stage nitrification process are: volumetric BOD loading = 0.05 to 0.14 kg/($m^3 \cdot$ d) (3 to 9 lb/1000 ($ft^3 \cdot$ d)), very low; F/M = 0.05 to 0.2 per day; MLSS = 1500 to 3500 mg/L; HRT = 3 to 6 h; $\theta_c$ = 15 to 100 days (very long); and $Q_r/Q$ = 0.50 to 2.00. More details on single-stage and separate-stage biological nitrification are given under advanced treatment for nitrogen removal in a later section.

## 21.8 Aeration and mixing systems

The treatment efficiency of an activated-sludge process depends on the oxygen dissolved in the aeration tank and the transfer of oxygen supply from the air and from pure oxygen. Oxygen is transferred to the mixed liquor in an aeration tank by dispersing compressed air bubbles through submerged diffusers with compressors or by entraining air into the mixed liquor by surface aerators. A combination of the above has also been used. An air diffusion system consists of air filters and other conditioning equipment, blowers, air piping, and diffusers, including diffuser cleaning devices.

A large number of combinations of mixing/aeration equipment and tank configuration have been developed.

**Oxygen transfer and utilization.** Oxygen transfer is a two-step process. Gaseous oxygen is first dissolved in the mixed liquor by diffusion or mechanical aeration or both. The dissolved oxygen is utilized by the microorganisms in the process of metabolism of the organic matter. When the rate of oxygen utilization exceeds the rate of oxygen dissolution, the dissolved oxygen (DO) in the mixed liquor will be depleted. This should be avoided. DO should be measured periodically, especially during the period of peak loading, and the air supply then adjusted accordingly.

The rate of oxygen transfer from air bubbles into solution is written as (Hammer, 1986)

$$r = k(\beta C_s - C_t) \tag{6.113}$$

where

$r$ = rate of oxygen transfer from air to liquid, mg/(L · h)

$k$ = transfer coefficient, per h

$\beta$ = oxygen saturation coefficient of the wastewater = 0.8 to 0.9

$C_s$ = DO concentration at saturation of top water (Table 1.2), mg/L

$C_t$ = DO concentration in mixed liquor, mg/L

$\beta C_s - C_t$ = DO deficit, mg/L

The greater the DO deficit, the higher the rate of oxygen transfer. The oxygen transfer coefficient $k$ depends on wastewater characteristics and, more importantly, on the physical features of the aeration, liquid depth of the aeration tank, mixing turbulence, and tank configuration.

The rate of DO utilization is essentially a function of F/M ratio and liquid temperature. The biological uptake of DO is approximately 30 mg/(L · h) for the conventional activated-sludge process, generally less than 10 mg/(L · h) for extended aeration, and as high as 100 mg/(L · h) for high-rate aeration. Aerobic biological reaction is independent of DO above a minimum critical concentration. However, if DO is below the critical value, the metabolism of microorganisms is limited by reduced oxygen supply. Critical concentrations reported for various activated-sludge systems range from 0.2 to 2.0 mg/L, the most commonly being 0.5 mg/L (Hummer, 1986). Ten States Standards stipulates that aeration equipment shall be capable of maintaining a minimum of 2.0 mg/L of DO in mixed liquor at all times and provide thorough mixing of the mixed liquor (GLUMRB, 1996).

Oxygen requirements generally depend on maximum diurnal organic loading (design peak hourly $BOD_5$), degree of treatment, and MLSS. The Ten States Standards (GLUMRS, 1996) further recommends that the normal design air requirement for all activated-sludge processes except extended aeration shall provide 1500 $ft^3$/lb of $BOD_5$ loading (94 $m^3$/kg $BOD_5$). For extended aeration the value shall be 2050 $ft^3$/lb of $BOD_5$ (128 $m^3$/kg of $BOD_5$). The design oxygen requirement for all activated-sludge processes shall be 1.1 lb oxygen/lb design peak hourly $BOD_5$ applied.

**Example 1:** Compare the rate of oxygen transfer when the existing DOs in the mixed liquor are 3.0 and 0.5 mg/L. Assume the transfer coefficient is 2.8/h and oxygen saturation coefficient is 0.88 for wastewater at 15.5°C.

**solution:**

Step 1. Find $C_s$ at 15.5°C

From Table 1.2

$$C_s = 9.92 \text{ mg/L}$$

Step 2. Calculate oxygen transfer rates, using Eq. (6.113)

$$\text{At DO} = 3.0 \text{ mg/L}$$

$$r = k(\beta C_s - C_t)$$

$$= (2.8/\text{h})\ (0.88 \times 9.92 - 3.0) \text{ mg/L}$$

$$= 16.0 \text{ mg/(L} \cdot \text{h)}$$

$$\text{At DO} = 0.5 \text{ mg/L}$$

$$r = (2.8/\text{h})\ (0.88 \times 9.92 - 0.5) \text{ mg/L}$$

$$= 23.0 \text{ mg/(L} \cdot \text{h)}$$

*Note*: The higher the DO deficit, the greater the oxygen transfer rate.

**Example 2:** Determine the oxygen transfer efficiency of a diffused aeration system providing 60 $m^3$/kg $BOD_5$ applied (960 $ft^3$/lb $BOD_5$). Assume that the aeration system is capable of transferring 1.1 kg of atmospheric oxygen to DO in the mixed liquor per kg $BOD_5$ applied, and that 1 $m^3$ of air at standard conditions (20°C, 760 mm Hg, and 36% relative humidity) contains 0.279 kg of oxygen.

**solution:**

Step 1. Calculate oxygen provided per kg $BOD_5$

$$\text{Oxygen provided} = 60 \text{ m}^3\text{/kg BOD} \times 0.279 \text{ kg/m}^3$$

$$= 16.7 \text{ kg/kg BOD}$$

Step 2. Determine oxygen transfer efficiency $e$

$$e = \text{oxygen transferred/oxygen provided}$$

$$= (1.1 \text{ kg/kg BOD})/(16.7 \text{ kg/kg BOD})$$

$$= 0.066$$

$$= 6.6\%$$

**Diffused air aeration.** A diffused air aeration system consists of diffusers submerged in the mixed liquor, header pipes, air mains, blowers (air compressors), and appurtenances through which the air passes. Ceramic diffusers (porous and nonporous) are manufactured as tubes, domes, or plates. Other diffusers, such as jets or U-tubes, may be used. Compressed air forced through the diffusers is released as fine bubbles. The oxygen is then introduced to the liquid.

The oxygen requirement for the aeration can be calculated using Eq. (6.94) or (6.95). In practice, the efficiency of oxygen transfer of a diffusion and sparging device seldom exceeds 8%. The oxygen transfer efficiency for various bubble sizes is available from the manufacturer's data. The diffuser system should be capable of providing for 200% of the designed average daily oxygen demand (GLUMRB, 1996).

**Air piping.** As compressed air flows through a pipe, its volume changes according to the pressure drop. The total headloss from the blower (or silencer outlet) to the diffuser inlet shall not exceed 0.5 lb/in$^2$ (3.4 kPa) under average operational conditions (GLUMRB, 1996). The pressure drop in a pipeline carrying compressed air can be calculated by the Darcy–Weisbach formula (Eq. (4.17)).

The value of friction factor $f$ can be determined from the Moody diagram. However, as an approximation, the friction factor of a steel pipe carrying compressed air is expressed as (Steel and McGhee, 1979)

$$f = 0.029D^{0.027}/Q^{0.148} \quad \text{(SI units)} \tag{6.114}$$

$$f = 0.013D^{0.027}/Q^{0.148} \quad \text{(US customary units)} \tag{6.115}$$

where $f$ = friction factor
$D$ = diameter of pipe, m or ft
$Q$ = flow of air, m$^3$/min or gal/min

The headloss in a straight pipe is calculated from

$$H = f(L/D)h_r \tag{6.116}$$

where $H$ = headloss in pipe, mm or in
$L$ = length of pipe, m or ft
$D$ = diameter of pipe, m or ft
$h_r$ = velocity head, mm or in

Another approximation of the estimation of headloss in a pipe can be expressed as

$$H = 9.82 \times 10^{-8}\,(fLTQ^2)/(PD^5) \tag{6.117}$$

where $T$ = temperature of air, K
$P$ = absolute pressure, atm
Other terms are as given above.

The absolute temperature may be estimated from the pressure rise using the following equation

$$T_2 = T_1(P_2/P_1)^{0.283} \tag{6.118}$$

Fittings in an air distribution system may be converted into the equivalent length of straight pipe as below

$$L = 55.4\ CD^{1.2} \qquad \text{(SI units)} \tag{6.119}$$

and

$$L = 230\ CD^{1.2} \qquad \text{(US customary units)} \tag{6.120}$$

where $L$ = equivalent length, m or ft
$D$ = diameter of pipe, m or ft
$C$ = resistance factor (0.25 to 2.0)

**Example:** Determine the headloss in 250 m (820 ft) of 0.30-m (12-in) steel pipeline carrying 42.5 $m^3$/min (1500 $ft^3$/min = 25 $ft^3$/s = 0.7 $m^3$/s) of air at a pressure of 0.66 atm (gauge). The ambient air temperature is 26°C (78°F).

**solution:**

Step 1. Estimate the temperature in the pipe using Eq. (6.118)

$$
\begin{aligned}
P_1 &= 1 \text{ atm} \\
P_2 &= 1 \text{ atm} + 0.66 \text{ atm} = 1.66 \text{ atm} \\
T_1 &= 26 + 273 = 299 \text{ (K)} \\
T_2 &= T_1(P_2/P_1)^{0.283} \\
&= 299 \text{ K } (1.66/1)^{0.283} \\
&= 345 \text{ K}
\end{aligned}
$$

Step 2. Estimate the friction factor using Eq. (6.114)

$$
\begin{aligned}
f &= 0.029\ D^{0.027}/Q^{0.148} \\
&= 0.029\ (0.3)^{0.027}/(42.5)^{0.148} \\
&= 0.0161
\end{aligned}
$$

Step 3. Estimate the head loss using Eq. (6.117)

$$
\begin{aligned}
H &= 9.82 \times 10^{8}(fLTQ^2)/(PD^5) \\
&= 9.82 \times 10^{-8} \times 0.0161 \times 250 \times 345 \times 42.5^2/(1.66 \times 0.3^5) \\
&= 61 \text{ (mm of water)}
\end{aligned}
$$

**Blower.** The blower (compressor) is used to convey air up to 103 kPa (15 lb/in$^2$). There are two types of blower commonly used for aeration, i.e. rotary position displacement (PD) and centrifugal units. Centrifugal blowers are usually used when the air supply is greater than 85 m$^3$/min (3000 ft$^3$/min or cfm) and are popular in Europe. For pressures up to 50 to 70 kPa (7 to 10 lb/in$^2$) and flows greater than 15 m$^3$/min (530 ft$^3$/min), centrifugal blowers are preferable. Rotary PD blowers are used primarily in small plants. Air flow is low when pressure is over 40 to 50 kPa (6 to 7 lb/in$^2$) and plant flows are less than 15 m$^3$/min (530 ft$^3$/min) (Steel and McGhee, 1979).

As an advantage, PD blowers are capable of operating over a wide range of discharge pressure, but it is difficult to adjust their air flow rate (only the speed is adjustable), they require more maintenance, and are noisy. In contrast, centrifugal blowers are less noisy in operation and take up less space. Their disadvantages include a limited range of operating pressure and reduced air flow when there is any increase in backpressure because of diffuser clogging. In many plants, single-stage centrifugal blowers can provide the head requirement (6 m or 20 ft of water). If higher pressures are needed, multiple units can provide a pressure of 28 m (90 ft) of water.

The power requirements for a blower may be estimated from the air flow, inlet and discharge pressures, and air temperature by the following equation which is based on the assumption of adiabatic compression conditions.

$$p = \frac{wRT_1}{29.7\ ne}\left[\left(\frac{P_2}{P_1}\right)^{0.283} - 1\right]$$

or

$$p = \frac{wRT_1}{8.41\ e}\left[\left(\frac{P_2}{P_1}\right)^{0.283} - 1\right] \quad \text{(SI units)} \tag{6.121}$$

$$p = \frac{wRT_1}{550\ ne}\left[\left(\frac{P_2}{P_1}\right)^{0.283} - 1\right]$$

or

$$p = \frac{wRT_1}{155.6\ e}\left[\left(\frac{P_2}{P_1}\right)^{0.283} - 1\right] \quad \text{(British system)} \tag{6.122}$$

where $p$ = power required, kW or hp
$w$ = weight of flow of air, kg/s or lb/s

$R$ = gas constant for air
= 8.314 kJ/(kmol · K) for SI units
= 53.3 ft · lb/(lb air · R)

(R = °F + 460 for British system)

$T_1$ = inlet absolute temperature, K or R
$P_1$ = inlet absolute pressure, atm or lb/in$^2$
$P_2$ = outlet absolute pressure, atm or lb/in$^2$
$n = (k - 1)/k = 0.283$ for air
$k = 1.395$ for air
550 = ft · lb/(s · hp)
29.7 = constant for conversion to SI units
$e$ = efficiency of the machine (usually 0.7 to 0.9)

As stated above, for higher discharge pressure application (>55 kPa, >8 lb/in$^2$) and for capacity smaller than 85 m$^3$/min (3000 ft$^3$/min) of free air per unit, rotary-lobe PD blowers are generally used. PD blowers are also used when significant water level variations are expected. The units cannot be throttled. Rugged inlet and discharge silencers are essential (Metcalf and Eddy, Inc. 1991).

**Example:** Estimate the power requirement of a blower providing 890 kg of oxygen per day through a diffused air system with an oxygen transfer efficiency of 7.2%. Assume the inlet temperature is 27°C, the discharge pressure is 6 m of water, and the efficiency of the blower is 80%

**solution:**

Step 1. Calculate weight of air flow $w$

$$\text{Oxygen required} = (890 \text{ kg/d})/0.072$$
$$= 12{,}360 \text{ kg/d}$$

In air, 23.2% is oxygen

$$w = (12{,}360 \text{ kg/d})/0.232$$
$$= 53{,}280 \text{ kg/d}$$
$$= 0.617 \text{ kg/s}$$

Step 2. Determine the power requirement for a blower $p$

$$T_1 = 27 + 273 = 300(\text{K})$$
$$P_1 = 1 \text{ atm}$$
$$P_2 = 1 \text{ atm} + 6 \text{ m}/10.345 \text{ m/atm}$$
$$= 1.58 \text{ atm}$$
$$e = 0.8$$

Using Eq. (6.121)

$$p = \frac{wRT_1}{8.41e}\left[\left(\frac{P_2}{P_1}\right)^{0.283} - 1\right]$$

$$= \frac{0.617 \times 8.314 \times 300}{8.41 \times 0.8}\left[\left(\frac{1.58}{1}\right)^{0.283} - 1\right]$$

$$= 31.6 \text{ (kW)}$$

**Mechanically aerated systems.** Mechanical aerators consist of electrical motors and propellers mounted on either a floating or a fixed support. The electrically driven propellers throw the bulk liquid through the air and oxygen transfer occurs both at the surface of the droplets and at the surface of the mixed liquor. Mechanical aerators may be mounted on either a horizontal or a vertical axis. Each group is divided into surface or submerged, high-speed (900–1800 rev/min) or low-speed (40 to 50 rev/min) (Steel and McGhee, 1979). Low-speed aerators are more expensive than high-speed ones but have fewer mechanical problems and are more desirable for biological floc formation. Selection of aerators is based on oxygen transfer efficiency and mixing requirements. Effective mixing is a function of liquid depth, unit design, and power supply.

Mechanical aerators are rated on the basis of oxygen transfer rate, expressed as kg of oxygen per kW · h (or lb/hp · h) under standard conditions, in which tap water with 0.0 mg/L DO (with sodium sulfite added) is tested at the temperature of 20°C. Commercially available surface reaerators range from 1.2 to 2.4 kg $O_2$/kW · h (2 to 4 lb $O_2$/hp · h) (Metcalf and Eddy, Inc. 1991).

Mechanical aerator requirements depend on the manufacturer's rating, the wastewater quality, the temperature, the altitude, and the desired DO level. The standard performance data must be adjusted to the anticipated field conditions, using the following equation (Eckenfelder, 1966)

$$N = N_0\left(\frac{\beta C_{\mathrm{w}} - C_{\mathrm{L}}}{C_{\mathrm{s20}}}\right)1.024^{T-20}\alpha \tag{6.123}$$

where $N$ = oxygen transfer rate under field conditions, kg/kW · h or lb/hp · h

$N_0$ = oxygen transfer rate provided by manufacturer, kg/kW · h or lb/hp · h

$\beta$ = salinity surface tension correction factor = 1 (usually)

$C_{\mathrm{w}}$ = oxygen saturation concentration for tap water at given altitude and temperature, mg/L

$C_L$ = operating DO concentration (2.0 mg/L)
$C_{s20}$ = oxygen saturation concentration in tap water at 20°C, mg/L
= 9.02
$T$ = temperature, °C
$\alpha$ = oxygen transfer correction factor for wastewater
= 0.8 to 0.9

**Example:** An activated-sludge plant is located at an elevation of 210 m. The desired DO level in the aeration tank is 2.0 mg/L. The range of operating temperature is from 8 to 32°C. The saturated DO values at 8, 20, and 32°C are 11.84, 9.02, and 7.29 mg/L, (from Table 1.2) respectively. The oxygen-transfer correction factor for the wastewater is 0.85. The manufacturer's rating for oxygen-transfer rate of the aerator under standard conditions is 2.0 kg oxygen/kW · h. Determine the power requirement for providing 780 kg oxygen per day to the aeration system.

**solution:**

Step 1. Determine DO saturation levels at altitude 210 m

For DO, after altitude correction (zero saturation at 9450 m)

$$C_w = C\left(1 - \frac{\text{altitude, m}}{9450\text{ m}}\right) = C\left(1 - \frac{210\text{ m}}{9450\text{ m}}\right)$$
$$= 0.98C$$

At 8°C

$$C_w = 0.98 \times 11.84 \text{ mg/L (from Table 1.2)}$$
$$= 11.6 \text{ mg/L}$$

At 32° C

$$C_w = 0.98 \times 7.29 \text{ mg/L [calculated by Eq. (1.4)]}$$
$$= 7.14 \text{ mg/L}$$

Step 2. Calculate oxygen transfer under field conditions

At 8°C, by Eq. (6.123)

$N_0$ = 2.0 kg oxygen/kW · h

$\beta$ = 1

$C_w$ = 11.6 mg/L

$C_L$ = 2.0 mg/L

$C_{s20}$ = 9.02 mg/L

$T$ = 8°C

$\alpha$ = 0.85

$$\begin{aligned} N &= N_0\left(\frac{\beta C_w - C_L}{C_{s20}}\right)1.024^{T-20}\alpha \\ &= (2\text{ kg/kW}\cdot\text{h})\left(\frac{1\times 11.6\text{ mg/L} - 2.0\text{ mg/L}}{9.02\text{ mg/L}}\right)(1.024^{8-20})0.85 \\ &= 1.36\text{ kg/kW}\cdot\text{h} \end{aligned}$$

At 32°C

$$\begin{aligned} N &= 2\left(\frac{7.14 - 2}{9.02}\right)(1.024^{32-20})\ 0.85 \\ &= 1.29\ (\text{kg/kW}\cdot\text{h}) \\ &= 31\text{ kg/kW}\cdot\text{d} \end{aligned}$$

Step 3. Calculate the power required per day, $p$

At 32°C

$$\begin{aligned} p &= (780\text{ kg/d})/(31\text{ kg/kW}\cdot\text{d}) \\ &= 25\text{ kW} \end{aligned}$$

*Note*: Transfer rate is lower with higher temperature.

**Aerated lagoon.** An aerated lagoon (pond) is a complete-mix flow-through system with or without solids recycle. Most systems operate without solids recycle. If the solids are returned to the lagoon, the system becomes a modified activated-sludge process.

The lagoons are 3 to 4 m (10 to 13 ft) deep. Solids in the complete-mix aerated pond are kept suspended at all times by aerators. Oxygen is usually supplied by means of surface aerators or diffused air devices. Depending on the hydraulic retention time, the effluent from an aerated pond will contain from one-third to one-half the concentration of the influent BOD in the form of cell tissue (Metcalf and Eddy, Inc. 1991). These solids must be removed by settling before the effluent is discharged. Settling can take place at a part of the aerated pond system separated with baffles or in a sedimentation basin.

The design factors for the process include BOD removal, temperature effects, oxygen requirements, mixing requirements, solids separation, and effluent characteristics. BOD removal and the effluent characteristics are generally estimated using a complete-mix hydraulic model and first-order reaction kinetics. The mathematical relationship for BOD removal in a complete-mix aerated lagoon is derived from the following equation

$$QS_0 - QS - kSV = 0 \tag{6.124}$$

Rearranging

$$\frac{S}{S_0} = \frac{1}{1 + k(V/Q)} = \frac{\text{effluent BOD}}{\text{influent BOD}} \tag{6.125}$$

$$= \frac{1}{1 + k\theta} \tag{6.126}$$

where $S$ = effluent $BOD_5$ concentration, mg/L
$S_0$ = influent $BOD_5$ concentration, mg/L
$k$ = overall first-order $BOD_5$ removal rate, per day
= 0.25 to 1.0, based on $e$
$Q$ = wastewater flow, $m^3$/d or Mgal/d
$\theta$ = total hydraulic retention time, day

Typical design values of $\theta$ for aerated ponds used for treating domestic wastewater vary from 3 to 6 days. The amounts of oxygen required for aerated lagoons range from 0.7 to 1.4 times the amount of $BOD_5$ removed (Metcalf and Eddy, Inc. 1991).

Mancini and Barnhart (1968) developed the resulting temperature in the aerated lagoon from the influent wastewater temperature, air temperature, surface area, and flow. The equation is

$$T_i - T_w = \frac{(T_w - T_a)fA}{Q} \tag{6.127}$$

where $T_i$ = influent wastewater temperature, °C or °F
$T_w$ = lagoon water temperature, °C or °F
$T_a$ = ambient air temperature, °C or °F
$f$ = proportionality factor
= $12 \times 10^{-6}$ (for British system)
= 0.5 (for SI units)
$A$ = surface area of lagoon, $m^2$ or $ft^2$
$Q$ = wastewater flow, $m^3$/d or Mgal/d

Rearranging Eq. (6.127), the lagoon water temperature is

$$T_w = \frac{AfT_a + QT_i}{Af + Q} \tag{6.128}$$

Aerated lagoons, usually followed by facultative lagoons, are used for first-stage treatment of high-strength domestic wastewaters and for pretreatment of industrial wastewaters. Their BOD removal efficiencies range from 60% to 70%. Low efficiency and foul odors may occur for improperly designed or poorly operated plants, especially if the aeration devices are inadequate. Wet weather flow, infiltration, and icing may cause the process to be upset.

**Example:** Design a complete-mix aerated lagoon system using the conditions given below.

| | |
|---|---|
| Wastewater flow | 3000 $m^3$/d (0.8 Mgal/d) |
| Influent soluble $BOD_5$ | 180 mg/L |
| Effluent soluble $BOD_5$ | 20 mg/L |
| Soluble $BOD_5$ first-order $k_{20}$ | 2.4 per day |
| Influent TSS (not biodegraded) | 190 mg/L |
| Final effluent TSS | 22 mg/L |
| MLVSS/MLSS | 0.8 |
| Kinetic coefficients: | |
| $k$ | 5 per day |
| $K_s$ | 60 mg/L BOD |
| $Y$ | 0.6 mg/mg |
| $k_d$ | 0.06 per day |
| Design depth of lagoon | 3 m (10 ft) |
| Design HRT, $\theta$ | 5 days |
| Detention time at settling basin | 2 days |
| Temperature coefficient | 1.07 |
| Wastewater temperature | 15°C (59°F) |
| Summer mean air temperature | 26.5°C (78°F) |
| Winter mean air temperature | 9°C (48°F) |
| Summer mean wastewater temperature | 20.2°C (68°F) |
| Winter mean wastewater temperature | 12.3°C (54°F) |
| Aeration constant $\alpha$ | 0.86 |
| Aeration constant $\beta$ | 1.0 |
| Plant site elevation | 210 m (640 ft) |

**solution:**

Step 1. Determine the surface area of the lagoon with a depth of 3 m

$$\text{Volume } V = Q\theta = 3000 \text{ m}^3\text{/d} \times 5 \text{ days}$$
$$= 15{,}000 \text{ m}^3$$
$$\text{Area } A = V/(3 \text{ m}) = 15{,}000 \text{ m}^3/3 \text{ m}$$
$$= 5000 \text{ m}^2$$
$$= 1.23 \text{ acres}$$

Step 2. Estimate lagoon wastewater temperature in summer and winter

Using Eq. (6.128), in summer ($T_a$ = 26.5°C)

$$T_w = \frac{AfT_a + QT_i}{Af + Q}$$

$$= \frac{5000 \times 0.5 \times 26.5 + 3000 \times 15}{5000 \times 0.5 + 3000}$$

$$= 20.2(°C)$$

$$= 68.4°F$$

In winter ($T_a = 9°C$)

$$T_w = \frac{5000 \times 0.5 \times 9 + 3000 \times 15}{5000 \times 0.5 + 3000}$$

$$= 12.3(°C)$$

$$= 54.1°F$$

Step 3. Calculate the soluble $BOD_5$ during the summer

Using Eq. (6.80)

$$S = \frac{K_s(1 + \theta k_d)}{\theta(Yk - k_d) - 1}$$

$$= \frac{60\,\text{mg/L}(1 + 5\text{ days} \times 0.06/\text{days})}{5\text{ days}(0.6 \times 5/\text{day} - 0.06/\text{day}) - 1}$$

$$= 5.7\text{ mg/L}$$

Step 4. Calculate first-order BOD removal rate constant for temperature effects

In summer at the mean wastewater temperature 20.2°C

$$k_t = k_{20}\,(1.07)^{T-20}$$

$$k_{20.2} = 2.4/\text{day}\,(1.07)^{20.2-20}$$

$$= 2.43/\text{day}$$

In winter mean wastewater temperature at 12.3°C

$$k_{12.3} = 2.4/\text{day}\,(1.07)^{12.3-20}$$

$$= 1.43/\text{day}$$

Step 5. Calculate the effluent $BOD_5$ using Eq. (6.126)

In summer mean wastewater temperature at 20.2°C

$$\frac{S}{S_0} = \frac{1}{1 + k\theta}$$

$$\frac{S}{180\text{ mg/L}} = \frac{1}{1 + 2.43 \times 5}$$

$$S = 13.7\text{ mg/L}$$

In winter at 12.3°C

$$S = \frac{180 \text{ mg/L}}{1 + 1.43 \times 5}$$

$$= 22.1 \text{ mg/L}$$

Ratio of winter to summer = 22.1 : 13.7 = 1.6 : 1

Step 6. Estimate the biological solids produced using Eq. (6.78), $Q = Q_c$

$$X = \frac{Y(S_0 - S)}{1 + k_d\theta} = \frac{0.6(180 - 5.7) \text{ mg/L}}{1 + 0.06 \times 5}$$

$$= 80 \text{ mg/L(of VSS)}$$

$$\text{TSS} = 80 \text{ mg/L} \div 0.8$$

$$= 100 \text{ mg/L}$$

Step 7. Calculate TSS in the lagoon effluent before settling

$$\text{TSS} = 190 \text{ mg/L} + 100 \text{ mg/L}$$

$$= 290 \text{ mg/L}$$

*Note:* With low overflow rate and a long detention time of 2 days, the final effluent can achieve 22 mg/L of TSS.

Step 8. Calculate the amount of biological solids wasted per day

Since $X = 80 \text{ mg/L} = 80 \text{ g/m}^3 = 0.08 \text{ kg/m}^3$ (from Step 6)

$$p_x = 3000 \text{ m}^3\text{/d} \times 0.08 \text{ kg/m}^3$$

$$= 240 \text{ kg/d}$$

Step 9. Calculate the oxygen required using Eq. (6.94a)

Using the conversion factor $f$ for $BOD_5$ to $BOD_L$, $f = 0.67$

$$\text{Oxygen} = \frac{Q(S_0 - S)}{(1000 \text{ g/kg}) f} - 1.42\, p_x$$

$$= \frac{3000 \text{ m}^3 \times (180 - 5.7) \text{ g/m}^3}{1000 \text{ g/kg} \times 0.67} - 1.42 \times 240 \text{ kg/d}$$

$$= 400 \text{ kg/d}$$

$$= 970 \text{ lb/d}$$

Step 10. Calculate the ratio of oxygen required to BOD removal

$$\text{BOD removal} = 3000\ \text{m}^3/\text{d} \times (180 - 5.7)\ \text{g/m}^3 \div 1000\ \text{g/kg}$$

$$= 522.9\ \text{kg/d}$$

$$= 1153\ \text{lb/d}$$

$$\frac{\text{Oxygen required}}{\text{BOD removed}} = \frac{440\ \text{kg/d}}{522.9\ \text{kg/d}}$$

$$= 0.84$$

Step 11. Determine the field transfer rate for the surface aerators

The example in the mechanical aerator shows that the oxygen transfer rate is lower in summer (higher temperature). From that example, the altitude correction factor is 0.98 for an elevation of 210 m. Select the surface aerator rating as 2.1 kg oxygen/kW · h (3.5 lb/hp · h). The solubility of tap water at 26.5 and 20.0°C is 7.95 and 9.02 mg/L, respectively. Let $C_L = 2.0$ g/L as GLUMRB (1996) recommended. The power requirements under field conditions can be estimated using Eq. (6.123).

$$N = N_0\left(\frac{\beta C_w - C_L}{C_{s20}}\right)1.024^{T-20}\alpha$$

$$= (2.1\,\text{kg/kW}\cdot\text{h})\left(\frac{1 \times 0.98 \times 7.95\ \text{mg/L} - 2\ \text{mg/L}}{9.02\ \text{mg/L}}\right)(1.024)^{26.5-20}\,(0.86)$$

$$= 1.35\ \text{kg/kW}\cdot\text{h}$$

$$= 32.4\ \text{kg/kW}\cdot\text{d}$$

Step 12. Determine the power requirements of the surface aerator

From Steps 9 and 11

$$\text{Power required} = \frac{\text{oxygen required}}{\text{field transfer rate}} = \frac{440\ \text{kg/d}}{32.4\ \text{kg/kW}\cdot\text{d}}$$

$$= 13.6\ \text{kW}$$

*Note*: This is the power required for oxygen transfer.

Step 13. Determine the total energy required for mixing the lagoon

Assume power required for complete-mix flow is 0.015 kW/m$^3$ (0.57 hp/1000 ft$^3$).

From Step 1

$$\text{Volume of the lagoon} = 15{,}000\ \text{m}^3$$

$$\text{Total power needed} = 0.015\ \text{kW/m}^3 \times 15{,}000\ \text{m}^3$$

$$= 225\ \text{kW}$$

Use 8 of 30 kW (40 hp) surface aerators providing 240 kW.

*Note*: This is the power required for mixing and is commonly the control factor in sizing the aerators for domestic wastewater treatment. For industrial wastewater treatment, the control factor is usually reversed.

## 22 Trickling Filter

A trickling filter is actually a unit process for introducing primary effluent into contact with biological growth and is a biological oxidation bed. The word "filter" does not mean any filtering or straining action; nevertheless, it is popularly and universally used.

### 22.1 Process description

The trickling filter is the most commonly used unit of the fixed-growth film-flow-type process. A trickling filter consists of: (1) a bed of coarse material, such as stone slates or plastic media, over which wastewater from primary effluent is sprayed; (2) an underdrain system; and (3) distributors. The underdrain is used to carry wastewater passing through the biological filter and drain to the subsequent treatment units and to provide ventilation of the filter and maintenance of the aerobic condition. Wastewater from the primary effluent is distributed to the surface of the filter bed by fixed spray nozzles (first developed) or rotary distributors. Sloughs of biomass from the media are settled in the secondary sedimentation tank.

Biological slime occurs on the surface of the support media while oxygen is supplied by air diffusion through the void spaces. It allows wastewater to trickle (usually in an intermittent fashion) downward through the bed media. Organic and inorganic nutrients are extracted from the liquid film by the microorganisms in the slime. The biological slime layer consists of aerobic, anaerobic, and facultative bacteria, algae, fungi, and protozoans. Higher animals such as sludge worms, filter-fly larvae, rotifers, and snails are also present. Facultative bacteria are the predominant microorganisms in the trickling filter. Nitrifying bacteria may occur in the lower part of a deep filter.

The biological activity of the trickling filter process can be described as shown in Fig. 6.27. The microbial layer on the filter is aerobic usually to a depth of only 0.1 to 0.2 mm. Most of the depth of the microbial film is anaerobic. As the wastewater flows over the slime layer, organic matter (nutrient) and dissolved oxygen are transferred to the aerobic zone by diffusion and extracted, and then metabolic end products such as carbon dioxide are released to the water.

Dissolved oxygen in the liquid is replenished by adsorption from air in the voids surrounding the support media. Microorganisms near the surface of the filter bed are in a rapid growth rate due to plenteous food

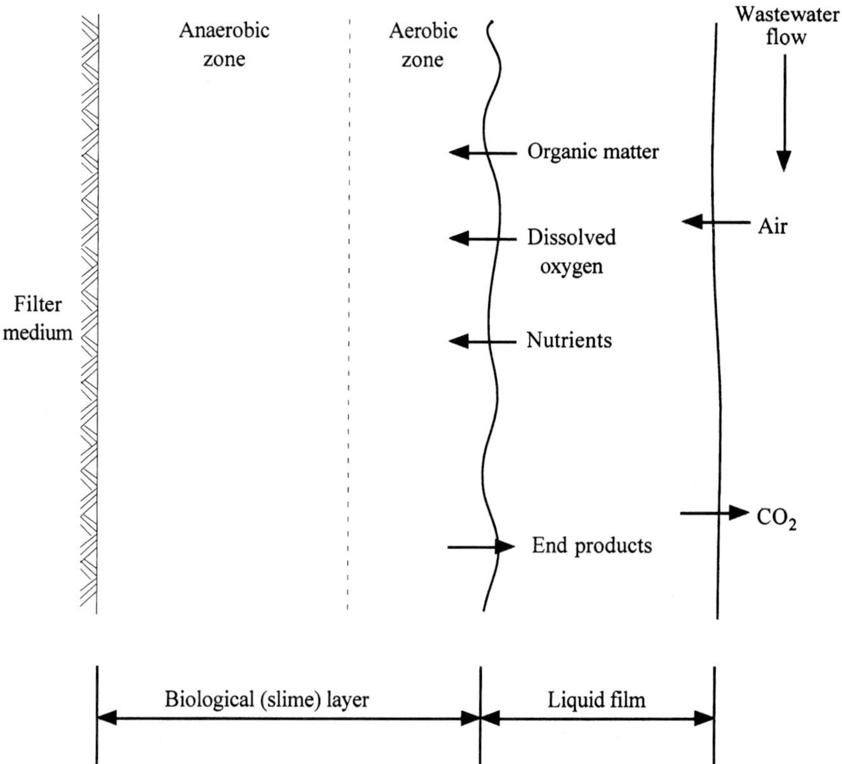


**Figure 6.27** Schematic diagram of attached-growth process.

supply, whereas microorganisms in the lower portion of the filter may be in a state of starvation. Overall, a trickling filter operation is considered to be in the endogenous growth phase. When slime becomes thicker and cells die and lyse, the slime layer will slough off and is subsequently removed by secondary settling.

For over a century, the trickling filter has been used as a secondary treatment unit. In a rock-fill conventional trickling filter, the rock size is 25 to 100 mm (1 to 4 in). The depth of the rock bed varies from 0.9 to 2.5 m (3 to 8 ft) with an average of 1.8 m (6 ft). The Ten States Standards recommends a minimum depth of trickling filter media of 1.8 m above the underdrains and that a rock and/or slag filter shall not exceed 3 m (10 ft) in depth (GLUMRB, 1996).

After declining in use in the late 1960s and early 1970s because of the development of RBC, trickling filters regained popularity in the late 1970s and 1980s, primarily due to new synthetic media. The new high-rate media were generally preferred over rocks because they are lighter, increased the surface area for biological growth, and improved treatment efficiency. High-rate media minimizes many of the common problems with rock media, such as uncontrolled sloughing, plugging, odors, and filter flies. Consequently, almost all trickling filters constructed in the late 1980s have been of high-rate media type (WPCF, 1988b).

Plastic media built in square, round, and other modules of corrugated shape have become popular. The depths of these plastic media range from 4 to 12 m (14 to 40 ft) (US EPA, 1980; Metcalf and Eddy, Inc. 1991). These materials increase void ratios and air flow. The plastics are 30% lighter than rock. A minimum clearance of 0.3 m (1 ft) between media and distribution arms shall be provided (GLUMRB, 1996).

### 22.2 Filter classification

Trickling filters are classified according to the applied hydraulic and organic loading rates. The hydraulic loading rate is expressed as the quantity of wastewater applied per day per unit area of bulk filter surface ($m^3/(d \cdot m^2)$, $gal/(d \cdot ft^2)$, or $Mgal/(d \cdot acre)$ or as depth of wastewater applied per unit of time. Organic loading rate is expressed as mass of $BOD_5$ applied per day per unit of bulk filter volume ($kg/(m^3 \cdot d)$, $lb/(1000\ ft^3 \cdot d)$). Common classifications, are low- or standard-rate, intermediate-rate, high-rate, super-high-rate, and roughing. Two-stage filters are frequently used, in which two trickling filters are connected in series. Various trickling filter classifications are summarized in Table 6.14.

### 22.3 Recirculation

Recirculation of a portion of the effluent to flow back through the filter is generally practised in modern trickling filter plants. The ratio of the return flow $Q_r$, to the influent flow $Q$ is called the recirculation ratio $r$. Techniques of recirculation vary widely, with a variety of configurations. The recirculation ratios range from 0 to 4 (Table 6.14) with usual ratios being 0.5 to 3.0.

The advantages of recirculation include an increase in biological solids in the system with continuous seeding of active biological material; elimination of shock load by diluting strong influent; maintenance of more uniform hydraulic and organic loads; an increase in the DO level of the influent; thinning of the biological slime layer; an improvement of treatment efficiency; reduction of filter clogging; and less nuisance problems.

### 22.4 Design formulas

Attempts have been made by numerous investigators to correlate operational data with the design parameters of trickling filters. The design of trickling filter plants is based on empirical, semiempirical, and mass balance concepts. Mathematical equations have been developed for

**TABLE 6.14 Typical Design Information for Trickling Filter Categories**

| Parameter | Low rate (Standard) | Intermediate rate | High rate | Super high rate | Roughing | Two stage |
|---|---|---|---|---|---|---|
| Filter media | Rock, slag | Rock, slag | Rock | Plastic | Plastic, redwood | Rock, plastic |
| Hydraulic loading | | | | | | |
| $m^3/(m^2 \cdot d)$ | 1–3.7 | 3.7–9.4 | 9.4–37 | 14–86 | 47–187 | 9.4–37 |
| $gal/(d \cdot ft^2)$ | 25–90 | 90–230 | 230–900 | 350–2100 | 1150–4600 | 230–900 |
| $Mgal/(d \cdot acre)$ | 1–4 | 4–10 | 10–39 | 15–91 | 50–200 | 10–39 |
| $BOD_5$ loading | | | | | | |
| $kg/(m^3 \cdot d)$ | 0.08–0.32 | 0.24–0.48 | 0.48–1.0 | 0.48–1.6 | 1.6–8.0 | 1.0–2.0 |
| $lb/(1000\ ft^3 \cdot d)$ | 5–20 | 15–30 | 30–60 | 30–100 | 100–500 | 60–120 |
| Recirculation rate | 0 | 0–1 | 1–2 | 1–2 | 1–4 | 0.5–2 |
| Filter flies | Many | some | few | few or none | few or none | few or none |
| Sloughing | intermittent | intermittent | continuous | continuous | continuous | continuous |
| Depth, m | 1.8–2.4 | 1.8–2.4 | 0.9–1.8 | 3–12 | 4.6–12 | 1.8–2.4 |
| ft | 6–8 | 6–8 | 3–6 | 10–40 | 15–40 | 6–8 |
| $BOD_5$ removal % (includes settling) | 80–90 | 50–70 | 65–85 | 65–80 | 40–65 | 85–95 |
| Nitrification | well | partial | little | little | no | well |
| Combined process | | | | | | |
| for secondary treatment | no | unlikely | likely | yes | yes | yes |
| for tertiary treatment | yes | yes | yes | yes | yes | yes |
| Type normally used* | SC/TF<br>ABF | SC/TF<br>ABF | SC/TF<br>RBC/TF | AS/TF<br>AS/BF | AS/TF<br>— | TF/TF<br>— |

* SC/TF=trickling filter after solid contactor
AS=activated sludge
ABF=activated biofilter
RBC=rotating biological contactor
BF=biofilter

SOURCES: US EPA (1974a), Metcalf and Eddy, Inc. (1991), WEF and ASCE (1991a, 1996a)

calculating the $BOD_5$ removal efficiency of biological filters on the basis of factors such as bed depth, types of media, recirculation, temperature, and loading rates. The design formulations of trickling filters of major interest include the NRC formula (1946), Velz formula (1948), Fairall (1956), Schulze formula (1960), Eckenfelder formula (1963), Galler and Gotaas formulas (1964, 1966), Germain formula (1966), Kincannon and Stover formula (1982), Logan formula (Logan *et al.*, 1987a, b), and British multiple regression analysis equation (1988). These are summarized elsewhere (WEF and ASCE, 1991a; McGhee, 1991).

**NRC formula.** The NRC (National Research Council) formula for trickling-filter performance is an empirical expression developed by the National Research Council from an extensive study of the operating data of trickling treatment plants at military bases within the United States during World War II in the early 1940s (NRC, 1946). It may be applied to single-stage and multistage rock filters with varying recirculation ratios. Graphic expressions for BOD removal efficiency are available. The equation for a single-stage or first-stage rock filter is

$$E_1 = \frac{100}{1 + 0.532\sqrt{W/VF}} \quad \text{(SI units)} \tag{6.129}$$

$$E_1 = \frac{100}{1 + 0.561\sqrt{W/VF}} \quad \text{(US customary units)} \tag{6.130}$$

where $E_1$ = efficiency of BOD removal for first stage at 20°C including recirculation and sedimentation, %
$W$ = BOD loading to filter, kg/d or lb/d
= flow times influent concentration
$V$ = volume of filter media, $m^3$ or 1000 $ft^3$
$F$ = recirculation factor

The recirculation factor is calculated by

$$F = \frac{1 + r}{(1 + 0.1r)^2} \tag{6.131}$$

where $r$ = recirculation ratio, $Q_r/Q$
$Q_r$ = recirculation flow, $m^3$/d or Mgal/d
$Q$ = wastewater flow, $m^3$/d or Mgal/d

The recirculation factor represents the average number of passes of the influent organic matter through the trickling filter. For the second-stage filter, the formula becomes

$$E_2 = \frac{100}{1 + \dfrac{0.0532}{1 - E_1}\sqrt{\dfrac{W'}{VF}}} \qquad \text{(SI units)} \qquad (6.132)$$

$$E_2 = \frac{100}{1 + \dfrac{0.0561}{1 - E_1}\sqrt{\dfrac{W'}{VF}}} \qquad \text{(British system)} \qquad (6.133)$$

where $E_2$ = efficiency of $BOD_5$ removal for second-stage filter, %
$W'$ = BOD loading applied to second-stage filter, kg/d or lb/d

Other terms are as described previously. Overall BOD removal efficiency of a two-stage filter system can be computed by

$$E = 100 - 100\left(1 - \frac{35}{100}\right)\left(1 - \frac{E_1}{100}\right)\left(1 - \frac{E_2}{100}\right) \qquad (6.134)$$

where the term 35 means that 35% of BOD of raw wastewater is removed by primary settling.

BOD removal efficiency in biological treatment process is significantly influenced by wastewater temperature. The effect of temperature can be calculated as

$$E_T = E_{20}\, 1.035^{T-20} \qquad (6.135)$$

**Example 1:** Estimate the BOD removal efficiency and effluent $BOD_5$ of a two-stage trickling filter using the NRC formula with the following given conditions.

| | |
|---|---|
| Wastewater temperature | 20°C |
| Plant flow $Q$ | 2 Mgal/d (7570 $m^3$) |
| $BOD_5$ in raw waste | 300 mg/L |
| Volume of filter (each) | 16,000 $ft^3$ (453 $m^3$) |
| Depth of filter | 7 ft (2.13 m) |
| Recirculation for filter 1 | = 1.5 $Q$ |
| Recirculation for filter 2 | = 0.8 $Q$ |

**solution:**

Step 1. Estimate BOD loading at the first stage

Influent BOD $C_1$ = 300 mg/L (1 – 0.35) = 195 mg/L

$$W = QC_1 = 2\text{ Mgal/d} \times 195\text{ mg/L} \times 8.34\text{ lb/(Mgal}\cdot\text{mg/L)}$$
$$= 3252\text{ lb/d}$$

Step 2. Calculate BOD removal efficiency of filter 1

Using Eqs. (6.131) and (6.130)

$$F = \frac{1 + r_1}{(1 + 0.1r_1)^2} = \frac{1 + 1.5}{(1 + 0.1 \times 1.5)^2}$$

$$= 1.89$$

$$E_1 = \frac{100}{1 + 0.0561\sqrt{W/VF}}$$

$$= \frac{100}{1 + 0.0561\sqrt{3252/(16 \times 1.89)}}$$

$$= 63.2(\%)$$

Step 3. Calculate effluent BOD concentration of filter 1

$$C_{1e} = 195 \text{ mg/L } (1 - 0.632)$$

$$= 71.8 \text{ mg/L}$$

Step 4. Calculate BOD removal efficiency of filter 2

$$F' = \frac{1 + 0.8}{(1 + 0.1 \times 0.8)^2} = 1.54$$

$$\text{Mass of influent BOD} = 2 \times 71.8 \times 8.34$$

$$= 1198 \text{ (lb/d)}$$

$$E_2 = \frac{100\%}{1 + 0.0561\sqrt{1198/(16 \times 1.54)}}$$

$$= 71.9\%$$

Step 5. Calculate effluent BOD concentration of filter 2

$$C_{2e} = 71.8 \text{ mg/L} \times (1 - 0.719)$$

$$= 20.1 \text{ mg/L}$$

Step 6. Calculate the overall efficiency

(a) Using Eq. (6.134)

$$E = 100 - 100[(1 - 0.35)(1 - 0.632)(1 - 0.719)]$$

$$= 93.3\%$$

(b)

$$E = (300 \text{ mg/L} - 20.1 \text{ mg/L})/(300 \text{ mg/L})$$

$$= 0.933$$

$$= 93.3\%$$

*Note*: At 20°C, no temperature correction is needed.

**Example 2:** Determine the size of a two-stage trickling filter using the NRC equations. Assume both filters have the same efficiency of $BOD_5$ removal and the same recirculation ratio. Other conditions are as follows.

| | |
|---|---|
| Wastewater temperature | 20°C |
| Wastewater flow $Q$ | 3785 $m^3$/d (1 Mgal/d) |
| Influent $BOD_5$ | 195 mg/L |
| Design effluent $BOD_5$ | 20 mg/L |
| Depth of each filter | 2 m (6.6 ft) |
| Recirculations for filters 1 and 2, $r_1 = r_2$ | 1.8 |

**solution:**

Step 1. Determine $E_1$ and $E_2$

$$\text{Overall efficiency} = (195 - 20)\ \text{mg/L} \times 100\%/195\ \text{mg/L}$$

$$= 89.7\%$$

$$E_1 + E_2(1 - E_1) = 0.897$$

$$E_{1r} + E_2 - E_1E_2 = 0.897$$

Since $E_1 = E_2$

Thus $$E_1^2 - 2E_1 + 0.897 = 0$$

$$E_1 = \frac{2 \pm \sqrt{4 - 4 \times 0.897}}{2}$$

$$E_1 = 0.68 = 68\%$$

Step 2. Calculate the recirculation factor $F$

$$F = \frac{1 + r}{(1 + 0.1r)^2} = \frac{1 + 1.8}{(1 + 0.1 \times 1.8)^2}$$

$$= 2.01$$

Step 3. Calculate mass BOD load to the first-stage filter, $W$

$$\text{Influent BOD} = 195\ \text{mg/L} = 195\ \text{g/m}^3$$

$$= 0.195\ \text{kg/m}^3$$

$$W = QC_1 = 3785\ \text{m}^3\text{/d} \times 0.195\ \text{kg/m}^3$$

$$= 738\ \text{kg/d}$$

Step 4. Calculate the volume $V$ of the first filter using Eq. (6.129)

$$E_1 = \frac{100}{1 + 0.532\sqrt{W/VF}}$$

From Step 1

$$68 = \frac{100}{1 + 0.532\sqrt{738/2.01V}}$$

$$\sqrt{367.2/V} = 0.885$$

$$V = 469(\text{m}^3)$$

Step 5. Calculate the diameter $d$ of the first filter

$$\text{Area} = V/\text{depth} = 469 \text{ m}^3/2 \text{ m}$$
$$= 234.5 \text{ m}^2$$
$$d^2\pi/4 = 234.5 \text{ m}^2$$
$$d = 17.3 \text{ m}$$

Step 6. Calculate the mass BOD loading $W'$ to the second-stage filter

$$W' = W(1 - E) = 738 \text{ kg/d } (1 - 0.68)$$
$$= 236.2 \text{ kg/d}$$

Step 7. Calculate the volume of the second filter using Eq. (6.132)

$$E_2 = \frac{100}{1 + \dfrac{0.532}{1 - E_1}\sqrt{\dfrac{W'}{VF}}}$$

$$68 = \frac{100}{1 + \dfrac{0.532}{1 - 0.68}\sqrt{\dfrac{236.2}{2.01V}}}$$

$$68 + 1225\sqrt{1/V} = 100$$

$$V = 1467(\text{m}^3)$$

Step 8. Calculate the diameter of the second-stage filter

$$A = 1467 \text{ m}^3/2 \text{ m}$$
$$= 733.5 \text{ m}^2$$
$$d^2 = 733.5 \text{m}^2 \times 4/\pi$$
$$d = 30.6 \text{ (m)}$$

Step 9. Check the BOD loading rate to each filter

(a) First-stage filter, from Steps 3 and 4

$$\text{BOD loading rate} = W \div V = 738 \text{ kg/d} \div 469 \text{ m}^3$$
$$= 1.57 \text{ kg/(m}^3 \cdot \text{d)}$$

*Note*: The rates are between 1.0 and 2.0 kg/(m$^3$ · d) (Table 6.14).

(b) Second-stage filter, from Steps 6 and 7

$$\text{BOD loading rate} = 236.2 \text{ kg/d} \div 1467 \text{ m}^3$$
$$= 0.161 \text{ kg/(m}^3 \cdot \text{d)}$$

Step 10. Check hydraulic loading rate to each filter

(a) For first-stage filter, recirculation ratio $r_1 = 1.8$, and from Step 5

$$\text{HLR} = \frac{(1 + 1.8) \times 3785 \text{ m}^3\text{/d}}{234.5 \text{ m}^2}$$
$$= 45.2 \text{ m}^3\text{/(m}^2 \cdot \text{d)}$$

(b) For second-stage filter, $r_2 = 1.8$, from Step 8

$$\text{HLR} = \frac{(1 + 1.8) \times 3785 \text{ m}^3\text{/d}}{733.5 \text{ m}^2}$$
$$= 14.4 \text{ m}^3\text{/(m}^2 \cdot \text{d)}$$

**Formulation for plastic media.** Numerous investigations have been undertaken to predict the performance of plastic media in the trickling filter process. The Eckenfelder formula (1963) and the Germain (1965) applied Schulze formulation (1960) are the ones most commonly used to describe the performance of plastic media packed trickling filters.

**Eckenfelder formula.** Eckenfelder (1963) and Eckenfelder and Barnhart (1963) developed an exponential formula based on the rate of waste removal for a pseudo-first-order reaction, as below:

$$S_e/S_i = \exp[-KA_s^{1+m}D/q^n] \qquad (6.136)$$

where
$S_e$ = effluent soluble $BOD_5$, mg/L
$S_i$ = influent soluble $BOD_5$, mg/L
$K$ = observed reaction rate constant, m/d or ft/d
$A_s$ = specific surface area
= surface area/volume, $m^2/m^3$ or $ft^2/ft^3$
$D$ = depth of media, m or ft
$q$ = influent volumetric flow rate
= $Q/A$
$Q$ = influent flow, $m^3$/d or $ft^3$/d
$A$ = cross-sectional area of filter, $m^2$ or $ft^2$
$m, n$ = empirical constants based on filter media

The mean time of contact $t$ of wastewater with the filter media is related to the filter depth, the hydraulic loading rate, and the nature of the filter packing. The relationship is expressed as

$$\frac{t}{D} = \frac{C}{q^n} = \frac{C}{(Q/A)^n} \tag{6.137}$$

and

$$C \cong 1/D^m \tag{6.138}$$

where $t$ = mean detention time
$C, n$ = constants related to the specific surface and configuration of the packing

Other terms are as in Eq. (6.136).

Eq. (6.136) may be simplified to the following form

$$S_e/S_i = \exp[-kD/q^n] \tag{6.139}$$

where $k$ is a new rate constant, per day.

**Example:** Estimate the effluent soluble $BOD_5$ from a 6-m (20-ft) plastic packed trickling filter with a diameter of 18 m (60 ft). The influent flow is 4540 $m^3$/d (1.2 Mgal/d) and the influent BOD is 140 mg/L. Assume that the rate constant $k$ = 1.95 per day and $n$ = 0.68.

**solution:**

Step 1. Calculate the area of the filter $A$

$$A = \pi(18\text{ m}/2)^2$$
$$= 254\text{ m}^2$$

Step 2. Calculate the hydraulic loading rate $q$

$$q = Q/A = 4540\text{ m}^3/\text{d} \div 254\text{ m}^2$$
$$= 17.9\text{ m}^3/(\text{m}^2 \cdot \text{d})$$
$$= 19.1\text{ (Mgal/d)/acre}$$

Step 3. Calculate the effluent soluble BOD using Eq. (6.139)

$$S_e = S_i \exp[-kD/q^n]$$
$$= 140\text{ mg/L} \exp[-1.95 \times 6/(17.9)^{0.68}]$$
$$= 27\text{ mg/L}$$

**Germain formula.** In 1966, Germain applied the Schultz (1960) formulation to a plastic media trickling filter and proposed a first-order equation as follows:

$$S_e/S_i = \exp[-k_{20}D/q^n] \tag{6.140}$$

where $S_e$ = total $BOD_5$ of settled effluent, mg/L
$S_i$ = total $BOD_5$ of wastewater applied to filter, mg/L
$k_{20}$ = treatability constant corresponding to depth of filter at 20°C
$D$ = depth of filter
$q$ = hydraulic loading rate, $m^3/(m^2 \cdot d)$ or gal/min/ft$^2$
$n$ = exponent constant of media, usually 0.5

The treatability constant at another depth of the filter must be corrected for depth when the $k_{20}$ value is determined at one depth. The relationship proposed by Albertson and Davis (1984) is

$$k_2 = k_1(D_1/D_2)^x \tag{6.141}$$

where $k_2$ = treatability constant corresponding to depth $D_2$ of filter 2
$k_1$ = treatability constant corresponding to depth $D_1$ of filter 1
$D_1$ = depth of filter 1, ft
$D_2$ = depth of filter 2, ft
$x$ = 0.5 for vertical and rock media filters
= 0.3 for crossflow plastic media filters

The values of $k_1$ and $k_2$ are a function of wastewater characteristics, the depth and configuration of the media, surface area of the filter, dosing cycle, and hydraulic loading rate. They are interdependent. Germain (1966) reported that the value of $k$ for a plastic media filter 6.6 m (21.6 ft) deep, treating domestic wastewater, was 0.24 $(L/s)^n/m^2$ and that $n$ is 0.5. This VFC media had a surface area of 88 $m^2/m^3$ (27 $ft^2/ft^3$). The ranges of $k$ values in $(L/s)^{0.5}/m^2$ for a 6-m (20-ft) tower trickling filter packed with plastic media at 20°C are 0.18 to 0.27 for domestic wastewater, 0.16 to 0.22 for domestic and food waste, 0.054 to 0.14 for fruit-canning wastes, 0.081 to 0.14 for meat packing wastes, 0.054 to 0.11 for paper mill wastes, 0.095 to 0.14 for potato processing, and 0.054 to 0.19 for a refinery. Multiplying the value in $(L/s)^{0.5}/m^2$ by 0.37 obtains the value in $(gal/min)^{0.5}/ft^2$.

**Distributor speed.** The dosing rate of $BOD_5$ is very important for treatment efficiency. The instantaneous dosing rate is a function of the distributor speed or the on–off times for a fixed distributor. The rotational speed of a rotary distributor is expressed as follows:

$$n = \frac{0.00044 q_t}{\alpha(\text{DR})} \qquad \text{(SI units)} \tag{6.142}$$

$$n = \frac{1.6 q_t}{\alpha(\text{DR})} \qquad \text{(US customary units)} \tag{6.143}$$

where $n$ = rotational speed of distributor, rev/min
$q_t$ = total applied hydraulic loading rate, $m^3/(m^2 \cdot d)$ or
= $q + q_r$ (gal/min)/ft$^2$
$q$ = influent wastewater hydraulic loading rate, $m^3/(m^2 \cdot d)$ or (gal/min)/ft$^2$
$q_r$ = recycle flow hydraulic loading rate, $m^3/(m^2 \cdot d)$ or (gal/min)/ft$^2$
$\alpha$ = number of arms in rotary distributor assembly
DR = dosing rate, cm or in per pass of distributor arm

The required dosing rates in inches per pass for trickling filters is determined approximately by multiplying the organic loading rate, expressed in lb $BOD_5$/1000 ft$^3$, by 0.12. For SI units, the dosing rate in cm per pass can be obtained by multiplying the loading rate in kg/m$^3$ by 0.30.

**Example:** Design a 8.0 m (26 ft) deep plastic packed trickling filter to treat domestic wastewater and seasonal (summer) food-process wastewater with given conditions as follows:

| | |
|---|---|
| Average year-round domestic flow $Q$ | 5590 m$^3$/d (1.48 Mgal/d) |
| Industrial wastewater flow | 4160 m$^3$/d (1.1 Mgal/d) |
| Influent domestic total $BOD_5$ | 240 mg/L |
| Influent domestic plus industrial $BOD_5$ | 520 mg/L |
| Final effluent $BOD_5$ | $\leq$ 24 mg/L |
| Value of $k$ at 26°C and at 6 m (from pilot plant study) | 0.27(L/s)$^{0.5}$/m$^2$ or 0.10 (gal/min)$^{0.5}$/ft$^2$ |
| Sustained low temperature in summer | 20°C |
| Sustained low temperature in winter | 10°C |

**solution:**

Step 1. Compute $k$ value at 20°C at 6 m

$$\begin{aligned} k_{20@6} &= k_{26@6}\theta^{T-26} \\ &= 0.27(\text{L/s})^{0.5}/\text{m}^2 \times 1.035^{20-26} \\ &= 0.22(\text{L/s})^{0.5}/\text{m}^2 \end{aligned}$$

Step 2. Correct the observed $k_{20}$ value for depth of 8 m, using Eq. (6.141)

$$k_2 = k_1(D_1/D_2)^x$$

At 8 m depth

$$\begin{aligned} k_{20@8} &= k_{20@6}(6/8)^{0.5} \\ &= 0.22(\text{L/s})^{0.5}/\text{m}^2 \times 0.866 \\ &= 0.19(\text{L/s})^{0.5}/\text{m}^2 \end{aligned}$$

Step 3. Compute the summer total flow

$$\begin{aligned} Q &= (5590 + 4160)\ \text{m}^3/\text{d} \\ &= 9750\ \text{m}^3/\text{d} \times 1000\ \text{L/m}^3 \div 86{,}400\ \text{s/d} \\ &= 112.8\ \text{L/s} \end{aligned}$$

Step 4. Determine the surface area required for an 8-m deep filter for summer, using Eq. (6.140)

$$S_e/S_i = \exp[-k_{20}D/q^n]$$

Substituting $Q/A$ for $q$ in Eq. (6.140) and rearranging yields

$$\begin{aligned} \ln S_e/S_i &= -k_{20}D/(Q/A)^n \\ A &= Q[-(\ln S_e/S_i)/k_{20}D]^{1/n} \\ &= 112.8[-(\ln 24/520)/(0.19 \times 8)]^{1/0.5} \\ &= 462\ (\text{m}^2) \end{aligned}$$

Step 5. Similarly, determine the surface area required for an 8-m deep filter during the winter at 10°C to meet the effluent requirements

(a) Determine $k_{10}$ for 6-m filter

$$\begin{aligned} k_{10@8} &= 0.27\ (\text{L/s})^{0.5}/\text{m}^2 \times 1.035^{10-26} \\ &= 0.156\ (\text{L/s})^{0.5}/\text{m}^2 \end{aligned}$$

(b) Correct $k_{10}$ value for 8-m filter

$$\begin{aligned} k_{10@8} &= 0.156\ [(\text{L/s})^{0.5}/\text{m}^2] \times (6/8)^{0.5} \\ &= 0.135\ (\text{L/s})^{0.5}/\text{m}^2 \end{aligned}$$

(c) Compute the area required (without industrial wastewaters in winter)

$$\begin{aligned} Q &= 5590\ \text{m}^3/\text{d} \times 1000\ \text{L/m}^3 \div 86{,}400\ \text{s/d} \\ &= 64.7\ \text{L/s} \\ A &= 64.7[-(\ln 24/240)/(0.135 \times 8)]^2 \\ &= 294\ (\text{m}^2) \end{aligned}$$

Step 6. Select the design area required

The required design area is controlled by the summer condition. Because the area required for the summer condition is larger (see Steps 4 and 5), the design area required is 462 $m^2$.

Step 7. Compute the hydraulic loading rates, HLR or $q_t$

(a) For summer

$$\begin{aligned}\text{HLR} &= 9750\ \text{m}^3/\text{d} \div 462\ \text{m}^2 \\ &= 21.1\ \text{m}^3/(\text{m}^2 \cdot \text{d}) \\ &= 518\ \text{gal/ft}^2 \cdot \text{d}\end{aligned}$$

(b) For winter

$$\begin{aligned}\text{HLR} &= 5590\ \text{m}^3/\text{d} \div 462\ \text{m}^2 \\ &= 12.1\ \text{m}^3/(\text{m}^2 \cdot \text{d}) \\ &= 297\ \text{gal/ft}^2 \cdot \text{d}\end{aligned}$$

Step 8. Check the organic (BOD) loading rates

(a) For summer

$$\begin{aligned}\text{Volume of filter} &= 8\ \text{m} \times 462\ \text{m}^2 \\ &= 3696\ \text{m}^3\end{aligned}$$

$$\begin{aligned}\text{BOD loading} &= \frac{9750\ \text{m}^3/\text{d} \times 520\ \text{g/m}^3}{3696\ \text{m}^3 \times 1000\ \text{g/kg}} \\ &= 1.37\ \text{kg}/(\text{m}^3 \cdot \text{d}) \\ &= 85.6\ \text{lb}/1000\ \text{ft}^3 \cdot \text{d}\end{aligned}$$

(b) For winter

$$\begin{aligned}\text{BOD loading} &= \frac{5590\ \text{m}^3/\text{d} \times 240\ \text{g/m}^3}{3696\ \text{m}^3 \times 1000\ \text{g/kg}} \\ &= 0.36\ \text{kg}/(\text{m}^3 \cdot \text{d}) \\ &= 22.5\ \text{lb}/1000\ \text{ft}^3 \cdot \text{d}\end{aligned}$$

Step 9. Determine rotation speed of rotary distributor, using Eq. (6.142)

The required dosing rates in cm/pass of arm for the trickling filter can be approximately estimated by multiplying the BOD loading rate in kg/($m^3 \cdot$ d) by 0.30.

(a) For summer, dose rate DR is, from Step 8a

$$\text{DR} = 1.37 \times 0.3 = 0.41\ \text{cm/pass}$$

Using 2 arms in the rotary distributor, $a = 2$; from Step 7a $= 21.1\ \text{m}^3/(\text{m}^2 \cdot \text{d})$

$$n = \frac{0.00044 q_t}{a(\text{DR})} = \frac{0.00044 \times 21.1}{2(0.41)}$$
$$= 0.0113\ (\text{rev/min})$$

or 1 revolution every 88.5 min

(b) For winter

From Step 8b

$$\text{DR} = 0.36 \times 0.3 = 0.108$$

From Step 7b

$$q_t = 12.1\ \text{m}^3/(\text{m}^2 \cdot \text{d})$$

Thus

$$n = \frac{0.00044 \times 12.1}{2 \times 0.108}$$
$$= 0.025(\text{rev/min})$$

or one revolution every 40 min

## 23 Rotating Biological Contactor

The use of plastic media to develop the rotating biological contactor (RBC) was commercialized in Germany in the late 1960s. The process is a fixed-film (attached growth) either aerobic or anaerobic biological treatment system for removal of carbonaceous and nitrogenous materials from domestic and industrial wastewaters. It was very popular during the late 1970s and early 1980s in the United States.

The RBC process can be used to modify and upgrade existing treatment systems as secondary or tertiary treatment. It has been successfully applied to all three steps of biological treatment, that is $\text{BOD}_5$ removal, nitrification, and denitrification.

The majority of RBC installations in the northern United States have been designed for removal of $\text{BOD}_5$ or ammonia nitrogen ($\text{NH}_3$-N), or both, from domestic wastewater. Currently in the United States there are approximately 600 installations for domestic wastewater treatment and more than 200 for industrial wastewater treatment. Over 1000 installations have been used in Europe, especially in Germany.

### 23.1 Hardware

The basic elements of the RBC system are media, shaft, bearings, drive, and cover. The RBC hardware consists of a large-diameter and closely spaced circular plastic media which is mounted on a horizontal shaft (Fig. 6.28). The shaft is supported by bearings and is slowly rotated by an electric motor. The plastic media are made of corrugated polyethylene or polystyrene material with various sizes and configurations and with various densities. The configuration designs are based on increasing stiffness and surface area, serving as spaces providing a tortuous wastewater flow path and stimulating air turbulence. As an exception, the Bio-Drum process consists of a drum filled with 38 mm plastic balls.

**Figure 6.28** (A) RBC units, (B) A view of an 84-unit RBCs for nitrification in Peoria, Illinois.

The diameters of the media range from 4 ft (1.22 m) to 12 ft (3.66 m), depending on the treatment capacity. The shaft lengths vary from 5 ft (1.52 m) to 27 ft (8.23 m), depending on the size of the RBC unit (Banerji, 1980). Commonly used RBC shafts are generally 25 to 27 ft (6.62 to 8.23 m) in length with a media diameter of 12 ft (3.66 m). Standard density media provide 100,000 ft$^2$ (9290 m$^2$) of surface area.

About 40% of the media is immersed in the wastewater at any time in trapezoidal, semicircular, or flat-bottom rectangular tanks with intermediate partitions in some cases. The shaft rotates at 1.5 to 1.7 rev/min for mechanical drive and 1.0 to 1.3 rev/min for air drive units (Fig. 6.29). The wastewater flows can be perpendicular to or parallel to the shafts.

### 23.2 Process description

In domestic wastewater treatment, the RBC does not require seeding to establish the biological growth. After RBC system startup, microorganisms naturally present in the wastewater begin to adhere to the rotating media surface and propagate until, in about 1 week, the entire

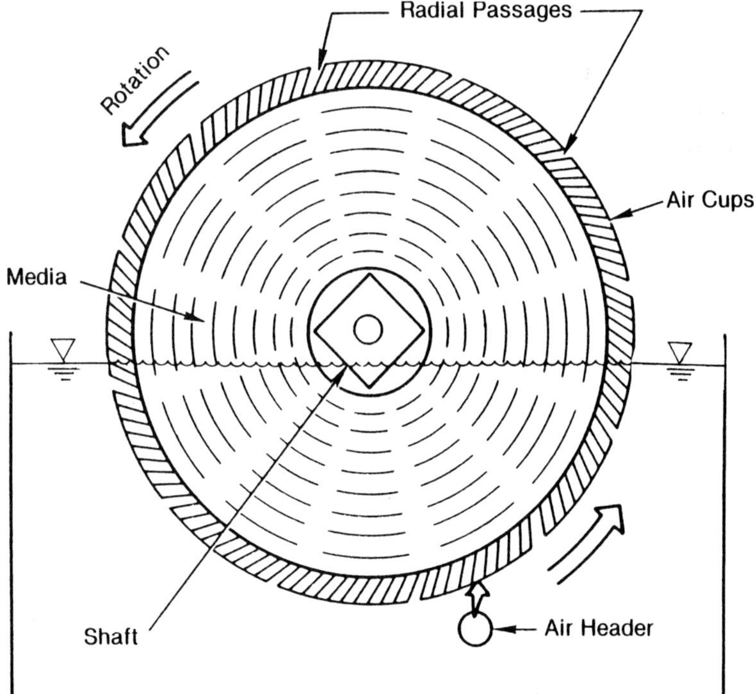


**Figure 6.29** Schematic diagram of air-drive RBC.

surface area will be covered with an approximately 1- to 4-mm thick layer of biological mass (biomass). The attached biomass contains about 50,000 to 100,000 mg/L suspended solids (Antonie, 1978). The microorganisms in the film of biomass (biofilm) on the media remove biodegradable organic matter, nitrogen, and dissolved oxygen in wastewater and convert the pollutants to more benign components (biomass and gaseous by-products).

In the first-stage RBC biofilm, the most commonly observed filamentous bacterium is *Sphaerotilus*, *Beggiatoa* (a sulfur bacterium); *Cladothrix, Nocardia, Oscillatoria,* and filamentous fungus, *Fusarium,* are also found, though less frequently. Nonfilamentous organisms in the first stage are *Zocloea* and *Zooglear filipendula, Aerobacter aerogen, Escherichia coli,* unicellular rods, spirilla and spirochaetes, and unicellular algae. The final stages harbor mostly the same forms of biota in addition to *Athrobotrys* and *Streptomyces* reported by investigators (Pretorius, 1971; Torpey *et al.,* 1971; Pescod and Nair, 1972; Sudo *et al.*, 1977; Clark *et al.,* 1978; Hitdlebaugh and Miller, 1980; Hoag *et al.*, 1980; Kinner *et al.*, 1982).

In continuous rotation, the media carry films of wastewater into the air, which then trickle down through the liquid film surface into the bulk liquid (Figs. 6.30 and 6.31). They also provide the surface area necessary for absorbing oxygen from air. Intimate contact between the wastewater and the biomass creates a constantly moving surface area for the bacteria-substrate-oxygen interface. The renewed liquid layer (wastewater film) on the biomass is rich in DO. Both substrates and DO

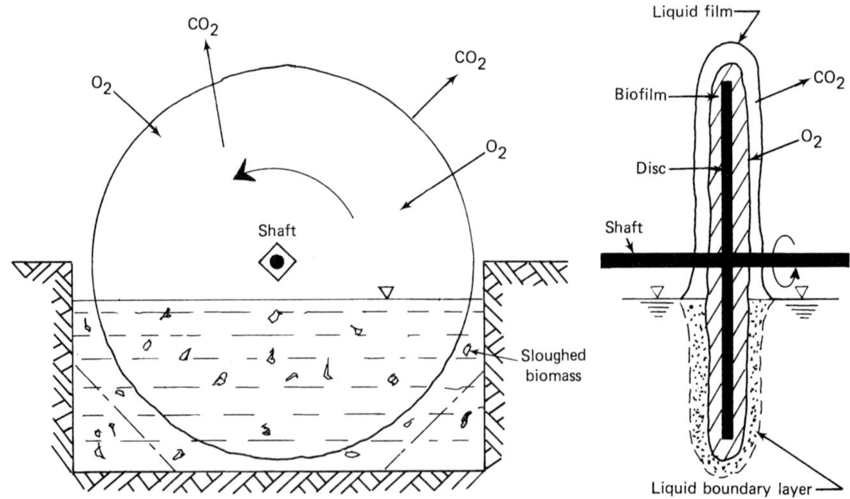


**Figure 6.30** Mechanism for attached growth media in RBC system.

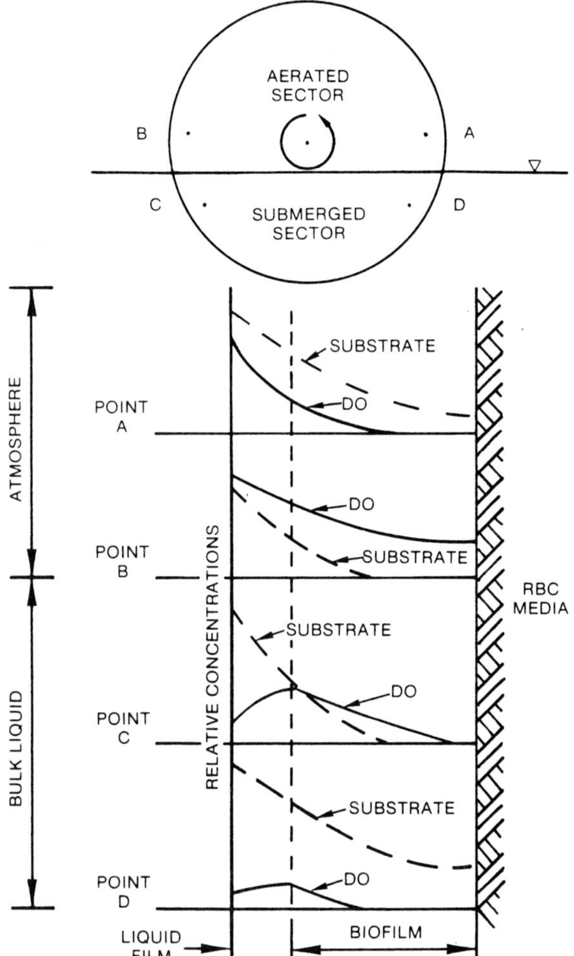


**Figure 6.31** Relative concentration of substrate and dissolved oxygen for a loading condition and RBC rotation speed as a function of media location *(source: US EPA 1984).*

penetrate the liquid film through mixing and diffusion into the biofilm for biological oxidation. Excess DO in the wastewater film is mixed with bulk wastewater in the tank and results in aeration of the wastewater. The rotating media are used for supporting growth of microorganisms and for providing contact between the microorganisms, the substrates, and DO.

A group of RBC units is usually separated by baffling into stages to avoid short circuiting in the tank. There can be one shaft or more in a stage. The hydraulic detention time in each stage is relatively short, on the order of 20 min under normal loading. Each RBC stage tends to operate as a complete-mix reactor. The density and species of microbial population in each stage can vary significantly, depending on wastewater loading conditions. If RBCs are designed for secondary treatment, heavy microbial growth, shaggy and gray in color, develops. Good carbonaceous removal usually occurs in the first and second stages. The succeeding stages can be used for nitrification, if designed, which will exhibit nitrifier growth, brown in color.

The shearing process from rotation exerted on the biomass periodically sloughs off the excess biomass from the media into the wastewater stream. This sloughing action prevents bridging and clogging between adjacent media. The mixing action of the rotating media keeps the sloughed biomass in suspension from settling to the RBC tank. The sloughed-off solids flow from stage to stage and finally into the clarifier following the RBC units. Intermediate clarification and sludge recycle are not necessary for the RBC process.

In comparison with other biological treatment process, the RBC process differs from the trickling filter process by having substantially longer retention time and dynamic rather than stationary media; and from the activated-sludge process by having attached (fixed) biomass rather than a suspended culture and sludge recycle.

The patented Surfact process was created and developed by the Philadelphia Wastewater Department (Nelson and Guarino, 1977). The process uses air-driven RBCs that are partially submerged in the aeration basins of an activated-sludge system. The RBCs provide fixed-film media for biological growth that are present in the recycled activated sludge in the aeration tanks. The results are more active biological coating on the fixed-film media than that which is found on such media when used as a separate secondary treatment. Surfact combines the advantages of both RBC (fixed-film growth) and activated-sludge systems in a single tank, producing additional biological solids in the system. The results can be either a higher treatment efficiency at the same flow rate or the same level of treatment at a higher flow rate.

### 23.3 Advantages

Whether used in a small or a large municipal wastewater treatment plant, the RBC process has provided 85% or more of $BOD_5$ and ammonia nitrogen removal from sewage. In addition to high treatment capacity, it gives good sludge separation. It has the advantages of smaller operation and maintenance costs, and of simplicity in operation. It can be retrofitted easily to existing plants.

The RBC process is similar to the trickling filter (fixed-film biological reactor) and to the activated-sludge process (suspended culture in the mixed liquor). However, the RBC has advantages over the trickling filter process, such as longer contact time (eight to ten times), relatively low land requirement (40% less), and less excavation, more surface area renewal for aeration, greater effectiveness for handling shock loadings, effective sloughing off of the excessive biomass, and without the nuisance of "filter flies." The RBC system may use less power than either the mechanical aeration (activated-sludge) or the trickling filter system of an equivalent capacity. It is anticipated that the RBC would exhibit a more consistent treatment efficiency during the winter months.

Over 95% of biological solids in the RBC units are attached to the media. These result in lower maintenance and power consumption of the RBC process over the activated-sludge process. In comparison with the activated-sludge process, there is no sludge or effluent recycle with a minimum process control requirement. Less skilled personnel are needed to operate the RBC process. A hydraulic surge or organic overloading will cause activated-sludge units to upset in process operation and thereby cause sludge bulking. This is not the case for the RBC process, which has a better process stability. Other advantages of the RBC process over the activated-sludge process are relatively low land requirement and less excavation, more flexibility for upgrading treatment facilities, less expense for nitrification, and better sludge settling without hindered settling.

### 23.4 Disadvantages

The problems of the first generation of RBC units are mainly caused by the failure of hardware and equipment. Significant effort has been made to correct these problems by the manufacturers. The second or third generation of RBC units may perform as designed.

Capital and installation costs for the RBC system, including an overhead structure, will be higher than that for an activated-sludge system of equal capacity. The land area requirement for the RBC process is about 30% to 40% about that for the activated-sludge process.

If low dissolved oxygen is coupled with available sulfide, the nuisance bacteria *Beggiatoa* may grow on the RBC media (Hitdlebaugh and Miller, 1980). The white biomass phenomenon is caused by the *Beggiatoa* propagation. These problems can be corrected by addition of hydrogen peroxide. Some minor disadvantages related to the RBC process as well as other biological treatment processes include that a large land area may be required for a very large facility, additional cost for enclosures, possible foul odor problem, shock loading recovery, extremes of wastewater, pH, *Thiotrix* or *Beggiatoa* growth, overloading, and controversy regarding the technological nature of RBC.

### 23.5 Soluble $BOD_5$

It is accepted that soluble $BOD_5$ (SBOD) is the control parameter for RBC performance. The SBOD can be determined by the BOD test using filtrate from the wastewater samples. For RBC design purposes, it can also be estimated from historical data on total BOD (TBOD) and suspended solids. The SBOD is suspended BOD subtracted from TBOD. Suspended BOD is directly correlated with total suspended solids (TSS). Their relationships are expressed as

$$\text{SBOD} = \text{TBOD} - \text{suspended BOD} \tag{6.144}$$

$$\text{Suspended BOD} = c\,(\text{TSS}) \tag{6.145}$$

$$\text{SBOD} = \text{TBOD} - c\,(\text{TSS}) \tag{6.146}$$

where $c$ = a coefficient
= 0.5 to 0.7 for domestic wastewater
= 0.5 for raw domestic wastewater (TSS > TBOD)
= 0.6 for raw wastewater (TSS ≅ TBOD)
(municipal with commercial and industrial wastewaters)
= 0.6 for primary effluents
= 0.5 for secondary effluents

**Example:** Historical data of the primary effluent show that the average TBOD is 145 mg/L and TSS is 130 mg/L. What is the influent SBOD concentration that can be used for the design of an RBC system? RBC is used as the secondary treatment unit.

**solution:**

For the primary effluent (RBC influent)

$$c = 0.6$$

Estimate SBOD concentration of RBC influent using Eq. (6.146)

$$\begin{aligned} \text{SBOD} &= \text{TBOD} - c\,(\text{TSS}) \\ &= 145\ \text{mg/L} - 0.6\,(130\ \text{mg/L}) \\ &= 67\ \text{mg/L} \end{aligned}$$

### 23.6 RBC process design

Many studies employ either the Monod growth kinetics or Michaelis—Menton enzyme kinetics for modeling organic and nitrogenous removals by the RBC process. Under steady-state conditions and a complete-mix chamber, based on the material mass balance expression for RBC stages, zero-, half-, and first-order kinetics, nonlinear second-order differential equations, conceptional models, and empirical models have been proposed. However, the usefulness of these models to predict RBC performance is not well established.

Factors affecting the RBC performance include wastewater temperature, influent substrate concentration, hydraulic retention time, tank volume to media surface area ratio, media rotational speed, and dissolved oxygen (Poon *et al.*, 1979).

**Manufacturer's empirical design approach.** In the United States, the design of an RBC system for most municipal wastewaters is normally based on empirical curves developed by various manufacturers. Unfortunately the empirical approach is often the least rational in its methodologies and omits many important performance parameters. The designers or the users have relied heavily on the manufacturers for planning and design assistance.

The RBC design aspects for organic removal vary considerably among the various manufacturers. For the design loading, Autotrol (Envirex) (1979) and Clow (1980) use applied $SBOD_5$ while Lyco (1982) uses applied $TBOD_5$. All can predict the water quality at intermediate points in the treatment process and at the effluent. However, the predicted performances are quite different among manufacturers.

The manufacturer's design curves are developed on the basis of observed municipal RBC wastewater treatment performances. For example, Figs. 6.32 and 6.33 show organic removal design curves. The manufacturer's empirical curves define effluent concentration (soluble or total $BOD_5$ or $NH_3$-N) or percent removal in terms of applied hydraulic loading rate or organic loading rate in conjunction with influent substrate concentration.

In design, starting with a desired effluent concentration or percent removal (① in Figs. 6.32 and 6.33) and given influent concentration ②, the hydraulic loading ③ can be selected. The RBC total media-surface area required is calculated by dividing the design flow rate by the hydraulic loading derived.

Work by Antonie (1978) demonstrated that, for any kinetics order higher than zero, overall $BOD_5$ removal for a given media surface area is enhanced by increasing the number of stages. Table 6.15 gives guidelines for stages recommended by three manufacturers.

One to four stages are recommended in most design manuals. A large number of stages are required to achieve higher removal efficiency. The relationship between design curves and stages has not been established and the reasons for recommendations are not generally given. The stage selection is an integral part of the design procedure and should be used intelligently. Based on substrate loading, the percent surface area for the first stage can be determined from the manufacturer's design curves

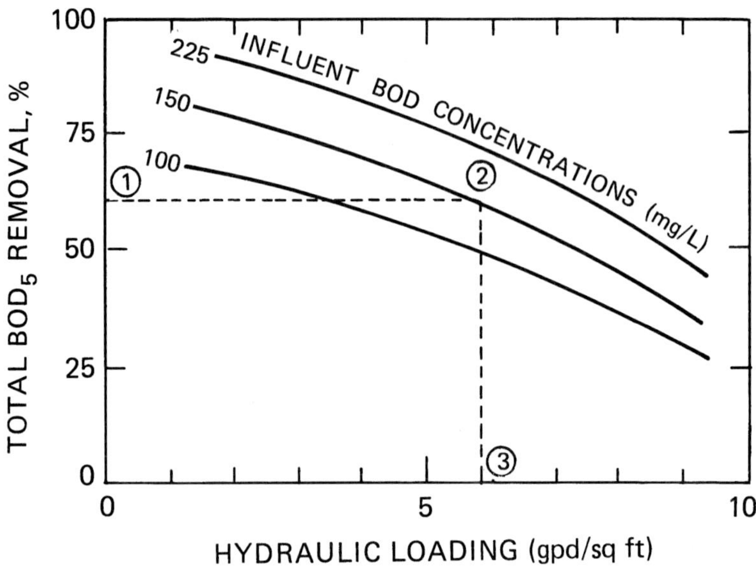


**Figure 6.32** RBC process (mechanical drive) design curves for percent total BOD removal (temperature > 13°C, 55°F).

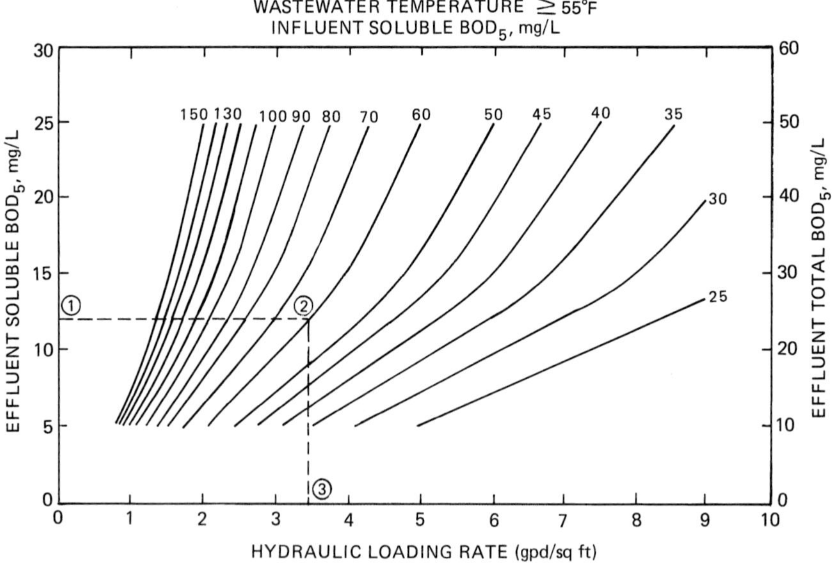


**Figure 6.33** RBC (mechanical drive) process design curves for total and soluble BOD removal ($T$ > 13°C, 55°F).

**TABLE 6.15 RBC Stage Recommendations**

| Autotrol (1979) | | | Lyco (1980) | |
|---|---|---|---|---|
| Target effluent $SBOD_5$, mg/L | Min. number of stages | Clow (1980) | Target $TBOD_5$ reduction | Number of stages |
| >25 | 1 | At least 4 stages per train | up to 40% | 1 |
| 15–25 | 1 or 2 | | 35–65% | 2 |
| 10–15 | 2 or 3 | | 60–85% | 3 |
| <10 | 3 or 4 | | 80–95% | 4 |
| | | | Minimum of four stages recommended for combined $BOD_5$ and $NH_3$-N removal | |

(not included). Typical design information for RBC used for various treatment levels is summarized in Table 6.16.

To avoid organic overloading and possible growth of nuisance organisms, and shaft overloads, limitation of organic loadings has been recommended as follows (Autotrol, 1979):

**TABLE 6.16 Typical Design Parameters for Rotating Biological Contactors Used for Various Treatment Levels**

| Parameter | Secondary | Combined nitrification | Separate nitrification |
|---|---|---|---|
| Hydraulic loading | | | |
| gal/(d · $ft^2$) | 2.0–40 | 0.75–2.0 | 1.0–2.5 |
| $m^3/(m^2 \cdot d)$ | 0.081–0.163 | 0.030–0.081 | 0.041–0.102 |
| Soluble $BOD_5$ loading | | | |
| lb/(1000 $ft^2$ · d) | 0.75–2.0 | 0.5–1.5 | 0.1–0.3 |
| Total (T) $BOD_5$ loading | | | |
| lb/(1000 $ft^2$ · d) | 2.0–3.5 | 1.5–3.0 | 0.2–0.6 |
| Maximum loading on first stage | | | |
| lb SBOD/4–6 (1000 $ft^2$ · d) | 4–6 | | |
| lb TBOD/ (1000 $ft^2$ · d) | 8–12 | 8–12 | |
| $NH_3$-N loading | | | |
| lb/(1000 $ft^2$ · d) | | 0.15–0.3 | 0.2–0.4 |
| Hydraulic retention time $\theta$, h | 0.7–1.5 | 1.5–4 | 1.2–2.9 |
| Effluent $BOD_5$, mg/L | 15–30 | 7–15 | 7–15 |
| Effluent $NH_3$-N, mg/L | | < 2 | 1–2 |

*Note*: lb/(1000 $ft^2$ · d) × 0.00488 = kg/($m^2$ · d)
SOURCE: Metcalf and Eddy Inc. (1991)

| Media | Organic ($SBOD_5$) loading, lb/(d · 1000 ft$^2$) Mechanical drive | Air drive |
|---|---|---|
| Standard density | 4.0 | 5.0 |
| High density | 1.7 | 2.5 |

*Note*: lb/(d · 1000 ft$^2$) × 0.00488 = kg/(d · m$^2$)

Then media distribution and choice of configurations can be made with engineering judgment or according to the manufacturer's design manual. Most RBC manufacturers have a standard ratio of tank volume to media surface area at 1.2 gal/ft$^2$ (48.9 L/m$^2$).

The manufacturers universally contend that wastewater temperatures above 55°F (12.8°C) do not affect the organic removal rate. Below 55°F varying degrees of decreased biological activity rate are predicted by the manufacturers. Surface area correction factors for wastewater treatment below 55°F are shown in Table 6.17. Most manufacturers require wastewater temperature not to exceed 90°F (32.2°C) which can damage the RBC media. Fortunately, this is normally not the case for municipal wastewaters.

In summary, an RBC system can be designed on the basis of information on total required media surface area, temperature correction, staging, and organic loading limits.

RBC performance histories relative to empirical design have been examined. Operational data indicated that the prediction results from manufacturer's design curves were considerably different from actual results

**TABLE 6.17 Factors for Required Surface Area Correction to Temperature**

| Temperature °F | Temperature °C | For soluble $BOD_5$ removal | For ammonia-nitrogen removal |
|---|---|---|---|
| 64 | 17.8 | 1.0 | 0.71 |
| 62 | 16.7 | 1.0 | 0.77 |
| 60 | 15.5 | 1.0 | 0.82 |
| 58 | 14.4 | 1.0 | 0.89 |
| 56 | 13.3 | 1.0 | 0.98 |
| 55 | 12.8 | 1.0 | 1.00 |
| 54 | 12.2 | 1.03 | 1.02 |
| 52 | 11.1 | 1.09 | 1.14 |
| 50 | 10.0 | 1.15 | 1.28 |
| 48 | 8.9 | 1.21 | 1.40 |
| 46 | 7.8 | 1.27 | 1.63 |
| 45 | 7.2 | 1.31 | 1.75 |
| 44 | 6.7 | 1.34 | 1.85 |
| 42 | 5.6 | 1.42 | 2.32 |
| 40 | 4.4 | 1.50 | – |

(Opatken 1982, Brenner *et al.*, 1984). They usually gave an overestimate of attainable removals. The manufacturer's design curves should be used with caution whenever first-stage loading exceeds the recommended values.

**Example 1:** At the first stage of the RBC system, what is the maximum hydraulic loading if the total $BOD_5$ for the influent and effluent are 150 and 60 mg/L, respectively?

**solution:**

Step 1. Calculate percent TBOD removal

$$\% \text{ removal} = \frac{(150 - 60) \text{ mg/L} \times 100\%}{150 \text{ mg/L}}$$

$$= 60\%$$

Step 2. Determine hydraulic loading limit

From Fig. 6.32, starting with 60% TBOD removal at point 1, draw a horizontal line to the intersect with the influent TBOD concentration of 150 mg/L at point 2. Next, starting at point 2, draw a vertical line to intersect the hydraulic loading axis at point 3. Read the value at point 3 as 5.8 gal/(d · $ft^2$). This is the answer.

**Example 2:** Compute the hydraulic and organic loading rates of an RBC system and of the first stage. Given:

Primary effluent (influent) flow = 1.5 Mgal/d

Influent total $BOD_5$ = 140 mg/L

Influent $SBOD_5$ = 75 mg/L

Area of each RBC shaft = 100,000 $ft^2$

Number of RBC shafts = 6

Number of trains = 2 (3 stages for each train)

No recirculation

**solution:**

Step 1. Compute the overall system hydraulic loading (HL)

$$\text{Surface area of total system} = 6 \times 100{,}000 \text{ ft}^2$$

$$= 600{,}000 \text{ ft}^2$$

$$\text{System HL} = \frac{1{,}500{,}000 \text{ gal/d}}{600{,}000 \text{ ft}^2}$$

$$= 2.5 \text{ gal/(d} \cdot \text{ft}^2)$$

*Note*: It is in the range of 2 to 4 gal/(d · $ft^2$).

Step 2. Compute the organic loadings of the overall RBC system

$$\text{Total BOD loading} = \frac{1.5\ \text{Mgal/d} \times 140\ \text{mg/L} \times 8.34\ \text{lb/(Mgal.mg/L)}}{600 \times 1000\ \text{ft}^2}$$

$$= 2.92/\text{lb}(1000\ \text{ft}^2 \cdot \text{d})$$

$$= 14.2\ \text{g/(m}^2 \cdot \text{d)}$$

*Note*: Design range = 2 to 3.5 lb/1000 $\text{ft}^2$.

$$\text{Soluble BOD loading} = \frac{1.5 \times 75 \times 8.34}{600 \times 1000}$$

$$= 1.56 \cdot \text{lb}(1000\ \text{ft}^2 \cdot \text{d})$$

$$= 7.6\ \text{g/(m}^2 \cdot \text{d)}$$

*Note*: Design range = 0.75 to 2.0 lb/1000 $\text{ft}^2$.

Step 3. Compute the organic loadings of the first stage

Two units are used as the first stage.

$$\text{Total BOD loading} = \frac{1.5 \times 140 \times 8.34}{2 \times 100{,}000}$$

$$= 8.76\ [\text{lb/(1000ft}^2 \cdot \text{d)}]$$

$$\text{Soluble BOD loading} = \frac{1.5 \times 75 \times 8.34}{200 \times 1000}$$

$$= 4.69\ [\text{lb/(1000ft}^2 \cdot \text{d)}]$$

**Example 3:** A municipal primary effluent with total and soluble $BOD_5$ of 140 and 60 mg/L, respectively, is to be treated with an RBC system. The RBC effluent TBOD and SBOD is to be 24 and 12 mg/L, respectively, or less. The temperature of wastewater is 60°F (above 55°F; no temperature correction needed). The design plant flow is 4.0 Mgal/d (15,140 $\text{m}^3$/d). The maximum hourly flow is 10.0 Mgal/d (37,850 $\text{m}^3$/d).

**solution:**

Step 1. Determine the allowable hydraulic loading rate

From Fig. 6.33, at effluent

$$\text{SBOD} = 12\ \text{mg/L with}$$

$$\text{Hydraulic loading} = 3.44\ \text{gal/(d} \cdot \text{ft}^2)$$

Step 2. Compute the required surface area of the RBC

$$\text{Area} = \frac{4{,}000{,}000 \text{ gal/d}}{3.44 \text{ gal/(d} \cdot \text{ft}^2)}$$

$$= 1{,}160{,}000 \text{ ft}^2$$

Step 3. Check the design for overall organic loading rate

$$\text{SBOD loading} = \frac{4 \text{ Mgal/d} \times 60 \text{ mg/L} \times 8.34 \text{ lb/(Mgal} \cdot \text{mg/L)}}{1.160{,}000 \text{ ft}^2}$$

$$= 1.73 \text{ [lb/(1000 ft}^2 \cdot \text{d)]}$$

It is under 2.0 [lb/(1000 ft$^2$ · d)], therefore OK.

Step 4. Size the first stage

The size factor for the area required for the first stage is 0.2 × SBOD loading. For this case, the size factor is 0.346 (0.2 × 1.73, or 34.6% of total surface area). This value can be obtained from the manufacturer's curves.

$$\text{Surface area for first stage} = 0.346 \times 1{,}160{,}000 \text{ ft}^2$$

$$\cong 400{,}000 \text{ ft}^2$$

Step 5. Determine configuration

(a) Choice standard media assemblies each having 100,000 ft$^2$ for the first stage

$$\text{Unit for first stage} = 400{,}000 \text{ ft}^2/100{,}000 \text{ ft}^2$$

$$= 4$$

Use 4 standard media; each is installed at the first stage of 4 trains.

(b) Choice Hi-Density media assemblies have 150,000 ft$^2$ for the following stages

$$\text{High-Density media area} = (1{,}160{,}000 - 400{,}000) \text{ ft}^2$$

$$= 760,\ 000 \text{ ft}^2$$

$$\text{No. of Hi-Density assemblies} = 760{,}000 \text{ ft}^2/150{,}000 \text{ ft}^2$$

$$= 5.1$$

Alternative 1: Use 4 standard media at stages 2 and 3.
Alternative 2: Use 4 Hi-density media at stages 2 and 4 standard density media at stage 3
Alternative 1 is less costly.
Alternative 2 has an advantage in preventing shock loadings.

(c) Possible configurations are designed using 4 trains with 3 stages each, as shown below:

(1) SSSS + SSSS + SSSS

or

(2) SSSS + HHHH + SSSS

*Note*: S = standard density media; H = high density media.

Step 6. Determine the required surface area for the secondary settling tank

(a) Compute surface area based on average (design) flow using an overflow rate of 600 gal/(d · ft$^2$)

$$\text{Area} = 4{,}000{,}000 \text{ (gal/d)}/600 \text{ (gal/(d} \cdot \text{ft}^2\text{)}$$
$$= 6{,}670 \text{ ft}^2$$

(b) Compute surface area based on peak flow using an overflow rate of 1200 gal/(d · ft$^2$)

$$\text{Area} = 10{,}000{,}000 \text{ (gal/d)}/1200 \text{ (gal/(d} \cdot \text{ft}^2\text{))}$$
$$= 8330 \text{ ft}^2$$

(c) On the basis of the above calculations, the size of the secondary settling tank is controlled by the maximum hourly flow rate.

## 24 Dual Biological Treatment

Dual biological treatment processes use a fixed film reactor in series with a suspended growth reactor. Dual processes include such as activated biofilter, trickling filter-solids contact, roughing filter-activated sludge, biofilter-activated sludge, trickling filter-activated sludge, roughing filter-RBC, roughing filter-aerated lagoon, roughing filter-facultative lagoon, roughing filter-pure oxygen activated sludge. Descriptions of these dual processes may be found elsewhere (Metcalf and Eddy, Inc. 1991; WEF and ASCE, 1991a).

## 25 Stabilization Ponds

The terms stabilization pond, lagoon, oxidation pond have been used synonymously. This is a relatively shallow earthen basin used as secondary or tertiary wastewater treatment, especially in rural areas. Stabilization ponds have been employed for treatment of wastewater for over 300 years. Ponds have become popular in small communities because of their low construction and operating costs. They are used to treat a variety of wastewaters from domestic and industrial wastes and functions under a wide range of weather conditions. Ponds can be employed alone or in combination with other treatment processes.

Stabilization ponds can be classified as facultative (aerobic-anaerobic), aerated, aerobic, and anaerobic ponds according to the dominant type of biological activity or reactions occurring in the pond. Other classifications can be based on the types of the influent (untreated, screened, settled effluent, or secondary (activated-sludge) effluent), the duration of discharge (nonexistent, intermittent, and continuous), and the method of oxygenation (photosynthesis, atmospheric surface reaeration, and mechanical aeration). Aerated ponds (aerated lagoons) have been discussed previously.

### 25.1 Facultative ponds

The most common type of stabilization pond is the facultative pond. It is also called the wastewater lagoon. Facultative ponds are usually 1.2 to 2.5 m (4 to 8 ft) in depth, with an aerobic layer overlying an anaerobic layer, often containing sludge deposits. The detention time is usually 5 to 30 days (US EPA, 1983b).

The ponds commonly receive no more pretreatment than screening (few with primary effluent). It can also be used to follow trickling filters, aerated ponds, or anaerobic ponds. They then store grit and heavy solids in the first or primary ponds to form an anaerobic layer. The system is a symbiotic relationship between heterotrophic bacteria and algae.

Bacteria found in an aerobic zone of a stabilization pond are primarily of the same type as those found in an activated-sludge process or in the zoogleal mass of a trickling filter. The most frequently isolated bacteria include *Beggiatoa alba, Sphaerotilus natans, Achromobacter, Alcaligenes, Flavobacterium, Pseudomonas,* and *Zoogloea* spp. (Lynch and Poole, 1979). These organisms decompose the organic materials present in the aerobic zone into oxidized end products.

Organic matter in wastewater is decomposed by bacterial activities, including both aerobic and anaerobic, which release phophorus, nitrogen, and carbon dioxide. Oxygen in the aerobic layer is supplied by surface reaeration and algal photosynthesis. Algae consume nutrient and carbon dioxide produced by bacteria and release oxygen to water. DO is used by bacteria, thus forming a symbiotic cycle. In the pond bottom, anaerobic breakdown of the solids in the sludge layer produces dissolved organics and gases such as methane, carbon dioxide, and hydrogen sulfide. Between the aerobic and anaerobic zones, there is a zone called the facultative zone. Temperature is a major factor for the biological symbiotic activities.

Organic loading rates on stabilization ponds are expressed in terms of kg $BOD_5$ applied per hectare of surface area per day (lb BOD/(acre · d)), or sometimes as BOD equivalent population per unit area. Typical organic loading rates are 22 to 67 kg BOD/(ha · d) (20 to 60 lb BOD/(a · d)). Typical detention times range from 25 to 180 days. Typical dimensions are 1.2 to 2.5 m (4 to 8 ft) deep with 4 to 60 ha (10 to 150 acres) of surface area (US EPA, 1983b).

Facultative ponds are commonly designed to reduce BOD to about 30 mg/L; but, in practice, it ranges from 30 to 40 mg/L or greater due to algae. Volatile organic removal is between 77% and 96%. Nitrogen removal achieves 40% to 95%. Less phosphorus removal is observed, being less than 40%. Effluent TSS levels range from 40 to 100 mg/L, mainly contributed by algae (WEF and ASCE, 1991b). The presence of algae in pond effluent is one of the most serious performance problems associated with facultative ponds. The ponds are effective in removal of fecal coliform (FC) due to their dying off. In most cases effluent FC densities are less than the limit of 200 FC/100 ml.

**Process design.** Several design formulas with operational data were presented in the design manual (US EPA, 1974b). Calculations of the size of a facultative pond are also illustrated for the areal loading rate procedure, Gloyna equation, Marais–Shaw equation, plug-flow model, and Wehner–Wilhelm equation. The design for partial-mix aerated lagoon is given elsewhere (WPCF, 1990). In this section, design methods of the areal loading rate method and the Wehner and Wilhelm model are discussed.

**Areal loading rate method.** The design procedure is usually based on organic loading rate and hydraulic residence time. Several empirical and rational models for the design of facultative ponds have been proposed. Several proposed design methods have been discussed elsewhere (US EPA, 1983b). The areal loading rate is the most conservative design method and can be adapted to specific standards. The recommended BOD loading rates based on average winter air temperature are given in Table 6.18 (US EPA , 1974b).

The surface area required for the facultative pond is determined by dividing the organic (BOD) load by the appropriate BOD loading rate listed in Table 6.18 or from the specific state standards. It can be

**TABLE 6.18 Recommended $BOD_5$ Loading Rates for Facultative Ponds**

| Average winter air temperature, °C | Water depth | | BOD loading rate | |
|---|---|---|---|---|
| | m | ft | kg/(ha · d) | lb/(acre · d) |
| <0 | 1.5–2.1 | 5–7 | 11–22 | 10–20 |
| 0–15 (59°F) | 1.2–1.8 | 4–6 | 22–45 | 20–40 |
| >15 | 1.1 | 3.6 | 45–90 | 40–80 |

SOURCE: US EPA (1974b)

expressed as the following equation (as used for other processes):

$$A = \frac{Q(\text{BOD})}{(\text{LR})(1000)} \qquad \text{(SI units)} \tag{6.147}$$

$$A = \frac{Q(\text{BOD})(8.34)}{(\text{LR})} \qquad \text{(British units)} \tag{6.148}$$

where $A$ = area required for facultative pond, ha or acre
BOD = BOD concentration in influent, mg/L
$Q$ = flow of influent, m$^3$/d or Mgal/d
LR = BOD loading rate for average winter air temperature, kg/(ha · d) or lb/(acre · d)
1000 = conversion factor, 1000 g = 1 kg
8.34 = conversion factor, 1 Mgal · mg/L = 8.34 lb

The BOD loading rate at the first cell in a series of cells (ponds) should not exceed 100 kg/(ha · d) (90 lb/(acre · d)) for warm climates, average winter air temperature greater than 15°C (59°F), and 40 kg/(ha · d) (36 lb/(acre · d)) for average winter air temperature less than 0°C (32°F).

**Example 1:** The design flow of facultative ponds for a small town is 1100 m$^3$/d (0.29 Mgal/d). The expected influent BOD is 210 mg/L. The average winter temperature is 10°C (50°F). Design a three-cell system with organic loading less than 80 kg/(ha · d) (72 lb/(acre · d)) in the primary cell. Also estimate the hydraulic detention time when average sludge depth is 0.5 m and there are seepage and evaporation losses of 2.0 mm of water per day.

**solution:**

Step 1. Determine area required for the total ponds

From Table 6.18, choose BOD loading rate LR = 38 kg/(ha · d) (33.9 lb/(acre · d)), because mean air temperature is 10°C. Using Eq. (6.147)

$$A = \frac{Q(\text{BOD})}{(\text{LR})(1000)} = \frac{1100 \text{ m}^3/\text{d} \times 210 \text{ g/m}^3}{38 \text{ kg/(ha} \cdot \text{d)} \times 1000 \text{ g/kg}}$$

$$= 6.08 \text{ ha} \,(16.6 \text{ areas})$$

$$= 60{,}800 \text{ m}^2$$

Step 2. Determine the area required for the primary cell (use LR = 80 kg/(ha · d))

$$\text{Area} = \frac{1100 \times 210}{80 \times 1000}$$

$$= 2.80 \,(\text{ha})$$

Step 3. Sizing for 3 cells

Referring to Table 6.18, choose water depth of 1.5 m (5 ft) for all cells.

(a) Primary cell:

$$\text{Area} = 2.88 \text{ ha} = 28{,}800 \text{ m}^2$$

Use 100-m (328-ft) wide, 288-m (945-ft) long and 1.5-m (5-ft) deep pond.

(b) Two other cells:

$$\text{Area for each} = (60{,}800 - 28{,}800) \text{ m}^2/2$$
$$= 16{,}000 \text{ m}^2$$

Choose 144 m in length, then the width = 111 m

The pond arrangements are as shown below.

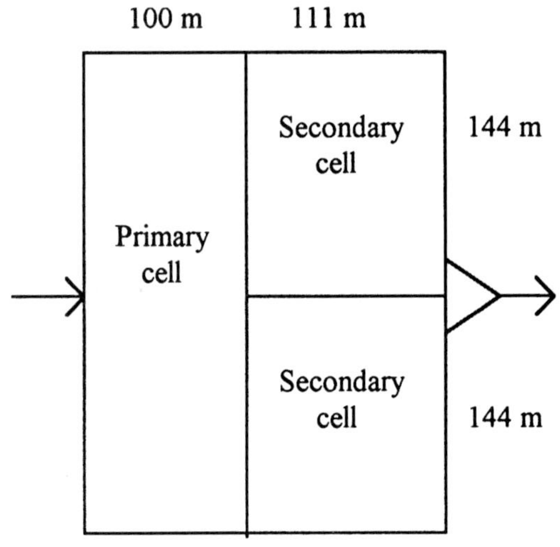


Step 4. Estimate the hydraulic retention time

(a) Calculate the storage volume $V$ (when average sludge depth = 0.5 m)

$$V = (1.5 \text{ m} - 0.5 \text{ m}) \times 60{,}800 \text{ m}^2$$
$$= 60{,}800 \text{ m}^3$$

(b) Calculate the water loss $V'$

$$V' = 0.002 \text{ m/d} \times 60{,}800 \text{ m}^2$$
$$= 122 \text{ m}^3\text{/d}$$

(c) Calculate the HRT storage time

$$\text{HRT} = \frac{V}{Q - V'} = \frac{60{,}800 \text{ m}^2}{1100 \text{ m}^3\text{/d} - 122 \text{ m}^3\text{/d}}$$
$$= 62 \text{ days}$$

**Example 2:** A lagoon system contains two cells in series. Each cell is 110 m × 220 m (360 ft × 720 ft) in size with a maximum operating depth of 1.64 m (5.4 ft) and a minimum operating depth of 0.55 m. (1.8 ft). The wastewater flow is 950 $m^3$/d (0.25 MGD) with an average $BOD_5$ of 105 mg/L. Determine the organic loading rate and the detention time of the lagoon. The temperature ranges 10 to 15°C.

**solution:**

Step 1. Calculate the $BOD_5$ loading rate

$$\text{BOD}_5 \text{ mass} = 950 \text{ m}^3/\text{d} \times 105 \text{ mg/L} \times 1000 \text{ L/m}^3 \times 10^{-6} \text{ kg/mg}$$
$$= 99.75 \text{ kg/d}$$

$$\text{Area of the primary cell} = 110 \text{ m} \times 220 \text{ m}$$
$$= 24{,}200 \text{ m}^2 \times 10^{-4} \text{ ha/m}^2$$
$$= 2.42 \text{ ha}$$

$$\text{BOD}_5 \text{ loading rate} = (99.75 \text{ kg/d})/(2.42 \text{ ha})$$
$$= 41.2 \text{ kg/(ha} \cdot \text{d)}$$

This loading rate is acceptable by checking with Table 6.18, at temperature >15°C, the recommended $BOD_5$ loading rate ranges 22 to 45 kg/(ha · d).

Step 2. Calculate the detention time

$$\text{The detention time} = 24{,}000 \text{ m}^2/\text{cell} \times (1.64 \text{ m} - 0.55 \text{ m}) \times 2 \text{ cells} \div 950 \text{ m}^3/\text{d}$$
$$= 55.5 \text{ days}$$

**Wehner and Wilhelm equation.** Wehner and Wilhelm (1958) used the first-order substrate removal rate equation for a reactor with an arbitrary flow-through pattern, which is between a plug-flow pattern and a complete-mix pattern. Their proposed equation is

$$\frac{C}{C_0} = \frac{4a\exp(1/2D)}{(1+a)^2\exp(a/2D) - (1-a)^2\exp(-a/2D)} \tag{6.149}$$

where $C$ = effluent substrate concentration, mg/L
$C_0$ = influent substrate concentration, mg/L
$a = \sqrt{1 + 4ktD}$
$k$ = first-order reaction constant, 1/h
$t$ = detention time, h
$D$ = dispersion factor $= H/uL$
$H$ = axial dispersion coefficient, $m^2$/h or $ft^2$/h
$u$ = fluid velocity, m/h or ft/h
$L$ = length of travel path of a typical particle, m or ft

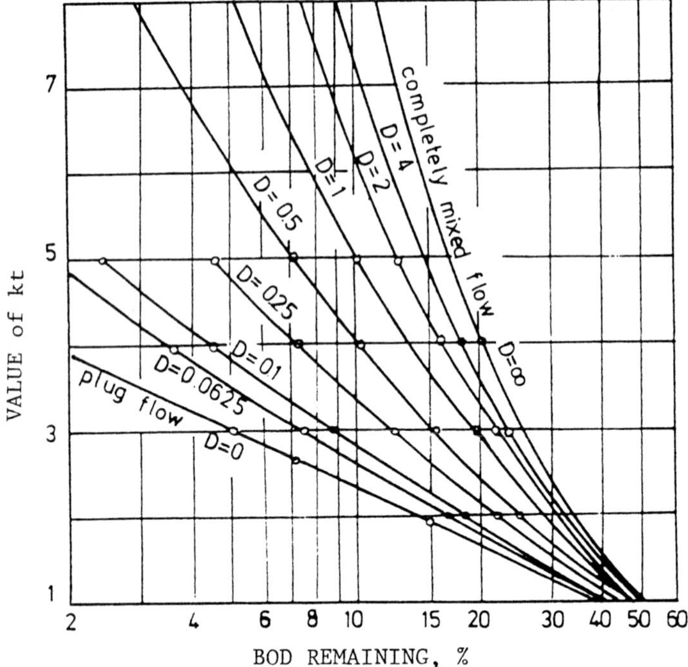


**Figure 6.34** Relationship between kt values and percent BOD remaining for various dispersions factors, by the Wehner-Wilhelm equation.

The Wehner-Wilhelm equation for arbitrary flow was proposed by Thirumurthi (1969) as a method of designing facultative pond systems. Thirumurthi developed a graph, shown in Fig. 6.34, to facilitate the use of the equation. In Fig. 6.34, the term $kt$ is plotted against the percent $BOD_5$ remaining ($C/C_0$) in the effluent for dispersion factors varying from zero for an ideal plug-flow reactor to infinity for a complete-mix reactor. Dispersion factors for stabilization ponds range from 0.1 to 2.0, with most values not exceeding 1.0 due to the mixing requirement. Typical values for the overall first-order $BOD_5$ removal rate constant $k$ vary from 0.05 to 1.0 per day, depending on the operating and hydraulic characteristics of the pond. The use of the arbitrary flow equation is complicated in selecting $k$ and $D$ values. A value of 0.15 per day is recommended for $k_{20}$ (US EPA, 1984). Temperature adjustment for $k$ can be determined by

$$k_T = k_{20}(1.09)^{T-20} \tag{6.150}$$

where $k_T$ = reaction rate at minimum operating water temperature T, per day

$k_{20}$ = reaction rate at 20°C, per day

$T$ = minimum operating temperature, °C

**Example:** Design a facultative pond system using the Wehner-Wilhelm model and Thirumurthi application with the following given data.

Design flow rate $Q$ = 1100 $m^3$/d (0.29 Mgal/d)

Influent TSS = 220 mg/L

Influent $BOD_5$ = 210 mg/L

Effluent $BOD_5$ = 30 mg/L

Overall first-order $k$ at 20°C = 0.22 per day

Pond dispersion factor $D$ = 0.5

Water temperature at critical period = 1°C

Pond depth = 2 m (6.6 ft)

Effective depth = 1.5 m (5 ft)

**solution:**

Step 1. Calculate the percentage of BOD remaining in the effluent

$$\begin{aligned} C/C_0 &= 30 \text{ mg/L} \times 100\%/(210 \text{ mg/L}) \\ &= 14.3\% \end{aligned}$$

Step 2. Calculate the temperature adjustment for $k_{20}$ using Eq. (6.150)

$$\begin{aligned} k_T &= k_{20}(1.09)^{T-20} = 0.22\,(1.09)^{1-20} \\ &= 0.043 \text{ (per day)} \end{aligned}$$

Step 3. Determine the value of $k_T t$ from Fig. 6.34

$$\text{At } C/C_0 = 14.3\% \text{ and } D = 0.5$$

$$k_T t = 3.1$$

Step 4. Calculate the detention time for the critical period of year

$$\begin{aligned} t &= 3.1/(0.043 \text{ d}^{-1}) \\ &= 72 \text{ days} \end{aligned}$$

Step 5. Determine the pond volume and surface area requirements

$$\begin{aligned} \text{Volume} &= Qt = 1100 \text{ m}^3\text{/d} \times 72 \text{ days} \\ &= 79{,}200 \text{ m}^3 \\ \text{Area} &= \text{volume/effective depth} = 79{,}200 \text{ m}^3/1.5 \text{ m} \\ &= 52{,}800 \text{ m}^2 \\ &= 5.28 \text{ ha} \\ &= 13.0 \text{ acres} \end{aligned}$$

Step 6. Check $BOD_5$ loading rate

$$\text{Loading} = \frac{1100 \text{ m}^3/\text{d} \times 210 \text{ g/m}^3}{5.28 \text{ ha} \times 1000 \text{ g/kg}}$$

$$= 43.8 \text{ kg/(ha} \cdot \text{d)}$$

$$= 40.0 \text{ lb(acre} \cdot \text{d)}$$

Step 7. Determine the power requirement for the surface aerators

Assume that the oxygen transfer capacity of the aerators is twice the value of the BOD applied per day and that a typical aerator has a transfer capacity of 22 kg $O_2$/(hp · d)

$$O_2 \text{ required} = 2 \times 1100 \text{ m}^3/\text{d} \times 210(\text{g/m}^3)/(1000 \text{ g/kg})$$

$$= 462 \text{ kg/d}$$

$$\text{Power} = (462 \text{ kg/d})/(22 \text{ kg/hp} \cdot \text{d})$$

$$= 21.0 \text{ hp}$$

$$= 15.7 \text{ kW}$$

Use seven 3-hp units

Step 8. Check the power input to determine the degree of mixing

$$\text{Power input} = 15.7 \text{ kW}/79.2 \times 1000 \text{ m}^3$$

$$= 0.20 \text{ kW}/1000 \text{ m}^3$$

$$= 0.0076 \text{ hp}/1000 \text{ ft}^3$$

*Note*: In practice, the minimum power input is about 28 to 54 kW/1000 $m^3$ (1.06 to 2.05 hp/1000 $ft^3$) for mixing (Metcalf and Eddy, Inc. 1991).

**Tertiary ponds.** Tertiary ponds, also called polishing or maturation ponds, serve as the third-stage process for effluent from activated-sludge or trickling filter secondary clarifier effluent. They are also used as a second stage following facultative ponds and are aerobic throughout their depth.

The water depth of the tertiary pond is usually 1 to1.5 m (3 to 4.5 ft). BOD loading rates are less than 17 kg/(ha · d) (15 lb/(acre · d)). Detention times are relatively short and vary from 4 to 15 days.

### 25.2 Aerobic ponds

Aerobic ponds, also referred to as high-rate aerobic ponds, are relatively shallow with usual depths ranging from 0.3 to 0.6 m (1 to 2 ft), allowing light to penetrate the full depth. The ponds maintain DO throughout their entire depth. Mixing is often provided to expose all algae to sunlight and to prevent deposition and subsequent anaerobic conditions. DO is supplied by algal photosynthesis and surface reaeration. Aerobic bacteria utilize and stabilize the organic waste. The HRT in the ponds is short—3 to 5 days.

The use of aerobic ponds is limited to warm and sunny climates, especially where a high degree of BOD removal is required but the land area is not limited. However, very little coliform die-off occurs.

### 25.3 Anaerobic ponds

Anaerobic ponds are usually deep and are subjected to heavy organic loading. There is no aerobic zone in an anaerobic pond. The depths of anaerobic ponds usually range from 2.5 to 5 m (8 to 16 ft). The detention times are 20 to 50 days (US EPA, 1983b).

Anaerobic bacteria decompose organic matter to carbon dioxide and methane. The principal biological reactions are acid formation and methane fermentation. These processes are similar to those of anaerobic digestion of sludge. Odorous compounds, such as organic acids and hydrogen sulfide, are also produced.

Anaerobic ponds are usually used to treat strong industrial and agricultural wastes. They have been used as pretreatment to facultative or aerobic ponds for strong industrial wastewaters and for rural communities with high organic load, such as food processing. They are not in wide application for municipal wastewater treatment.

The advantages of anaerobic ponds compared with an aerobic treatment process are low production of waste biological sludge and no need for aeration equipment. An important disadvantage of the anaerobic pond is the generation of odorous compounds. Its incomplete stabilization of waste requires a second-stage aerobic process. It requires a relatively high temperature for anaerobic decomposition of wastes.

Normal operation, to achieve a BOD removal efficiency of at least 75%, entails a loading rate of 0.32 kg B0D/($m^3 \cdot d$) (20 lb/(1000 $ft^3 \cdot d$)), a minimum detention time of 4 days, and a minimum operating temperature of 24°C (75°F) (Hummer, 1986).

## 26 Secondary Clarifier

The secondary settling tank is an integral part of both suspended- and attached-growth biological treatment processes. It is vital to the operation and performance of the activated-sludge process. The clarifier separates MLSS from the treated wastewater prior to discharge. It thickens the MLSS before their turn in the aeration tank or their wasting.

Secondary clarifiers are generally classified as Type 3 settling. Types 1 and 2 may also occur. The sloughed-off solids are commonly well-oxidized particles that settle readily. On the other hand, greater viability of activated sludge results in lighter, more buoyant flocs, with reduced settling velocities. In addition, this is the result of microbial production of gas bubbles that buoy up the tiny biological clusters.

### 26.1 Basin sizing for attached-growth biological treatment effluent

The Ten States Standards (GLUMRB, 1996) recommends that surface overflow rate for the settling tank following trickling filter should not exceed 1200 gal/(d · ft$^2$) (49 m$^3$/(m$^3$ · d)) based on design peak hourly flow. In practice, typical overflow rates are 600 gal/(d · ft$^2$) (24.4 m$^3$/(m$^2$ · d)) for plants smaller than 1 Mgal/d (3785 m$^3$/d) and 800 gal/(d · ft$^2$) (33 m$^3$/(m$^2$ · d)) for larger plants. The minimum side water depth is 10 ft (3 m) with greater depths for larger diameter basins. The retention time in the secondary settling tank is in the range of 2 to 3 h. The maximum recommended weir loading rate is 20,000 gal/(d · ft) (250 m$^3$/(d · m)) for plants equal or less than 1 Mgal/d and 30,000 gal/(d · ft) (373 m$^3$/(d · m)) for plants greater than 1 Mgal/d (3785 m$^3$/d).

Secondary clarifiers following trickling filters are usually sized on the basis of the hydraulic loading rate. Typical design parameters for secondary clarifiers are listed in Table 6.19 (US EPA, 1975a). Referring to the hydraulic loading rates from Table 6.20, sizing should be calculated for both peak and design average flow conditions and the largest value calculated should be used. At the selected hydraulic loading rates, settled effluent quality is limited primarily by the performance of the biological reactor, not of the settling basins. Solids loading limits are not involved in the size of clarifiers following trickling filters. Where further treatment follows the clarifier, cost optimization may be considered in sizing the settling basins.

**TABLE 6.19 Typical Design Parameters for Secondary Sedimentation Tanks**

| | Hydraulic loading gal/(d · ft$^2$)† | | Solids loading,* lb solids/(d · ft$^2$)† | | |
|---|---|---|---|---|---|
| Type of treatment | Average | Peak | Average | Peak | Depth, ft |
| Settling following tracking filtration | 400–600 | 1000–2000 | – | – | 10–12 |
| Settling following air activated sludge (excluding extended aeration) | 400–800 | 1000–1200 | 20–30 | 50 | 12–15 |
| Settling following extended aeration | 200–400 | 800 | 20–30 | 50 | 12–15 |
| Settling following oxygen activated sludge with primary settling | 400–800 | 1000–1200 | 25–35 | 50 | 12–15 |

* Allowable solids loading are generally governed by sludge thickening characteristics associated with cold weather operations.

† gal/(d · ft$^2$) × 0.0407 = m$^3$/(m$^2$ · d)

lb/(d · ft$^2$) × 4.8827 = kg/(d · m$^2$)

SOURCE: US EPA (1975a)

**TABLE 6.20 Recommended Design Overflow Rate and Peak Solids Loading Rate for Secondary Settling Tanks Following Activated-Sludge Processes**

| Treatment process | Surface loading at design peak hourly flow,* gal/(d · ft$^2$) (m$^3$ · m$^2$ · d) | Peak solids loading rate,‡ lb/(d · ft$^2$) (kg/(d · m$^2$) |
|---|---|---|
| Conventional<br>Step aeration<br>Complete mix<br>Contact stabilization<br>Carbonaceous state of separate-stage nitrification | 1200 (49)<br>or<br>1000 (41)† | 50 (244) |
| Extended aeration<br>Single-stage nitrification | 1000 (41) | 35 (171) |
| Two-stage nitrification | 800 (33) | 35 (171) |
| Activated sludge with chemical addition to mixed liquor for phosphorus removal | 900 (37)§ | as above |

*Based on influent flow only.
†Computed on the basis of design maximum daily flow rate plus design maximum return sludge rate requirement, and the design MLSS under aeration.
‡For plant effluent ≤ 20 mg/L.
§When effluent P concentration of 1.0 mg/L or less is required.
SOURCE: GLUMRB (1996)

**Example 1:** A trickling filter plant treats a flow of 1.0 Mgal/d (3785 m$^3$/d) with a raw wastewater $BOD_5$ of 200 mg/L and total suspended solids of 240 mg/L. Estimate the daily solids production of the plant assuming that the primary clarifiers remove 32% of the $BOD_5$ and 65% of TSS.

**solution:**

Step 1. Calculate the $BOD_5$ loading to the secondary clarifiers

$$BOD_5 = (1 - 0.32) \times (200 \text{ mg/L}) \times 1.0 \text{ Mgal/d} \times 8.34 \text{ lb/(Mgal} \cdot \text{mg/L)}$$
$$= 1134 \text{ lb/d } (=515 \text{ kg/d})$$

Step 2. Estimate solids ($S_2$) production of the secondary clarifiers

*Note*: The biological solids generated in secondary treatment range from 0.4 to 0.5 lb per lb (kg/kg) $BOD_5$ applied. Use 0.45 lb/lb in this example.

$$S_2 = 0.45 \text{ lb/lb} \times 1134 \text{ lb/d}$$
$$= 510 \text{ lb/d } (= 32 \text{ kg/d})$$

Step 3. Estimate solids ($S_1$) generated from the primary clarifiers

$$S_1 = 0.65 \times (240 \text{ mg/L}) \times 1.0 \text{ Mgal/d} \times 8.34 \text{ lb/(Mgal} \cdot \text{mg/L)}$$
$$= 1300 \text{ lb/d } (= 590 \text{ kg/d})$$

Step 4. Calculate the total daily solids production ($S$)

$$S = S_1 + S_2 = 1300 \text{ lb/d} + 510 \text{ lb/d}$$
$$= 1810 \text{ lb/d } (= 822 \text{ kg/d})$$

*Or solve this problem using SI units*

Step 1. Calculate the $BOD_5$ loading to the secondary clarifiers

$$BOD_5 = (1 - 0.32) \times (200 \text{ mg/L}) \times (3785 \text{ m}^3\text{/d}) \times 0.001 \text{ [(kg/m}^3\text{)/(mg/L)]}$$
$$= 515 \text{ kg/d } (= 1134 \text{ lb/d})$$

Step 2. Estimate solids ($S_2$) production of the secondary clarifiers

*Note*: The biological solids generated in secondary treatment range from 0.4 to 0.5 lb per lb (kg/kg) $BOD_5$ applied. Use 0.45 kg/kg in this example.

$$S_2 = 0.45 \text{ kg/kg} \times 515 \text{ kg/d}$$
$$= 232 \text{ kg/d } (= 510 \text{ lb/d})$$

Step 3. Estimate solids ($S_1$) generated from the primary clarifiers

$$S_1 = 0.65 \times (240 \text{ mg/L}) \times 3785 \text{ m}^3\text{/d} \times 0.001 \text{ [(kg/m}^3\text{)/(mg/L)]}$$
$$= 590 \text{ kg/d} = (1300 \text{ lb/d})$$

Step 4. Calculate the total daily solids production ($S$)

$$S = S_1 + S_2 = 590 \text{ kg/d} + 232 \text{ kg/d}$$
$$= 822 \text{ kg/d } (= 1810 \text{ lb/d})$$

**Example 2:** Design a two-stage trickling filter system with a design average flow of 5680 $m^3$/d (1.5 Mgal/d) and intermediate and secondary settling tanks under conditions given as follows. The primary effluent (system influent)

BOD = 190 mg/L. The design BOD loading rate is 1.5 kg/($m^3 \cdot$ d) (93.8 lb/(1000 $ft^3 \cdot$ d)).

The recycle ratio of both filters is 0.8. Both clarifiers have 20% recycled to the influent.

**solution:**

Step 1. Determine volume required for the filters (plastic media)

$$\text{Volume} = \frac{5680 \text{ m}^3\text{/d} \times 0.19 \text{ kg/m}^3}{1.5 \text{ kg/(m}^3 \cdot \text{d)}}$$
$$= 719 \text{ m}^3$$

Volume of each filter = 360 $m^3$

Step 2. Determine surface area of the filter

Use side water depth of 4 m (13 ft) since minimum depth is 3 m.

$$\text{Area} = 360\ \text{m}^3/4\ \text{m} = 90\ \text{m}^2$$

$$\begin{aligned}\text{Diameter} &= (4 \times 90\ \text{m}^2/3.14)^{0.5}\\ &= 10.7\ \text{m}\\ &= 35\ \text{ft}\end{aligned}$$

Step 3. Check hydraulic loading rate

$$\begin{aligned}\text{HRT} &= (5680\ \text{m}^3/\text{d})(1 + 0.8)/90\ \text{m}^2\\ &= 113.6\ \text{m}^3/(\text{m}^2 \cdot \text{d}) \qquad (\text{OK}, <375\text{m}^3/(\text{m}^2 \cdot \text{d}))\end{aligned}$$

Use 4 m (13 ft) deep filters with 90 $\text{m}^2$ (970 $\text{ft}^2$) area.

Step 4. Sizing for the intermediate clarifier

Use HLR = 41 $\text{m}^3/(\text{m}^2 \cdot \text{d})$ (1000 gal/(d · $\text{ft}^2$)) and minimum depth of 3 m (10 ft)

$$\text{Area required} = \frac{5680\ \text{m}^3/\text{d} \times 1.2}{41\ \text{m}^3/(\text{m}^2 \cdot \text{d})}$$

$$= 166\ \text{m}^2$$

$$\begin{aligned}\text{Diameter} &= (4 \times 166\ \text{m}^2/3.14)^{1/2}\\ &= 14.6\ \text{m}\\ &= 48\ \text{ft}\end{aligned}$$

Step 5. Sizing for the secondary clarifier

Using HLR = 31 $\text{m}^3/(\text{m}^2 \cdot \text{d})$ (760 gal/(d · $\text{ft}^2$)) and minimum depth of 3 m (10 ft)

$$\text{Area required} = \frac{5680\ \text{m}^3/\text{d} \times 1.2}{31\ \text{m}^3/(\text{m}^2 \cdot \text{d})}$$

$$= 220\ \text{m}^2$$

$$\begin{aligned}\text{Diameter} &= (4 \times 220\ \text{m}^2/3.14)^{1/2}\\ &= 16.7\ \text{m}\\ &= 55\ \text{ft}\end{aligned}$$

## 26.2 Basin sizing for suspended-growth biological treatment

In order to produce the proper concentration of return sludge, activated-sludge settling tanks must be designed to meet thickening as well as separation requirements. Since the rate of recirculation of RAS from the secondary settling tanks to the aeration or reaeration basins is quite high in activated-sludge processes, their surface overflow rate and weir overflow rate should be adjusted for the various modified processes to minimize the problems with sludge loadings, density currents, inlet hydraulic turbulence, and occasional poor sludge settle-ability. The size of the secondary settling tank must be based on the large surface area determined for surface overflow rate and solids loading rate. Table 6.20 presents the design criteria for secondary clarifiers following activated-sludge processes (GLUMRB, 1996). The values given in Tables 6.19 and 6.20 are comparable.

Solids loading rate is of primary importance to insure adequate function in the secondary settling tanks following aeration basins. In practice, most domestic wastewater plants have values of solids volume index in the range of 100 to 250 mg/L (WEF and ASCE, 1991a). Detail discussions of the secondary clarifier can be found in this manual. Most design engineers prefer to keep the maximum solids loading rates in the range of 4 to 6 kg/($m^2 \cdot$ h) (20 to 30 lb/(d $\cdot$ $ft^2$)). Solids loadings rates of 10 kg/($m^2 \cdot$ h) (50 lb/(d $\cdot$ $ft^2$) or more have been found in some well-operating plants.

The maximum allowable hydraulic loading rate HLR as a function of the initial settling velocity ISV at the design MLSS concentration was proposed by Wilson and Lee (1982). The equation is expressed as follows:

$$\text{HRT} = Q/A = 24 \times \text{ISV/CSF} \qquad (6.151)$$

where HRT = hydraulic retention time, h
$Q$ = limiting hydraulic capacity, $m^3$/d
$A$ = area of the clarifier, $m^2$
24 = unit conversion factor, 24 h/d
ISV = initial settling velocity at the design MLSS concentration, m/h
CSF = clarifier safety factor, 1.5 to 3, typically 2

The values of ISV change with MLSS concentrations and other conditions. Batch-settling analyses should be performed. The maximum anticipated operational MLSS or the corresponding minimum ISV should be used in Eq. (6.151).

Numerous state regulations limit the maximum allowable weir loading rates to 124 $m^3$/(d $\cdot$ m) (10,000 gal/(d $\cdot$ ft)) for small plants (less than 3785 $m^3$/d or 1 Mgal/d) and to 186 $m^3$/(d $\cdot$ m) (15,000 gal/(d $\cdot$ ft)) for larger treatment plants (WEF and ASCE, 1991a). It is a general consensus that substantially high weir loading rates will not impair performance.

The depth of secondary clarifiers is commonly designed as 4 to 5 m (13 to 16 ft). The deeper tanks increase TSS removal and RAS concentration as well as costs. Typically, the larger the tank diameter, the deeper the sidewall depth. The shapes of settling tanks include rectangular, circular, and square. A design example of a secondary clarifier following the activated-sludge process is given previously in section 21.5, Example 1B.

**Example:** Determine the size of three identical secondary settling tanks for an activated-sludge process with a recycle rate of 25%, an MLSS concentration of 3600 mg/L, an average design flow of 22,710 $m^3/d$ (6 Mgal/d), and an anticipated peak hourly flow of 53,000 $m^3/d$ (14 Mgal/d). Use solids loading rates of 4.0 and 10.0 $kg/(m^2 \cdot h)$ for average and peak flow respectively.

**solution:**

Step 1. Compute the peak solids loading

$$3600 \text{ mg/L} = 3600 \text{ g/m}^3 = 3.6 \text{ kg/m}^3$$

$$\text{Loading} = 53{,}000 \text{ m}^3\text{/d} \times 3.6 \text{ kg/m}^3 \times 1.25$$
$$= 238{,}500 \text{ kg/d}$$

Step 2. Compute the design average solids loading

$$\text{Loading} = 22{,}710 \text{ m}^3\text{/d} \times 3.6 \text{ kg/m}^3 \times 1.25$$
$$= 102{,}200 \text{ kg/d}$$

Step 3. Compute the surface area required (each of three)

(a) At design flow

$$A = \frac{102{,}200 \text{ kg/d}}{4.0 \text{ kg/(m}^2 \cdot \text{h)} \times 24 \text{ h/d} \times 3 \text{ (units)}}$$
$$= 355 \text{ m}^2$$
$$= 3820 \text{ ft}^2$$

(b) At peak flow

$$A = 238{,}500/(10.0 \times 24 \times 3)$$
$$= 331 \text{ m}^2$$
$$= 3563 \text{ ft}^2$$

The design flow is the controlling factor; and the required surface area $A$ = 355 $m^2$.

Step 3. Determine the diameter of circular clarifiers

$$d^2 = 4A/\pi = 4 \times 355 \text{ m}^2/3.14$$
$$d = 21.3 \text{ m}$$
$$= 70 \text{ ft}$$

Use sidewall depth of 4 m (13 ft) for the tank diameter of 21 to 30 m (70 to 100 ft).

Step 4. Check hydraulic loading rate at the design flow

$$\text{HLR} = (22{,}710 \text{ m}^3/\text{d})/(3 \times 355 \text{ m}^2)$$
$$= 21.3 \text{ m}^3/(\text{m}^2 \cdot \text{d})$$

Step 5. Check HLR at the peak flow

$$\text{HLR} = (53{,}000 \text{ m}^3/\text{d})/(3 \times 355 \text{ m}^2)$$
$$= 49.8 \text{ m}^3/(\text{m}^2 \cdot \text{d})$$

(It is slightly over the limit of 49 $\text{m}^3/(\text{m}^2 \cdot \text{d})$; see Table 6.20.)

Step 6. Compute the weir loading rate

$$\text{Perimeter} = \pi d = 3.14 \times 21.3 \text{ m}$$
$$= 66.9 \text{ m}$$

At average design flow

$$\text{Weir loading} = (22{,}710 \text{ m}^3/\text{d})/(66.9 \text{ m} \times 3)$$
$$= 113 \text{ m}^3/(\text{m} \cdot \text{d}) \qquad (\text{OK}, < 125 \text{ m}^3/\text{m} \cdot \text{d limit})$$

At the peak flow

$$\text{Weir loading} = (53{,}000 \text{ m}^3/\text{d})/(66.9 \text{ m} \times 3)$$
$$= 264 \text{ m}^3/(\text{m} \cdot \text{d})$$

## 27 Effluent Disinfection

Effluent disinfection is the last treatment step of a secondary or tertiary treatment process. Disinfection is a chemical treatment method carried out by adding the selected disinfectant to an effluent to destroy or inactivate the disease-causing organisms. The purposes of effluent disinfection are to protect public health by killing or inactivating pathogenic organisms such as enteric bacteria, viruses, and protozoans, and to meet the effluent discharge standards.

The disinfection agents (chemicals) include chlorine, ozone, ultraviolet (UV) radiation, chlorine dioxide, and bromine. Design of UV irradiation can be referred to the manufacturer or elsewhere (WEF and ASCE, 1996a).

The chlorination–dechlorination process is currently widely practiced in the United States. Chlorine is added to a secondary effluent for a certain contact time (20 to 45 min for average flow and 15 min at peak flow), then the effluent is dechlorinated before discharge only during warm weather when people use water as primary contact. In the United States most states adopt a coliform limitation of 200 fecal coliform/100 ml.

Chlorination of effluents is usually accomplished with liquid chlorine. Alternative methods use calcium or sodium hypochlorite or chlorine dioxide. Disinfection kinetics and chemistry of chlorination are discussed in Chapter 5, numerous literature, and textbooks. An excellent review of effluent disinfection can be found in Design Manual (US EPA, 1986).

## 27.1 Chlorine dosage

If a small quantity of chlorine is added to wastewater or effluent, it will react rapidly with reducing substances such as hydrogen sulfide and ferrous iron, and be destroyed. Under these conditions, there are no disinfection effects. If enough chlorine is added to react with all reducing compounds, then a little more added chlorine will react with organic materials present in wastewater and form chlororganic compounds, which have slight disinfection activities. Again, if enough chlorine is introduced to react with all reducing compounds and all organic materials, then a little more chlorine added will react with ammonia or other nitrogenous compounds to produce chloramines or other combined forms of chlorine, which do have disinfection capabilities. Therefore chlorine dosage and residual chlorine are very important factors of disinfection operation. In addition to its disinfection purpose, chlorination is also applied for prevention of wastewater decomposition, prechlorination of primary influent, control of activated sludge bulking, and reduction of BOD.

Chlorinators are designed to have a capacity adequate to produce an effluent to meet coliform density limits specified by the regulatory agency. Usually, multiple units are installed for adequate capacity and to prevent excessive chlorine residuals in the effluent. Table 6.21 shows the recommended chlorine dosing capacity for treating normal

**TABLE 6.21 Recommended Chlorine Dosing Capacity for Various Types of Treatment Based on Design Average Flow**

| Type of treatment | Illinois EPA dosage, mg/L | GLUMRB dosage, mg/L |
|---|---|---|
| Primary settled effluent | 20 | |
| Lagoon effluent (unfiltered) | 20 | |
| Lagoon effluent (filtered) | 10 | |
| Trickling filter plant effluent | 10 | 10 |
| Activated sludge plant effluent | 6 | 8 |
| Activated sludge plant with chemical addition | 4 | |
| Nitrified effluent | | 6 |
| Filtered effluent following mechanical biological treatment | 4 | 6 |

SOURCE: Illinois EPA (1997), GLUMRBS (1996)

domestic wastewater, based on design average flow (Illinois EPA, 1997; GLUMRB, 1996).

For small applications, 150-lb (68-kg) chlorine cylinders are typically used where chlorine gas consumption is less than 150 lb/d. Chlorine cylinders are stored in an upright position with adequate support brackets and chains at 2/3 of cylinder height for each cylinder. For larger applications where the average daily chlorine gas consumption is greater than 150 lb, 1-ton (907-kg) containers are employed. Tank cars, usually accompanied by evaporators, are used for large installations (>10 Mgal/d, 0.44 $m^3$/s). In this case, area-wide public safety should be evaluated as part of the design consideration.

**Example 1:** Estimate a monthly supply of liquid chlorine for trickling filter plant effluent disinfection. The design average flow of the plant is 3.0 Mgal/d (11,360 $m^3$/d).

**solution:**

Step 1. Find the recommended dosage

From Table 6.21, the recommended dosage for trickling filter plant effluent is 10 mg/L.

Step 2. Compute the daily consumption

$$\begin{aligned}\text{Chlorine} &= 3.0\ \text{Mgal/d} \times 10\ \text{mg/L} \times 8.34\ \text{lb/(Mgal} \cdot \text{(mg/L)} \\ &= 250\ \text{lb/d}\end{aligned}$$

The daily consumption is over 150 lb/d (68 kg/d); thus choose 1-ton (2000-lb, 907-kg) containers.

Step 3. Compute the number of 1-ton containers required for 1 month's (M) supply

$$\begin{aligned}\text{Monthly need} &= 250\ \text{lb/d} \times 30\ \text{d/M} \\ &= 7500\ \text{lb/M}\end{aligned}$$

The plant needs four 1-ton containers, which is enough for 1 month's consumption.

**Example 2:** Determine the feeding rate in gal/min of sodium hypochlorite (NaOCl) solution containing 10% available chlorine. The daily chlorine dosage for the plant is 480 kg/d (1060 lb/d).

**solution:**

Step 1. Calculate chlorine concentration of the solution

$$10\% = 100{,}000\ \text{mg/L} = 100\ \text{g/L}$$

Step 2. Calculate the solution feed rate

$$\text{Feeding rate} = \frac{480{,}000 \text{ g/d}}{100 \text{ g/L} \times 1440 \text{ min/d}}$$

$$= 3.333 \text{ L/min}$$

$$= 0.88 \text{ gal/min}$$

**Example 3:** A wastewater treatment plant having an average flow of 28,400 $m^3$/d (7.5 Mgal/d) needs an average chlorine dosage of 8 mg/L. Chlorine is applied daily via 10% NaOCl solution. The time of shipment from the vendor to the plant is 2 days. A minimum 10-day supply is required for reserve purposes. Estimate the storage tank capacity required for NaOCl solution with a 0.03% per day decay rate.

**solution:**

Step 1. Determine daily chlorine requirement

$$\text{Chlorine} = \frac{28{,}000 \text{ m}^3\text{/d} \times 8 \text{ g/m}^3}{1000 \text{ g/kg}}$$

$$= 227.2 \text{ kg/d}$$

Step 2. Calculate daily volume of NaOCl solution used

$$10\% \text{ solution} = 100 \text{ g/L} = 100 \text{ kg/m}^3$$

$$\text{Volume} = \frac{227.2 \text{ kg/d}}{100 \text{ kg/m}^3}$$

$$= 2.27 \text{ m}^3\text{/d}$$

Step 3. Calculate the storage tank for NaOCl solution

$$\text{Tank volume} = (2.27 \text{ m}^3\text{/d})(2 \text{ days} + 10 \text{ days})$$

$$= 27.24 \text{ m}^3$$

Step 4. Calculate the tank volume with NaOCl decay correction.

$$\text{Correction factor} = (0.03\%\text{/day}) \times 12 \text{ days}$$

$$= 0.36\%$$

$$\text{The volume requirement} = \frac{27.24 \text{ m}^3 \times 10\%}{10\% - 0.36\%}$$

$$= 28.3 \text{ m}^3$$

**Example 4:** Calculate the residual chlorine concentration that must be maintained to achieve a fecal coliform (FC) density equal or less than 200/ 100 mL in the chlorinated secondary effluent. The influent of chlorine contact tank contains 4 3 106 FC/100 mL. The contact time is 30 min.

**solution:**

Step 1. Calculate the residual chlorine needed during the average flow using equation (White, 1986) modified from Eq. (5.180a)

$$N/N_0 = (1 + 0.23C_t t)^{-3} \quad \text{where } C_t \text{ is chlorine residual at time } t \text{ in mg/L}$$

$$200/(4 \times 10^6) = (1 + 0.23\ C_t t)^{-3}$$

$$5 \times 10^{-5} = (1 + 0.23\ C_t t)^{-3}$$

$$1 + 0.23\ C_t t = (0.2 \times 10^5)^{1/3} = 27.14$$

$$C_t t = 113.7$$

$$\text{When } t = 30 \text{ min, } C_t = 113.7/30 = 3.8 \text{ (mg/L)}$$

Step 2. Calculate the residual chlorine needed during the peak hourly flow Assume the ratio of the peak hourly flow to the average flow is 3.0

$$C_t = 3.8 \text{ mg/L} \times 3 = 11.4 \text{ mg/L}$$

## 27.2 Dechlorination

There have been concerns that wastewater disinfection may do more harm than good due to the toxicity of disinfection by-products. Many states have adopted seasonal chlorination (warm weather periods) in addition to dechlorination. Dechlorination is important in cases where chlorinated wastewater comes into contact with fish and other aquatic animals. Sulfur compounds, activated carbon, hydrogen peroxide, and ammonia can be used to reduce the residual chlorine in a disinfected wastewater prior to discharge. The first two are the most widely used. Sulfur compounds include sulfur dioxide ($SO_2$), sodium metabisulfite ($NaS_2O_5$), sodium bisulfite ($NaHSO_3$), and sodium sulfite ($Na_2SO_3$). Dechlorination reactions with those compounds are shown below.

Free chlorine:

$$SO_2 + 2H_2O + Cl_2 \rightarrow H_2SO_4 + 2HCl \tag{6.152}$$

$$SO_2 + H_2O + HOCl \rightarrow 3H^+ + Cl^- + SO_4^{2-} \tag{6.153}$$

$$Na_2S_2O_5 + 2Cl_2 + 3H_2O \rightarrow 2NaHSO_4 + 4HCl \tag{6.154}$$

$$NaHSO_3 + H_2O + Cl_2 \rightarrow NaHSO_4 + 2HCl \tag{6.155}$$

Chloramine:

$$SO_2 + 2H_2O + NH_2Cl \rightarrow NH_4^+ + 2H^+ + Cl^- + SO_4^{2-} \tag{6.156}$$

$$3Na_2S_2O_5 + 9H_2O + 2NH_3 + 6Cl_2 \rightarrow 6NaHSO_4 + 10HCl + 2NH_4Cl \tag{6.157}$$

$$3NaHSO_3 + 3H_2O + NH_3 + 3Cl_2 \rightarrow 3NaHSO_4 + 5HCl + NH_4Cl \tag{6.158}$$

$$C^* + 2H_2O + 2Cl_2 \rightarrow 4HCl + C^*O_2 \quad (6.159)$$

$$C^* + H_2O + NH_2Cl \rightarrow NH_4^+ + Cl^- + C^*O \quad (6.160)$$

$$C^*O + 2NH_2Cl \rightarrow N_2 + 2HCl + H_2O + C^* \quad (6.161)$$

$$C^* + H_2O + 2NHCl_2 \rightarrow N_2 + 4HCl + C^*O \quad (6.162)$$

The most common dechlorination chemicals are sulfur compounds, particularly sulfur dioxide gas, or aqueous solutions of bisulfite or sulfite. A pellet dechlorination system can be used for small plants. Liquid sulfur dioxide gas cylinders are in 50 gal (190 L) drums.

Sulfur dioxide is a deadly gas that affects the central nervous system. It is colorless and nonflammable. When sulfur dioxide dissolves in water, it forms sulfurous acid which is a strong reducing chemical. As sulfur dioxide is introduced to chlorinated effluent, it reduces all forms of chlorine to chloride and converts sulfur to sulfate. Both by-products, chlorides and sulfates, occur commonly in natural waters.

Sodium metabisulfite is a cream-colored powder readily soluble in water at various strengths. Sodium bisulfite is a white powder, or in granular form, and is available in solution with strengths up to 44%. Both sodium metabisulfite and sodium bisulfite are safe chemicals. These are most commonly used in small plants and in a few larger plants.

The dosage of dechlorination chemical depends on the chlorine residual in the effluent, the final residual chlorine standards, and the type of dechlorinating chemical used. Theoretical requirements to neutralize 1 mg/L of chlorine for sulfur dioxide (gas), sodium bisulfite, and sodium metabisulfite are 0.90, 1.46, and 1.34 mg/L, respectively. Theoretical values may be employed for initial estimation for sizing dechlorinating equipment under good mixing conditions. An extra 10% of dechlorination is designed above theoretical values. However, excess sulfur dioxide may consume dissolved oxygen at a maximum of 0.25 mg DO for every 1 mg of sulfur dioxide. Since the sulfur dioxide reacts very rapidly with the chlorine residual, no additional contact time is necessary. Dechlorination may remove all chlorine-induced toxicity from the effluents.

### 27.3 Process design

The design of an ozonation or ultraviolet irradiation system can usually be provided by the manufacturers. Only chlorination is discussed in this section.

**Traditional design.** The disinfection efficiency of effluent chlorination depends on chlorine dosage, contact time, temperature, pH, and characteristics of wastewater such as TSS, nitrogen concentrations, and type and density of organisms. The first two factors can be designed for the best performance for chlorination of water and wastewater effluent. In general, residual chlorine at 0.5 mg/L after 20 to 30 min of contact time is required. The State of California requires a minimum of 30 min contact time. The chlorine contact tank can be sized on the basis of contact time and plant flow. However, many engineers disagree with this method.

Contact time is related to the configuration of the chlorine contact tank. The drag on the sides of a deep, narrow tank causes relatively poor dispersion characteristics. Sepp (1977) observed that the dispersion number usually decreases with increasing length-to-width ($L/W$) ratios. The depth-to-width ($D/W$) ratio should be 1.0 or less in chlorine tanks. Adequate baffling, reduction of side drag, and elimination of dead spaces can provide a reasonably long tank with adequate plug-flow hydraulics (Sepp, 1981). Marske and Boyle (1973) claimed that adequate plug-flow tanks can be achieved by $L/W$ ratios of 40–70 to 1. In design, in addition to $L/W$ ratios, the usual dispersion parameters and geometric configurations must be considered.

**Example:** Determine the size of a chlorine contact tank for a design average flow of 0.131 m$^3$/s (3 Mgal/d) and a peak flow of 0.329 m$^3$/s (7.5 Mgal/d).

**solution:**

Step 1. Calculate the tank volume required at peak design flow. Select contact time of 20 min at the peak flow

$$\begin{aligned}\text{Volume} &= 0.329\ \text{m}^3/\text{s} \times 60\ \text{s/min} \times 20\ \text{min}\\ &= 395\ \text{m}^3\end{aligned}$$

Step 2. Calculate the volume required at the average flow

Use contact time of 30 min at the design flow

$$\begin{aligned}\text{Volume} &= 0.131\ \text{m}^3/\text{s} \times 60\ \text{s/min} \times 30\ \text{min}\\ &= 236\ \text{m}^3\end{aligned}$$

Peak flow is the control factor.

Step 3. Determine chlorine contact tank configuration and dimensions

Provide one basin with three-pass-around-the-end baffled arrangement. Select channel width $W$ of 2.2 m (6.1 ft) and water depth $D$ of 1.8 m (5.9 ft) with 0.6 m (2.0 ft) of freeboard; thus

$$\begin{aligned}D/W &= 1.8\ \text{m}/2.2\ \text{m}\\ &= 0.82\ (\text{OK if} < 1.0)\end{aligned}$$

$$\begin{aligned}\text{Cross-sectional area} &= 1.8\ \text{m} \times 2.2\ \text{m}\\ &= 3.96\ \text{m}^2\end{aligned}$$

$$\text{Total length } L = 395\ \text{m}^3/3.96\ \text{m}^2 = 99.7\ \text{m (say 100 m)}$$

$$\text{Length of each pass} = 100\ \text{m}/3 = 33.3\ \text{m}$$

$$\text{Use each pass} = 33.5 \text{ m (110 ft)}$$

$$\text{Check } L/W = 100 \text{ m}/2.2 \text{ m}$$

$$= 45.5 \text{ (OK if in the range of 40 to 70)}$$

$$\text{Tank size: } D \times W \times L = (1.8 \text{ m} + 0.6 \text{ m}) \times (2.2 \text{ m} \times 3) \times 33.5 \text{ m}$$

$$\text{Actual liquid volume} = 1.8 \text{ m} \times 2.2 \text{ m} \times 100 \text{ m}$$

$$= 396 \text{ m}^3$$

Step 4. Check the contact time $t$ at peak flow

$$t = \frac{396 \text{ m}^3}{0.329 \text{ m}^3/\text{s} \times 60 \text{ s/min}}$$

$$= 20.1 \text{ min}$$

**Collins–Selleck model.** Traditionally, chlorination has been considered as a first-order reaction that follows the theoretical Chick or Chick–Watson model. However, in practice, it has been observed not to be the case for wastewater chlorination. Collins *et al.* (1971) found that initial mixing of the chlorine solution and wastewater has a profound effect on coliform bacteria reduction. The efficiency of wastewater chlorination, as measured by coliform reduction, was related to the residence time. It was recommended that the chlorine contact tank be designed to approach plug-flow reactors with rapid initial mixing. Collins *et al.* (1971) suggested that amperometric chlorine residuals and contact time could be used to predict the effluent coliform density.

Collins and Selleck (1972) found that the coliform bacteria survival ratio often produces an initial log period and a declining rate of inactivation as a plot of the log survival ratio versus time (Fig. 6.35a). After the initial log period, there is a straight-line relationship between log survival and log of the product of chlorine residual concentration $C$ and contact time $t$ (Fig. 6.35b).

The rate equation of the Collins and Selleck model is

$$\frac{dN}{dt} = -kN \tag{6.163}$$

where $N$ = bacteria density at time $t$

$t$ = time

$k$ = rate constant characteristic of type of disinfectant, microorganism, and water quality

= 0 for $Ct < b$

= $K$ for $Ct = b$

= $K/b(Ct)$ for $Ct > b$

$b$ = log time

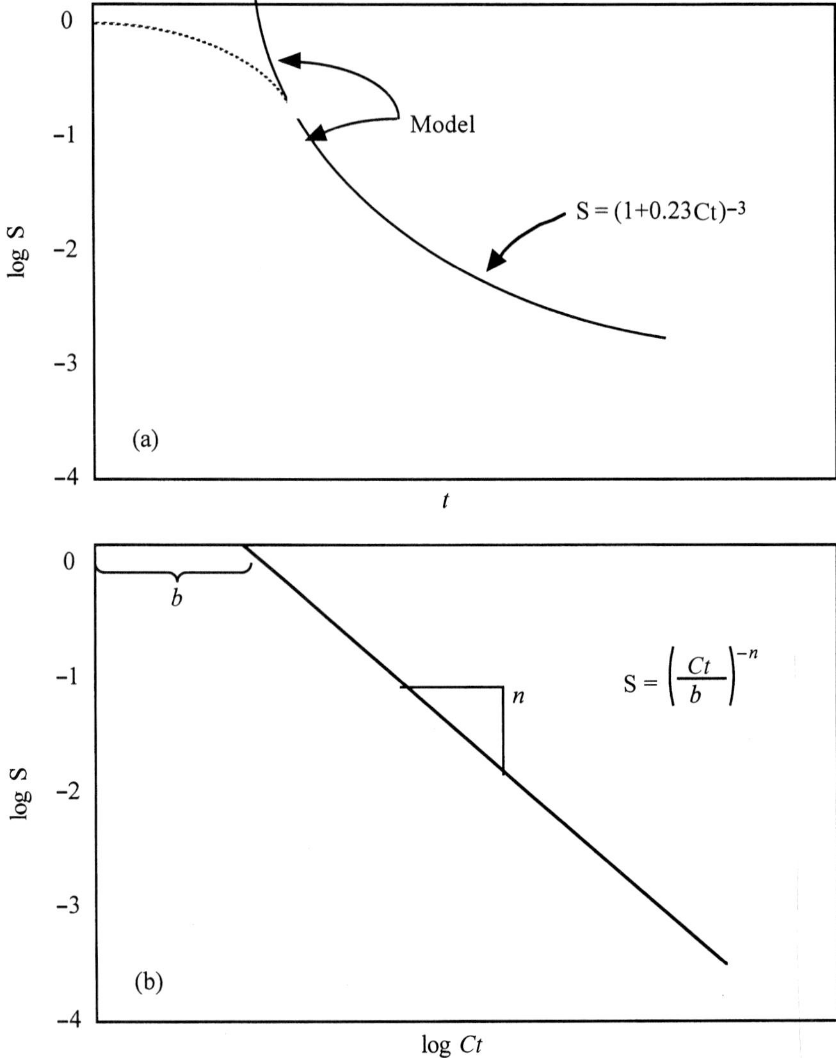


**Figure 6.35** Form of plots after Collins–Selleck model: (a) log $S$ versus time and (b) log $S$ versus log $Ct$.

After integration and application of the boundary conditions, the rate equation for Collins and Selleck becomes

$$\frac{N}{N_0} = 1 \qquad \text{for } Ct < b \tag{6.164}$$

$$\frac{N}{N_0} = \left(\frac{Ct}{b}\right)^n \qquad \text{for } Ct > b \tag{6.165}$$

where $N$ = bacteria density at time $t$ after chlorine contact
$N_0$ = bacteria density at time zero in unchlorinated effluent
$C$ = chlorine residual after contact, mg/L
$t$ = theoretical contact time, min
$b$ = an intercept when $N/N_0 = 1.0$, min · mg/L
= a product of concentration and time
$n$ = slope of the curve

Equation (6.164) applies to the lag period. Equation (6.165) plots as a straight line on log–log paper and expresses the declining rate. The values of $b$ and $n$, obtained from several researchers for various types of chlorine applied to various stages of wastewaters with different bacteria species, are summarized by J.M. Montgomery Engineers (1985). Wide variations of those values were observed. Collins and Selleck (1972) found that the slope $n = -3$ for ideal plug flow, $n = -1.5$ for complete-mix flow, and $b$ was about 4 in a primary effluent.

Robert *et al.* (1980) stated that the Collins–Selleck model is a useful empirical design tool although it has no rational mechanistic basis in describing chemical disinfection. Sepp (1977) recommended that the design of chlorine disinfection systems should incorporate rapid initial mixing, reliable automatic chlorine residual control, and adequate chlorine contact in well-designed contact tanks.

**Example:** A pilot study was conducted to determine the efficiency of combined chlorine to disinfect a secondary effluent. The results show that when $b = 2.87$ min · mg/L and $n = -3.38$, the geometric mean density of fecal coliform (FC) in the effluent is 14,000 FC/100 mL. Estimate FC density in the chlorinated effluent when contact times are 10, 20, and 30 min with the same chlorine residual of 1.0 mg/L.

**solution:**

Step 1. Determine mean FC survival ratios

$$Ct = 1.0\ \text{mg/L} \times 10\ \text{min}$$
$$= 10\ \text{min} \cdot \text{mg/L}\ (>b = 2.87)$$

Using Eq. (6.165)

(1) At $t = 10$ min

$$\frac{N}{N_0} = \left(\frac{1.0\,\text{mg/L} \times 10\ \text{min}}{2.87\,\text{min} \cdot \text{mg/L}}\right)^{-3.38}$$
$$= 0.0147$$

(2) At $t = 20$ min

$$\frac{N}{N_0} = \left(\frac{1.0 \times 20}{2.87}\right)^{-3.38}$$

$$= 0.00141$$

(3) At $t = 30$ min

$$\frac{N}{N_0} = \left(\frac{1.0 \times 30}{2.87}\right)^{-3.38}$$

$$= 0.000359$$

Step 2. Calculate FC densities in chlorinated effluent

(1) At $t = 10$ min

$$N = 0.0147 \times N_0 = 0.0147 \times 14{,}000 \text{ FC/100 mL}$$

$$= 210 \text{ FC/100 mL}$$

(2) At $t = 20$ min

$$N = 0.00141 \times 14{,}000$$

$$= 20(\text{FC/100 mL})$$

(3) At $t = 30$ min

$$N = 0.000359 \times 14{,}000$$

$$= 5(\text{FC/100 mL})$$

## 28 Advanced Wastewater Treatment

Advanced wastewater treatment (AWT) refers to those additional treatment techniques needed to further reduce suspended and dissolved substances remaining after secondary treatment. It is also called tertiary treatment. Secondary treatment removes 85% to 95% of BOD and TSS and minor portions of nitrogen, phosphorus, and heavy metals. The purposes of AWT are to improve the effluent quality to meet stringent effluent standards and to reclaim waste-water for reuse as a valuable water resource.

The targets for removal by the AWT process include suspended solids, organic matter, nutrients, dissolved solids, refractory organics, and specific toxic compounds. More specifically, the common AWT processes are suspended solids removal, nitrogen control, and phosphorus removal. Suspended solids are removed by chemical coagulation and filtration. Phosphorus removal is done to reduce eutrophication of receiving waters from wastewater discharge. Ammonia is oxidized to nitrate to reduce its toxicity to aquatic life and nitrogenous oxygen demand in the receiving water bodies.

## 28.1 Suspended solids removal

Treatment techniques for the reduction of suspended solids in secondary effluents include chemical coagulation followed by gravity sedimentation or dissolved air flotation, and physical straining or filtration such as wedge-wire screens, microscreens, other screening devices, diatomaceous earth filters, ultrafiltration, and granular media filters. These treatment processes are discussed in Chapter 5. More detailed descriptions and practical applications for SS removal are given in the US EPA's manual (1975a). Chemical coagulants used for SS removal include aluminum compounds, iron compounds, soda ash, caustic soda, carbon dioxide, and polymers. Some of the above processes are described in Chapter 5 and in other textbooks.

The use of a filtration process similar to that employed in drinking water treatment plant can remove the residual SS, BOD, and microorganisms from secondary effluent. Conventional sand filters or granular-media filters for water treatment may serve as advanced wastewater treatment and are discussed by Metcalf and Eddy, Inc. (1991). However, the filters may require more frequent backwashing. Design surface loading rates for wastewater filtration usually range from 2.0 to 2.7 L/($m^2 \cdot s$) (3 to 4 gal/(min · $ft^2$)). Filtration lasts about 24 h, and the effluent quality expected is 5 to 10 mg/L of suspended solids.

## 28.2 Phosphorus removal

The typical forms of phosphorus found in wastewater include the orthophosphates, polyphosphates (molecularly dehydrated phosphates), and organic phosphates. Orthophosphates such as $PO_4^{3-}$, $HPO_4^{2-}$, $H_2PO_4^-$, $H_3PO_4$ are available for biological uptake without further breakdown. The polyphosphates undergo hydrolysis in aqueous solutions and revert to the orthophosphate forms. This hydrolysis is a very slow process. The organic phosphorus is an important constituent of industrial wastes and less important in most domestic wastewaters.

The total domestic phosphorus contribution to wastewater is about 1.6 kg per person per year (3.5 lb per capita per year). The average total phosphorus concentration in domestic raw wastewater is about 10 mg/L, expressed as elemental phosphorus, P (US EPA, 1976). Approximately 30% to 50% of the phosphorus is from sanitary wastes, while the remaining 70% to 50% is from phosphate builders in detergents.

Phosphorus is one of major contributors to eutrophication of receiving waters. Removal of phosphorus is a necessary part of pollution prevention to reduce eutrophication. In most cases, the effluent standards range from 0.1 to 2.0 mg/L as P, with many established at 1.0 mg/L. Percentage reduction requirements range from 80% to 95%.

**Chemical precipitation.** Phosphorus removal can be achieved by chemical precipitation or a biological method. In practice, phosphorus in wastewater is removed by chemical precipitation with metal salt (aluminum or iron) or/and polymer, or lime. The chemical can be dosed before primary sedimentation, in the trickling filter, in the activated sludge process, and in the secondary classifier. Aluminum ions can flocculate phosphate ions to form aluminum phosphate which then precipitates:

$$Al^{3+} + H_nPO_4^{3-,n+} \rightarrow AlPO_4 \downarrow + nH^+ \tag{6.166}$$

Alum, $Al_2(SO_4)_3 \cdot 14H_2O$, is most commonly used as a source of aluminum. The precipitation reactions for phosphorus removal by three chemical additions are as follows:

Alum:

$$Al_2(SO_4)_3 \cdot 14H_2O + 2HPO_4^{2-} \rightarrow 2AlPO_4\downarrow + 2H^+ + 3SO_4^{2-} + 14H_2O \tag{6.167}$$

Sodium aluminate:

$$Na_2O \cdot Al_2O_3 + 2HPO_4^{2-} + 4H_2O \rightarrow 2AlPO_4\downarrow + 2NaOH + 6OH^- + 2H^+ \tag{6.168}$$

Ferric chloride, $FeCl_3 \cdot 6H_2O$:

$$FeCl_3 + HPO_4^{2-} \rightarrow FePO_4\downarrow + H^+ + 3Cl^- \tag{6.169}$$

Lime CaO:

$$CaO + H_2O \rightarrow Ca(OH)_2 \tag{6.170}$$

$$5Ca(OH)_2 + 3HPO_4^{2-} \rightarrow Ca_5(PO_4)_3OH\downarrow + 3H_2O + 6OH^- \tag{6.171}$$

Alum and ferric chloride decrease the pH while lime increases pH. The optimum pH for alum and ferric chloride is between 5.5 and 6.0. The effective pH for lime is above 10.0. Theoretically, 9.6 g of alum is required to remove 1 g of phosphorus. However, there is an excess alum requirement due to the competing reactions with natural alkalinity. The alkalinity reaction is

$$Al_2(SO_4)_3 \cdot 18H_2O + 6HCO_3^- \rightarrow 2Al(OH)_3\downarrow + 3SO_4^{2-} + 6CO_2 + 18H_2O \tag{6.172}$$

As a result, the requirement of alum to remove phosphorus demands weight ratios of approximately 13:1, 16:1, and 22:1 for 75%, 85%, and 95% phosphorus removal, respectively (from Eqs. (6.167) and (6.172)).

The molar ratio of ferric ion to orthophosphate is 1 to 1. Similarly to alum, a greater amount of iron is required to precipitate phosphorus and to react with alkalinity in wastewater. Its competing reaction with natural alkalinity is

$$FeCl_2 \cdot 6H_2O + 3HCO_3^- \rightarrow Fe(OH)_3 \downarrow + 3Cl^- + 3CO_2 + 6H_2O \qquad (6.173)$$

The stoichiometric weight ratios of Fe:P is 1.8:1. Weight ratio of lime (CaO) to phosphorus is 2.2:1. The reaction of Eq. (6.173) is slow. Therefore, lime or some other alkali may be added to raise the pH and supply hydroxyl ion for better coagulation. The reaction is

$$2FeCl_3 \cdot 6H_2O + 3Ca(OH)_2 \rightarrow 2Fe(OH)_3 \downarrow + 3CaCl_2 + 12H_2O \quad (6.174)$$

Both ferric ($Fe^{3+}$) and ferrous ($Fe^{2+}$) ions are used in the precipitation of phosphorus. Ferrous sulfate application for phosphorus removal is similar to that of ferric sulfate.

Calcium ion reacts with phosphate ion in the presence of hydroxyl ion to form hydroxyapatite. Lime usually comes in a dry form, calcium oxide. It must be mixed with water to form a slurry (calcium hydroxide, $Ca(OH)_2$) in order to be fed to a wastewater. Equipment required for lime precipitation includes a lime-feed device, mixing chamber, settling tank, and pumps and piping.

**Phosphorus removal in primary and secondary plants.** In conventional primary and secondary treatment of wastewater, phosphorus removal is only sparingly undertaken because the majority of phosphorus in wastewater is in soluble form. Primary sedimentation removes only 5% to 10% of phosphorus. Secondary biological treatment removes 10% to 20% of phosphorus by biological uptake. The effectiveness of primary and secondary treatment without (as in most conventional plants) and with chemical addition for phosphorus removal as well as for SS and BOD removals is presented in Table 6.22. Generally, the

**TABLE 6.22 Efficiencies of Primary and Secondary Treatment Without and with Mineral Addition for Phosphorus and Other Constituents Removal**

| | Phosphorus removal % | | Suspended solids removal, % | | BOD removal, % | |
|---|---|---|---|---|---|---|
| Treatment process | Without | With | Without | With | Without | With |
| Primary | 5–10 | 70–90 | 40–70 | 60–75 | 25–40 | 40–50 |
| Secondary | | | | | | |
| Activated sludge | 10–20 | 80–95 | 85–95 | 85–95 | 85–95 | 85–95 |
| Trickling filter | 10–20 | 80–95 | 70–92 | 85–95 | 80–90 | 80–95 |
| RBC* | 8–12 | | | | | |

*RBC = rotating biological contactor.
SOURCE: US EPA 1976

total phosphorus concentration of 10 mg/L in the raw wastewater is reduced to about 9 mg/L in the primary effluent and to 8 mg/L in the secondary effluent.

**Example 1:** Alum is added in the aeration basin of a conventional activated-sludge process for phosphorus removal. The effluent limit for total phosphorus is 0.5 mg/L. The alum is dosed at 140 mg/L. The average total phosphorus concentrations are 10.0 and 9.0 mg/L respectively for the influent and effluent (influent of the aeration basin) of the primary clarifier. Determine the molar ratio of aluminum to phosphorus and the weight ratio of alum dosed to the phosphorus content in the wastewater. Estimate the amount of sludge generated, assuming 0.5 mg/L of biological solids are produced by 1 mg/L of BOD reduction and using the following given data for the aeration basin (with alum dosage).

$$\begin{aligned} \text{Influent BOD} &= 148 \text{ mg/L} \\ \text{Effluent BOD} &= 10 \text{ mg/L} \\ \text{Influent TSS} &= 140 \text{ mg/L Effluent} \\ \text{TSS} &= 12 \text{ mg/L} \end{aligned}$$

Also estimate the increase of sludge production of chemical–biological treatment compared to biological treatment, assuming 30 mg/L for both TSS and BOD in the conventional system.

**solution:**

Step 1. Compute the molar ratio of Al to P

$$\begin{aligned} \text{Molecular wt of } Al_2(SO_4)_3 \cdot 18H_2O &= 27 \times 2 + 3(32 + 16 \times 4) + 14(2 + 16) \\ &= 666 \\ \text{Alum dosage} &= 140 \text{ mg/L} \\ \text{Aluminum dosage} &= 140 \text{ mg/L } (2 \times 27/594) \\ &= 12.7 \text{ mg/L} \\ \text{Molar ratio of Al:P} &= 12.7 \text{ mg/L} : 9.0 \text{ mg/L} \\ &= 1.4 : 1 \end{aligned}$$

Step 2. Compute the weight ratio of alum dosed to P in the influent

$$\frac{\text{Alum dosed}}{\text{P in the influent}} = \frac{140 \text{ mg/L}}{9.0 \text{ mg/L}} = \frac{15.6}{1} \tag{6.174}$$

Step 3. Estimate sludge residue for TSS removal

$$\text{TSS removed} = 140 \text{ mg/L} - 12 \text{ mg/L} = 128 \text{ mg/L}$$

Step 4. Estimate biological solids from BOD removal

$$\begin{aligned} \text{Biological solids} &= 0.5\ (148 \text{ mg/L} - 10 \text{ mg/L}) \\ &= 69 \text{ mg/L} \end{aligned}$$

Step 5. Estimate organic P in biological solids

Assuming organic P in biological solids removed is 2% by weight,

$$\text{P in biological solids} = 69 \text{ mg/L} \times 0.02$$
$$= 1.4 \text{ mg/L}$$

Step 6. Compute P removed by alum precipitation

$$\text{P removed by alum} = \text{P in (influent} - \text{biological solids} - \text{effluent)}$$
$$= (9.0 - 1.4 - 0.5) \text{ mg/L}$$
$$= 7.1 \text{ mg/L}$$

Step 7. Compute $AlPO_4$ precipitate

$$AlPO_4 \text{ precipitate} = \frac{\text{(P removed by alum) (M.wt. of } AlPO_4)}{\text{(M.wt. of P)}}$$
$$= \frac{7.1 \text{ mg/L} (27 + 31 + 16 \times 4)}{31}$$
$$= 27.9 \text{ mg/L}$$

*Note*: M.wt. = molecular weight

Step 8. Compute unused alum

Refer to Eq. (6.167) to find alum used

$$\text{Alum used} = \frac{\text{(P removed)(M.wt. of alum)}}{2 \times \text{M.wt. of P}}$$
$$= \frac{7.1 \text{ mg/L} \times 666}{2 \times 31}$$
$$= 76.3 \text{ mg/L}$$

$$\text{Unused alum} = \text{alum dosed} - \text{alum used} = 140 \text{ mg/L} - 76.3 \text{ mg/L}$$
$$= 63.7 \text{ mg/L}$$

Step 9. Compute $Al(OH)_3$ precipitate

Referring to Eq. (6.172), the unused alum reacts with natural alkalinity.

$$Al(OH)_3 \text{ precipitate} = \frac{\text{(unused alum)(2} \times \text{m.wt. of } Al(OH)_3)}{\text{M.wt. of alum}}$$
$$= \frac{63.7 \text{ mg/L}(2 \times (27 + 17 \times 3))}{666}$$
$$= 14.9 \text{ mg/L}$$

Step 10. Compute total sludge produced in the secondary clarifier

$$\text{Total sludge} = \text{TSS removed} + \text{biological solids} + AlPO_4 \downarrow + Al(OH)_3 \downarrow$$
$$= (128 + 69 + 27.9 + 14.9)\ \text{mg/L}$$
$$= 239.8\ \text{mg/L}$$

Step 11. Estimate sludge generated from the conventional activated-sludge system

$$\text{TSS removal} = 140\ \text{mg/L} - 30\ \text{mg/L} = 110\ \text{mg/L}$$
$$\text{Biological sludge} = 0.5\ (148\ \text{mg/L} - 30\ \text{mg/L})$$
$$= 59\ \text{mg/L}$$
$$\text{Total sludge} = 110\ \text{mg/L} + 59\ \text{mg/L}$$
$$= 169\ \text{mg/L}$$

Step 12. Increase of sludge production by chemical–biological process over biological process

$$\text{Increase} = 243.8\ \text{mg/L} - 169\ \text{mg/L}$$
$$= 74.8\ \text{mg/L}$$
$$\%\ \text{increase} = 74.8\ \text{mg/L} \times 100\%/169\ \text{mg/L}$$
$$= 44.3\%$$

**Example 2:** A wastewater has a soluble phosphorus concentration of 7.0 mg/L as P. Determine the stoichiometric quantity of ferric chloride required to remove P completely.

**solution:**

Step 1. Calculate the gram molecular weights

Equation (6.169) shows that 1 mole of ferric chloride reacts with 1 mole of orthophosphate to be removed. The pertinent gram molecular weights for $FeCl_3$ and P are:

$$FeCl_3 = 55.845 + 35.453 \times 3 = 162.204\ (\text{g})$$
$$P = 30.975\ \text{g}$$

Step 2. Calculate the stoichiometric amount ($Y$) of $FeCl_3$ required

$$Y = 7.0\ \text{mg/L}\ (162.204\ \text{g}/30.974\ \text{g})$$
$$= 36.66\ \text{mg/L}$$

*Notes*: In practice, the actual $FeCl_3$ dose would be 1.5 to 3.0 times the stoichiometric amount. Likewise, the alum dose would be 1.3 to 2.5 times the stoichiometric amount due to side reaction, solubility products limitations, and wastewater quality variations. Jar tests on wastewater are needed for determining the actual amount.

**Phosphorus removal by mineral addition to secondary effluent.** Generally, an alum dosage of about 200 mg/L is required for phosphorus removal from typical municipal raw wastewater and a dosage of 50 to 100 mg/L is sufficient for secondary effluent. Iron salts have little application because of residual iron remaining in the treated wastewater. An Al : P molar ratio of 1 : 1 to 2 : 1 is required. The optimum pH for alum treatment is near 6.0, whereas that for iron is near 5.0 (Recht and Ghassemi, 1970). The pH of high alkalinity waters may be reduced either by using high dosages of alum or by adding supplementary dosages of sulfuric acid. Anionic polyelectrolytes (coagulant aids) can be used to enhance P removal. If higher phosphorus removal is required, a filtration process must be used to achieve 0.1 mg/L of residual P.

The surface overflow rates for settling tanks range from 24 to 58 $m^3/(m^2 \cdot d)$ (580 to 1440 $gal/(d \cdot ft^2)$). Filtration rates are 0.08 to 0.20 $m^3/(m^2 \cdot min)$ (2 to 5 $gal/(min \cdot ft^2)$) (US EPA, 1976).

**Example:** Estimate the daily liquid alum requirement for phosphorus removal from the secondary effluent which has an average design flow of 1 Mgal/d (3785 $m^3/d$) and an average phosphorus concentration of 8 mg/L. Assume that the specific weight of alum is 11.1 lb/gal and alum contains 4.37% of aluminum.

**solution:**

Step 1. Calculate the mass of the incoming P loading rate

$$\text{Load} = 1\ \text{Mgal/d} \times 8\ \text{mg/L} \times 8.34\ \text{lb/(Mgal} \cdot \text{mg/L)}$$
$$= 66.7\ \text{lb/d}$$

Step 2. Determine aluminum in mol/d required

Atomic weights of phosphorus and aluminum are 31 and 27, respectively.

$$\text{P load} = 66.7\ \text{lb/d} \div 31\ \text{lb/mol}$$
$$= 2.15\ \text{mol/d}$$

Usually a molar ratio of Al : P = 2 : 1 is used, refering to Eq. (6.167)

$$\text{Al required} = 2.15\ \text{mol/d} \times 2$$
$$= 4.3\ \text{mol/d}$$

Step 3. Determine liquid alum required

$$\text{Mass of Al} = 4.3\ \text{mol/d} \times 27\ \text{lb/mol}$$
$$= 116.1\ \text{lb/d}$$

*Note*: Or this can be solved simply as:

$$\frac{\text{Al required}}{\text{P load}} = \frac{2 \times 27}{31}$$

$$\text{Al required} = 66.7 \text{ lb/d} \times 54/31$$

$$= 116.1 \text{ lb/d}$$

Using liquid alum having 4.37% of Al, with 11.1 lb/gal

$$\text{Liquid alum required} = \frac{116.1 \text{ lb/d}}{0.0437(11.1 \text{ lb/gal})}$$

$$= 239 \text{ gal/d} = 0.166 \text{ gal/min}$$

$$= 0.628 \text{ L/min}$$

**Phosphorus removal by lime treatment of secondary effluent.** For phosphorus removal, lime can be added to the primary sedimentation tank and to the secondary effluent. Lime treatment of wastewater is essentially the same process as is used for lime softening of drinking water, but with a different purpose. While softening may occur, the primary objective is to remove phosphorus by precipitation as hydroxyapatite (Eq. (6.171)).

During phosphorus precipitation, a competing reaction of lime with alkalinity will occur. This reaction results in calcium removal, an action of softening which has a very important effect on the phosphorus removal efficiency of the process. This reaction may occur in two ways:

$$Ca(OH)_2 + Ca(HCO_3)_2 \rightarrow 2CaCO_3 \downarrow + 2H_2O \qquad (6.175)$$

$$Ca(OH)_2 + NaHCO_3 \rightarrow CaCO_3 \downarrow + NaOH + H_2O \qquad (6.176)$$

Another reaction may occur to precipitate magnesium hydroxide:

$$Mg^{2+} + 2OH^- \rightarrow Mg(OH)_2 \downarrow \qquad (6.177)$$

In the two-stage lime treatment process, recarbonation after the first stage is required to reduce the pH and to precipitate excess lime as calcium carbonate by adding carbon dioxide. The chemical reaction is

$$Ca^{2+} + CO_2 + 2OH^- \rightarrow CaCO_3 + H_2O \qquad (6.178)$$

Lime can be added to secondary effluent by single-stage or two-stage treatment. For the single-stage process (Fig. 6.36), lime is mixed with feed wastewater to raise the pH (10–11); this is then followed by flocculation and sedimentation. The clarified water is adjusted to lower pH with carbon dioxide and is filtered through a multimedia filter to prevent postprecipitation of calcium carbonate before discharge. The settled lime sediment may be disposed of to a landfill or may be recalcined for recovery of lime. In recovery, the sediment is thickened, dewatered by centrifuge or vacuum filter, calcined, and then reused.

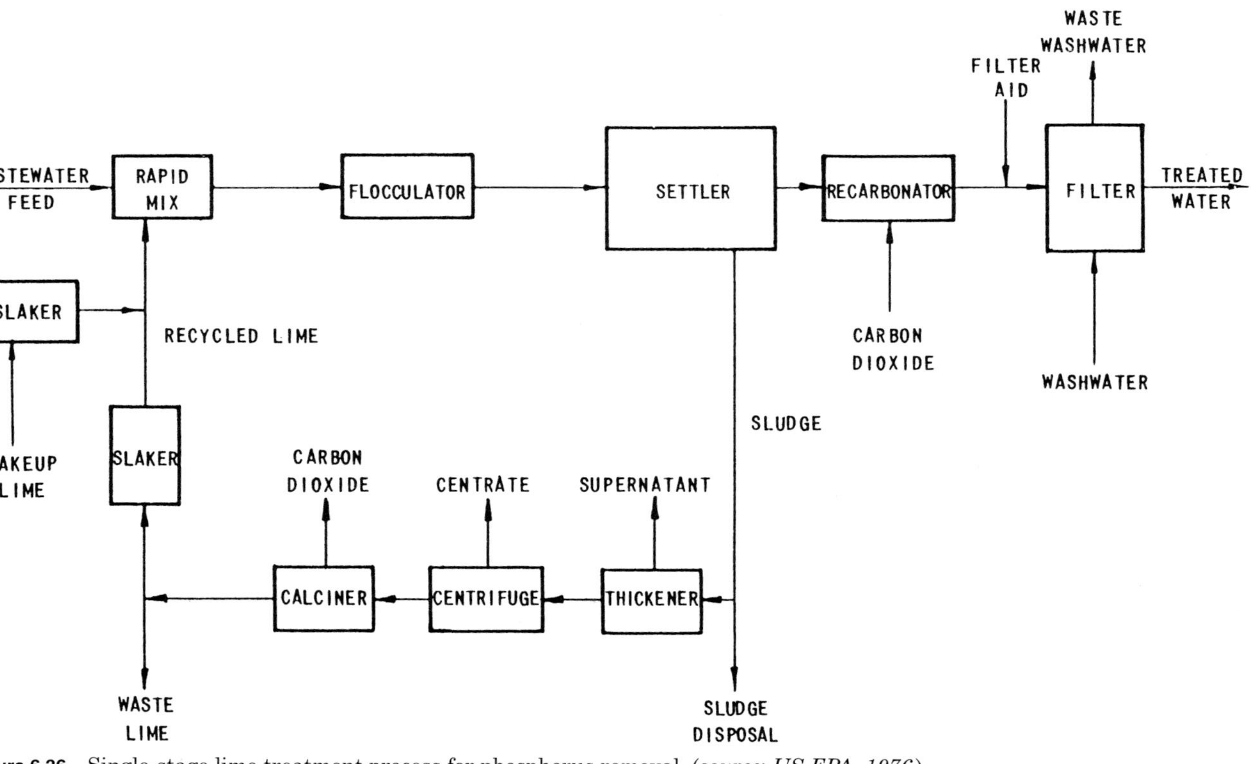


**Figure 6.36** Single-stage lime treatment process for phosphorus removal (*source*: *US EPA, 1976*).

For a typical two-stage process (Fig. 6.37), in the first-stage clarifier of the two-stage process, sufficient lime is dosed to raise the pH above 11 to precipitate the soluble phosphorus as hydroxyapatite, calcium carbonate, and magnesium hydroxide.

The calcium carbonate precipitate formed in the process acts as a coagulant for suspended solids removal. The excess soluble calcium is removed in the second clarifier as a calcium carbonate precipitate by adding carbon dioxide gas to reduce the pH to about 10. The calcium carbonate settles out in the secondary (or tertiary) clarifier. To remove the residual levels of phosphorus and suspended solids, the secondary (or tertiary) clarified effluent is filtered, then discharged. Usually pH is reduced to about 8.0 and 8.5 in first-stage recarbonation and further reduced to 7.0 in second-stage recarbonation. Fifteen minutes' contact time in the recarbonation basin is recommended.

Typical hydraulic loading rates for chemical treatment clarifiers are 32 to 61 $m^3/(m^2 \cdot d)$ (800 to 1500° $gal/(d \cdot ft^2)$).

**Example 1:** From the results of jar tests, calcium concentration drops from 56 to 38 mg/L when the pH is reduced from 10 to 7 by addition of carbon dioxide. Determine the carbon dioxide dose rate required for recarbonation.

**solution:**

Step 1. Determine theoretical Ca : $CO_2$ ratio, using Eq. (6.178)

$$Ca^{2+} + CO_2 + 2OH^- \rightarrow CaCO_3 \downarrow + H_2O$$

then

$$\begin{aligned} CO_2 : Ca &= (12 + 2 \times 16) : 40 \\ &= 1.1 : 1 \end{aligned}$$

Step 2. Calculate $CO_2$ theoretical requirement

$$CO_2 \text{ required} : \text{Ca reduction} = 1.1 : 1$$

$$\begin{aligned} CO_2 &= (56 - 38) \text{ mg/L} \times 1.1 \\ &= 19.8 \text{ mg/L} \\ &\cong 20 \text{ mg/L} \end{aligned}$$

Step 3. Calculate field requirement of $CO_2$

A safe factor of 20% should be added to calculated dosage to compensate for inefficiency in absorption.

$$\begin{aligned} CO_2 \text{ required} &= 20 \text{ mg/L} \times 1.2 \\ &= 24 \text{ mg/L} \end{aligned}$$

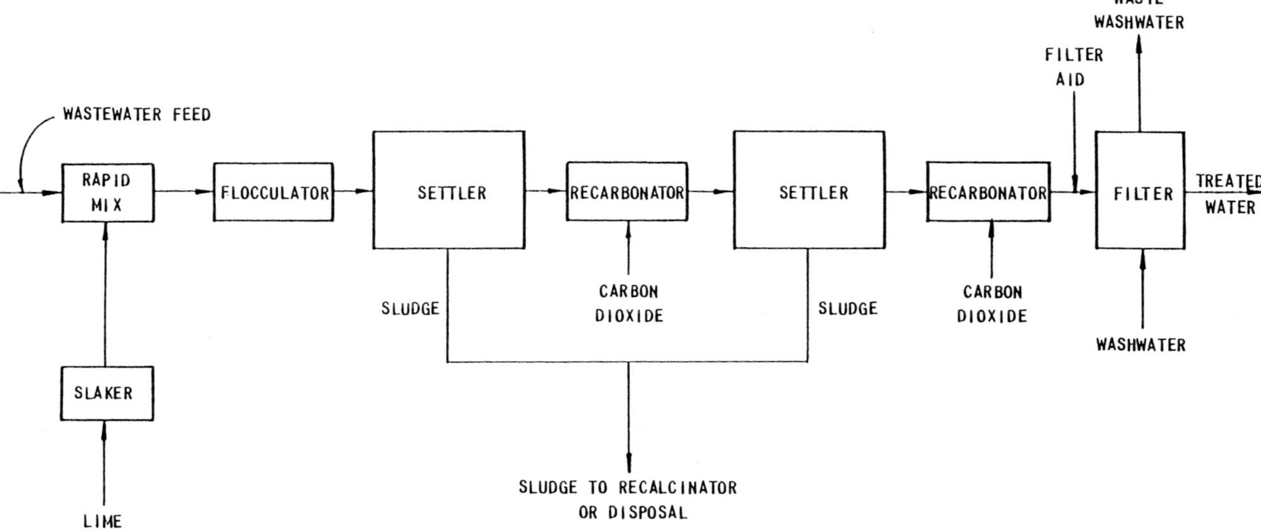


**Figure 6.37** Two-stage lime treatment process for phosphorus removal (*source*: *US EPA, 1976*).

**Example 2:** A single-stage lime treatment system is used as advanced treatment for phosphorus removal. Estimate the residue mass and volume generated in the clarifier under the following conditions.

Influent design flow = 37,850 $m^3$/d (10 Mgal/d)

Lime dosage as $Ca(OH)_2$ = 280 mg/L

Influent $PO_4^{3-}$ as P = 8 mg/L

Influent $Ca^{2+}$ = 62 mg/L

Influent $Mg^{2+}$ = 3 mg/L

Effluent $PO_4^{3-}$ as P = 0.6 mg/L

Effluent $Ca^{2+}$ = 20 mg/L

Effluent Mg = 0 mg/L

Influent suspended solids = 20 mg/L

Effluent suspended solids = 2 mg/L

Specific gravity of residue = 1.07

Moisture content of residue = 92%

**solution:**

Step 1. Compute the mass of $Ca_5(PO_4)_3$ (OH) formed

(a) Calculate the moles of P removed

$$\text{Mole of P removed} = \frac{8\,\text{mg/L} - 0.6\,\text{mg/L}}{30.974\,\text{g/mol} \times 1000\,\text{mg/g}}$$

$$= 0.239 \times 10^{-3}\ \text{mol/L}$$

$$= 0.239\ \text{mM/L}$$

where mM = millimol

(b) Calculate the moles of $Ca_5(PO_4)_3(OH)$ formed

From Eq. (6.171), 3 moles of P removed will form 1 mole of $Ca_5(PO_4)_3(OH)$

$$\text{Mole of } Ca_5(PO_4)_3(OH) \text{ formed} = 1/3 \times 0.239\ \text{mM/L}$$

$$= 0.080\ \text{mM/L}$$

(c) Calculate the mass of $Ca_5(PO_4)_3(OH)$ formed

$$\text{Mass of 1 mole } Ca_5(PO_4)_3(OH) = 40 \times 5 + 30.97 \times 3 + 16 \times 13 + 1\ \text{g/mol}$$

$$= 501.9\ \text{g/mol}$$

$$= 501.9\ \text{mg/mM}$$

$$\text{Mass of } Ca_5(PO_4)_3(OH) \text{ formed} = 0.08\ \text{mM/L} \times 501.9\ \text{mg/mM}$$

$$= 40.2\ \text{mg/L}$$

Step 2. Calculate the mass of calcium carbonate formed

(a) Calculate the mass of Ca added in the lime dosage

$$\text{Mass of Ca in Ca(OH)}_2 = 280\ \text{mg/L} \times 40/(40 + 17 \times 2)$$
$$= 151.4\ \text{mg/L}$$

(b) Determine the mass of $Ca^{2+}$ in $Ca_5(PO_4)_3(OH)$

$$\text{Mass of Ca in Ca}_5\text{(PO}_4\text{)}_3\text{(OH)} = 5\ (40\ \text{mg/mM}) \times 0.08\ \text{mM/L}$$
$$= 16\ \text{mg/L}$$

(c) Calculate the mass of total Ca present in $CaCO_3$

$$\text{Ca in CaCO}_3 = \text{Ca in Ca(OH)}_2 + \text{Ca in influent}$$
$$- \text{Ca in Ca}_5\text{(PO}_4\text{)}_3\text{(OH)} - \text{Ca in effluent}$$
$$= (151.4 + 62 - 16 - 20)\ \text{mg/L}$$
$$= 177.4\ \text{mg/L}$$

(d) Convert the mass of Ca as expressed in $CaCO_3$

$$\text{Mass of Ca as CaCO}_3 = 177.4\ \text{mg/L} \times (40 + 12 + 48)/40$$
$$= 443.5\ \text{mg/L}$$

Step 3. Determine the mass of $Mg(OH)_2$ formed

(a) Calculate the mass of Mg removed

$$\text{Mole of Mg}^{2+}\ \text{removed} = 3\ \text{mg/L}/(24.3\ \text{g/mol} \times 1000\ \text{mg/g})$$
$$= 0.123 \times 10^{-3}\ \text{mol/L}$$
$$= 0.123\ \text{mM/L}$$

(b) Calculate the mass of $Mg(OH)_2$ formed

$$\text{Mass of Mg(OH)}_2 = 0.123\ \text{mM/L} \times (24.3 + 17 \times 2)\ \text{mg/mM}$$
$$= 7.2\ \text{mg/L}$$

Step 4. Calculate the total mass of residue removed as a result of the lime application

(a) Mass of residue due to lime dosage

$$\text{Chemical mass} = \text{sum of mass from Ca}_5\text{(PO}_4\text{)}_3\text{(OH), CaCO}_3\text{, and Mg (OH)}_2$$
$$= (40.2 + 443.5 + 7.2)\ \text{mg/L} \times 37{,}850\ \text{m}^3\text{/d}$$
$$= 490.9\ \text{g/m}^3 \times 37{,}850\ \text{m}^3\text{/d}$$
$$= 18{,}580{,}000\ \text{g/d}$$
$$= 18{,}580\ \text{kg/d}$$

(b) Calculate the mass of SS removed

$$\text{Mass of SS} = (20 - 2)\ \text{mg/L} \times 37{,}850\ \text{m}^3\text{/d}$$
$$= 681\ \text{kg/d}$$

(c) Calculate the total mass of residues

$$\text{Total mass} = \text{mass of chemical residue} + \text{mass of SS}$$
$$= 18{,}580\ \text{kg/d} + 681\ \text{kg/d}$$
$$= 19{,}261\ \text{kg/d}$$

Step 5. Determine total volume $V$ of residue from lime precipitation process

Given: specific gravity of residue = 1.07, moisture content of residue = 92%, then

$$\text{Solid content} = 8\%$$

$$V = \frac{19{,}261\ \text{kg/d}}{1000\ \text{kg/m}^3 \times 1.07 \times 0.08}$$
$$= 225\ \text{m}^3\text{/d}$$
$$= 7945\ \text{ft}^3\text{/d}$$

**Phosphorus removal by biological processes.** Numerous biological methods of removing phosphorus have been developed. Integrated biological processes for nutrient removal from wastewater use a combination of biological and chemical methods to bring phosphorus and nitrogen concentrations to below the effluent standards. Some patented (17 years) integrated biological processes have been installed in the United States. The A/O process (Fig. 6.38) for mainstream P removal, Phostrip process (Fig. 6.39) for sidestream P removal, sequencing batch reactor (SBR, Fig. 6.40), sidestream fermentation process (OWASA nitrification, Fig. 6.41), and chemical polishing are used for phosphorus removal. Combined removal of phosphorus and nitrogen by biological methods includes the

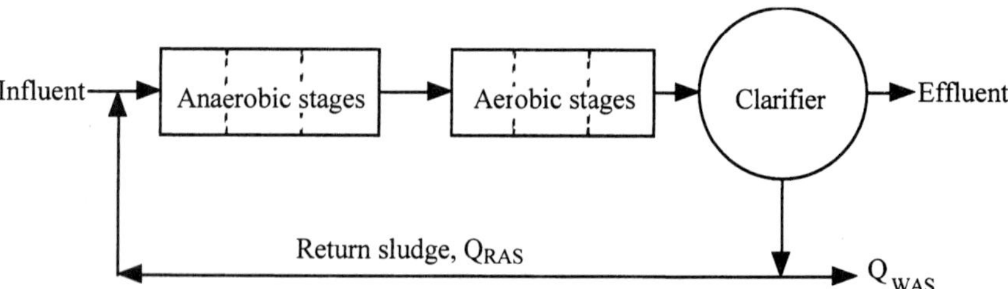


**Figure 6.38** A/O process.

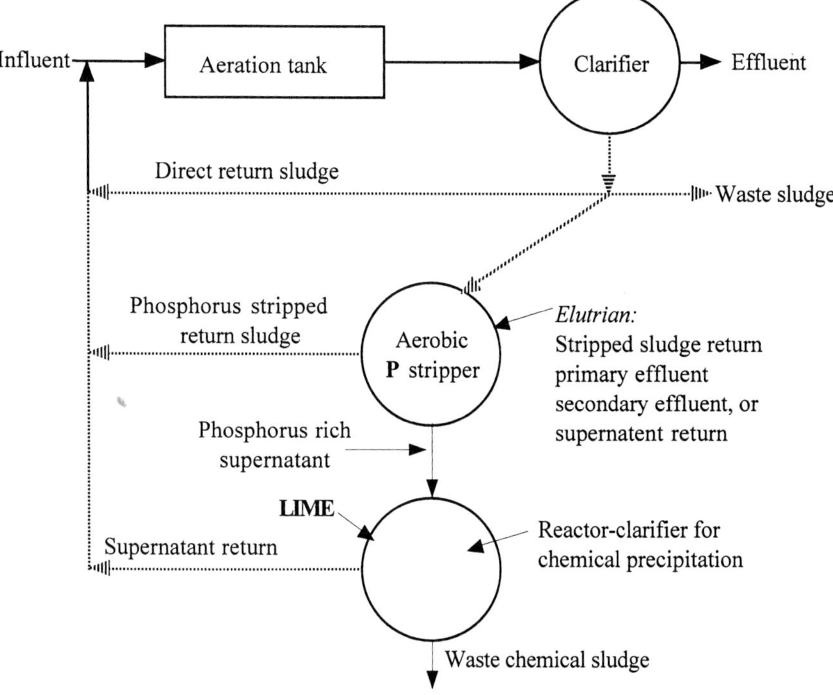


**Figure 6.39** PhoStrip process.

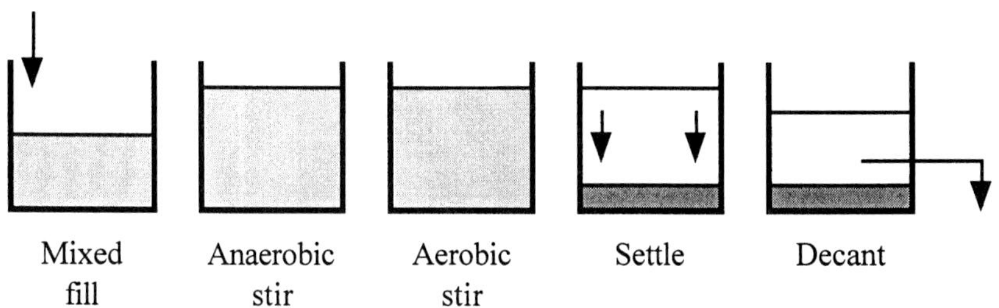


**Figure 6.40** Sequencing batch reactor for removal of phosphorus and carbonaceous BOD.

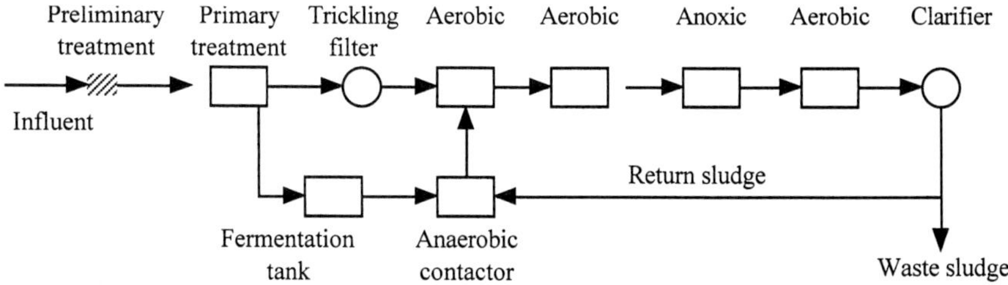


**Figure 6.41** OWASA nitrification process.

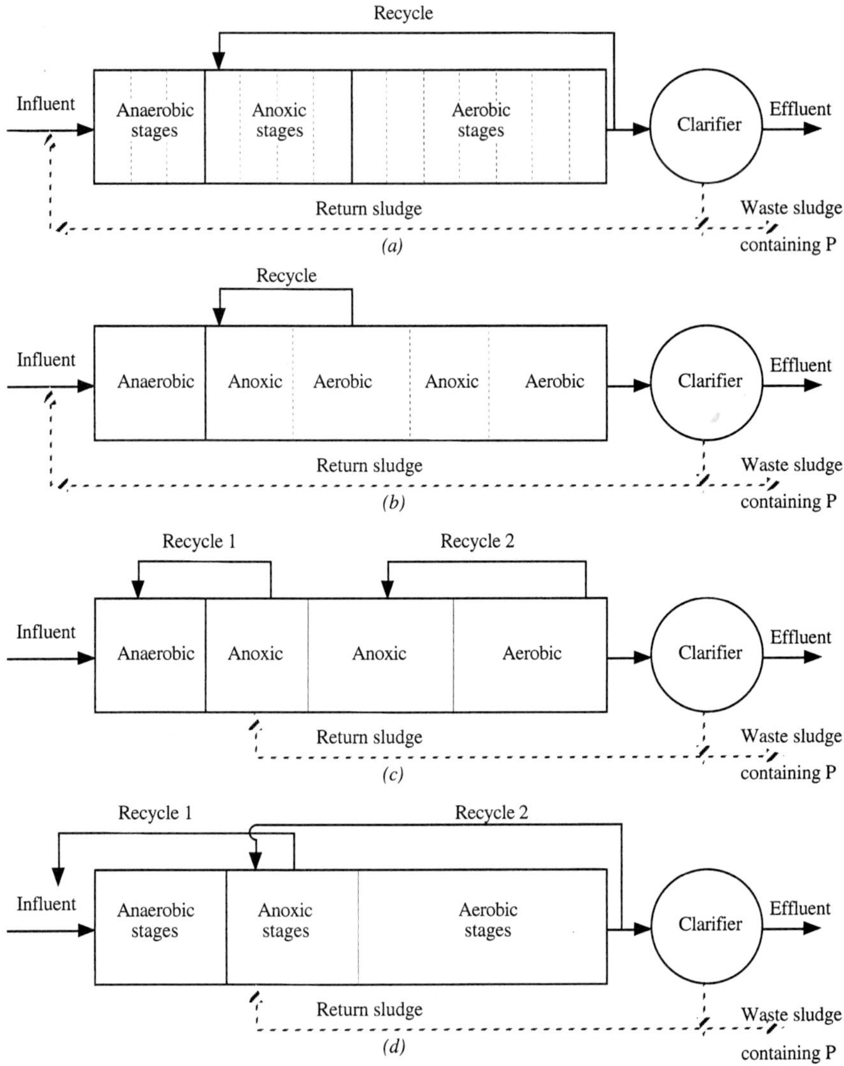


**Figure 6.42** Combined biological nitrogen and phosphorus removal processes: (a) $A^2/O$ process; (b) five-stage Bardenpho process; (c) UCT process; and (d) VIP process (*source: Metcalfe and Eddy, Inc. 1991*).

$A^2/O$ process (Fig. 6.42a), the modified (5-stage) Bardenpho process (Fig. 6.42b), the University of Cape Town (UCT) process (Fig. 6.42c), the VIP process (Virginia Institute Plant in Norfolk, Virginia) (Fig. 6.42d), PhoStrip II process (Fig. 6.43), SBR (Fig. 6.44), and phased isolation ditch. For nitrogen control, biological denitrification includes the Wuhrmann process (Fig. 6.45), Ludzack–Ettinger process (Figs. 6.46

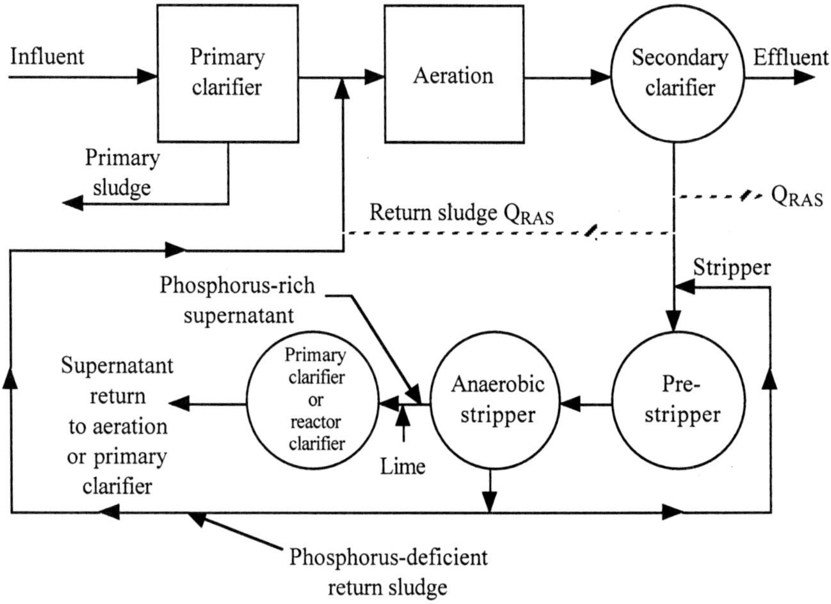

Influent
Primary clarifier
Aeration
Secondary clarifier
Effluent
Primary sludge
Return sludge $Q_{RAS}$
$Q_{RAS}$
Stripper
Phosphorus-rich supernatant
Supernatant return to aeration or primary clarifier
Primary clarifier or reactor clarifier
Anaerobic stripper
Pre-stripper
Lime
Phosphorus-deficient return sludge

**Figure 6.43** PhoStrip II process for phosphorus and nitrogen removal.

Mixed fill
Anaerobic stir
Aerobic stir
Anoxic stir
Settle
Decant

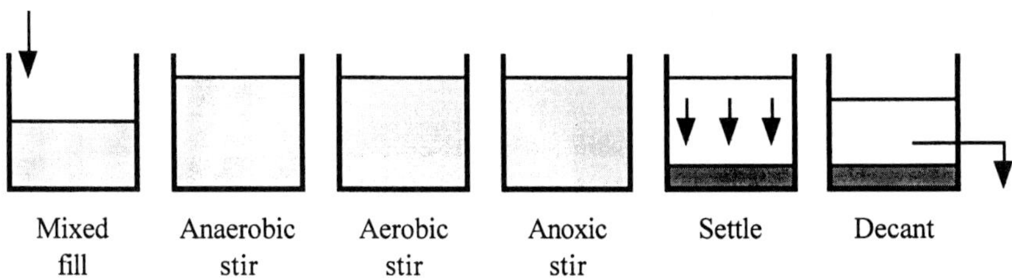

**Figure 6.44** Sequencing batch reactor for carbon oxidation plus phosphorus and nitrogen removal.

Influent
Aerobic
Anoxic
Clarifier
Effluent
Return sludge, $Q_{RAS}$
Waste sludge
$Q_{WAS}$

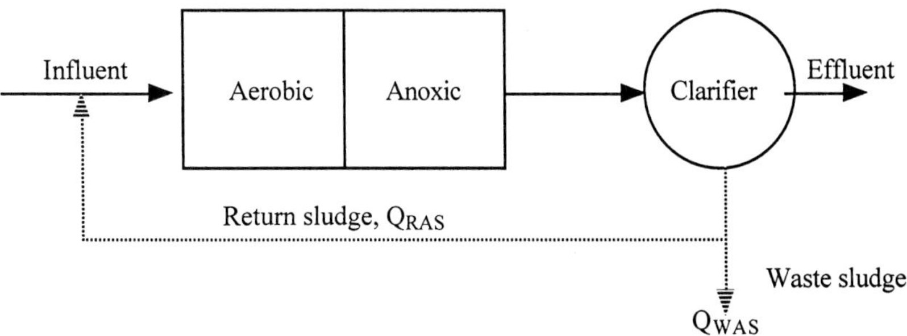

**Figure 6.45** Wuhrmann process for nitrogen removal.

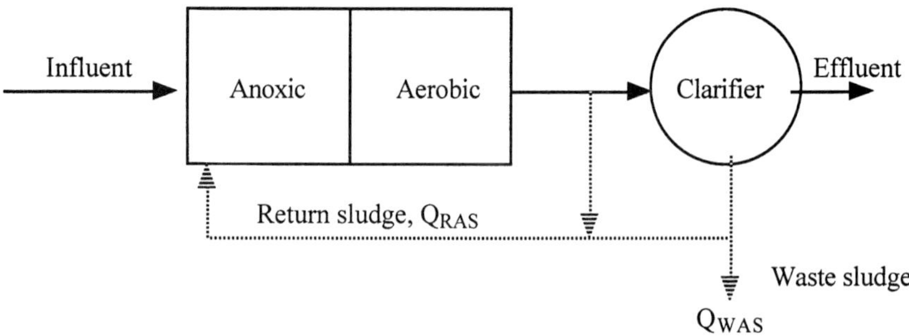


**Figure 6.46** Ludzack–Ettinger process for nitrogen removal.

and 6.47), Bordenpho (4-stage) process, oxidation ditch, phase isolation ditch, dual-sludge process (Fig. 6.48), triple-stage process (Fig. 6.49), denitrification filter, RBC, and fluidized bed process. The basic theory, stoichiometry, kinetics, design considerations, and practice of integrated systems in the US are discussed in detail elsewhere (WEF and ASCE 1991b, Metcalf and Eddy, Inc. 1991).

Biological processes for nutrient removal feature the exposure of alternate anaerobic and aerobic conditions to the microorganisms so that their uptake of phosphorus will be above normal levels. Phosphorus is not only utilized for cell synthesis, maintenance, and energy transport, but is also stored for subsequent use by the microorganisms. The sludge produced containing phosphorus is either wasted or removed through a sidestream (return sludge stream). The alternate exposure of the microorganisms to anaerobic and aerobic (oxic) conditions in the main biological treatment is called the mainstream process. A design example (Ex. 15.2) is illustrated

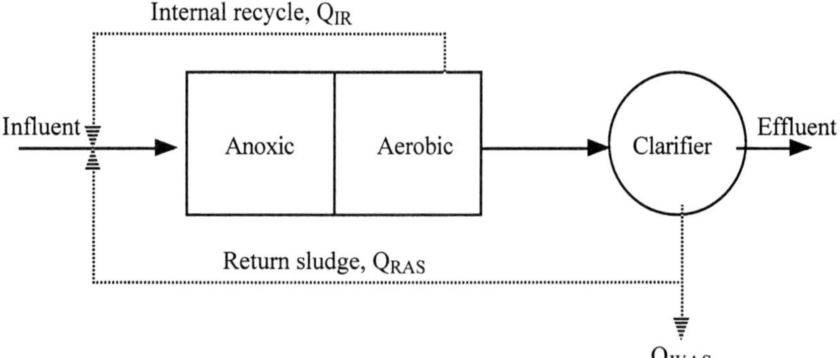


**Figure 6.47** Modified Ludzack–Ettinger process for nitrogen removal.

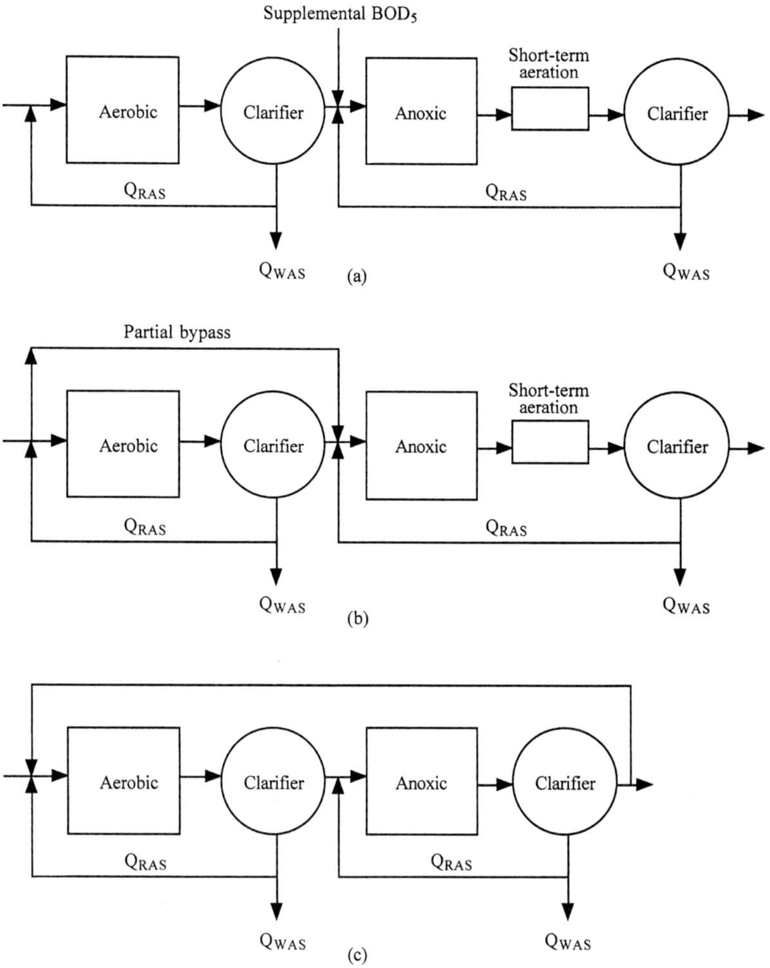

**Figure 6.48** Dual-sludge processes for nitrogen removal (*source*: *Grady and Lim, 1980*).

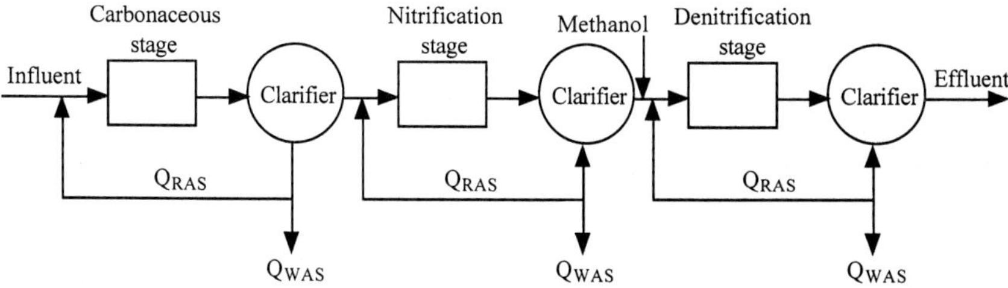


**Figure 6.49** Triple-sludge process for nitrogen removal.

by WEF and ASCE (1991b); this uses the activated-sludge kinetic model developed by Lawrence and McCarty (1970), combined with empirical equations for complete denitrification using a four-stage Bardenpho process.

### 28.3 Nitrogen control

Nitrogen (N) in wastewater exists commonly in the form of organic, ammonia, nitrite, nitrate, and gaseous nitrogen. The organic nitrogen includes both soluble and particulate forms. The soluble organic nitrogen is mainly in the form of urea and amino acids. The main sources are from human excreta, kitchen garbage, and industrial (food processing) wastes. Typical domestic wastewater contains 20 mg/L of organic nitrogen and 15 mg/L of inorganic nitrogen.

**Environmental effects.** Nitrogen compounds, particularly ammonia, will exert a significant oxygen demand through biological nitrification and may cause eutrophication in receiving waters. Ammonia (unionized) can be toxic to aquatic organisms and it readily reacts with chlorine (affecting disinfection efficiency). A high nitrate ($NO_3$) level in water supplies has been reported to cause methemoglobinemia in infants. The need for nitrogen control in wastewater effluents has generally been recognized.

**Treatment process.** Many treatment processes have been developed with the specific purpose of transforming nitrogen compounds or removing nitrogen from the wastewater stream.

**Conventional processes.** In conventional treatment processes, the concentration of inorganic nitrogen is not affected by primary sedimentation and is increased more than 50% to 24 mg/L after secondary (biological) treatment. The overall primary and secondary treatment removes 25% to 75% (5 to 15 mg/L) of organic nitrogen. Typically, about 14% and 26% of total nitrogen from raw wastewater are removed by conventional primary and secondary treatment processes, respectively.

Biological processes remove particulate organic nitrogen and transform some to ammonium and other inorganic forms. A fraction of the ammonium present will be assimilated into organic materials of cells. Soluble organic nitrogen is partially transformed to ammonium by microorganisms, with 1 to 3 mg/L of organic nitrogen remaining insoluble in the secondary effluent.

**Advanced processes.** Advanced wastewater treatment processes designed to remove wastewater constituents other than nitrogen often remove some nitrogen compounds as well. Removal is usually limited to particulate forms, and the removal efficiency is not high.

Tertiary filtration removes the suspended organic nitrogen from the secondary effluent. However the majority of nitrogen is inorganic (ammonium). Reverse osmosis and electrodialysis can be used as a tertiary process for ammonium removal. Their effectiveness is 80% and 40%, respectively. In practice, RO and electrodialysis are not used for wastewater treatment.

Chemical coagulation for phosphorus removal also removes particulate organic nitrogen. The process for nitrogen control may be divided into two categories, i.e. nitrification and nitrification–denitrification, depending on the quality requirements of wastewater effluent. The nitrification process is the oxidation of organic and ammonia nitrogen ($NH_3$-N) to nitrate, a less objectionable form; it is merely the conversion of nitrogen from one form to another form in the wastewater. Denitrification is the reduction of nitrate to nitrogen gas, constituting a removal of nitrogen from wastewater. Nitrification is used only to control $NH_3$-N concentration in the wastewater. Nitrification–denitrification is employed to reduce the total level of nitrogen in the effluent.

There are several methods of nitrogen control. They are biological nitrification–denitrification, break-point chlorination, selective ion exchange, and air (ammonia) stripping. This section is mainly devoted to biological nitrification or biological technologies, since most processes have been discussed previously in Chapter 5.

Biological nitrification can be achieved in separate stage processes following secondary treatment (most cases) or in combination for carbon oxidation–nitrification and for carbon oxidation–nitrification–denitrification. The secondary effluent with its high ammonia content and low BOD provides greater growth potential for the nitrifiers relative to the heterotrophic bacteria. The nitrification process is operated at an increased sludge age to compensate for lower temperature. After the nitrifiers have oxidized the ammonia in the aeration tank, the activated sludge containing a large fraction of nitrifiers is settled in the final clarifier for return to the aeration tank and for waste.

**Nitrification reaction.** Biological nitrification is an aerobic autotrophic process in which the energy for bacterial growth is derived from the oxidation of inorganic compounds, primarily ammonia nitrogen. Autotrophic nitrifiers, in contrast to heterotrophs, use inorganic carbon dioxide instead of organic carbon for cell synthesis. The yield of nitrifier cells per unit of substrate metabolized is many times smaller than that for heterotrophic bacteria.

Although a variety of nitrifying bacteria exist in nature, the two genera associated with biological nitrification are *Nitrosomonas* and *Nitrobacter.* The oxidation of ammonia to nitrate is a two-step process requiring both nitrifiers for the conversion. *Nitrosomonas* oxidizes ammonia to nitrite, while *Nitrobacter* subsequently transforms nitrite to nitrate. The respective oxidation reactions are as follows:

*Ammonia oxidation:*

$$NH_4^+ + 1.5O_2 + 2HCO_3^- \xrightarrow{Nitrosomonas} NO_2^- + 2H_2CO_3 + H_2O \quad (6.179)$$

*Nitrite oxidation:*

$$NO_2^- + 0.5O_2 \xrightarrow{Nitrobacter} NO_3^- \quad (6.180)$$

*Overall reaction:*

$$NH_4^+ + 2O_2 + 2HCO_3^- \xrightarrow{Nitrifiers} NO_3^- + 2H_2CO_3 + H_2O \quad (6.181)$$

*Note*: The above three equations are essentially the same as Eqs. (1.87), (1.88), and (1.89), respectively.

To oxidize 1 mg/L of $NH_3$-N, theoretically 4.56 mg/L of oxygen is required when synthesis of nitrifiers is neglected. As ammonia is oxidized and bicarbonate is utilized; nitrate is formed and carbonic acid is produced. The carbonic acid will further depress the pH of the wastewater. Theoretically, alkalinity of 7.14 mg/L $CaCO_3$ is consumed for each mg/L of $NH_3$-N oxidized. The destruction of alkalinity and increase of carbonic acid results in a drop in pH of the wastewater.

Reviews of the synthesis and energy relationship associated with biological nitrification are available elsewhere (Painter, 1970, 1975; Haug and McCarty, 1972; US EPA, 1975). Overall oxidation and synthesis–oxidation reactions of ammonia are presented below (US EPA, 1975c):

$$55NH_4^+ + 76O_2 + 109HCO_3^- \rightarrow C_5H_7NO_2 + 54NO_2^- + 57H_2O + 104H_2CO_3$$

*Nitrosomonas* (6.182)

and

$$400NO_2^- + NH_4^+ + 4H_2CO_3 + HCO_3^- + 195O_2 \rightarrow C_5H_7NO_2 + 3H_2O + 400NO_2^-$$

*Nitrobacter* (6.183)

The overall synthesis and oxidation reaction is

$$NH_4^+ + 1.83O_2 + 1.98HCO_3^- \rightarrow 0.021C_5H_7NO_2 + 1.041H_2O + 0.98NO_3^- + 1.88H_2CO_3 \quad (6.184)$$

**Nitrifying biofilm properties.** Although the principal genera *Nitrosomonas* and *Nitrobacter* are responsible for biological nitrification, heterotrophic nitrification can also occur when nitrite and/or nitrate are produced from organic or inorganic compounds by heterotrophic organisms (>100 species, including fungi). However, the amount of oxidation nitrogen formed by heterotrophic organisms is relatively small (Painter, 1970).

The growth rate for nitrifying bacteria is much less than of heterotrophic bacteria. Nitrifying bacteria have a longer generation time of at least 10 to 30 h (Painter, 1970, 1975). They are also much more sensitive to environmental conditions as well as to growth inhibitors. The growth rate for nitrite oxidizers is much greater than that for the ammonia oxidizers.

According to Painter (1970), experimental yield values of *Nitrosomonas* lay between 0.04 and 0.13 lb volatile suspended solids (VSS) grown per pound of ammonia nitrogen oxidized; and for *Nitrobacter* the yield ranged from 0.02 to 0.07 lb VSS per pound of nitrite nitrogen oxidized. Yield values based on thermodynamic theory are 0.29 and 0.084, respectively for *Nitrosomonas* and *Nitrobacter* (Haug and McCarty, 1972). The lower value of the experimental yield is due to the diversion of a portion of the free energy released by oxidation to microorganism maintenance functions. Based on Eqs. (6.182) and (6.183), the yield for *Nitrosomonas* and *Nitrobacter* should be 0.15 mg cells/mg $NH_4^+$-N and 0.02 mg cells/mg $NO_2$-N, respectively.

**Kinetics of nitrification.** The kinetics of biological nitrification, using mathematical expressions for the oxidation of ammonia and nitrite, have been proposed by many investigators. A variety of environmental factors, such as ammonia concentration, pH, temperature, and dissolved oxygen level, affect the kinetics. For the combined carbon oxidation–nitrification processes, BOD to TKN ratio is also an important factor. The reaction is enhanced with higher pH, higher temperature, and higher DO concentration.

**Effect of ammonia concentration on kinetics.** Numerous investigators have developed growth kinetics and substrate utilization equations for nitrifiers. However, the most popular method is the Monod model (1949) of population dynamics. The growth rate of nitrifiers is a function of ammonia concentration dynamics, as follows:

$$\mu = \frac{1}{X}\frac{dS}{dt} = \hat{\mu}\frac{S}{K_s + S} \tag{6.185}$$

where $\mu$ = growth rate of nitrifier, per day
$\hat{\mu}$ = maximum specific growth rate of nitrifier, per day
$X$ = concentration of bacteria
$S$ = concentration of substrate ($NH_3$-N), mg/L
$t$ = time, days
$K_s$ = half velocity constant, i.e. substrate concentration (mg/L) at half of the maximum growth rate

Equation (6.185) assumes no mass transfer or oxygen transfer limitations.

Values of $K_s$ for *Nitrosomonas* and *Nitrobacter* are generally equal to or less than 1 mg/L (0.18 to 1.0 mg/L mostly) at liquid temperatures of 20°C or less (Haug and McCarty, 1972). The Monod expression for biological nitrification provides for a continuous transition between first- and zero-order kinetics based on substrate concentration. Since $K_s$ values for nitrification are much less than ammonia concentrations found in wastewater, many investigators (Huang and Hopson, 1974, Wild et al., 1971; Kiff, 1972) found that the growth kinetic model reduces to a zero-order expression

$$\mu = \frac{1}{X}\frac{dS}{dt} = \hat{\mu} \tag{6.186}$$

Equation (6.186) indicates that the nitrification rate is independent of the initial substrate concentration and the mixing regime. This suggests that nitrifiers are reproducing at or near their maximum growth rate. As substrate is removed, and as $S$ approaches the value of $K_s$ or less, further substrate removal will begin to approximate a first-order reaction.

Estimates of the maximum growth rates ($\hat{\mu}$) of *Nitrosomonas* and *Nitrobacter* under various environmental conditions, found by other investigators, are summarized elsewhere (US EPA, 1975c; Brenner *et al.*, 1984). Both $\hat{\mu}$ and $K_s$ are found to increase with increasing temperature. All nitrification studies were conducted with activated-sludge processes or river waters. Maximum growth rates might be expected to approximate those of suspended growth process in which ammonia mass transfer and DO are not limiting.

**TABLE 6.23 Values of Kinetic Coefficients for the Suspended Growth Nitrification Process (Pure Culture Values)**

| Kinetic coefficient | Unit | Value (20°C) | |
|---|---|---|---|
| | | Range | Typical |
| *Nitrosomonas* | | | |
| $\mu_m$ | per day | 0.3–2.0 | 0.7 |
| $K_s$ | $NH_4^+$-N, mg/L | 0.2–2.0 | 0.6 |
| *Nitrobacter* | | | |
| $\mu_m$ | per day | 0.4–3.0 | 1.0 |
| $K_s$ | $NO_2^-$-N, mg/L | 0.2–5.0 | 1.4 |
| *Overall* | | | |
| $\mu_m$ | per day | 0.3–3.0 | 1.0 |
| $K_s$ | $NH_4^+$-N, mg/L | 0.2–5.0 | 1.4 |
| $Y$ | mg VSS/mg $NH_4^+$-N | 0.1–0.3 | 0.2 |
| $k_d$ | per day | 0.03–0.06 | 0.05 |

SOURCES: US EPA, 1975b and 1975c; Schroeder, 1977

The ranges and typical values of kinetic coefficients for suspended growth nitrification processes using pure cultures are summarized in Table 6.23. In practice, the values of those coefficients for nitrifiers in activated-sludge processes will be considerably less than the values listed in the table.

The reported data indicate that the maximum growth rate of *Nitrobacter* is considerably greater than that of *Nitrosomonas*. Therefore the oxidation of ammonia to nitrite is the rate-limiting reaction in nitrification. For this reason, nitrite does not accumulate in large amounts in mature nitrification systems for municipal wastewaters.

**Oxidation rate.** The ammonia oxidation rate is related to the *Nitrosomonas* growth rate. Its maximum oxidation rate can be expressed as

$$r_N = \frac{\mu_N}{Y_N} = \hat{r}_N \frac{N}{K_N + N} \tag{6.187}$$

where $r_N$ = ammonia oxidation rate, lb $NH_4^+$-N oxidized/(lb VSS · d)

$\mu_N$ = *Nitrosomonas* growth rate, $day^{-1}$

$Y_N$ = organism yield coefficient, lb *Nitrosomonas* grown(VSS)/lb $NH_4^+$-N removed

$\hat{r}_N$ = $\hat{\mu}_N/Y_N$ = peak ammonia oxidation rate, lb $NH_4^+$-oxidized/(lb VSS · d)

$\hat{\mu}_N$ = peak *Nitrosomonas* growth rate, $day^{-1}$

$N$ = $NH_4^+$-N concentration, mg/L

$K_N$ = half saturation constant, mg/L $NH_4^+$-N, mg/L

**Loading rate.** Referring to Eqs. (6.185) and (6.187), the growth rate of nitrifier is proportional to substrate ($NH_3$-N) concentration. The ammonia loading rates applied to the biological nitrification unit (aeration tank) are 160 to 320 g/($m^3 \cdot$ d) (10 to 20 lb/(1000 $ft^3 \cdot$ d)) with corresponding wastewater temperature of 10 to 20°C, respectively. The aeration periods are 4 to 6 h for an average wastewater secondary effluent (Hammer, 1986).

**BOD concentration.** The combined process oxidizes a high proportion of the influent organics ($SBOD_5$) relative to the $NH_3$-N concentration, while less nitrifiers are present in the biofilm. Separate nitrification processes have relatively low $SBOD_5$ values, relative to the $NH_3$-N content, when a high level of $SBOD_5$ removal is provided prior to the nitrification stages. High populations of nitrifiers occur in the separate-stage nitrification process.

In combined carbon oxidation–nitrification processes the ratio of BOD to TKN is greater than 5, whereas in separate processes the BOD to TKN ratio in the second stage is greater than 1 and less than 3. Reducing the ratio to 3 does not require a high degree of treatment in the first stage (US EPA, 1975b; McGhee, 1991).

Biological nitrification is sensitive to organic loading, depending on the carbon removal sections, because of the nitrifiers having a much slower growth rate. Nitrification will take place when organic content is reduced to a certain level. There are discrepancies about what is the critical concentration for organic matter. The values cited in the literature are as follows: $TBOD_5$ approaching 30 mg/L (Antonie, 1978; Khan and Raman, 1980); $TBOD_5$ reaching 20 mg/L (Banerji, 1980); $SBOD_5$ less than 20 mg/L (Autotrol Corp, 1979); $SBOD_5$ reduced to 15 mg/L; $SBOD_5$ around 10 mg/L (Miller *et al.*, 1980).

**Temperature.** The optimum range for nitrification is 30 to 36°C (Haug and McCarty, 1972; Ford *et al.*, 1980). Nitrifiers have been found not to grow at temperatures below 4°C or above 45°C (Ford *et al.*, 1980).

In a suspended-growth activated-sludge system, Downing and Hopwood (1964) found that the maximum growth $\hat{\mu}$ and the half-velocity constant $K_N$ for both *Nitrosomonas* and *Nitrobacter* were markedly influenced by temperature. The relationships developed for *Nitrosomonas* are as follows:

$$\hat{\mu} = 0.47e^{0.098(T-15)}, \text{ per day} \tag{6.188}$$

and

$$K_N = 10^{0.051T-1.158}, \text{ mg/L as N} \tag{6.189}$$

where $T$ is the wastewater temperature, °C.

For attached-growth systems, different temperature effects have been reported. Attached-growth systems have an advantage in withstanding low temperature (<15°C) without as severe a loss of nitrification rates as suspended-growth systems.

The following equation describes temperature effects on nitrification rate at a temperature *T*, °C

$$\mu_T = \mu_{20}\theta^{T-20} \tag{6.190}$$

where *T* is wastewater temperature, °C and $\theta$ is a temperature correction coefficient. From laboratory and pilot RBC studies, the value of $\theta$ is found to be 1.10 (Mueller *et al.,* 1980).

**Dissolved oxygen.** When dissolved oxygen concentration is a limiting factor, the growth rate of nitrifiers can be expressed as following the Monod-type relationships as follows:

$$\mu_{\mathrm{N}} = \hat{\mu}_{\mathrm{N}}\left(\frac{\mathrm{DO}}{K_{\mathrm{O_2}} + \mathrm{DO}}\right) \tag{6.191}$$

where $K_{O_2}$ is the half-saturation constant for oxygen in mg/L. Values of $K_{O_2}$ have been reported to vary from 0.15 mg/L at 15°C to 2.0 mg/L at 20°C (US EPA, 1975b).

**Alkalinity and pH.** From stoichiometry (Eq. (6.181)), 7.14 mg/L of alkalinity as $CaCO_3$ will be destroyed for every one mg/L of $NH_3$-N oxidized. However, Sherrard (1976) claimed this value to be in error and suggested that it should be less for biological nitrification, because ammonia is incorporated into the biomass resulting in a lesser quantity of $NH_3$-N available for oxidation to nitrate. The alkalinity destruction in a nitrifying activated sludge process is a function of solids retention time and influent wastewater $BOD_5$:N:P ratio. Alleman and Irving (1980) observed a considerably low value (2.73 mg/L of alkalinity as $CaCO_3$ per mg/L of $NH_3$-N oxidized) for nitrification in a sequential batch reactor.

With the RBC system at Cadillac, Michigan, alkalinity declined 8.1 mg/L for each mg/L of $NH_3$-N oxidized (Singhal, 1980; Chou *et al.,* 1980). At Princeton, Illinois, the alkalinity reduction ratios were respectively 6.8 and 9.1 for stages 3 and 4 in a combined $BOD_5$-nitrification 5-stage RBC system (Lin *et al.*, 1982). Therefore it is important to have sufficient alkalinity buffering capacity for nitrifiers to maintain the acceptable pH range.

Reported data show a wide range of optimum pH of 7.0 to 9.5 with maximum activity at approximately pH 8.5 (Painter, 1970, 1975; Haug and McCarty, 1972; Antonie, 1978). Below pH 7.0, adverse effects on ammonia oxidation become pronounced.

Full-scale kinetic studies conducted at the Autotrol Corporation (1979) showed that the nitrification rate declined from 0.31 lb $NH_3$-N/(d · 1000 ft$^2$) (1.52 g $NH_3$-N/(m$^2$ · d)) at pH 7.0 to 0.17 lb/(d · 1000 ft$^2$) (0.83 g/(m$^2$ · d)) at pH 6.5.

Downing and Hopwood (1964) proposed the following relationship for the growth rate for nitrifiers and pH values up to 7.2 in a combined carbon oxidation and nitrification system.

$$\mu_N = \hat{\mu}_N[1 - 0.833(6.2 - \text{pH})] \tag{6.192}$$

It is assumed that the growth rate is constant for pH levels between 7.2 and 8.0. Eq. (6.192), when applied to separate-stage nitrification systems, is probably conservative.

**Combined kinetics expression.** As previously presented, the major factors which affect the nitrification rate are ammonia nitrogen concentration, temperature, DO, and pH. In the absence of toxic or inhibitory substance in the wastewater, US EPA (1975b) applied Chen's model (1970) which combined Monod's expression for ammonia-N concentration (Eq. (6.185)), DO (Eq. (6.191)), and pH (Eq. (6.192)) effects on nitrifier growth as follows:

$$\mu_N = \hat{\mu}_N\left(\frac{N}{K_N + N}\right)\left(\frac{\text{DO}}{K_{O_2} + \text{DO}}\right)[1 - 0.833(6.2 - \text{pH})] \tag{6.193a}$$

Substituting the effects of temperature (Eq. (6.188)), Eq. (6.189) for $K_N$, 1.3 mg/L for $K_{O_2}$, the following equation is valid for *Nitrosomonas* for pH $< 7.2$ and wastewater temperature between 8 and 30°C

$$\mu_N = 0.47\left[e^{0.098(T-15)}\right]\left[\frac{N}{10^{(0.051T-1.158)} + N}\right]\left[\frac{\text{DO}}{1.3 + \text{DO}}\right] \tag{6.193b}$$

$$[1 - 0.833(6.2 - \text{pH})]$$

The terms in the first bracket represent the effect of temperature. The second bracket terms are the Monod expression for the effect of $NH_3$-N concentration. The terms in the third bracket take into account the effect of DO. The terms in the last bracket account for the effect of pH: it will be unity for pH $\geq 7.2$. From Eq. (6.193) it can be seen that if any one factor becomes limiting, even if all the others are non-limiting, the nitrification rate will be much lower, perhaps even approaching zero.

**Example:** Design an activated-sludge process for carbon oxidation–nitrification using the following given data. Determine the volume of the aeration tank, daily oxygen requirement, and the mass of organisms removed daily from the system.

| | |
|---|---|
| Design average flow | 3785 $m^3$/d (1 Mgal/d) |
| BOD of influent (primary effluent) | 160 mg/L |
| TKN of influent (primary effluent) | 30 mg/L |
| $NH_3$-N of influent (primary effluent) | 15 mg/L |
| Minimum DO in aeration tank | 2.0 mg/L |
| Temperature (minimum) | 16°C |
| Maximum growth rate, $\hat{\mu}$ | 1.0 per day |
| Minimum pH | 7.2 |
| Assume a safety factor, SF (required due to transient loading conditions) | 2.5 |
| MLSS | 2500 mg/L |
| Maximum effluent SS or BOD | 15 mg/L |
| Total alkalinity, as $CaCO_3$ | 190 mg/L |

**solution:**

Step 1. Compute the maximum growth rate of nitrifiers under the stated operating conditions

$$T = 16°\text{C}$$
$$\text{DO} = 2 \text{ mg/L}$$
$$\text{pH} \geq 7.2$$

Additional required data for Eqs. (6.193a) and (6.193b)

$$\hat{\mu}_\text{N} = 0.47 \text{ d}^{-1}$$
$$K_{\text{O}_2} = 1.3 \text{ mg/L}$$
$$K_\text{N} = 10^{0.051T-1.158} = 10^{0.051\times 16-1.158}$$
$$= 0.455 \text{ mg/L as N}$$
$$N = 15 \text{ mg/L as N}$$

$$\text{Temperature correction factor} = e^{0.098(16-15)} = 1.10$$

$$\text{pH correction factor} = 1 \text{ for pH} \geq 7.2$$

Eqs. (6.193a) and (6.193b)

$$\mu_\text{N} = \hat{\mu}_\text{N}\left(\frac{N}{K_\text{N} + N}\right)\left(\frac{\text{DO}}{K_{\text{O}_2} + \text{DO}}\right)[1 - 0.833(6.2 - \text{pH})]$$

$$= 0.47\text{d}^{-1}\left[e^{0.098(T-15)}\right]\left(\frac{15}{0.455 + 15}\right)\left(\frac{2}{1.3 + 2}\right)(1)$$

$$= (0.47 \text{ d}^{-1})(1.1)(0.97)(0.61)(1)$$

$$= 0.30 \text{ d}^{-1}$$

Step 2. Compute the maximum ammonia oxidation rate under the environmental conditions of temperature, pH, and DO

Using Eq. (6.187)

$$r_N = \frac{\mu_N}{Y_N}$$

Referring to Table 6.23

$$Y_N = 0.2 \text{ mg VSS/mg } NH_4^+\text{-N}$$

$$\text{Max. } \hat{r}_N = (0.30 \text{ d}^{-1})/0.2$$

$$= 1.5 \text{ d}^{-1}$$

Step 3. Compute the minimum cell residence time

Using Eq. (6.76)

$$1/\theta_c \approx Yk' - k_d$$

$$Y = Y_N = 0.2 \text{ (from Table 6.23)}$$

$$k' = \hat{r}_N = 1.5 \text{ d}^{-1} \text{ (from Step 2)}$$

$$k_d = 0.05 \text{ (from Table 6.23)}$$

$$1/\theta_c = 0.2(1.5 \text{ d}^{-1}) - 0.05 \text{ d}^{-1}$$

$$= 0.25 \text{ d}^{-1}$$

$$\text{Minimum } \theta_{c-\min} = 1/0.25 \text{ d}^{-1}$$

$$= 4 \text{ days}$$

*Note*: Another rough estimate of $\theta_c$ (solids retention time) can be determined by

$$\theta_c = 1/\mu_N = 1/0.30 \text{ d}^{-1} = 3.33 \text{ days}$$

Step 4. Compute the design cell residence time

$$\text{Design } \theta_{c-d} = \text{SF} \times \text{min.}\, \theta_{c-\min} = 2.5 \times 4 \text{ days}$$

$$= 10 \text{ days}$$

Step 5. Compute the design specific substrate utilization rate $U$ for ammonia oxidation

Using Eq. (6.84)

$$\frac{1}{\theta_c} = YU - k_d$$

or

$$U = \frac{1}{Y}\left(\frac{1}{\theta_c} + k_d\right)$$

Applying $\theta_{c-d}$ as $\theta_c$

$$= \frac{1}{0.2}\left(\frac{1}{10 \text{ days}} + 0.05 \text{ d}^{-1}\right)$$

$$= 0.75 \text{ d}^{-1}$$

*Note*: $U = r_N$ for Eq. (6.187).

Step 6. Compute the steady state ammonia concentration of the effluent, $N$

Using Eq. (6.187)

$$r_N = \hat{r}_N \frac{N}{K_N + N}$$

where $U = 0.75 \text{ d}^{-1}$ (Step 5)
$K_N = 0.455$ mg/L for $NH_4^+$-N (Step 1)
$\hat{r}_N = 1.50 \text{ d}^{-1}$ = maximum growth rate (Step 2)

$$0.75 \text{ d}^{-1} = \frac{(1.50 \text{ d}^{-1})N}{(0.455 \text{ mg/L}) + N}$$

$$N = 0.455 \text{ mg/L}$$

Step 7. Compute organic (BOD) removal rate $U$

The design cell residence time $\theta_{c-d}$applies to both the nitrifiers and heterotrophic bacteria.

$$\frac{1}{\theta_{c-d}} = YU - k_d$$

$$\theta_{c-d} = 10 \text{ days}$$

Referring to Table 6.12, for heterotrophic kinetics

$$Y = 0.6 \text{ kg VSS/kg BOD}_5$$

$$k_d = 0.06 \text{ day}^{-1}$$

Then

$$\frac{1}{10 \text{ d}} = 0.6U - (0.06 \text{ day}^{-1})$$

$$U = 0.27 \text{ kg BOD}_5 \text{ removed/(kg MLVSS} \cdot \text{d)}$$

assuming there is 90% $BOD_5$ removal efficiency.

The food to microorganism ratio is

$$F/M = 0.27/0.9$$
$$= 0.30 \text{ kg BOD}_5 \text{ applied/(kg MLVSS} \cdot \text{d)}$$

Step 8. Compute the hydraulic retention time $\theta$ required for organic and ammonia oxidations

Using Eq. (6.83)

$$U = \frac{S_0 - S}{\theta X} \qquad \text{or} \qquad \theta = \frac{S_0 - S}{UX}$$

(a) For organic oxidation

$$S_0 = 160 \text{ mg/L} \quad \text{(given)}$$
$$S = 15 \text{ mg/L} \quad \text{(regulated)}$$
$$U = 0.27 \text{ day}^{-1} \quad \text{(Step 7)}$$
$$X = \text{MLVSS} = 0.8 \text{ MLSS} = 0.8 \times 2500 \text{ mg/L}$$
$$= 2000 \text{ mg/L}$$

Therefore
$$\theta = \frac{(160 - 15) \text{ mg/L}}{(0.27 \text{ day}^{-1})(2000 \text{ mg/L})}$$
$$= 0.269 \text{ day}$$
$$= 6.4 \text{ h}$$

(b) For nitrification

$$N_0 = \text{TKN} = 30 \text{ mg/L} \quad \text{(given)}$$
$$N = 0.45 \text{ mg/L} \quad \text{(Step 6)}$$
$$U = 0.75 \text{ d}^{-1} \quad \text{(Step 5)}$$
$$X = 2000 \text{ mg/L} \times 0.08 \quad \text{(assuming 8\% of VSS is nitrifiers)}$$
$$= 160 \text{ mg/L}$$
$$\theta = \frac{(30 - 0.45) \text{ mg/L}}{(0.75 \text{ d}^{-1})(160 \text{ mg/L})}$$
$$= 0.246 \text{ days}$$
$$= 5.9 \text{ h}$$

*Note*: Organic oxidation controls HRT; $\theta = 6.4$ h

Step 9. Determine the volume $V$ of aeration tank required, based on organic removal process

$$V = Q\theta = (3785\ \text{m}^3/\text{d})\ (0.269\ \text{day})$$
$$= 1020\ \text{m}^3$$
$$= 36{,}000\ \text{ft}^3$$

Step 10. Compute $BOD_5$ loading rate

$$\text{BOD loading} = (3785\ \text{m}^3/\text{d})(160\ \text{g/m}^3)/(1000\ \text{g/kg})$$
$$= 606\ \text{kg/d}$$
$$\text{BOD loading rate} = (606\ \text{kg/d})/(1020\ \text{m}^3)$$
$$= 0.59\ \text{kg/(m}^3 \cdot \text{d)}$$
$$= 37.1\ \text{lb/(1000 ft}^3 \cdot \text{d) (OK)}$$

*Note*: This rate is below the range of 50 to 120 lb/(1000 $\text{ft}^3 \cdot$ d) for the complete-mix activated-sludge process.

Step 11. Estimate the total quantity of oxygen supply needed

The total quantity of oxygen required can be estimated on the basis of oxygen demand for $BOD_5$ and TKN removal and net mass of volatile solids (cells) produced with conversion factors. A procedure similar to the one illustrated in the section on the activated-sludge process may be used.

Alternatively, a rough estimation can be calculated by the following equation from the influent $BOD_5$ and TKN concentrations (Metcalf and Eddy, Inc. 1991)

$$O_2,\ \text{lb/d} = Q(kS_0 + 4.57\ \text{TKN})\ 8.34$$

where $Q$ = influent flow, Mgal/d
$k$ = conversion factor for BOD loading on nitrification system
= 1.1 to 1.25
4.57 = conversion factor for complete oxidation of TKN
8.34 = a conversion factor, lb/((Mgal · mg/L)

For this example, let $k$ = 1.18 and SF = 2.5. Then

$$\text{Required } O_2,\ \text{lb/d} = 1\ \text{Mgal/d}\ [1.18(160\ \text{mg/L}) + 4.57(30\ \text{mg/L})]$$
$$\times [8.34\ \text{lb/(Mgal} \cdot \text{mg/L)}] \times 2.5$$
$$= 6795\ \text{lb/d}$$
$$= 3082\ \text{kg/d}$$

Step 12. Determine the sludge wasting schedule

Sludge wasting includes solids contained in the effluent from secondary clarifier effluent and sludge waste in the return activated sludge or mixed liquor.

The sludge to be wasted under steady-state conditions is the denominator of Eq. (6.71)

$$\theta_c = \frac{VX}{Q_{wa}X + Q_eX_e}$$

Total solids wasted per day is

$$Q_{wa}X + Q_eX_e = VX/\theta_c$$

$$V = 1020 \text{ m}^3$$

$$X = \text{MLVSS} = 0.8 \times \text{MLSS} = 0.8 \times 2500 \text{ mg/L}$$

$$= 2000 \text{ mg/L}$$

$$VX = 1020 \text{ m}^3 \times 2000 \text{ (g/m}^3\text{)/(1000 g/kg)}$$

$$= 2040 \text{ kg}$$

$$\theta_c = 10 \text{ days (use design } \theta_{c-d}\text{, Step 4)}$$

$$VX/\theta_c = 2040 \text{ kg/10 days}$$

$$= 204 \text{ kg/d}$$

The solids contained in the clarifier effluent at 3785 $m^3$/d (1 Mgal/d) are calculated from the VSS in the effluent

$$\text{Effluent VSS} = 0.8 \times 15 \text{ mg/L} = 12 \text{ mg/L} = 12 \text{ g/m}^3$$

$$Q_eX_e = 3785 \text{ m}^3\text{/d} \times 12\text{(g/m}^3\text{)/(1000 g/kg)}$$

$$= 45.4 \text{ kg/d}$$

The VSS (microorganism concentration) to be wasted from mixed liquor or return sludge is

$$Q_{wa}X = VX/\theta_c - Q_eX_e = (204 - 45) \text{ kg/d}$$

$$= 159 \text{ kg/d}$$

Step 13. Check the buffering capacity of the wastewater

Theoretically 7.14 mg/L of alkalinity as $CaCO_3$ is destroyed per mg/L of $NM_4^+$-N oxidized. The alkalinity remaining after nitrification would be at least

$$\text{Alk.} = 190 \text{ mg/L} - 7.14 \text{ (15 mg/L)}$$

$$= 83 \text{ mg/L as } CaCO_3$$

This should be sufficient to maintain the pH value in the aeration tanks above 7.2.

**Combined carbon oxidation–nitrification in attached growth reactors.** Two attached growth reactors, the trickling filter process and the RBC process, can be used for combined carbon oxidation–nitrification.

Detailed design procedures for these two processes are described elsewhere (US EPA, 1975c) and in the manufacturers' design manual.

In the design of nitrification with trickling filters, the total surface area required is determined on the basis of empirical unit surface area for unit $NH_4^+$-N oxidized per day under the BOD/TKN and the sustained temperature conditions. The choice of filter media is based on the effluent $BOD_5$ and SS requirements. A circulation rate of 1:1 is usually adequate.

**Nitrification with RBC system.** For a combined oxidation–nitrification process with the RBC system, a two-step design procedure is needed. Significant nitrification will not occur in the RBC process until the soluble BOD concentration is reduced to 15 mg/L or less (Autotrol, 1979). In the first design the media surface area required to reduce SBOD is determined as shown in the example illustrated in the previous section. For influent ammonia nitrogen concentrations of 15 mg/L or above, it is necessary to reduce SBOD concentrations to less than 15 mg/L, i.e. to about the same value as the ammonia nitrogen concentration. The second design uses the nitrification design curves (Fig. 6.50) to determine the RBC area to reduce the influent ammonia nitrogen level to the required effluent concentration. The sum of the two RBC surface areas is determined as the total surface area required for the combined

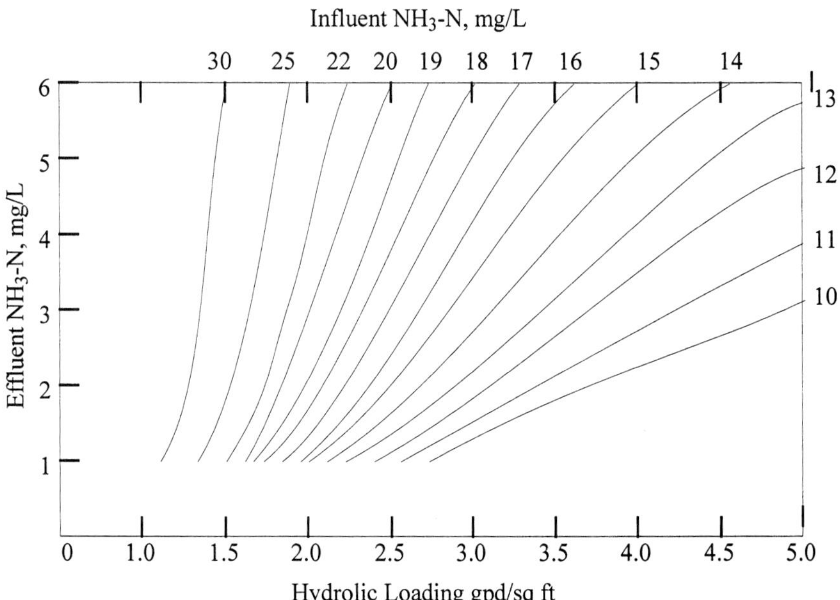


**Figure 6.50** RBC process design curve for nitrification of domestic wastewater *(source: Autotrol, 1979).*

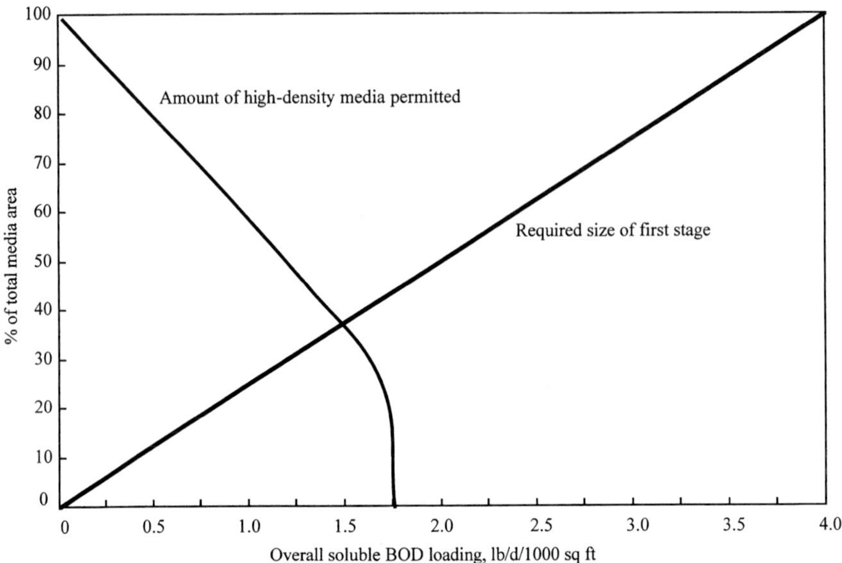


**Figure 6.51** Design curves for RBC carbon oxidation–nitrification of domestic wastewater.

carbon oxidation–nitrification system. Figure 6.51 shows the design curves for RBC carbon oxidization–nitrification of municipal wastewater. If the wastewater temperature is less than 55°F (12.8°C), a separate temperature correction should be made to each of the surface areas determined.

**Example:** Design an RBC system for organic and nitrogen removal with the following data. The effluent SBOD and $NH_3$-N are to be 6 and 2 mg/L, respectively. Total BOD = 12 mg/L.

| | |
|---|---|
| Design inflow | 4.0 Mgal/d (15,140 $m^3$/d) |
| Influent soluble BOD | 70 mg/L |
| Effluent soluble BOD | 6 mg/L |
| Influent $NH_3$-N | 18 mg/L |
| Effluent $NH_3$-N | 2 mg/L |
| Wastewater temperature | ≥ 55°F (12.8°C) |

**solution:**

Step 1. Determine the required surface area

(a) Determine hydraulic loading HL to reduce SBOD from 70 mg/L to 15 mg/L for starting nitrification

From Fig. 6.33

$$HL_b = 3.32 \text{ gal/(d} \cdot \text{ft}^2) \text{ for the influent SBOD} = 70 \text{ mg/L}$$

(b) Determine HL to reduce $NH_3$-N from 18 mg/L to 2 mg/L

From Fig. 6.50

$$HL_n = 2.1 \text{ gal/(d} \cdot \text{ft}^2)$$

(c) Determine overall hydraulic loading $HL_o$

$$\frac{1}{HL_o} = \frac{1}{HL_b} + \frac{1}{HL_n} = \frac{1}{3.32} + \frac{1}{2.1}$$

$$HL_o = 1.29 \text{ gal/(d} \cdot \text{ft}^2)$$

(d) Determine HL for reducing SBOD to 6 mg/L

From Fig. 6.33

$$HL_b = 2.0 \text{ gal/(d} \cdot \text{ft}^2)$$

This is greater than $HL_o$; thus, nitrification controls the overall design HL. Calculate the surface area requirement

$$\text{Area} = \frac{4{,}000{,}000 \text{ gal/d}}{1.29 \text{ gal/(d} \cdot \text{ft}^2)}$$

$$= 3{,}100{,}000 \text{ ft}^2$$

Step 2. Determine the size of first stage

(a) Calculate overall SBOD loading area

$$\text{SBOD loading} = \frac{4 \text{ Mgal/d} \times 70 \text{ mg/L} \times 8.34 \text{ lb/(Mgal} \cdot \text{mg/L)}}{3{,}100{,}000 \text{ ft}^2}$$

$$= 0.75 \text{ lb/d} \cdot 1000 \text{ ft}^2$$

(b) Sizing of first stage

The size of the first stage can be found from the manufacturer's design curve (Fig. 6.51) or by proportional calculation. An SBOD loading rate of 4.00 lb/(d · 1000 ft²) uses 100% of the total media area. Therefore

$$\frac{X\%}{100\%} = \frac{0.75 \text{ lb/(d} \cdot 1000 \text{ ft}^2)}{4.00 \text{ lb/(d} \cdot 1000 \text{ ft}^2)}$$

$$X = 18.8\% \text{ or } 19\%$$

(c) Calculate surface area of first stage

$$\text{Area} = 0.19\ (3{,}100{,}000 \text{ ft}^2)$$

$$= 589{,}000 \text{ ft}^2$$

Step 3. Determine media distribution

(a) Using high-density media, from Fig. 6.51,

maximum % of total media area =67% for 0.75 lb SBOD/(d · 1000 ft$^2$)

(b) High-density surface area

$$= 0.67\ (3{,}100{,}000\ \text{ft}^2)$$
$$= 2{,}077{,}000\ \text{ft}^2$$

(c) Standard media surface area

$$= (1 - 0.67)(3{,}100{,}000\ \text{ft}^2)$$
$$= 1{,}023{,}000\ \text{ft}^2$$

Step 4. Select configuration

(a) Standard media, S, (100,000 ft$^2$ per unit) assemblies in first stage

$$= \frac{589{,}000\ \text{ft}^2}{100{,}000\ \text{ft}^2/\text{unit}}$$

$$= 5.9\ \text{units (use 6 units with 3 trains)}$$

(b) Standard media (total)

$$= \frac{1{,}023{,}000}{100{,}000}$$

$$= 10.23\ \text{(use 12 units)}$$

(c) High-density media, H, (150,000 ft$^2$ per unit) assemblies

$$= \frac{2{,}077{,}000}{150{,}000}$$

$$= 13.8\ \text{(use 15 units)}$$

(d) Choose between three trains of five-stage operation as shown below

3 (SS + SS + HH + HH + H) = 27 (units with 5 stages in 3 trains)

or three trains of four-stage operation

3 (SS + SS + HHH + HH) = 27 (units with 4 stages in 3 trains)

Step 5. No temperature correction is needed, since the wastewater temperature is above 12.8°C (55°F).

Step 6. Determine power consumption

A rough estimation of power consumption is 2.5 kW per shaft

$$\text{Power consumption} = 2.5 \text{ kW/shaft} \times 27 \text{ shafts}$$
$$= 67.5 \text{ kW}$$

**Denitrification.** Biological denitrification is the conversion of nitrate to gaseous nitrogen species and to cell material by the ubiquitous heterotrophic facultative aerobic bacteria and some fungi. Denitrifiers include a broad group of bacteria such as *Pseudomonas, Micrococcus, Archromobacter,* and *Bacillus* (US EPA, 1975b). These groups of organisms can use either nitrate or oxygen as electron acceptor (hydrogen donor) for conversion of nitrate to nitrogen gas. Denitrification occurs in both aerobic and anoxic conditions. An anaerobic condition in the liquid is not necessary for denitrification, and 1 to 2 mg/L of DO does not influence denitrification. The term anoxic is preferred over anaerobic when describing the process of denitrification.

The conversion of nitrate to gaseous end products is a two-step process, called dissimilatory denitrification, with a series of enzymatic reactions. The first step is a conversion of nitrate to nitrite, and the second step converts the nitrite to nitrogen gas. The nitrate dissimilations are expressed as follows (Stanier *et al.,* 1963)

$$2NO_3^- \; 4e^- + 4H^+ \rightarrow 2NO_2^- + 2H_2O \tag{6.194}$$

$$2NO_2^- + 6e^- + 8H^+ \rightarrow N_2 + 4H_2O \tag{6.195}$$

Overall transformation yields

$$2NO_3^- + 10e^- + 12H^+ \rightarrow N_2 + 6H_2O \tag{6.196}$$

For each molecule of nitrate reduced, five electrons can be accepted. When methanol ($CH_3OH$) is used on the organic carbon sources, the dissimilatory denitrification can be expressed as follows:

*First step*

$$6NO_3^- + 2CH_3OH \rightarrow 6NO_2^- + 4H_2O + 2CO_2 \tag{6.197}$$

*Second step*

$$6NO_2^- + 3CH_3OH \rightarrow 3N_2 + 3H_2O + 3CO_2 + 6OH^- \tag{6.198}$$

*Overall transformation*

$$6NO_3^- + 5CH_3OH \rightarrow 3N_2 + 7H_2O + 5CO_2 + 6OH^- \tag{6.199}$$

In Eq. (6.199), nitrate serves as the electron acceptor and methanol as the electron donor.

When methanol is used as an organic carbon source in the conversion of nitrate to cell material, a process termed assimilatory (or synthesis) denitrification, the stoichiometric equation is

$$3NO_3^- + 14CH_3OH + 4H_2CO_3 \rightarrow 3C_5H_7O_2N + 20H_2O + 3HCO_3^- \quad (6.200)$$

From Eq. (6.200), neglecting cell synthesis, 1.9 mg of methanol is required for each mg of $NO_3$-N reduction (M/N ratio). Including synthesis results in an increase in the methanol requirement to 2.47 mg. The total methanol requirement can be calculated from that required for nitrate and nitrite reductions and deoxygenation as follows (McCarty *et al.*, 1969):

$$C_m = 2.47\,(NO_3\text{-N}) + 1.53\,(NO_2\text{-N}) + 0.87DO \quad (6.201)$$

where $C_m$ = methanol required, mg/L
$NO_3$-N = nitrate nitrogen removed, mg/L
$NO_2$-N = nitrite nitrogen removed, mg/L
DO = dissolved oxygen removed, mg/L

Biomass production can be computed similarly

$$C_b = 0.53\,(NO_3\text{-N}) + 0.32\,(NO_2\text{-N}) + 0.19DO \quad (6.202)$$

where $C_b$ = biomass production, mg/L

In general, an M/N ratio of 2.5 to 3.0 is sufficient for complete denitrification (US EPA, 1975c). A commonly used design value for the required methanol dosage is 3 mg/L per mg/L of $NO_3$-N to be reduced.

**Example:** Determine the methanol dosage requirement, the M/N ratio, and biomass generated for complete denitrification of an influent with a nitrate-N of 24 mg/L, nitrite-N of 0.5 mg/L, and DO of 2.5 mg/L.

**solution:**

Step 1. Determine methanol required, using Eq. (6.201)

$$\begin{aligned} C_m &= 2.47\,(NO_3\text{-N}) + 1.53\,(NO_2\text{-N}) + 0.87\,DO \\ &= 2.47\,(24\text{ mg/L}) + 1.53\,(0.5\text{ mg/L}) + 0.87\,(2.5\text{ mg/L}) \\ &= 62.2\text{ mg/L} \end{aligned}$$

Step 2. Calculate M/N ratio

$$\begin{aligned} M/N &= (62.2\text{ mg/L})/(24\text{ mg/L}) \\ &= 2.6 \end{aligned}$$

Step 3. Calculate biomass generated, using Eq. (6.202)

$$C_b = 0.53\ (\text{NO}_3\text{-N}) + 0.32\ (\text{NO}_2\text{-N}) + 0.19\ \text{DO}$$

$$= 0.53\ (24\ \text{mg/L}) + 0.32\ (0.5\ \text{mg/L}) + 0.19\ (2.5\ \text{mg/L})$$

$$= 13.4\ \text{mg/L}$$

**Kinetics of denitrification.** Similarly to nitrification, environmental factors affect the kinetic rate of denitrifier growth and nitrate removal. These factors are mentioned in the previous section.

**Effect of nitrate on kinetics.** The effect of nitrate on denitrifier growth rate can be expressed by the Monod equation

$$\mu_D = \hat{\mu}_D \frac{D}{K_D + D} \tag{6.203}$$

where $\mu_D$ = growth rate of denitrifier, per day
$\hat{\mu}_D$ = maximum growth rate of denitrifier, per day
$D$ = nitrate-N concentration, mg/L
$K_D$ = half saturation constant, mg/L nitrate-N
= 0.08 mg/L $NO_3$-N, for suspended-growth systems without solids recycle at 20°C
= 0.16 mg/L $NO_3$-N, for suspended-growth systems with solids recycle at 20°C
= 0.06 mg/L $NO_3$-N, for attached-growth systems at 25°C

With the effects of wastewater temperature ($T$, °C) and dissolved oxygen (DO, mg/L), the overall rate of denitrification can be described as follows:

$$\mu_D = \hat{\mu}_D \times \frac{D}{K_D + D} \times 1.09^{(T-20)} \times (1 - \text{DO}) \tag{6.203a}$$

Denitrification rates can be related to denitrifier growth rates by the following relationship (US EPA, 1975c)

$$r_D = \mu_D / Y_D \tag{6.204}$$

where $r_D$ = nitrate removal rate, lb ($NO_3$-N)/lb VSS/d
$Y_D$ = gross yield of denitrifier, lb VSS/lb ($NO_3$-N)

Similarly, peak denitrification rates are related to maximum denitrifier growth rates as follows:

$$\hat{r}_D = \hat{\mu}_D / Y_D \tag{6.205}$$

**Solids retention time.** Consideration of solids production and solids retention time is an important design consideration for the system. Similar to Eq. (6.84), a mass balance of the biomass in a completely mixed reactor yields the relationship (Lawrence and McCarty, 1970)

$$\frac{1}{\theta_c} = Y_D r_D - K_d \tag{6.206}$$

where $\theta_c$ = solids retention time, day
$K_d$ = decay coefficient, day$^{-1}$

**Denitrification with RBC process.** The RBC process has been applied to biological denitrification by completely submerging the rotating media and by adding an appropriate source of organic carbon. Figure 6.52 is a schematic process flow diagram of carbon oxidation–nitrification–denitrification. Methanol is added to the denitrification stage, in which the rotational speed is reduced. Methanol requirements are a significant portion of the operating cost. In a completely submerged mode, RBC will remove nitrate nitrogen at a rate of approximately 1 lb/(d · 1000 ft$^2$) while treating influent nitrate concentrations up to 25 mg/L and producing effluent nitrate nitrogen concentrations below 5 mg/L. The design curves for RBC denitrification of municipal wastewater are presented in Fig. 6.53 (Autotrol, 1979).

Denitrification is a relatively rapid reaction compared to nitrification. It is generally more economical to reduce nitrate nitrogen to as low a level as can be achieved by the RBC process, i.e. ≤ 1.0 mg/L.

**Example:** Design a denitrification RBC system following organic oxidation–nitrification, using data for the example in the section on nitrification with the RBC system. The treatment plant is designed to produce a final effluent of 4 mg/L total nitrogen (2 mg/L of $NH_3$-N, 3 mg/L of TKN, and 1 mg/L of $NO_3$-N).

*Given conditions:*

Design flow = 4.0 Mgal/d

Influent $NH_3$-N (nitrification) = 18 mg/L

Effluent $NH_3$-N = 2 mg/L

Wastewater temperature ≥ 55°F

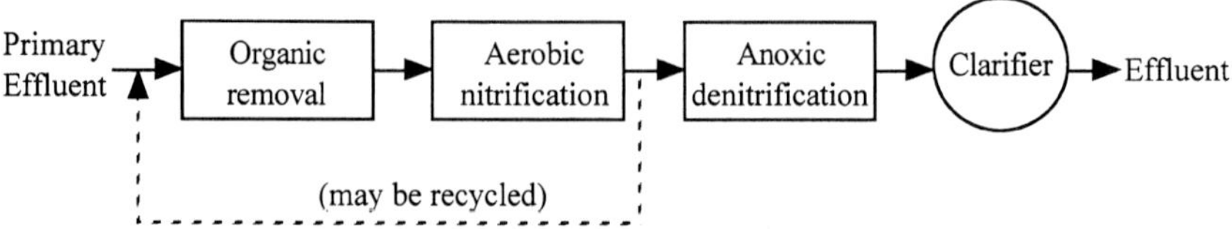


**Figure 6.52** Schematic RBC process for BOD removal, nitrification, and denitrification.

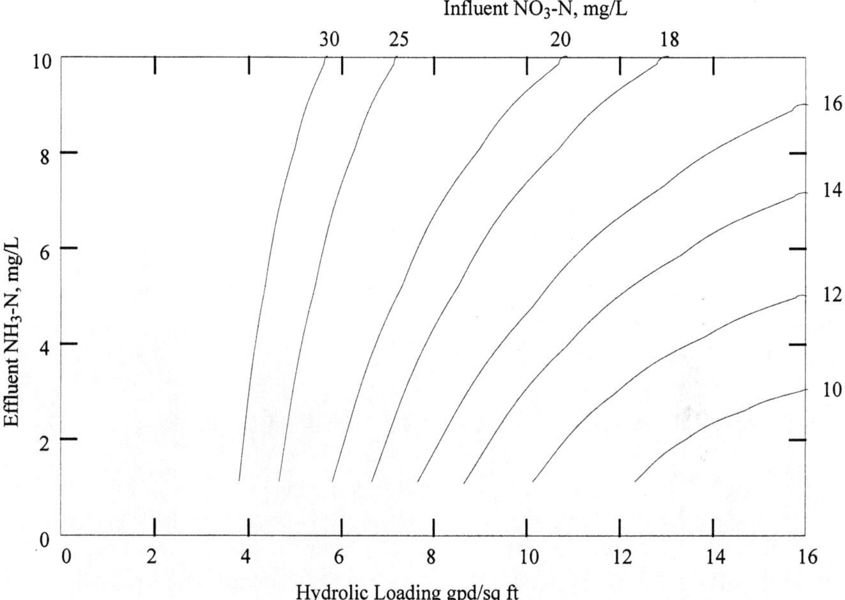


**Figure 6.53** Design curves for RBC denitrification of domestic wastewater.

**solution:**

Step 1. Determine surface area for denitrification shafts

(a) As in the previous RBC example for BOD removal and nitrification, the overall hydraulic loading $HL_o = 1.29\ gal/(d \cdot ft^2)$

$$\text{The surface area required} = 3{,}100{,}000\ ft^2$$

(b) Influent $NO_3$-N of denitrification

$$= (\text{influent} - \text{effluent})\ NH_3\text{-N}$$
$$= (18 - 2)\ mg/L$$
$$= 16\ mg/L$$

*Note*: Assume that all $NH_3$-N removed is stoichiometrically converted to $NO_3$-N, even though field practice often shows a loss of $NO_3$-N.

(c) Referring to Fig. 6.53, the hydraulic loading rate to reduce $NO_3$-N from 16 mg/L to 1 mg/L

$$HL = 7.45\ gal/(d \cdot ft^2)$$

(d) Calculate the surface area for denitrification

$$\text{Surface area} = \frac{4{,}000{,}000 \text{ gal/d}}{7.45 \text{ gal/(d} \cdot \text{ft}^2)}$$

$$= 540{,}000 \text{ ft}^2$$

Step 2. Select configuration

(a) Use standard media (100,000 ft$^2$) for denitrification

(b) Number of shafts

$$= \frac{540{,}000 \text{ ft}^2}{100{,}000 \text{ ft}^2}$$

$$= 5.4 \text{ (use 6 shafts)}$$

(c) A two-stage operation with three trains is recommended. Install two assemblies in a separate basin with baffles between adjacent shafts.

## 29 Sludge (Residuals) Treatment and Management

Residuals is a term currently used to refer to "sludge." The bulk of residuals (sludge) generated from wastewater by physical primary and biological (secondary) and advanced (tertiary) treatment processes must be treated and properly disposed of. The higher the degree of wastewater treatment, the larger the quantity of sludge to be treated and handled. With the advent of strict rules and regulations involving the handling and disposal of sludge, the need for reducing the volume of sludge has become increasingly important in order to reduce the operating costs (approximately 50% of the plant costs) of wastewater treatment plants. Sludge treatment and disposal is a complex problem facing wastewater treatment professionals. A properly designed and efficiently operated sludge processing and disposal system is essential to the overall success of the wastewater treatment effort.

### 29.1 Quantity and characteristics of sludge

The quantity and characteristics of the sludge produced depend on the character of raw wastewater and the wastewater treatment processes. In the United States, approximately 12,750 public owned treatment works (POTW) generate 5.4 million dry metric tons of sludge annually, or 21 kg (47 lb) of dry sewage sludge (biosolids) per person (Federal Register, 1993).

Some estimates of the amounts of solids generated by the treatment unit processes may be inferred from previous examples. Grit collected from the preliminary treatment units is not biodegradable. It is usually transported to a sanitary landfill without further treatment.

Characteristics of wastewater sludges, including total solids and volatile solids contents, pH, nutrients, organic matter, pathogens, metals, organic chemicals, and hazardous pollutants, are discussed in detail elsewhere (Federal Register, 1993; US EPA, 1995).

Sludge from primary settling tanks contains from 3% to 7% solids which are approximately 60% to 80% organic (Davis and Cornwell, 1991; Federal Register, 1993). Primary sludge solids are usually gray in color, slimy, fairly coarse, and with highly obnoxious odors. Primary sludge is readily digested under suitable operational conditions (organics are decomposed by bacteria). Table 6.24 provides the solids concentrations of primary sludge and sludge produced in different biological treatment systems.

Sludge from secondary settling tanks has commonly a brownish, flocculant appearance and an earthy odor. If the color is dark, the sludge may be approaching a septic condition. Secondary sludge consists mainly of microorganisms (75% to 90% organic) and inert materials. The organic matter may be assumed to have a specific gravity of 1.01 to 1.06, depending on its source, whereas the inorganic particles have a specific gravity of 2.5 (McGhee, 1991).

In general, secondary sludges are more flocculant than primary sludge solids, less fibrous. Waste activated sludge usually contains 0.5% to 2% solids, whereas trickling filter sludge has 2% to 5% solids (Davis and Cornwell, 1991; Hammer, 1986). Activated sludge and trickling filter sludge can be digested readily, either alone or when mixed with primary sludge.

**TABLE 6.24 Solids Concentrations and Other Characteristics of Various Types of Sludge**

| Wastewater treatment | Primary, gravity | Secondary, biological | Advanced (tertiary), chemical precipitation, filtration |
|---|---|---|---|
| Sludge | | | |
| Amounts generated, L/m$^3$ of wastewater | 2.5–3.5 | 15–20 | 25–30 |
| Solids content, % | 3–7 | 0.5–2 | 0.2–1.5 |
| Organic content, % | 60–80 | 50–60 | 35–50 |
| Treatability, relative | easy | difficult | difficult |
| Dewatered by belt filter | | | |
| Feed solids, % | 3–7 | 3–6 | |
| Cake solids, % | 28–44 | 20–35 | |

SOURCES: WPCF (1988a), WEF and ASCE (1991b), US EPA (1991)

Sludge from chemical (metal salts) precipitation is generally dark in color or red (with iron) and slimy. Lime sludge is grayish brown. Chemical sludge may create odor. Decomposition of chemical sludge occurs at a slower rate.

The nature of sludge from the tertiary (advanced) treatment process depends on the unit process. Chemical sludge from phosphorus removal is difficult to handle and treat. Tertiary sludge combined with biological nitrification and denitrification is similar to waste activated sludge.

**Example:** Estimate the solids generated in the primary and secondary clarifiers at a secondary (activated sludge) treatment plant. Assume that the primary settling tank removes 65% of the TSS and 33% of the $BOD_5$. Also determine the volume of each sludge, assuming 6% and 1.2% of solids in the primary and secondary effluents, respectively.

Average plant flow = 3785 $m^3$/d (1 Mgal/d)

Primary influent TSS = 240 mg/L

Primary influent BOD = 200 mg/L

Secondary effluent BOD = 30 mg/L

Secondary effluent TSS = 24 mg/L

Bacteria growth rate $Y = 0.23$ kg (0.5 lb) sludge solids per kg (lb) BOD removed

**solution:**

Step 1. Calculate the quantity of dry primary solids produced daily

$$1\ \text{mg/L} = 1\ \text{g/m}^3$$

$$\begin{aligned}\text{Primary sludge} &= 3785\ \text{m}^3/\text{d} \times 240\ \text{g/m}^3 \times 0.65/(1000\ \text{g/kg}) \\ &= 590\ \text{kg/d} \\ &= 1300\ \text{lb/d}\end{aligned}$$

Step 2. Calculate the primary effluent TSS and BOD concentrations

$$\text{TSS} = 240\ \text{mg/L} \times (1 - 0.65) = 84\ \text{mg/L}$$

$$\text{BOD} = 200\ \text{mg/L} \times (1 - 0.33) = 134\ \text{mg/L}$$

Step 3. Calculate TSS removed in the secondary clarifier

$$\begin{aligned}\text{Secondary (TSS) solids} &= 3785\ \text{m}^3/\text{d} \times (84 - 24)\ (\text{g/m}^3/(1000\ \text{g/kg}) \\ &= 227\ \text{kg/d} \\ &= 500\ \text{lb/d}\end{aligned}$$

Step 4. Calculate biological solids produced due to BOD removal

$$\begin{aligned}\text{BOD removed} &= 3785\ \text{m}^3/\text{d}\ (134 - 30)\ \text{g/m}^3/(1000\ \text{g/kg}) \\ &= 394\ \text{kg/d} \\ &= 869\ \text{lb/d}\end{aligned}$$

$$\text{Biological solids} = 394 \text{ kg/d} \times Y = 394 \text{ kg/d} \times 0.23 \text{ kg/kg}$$
$$= 90 \text{ kg/d}$$
$$= 198 \text{ lb/d}$$

Step 5. Calculate total amount of solids produced from the secondary clarifier and from the whole plant (from Steps 3 and 4)

$$\text{Secondary solids} = (227 + 90) \text{ kg/d}$$
$$= 317 \text{ kg/d}$$
$$= 698 \text{ lb/d}$$
$$\text{Solids of the plant} = \text{Step 1} + \text{Step 5}$$
$$= (590 + 317) \text{ kg/d}$$
$$= 907 \text{ kg/d}$$
$$= 1998 \text{ lb/d}$$

Step 6. Determine the volume of each type of sludge

Assuming sp. gr. of sludge = 1.0

$$\text{Primary sludge volume } V_1 = \frac{590 \text{ kg/d}}{0.06 \times 1000 \text{ kg/m}^3}$$
$$= 9.8 \text{ m}^3\text{/d}$$
$$= 2590 \text{ gal/d}$$

$$\text{Secondary sludge volume } V_2 = \frac{317 \text{ kg/d}}{0.012 \times 1000 \text{ kg/m}^3}$$
$$= 26.4 \text{ m}^3\text{/d}$$
$$= 6980 \text{ gal/d}$$

**Mass–volume relation.** The mass of solids in a slurry is related to volatile and fixed suspended solids contents. The specific gravity (sp. gr.) of a slurry is

$$S_s = \frac{m_w + m_v + m_f}{V_s} \tag{6.207}$$

where $S_s$ = specific gravity of slurry, g/cm$^3$ or lb/ft$^3$
$m_w$ = mass of water, kg or lb
$m_v$ = mass of VSS, kg or lb
$m_f$ = mass of FSS, kg or lb
$V_s$ = volume of sludge slurry, m$^3$ or ft$^3$

Since

$$V_s = V_w + V_v + V_f \tag{6.208}$$

where

$V_w, V_v, V_f$ = volume of water, VSS, FSS, $m^3$ or $ft^3$

then

$$\frac{m_s}{S_s} = \frac{m_w}{S_w} + \frac{m_v}{S_v} + \frac{m_f}{S_f} \tag{6.209}$$

where

$m_s$ = mass of slurry, kg or lb

**Moisture content.** The moisture (water) $\rho_w$ or total solids $\rho_s$ content of a sludge, expressed on a percentage basis, can be computed as

$$\rho_w = \frac{100 m_w}{m_w + m_s} = \frac{100 m_w}{m_w + m_v + m_f} \tag{6.210}$$

$$p_s = 100 - \rho_w \tag{6.211}$$

The volume of a sludge, related to its total solids content, is

$$V_s = \frac{m_s}{(\rho_s/100)S_s} \tag{6.212}$$

The specific gravity of organic matter (VSS) is close to that of water (1.00). The sp. gr. of activated sludge is 1.01 to 1.10, the sp. gr. of FSS is 2.5, and that for chemical sludge ranges from 1.5 to 2.5 (Droste, 1997).

**Example:** Determine the sp. gr. of waste activated sludge that contains 80% VSS and has a solids concentration of 1.5%. Also, determine the volume of 1 kg of sludge.

**solution:**

Step 1. Calculate mass of VSS, FSS, and water in 1000 g of sludge

1.5% sludge = 15,000 mg/L of solids

= 15 g/L

VSS = 15 g/L × 0.8 = 12 g/L $= m_v$

FSS = 15 g/L × 0.2 = 3 g/L $= m_f$

If $m_s$ = 1000 g

Then $m_w$ = (1000 − 12 − 3) g = 985 g

Step 2. Calculate sp. gr. of the sludge, using Eq. (6.209)

$$\frac{m_s}{S_s} = \frac{m_w}{S_w} + \frac{m_v}{S_v} + \frac{m_f}{S_f}$$

$$\frac{1000}{S_s} = \frac{985}{1.0} + \frac{12}{1.0} + \frac{3}{2.5}$$

$$S_s = \frac{1}{0.9982} = 1.0018\ (\text{g/cm}^3)$$

Step 3. Calculate the volume per kg of sludge

$$m_s = 1000\ \text{g} \times 0.015 = 15\ \text{g}$$

Using Eq. (6.212)

$$V_s = \frac{m_s}{(\rho_s/100)S_s} = \frac{15\ \text{g}}{(1.5/100)S_s}$$

$$= \frac{15\ \text{g}}{(1.5/100)(1.0018\ \text{g/cm}^3)(1000\ \text{cm}^3/\text{L})}$$

$$= 0.998\ \text{L}$$

## 29.2 Sludge treatment alternatives

A variety of treatment processes and overall sludge management options can be established, depending on the type and quantity of sludge generated. Figure 6.54 illustrates schematically the wastewater sludge treatment alternatives. The basic processes for sludge treatment include thickening, stabilization, conditioning, dewatering, and volume reduction. More details of sludge treatments are available from manufacturers' manuals and elsewhere (US EPA,1979; 1991; Metcalf and Eddy, Inc. 1991; WEF and ASCE, 1991b, 1996b).

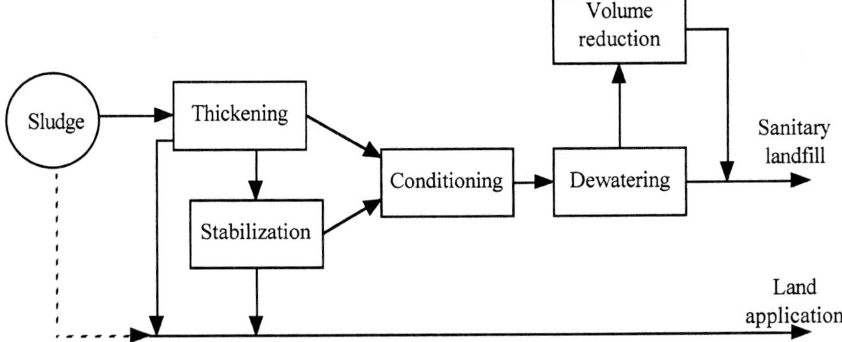


**Figure 6.54** Sludge processing alternatives.

**Sludge thickening.** As stated previously, all types of sludge contain a large volume of water (Table 6.24). The purposes of sludge thickening are to reduce the sludge volume to be handled in the subsequent sludge processing units (pump, digester, dewatering equipment) and to reduce the construction and operating costs of subsequent processes. Sludge thickening is a procedure used to remove water and increase the solids content. For example, if waste activated sludge with 0.6% solids is thickened to a content of 3.0% solids, a five-fold decrease in sludge volume is achieved.

Thickening a sludge with 3% to 8% solids may reduce its volume by 50%. Sludge thickening mainly involves physical processes such as gravity settling, flotation, centrifugation, and gravity belts.

**Example:** Estimate the sludge volume reduction when the sludge is thickened from 4% to 7% solids concentration. The daily sludge production is 100 $m^3$ (26,420 gal).

**solution:**

Step 1. Calculate amount of dry sludge produced

$$\text{Dry solids} = 100\ \text{m}^3 \cdot \text{d} \times 1000\ \text{kg/m}^3 \times 0.04$$
$$= 4000\ \text{kg/d}$$

Step 2. Calculate volume in 7% solids content

$$\text{Volume} = (4000\ \text{kg/d})/[0.07(1000\ \text{kg/m}^3)]$$
$$= 57.1\ \text{m}^3/\text{d}$$
$$= 15{,}100\ \text{gal/d}$$

*Note*: Steps 1 and 2 can be solved by the mass balance method:

$$\text{Volume} = 100\ \text{m}^3/\text{d} \times 4\%/7\%$$
$$= 57.1\ \text{m}^3/\text{d}$$

Step 3. Calculate percentage sludge volume reduction

$$\text{Percent volume reduction} = \frac{(100 - 57.1)\text{m}^3 \times 100\%}{100\ \text{m}^3}$$
$$= 42.9\%$$

**Gravity thickening.** Gravity thickening (Fig. 6.55) uses gravity forces to separate solids from the sludge. The equipment is similar in design to a conventional sedimentation basin. Sludge withdrawn from primary clarifiers or sludge blending tanks is applied to the gravity thickener through a central inlet wall. The normal solids loading rates range from 30 to 60 kg solids per $m^2$ of tank bottom per day (6 to 12 lb/($ft^2 \cdot d$)) (Hammer, 1986). Coagulant is sometimes added for improving the settling.

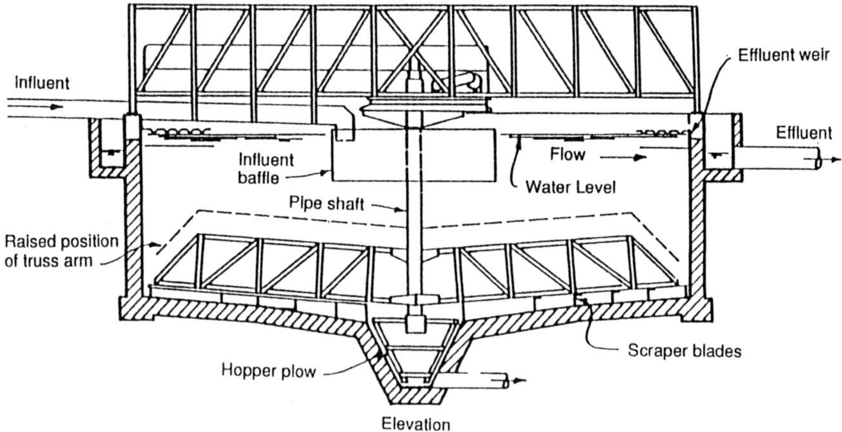


**Figure 6.55** Schematic of gravity thickener *(source: US EPA, 1991).*

Typical hydraulic loading rates are from 16 to 32 $m^3/(m^2 \cdot d)$ (390 to 785 gal/(d · $ft^2$)). For waste activated sludge or for a very thin mixture, hydraulic loading rates range from 4 to 8 $m^3/(m^2 \cdot d)$ (100 to 200 gal/ (d · $ft^2$)) for secondary sludges, and 16 to 32 $m^3/(m^2 \cdot d)$ (390 to 785 gal/(d · $ft^2$)) for primary sludges (US EPA, 1979). For activated sludge, residence time in the thickener needs to be more than 18 h to reduce gas production and other undesirable effects (WEF and ASCE, 1991b). A typical design is a circular tank with a side depth of 3 to 4 m and a floor sloping at 1:4 to 1:6.

**Example 1:** A residual with 4% solids is thickened to a 9% solids content. What is the concentration factor?

**solution:**

$$\text{Concentration factor} = \frac{\text{thickened solids concentraiton, \%}}{\text{solids content in influent, \%}}$$

$$= \frac{9\%}{4\%}$$

$$= 2.25$$

*Note*: Concentration factor should be 2 or more for primary sludges, and 3 or more for secondary sludges.

**Example 2:** A 9 m (30 ft) diameter by 3 m (10 ft) side wall depth gravity thickener is concentrating a primary sludge. The sludge flow is 303 L/min (80 gal/min) with average solids content of 4.4%. The thickened sludge is withdrawn at 125 L/min (33 gal/min) with 6.8% solids. The sludge blanket is 1 m (3.3 ft) thick. The effluent of the thickener has a TSS level of 660 mg/L. Determine: (1) the sludge detention time; (2) whether the blanket will increase or decrease in depth under the stated conditions.

**solution:**

Step 1. Determine the sludge detention time

(a) Compute the volume of the sludge blanket

$$\begin{aligned}\text{Volume} &= \pi\,(9\text{ m}/2)^2 \times 1\text{ m}\\ &= 63.6\text{ m}^3\\ &= 63{,}600\text{ L}\end{aligned}$$

(b) Compute the daily sludge pumped

$$\begin{aligned}\text{Pumpage} &= 125\text{ L/min} \times 1440\text{ min/d}\\ &= 180{,}000\text{ L/d}\end{aligned}$$

(c) Compute the sludge detention time (DT)

$$\begin{aligned}\text{DT} &= \text{sludge volume/sludge pumpage}\\ &= 63{,}600\text{ L/(180,000 L/d)}\\ &= 0.353\text{ day}\\ &= 8.48\text{ h}\end{aligned}$$

Step 2. Determine sludge blanket increase or otherwise

(a) Estimate the amount of solids entering the thickener

Assume the sludge = 1 kg/L

$$\begin{aligned}\text{Solids in} &= 303\text{ L/min} \times 1440\text{ min/d} \times 1\text{ kg/L} \times 0.044\\ &= 19{,}200\text{ kg/d}\end{aligned}$$

(b) Compute the sludge withdrawal rate

$$\begin{aligned}\text{Withdrawal rate} &= 125\text{ L/min} \times 1440\text{ min/d} \times 1\text{ kg/L} \times 0.068\\ &= 12{,}240\text{ kg/d}\end{aligned}$$

(c) Compute solids discharge in the effluent of the thickener

$$\begin{aligned}\text{Effluent TSS} &= 660\text{ mg/L} = 0.066\%\\ \text{Solids lost} &= 303 \times 1440 \times 1 \times (0.066\%/100\%)\\ &= 288\text{ kg/d}\\ \text{say} &= 290\text{ kg/d}\end{aligned}$$

(d) Compute total solids out

$$\text{Solids out} = (\text{Step 2b}) + (\text{Step 2c}) = (12{,}240 + 290)\ \text{kg/d}$$
$$= 12{,}530\ \text{kg/d}$$

(e) Compare Steps 2a and 2d

$$\text{Solids in } 19{,}200\ \text{kg/d} > \text{solids out, } 12{,}530\ \text{kg/d}$$

*Answer*: The sludge blanket will increase in depth.

*Note*: the withdrawal rate also can be expressed as per minute

**Dissolved air flotation thickening.** Flotation thickeners include dissolved air flotation (DAF), vacuum flotation, and depressed-air flotation. Only DAF is used for wastewater sludge thickening in the US. It offers significant advantages in thickening light sludges such as activated sludge. The DAF thickener (Fig. 6.56) separates solids from the liquid phase in

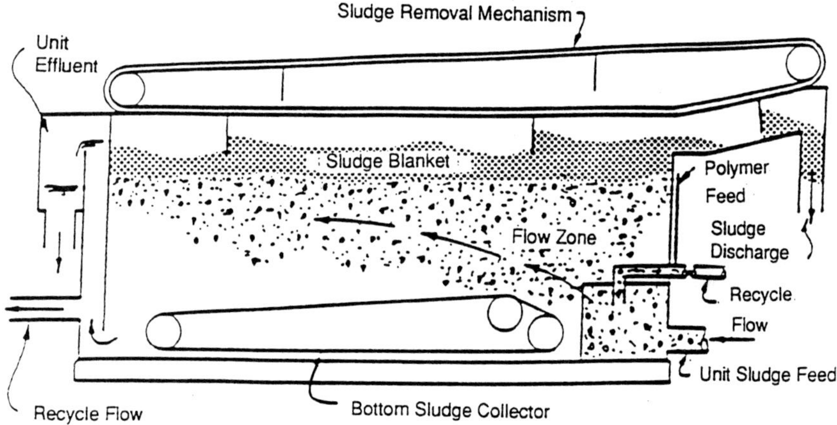


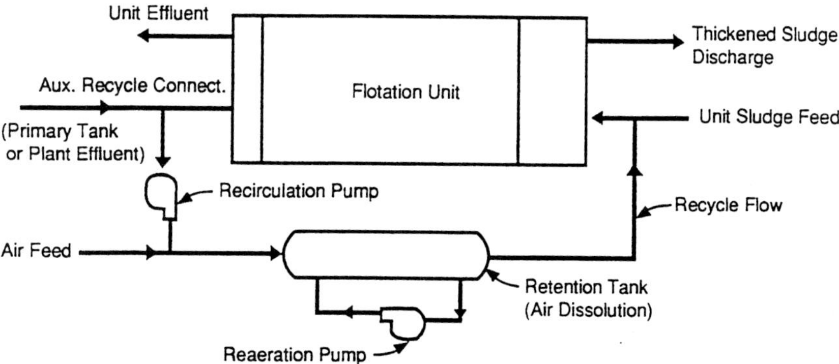


**Figure 6.56** Dissolved air flotation thickener (*source: US EPA, 1991*).

an upward direction by attaching fine air bubbles (60 to 100 $\mu$m) to particles of suspended solids which then float. The influent stream at the tank bottom is saturated with air, pressurized (280 to 550 kPa), then released to the inlet distributor. The retention tank is maintained at a pressure of 3.2 to 4.9 kg/cm$^2$ (45 to 70 lb/in$^2$) (US EPA, 1987b).

The ratio of the quantity of air supplied and dissolved into the recycle or waste stream to that of solids (the air-to-solids, A/S ratio) is probably the most important factor affecting the performance of the flotation thickener. Normal loading rates for waste activated sludge range from 10 to 20 kg solids/(m$^2$ · d) (2 to 4 lb/(ft$^2$ · d)). DAF thickening produces about 4% solids with a solids recovery of 85% (Hammer, 1986). The sludge volume index, SVI, is also an important factor for DAF operation.

**Example 1:** Determine hydraulic and solids loading rates for a DAF thickener. The thickener, of 9 m (30 ft) diameter, treats 303 L/min (80 gal/min) of waste activated sludge with a TSS concentration of 7800 mg/L.

**solution:**

Step 1. Calculate liquid surface area

$$\text{Area} = \pi\ (9\ \text{m}/2)^2 = 63.6\ \text{m}^2$$

$$\text{Hydraulic loading} = (303\ \text{L/min})/63.6\ \text{m}^2$$

$$= 4.76\ \text{L/(min} \cdot \text{m}^2)$$

Step 2. Calculate solids loading rate

$$\text{TSS in WAS} = 7800\ \text{mg/L} = 0.78\%$$

$$\text{Solids loaded} = 303\ \text{L/min} \times 60\ \text{min/h} \times 1\ \text{kg/L} \times (0.78\%/100\%)$$

$$= 141.8\ \text{kg/h}$$

$$\text{Solids loading rate} = (141.8\ \text{kg/h})/(63.6\ \text{m}^2)$$

$$= 2.23\ \text{kg/(m}^2 \cdot \text{h)}$$

$$= 53.3\ \text{kg/(m}^2 \cdot \text{d)}$$

$$= 10.9\ \text{lb/(ft}^2 \cdot \text{h)}$$

**Example 2:** A DAF thickener treats 303 L/min (80 ft/min) of waste activated sludge at 8600 mg/L. Air is added at a rate of 170 L/min (6.0 ft$^3$/min). Determine the air-to-solids ratio. Use 1.2 g of air per liter of air (0.075 lb/ft$^3$) under the plant conditions.

**solution:**

Step 1. Compute mass of air added

$$\text{Air} = 170\ \text{L/min} \times 1.2\ \text{g/L}$$

$$= 204\ \text{g/min}$$

Step 2. Compute mass of solids treated

$$\text{Solids} = 8.6 \text{ g/L} \times 303 \text{ L/min}$$
$$= 2606 \text{ g/min}$$

Step 3. Compute A/S ratio

$$\text{A/S} = (204 \text{ g/min})/(2606 \text{ g/min})$$
$$= 0.078 \text{ g air/g solids}$$
$$\text{or} = 0.078 \text{ lb air/lb solids}$$

**Example 3:** Determine the concentration factor and the solids removal efficiency of a DAF thickener with conditions the same as in Example 2. The thickened sludge or float has 3.7% solids and the effluent TSS concentration is 166 mg/L.

**solution:**

Step 1. Compute the concentration factor CF

From Example 2,

The influent sludge solids content = 8600 mg/L

= 0.86%

$$\text{CF} = \frac{\text{float solids content}}{\text{influent solids content}} = \frac{3.7\%}{0.86\%}$$
$$= 4.3$$

Step 2. Compute the solids removal efficiency

$$\text{Efficiency} = \frac{\text{influent TSS} - \text{effluent TSS}}{\text{influent TSS}}$$
$$= \frac{(8600 - 166) \text{ mg/L} \times 100\%}{8600 \text{ mg/L}}$$
$$= 98.1\%$$

**Centrifuge thickening.** A centrifuge acts both to thicken and to dewater sludge. The centrifuge process separates liquid and solids by the influence of centrifugal force which is typically 50 to 300 times that of gravity (WEF and ASCE, 1996b). Centrifuges may be used for thickening waste activated sludge or as dewatering devices for digested or conditioned sludges. Three basic types (solid bowl, imperforate basket, and disc-nozzle) are commonly installed to thicken or dewater wastewater sludge.

The solid bowl scroll centrifuge (Fig. 6.57) is the most widely used type. It rotates along a horizontal axis and operates in a continuous-feed manner. It consists of a rotating bowl having a cylindrical-conical shape

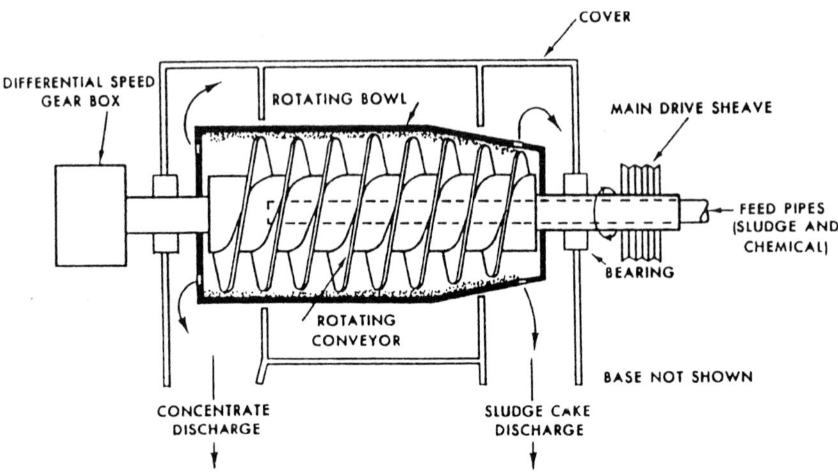


**Figure 6.57** Solid bowl scroll centrifuge (*source: US EPA, 1991*).

and a screw conveyer. Sludge is introduced into the rotating bowl through a stationary feed pipe continuously and the solids concentrate on the periphery. The gravitational force causes the solids to settle out on the inner surface of the rotating bowl. A helical scroll, spinning at a slightly different speed, moves the settled solids toward the tapered end (outlet ports) and then discharges them. The light liquid pools above the sludge layer flows toward the concentrate outlet ports. The unit has a low cost/capacity ratio.

The basket centrifuge, also called the imperforate bowl (Fig. 6.58), is a knife-discharge type and operates on a batch basis. Liquid sludge is transported by a pipe through the top and is fed to the bottom of a vertically mounted spinning bowl. Solids accumulate against the wall of the bowl by centrifugal force and the concentrate is decanted. The duration of the feed time is controlled by a preset timer or a concentrate monitor (usually 60% to 85% of maximum depth). When the feed is stopped, the bowl begins to decelerate. As a certain point a nozzle skimmer (plow or knife) enters the bowl to remove the retained solids. The solids fall through the bottom of the bowl into a hopper. The plow retracts and the bowl accelerates, starting a new cycle. The units needs a skilled operator.

The disc-nozzle centrifuge (Fig. 6.59) rotates along a vertical axis and operates in a continuous manner. The liquid sludge is fed normally through the top of the unit (bottom feed is also possible) and flows through a feedwell in the center of the rotor and to a set of some 50 conical discs. An impeller within the rotor accelerates and distributes the feed slurry, filling the rotor interior. The centrifugal force is applied to

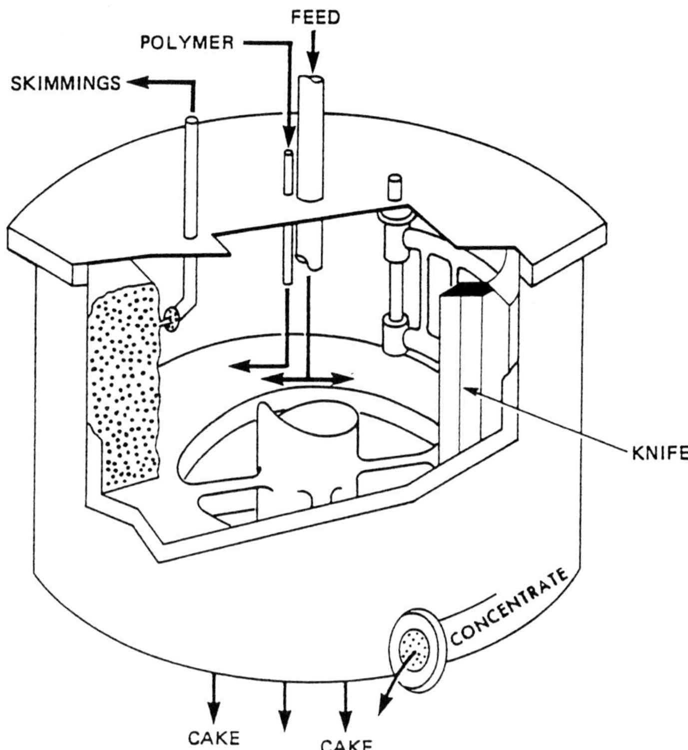


**Figure 6.58** Schematic of imperforate basket centrifuge (*source: US EPA, 1991.*)

the relatively thin film of liquid and solid between the discs. The force throws the denser solid materials to the wall of the rotor bowl, where it is subjected to additional centrifugal force and concentrated before it is discharged through nozzles located on the periphery. The clarified liquid passes on through the disc stack into the weir at the top of the bowl, and is then discharged.

The performance of a centrifuge is usually determined by the percentage of capture. It can be calculated as follows:

$$\text{Percent capture} = \left[1 - \frac{C_r(C_c - C_s)}{C_s(C_c - C_r)}\right] \times 100 \qquad (6.213)$$

where $C_r$ = concentration of solids in rejected wastewater (concentrate), mg/L, %

$C_c$ = concentration of solids in sludge cake, mg/L, %

$C_s$ = concentration of solids in sludge feed, mg/L, %

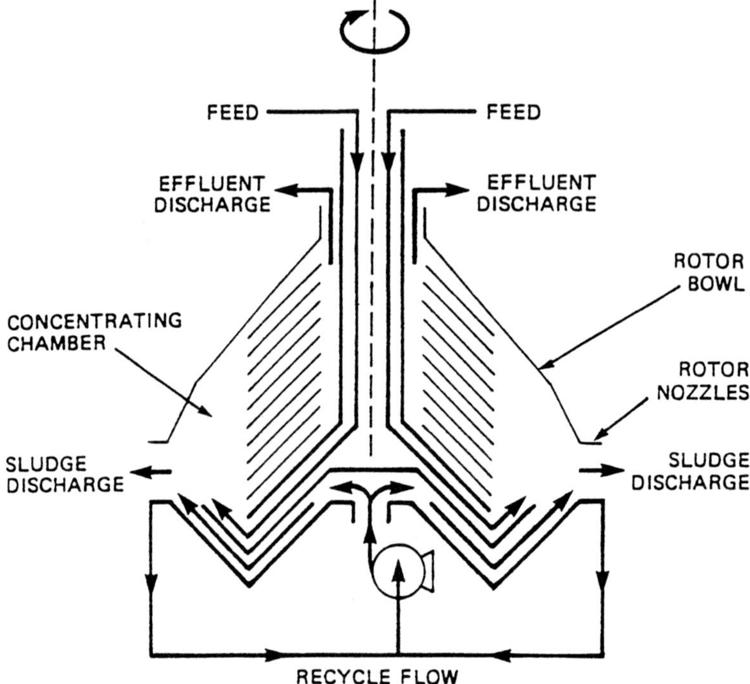


**Figure 6.59** Schematic of a disc-nozzle centrifuge (*source: US EPA, 1991*).

**Example:** A basket centrifuge is applied with 378 L/min (100 gal/min) of waste activated sludge at a solids content of 0.77%. The basket run time for 80% depth is 18 min with a skimming operating time of 2 min. The average solids content in the thickened sludge is 7.2%. The solids concentration in the effluent is 950 mg/L. Determine hourly hydraulic loading and solids loading, dry solids produced per cycle and per day, and the efficiency of solids capture.

**solution:**

Step 1. Calculate hydraulic loading (HL) rate

$$\begin{aligned} \text{HL} &= 378 \text{ L/min} \times 60 \text{ min/h} \\ &= 22{,}680 \text{ L/h } (=22.7 \text{ m}^3\text{/h}) \\ &= 5990 \text{ gal/h} \end{aligned}$$

Step 2. Calculate solids loading SL

$$\begin{aligned} \text{SL} &= 378 \text{ L/min} \times 60 \text{ min/h} \times (0.77\%/100\%) \times 1 \text{ kg/L} \\ &= 175 \text{ kg/h} \\ &= 385 \text{ lb/h} \end{aligned}$$

Step 3. Calculate the solids production per cycle and per day

$$\text{Solids produced per cycle} = \frac{175 \text{ kg/h} \times 18 \text{ min/cycle}}{60 \text{ min/h}}$$

$$= 52.5 \text{ kg/cycle}$$

$$\text{Total time for each cycle} = 18 \text{ min} + 2 \text{ min} = 20 \text{ min/cycle}$$

$$\text{Solids produced per day} = \frac{52.5 \text{ kg/cycle} \times 1440 \text{ min/d}}{20 \text{ min/cycle}}$$

$$= 3780 \text{ kg/d}$$

$$= 8330 \text{ lb/d}$$

Step 4. Calculate the efficiency of solids capture

Given: $C_r = 950 \text{ mg/L} = 0.095\%$

$C_c = 7.2\%$

$C_s = 0.77\%$

Using Eq. (6.213)

$$\text{Efficiency} = \left[1 - \frac{C_r(C_c - C_s)}{C_s(C_c - C_r)}\right] \times 100\%$$

$$= \left[1 - \frac{0.095(6.2 - 0.77)}{0.77(6.2 - 0.095)}\right] \times 100\%$$

$$= 88.8\%$$

**Gravity belt thickening.** The gravity belt thickener is a relatively new sludge thickening technique, stemming from the application of belt presses for sludge dewatering. It is effective for raw sludge and digested sludge with less than 2% solids content. A gravity belt moves over rollers driven by a variable-speed drive unit. The sludge is conditioned with polymer and applied into a feed/distribution box at one end. The sludge is distributed evenly across the width of the moving belt as the liquid drains through, and the solids are carried toward the discharge end of the thickener and removed. The belt travels through a wash cycle. Additional material on gravity belt thickeners and rotary drum thickeners may be found in WEF and ASCE (1991b, 1996b).

**Example 1:** An operator adds 2 lb (0.91 kg) of a dry polymer to 50 gal (189 L) of water. What is the polymer strength?

**solution:**

Step 1. Calculate the mass of water plus polymer, $W$

$$W = 50 \text{ gal} \times 8.34 \text{ lb/gal} + 2 \text{ lb}$$
$$= 419 \text{ lb}$$

Step 2. Calculate the strength of polymer solution

$$\text{Polymer strength} = \frac{\text{mass of polymer}}{\text{total mass}} = \frac{2 \text{ lb}}{419 \text{ lb}} \times 100\% = 0.48\%$$

**Example 2:** Determine the amount of dry polymer needed to prepare a 0.12% solution in a 1000-gal (3.78 $m^3$) tank. If 8 gal/min of this 0.12% polymer solution is added to a waste activated sludge flow of 88 gal/min (333 L/min) with 8600 mg/L solids concentration, compute the dosage of polymer per ton of sludge.

**solution:**

Step 1. Compute dry polymer required for 1000-gal solution

$$\text{Polymer} = 1000 \text{ gal} \times 8.34 \text{ lb/gal} \times (0.12\%/100\%)$$
$$= 10.0 \text{ lb}$$

Step 2. Compute the dosage in lb/ton of sludge

$$\text{Solids at 8600 mg/L} = 0.86\%$$

$$\text{Dosage} = \frac{\text{lb polymer added}}{\text{lb dry solid}} \times \frac{2000 \text{ lb}}{\text{ton}}$$

$$= \frac{8 \text{ gal/min} \times 8.34 \text{ lb/gal} \times 0.12\% \times 2000 \text{ lb/ton}}{88 \text{ gal/min} \times 8.34 \text{ lb/gal} \times 0.86\%}$$

$$= 25.4 \text{ lb polymer/ton dry sluge solid}$$

**Sludge stabilization.** After sludge has thickened, it requires stabilization to convert the organic solids to a more refractory or inert form. Thus the sludge can be handled or used as a soil conditioner without causing a nuisance or health hazard. The purposes of sludge stabilization are to reduce pathogens, eliminate odor-causing materials, and to inhibit, reduce, and eliminate the potential for putrefaction.

Treatment processes commonly used for stabilization of wastewater sludges include anaerobic digestion, aerobic digestion, chemical (lime, disinfectants) stabilization, and composting.

**Anaerobic digestion.** Anaerobic sludge digestion is the biochemical degradation (oxidation) of complex organic substances in the absence of free oxygen. The process is one of the oldest and most widely used methods.

During anaerobic digestion, energy is released, and much of the volatile organic matter is converted to methane, carbon dioxide, and water. Thus little carbon and energy are available to sustain further biological activity, and the remaining residuals are rendered stable.

Anaerobic digestion involves three basic successive phases of fermentation: hydrolysis, acid formation, and methane formation (WEF and ASCE, 1996b; US EPA, 1991). In the first phase of digestion, extracellular enzymes (enzymes operating outside the cells) break down complex organic substances (proteins, cellulose, lignins, lipids) into soluble organic fatty acids, alcohols, carbon dioxide, and ammonia.

In the second phase, acid-forming bacteria, including facultative bacteria, convert the products of the first stage into short-chain organic acids such as acetic acid, propionic acid, other low molecular weight organic acids, carbon dioxide, and hydrogen. These volatile organic acids tend to reduce the pH, although alkalinity buffering materials are also produced. Organic matter is converted into a form suitable for breakdown by the second group of bacteria.

The third phase, strictly anaerobic, involves two groups of methane-forming bacteria (methanogens). One group converts carbon and hydrogen to methane. The other group converts acetate to methane, carbon dioxide, and other trace gases. Both groups of bacteria are anaerobic. Closed digesters are used for anaerobic digestion. It should be noted that many authors consider only two phases (excluding phase 1 above).

The most important factors affecting the performance of anaerobic digesters are solids residence time, hydraulic residence time, temperature, pH, and toxic materials. Methanogens are very sensitive to environmental conditions. Anaerobic digesters are usually heated to maintain a temperature of 34 to 36°C (94 to 97°F). The methane bacteria are active in the mesophilic range (27 to 43°C, 80 to 110°F). They have a slower growth rate than the acid formers and are very specific in food supply requirements. The anaerobic digester (two-stage) usually provides 10 to 20 days' detention of sludge (WEF and ASCE, 1996b). Optimum methane production typically occurs when the pH is maintained between 6.8 and 6.2. If the temperature falls below the operating range and/or the digestion times falls below 15 days, the digester may become upset and require close monitoring and attention.

If concentrations of certain materials such as ammonia, sulfide, light metal cations, and heavy metals increase significantly in anaerobic digesters, they can inhibit or upset the process of performance.

The key parameter for digester sizing is solids residence time. For digester systems without recycle, there is no difference between SRT and HRT. Volatile solids loading rate is also used frequently as a basis for design. Typically, design SRT values are from 30 to 60 days for low-rate digesters and 10 to 20 days for high-rate digesters (WEF and ASCE, 1991b).

Volatile solids loading criteria are generally based on sustained loading conditions (typically peak month or peak week solids production), with provisions for avoiding excessive loading during shorter time periods. A typical design sustained peak volatile solids loading rate is 1.9 to 2.5 kg VS/($m^3 \cdot d$) (0.12 to 0.16 lb VS/($ft^3 \cdot d$)). A maximum limit of 3.2 kg VS/($m^3 \cdot d$) (0.20 lb VS/($ft^3 \cdot d$)) is often used (WEF and ASCE, 1991b).

The configuration of an anaerobic digester is typically a single-stage or two-stage process (Fig. 6.60). For the low rate, single-stage digester, three separate layers (scum, supernatant, and sludge layers) form as decomposition occurs. The stabilized (digested) sludge settles at the bottom of the digester. The supernatant is usually returned to the plant influent. In a single-stage high rate process, the digester is heated and

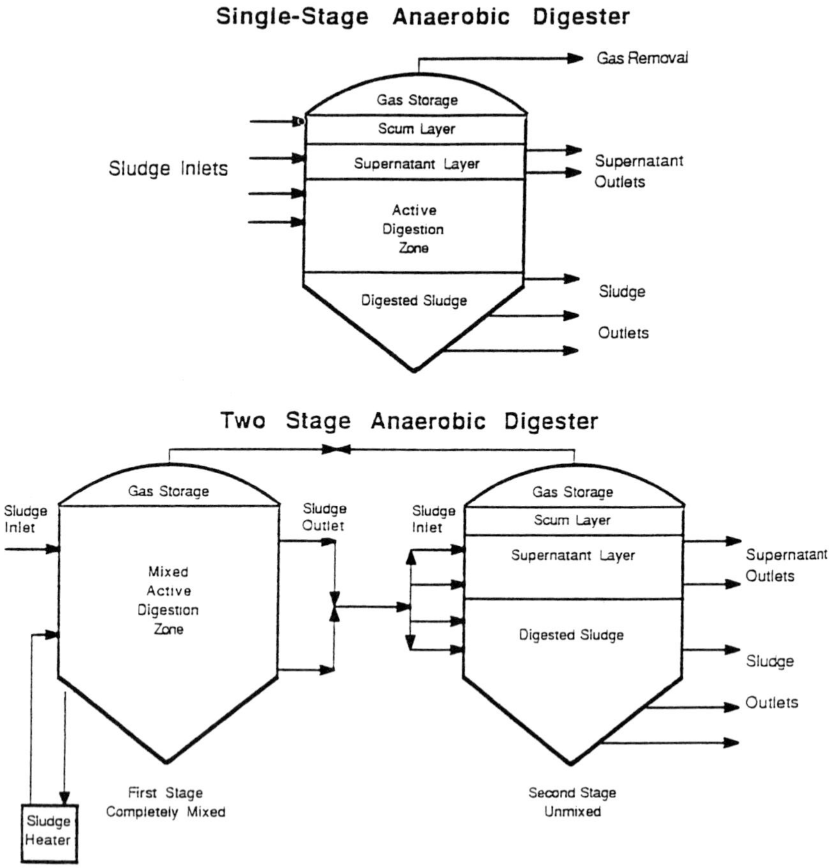

**Figure 6.60** Configuration of (a) single-stage and (b) two-stage anaerobic digesters (*source: US EPA, 1991*).

mixed, and supernatant is not withdrawn. In a two-stage system, sludge is stabilized in the first stage, whereas the second stage provides settling and thickening. The digester is heated to 34 to 36°C.

An anaerobic digester is designed to provide warm, oxygen-free, and well-mixed conditions to digest organic matter and to reduce pathogenic organisms. The hardware include covers, heaters, and mixers. Operation of the process demands control of food supply, temperature (31 to 36°C, 88 to 97°F), pH (6.8 to 7.2), alkalinity (2000 to 3500 mg/L as $CaCO_3$), and detention time (60 days at 20°C to 15 days at 35°C).

**Gas production.** Gas production is one of the important parameters for measuring the performance of the digester. Typically, gas production ranges from 810 to 1120 L of digester gas per kg volatile solids (13 to 18 $ft^3$ gas/lb VS) destroyed. Gas produced from a properly operated digester contains approximately 65% to 69% methane and 31% to 35% carbon dioxide. If more than 35% of gas is carbon dioxide, there is probably something wrong with the digestion system (US EPA, 1991). The quantity of methane gas produced can be computed by the following equation (McCarty, 1964)

$$V = 350[Q(S_0 - S)/(1000) - 1.42\, P_x] \quad \text{(SI units)} \tag{6.214}$$

$$V = 5.62[Q(S_0 - S)8.34 - 1.42\, P_x] \quad \text{(British system)} \tag{6.215}$$

where
$V$ = volume of methane produced at standard conditions (0°C, 32°F and 1 atm), L/d or $ft^3$/d
350, 5.62 = theoretical conversion factor for the amount of methane produced per kg (lb) of ultimate BOD oxidized (see Example 1), 350 L/kg or 5.62 $ft^3$/lb
1000 = 1000 g/kg
$Q$ = flow rate, $m^3$/d or Mgal/d
$S_0$ = influent ultimate BOD, mg/L
$S$ = effluent ultimate BOD, mg/L
8.34 = conversion factor, lb/(Mgal · mg/L)
$P_x$ = net mass of cell tissue produced, kg/d or lb/d

For a complete-mix high-rate two-stage anaerobic digester without recycle, the mass of biological solids synthesized daily, $P_x$, can be estimated by the equation below

$$P_x = \frac{Y[Q(S_0 - S)]}{1 + k_d\theta_c} \quad \text{(SI units)} \tag{6.216}$$

$$P_x = \frac{Y[Q(S_0 - S)8.34]}{1 + k_d\theta_c} \quad \text{(British system)} \tag{6.217}$$

where $Y$ = yield coefficient, kg/kg or lb/lb
$k_d$ = endogenous coefficient, per day
$\theta_c$ = mean cell residence time, day
Other terms are as defined previously.

**Example 1:** Determine the amount of methane generated per kg of ultimate BOD stabilized. Use glucose, $C_6H_{12}O_6$, as BOD.

**solution:**

Step 1. Use the balanced equation for converting glucose to methane and carbon dioxide under anaerobic conditions

$$C_6H_{12}O_6 \longrightarrow 3CH_4 + 3CO_2$$

$$\text{M.W.} \quad 180 \qquad\qquad 48$$

$$1 \qquad\qquad x$$

$$x = 48/180 = 0.267$$

Step 2. Determine oxygen requirement for methane

Using a balanced equation of oxidation of methane to carbon dioxide and water

$$3CH_4 + 6O_2 \rightarrow 3CO_2 + 6H_2O$$

$$\text{M.W.} \quad 48 \qquad 192$$

$$\text{Ultimate BOD per kg of glucose} = (192/180)\ \text{kg} = 1.07\ \text{kg}$$

Step 3. Calculate the rate of the amount of methane generated per kg of $BOD_L$ converted

$$\frac{\text{kg CH}_4}{\text{kg BOD}_L} = \frac{0.267}{1.07} = \frac{0.25}{1.0}$$

$$\text{or simply } 48/192 = 0.25/1.0$$

*Note*: 0.25 (or simply 48/192 = 0.25) kg of methane is produced by each kg of $BOD_L$ stabilized.

Step 4. Calculate the volume equivalent of 0.25 kg of methane at the standard conditions (0°C and 1 atm)

$$\text{Volume} = (0.25\ \text{kg} \times 1000\ \text{g/kg})\left(\frac{1\ \text{mole}}{16\ \text{g}}\right)\left(\frac{22.4\ \text{L}}{\text{mole}}\right)$$

$$= 350\ \text{L}$$

$$= 12.36\ \text{ft}^3$$

*Note*: 350 L of methane is produced per kg of ultimate BOD stabilized, or 5.60 $ft^3$ of methane is produced per lb of ultimate BOD stabilized (by a conversion factor, L/kg × 0.016 = $ft^3$/lb).

**Example 2:** Determine the size of anaerobic digester required to treat primary sludge under the conditions given below, using a complete-mix reactor. Also check loading rate and estimate methane and total gas production.

| | | |
|---|---|---|
| Average design flow | 3785 $m^3/d$ | (1 Mgal/d) |
| Dry solid removed | 0.15 $kg/m^3$ | (1250 lb/Mgal) |
| Ultimate $BOD_L$ removed | 0.14 $kg/m^3$ | (1170 lb/Mgal) |
| Sludge solids content | 5% | (95% moisture) |
| Specific gravity of solids | 1.01 | |
| $\theta_c$ | 15 days | |
| Temperature | 35°C | |
| $Y$ | 0.5 kg cells/kg $BOD_L$ | |
| $K_d$ | 0.04 per day | |
| Efficiency of waste utilization | 0.66 | |

**solution:**

Step 1. Calculate the sludge volume produced per day

$$\text{Sludge volume per day} = \frac{(3785\ m^3/d)(0.15\ kg/m^3)}{1.01(1000\ kg/m^3)(0.05\ kg/kg)}$$

$$Q = 11.24\ m^3/d$$

$$= 397\ ft^3/d$$

Step 2. Calculate the total $BOD_L$ loading rate

$$BOD_L\ \text{loading} = 0.14\ kg/m^3 \times 3785\ m^3/d$$

$$= 530\ kg/d$$

$$= 1.170\ lb/d$$

Step 3. Calculate the volume of the digester

From Step 1

$$Q = 11.24\ m^3/d$$

$$V = Q\theta_c = (11.24\ m^3/d)\ (15\ days)$$

$$= 169\ m^3$$

$$= 5968\ ft^3$$

Use the same volume for both the first and second stages.

Step 4. Check $BOD_L$ volumetric loading rate

$$BOD_L = \frac{530\ kg/d}{169\ m^3}$$

$$= 3.14\ kg/(m^3 \cdot d)$$

Step 5. Estimate the mass of volatile solids produced per day, using Eq. (6.216)

$$QS_0 = 530 \text{ kg/d (from Step 2)}$$

$$QS = 530 \text{ kg/d } (1 - 0.66) = 180 \text{ kg/d}$$

$$P_x = \frac{YQ(S_0 - S)}{1 + k_d\theta_c}$$

$$= \frac{0.05(530 - 180) \text{ kg/d}}{1 + (0.04 \text{ day}^{-1})15 \text{ days}}$$

$$= 10.9 \text{ kg/d}$$

Step 6. Calculate the percentage of stabilization

$$\% \text{ stabilization} = \frac{[(530 - 180) - 1.42(10.9)] \text{ kg/d} \times 100\%}{530 \text{ kg/d}}$$

$$= 63.1\%$$

Step 7. Calculate the volume of methane produced per day, using Eq. (6.214)

$$V = 350 \text{ L/kg } [(530 - 180) - 1.42\ (10.9)] \text{ kg/d}$$

$$= 117{,}000 \text{ L/d}$$

$$= 117 \text{ m}^3\text{/d}$$

$$= 4130 \text{ ft}^3\text{/d}$$

Step 8. Estimate total gas ($CH_4$ + $CO_2$) production

Normally 65% to 69% of gas is methane. Use 67%.

$$\text{Total gas volume} = (117 \text{ m}^3\text{/d})/0.67$$

$$= 175 \text{ m}^3\text{/d}$$

$$= 6170 \text{ ft}^3\text{/d}$$

**Example 3:** Estimate the gas produced in a anaerobic digester treating the secondary sludge from 1.0 MGD (3785 $m^3$/d) of wastewater containing 240 mg/L of total suspended solids (TSS).

**solution:**

Step 1. Estimate TSS removed, assuming the over-all secondary treatment process has TSS removal of 90%

$$\text{TSS removed} = 240 \text{ mgL} \times 0.9 = 216 \text{ mg/L}$$

Step 2. Estimate volatile suspended solids (VSS) in raw sewage removed, assuming VSS/TSS = 0.75

$$\text{VSS removed} = 216 \text{ mg/L} \times 0.75 = 162 \text{ mg/L}$$

Step 3. Estimate volatile suspended solids (VSS) in the sludge reduced 65%

$$\text{VSS reduced} = 162 \text{ mg/L} \times 0.65 = 105.3 \text{ mg/L}$$

Step 4. Calculate volatile solids (VS) reduced in 1 Mgal/d of sewage

$$\text{VS reduced} = (1 \text{ Mgal/d}) \times 8.34 \text{ lb/(Mgal} \cdot \text{mg/L)} \times 105.3 \text{ mg/L} = 878 \text{ lb/d}$$

Step 5. Estimate daily gas production for 1 MGD of sewage, assuming 15.0 $ft^3$/lb of VS

$$\text{Gas produced} = 15.0 \text{ ft}^3\text{/lb} \times 878 \text{ lb/d}$$

$$= 13{,}170 \text{ ft}^3\text{/d}$$

**Example 4:** The desired digester temperature is 91°F (32.8°C). The average raw sludge temperature is 45°F (7.2°C). The volume of raw sludge is 1,800 gal/d (6,810 L/d) with a specific weight of 1.20. Determine the quantity of heat needed for the digest, if SRT = 50 days and radiation heat loss of the inner surface is 7 Btu/h/$ft^2$.

**solution:**

Step 1. Calculate the heat required to raise the temperature of the sludge

$$\text{Heat} = 1800 \text{ gal/d} \times 8.34 \text{ lb/gal} \times (91 - 45) \text{ °F}$$

$$= 690{,}600 \text{ Btu/d}$$

Step 2. Determine the capacity of the digester with SRT = 50 days

$$\text{Digester} = 1800 \text{ gal/d} \times 0.1337 \text{ ft}^3\text{/gal} \times 50 \text{ days}$$

$$= 12{,}030 \text{ ft}^3$$

Use one digester 15 ft deep, thus the diameter ($D$) is

$$15 \times (3.14/4)D^2 = 12{,}030$$

$$D = 32.0 \text{ ft}$$

Step 3. Calculate radiation heat loss, assuming 7 Btu/h/$ft^2$ from the inner surface

$$\text{Area of the tank floor or roof} = 3.14 \times 16 \text{ ft} \times 16 \text{ ft} = 804 \text{ ft}^2$$

$$\text{Area of the wall} = 32 \text{ ft} \times 3.14 \times 15 \text{ ft} = 1507 \text{ ft}^2$$

$$\text{Total area} = 804 \text{ ft}^2 + 804 \text{ ft}^2 + 1507 \text{ ft}^2 = 3115 \text{ ft}^2$$

$$\text{Radiation loss} = 7 \text{ Btu/h/ft}^2 \times 24 \text{ h/d} \times 3115 \text{ ft}^2 = 523{,}300 \text{ Btu/d}$$

Step 4. Calculate the total heat required

$$\text{Sum of heat to raise sludge and radiation loss} = 690{,}600 \text{ Btu/d}$$

$$+ 523{,}300 \text{ Btu/d}$$

$$= 1{,}213{,}900 \text{ Btu/d}$$

Adding extra heat (25%) for other losses

Total heat required = 1,213,900 Btu/d × 1.25 = 1,517,400 Btu/d

**Egg-shaped digesters.** The egg-shaped digester looks similar to an upright egg and is different from the conventional American digester and the conventional German digester. The American digester typically is a relatively shallow cylindrical vessel with moderate floor and roof slopes. The typical conventional German digester is a deep cylindrical tank with steeply sloped top and bottom cones. The egg-shaped digester was first installed in Germany in the 1950s.

The advantages of the egg-shaped digester over conventional digesters include the elimination of abrupt change in the vessel geometry, reduced inactive volume of the digester, more cost effectiveness and less energy costs. The egg shape of the vessel provides unique advantages. Its mechanisms are as follows. Heavy solids cannot settle out over a large unmanageable bottom area; they must concentrate in the steep-sided bottom cone. Light solids do not accumulate across a large diameter surface area; they concentrate at the top of a steep-sided converging cone. The settleable materials can be pumped to the top of the vessel to sink light material into the digesting main mass. Likewise, light materials can be pumped to the bottom zone to stir the heavier materials. Small amounts of energy are required to manage the top and bottom zones. Moving materials through the main digesting mass to assure homogeneous digester conditions can be done more effectively (CBI Walker, 1998).

The egg-shaped digester is currently constructed of steel. It has become popular in the United States and other countries. More detailed description and design information may be found elsewhere (CBI Walker, 1998; Stukenberg *et al.*, 1990).

**Aerobic digestion.** Aerobic digestion is used to stabilize primary sludge, secondary sludge, or a combination of these by long-term aeration. The process converts organic sludge solids to carbon dioxide, ammonia, and water by aerobic bacteria with reduction of volatile solids, pathogens, and offensive odor.

In a conventional aerobic digester, concentration of influent VSS must be no more than 3 % for retention times of 15 to 20 days. High-purity oxygen may be used for oxygen supply.

Sludge is introduced to the aerobic digester on a batch (mostly), semibatch, or continuous basis. Aerobic digesters are typically a single-stage open tank like the activated-sludge aeration tank. In the batch basis, the digester is filled with raw sludge and aerated for 2 to 3 weeks, then stopped. The supernatant is decanted and the settled solids are removed (US EPA, 1991). For the semibatch basis, raw sludge is added every couple of days; the supernatant is decanted periodically, and the settled solids are held in the digester for a long time before being removed.

The design standard for aerobic digesters varies with sludge characteristics and method of ultimate sludge disposal. Sizing of the digester is commonly based on volatile solids loading of the digester or the volatile solids loading rate. It is generally determined by pilot and/or full-scale study. In general, volatile suspended solids loading rates for aerobic digesters vary from 1.1 to 3.2 kg VSS/($m^3 \cdot$ d) (0.07 to 0.20 lb VSS/($ft^3 \cdot$ d)), depending on the temperature and type of sludge. Solids retention time is 15 to 20 days for activated sludge only, and that for primary plus activated sludge is 20 to 25 days (US EPA, 1991). An aeration period ranges from 200 to 300 degree-days which is computed by multiplying the digester's temperature in °C by the sludge age. According to Rule 503 (Federal Register, 1993), if the sludge is for land application, the residence time requirements range from 60 days at 15°C (900° · d) to 40 days at 20°C (800° · d). Desirable aerobic digestion temperatures are approximately 18 to 27°C (65 to 80°F). A properly operated aerobic digester is capable of achieving a volatile solids reduction of 40% to 50% and a pathogens reduction of 90%.

GLUMRB (1996) recommended volume requirements based on population equivalent (PE). Table 6.25 presents the recommended digestion tank capacity based on a solids content of 2% with supernatant separation performed in a separate tank. If supernatant separation is performed in the digestion tank, a minimum of 25% additional volume is required. If high solids content (>2%) is applied, the digestion tank volume may be reduced proportionally. A digestion temperature of 15°C (59°F) and a solids retention time of 27 days (405° · d) is recommended. Cover and heating are required for cold temperature climate areas.

The air requirements of aerobic digestion are based on DO (1 to 2 mg/L) and mixing to keep solids in suspension. If DO in the digestion tank falls below 1.0 mg/L, the aerobic digestion process will be negatively

**TABLE 6.25 Recommended Volume Required for Aerobic Digester**

| | Volume per population equivalent | |
|---|---|---|
| Type of sludge | $m^3$ | $ft^3$ |
| Waste activated sludge—no primary settling | 0.13* | 4.5 |
| Primary plus activated sludge | 0.11* | 4.0 |
| Waste activated sludge exclusive of primary sludge | 0.06* | 2.0 |
| Extended aeration activated sludge | 0.09 | 3.0 |
| Primary plus fixed film reactor sludge | 0.09 | 3.0 |

*These volumes also apply to waste activated sludge from single-stage nitrification facilities with less than 24 h detention time based on design average flow.

SOURCE: Greater Lakes Upper Mississippi River Board, 1996

impacted. The diffused air requirements for waste activated sludge only is 20 to 35 $ft^3$/min per 1000 $ft^3$ (20 to 35 L/(min · $m^3$)) (US EPA, 1991). The oxygen requirement is 2.3 kg oxygen per kg VSS destroyed (Metcalf and Eddy, Inc. 1991).

The aerobic tank volume can be computed by the following equation (Water Pollution Control Federation, 1985b)

$$V = \frac{Q_i(X_i + YS_i)}{X(K_d P_v + 1/\theta_c)} \tag{6.218}$$

where $V$ = volume of aerobic digester, $ft^3$
$Q_i$ = influent average flow rate to digester, $ft^3$ /d
$X_i$ = influent suspended solids concentration, mg/L
$Y$ = fraction of the influent $BOD_5$ consisting of raw primary sludge, in decimals
$S_i$ = influent $BOD_5$, mg/L
$X$ = digester suspended solids concentration, mg/L
$K_d$ = reaction-rate constant, $day^{-1}$
$P_v$ = volatile fraction of digester suspended solids, in decimals
$\theta$ = solids retention time, day

The term $YS_i$ can be eliminated if no primary sludge is included in the influent of the aerobic digester. Equation (6.218) is not to be used for a system where significant nitrification will occur.

**Example 1:** The pH of an aerobic digester is found to have declined to 6.3. How much sodium hydroxide must be added to raise the pH to 7.0? The volume of the digester is 378 $m^3$ (0.1 Mgal). Results from jar tests show that 36 mg of caustic soda will raise the pH to 7.0 in a 2-L jar.

**solution:**

$$\begin{aligned}\text{NaOH required per m}^3 &= 36\text{ mg/2 L} = 18\text{ mg/L}\\ &= 18\text{ g/m}^3\\ \text{NaOH to be added} &= 18\text{ g/m}^3 \times 378\text{ m}^3\\ &= 6804\text{ g}\\ &= 6.8\text{ kg} = 15\text{ lb}\end{aligned}$$

**Example 2:** A 50 ft (15 m) diameter aerobic digester with 10 ft (3 m) side-wall depth treats 13,000 gal/d (49.2 $m^3$/d) of thickened secondary sludge. The sludge has 3.0% of solids content and is 80% volatile matter. Determine the hydraulic digestion time and volatile solids loading rate.

**solution:**

Step 1. Compute the effective volume of the digester

$$V = 3.14\ (50\ \text{ft}/2)^2 \times 10\ \text{ft} \times 7.48\ \text{gal/ft}^3$$
$$= 147{,}000\ \text{gal}$$
$$= 556\ \text{m}^3$$

Step 2. Compute the hydraulic digestion time

$$\text{Digestion time} = V/Q = 147{,}000\ \text{gal}/13{,}000\ \text{gal/d}$$
$$= 11.3\ \text{days}$$

Step 3. Compute the volatile solids loading rate

$$\text{Solids} = 3.0\% = 30{,}000\ \text{mg/L}$$
$$\text{VSS} = 30{,}000\ \text{mg/L} \times 0.8 = 24{,}000\ \text{mg/L}$$
$$\text{Volume} = 147{,}000\ \text{gal}/(6.48\ \text{ft}^3/\text{gal}) = 19{,}650\ \text{ft}^3$$

$$\text{VSS loading} = \frac{\text{VSS applied, lb/d}}{\text{Volume of digester, ft}^3}$$

$$= \frac{0.013\ \text{Mgal/d} \times 24{,}000\ \text{mg/L} \times 8.34\ \text{lb/(Mgal/d)(mg/L)}}{19{,}650\ \text{ft}^3}$$

$$= 0.132\ \text{lb/(d} \cdot \text{ft}^3)$$
$$= 2.11\ \text{kg/(d} \cdot \text{m}^3)$$

**Example 3:** Design an aerobic digester to receive thickened waste activated sludge at 1500 kg/d (3300 lb/d) with 2.5% solids content and a specific gravity of 1.02. Assume that the following conditions apply:

| | |
|---|---|
| Minimum winter temperature | 15°C |
| Maximum summer temperature | 25°C |
| Residence time (land application, Rule 503) | 900° · d at 15°C<br>700° · d at 25°C |
| VSS/TSS | 0.80 |
| SS concentration in digester | 75% of influent SS |
| $K_d$ at 15°C | 0.05 day$^{-1}$ |
| Oxygen supply and mixing | diffused air |

**solution:**

Step 1. Calculate the quantity of sludge to be treated per day, $Q_i$

$$Q_i = \frac{1500\ \text{kg/d}}{1000\ \text{kg/m}^3(1.02)(0.025)}$$

$$= 58.8\ \text{m}^3/\text{d}$$

Step 2. Calculate sludge age required for winter and summer conditions

During the winter:

$$\text{Sludge age} = 900°\text{d}/15° = 60 \text{ days}$$

During the summer:

$$\text{Sludge age} = 700°\text{d}/25° = 28 \text{ days}$$

The winter period is the controlling factor.
Under these conditions, assume the volatile solids reduction is 45% (normally 40% to 50%).

*Note*: A curve for determining volatile solids reduction in an aerobic digester as a function of digester liquid temperature and sludge age is available (WPCF, 1985b). It can be used to compare winter and summer conditions.

Step 3. Calculate volatile solids reduction

$$\begin{aligned}\text{Total mass of VSS} &= 1500 \text{ kg/d} \times 0.8 \\ &= 1200 \text{ kg/d} \\ \text{VSS reduction} &= 1200 \text{ kg/d} \times 0.45 \\ &= 540 \text{ kg/d}\end{aligned}$$

Step 4. Calculate the oxygen required

Using 2.3 kg $O_2$/kg VSS reduction

$$\begin{aligned}O_2 &= 540 \text{ kg/d} \times 2.3 \\ &= 1242 \text{ kg/d}\end{aligned}$$

Convert to the volume of air required under standard conditions, taking the mass of air as 1.2 kg/m$^3$ (0.75 lb/ft$^3$) with 23.2% of oxygen.

$$\text{Air} = \frac{1242 \text{ kg/d}}{1.2 \text{ kg/m}^3 \times 0.232}$$

$$= 4460 \text{ m}^3/\text{d}$$

Assume an oxygen transfer efficiency of 9%. Total air required is

$$\begin{aligned}\text{Air} &= (4460 \text{ m}^3/\text{d})/0.09 \\ &= 49{,}600 \text{ m}^3/\text{d} \\ &= 34.4 \text{ m}^3/\text{min} \\ &= 1210 \text{ ft}^3/\text{min}\end{aligned}$$

Step 5. Calculate the volume of the digester under winter conditions

Using Eq. (6.218)

$$X_i = 2.5\% = 25{,}000 \text{ mg/L}$$

$$X = 25{,}000 \text{ mg/L} \times 0.75 = 18{,}750 \text{ mg/L}$$

$$K_d = 0.05 \text{ day}^{-1}$$

$$P_v = 0.8$$

$$\theta_c = 60 \text{ days}$$

The terms $YS_i$ is neglected, since no primary sludge is applied.

$$V = \frac{Q_i(X_i + YS_i)}{X(K_d P_v + 1/\theta_c)}$$

$$= \frac{58.8 \text{ m}^3/\text{d} \times 25{,}000 \text{ mg/L}}{18{,}750 \text{ mg/L}(0.05 \text{ day}^{-1} \times 0.8 + 1/60 \text{ days})}$$

$$= 1383 \text{ m}^3$$

$$= 48{,}860 \text{ ft}^3$$

Step 6. Check the diffused air requirement per $m^3$ of digester volume, from Step 4

$$\text{Air} = \frac{34.4 \text{ m}^3/\text{min}}{1{,}383 \text{ m}^3}$$

$$= 0.025 \text{ m}^3/(\text{min} \cdot \text{m}^3)$$

$$= 25 \text{ L/min} \cdot \text{m}^3$$

$$= 25 \text{ ft}^3/\text{min}/1000 \text{ ft}^3$$

This value is at the lowest range of regulatory requirement, 25 to 35 L/min · $m^3$.

**Example 4:** Determine the size of an aerobic digester to treat the primary sludge of 590 kg/d (1300 lb/d) plus activated sludge of 380 kg/d (840 kg/d). Assume the mixed sludge has a solids content of 3.8% and a retention time of 18 days.

**solution:**

Step 1. Calculate the mass of mixed sludge

$$\text{Mass} = (590 \text{ kg/d} + 380 \text{ kg/d})/0.038$$

$$= 25{,}530 \text{ kg/d} \cong 25.5 \text{ m}^3/\text{d}$$

Step 2. Calculate the volume ($V$) of the aerobic digester

The volume of daily sludge production is approximately 25.5 $m^3$/d, thus

$$V = 25.5 \text{ m}^3/\text{d} \times 18 \text{ days}$$

$$= 459 \text{ m}^3$$

Step 3. Check the solids loading rate assuming the solids are 80% volatile

$$\text{VSS} = (590 + 380) \text{ kg/d} \times 0.8 = 776 \text{ kg/d}$$

$$\text{Loading rate} = (776 \text{ kg VSS/d})/459 \text{ m}^3$$
$$= 1.69 \text{ kg VSS(m}^3 \cdot \text{d) [in the range of 1.1 to 3.2 kg VSS/(m}^3 \cdot \text{d)]}$$

**Lime stabilization.** Chemical stabilization of wastewater sludge is an alternative to biological stabilization. It involves chemical oxidation (commonly using chlorine) and pH adjustment under basic conditions, which is achieved by addition of lime or lime-containing matter. Chemical oxidation is achieved by dosing the sludge with chlorine (or ozone, hydrogen peroxide). The sludge is deodorized and microbiological activities slowed down. The sludge can then be dewatered and disposed of.

In the lime stabilization process, lime is added to sludge in sufficient quantity to raise the pH to 12 or higher for a minimum of 2 h contact. The highly alkaline environment will inactivate biological growth and destroy pathogens. The sludge does not putrefy, create odors, or pose a health hazard. However, if the pH drops below 11, renewed bacteria and pathogen growth can reoccur. Since the addition of lime does not reduce the volatile organics, the sludge must be further treated and disposed of before the organic matter starts to putrefy again.

With 40 CFR Part 503 regulations (Federal Register, 1993) to reduce pathogens, a supplemental heat system can be used to reduce the quantity of lime required to obtain the proper temperature. Additional heating reduces the lime dosage and operating costs.

**Composting.** Composting is an aerobic biological decomposition of organic material to a stable end product at an elevated temperature. Some facilities have utilized anaerobic composting in a sanitary, nuisance-free environment to create a stable humus-like material suitable for plant growth. Approximately 20% to 30% of volatile solids are converted to carbon dioxide and water at temperatures in the pasteurization range of 50 to 70°C (122 to 158°F). At these temperatures most enteric pathogens are destroyed. Thermophilic bacteria are responsible for decomposing organic matter. The finished compost has a moisture content of 40% to 50% and a volatile solids content of 40% or less (US EPA, 1991).

The optimum moisture content for composting is 50% to 60% water. Dewatered wastewater sludges are generally too wet to meet optimum composing conditions. If the moisture content is over 60% water, proper structural integrity must be obtained. The dewatered sludge and bulking agent must be uniformly mixed. Sufficient air must be supplied to the composting pile, either by forced aeration or windrow turning to maintain oxygen levels between 5% and 15%. Odor control may be required on some sites.

In order to comply with the "process to significantly reduce pathogens" (PSRP) requirements of 40 CFR Part 257 (Federal Register, 1993), the compost pile must be maintained at a minimum operating temperature of 40°C (104°F) for at least 5 days. During this time, the temperature must be allowed to increase above 55°C (131°F) for at least 4 h to ensure pathogen destruction. In order to comply with the "process to further reduce pathogens" (PFRP) requirements of 40 CFR Part 257 (Federal Register, 1993), the in-vessel and static aerated compost piles must be maintained at a minimum operating temperature of 55°C for at least 3 days. The windrow pile must be maintained at a minimum operating temperature of 55°C for 15 days. In addition, there must be at least three turnings of the compost pile during this period.

**Sludge conditioning.** Mechanically concentrated (thickened) sludges and biologically and chemically stabilized sludges still require some conditioning steps. Sludge conditioning can involve chemical and/or physical treatment to enhance water removal. Sludge conditioning is undertaken before sludge dewatering.

Details of process chemistry, design considerations, mechanical components, system layout, operation, and costs of conditional methods are presented in *Sludge Conditioning Manual* (WAPCF, 1988a). Some additional sludge conditioning processes disinfect sludge, control odors, alter the nature of solids, provide limited solids destruction, and increase solids recovery.

**Chemical conditioning.** Chemical conditioning can reduce the 90% to 99% incoming sludge moisture content to 65% to 80%, depending on the nature of the sludge to be treated (WPCF, 1988a). Chemical conditioning results in coagulation of the solids and release of the absorbed water. Chemicals used for sludge conditioning include inorganic compounds such as lime, pebble quicklime, ferric chloride, alum, and organic polymers (polyelectrolytes). Addition of conditioning chemicals may increase the dry solids of the sludge. Inorganic chemicals can increase the dry solids by 20% to 30%, but polymers do not increase the dry solids significantly.

**Physical conditioning.** The physical process includes using hot and cold temperatures to change sludge characteristics. The commonly used physical conditioning methods are thermal conditioning and elutriation. Less commonly used methods include freeze–thaw, solvent extraction, irradiation, and ultrasonic vibration.

The thermal conditioning (heat treatment) process involves heating the sludge to a temperature of 177 to 240°C (350 to 464°F) in a reaction vessel under pressure of 1720 to 2760 kN/m$^2$ (250 to 400 lb/in$^2$ [psig]) for a period of 15 to 40 min (US EPA, 1991). One modification of the process involves the addition of a small amount of air. Heat coagulates solids, breaks down the structure of microbial cells in waste activated sludge, and releases the water bound in the cell. The heat-treated sludge is sterilized and practically deodorized. It has excellent dewatering characteristics and does not normally require chemical conditioning to dewater well on mechanical equipment, Yielding cake solids concentrations of 40% to 50%.

Heat treatment is suitable for many types of sludge that cannot be stabilized biologically due to the presence of toxic materials and is relatively insensitive to change in sludge composition. However, the process produces liquid sidestreams with high concentrations of organics, ammonia nitrogen, and color. Significant odorous off-gases are also generated which must be collected and treated before release. The process has high capital cost because of its mechanical complexity and the use of corrosion-resistant material. It also requires close supervision, skilled operators, and O and M programs. More detailed description of thermal processing of sludge can be found elsewhere (WPCF, 1988a; WEF and ASCE, 1991b).

A physical conditioning method used in the past is elutriation. In elutriation, a washing process, sludge is mixed with a liquid for the purpose of transferring certain soluble organic or inorganic components to the liquid. Wastewater effluent is usually used for elutriation. The volume of wastewater is two to six times the volume of sludge. The elutriation of the sludge produces large volumes of liquid that contain a high concentration of suspended solids. This liquid, when returned to the treatment plant, increases the solids and organic loadings. Elutriation tanks are designed to act as gravity thickeners with solids loading rates of 39 to 49 $kg/(m^2 \cdot d)$ (8 to 10 $lb/(ft^2 \cdot d)$) (Qasim, 1985).

**Sludge dewatering.** Following stabilization and/or conditioning, wastewater sludge can be ultimately disposed of, or can be dewatered prior to further treatment and/or ultimate disposal. It is generally more economical to dewater before disposal. The primary objective of dewatering is to reduce sludge moisture. Subsequently, it reduces the costs of pumping and hauling to the disposal site. Dewatered sludge is easier to handle than thickened or liquid sludge. Dewatering is also required before composting, prior to sludge incineration, and prior to landfill. An advantage of dewatering is that it makes the sludge odorless and nonputrescible.

Sludge can be dewatered by slow natural evaporation and percolation (drying beds, drying lagoons) or by mechanical devices, mechanically assisted physical means, such as vacuum filtration, pressure filtration, centrifugation, or recessed plate filtration.

**Vacuum filtration.** Vacuum filtration has been used for wastewater sludge dewatering for almost seven decades. Its use has declined owing to competition from the belt press. A rotary vacuum filter consists of a cylindrical drum covered with cloth of natural or synthetic fabric, coil springs, or woven stainless steel mesh. The drum is partly submerged (20% to 40%) in a vat containing the sludge to be dewatered.

The filter drum is divided into compartments. In sequence, each compartment is subject to vacuums ranging from 38 to 75 cm (18 to 30 in) of mercury. As it slowly rotates, vacuum is applied immediately under the mat formation or sludge pick-up zone. Suction continues to dewater the solids adhering to the filter medium as it rotates out of the liquid. This is called the drying zone of the cycle. The cake drying zone represents from 40% to 60% of the drum surface. The vacuum is then stopped, and the solids cake is removed to a sludge hopper or a conveyor. The filter medium is washed by water sprays before reentering the vat.

For small plants, 35 h operation per week is often designed. This allows daily 7 h operation with 1 h for start-up and wash-down. For larger plants, vacuum filters are often operated for a period of 16 to 20 h/d (Hammer, 1986).

**Example:** A 3 m (10 ft) diameter by 4.5 m (15 ft) long vacuum filter dewaters 5400 kg/d primary plus secondary sludge solids. The filter is operated 7 h/d with a drum cycle time of 4 min. The dewatered sludge has 25% solids content. The filter yield is 17.3 kg/(m$^2$ · h). Determine the filter loading rate and percent solids recovery. What is the daily operating hours required if the dewatered sludge is 30% solids with the same percentage of solids recovery and a filter yield of 9.8 kg/(m$^2$ · h)?

**solution:**

Step 1. Compute the surface area of the filter

$$\text{Area} = \pi(3\ \text{m})(4.5\ \text{m}) = 42.4\ \text{m}^2$$

Step 2. Compute the filter loading rate

$$\begin{aligned}\text{Loading} &= \frac{\text{solids loading}}{\text{area} \times \text{operating time}} = \frac{5400\ \text{kg/d}}{42.4\ \text{m}^2 \times 7\ \text{h/d}} \\ &= 18.2\ \text{kg/(m}^2 \cdot \text{h)} \\ &= 3.7\ \text{lb/(ft}^2 \cdot \text{h)}\end{aligned}$$

Step 3. Compute the percent solids recovery

$$\begin{aligned}\% &= \frac{(17.3\ \text{kg/(m}^2 \cdot \text{h))} \times 100\%}{18.2\ \text{kg/(m}^2 \cdot \text{h)}} \\ &= 95\%\end{aligned}$$

Step 4. Compute the daily filter operation required with 30% solids

Since filter yield is determined by

$$\text{Filter yield} = \frac{\text{solids loading} \times \text{factor of recovery}}{\text{h/d operation} \times \text{filter area}}$$

rearranging the above:

$$\text{Operation, h/d} = \frac{\text{solids loading} \times \text{factor of recovery}}{\text{filter yield} \times \text{filter area}}$$

$$= \frac{5400 \text{ kg/d} \times 0.95}{9.8 \text{ kg/(m}^2 \cdot \text{h)} \times 42.4 \text{ m}^2}$$

$$= 12.3 \text{ h/d}$$

**Pressure filtration.** There are two types of pressure filtration process as used for sludge dewatering. Normally, they are the belt filter press and the plate and frame filter press. The operating mechanics of these two filter press types are completely different. The belt filter press is popular due to availability of smaller sizes and uses polymer for chemical flocculation of the sludge. It consumes much less energy than the vacuum filter. The plate and frame filter press is used primarily to dewater chemical sludges. It is a large machine and uses lime and ferric chloride for conditioning of organic sludge prior to dewatering. The filtered cake is very compact and dry.

*Belt filter press.* A belt filter press consists of two endless, tensioned, porous belts. The belts travel continuously over a series of rollers of varied diameter that squeeze out the water from the sludge and produce a dried cake that can be easily removed from the belts. Variations in belt filter press design are available from different manufacturers.

Chemical conditioning of sludge is vital to the efficiency of a belt filter press. Polymers are used as a conditioning agent, especially cationic polymers. The operation of the belt filter press comprises three zones, i.e. gravity drainage, low-pressure and high-pressure zones. Conditioned sludge is first introduced uniformly to the gravity drainage zone and onto the moving belt where it is allowed to thicken. The majority of free water is removed from the sludge by gravity. Free water readily separates from the slurry and is recycled back through the treatment system. The efficiency of gravity drainage depends on the type of sludge, the quality of the sludge, sludge conditioning, the belt screen mesh, and design of the drainage section. Typically, gravity drainage occurs on a flat or slightly inclined belt for a period of 1 to 2 min. A 5% to 10% increase in solids content is achieved in this zone; that is, 1% to 5% sludge feed produces 6% to 15% solids prior to compression (US EPA, 1991).

Following the gravity drainage zone, pressure is applied in the low-pressure (wedge) zone. The pressure can come from the compression of the sludge between two belts or from the application of a vacuum on the lower belt. In the wedge zone, the thickened sludge is subjected to low pressure to further remove water from the sludge matrix. This is to prepare an even firmer sludge cake that can withstand the shear forces to which it is subjected by rollers.

On some units, the low-pressure zone is followed by a high-pressure zone, where the sludge is sandwiched between porous belts and subjected to shearing forces as the belts pass through a series of rollers of decreasing diameter. The roller arrangement progressively increases the pressure. The squeezing and shearing forces induce the further release of additional amounts of water from the sludge cake.

The final dewatered filter cake is removed from the belts by scraper blades into a hopper or conveyor belt for transfer to the sludge management area. After the cake is removed, a spray of water is applied to wash and rinse the belts. The spray rinse water, together with the filtrate, is recycled back to primary or secondary treatment.

Belt filter presses are commercially available in metric sizes from 0.5 to 3.5 m in belt width, with the most common size between 1.0 and 2.5 m. Hydraulic loading based on belt width varies from 25 to 100 gal/(min · m) (1.6 to 6.3 L/(m · s)). Sludge loading rates range from 200 to 1500 lb/(m · h) (90 to 680 kg/(m · h)) (Metcalf and Eddy, Inc. 1991). Typical operating parameters for belt filter press dewatering of polymer-conditioned wastewater sludges are presented in Table 6.26.

Odor is often a problem with the belt filter press. Adequate ventilation to remove hydrogen sulfide or other gases is also a safety consideration.

**TABLE 6.26 Typical Operating Parameters for Belt Filter Press Dewatering of Polymer Conditioned Sludge**

| Type of sludge | Feed solids, percent | Hydraulic loading, gal/(min · m*) | Solids loading, lb/(m* · h) | Cake solids, percent | Polymer dosage, lb/ton |
|---|---|---|---|---|---|
| Anaerobically digested | | | | | |
| Primary | 4–7 | 40–50 | 1000–1600 | 25–44 | 3–6 |
| Primary plus WAS | 2–6 | 40–50 | 500–1000 | 15–35 | 6–12 |
| Aerobically digested without primary | 1–3 | 30–45 | 200–500 | 12–20 | 8–14 |
| Raw primary and waste activated | 3–6 | 40–50 | 800–1200 | 20–35 | 4–10 |
| Thickened waste activated | 3–5 | 40–50 | 800–1000 | 14–20 | 6–8 |
| Extended aeration waste activated | 1–3 | 30–45 | 200–500 | 11–22 | 8–14 |
| Heat-treated primary plus waste activated | 4–8 | 35–50 | 1000–1800 | 25–50 | 1–2 |

*Notes*: 1.0 gal/(min · m) = 0.225 $m^3$/(m · h)
1.0 lb/(m · h) = 0.454 kg/(m · h)
1.1 lb/ton = 0.500 kg/tonne
*Loading per meter belt width

SOURCES: WPCF, 1983; Viessman and Hammer, 1993; WEF, 1996b

Hydrogen peroxide or potassium permanganate can be used to oxidize the odor-causing chemical ($H_2S$). Potassium permanganate can improve the dewatering efficiency of the sludge and can reduce the quantity of polymer required.

**Example:** A belt filter press (BFP) is designed to dewater anaerobically digested primary plus waste activated sludge for 7 h/d and 5 d/wk. The effective belt width is 2.0 m (commonly used). Estimate hydraulic and solids loading rates, polymer dosage, and solids capture (recovery) with the following given data.

| | |
|---|---|
| Sludge production | 20 gal/min |
| Total solids in sludge feed | 3.5% |
| Total solids in cake | 25% |
| Sp. gr. of sludge feed | 1.03 |
| Sp. gr. of dewatered cake | 1.08 |
| Sp. gr. of filtrate | 1.01 |
| Polymer dose (0.2% by weight) | 7.2 gal/min |
| TSS in wastewater of BFP | 1900 mg/L |
| Wash water | 40 gal/min |

**solution:**

Step 1. Calculate the average weekly sludge production

$$\text{Wet sludge} = 20\ \text{gal/min} \times 1440\ \text{min/d} \times 7\ \text{d/wk} \times 8.34\ \text{lb/gal} \times 1.03$$
$$= 1{,}731{,}780\ \text{lb/wk}$$
$$\text{Dry solids} = 1{,}731{,}780\ \text{lb/wk} \times 0.035$$
$$= 60{,}610\ \text{lb/wk}$$

Step 2. Calculate daily and hourly dry solids dewatering requirement based on the designed operation schedule: 7 h/d and 5 d/wk

$$\text{Daily rate} = (60{,}610\ \text{lb/wk})/(5\ \text{d/wk})$$
$$= 12{,}120\ \text{lb/d}$$
$$\text{Hourly rate} = (12{,}120\ \text{lb/d})/(7\ \text{h/d})$$
$$= 1730\ \text{lb/h}$$

Step 3. Select belt filter press size and calculate solids loading rate

Use one 2.0 m belt (most common size) and one more identical size for standby. Calculate solids loading rate

$$\text{Solids loading} = (1730\ \text{lb/h})/2\ \text{m}$$
$$= 865\ \text{lb/(h} \cdot \text{m)}\ \text{(OK, within 500 to 1000 lb/(h} \cdot \text{m) range)}$$

Step 4. Calculate hydraulic loading rate

$$\text{Flow to the press} = 20\ \text{gal/min} \times (7\ \text{days/5 days})(24\ \text{h/7 h})$$
$$= 96\ \text{gal/min}$$

$$\begin{aligned} \text{HL} &= 96\ (\text{gal/min})/2\ \text{m} \\ &= 48\ \text{gal/(min} \cdot \text{m)} \\ &\qquad (\text{OK, within 40 to 50 gal/(min} \cdot \text{m) range}) \end{aligned}$$

Step 5. Calculate dosage of polymer (0.2% by weight) with a 7.2 gal/min rate

$$\text{Dosage} = \frac{(7.2\ \text{gal/min} \times 60\ \text{min/h} \times 0.002) \times 8.34\ \text{lb/gal}}{2.0\ \text{m} \times 865\ \text{lb/(h} \cdot \text{m)} \times (1\ \text{ton/2000 lb})}$$

$$= 8.33\ \text{lb/ton}$$

$$= 4.16\ \text{kg/tonne}$$

Step 6. Calculate solids recovery rate

(a) Estimate the volumetric flow of cake, $q$ (from Step 4)

$$q = \frac{96\ \text{gal/min} \times 0.035}{(25\%/100\%) \times 1.08}$$

$$= 12.4\ \text{gal/min}$$

(b) Calculate flow rate of filtrate

$$\begin{aligned} \text{Flow of filtrate} &= \text{sludge feed flow} - \text{cake flow} \\ &= 96\ \text{gal/min} - 12.4\ \text{gal/min} \\ &= 83.6\ \text{gal/min} \end{aligned}$$

(c) Calculate total solids in the filtrate of wastewater from the press

$$\text{Solids} = \frac{83.6\ \text{gal/min} \times 60\ \text{min/h} \times 1900\ \text{mg/L} \times 8.34\ \text{lb/gal}}{2.0\ \text{m} \times 1{,}000{,}000\ \text{mg/L}}$$

$$= 40\text{lb/(h} \cdot \text{m)}$$

(d) Calculate solids capture (recovery)

$$\text{Solids caputre} = \frac{\text{solids in feed} - \text{solids in filtrate}}{\text{solids in feed}} \times 100\%$$

$$= \frac{(865 - 40)\ \text{lb/(h} \cdot \text{m)} \times 100\%}{865\ \text{lb/(h} \cdot \text{m)}}$$

$$= 95.4\%$$

*Plate and frame filter press.* There are several types of filter press available. The most common type, the plate and frame filter press, consists of vertical plates that are held rigidly in a frame and pressed together between fixed and moving ends (Fig. 6.61a). As shown in Fig. 6.61b, each chamber is formed by paired recessed plates. A series of individual

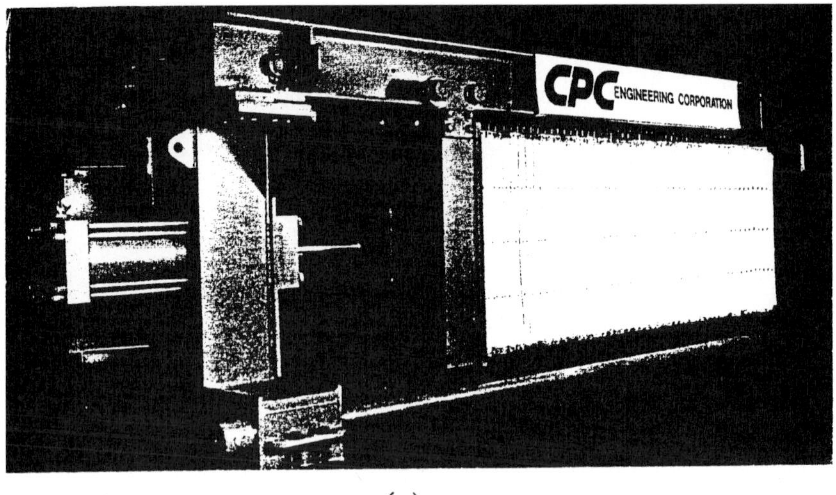


(a)



(b)

**Figure 6.61** Plate and frame filter press: (a) side view of a filter press; (b) schematic cross section of chamber area during fill cycle.

chambers have a common feed port and two or more common filtrate channels. Fabric filter media are mounted on the face of each individual plate, thus providing initial solids capture.

The filter press operates in batch manner to dewater sludge. Despite its name, the filter press does not press or squeeze sludge. Instead, when the filter is closed, the recessed faces of adjacent plates form a pressure-tight chamber; the sludge is pumped into the press at pressure up to 225 lb/in$^2$ (15 atm, 1550 kN/m$^2$) and maintained for less than 2 h (CPC Eng., 1980). The water passes through the filter cloth, while the solids are retained and form a cake on the cloth. During the initial phase of operation the fabric cloth retains the solids which then become the filter media. This action results in high solids capture and a filtrate of maximum clarity. At the end of a filtration cycle, the feed pump system is stopped and the chamber plates are opened, allowing the cake to fall into appropriate hoppers or conveyors. Filter cloth usually requires a precoat (ash or diatomaceous earth) to aid retention of solids on the cloth and release of cake. Drainage ports are provided at both the top and the bottom of each chamber (Figure 6.61b). This figure shows a fixed volume, recessed plate filter press. Another type, the variable volume recessed plate filter press, is also used for sludge dewatering.

The advantages of the plate filter press include production of higher solids cake, 35% to 50% versus 18% for both belt press (US EPA, 1991) and vacuum filters, higher clarity filtrate, lower energy consumption, lower o & M costs, longer equipment service life, and relatively simple operation. However, sufficient washing and air drying time between cycles is required.

**Example:** A plate and frame filter press has a plate surface of 400 ft$^2$ and is used to dewater conditioned waste activated sludge with a solids concentration of 2.5%. The filtration time is 110 min at 225 lb/in$^2$ (gage) and the time to discharge the sludge and restart the feed is 10 min. The total volume of sludge dewatered is 3450 gal. (1) Determine the solids loading rate and the net filter yield in lb/(h · ft$^2$). (2) What will be the time of filtration required to produce the same net yield if the feed solids concentration is reduced to 2%?

**solution:**

Step 1. Compute solids loading rate

$$\text{Solids loading} = \frac{\text{mass of dry solids applied}}{\text{filtration time} \times \text{plate area}}$$

$$= \frac{3450 \text{ gal} \times 8.34 \text{ lb/gal} \times 0.025}{110 \text{ min/(60 min/h)} \times 400 \text{ ft}^2}$$

$$= 0.98 \text{ lb/(h} \cdot \text{ft}^2)$$

Step 2. Compute the net filter yield

$$\text{Net yield} = \text{solids loading} \times \frac{\text{filtration time}}{\text{time per cycle}}$$

$$= 0.98 \text{ lb/(h} \cdot \text{ft}^2) \times \frac{110 \text{ min}}{(110 + 10) \text{ min}}$$

$$= 0.90 \text{ lb/(h} \cdot \text{ft}^2)$$

Step 3. Determine the filtration time required at 2% solids feed

$$\text{Time} = 110 \text{ min } (2.5\%/2\%)$$
$$= 137.5 \text{ min}$$

**Centrifugation.** The centrifugation process is also used for dewatering raw, digested, and waste activated sludge. The process is discussed in more detail in the section on sludge thickening. Typically, a centrifuge produces a cake with 15% to 30% dry solids. The solids capture ranges from 50% to 80% and 80% to 95%, without and with proper chemical sludge conditioning, respectively (US EPA, 1987b).

**Sludge drying beds.** Sludge drying beds remove moisture by natural evaporation, gravity, and/or induced drainage, and are the most widely used method of dewatering municipal sludge in the United States. They are usually used for dewatering well-digested sludge. Drying beds include conventional sand beds, paved beds, unpaved beds, wedge-wire beds, and vacuum-assisted beds.

Drying beds are less complex, easier to operate, require less operating energy, and produce higher solids cake than mechanical dewatering systems. However, more land is needed for them. Drying beds are usually used for small- and mid-size wastewater treatment plants with design flow of less than 7500 m$^3$/d (2 Mgal/d), due to land restrictions (WEF, 1996b).

*Sand drying beds.* Sand drying beds generally consist of 10 to 23 cm (4 to 9 in) of sand placed over a 20 to 50 cm (8 to 20 in) layer of gravel. The diameters of sand and gravel range from 0.3 to 1.22 mm and from 0.3 to 2.5 cm, respectively. The water drains to an underdrain system that consists of perforated pipe at least 10 cm (4 in) in diameter and spaced 2.4 to 6 m (8 to 20 ft) apart. Drying beds usually consist of a 0.3 to 1 m (1 to 3 ft) high retaining wall enclosing process drainage media (US EPA, 1989, 1991). The drying area is partitioned into individual beds, 6 m wide by 6 to 30 m long (20 ft wide by 20 to 100 ft long), or of a convenient size so that one or two beds will be filled in a normal cycle (Metcalf and Eddy, Inc. 1991).

In a typical sand drying bed digested and/or conditioned sludge is discharged (at least 0.75 m/s velocity) on the bed in a 30 to 45 cm (12 to 18 in) layer and allowed to dewater by drainage through the sludge mass and supporting sand and by evaporation from the surface exposed to air. Dissolved gases are released and rise to the surface. The water drains through the sand and is collected in the underdrain system and generally returned to the plant for further treatment.

Design area requirements for open sludge drying beds are based on sludge loading rate, which is calculated on a per capita basis or on a unit load of mass of dry solids per unit area per year. On a per capita basis, surface areas range from 1.0 to 2.5 $ft^2$/person (0.09 to 0.23 $m^2$/person), depending on sludge type. Sludge loading rates are between 12 and 23 lb dry solids/($ft^2 \cdot yr$) (59 and 113 kg/($m^2 \cdot yr$) (Metcalf and Eddy, Inc. 1991).

The design, use, and performance of the drying bed are affected by the type of sludge, sludge conditioning, sludge application rates and depth, dewatered sludge removal techniques, and climatic conditions.

**Example:** A sand drying bed, 6 m by 30 m (20 by 100 ft), receives conditioned sludge to a depth of 30 cm (12 in). The sludge feed contains 3% solids. The sand bed requires 29 days to dry and one day to remove the sludge from the bed for another application. Determine the amount of sludge applied per application and the annual solids loading rate. The sp. gr. of sludge feed is 1.02.

**solution:**

Step 1. Compute the volume of sludge applied per application, $V$

$$\begin{aligned} V &= 6\ \text{m} \times 30\ \text{m} \times 0.3\ \text{m/app.} \\ &= 54\ \text{m}^3\text{/app.} \\ &= 1907\ \text{ft}^3\text{/app.} \\ &= 14{,}300\ \text{gal/app.} \end{aligned}$$

Step 2. Compute the solids applied per application

$$\begin{aligned} \text{Solids} &= 54\ \text{m}^3\text{/app} \times 1000\ \text{kg/m}^3 \times 1.02 \times 0.03 \\ &= 1650\ \text{kg/app.} \\ &= 3640\ \text{lb/app.} \end{aligned}$$

Step 3. Compute annual solids loading rate

$$\text{Surface area of the bed} = 6\ \text{m} \times 30\ \text{m} = 180\ \text{m}^2$$

$$\begin{aligned} \text{Loading rate} &= \frac{\text{solids applied} \times 365\ \text{d/yr}}{\text{surface area} \times \text{cycle}} \\ &= \frac{1650\ \text{kg/app} \times 365\ \text{d/yr}}{180\ \text{m}^2 \times 30\ \text{d/app.}} \\ &= 111\ \text{kg/(m}^2 \cdot \text{yr) (OK within 59–113 kg/(m}^2 \cdot \text{yr) range)} \\ &= 22.8\ \text{lb/(ft}^2 \cdot \text{yr)} \end{aligned}$$

*Paved beds.* Paved beds consist of a concrete or asphalt pavement above a porous gravel subbase with a slope of at least 1.5%. There are two types of drainage, i.e. a drainage type and a decant type. Unpaved areas, constructed as sand drains, are placed around the perimeter or along the center of the bed to collect and convey drainage water. The advantage of the paved bed is that sludge can be removed by a front-end loader. The bed areas are larger and typically rectangular in shape, being 6 to 15 m (20 to 50 ft) wide by 21 to 46 m (70 to 150 ft) long with vertical side walls as sand beds. Cake solids contents as much as 50% can be achieved in an arid climate.

Well designed paved beds may remove about 20% to 30% of the water with good settling solids. Solids concentration may range from 40% to 50% for a 30 to 40 days drying time in an arid climate for a 30 cm (1 ft) sludge layer (Metcalf and Eddy, Inc. 1991).

*Unpaved beds.* Unpaved beds may be used in warm and dry climate areas where groundwater is not a concern. The beds are similar to paved beds with decanting. Sufficient storage area is required during a wet weather period, because access to the beds is restricted during wet weather.

*Vacuum-assisted drying beds.* The vacuum-assisted drying bed consists of a reinforced concrete ground slab, a layer of supporting aggregate, and a rigid porous media plate on top. The space between the concrete slab and the multimedia plate is the vacuum chamber, connected to a vacuum pump. This is to accelerate sludge dewatering and drying.

Polymer-preconditioned sludge is applied to the surface of the media plates until it is entirely covered. The vacuum is then applied to remove the free water from the sludge. The sludge is air dried for 24 to 48 h. Essentially all of the solids remain on top of the media plates form a cake of fairly uniform thickness. The cake has solids concentration from 9% to 35%, depending on sludge feed (WEF, 1996b).

The dewatered sludge is removed by a front-end loader. The porous multimedia plates are washed with a high-pressure hose to remove the remaining sludge residue before another application.

The vacuum-assisted drying bed has a short cycle time and needs less footage. However, it is labor intensive and is expensive to operate. Drying beds are sometimes enclosed in a greenhouse-type glass structure to increase drying efficiency in wet or colder climates. The enclosure also serves for odor and insect control, and improves the overall appearance of the treatment plant. Good ventilation is required to allow moisture to escape.

*Wedge-wire drying beds.* The wedge-wire (wedgewater) drying bed was developed in England and there exist a few installations in the United States. The material for the drying bed consists of stainless steel wedge wire or high-density polyurethane. It is a physical process similar to the vacuum-assisted drying bed.

The bed consists of a shallow rectangular watertight basin fitted with a false floor of artificial media, wedge-wire panels (Fig. 6.62). The panels

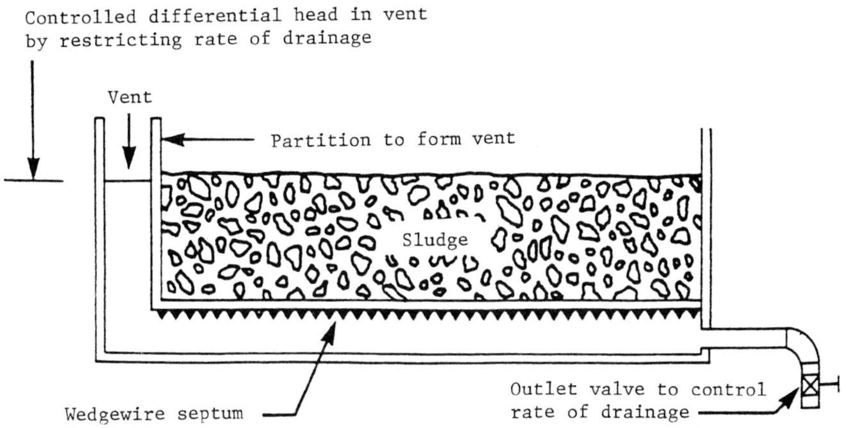

Controlled differential head in vent by restricting rate of drainage

Vent

Partition to form vent

Sludge

Outlet valve to control rate of drainage

Wedgewire septum

**Figure 6.62** Crosssection of a wedgewire drying bed (*source: US EPA, 1991*).

(wedge-wire septum) have wedge-shaped slotted openings of 0.25 mm (0.01 in). The false floor is made watertight with caulking where the panels abut the walls. An outlet valve to control the drainage rate is located underneath the false floor. Water or plant effluent is introduced to the bed from beneath the wedge-wire septum up to a depth of 2.5 cm (1 in) over the septum. This water serves as a cushion and prevents compression or other disturbance of the colloidal particles. The sludge is introduced slowly onto a horizontal water level to float without causing upward or downward pressure across the wedgewire surface. After the bed is filled with sludge, the initially separate water layer and drainage water are allowed to percolate away at a controlled flow rate through the outlet valve. After the free water has been drained, the sludge further concentrates by drainage and evaporation until there is a requirement for sludge removal (US EPA, 1987b).

The advantages of this process include no clogging in the wedgewire, treatability of aerobically digested sludge, constant and rapid drainage, higher throughput than the sand bed, and ease of operation and maintenance. However, the capital cost is higher than that of other drying beds.

*Sludge drying lagoons.* A sludge drying lagoon is similar to a sand bed in that the digested sludge is introduced to a lagoon and removed after a period of drying. Unlike the sand drying bed, the lagoon does not have an underdrain system to drain water. Sludge drying lagoons are not suitable for dewatering untreated sludges, lime treated sludges, or sludge with a high strength of supernatant, due to odor and nuisance potential.

Sludge drying lagoons are operated by periodically decanting the supernatant back to the plant and by evaporation. They are periodically dredged to remove sludge. Unconditioned digested sludge is introduced to the lagoon to a depth of 0.75 to 1.25 m (2.5 to 4 ft). It will dry mainly by evaporation in 3 to 5 months, depending on climate. Dried sludge is removed mechanically at a solids content of 20% to 30% percent. If sludge is to be used for soil conditioning, it needs to be stored for further drying. A 3-year cycle may be applied: lagooning for 1 year, drying for 18 months, and cleaning and resting for 6 months.

Solids loading rates are 36 to 39 kg/($m^3 \cdot$ y) (2.2 to 2.4 lb/($ft^3 \cdot$ y)) of lagoon capacity (US EPA, 1987b). The lagoons are operated in parallel; at least two units are essential. Very little process control is needed. Sludge lagoons are the most basic treatment units.

*Other sludge volume reduction processes.* Processes such as heat treatment and thermal combustion can reduce the moisture content and volume of the sludge. Detailed discussions are presented in the section on solids treatment.

**Example:** A POTW generates 1800 kg/d (3970 lb/d) of dewatered sludge at 33% solids concentration. The moisture content of composted material and compost mixture are 35% and 55%, respectively. What is the mass of composted material that must be blended daily with the dewatered sludge?

**solution:**

Step 1. Compute moisture of dewatered sludge

$$\begin{aligned}\text{Sludge moisture} &= 100\% - \text{solids content} = 100\% - 33\% \\ &= 67\%\end{aligned}$$

Step 2. Compute composted material required daily

Let $X$ and $Y$ be the mass of dewatered sludge and composted material, respectively. Then $X = 1800$ kg/d and

$$\begin{aligned}(X + Y)\text{ (mixture moisture)} &= X\text{ (s1. mois.)} + Y\text{ (comp. mois.)} \\ (X + Y)\,55\% &= X\,(67\%) + Y\,(35\%) \\ Y\,(55 - 35) &= X\,(67 - 55) \\ Y &= (1800\text{ kg/d})\,(12/20) \\ &= 1080\text{ kg/d} \\ &= 2380\text{ lb/d}\end{aligned}$$

### 29.3 Sewage sludge biosolids

Sewage sludge processing (i.e. thickening, stabilization, conditioning, and dewatering) produces a volume reduction, and this reduction in volume decreases the capital and operating costs. Digestion or composting of the sludge reduces the level of pathogens and odors. The degree of the sludge treatment process is very important to eliminate pathogens when considering land application of the sludge, when distributing and marketing it, and when placing it in monofills or on a surface disposal site.

The end product of wastewater sludge treatment processes is referred to as "biosolids." Webster's Collegiate Dictionary (1997) defines biosolids as solid organic matter recovered from a sewage treatment process and used especially as fertilizer for plants. The *McGraw-Hill Dictionary of Scientific and Technical Terms* (5th ed., 1994) defines biosolid as a recyclable, primarily organic solid material produced by wastewater treatment processes.

The term "biosolids" has gained recently in popularity as a synonym for sewage sludge because it perhaps has more reuse potential than the term "treated wastewater sludge." The name was chosen by the Water Environmental Federation (WEF) in early 1990s.

### 29.4 Use and disposal of sewage sludge biosolids

Biosolids are commonly used and disposed of in many ways. The beneficial uses of biosolids include land application to agricultural lands, land application to nonagricultural lands, and the sale or give-away of biosolids for use on home gardens. Nonagricultural areas may include compost, forests, public contact (parks, highways, recreational areas, golf courses, etc.). Case histories of beneficial use programs for biosolids use and management can be found elsewhere (WEF, 1994).

Disposal methods of biosolids include disposal in municipal landfills, disposal on delicate sites, surface disposal, and incineration. Surface disposal includes piles of biosolids left on the land surface and land application to dedicated nonagricultural land, and disposal in sludge-only landfills (monofills).

### 29.5 Regulatory requirements

In 1993, the US EPA promulgated "the standard for the use and disposal of sewage sludge, final rule, title 40 of the Code of Federation Regulations (CFR), Parts 257, 403, and 503" (Federal Register, 1993). All the above three plus "Phase-in submission of sewage sludge permit application: final rule, revisions to 40 CFR Parts 122, 123, and 501" are reprinted by the Water Environmental Federation, stock #P0100 (1993b). "The Part 503 rule" or "Part 503" was developed to protect public health and the environment from any reasonably anticipated adverse effects of using or disposing of sewage sludge biosolids. Two easy-to-read versions of the Part 503 rule were published by US EPA (1994, 1995).

The CFR 40 Part 503 includes five subparts. They are general provisions, requirement for land application, surface disposal, pathogen and vector attraction (flies, mosquitoes, and other potential disease-carrying organisms) reduction, and incineration. For each of the regulated use or disposal methods, the rule covers general requirements, pollutant limits, operational standards (pathogen and vector attraction reduction for land application and surface disposal; total hydrocarbons or carbon dioxide for incineration), management practices, and requirement for the frequency of monitoring, record keeping, and reporting. The requirements of the Part 503 are self-implementing and must be followed even without the insurance of a permit. State regulatory agencies may have their own rules governing the use or disposal of sewage sludge biosolids or domestic septage.

Part 503 applies to any person who applies biosolids to the land, or burns the biosolids in an incinerator, and to the operator/owner of a surface disposal site, or to any person preparing to use, dispose of, or incinerate biosolids. A person is defined as an individual, association, partnership, corporation, municipality, state, or federal agency, or an agent or employee thereof.

A person must apply for a permit covering biosolids use or disposal standards if they own or operate a treatment works treating domestic sewage (TWTDS). In most cases, Part 503 requirements will be incorporated over a time period into National Pollutant Discharge Elimination System (NPDES) permits issued to Public Owned Treatment Works (POTWs) and TWTDSs. Application for a federal biosolids permit must be submitted to the appropriate US EPA Regional Office, not the state. Until the biosolids management programs of individual states are approved by the US EPA, the US EPA will remain the permitting authority.

Most sewage sludge biosolids currently generated by POTWs in the United States meet the minimum pollutant limits and pathogen and vector attraction reduction requirements set forth in the Part 503. Some biosolids already meet the most stringent Part 503 pollutant limit standards and requirements of pathogen and vector attraction reduction.

**Pathogen reduction requirements.** The Part 503 pathogen reduction requirements for land application of biosolids are divided into two categories: Class A and Class B biosolids. In addition to meeting the requirement in one of the six treatment alternatives, the Class A requirement is to reduce the pathogens in biosolids (fecal coliform, or *Salmonella* sp. bacteria, enteric viruses, parasites, and viable helminth ova) in biosolids to below the detectable levels. When this goal is reached, Class A biosolids can be applied without any pathogen-related restriction on the site. The six treatment alternatives for biosolids include: alternative 1, thermally treated; alternative 2, high pH–high temperature treatment; alternative 3, other processes treatment; alternative 4, unknown processes; alternative 5, use of the processes to further reduce pathogens (PFRP); and alternative 6, use of a process equivalent to PFRP.

**Class A biosolids.** The pathogen reduction requirements must be met for all six alternatives (503.32a) and vector attraction reduction (503.33) for Class A with respect to pathogens. Regardless of the alternative chosen, either the FC density in the biosolids must be less than 1000, the most probable number (MPN) per gram dry total solids; or the density of *Salmonella* sp. in the biosolids must be less than 3 MPN per 4 g of dry total solids. Either of these requirements must be met at one of the following times: (1) when the biosolids are used or disposed of; (2) when the biosolids are prepared for sale or give away in a bag or other container for land application; or (3) when the biosolids or derived materials are prepared to meet the requirements for excellent quality biosolids.

Table 6.27 lists the four time–temperature regimes for Class A pathogen reduction under alternative 1 (time and temperature). Alternative 2 (pH and time) raises and maintains sludge pH above 12 for 72 h and keeps it at 52°C for 12 h. Alternative 3 is an analysis of the

**TABLE 6.27 The Four Time-Temperature Regimes for Class A Pathogen Reduction Under Alternative 1**

| Regime | Applies to | Requirement | Time-temperature relationships* |
|---|---|---|---|
| A | Biosolids with 7% solids or greater (except those covered by Regime B) | Temperature of biosolids must be 50°C or higher for 20 min or longer | $D = \frac{131{,}700{,}000}{10^{0.14T}}$ (Eq. 3 of section 503.32) |
| B | Biosolids with 7% solids or greater in the form of small particles and heated by contact with either warmed gases or an immiscible liquid | Temperature of biosolids must be 50°C or higher for 15 s or longer | $D = \frac{131{,}700{,}000}{10^{0.14T}}$ |
| C | Biosolids with less than 7% solids | Heated for at least 15 s but less than 30 min | $D = \frac{131{,}700{,}000}{10^{0.14T}}$ |
| D | Biosolids with less than 7% solids | Temperature of sludge is 50°C or higher with at least 30 min or longer contact time | $D = \frac{50{,}070{,}000}{10^{0.14T}}$ (Eq. 4 of section 503.32) |

*$D$ = time in days; $T$ = temperature in degrees Celsius.
SOURCE: US EPA (1994, 1995)

treatment process effectiveness of the Enteric viruses and helminth ova are to be determined prior to and after treatment. The final density is to be less than one plaque-forming unit per 4 g of dry solids. Alternative 4 (analysis with unidentified treatment process) analyzes for fecal coliform, *Salmonella,* Enteric viruses, and helminth ova at the time of use or disposal. Alternative 5 (PFRP process) treats sludge with one of composting and Class B alternative 2 processes (aerobic digestion, air drying, anaerobic digestion, heat drying, heat treatment, thermophilic aerobic digestion, beta ray irradiation, gamma ray irradiation, and pasteurization). Alternative 6 (PFRP equivalent process) allows the permitting authority to approve a process not currently identified as a PFRP process through a review by the Pathogen Equivalency Committee (Federal Register, 1993).

**Example 1:** What is the required minimum time to achieve Class A pathogen requirement when a biosolid with 15% solids content is heated at 60°C (140°F)?

**solution:**

Step 1. Select the regime under the given condition

Referring to Table 6.27, Regime A is chosen.

Step 2. Compute minimum time required

$$D = \frac{131{,}700{,}000}{10^{0.14T}} = \frac{131{,}700{,}000}{10^{0.14(60)}}$$

$$= 0.5243 \text{ days}$$

$$= 12.58 \text{ h}$$

**Example 2:** Determine the required minimum temperature to treat a biosolid containing 15% solids with a heating time of 30 min.

**solution:**

$$D = 30 \text{ min} = 0.0208 \text{ day}$$

$$\mathrm{D}(10^{0.14T}) = 131{,}700{,}000$$

$$10^{0.14T} = 131{,}700{,}000/0.0208 = 6{,}331{,}700{,}000$$

$$0.14T = 9.80$$

$$T = 70.0\ (^\circ\mathrm{C})$$

**Class B biosolids.** Class B requirements ensure that pathogens have been reduced to levels that are unlikely to pose a threat to public health and the environment under specific conditions of use. Class B biosolids also must meet one of the three alternatives: (1) monitoring of indicator organisms (fecal coliform) density; (2) biosolids are treated in one of the processes to significantly reduce pathogens (PSRP); and (3) use of the processes equivalent to PSRP.

**Land application.** Biosolids are applied to land, either to condition the soil or to fertilize crops or other vegetation grown in the soil. Biosolids can be applied to land in bulk or sold or given away in bags or other containers. The application sites may be categorized as nonpublic contact sites (agricultural lands, forests, reclamation sites) and public contact sites (public parks, roadsides, golf courses, nurseries, lawns, and home gardens). In the United States, about half of the biosolids produced are ultimately disposed of through land application.

Approximately one-third of the 5.4 million dry metric tons of wastewater sludge produced annually in the US at POTWs is used for land application. Of that, about two-thirds is applied on agricultural lands (US EPA, 1995). Land application of wastewater sludge has been practiced for centuries in many countries, because the nutrients (nitrogen and phosphorus) and organic matter in sludge can be beneficially utilized to grow crops and vegetation. However, microorganisms (bacteria, viruses, protozoa, and other pathogens), heavy metals, and toxic organic chemicals are major public health concerns for land application of biosolids. Proper management of biosolids utilization is required.

The application rate of biosolids applied to agricultural land must be equal to or less than the "agronomic rate". The agronomic rate is defined in the Part 503 as the rate designed to provide the amount of nitrogen needed by the crop or vegetation while minimizing the nitrogen in the biosolids passing below the root zone of the crop or vegetation and flowing to the groundwater.

Biosolids may be sprayed or spread on the soil surface and left on the surface (pasture, range land, lawn, forest), tilled into the soil after being applied, or injected directly below the surface. Land application of biosolids must meet risk-based pollutant limits specified in the Part 503. Operation standards are to control pathogens and to reduce the attraction of vectors. In addition, the application must meet the general requirements, management practices, and the monitoring, record keeping, and reporting requirements.

All land application of biosolids must meet the ceiling concentration limits for 10 heavy metals, listed in the second column of Table 6.28. If a limit for any one of the pollutants is exceeded, the biosolids cannot be applied to the land until such time that the ceiling concentration limits are no longer exceeded.

Biosolids applied to the land also must meet either pollutant concentration (PC) limits, cumulative pollutant loading rate (CPLR) limits, or annual pollutant loading rate limits for these 10 heavy metals. Either

**TABLE 6.28 Pollutant Limits for Land Application of Sewage Biosolids**

| Pollutant | Ceiling concentration limits for all biosolids applied to land, mg/kg* | Pollutant concentration (PC) limits, mg/kg*† | Cumulative pollutant loading rate limits (CPLR), kg/ha | Annual pollutant loading rate limits (APLR), kg/ha over 365-day period |
|---|---|---|---|---|
| Arsenic | 75 | 41 | 41 | 2.0 |
| Cadmium | 85 | 39 | 39 | 1.9 |
| Chromium | 3000 | 1200 | 3000 | 150 |
| Copper | 4300 | 1500 | 1500 | 75 |
| Lead | 840 | 300 | 300 | 15 |
| Mercury | 57 | 17 | 17 | 0.85 |
| Molybdenum† | 75 | –‡ | –‡ | –‡ |
| Nickel | 420 | 420 | 420 | 21 |
| Selenium | 100 | 36 | 100 | 5.0 |
| Zinc | 7500 | 2800 | 2800 | 140 |
| Applies to: | All biosolids that are land applied | Bulk biosolids and bagged biosolids§ | Bulk biosolids | Bagged biosolids§ |
| From Part 503, Section 503.13 | Table 1 | Table 3 | Table 2 | Table 4 |

* Dry-weight basis.
† Monthly average: also include exceptional quality (EQ) biosolids.
‡ EPA is re-examining these limits.
§ Bagged biosolids are sold or given away in a bag or other container.
SOURCE: US EPA, 1994, 1995

Class A or Class B pathogen requirements and site restrictions must be met. Finally, one of 10 options for vector attraction reduction must also be met.

**Annual whole sludge application rate.** The annual whole sludge application rate (AWSAR) for biosolids sold or given away in a bag or other container for application to land can be determined by dividing the annual pollutant loading rate by pollutant concentration. It is expressed as

$$\text{AWSAR} = \frac{\text{APLR}}{0.001\ \text{PC}} \tag{6.219}$$

where AWSAR = annual whole sludge application rate, metric ton/(ha · yr)

APLR = annual pollutant loading rate, kg/(ha · yr) (Table 6.28)

0.001 = a conversion factor, kg/metric ton or mg/kg

PC = pollutant concentration, mg/kg, dry weight

**Example:** Ten heavy metals listed in Table 6.28 were analyzed in biosolids to be sold. The chromium concentration is 240 mg/kg of biosolids. Determine the AWSAR of chromium for the biosolids.

**solution:**

Step 1. Find APLR for chromium from Table 6.28

$$\text{APLR} = 150 \text{ kg/(ha} \cdot \text{yr)}$$

Step 2. Compute AWSAR for chromium, using Eq. (6.219)

$$\text{PC} = 1200 \text{ mg/kg}$$

$$\text{AWSAR} = \frac{\text{APLR}}{0.001 \text{ PC}}$$

$$= \frac{150 \text{ kg/(ha} \cdot \text{y)}}{0.001 \text{ kg/(metric ton} \cdot \text{(mg/kg))} \times 1200 \text{ mg/kg}}$$

$$= 125 \text{ metric ton/(ha} \cdot \text{y)}$$

$$= \frac{(125 \text{ metric ton/(ha} \cdot \text{y)} \times (2205 \text{ lb/metric ton})}{107{,}600 \text{ ft}^2\text{/ha}}$$

$$= 2560 \text{ lb/1000 ft}^2\text{/y}$$

$$= 1.28 \text{ ton/1000 ft}^2\text{/y}$$

*Note*: 1 lb/1000 $\text{ft}^2$/y = 48.83 kg/ha/y; 1 ton/1000 $\text{ft}^2$/y = 97.65 metric ton/ha/y

**Site evaluation and selection.** Details of site evaluation and the selection process are discussed by Federal Register (1993) and by the US EPA (1995). The discussion covers Part 503 requirements, preliminary planning, phases 1 and 2, site evaluation, and site screening.

**Calculation of annual biosolids application rate on agricultural land.** Sewage sludge has been applied to agricultural land for a long time. The use of biosolids in agricultural land can partially replace costly commercial fertilizers. Generally, the intention is to optimize crop yields with application of biosolids and if required, supplemental fertilizers. The annual application rates of biosolids should not exceed the nutrients (N and P) requirements of the crop grown on an agricultural soil. In addition, land application of biosolids must meet the Part 503 requirements stated above.

**Calculation based on nitrogen.** In order to prevent groundwater contamination by ammonia nitrogen due to excess land application of biosolids, Part 503 requires that bulk biosolids be applied to a site at a rate equal to or less than the agronomic rate for nitrogen at the site.

Nitrogen may be present in biosolids in inorganic forms such as ammonium ($NH_4$) or nitrate ($NO_3$), or in organic forms. $NO_3$-N is the most water-soluble form of N, and is of most concern for groundwater contamination. $NH_4$-N can readily be volatilized as ammonia. Inorganic N is the plant available nitrogen (PAN). Not all the N in biosolids is immediately available for crop use, because some N is present as difficulty decomposable (refratory) organic N (Org-N), Org-N = total N − ($NO_3$-N) − ($NH_4$-N) which is in microbial cell tissue and other organic compounds. Organic N must be decomposed into mineral or inorganic forms, such as $NH_4$-N and $NO_3$, before it can be used by plants. Thus, the availability of Org-N for crops depends on the microbial breakdown of organic materials (biosolids, manure, crop residuals, soil organics, etc.) in soil.

When calculating the agronomic N rate for biosolids, the residual N from previously applied biosolids that will be mineralized and released as PAN must be accounted for as part of the overall budget for the total PAN. The residual N credit can be estimated for some sites undergoing soil nitrate tests, but the PAN credit is commonly estimated by multiplying a mineralization factor ($K_{min}$) by the amount of biosolids Org-N still remaining in the soil 1 or 2 years after biosolids application. Mineralization factors for different types of biosolids may be found in Table 6.29.

**Example 1:** An aerobic digested biosolid with 2.8% of organic nitrogen was applied at a rate of 4.5 dry ton per acre for the 1997 growing season. No other biosolids were applied to the same land in 1998. Determine the amount of plant-available nitrogen that will be mineralized from the biosolids Org-N in 1997 for the 1999 growing season.

**solution:**

Step 1. Calculate the biosolids Org-N applied in 1997

$$\text{Org-N} = 4.5 \text{ ton/acre} \times (2.8\%/100\%) \times 2000 \text{ lb/ton}$$

$$= 252 \text{ lb/acre}$$

**TABLE 6.29 Estimated Mineralization Rate for Various Type of Sewage Sludge**

| Year of growing season | Unstablized primary and WAS | Aerobically digested | Anaerobically digested | Composted |
|---|---|---|---|---|
| 0–1 (year of application) | 0.40 | 0.30 | 0.20 | 0.1 |
| 1–2 | 0.20 | 0.15 | 0.10 | 0.05 |
| 2–3 | 0.10 | 0.08 | 0.05 | – |
| 3–4 | 0.05 | 0.04 | – | – |

SOURCE: Sommers *et al.* (1981)

Step 2. Determine Org-N released as PAN in 1999 growing season

Construct a PAN credit for the application year, 1 year, and 2 years later for growing seasons as follows:

| | | Org-N, lb/acre | | |
|---|---|---|---|---|
| Year of growing season (1) | Mineralization rate (2) | Starting (3) | Mineralized (4) | Remaining (5) |
| 0–1 (1997 application) | 0.30 | 252 | 76 | 176 |
| 1–2 (1998) | 0.15 | 176 | 26 | 150 |
| 2–3 (1999) | 0.08 | 150 | **12** | 138 |

*Notes*: 1. Values of col. 2 are from Table 6.29
2. Col. 4 = col. 2 × col. 3
3. Col. 5 = col. 3 − col.4
4. 1 lb/acre = 1.12 kg/ha

*Answer:* Org-N released as PAN in 1999 = 12 lb/acre (*as bold shown above*).

The $NH_4$-N and $NO_3$-N added by biosolids is considered to be available for plant use. Plant-available nitrogen comprises $NH_4$-N and $NO_3$-N provided by biosolids and by fertilizer salts or other sources of these mineral forms of N.

When biosolids or animal manure are applied on a soil surface, the amount of PAN is reduced by the amount of $NH_4$-N lost by volatilization of ammonia. The volatilization factor, $K_{vol}$ is used for estimating the amount of $NH_4$-N.

The PAN of biosolids for the first year of application may be determined as

$$PAN = (NO_3\text{-N}) + K_{vol}(NH_4\text{-N}) + K_{min}(\text{Org-N}) \qquad (6.220)$$

where PAN = plant available N in biosolids, lb/dry ton
$NO_3$-N = nitrate N content in biosolids, lb/dry ton
$K_{vol}$ = volatilization factor, or fraction of $NH_4$-N not lost as $NH_3$ gas
= 0.5 for liquid and surface applied
= 1.0 for liquid and injection into soil
= 0.5 for dewatered and surface applied
$NH_4$-N = ammonium N content in biosolids, lb/dry ton
$K_{min}$ = mineralization factor, or fraction of Org-N converted to PAN, Table 6.29
Org-N = organic N content in biosolids, lb/dry ton
= Total N – ($NO_3$-N) – ($NH_4$-N)

**Example 2:** An anaerobically digested and dewatered biosolid is to be surface applied on an agricultural land. The biosolids analysis of N content is: $NO_3 - N = 1500$ mg/kg, $NH_4$-N = 1.2%, and total N = 3.7%, based on the dry weight. The solids content of the biosolids is 4.3%. Determine PAN.

**solution:**

Step 1. Convert N forms into lb/dry ton

$$\begin{aligned} NO_3\text{-N} &= 1500 \text{ mg/kg} = 1.500 \times 10^{-3} \text{ kg/kg} \\ &= 1.5 \times 10^{-3} \text{ lb/lb} \\ &= 3 \text{ lb/ton} \\ NH_4\text{-N} &= 2000 \text{ lb/ton} \times 1.2\%/100\% \\ &= 24 \text{ lb/ton} \\ \text{Total N} &= 2000 \text{ lb/ton} \times 0.037 = 74 \text{ lb/ton} \\ \text{Org-N} &= \text{Total N} - (NO_3\text{-N}) - (NH_4\text{-N}) \\ &= (74 - 3 - 24) \text{ lb/ton} \\ &= 47 \text{ lb/ton} \end{aligned}$$

Step 2. Compute PAN

For dewatered biosolids and surface applied, $K_{vol} = 0.5$.
Referring to Table 6.29, $K_{min} = 0.20$ for the first year.
Using Eq. (6.220),

$$\begin{aligned} \text{PAN} &= NO_3\text{-N} + K_{vol}(NH_4\text{-N}) + K_{min}\,(\text{Org-N}) \\ &= 3 \text{ lb/ton} + 0.5\,(24 \text{ lb/ton}) + 0.2\,(47 \text{ lb/ton}) \\ &= 24.4 \text{ lb/ton} \end{aligned}$$

*Note*: 1 lb/ton = 0.5 kg/metric ton = 0.5 mg/kg

It is necessary to determine the adjusted fertilizer N rate by subtracting "total N available from existing, anticipated, and planned sources" from the total N requirement. The procedures to compute the adjusted fertilizer N requirement for the crop to be grown may be summarized as follows (US EPA, 1995).

1. Determine the total N requirement (lb/ton) of crop to be grown. This value can be obtained from the Cooperative Extension Service of agricultural agents or universities, USDA-Natural Resources Conservation Service, or other agronomy professionals.
2. Determine N provided from other N sources added or mineralized in the soil

   a. N from a previous legume crop (legume credit) or green manure crop
   b. N from supplemental fertilizers already added or expected to be added

c. N that will be added by irrigation water
d. Estimate of available N from previous biosolids application (Example 1)
e. Estimate of available N from previous manure application
f. Soil nitrate test of available N present in soil [this quantity can be substituted in place of (a + d + e) if test is conducted properly; do not use this test value if estimates for a, d, and e are used]

Total N available from existing, expected, and planned sources of N = a + b + c + d + e or = b + c + f

3. Determine the loss of available N by denitrification, immobilization, or $NH_4$ fixation

Check with state regulatory agencies for approval before using this site-specific factor.

4. Compute the adjusted fertilizer N required for the crop to be grown

$$\text{N required} = \text{step 1} - \text{step 2} + \text{step 3}$$

Finally, the agronomic N application rate of biosolids can be calculated by dividing the adjusted fertilizer N required (step 4 above) by PAN (using Eq. (6.220)).

**Example 3:** Determine the agronomic N rate from Example 2, assuming the adjusted fertilizer N rate is 61 lb/acre (68.3 kg/ha). The solids content of the biosolids is 4.3%.

**solution:**

Step 1. Compute agronomic N rate in dry ton/acre

$$\begin{aligned} \text{Rate} &= (\text{adj. fert. N rate}) \div \text{PAN} \\ &= (61\ \text{lb/acre}) \div (24.4\ \text{lb/ton}) \\ &= 2.5\ \text{dry ton/acre} \end{aligned}$$

Step 2. Convert agronomic N rate in dry ton/acre to wet gal/acre

$$\begin{aligned} \text{Rate} &= (2.5\ \text{dry ton/acre}) \div (4.3\ \text{dry ton/100 wet ton}) \\ &= 58\ \text{wet ton/acre} \\ &= (58\ \text{wet ton/acre})\ (2000\ \text{lb/wet ton})\ (1\ \text{gal/8.34 lb}) \\ &= 13{,}900\ \text{gal/acre} \end{aligned}$$

*Note*: 1 gal/acre = 9.353 L/ha

**Calculation based on phosphorus.** The majority of phosphorus (P) in biosolids is present as inorganic forms, resulting from mineralization of biosolids organic matter. Generally, the P concentration in biosolids is considered to be about 50% available for plant uptake, as is the P normally applied to soils with commercial fertilizers. The P fertilizer required for the crop to be grown is determined from the soil fertility test for available P and the crop yield. The agronomic P rate of biosolids for land application is expressed as (US EPA, 1995)

$$\text{Agronomic P rate} = P_{req} \div (\text{available } P_2O_5/\text{dry ton}) \quad (6.221)$$

where $P_{req}$ = P fertilizer recommended for harvested crop, or the quantity of P removed by the crop

$$\text{Available } P_2O_5 = 0.5\,(\text{total } P_2O_5/\text{dry ton}) \quad (6.222)$$

$$\text{Total } P_2O_5/\text{dry ton} = \%\text{ of P in biosolids} \times 20 \times 2.3 \quad (6.223)$$

where 20 = 0.01 (= 1%) × 2000 lb/ton
2.3 = MW ratio of $P_2O_5 : P_2 = 142 : 62 = 2.3 : 1$

If biosolids application rates are based on the crop's P requirement, supplemental N fertilization is needed to optimize crop yield for nearly all biosolids applications.

**Calculation based on pollutant limitation.** Most biosolids are likely to contain heavy metals concentrations that do not exceed the Part 503 pollutant concentration limits. Therefore, pollutant loading limits are not a limiting factor for determining annual biosolids application rates on agricultural lands. Biosolids meeting pollutant concentration limits, as well as certain pathogen and vector attraction reduction requirements, generally are subject to meet the the CPLRs requirement. A CPLR is the maximum amount of a pollutant that can be applied to a site by all bulk biosolids applications after July 20, 1993. When the maximum CPLR is reached at the application site for any one of the 10 metals regulated by Part 503 (molybdenum was deleted with effect from February 25, 1994), no more additional biosolids are allowed to be applied to the site.

For some biosolids with one or more of the pollutant concentrations exceeding the Part 503 pollutant concentration limits, the CPLRs as shown in Table 6.30 must be met. In these cases, the CPLRs could eventually be the limiting factor for annual biosolids applications rather than the agronomic (N or P) rate of application.

For biosolids meeting the Part 503 CPLRs, the maximum total quantity of biosolids allowed to be applied to a site can be calculated on the basis of the CPLR and the pollutant concentration in the biosolids as follows:

Maximum allowed in dry ton/acre

$$= (\text{CPLR in lb/acre}) \div [0.002\,(\text{mg pollutant/kg dry biosolids})] \quad (6.224)$$

**TABLE 6.30 Part 503 Cumulative Pollutant Loading Rate Limits**

| Pollutant | CPLR limits | |
|---|---|---|
| | kg/ha | lb/acre |
| Arsenic | 41 | 37 |
| Cadmium | 39 | 35 |
| Chromium* | 3000 | 2700 |
| Copper | 1500 | 1300 |
| Lead | 300 | 270 |
| Mercury | 17 | 15 |
| Molybdenum† | – | – |
| Nickel | 420 | 380 |
| Selenium | 100 | 90 |
| Zinc | 2800 | 2500 |

*The chromium limit will most likely be deleted from the Part 503 rule.

†The CRLR for Mo was deleted from Part 503 with effect from February 25, 1994.

After computing for each of the eight or nine pollutants regulated by the Part 503, the lowest total biosolids value will be used as the maximum quantity of biosolids permitted to be applied to the site. The individual pollutant loading applied by each biosolids application can be calculated by

$$\text{lb of pollutant/acre} = \text{biosolids application rate in dry ton/acre} \times (0.002\ \text{mg/kg}) \qquad (6.225)$$

The pollutant loading for each individual biosolid application must be calculated and recorded to keep a cumulative summation of the total quantity of each pollutant that has been applied to each site receiving the biosolids.

**Supplemental K fertilizer.** After the agronomic application rate of biosolids has been determined, the amounts of plant available N, P, and K added by the biosolids must be computed and compared to the fertilizer recommendation for the crop grown at a specified yield level. If one or more of these three nutrients provided by the biosolids is less than the amount recommended, supplemental fertilizers are needed to achieve crop yield.

Potassium, K, is a soluble nutrient. Most of the concentration of K in raw wastewater is discharged with the treatment plant effluents. Generally, biosolids contain low concentrations of potassium, which is one of the major plant nutrients. Fertilizer potash ($K_2O$) or other sources of K usually are needed to supplement the amounts of $K_2O$ added by biosolids application.

Because K is readily soluble, all the K in biosolids is considered to be available for crop growth. The quantity of $K_2O$ provided (credited) by biosolids application can be calculated as

$$K_2O \text{ added by biosolids} = \text{applied rate in dry ton/acre} \times \text{avail. } K_2O \text{ in lb/dry ton} \quad (6.226)$$

where

$$\text{Available } K_2O = \% \text{ of K in biosolids} \times 20 \times 1.2 \quad (6.227)$$

where $20 = 2000 \text{ lb/dry ton} \cdot \%$
$1.2 = K_2O/K_2 = 94/78$

**Example 4**: A K fertilizer recommendation for a soybean field is 150 lb $K_2O$/acre. The agronomic N rate of biosolid application is 1.5 dry ton/acre. The biosolid contains 0.5% K. Compute the supplemental (additional) $K_2O$ required.

**solution:**

Step 1. Compute the available $K_2O$, using Eq. (6.227)

$$\text{Available. } K_2O = 0.5 \times 20 \text{ lb/dry ton} \times 1.2$$
$$= 12 \text{ lb/dry ton}$$

Step 2. Compute $K_2O$ added by biosolids, using Eq. (6.226)

$$\text{Biosolids } K_2O \text{ applied} = \text{biosolids rate} \times \text{available } K_2O$$
$$= 1.5 \text{ dry ton/acre} \times 12 \text{ lb/dry ton}$$
$$= 18 \text{ lb/acre}$$

Step 3. Compute supplemental $K_2O$ needed

$$\text{Additional } K_2O \text{ needed} = \text{recommended} - \text{biosolids added}$$
$$= 150 \text{ lb/acre} - 18 \text{ lb/acre}$$
$$= 132 \text{ lb/acre}$$

*Note*: 1 lb/acre = 1.12 kg/ha

**Example 5:** (This example combines Examples 1 to 4 and more) Determine biosolid annual application for the 1999 growing season for an agricultural site on the basis of the agronomic N rate, P rate, and long-term pollutant limitations required by Part 503. Also determine the supplemental nutrients to be added. An anaerobically digested liquid biosolid is designed to be applied to farmland at 10 dry ton/acre. The results of laboratory analyses for the biosolids are shown below:

| | Heavy metals, mg/kg |
|---|---|
| Total solids = 5.5% | As = 10 |
| Total N = 4.3% | Cd = 7 |
| $NH_4$-N = 1.1% | Cr = 120 |
| $NO_3$-N = 500 mg/kg | Cu = 1100 |
| Total P = 2.2% | Pb = 140 |
| Total K = 0.55% | Hg = 5 |
| pH = 7.0 | Mo = 10 |
| | Ni = 120 |
| | Se = 6 |
| | Zn = 3400 |

If plant nutrients added by biosolids application are not sufficient, supplemental fertilizer nutrients must be provided. Routine soil fertility tests, monitoring, and records must meet Part 503 and state agency requirements. The field is divided into two portions for rotating cropping of corn, soybeans, and wheat. The same biosolids have been applied to the field during the previous 2 years. The biosolids application rate and its Org-N content are as follows:

| | Application rate, ton/acre | | Org-N in biosolids, % | |
|---|---|---|---|---|
| Crop | 1997 | 1998 | 1997 | 1998 |
| Corn | 3.5 | 4.4 | 3.0 | 3.3 |
| Wheat | 5.0 | 0 | 3.1 | – |

According to the County Farm Bureau (Peoria, Illinois) for the 1999 growing season (personal communication), the crop yield and required nutrients in lb/(acre · yr) are given below:

| | | lb/(acre · yr) | | |
|---|---|---|---|---|
| Crop | Yield | N | $P_2O_5$ | $K_2O$ |
| Corn | 160 | 150 | 70 | 140 |
| Wheat | 70 | 70 | 80 | 125 |
| Soybeans | 45 | 0 | 60 | 150 |

For the 1999 growing season, one-half of the field will be planted in wheat and liquid biosolids will be injected to the soil in the fall (1998) after soybeans are harvested and before the winter wheat is planted in the fall of 1998. For the corn fields, biosolids will be surface applied and tilled in the spring of 1999 before corn is planted. There will be no other source of N for the corn field, except for residual N from 1997 and 1998 biosolid applications.

The wheat field will have 23 lb of N per acre from the preceding soybean crop; and a residual N credit for the 1997 biosolids application. No manure will be applied and no irrigation will be made to either field.

**solution:**

Step 1. Compute agronomic rate for each field

(a) Convert N from percent and mg/kg to lb/ton

$$\begin{aligned}\text{Total N} &= 2000\ \text{lb/ton} \times 0.043\\ &= 86\ \text{lb/ton}\\ \text{NH}_4\text{-N} &= 2000\ \text{lb/ton} \times 0.011 = 22\ \text{lb/ton}\\ \text{NO}_3\text{-N} &= 500\ \text{mg/kg} = 500 \times \frac{1\ \text{lb}}{10^6\ \text{lb}} = 500 \times \frac{1\ \text{lb}}{500\ \text{ton}}\\ &= 500 \times 0.002\ \text{lb/ton}\\ &= 1\ \text{lb/ton}\end{aligned}$$

(b) Compute Org-N

$$\begin{aligned}\text{Org-N} &= (86 - 22 - 1)\ \text{lb/ton}\\ &= 63\ \text{lb/ton}\end{aligned}$$

(c) Compute PAN for surface applied biosolids, using Eq. (6.220)
From Table 6.29

$$\begin{aligned}K_{\text{min}} &= 0.20\ \text{for the first year}\\ &= 0.10\ \text{for the second year}\\ &= 0.05\ \text{for the third year}\\ K_{\text{vol}} &= 0.7\ \text{for the corn field, surface applied and tilled}\\ &= 1.0\ \text{for the wheat field, injected}\end{aligned}$$

For the corn field

$$\begin{aligned}\text{PAN} &= (\text{NO}_3\text{-N}) + K_{\text{vol}}(\text{NH}_4\text{-N}) + K_{\text{min}}(\text{Org-N})\\ &= 1\ \text{lb/ton} + 0.7\ (22\ \text{lb/ton}) + 0.2\ (63\ \text{lb/ton})\\ &= 29\ \text{lb/ton}\end{aligned}$$

For the wheat field

$$\begin{aligned}\text{PAN} &= 1\ \text{lb/ton} + 1.0\ (22\ \text{lb/ton}) + 0.2\ (63\ \text{lb/ton})\\ &= 36\ \text{lb/ton}\end{aligned}$$

(d) Compute the Org-N applied in previous years using the given data
For the corn field, Org-N applied in 1997 and 1998 was:

$$\begin{aligned}\text{In 1997: Org-N} &= 3.5\ \text{ton/acre} \times 2000\ \text{lb/ton} \times 3\%/100\%\\ &= 210\ \text{lb/acre}\end{aligned}$$

$$\text{In 1998: Org-N} = 4.4 \text{ ton/acre} \times 2000 \text{ lb/ton} \times 0.033$$
$$= 290 \text{ lb/acre}$$

For the wheat field, Org-N originally applied in 1997 was

$$\text{In 1997: Org-N} = 5.0 \text{ ton/acre} \times 2000 \text{ lb/ton} \times 0.031$$
$$= 310 \text{ lb/acre}$$

(e) Compute the residual N mineralized from previous years' biosolids applications

Construct PAN credits for the application year, 1 and 2 years later than the growing season, as for Example 1. Take values of starting Org-N from Step 1 d.

| Year of growing season | Mineralization rate $K_{min}$ | Org-N, lb/acre | | |
|---|---|---|---|---|
| | | Starting | Mineralized | Remaining |
| Corn field: 1997 application | | | | |
| 0–1 (1997) | 0.20 | 210 | 42 | 168 |
| 1–2 (1998) | 0.10 | 168 | 17 | 151 |
| 2–3 (1999) | 0.05 | 151 | **8** | 143 |
| Corn field: 1998 application | | | | |
| 0–1 (1998) | 0.20 | 290 | 58 | 232 |
| 1–2 (1999) | 0.10 | 232 | **23** | 209 |
| 2–3 (2000) | 0.05 | 209 | 10 | 199 |
| Wheat field: 1997 application | | | | |
| 0–1 (1997) | 0.20 | 310 | 62 | 248 |
| 1–2 (1998) | 0.10 | 248 | 25 | 223 |
| 2–3 (1999) | 0.05 | 223 | **11** | 212 |

The PAN credit for the 1999 growing season on the corn field due to biosolid applications in 1997 and 1998 is **8** + **23** (in bold in the above table) = 31 lb of N per acre, while that on the wheat field due to 1997 application is **11** lb of N per acre.

(f) Determine agronomic N rate for both fields
For the corn field

1. Total N required for corn grown (given) = 150 lb/acre
2. N provided from other sources

2a. Estimate of available N from previous applications (from Step 1 e ) = 31 lb/acre

3. Loss of available N = 0
4. Adjusted fertilizer N required = (1) − (2a) − (3) = 119 lb/acre
5. The PAN dry ton biosolid to be applied, from Step 1 (c) = 29 lb/ton
6. The agronomic N rate of biosolid, (4)/(5) = 4.1 dry ton/acre
7. Convert biosolids rate into wet ton/acre, 4.1/0.055 = 74.6 wet ton/acre
8. Convert biosolid rate into gal/acre = 17,900 gal/acre
   = (74.6 wet ton/acre × 2000 lb/ton ÷ 8.34 lb/gal)

For the wheat field

| | | |
|---|---|---|
| 1. | Total N required for growing wheat | = 70 lb/acre |
| 2. | N provided from other sources | |
| 2a. | N from previous legume credit (given) | = 23 lb/acre |
| 2b. | N from previous biosolids applications, from Step 1 (e) | =11 lb/acre |
| 3. | Loss of available N | =0 |
| 4. | Adjusted fertilizer N required for wheat (1) − (2a) − (2b) − (3) | = 36 lb/acre |
| 5. | PAN/dry ton biosolids to be applied, from Step 1 c | = 36 lb/ton |
| 6. | Agronomic N rate of biosolids, (4)/(5) | =1 dry ton/acre |
| 7. | Convert biosolids rate into wet ton/acre | =18 wet ton/acre |
| 8. | Convert biosolids rate into gal/acre | = 4320 gal/acre |

Step 2. Determine cumulative pollution loading and maximum permitted application rate

(a) Comparing the pollutants concentration in biosolid with the Part 503 pollutant concentration limits, and comparing the concentrations of the 10 heavy metals in the biosolid with the Part 503 "CPLR limits" (Table 6.28), the results show that all heavy metals concentrations in biosolids are less than the limits, except for zinc; therefore, the CPLR limits must be met for the biosolids.

(b) Compute the maximum permitted biosolids application rate for each pollutant. The maximum total amount of biosolids permitted to be applied to soil can be calculated from the CPLR limits from Table 6.30 and the concentrations of pollutants, using Eq. (6.224).

The maximum biosolids application rate allows for

As = 37 lb/acre ÷ [0.002 (10 mg/kg)] = 1850 dry ton/acre

Cd = 35 lb/acre ÷ [0.002 (7 mg/kg)] =2550 dry ton/acre

Cr will be deleted from Part 503

Cu = 1300 lb/acre ÷ [0.002 (1100 mg/kg)] = 590 dry ton/acre

Pb = 270 lb/acre ÷ [0.002 (140 mg/kg)] = 964 dry ton/acre

Hg = 15 lb/acre ÷ [0.002 (5 mg/kg)] = 1500 dry ton/acre

Mo deleted

Ni = 380 lb/acre ÷ [0.002 (120 mg/kg)] = 1583 dry ton/acre

Se = 90 lb/acre ÷ [0.002 (6 mg/kg)] = 7500 dry ton/acre

Zn = 2500 lb/acre ÷ [0.002 (3400 mg/kg)] = 367 dry ton/acre

*Note*: Zn is the lowest allowed application rate, and it is much higher than the agronomic N rate. Thus heavy metal pollutants are not the limit factor.

(c) Compute the amount of each pollutant in the 1999 growing season

According to Part 503, the pollutant loading for each biosolid application must be calculated and recorded to keep a cumulative summation of the total amount of each pollution. The amount of each pollutant can be calculated, using Eq. (6.225), from the given pollutant concentrations for corn and wheat fields. The results are shown below.

| | | Amount applied, lb/acre | |
|---|---|---|---|
| Pollutant | Concentration in biosolids, mg/kg | Corn field (1.4 ton/acre)[†] | Wheat field (1.0 ton/acre) |
| As | 10 | 0.028[*] | 0.020 |
| Cd | 7 | 0.020 | 0.014 |
| Cr (deleted) | – | – | – |
| Cu | 1100 | 3.08 | 2.2 |
| Pb | 140 | 0.39 | 0.28 |
| Hg | 5 | 0.014 | 0.010 |
| Mo (deleted) | – | – | – |
| Ni | 120 | 0.34 | 0.24 |
| Se | 6 | 0.017 | 0.012 |
| Zn | 2400 | 6.72 | 6.8 |

[†] see Step 4 (the agronomic P rate for the corn field is 1.4 dry ton/acre only)
[*] lb pollutant/acre: 1.4 ton/acre × 0.002 (10 mg/kg) = 0.028 lb/acre

(d) Estimate the number of years biosolids can be applied

Assume that biosolids continue to have the same quality over time, and zinc would continue to be the limiting pollutant (see Step 2 b, 367 dry ton/acre is the maximum biosolids application rate for zinc). For this example, it is assumed that an average rate of 1.4 dry ton (acre · yr) would be applied over time. The number of years biosolid could be applied before reaching the CPLR can be computed from maximum biosolid allowed, divided by the average annual application rate, as follows:

$$\text{Number of years} = 367 \text{ dry ton/acre} \div 1.4 \text{ dry ton/(acre} \cdot \text{yr)}$$
$$= 262 \text{ years}$$

Step 3. Compute the agronomic P rate for each field

(a) The total $P_2O_5$ can be computed by Eq. (6.223). Given that anaerobic biosolids have 2.2% total P

$$\text{Total } P_2O_5 = 2.2\% \times 20 \times 2.3 \text{ lb/(ton} \cdot \%)$$
$$= 101 \text{ lb/dry ton}$$

(b) The plant available $P_2O_5$ is computed by Eq. (6.222)

$$\text{Available } P_2O_5 = 0.5 \text{ (101 lb/dry ton)}$$
$$= 50 \text{ lb/dry ton}$$

(c) For the corn field

$$P_{req} = 70 \text{ lb } P_2O_5\text{/acre (given)}$$

For the wheat field

$$P_{req} = 80 \text{ lb } P_2O_5\text{/acre (given)}$$

(d) Calculate the agronomic P rate using Eq. (6.221)

For the corn field

$$\begin{aligned}\text{Agronomic P rate} &= P_{reg} \div (\text{available } P_2O_5\text{/dry ton})\\ &= (70 \text{ lb/acre}) \div (50 \text{ lb/dry ton})\\ &= 1.4 \text{ dry ton/acre}\end{aligned}$$

For the wheat field

$$\begin{aligned}\text{Agronomic P rate} &= 80 \div 50\\ &= 1.6 \text{ dry ton/acre}\end{aligned}$$

Step 4. Determine the application rate of biosolid on the corn field

Because the agronomic P rate for the corn field (1.4 dry ton/acre) is less than the agronomic N rate (4.1 dry ton/acre, Step 1 f-6), this rate, 1.4 dry ton/acre, is selected to be used for the 1999 growing season. A supplemental N fertilizer must be added to fulfill the remaining N needs for corn not supplied by the biosolid.

Step 5. Determine the supplemental N fertilizer for the corn field

The amount of additional N needed for the corn can be computed by multiplying the PAN in lb/dry ton by the rate of biosolids application, then subtracting this PAN in lb/acre from the adjusted fertilizer N requirement. Referring to Step 1 c and Step 1 f-4, supplemental N can be determined as

$$\begin{aligned}\text{Biosolids N credit} &= 29 \text{ lb PAN/dry ton} \times 1.4 \text{ dry ton/acre}\\ &= 41.0 \text{ lb PAN/acre}\\ \text{Supplemental N} &= 119 \text{ lb/acre} - 41 \text{ lb/acre}\\ &= 78 \text{ lb/acre}\end{aligned}$$

Step 6. Determine the supplemental K fertilizer for the corn field

(a) Calculate the available $K_2O$ added by biosolids, using Eq. (6.227)
Given: % K in the biosolid = 0.55

$$\begin{aligned}\text{Available } K_2O &= \%\text{ K in biosolids} \times 20 \text{ lb/dry ton} \cdot \% \times 1.2\\ &= 0.55 \times 24 \text{ lb/dry ton}\\ &= 13 \text{ lb/dry ton}\end{aligned}$$

(b) Compute the amount of biosolid $K_2O$ applied, using Eq. (6.226)

$$\begin{aligned}\text{Biosolid } K_2O \text{ added} &= 1.4 \text{ dry ton/acre (see Step 4)} \times 13 \text{ lb/dry ton}\\ &= 18.2 \text{ lb/acre (say 18 lb/acre)}\end{aligned}$$

(c) Compute supplemental $K_2O$ needed

$$\begin{aligned}\text{Additional } K_2O \text{ needed} &= \text{required (given)} - \text{biosolids } K_2O \\ &= 140 \text{ lb/acre} - 18 \text{ lb/acre} \\ &= 122 \text{ lb/acre}\end{aligned}$$

Step 7. Determine the application rate of biosolid for the wheat field

For the wheat field, the agronomic N rate of 1.0 dry ton/acre of biosolids application is the choice since it is less than the agronomic P rate (1.6 dry ton/acre).

Step 8. Determine the supplemental P for the wheat field, referring to Steps 3 and 5

$$\begin{aligned}\text{Biosolids P credit} &= 50 \text{ lb } P_2O_5/\text{dry ton} \times 1.0 \text{ dry ton/acre} \\ &= 50 \text{ lb } P_2O_5/\text{acre} \\ \text{Additional P needed} &= (80 - 50) \text{ lb } P_2O_5/\text{acre} \\ &= 30 \text{ lb } P_2O_5/\text{acre}\end{aligned}$$

Step 9. Determine the supplemental K for the wheat field

(a) Compute the biosolid $K_2O$ added

From Step 6 (a)

$$\begin{aligned}\text{available } K_2O &= 13 \text{ lb/dry ton (from step 6 a)} \\ \text{Biosolid } K_2O \text{ added} &= 1.0 \text{ dry ton/acre} \times 13 \text{ lb/dry ton} \\ &= 13 \text{ lb/acre}\end{aligned}$$

(b) Compute additional $K_2O$ needed

$$\begin{aligned}&= 125 \text{ lb/acre} - 13 \text{ lb/acre} \\ &= 112 \text{ lb/acre}\end{aligned}$$

## 30 Wetlands

Wetlands are defined in Federal Register (40 CFR 230.3(t)) as "those areas that are inundated or saturated by surface or groundwater at the frequency and duration sufficient to support, and that under normal circumstances do support, a prevalence of vegetation typically adapted for life in saturated soil conditions."

Since the early 1950s, researches have been conducted on treatment capacity and ecological impacts of wetlands for wastewater treatment. The results of researches indicated that natural and man-made (constructed) wetlands can treat domestic wastewater and plant effluent (for polishing) effectively to remove $BOD_5$, TSS, nutrients (N and P), and color. The wetlands treatments included for domestic, industrial, and agricultural wastewaters.

In the early 1980s, the US EPA published a comprehensive literature review on the effects of wastewater treatment on wetlands (US EPA, 1983), the ecological impacts of wastewater on wetlands—an annotated bibliography (US EPA, 1984a) and the wetland evaluation methodologies (US EPA, 1984b).

The wetland treatment of wastewater had gained acceptance, especially for small communities with strained capital and O & M costs with available large land available. In a wetland, the aquatic system treats wastewater by means of both physical sedimentation and bacterial metabolism that are similar to the conventional wastewater treatment.

The first design manual on wetlands and aquatic plant systems for municipal wastewater treatments was published by the US EPA (1988). Later a more comprehensive manual of constructed wetlands treatment of municipal wastewaters was published (US EPA, 2000). In its 1994 report, the US EPA was considering wetlands at CERCLA (the Comprehensive Environmental Response, Compensation and Liability Act, or Superfund) sites (US EPA, 1994).

There are three categories of aquatic wastewater treatment systems (US EPA, 1988):

1. Natural wetlands
2. Aquatic plant systems
3. Constructed wetlands

## 30.1 Natural wetlands

Natural wetlands are referred to as marshes, swamps, bogs, and similar areas with native vegetation. They are existing wetlands that had little or no modification by humans. Grasses or forbs (nonwoody plants) are generally dominant in marshes; trees and shrubs grow in swamp area; and sedge and peat vegetations occur in various bogs. The modification or direct use of natural wetlands for wastewater treatment is discouraged. Natural wetlands are effective as wastewater treatment. The percent removals of some parameters by natural wetlands are (Reed *et al.*, 1979):

$BOD_5$: 70 to 96

TSS: 60 to 90

Nitrogen: 40 to 90

Phosphorus: 19 to 98 (seasonal)

## 30.2 Aquatic plant systems

Aquatic plant systems are shallow ponds with floating or submerged aquatic plants. Water hyacinth and duckweed are the most thoroughly studied floating plant systems. These systems are usually used for advanced treatment of secondary effluent or in oxidation ponds for polishing purposes. The plants derive the needed carbon dioxide from the atmosphere and use nutrients in the water. The water hyacinth has rapid growth with an extensive root system to support media for bacteria. It is temperature sensitive and is rapidly killed by frost. Removal of dead plants is cumbersome.

Duckweed occurs in natural wetlands and in lagoons (common at the highway rest areas). It is less sensitive to cold weather. However, it has a shallow root system, is sensitive to wind, and is easily flushed away by high wind.

The second type of aquatic system is the submerged plant system. Submerged plants are either rooted in the bottom sediments or suspended in the water column. Typically, photosynthesis in these plants occurs in the water column. Submerged plants are relatively easily inhibited by high turbidity in the water. These plants tend to be shaded by algal growths and are killed by anaerobic conditions.

### 30.3 Constructed wetlands

Artificial (constructed) wetlands are wetlands that have been built or extensively modified by humans. Modification examples are filling, draining, or altering the flow patterns or physical properties of the wetland. They are typically constructed with uniform depths and regular shapes near the source of the wastewater and are often located in upland areas where no wetlands have historically existed. Constructed wetlands are almost always regulated as wastewater treatment facilities and cannot be used for compensatory mitigation. Constructed wetlands that provide advanced treatment to wastewater that has been pretreated to secondary levels, and also other benefits such as wildlife habitat, research laboratories, or recreational uses, are sometimes called enhancement wetlands.

Constructed wetlands have been classified into two types (US EPA, 2000); namely, free water surface (FWS) wetlands and vegetated submerged bed (VSB) wetland systems. The FWS wetlands (Fig. 6.63) also known as surface flow wetlands, closely resembling natural wetlands in appearance because they contain aquatic plants that are rooted in a soil layer on the bottom of the wetland with water flowing through the leaves and stems of plants.

Vegetated submerged bed systems (Fig. 6.64), also known as subsurface flow wetlands) do not resemble natural wetlands because they have no standing water. They contain a bed of media (such as crushed rock, small stones, gravel, sand, or soil) that has been planted with aquatic plants. When properly designed and operated, wastewater stays beneath

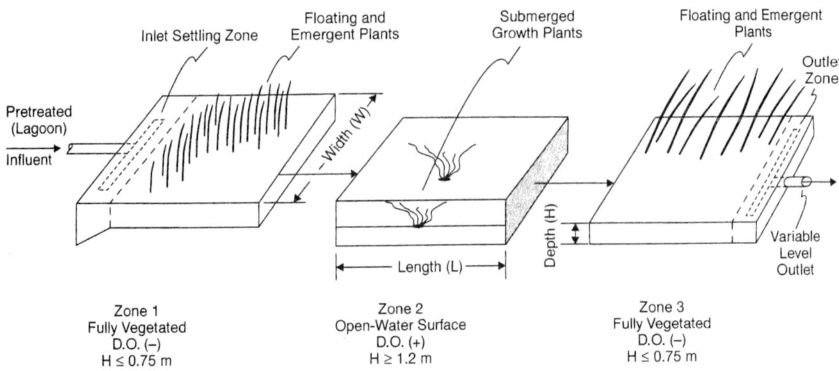


**Figure 6.63** Elements of a free water surface (FWS) constructed wetland (*source: US EPA 2001*).

the surface of the media, flows in contact with the roots and rhizomes of the plants, and is not visible or available to wildlife.

Unlike natural wetlands, constructed wetlands are designed and operated to meet certain performance standards. The constructed wetland, once designed and operated, requires regular monitoring to ensure proper operation and routine management to maintain optimum performance. On the basis of monitoring results, the system may need minor modifications.

**Environmental and health considerations.** Since wetland systems affect aesthetic, ecological, and recreational impacts in the area, during the planning of constructed wetland, some technical, environmental, and

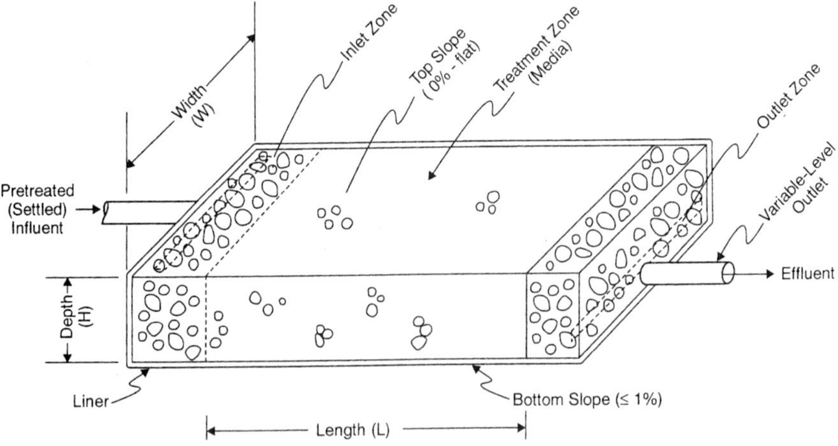


**Figure 6.64** Elements of a vegetated submerged bed (VSB) system (*source*: *US EPA 2001*).

health factors have to be considered and evaluated. Those factors are (US EPA, 1983, 1984b, 1988, 2000):

- regulatory issues,
- uses of constructed wetland,
- changes in hydrologic regime,
- cumulative impacts of inorganic and organic chemicals,
- long-term biological effects,
- mitigation/management issues, and
- health/disease considerations.

Constructed wetland is relatively less regulated than a conventional wastewater treatment system. It is a reliable, affordable, and operable process for a small (less than 1 MGD or 3800 $m^3$/d) wastewater treatment plant. Constructed wetlands are also versatile, to be used either as the primary means of secondary treatment or as a means of final polishing. Nowadays, the use of a constructed wetland is not uniformly accepted as a proven technology or an emerging treatment technology depending upon regulatory agencies. It has some degree of risks to ecosystem and toxic compounds (still unknown). Also, the design process is still empirical.

Sources of surface water and groundwater supplies have to be studied, including charge and discharge. Sequential changes of the hydrologic regime in the wetland affect all aspects of the ecosystem. These changes must be understood to determine nutrient cycle, sedimentation rates, erosion patterns, plant and animal community compositions, and hydrological budgets. These effects ultimately depend on the loading rate, discharge quantity and quality, and size and type of receiving wetland.

Wetlands achieve good removal of nitrogen via denitrification. Phosphorus removal is more variable. Wetlands have some ability to assimilate organic and other compounds, and change soil chemistry. These may cause the shift of dominant animal and plant species composition, especially benthic invertebrate communities. Some metals do accumulate within wetlands. Various physical and chemical parameters affect bioaccumulation. Refractory chemicals, such as surfactants, phenols, pesticides, etc., are typically accumulated in wetlands. The long-term ecological effects of these accumulation are less known.

Shifts in plant species composition and area distribution will change in biomass production, detrital cycle habitat, aquatic food chain (web), wildlife, animal, and fish species and population. Eventually these may reduce water quality. The pathogens of concern in wetland treatment system are bacteria, parasites, and viruses. Disease-carrying insect vectors are potential risks.

Wetland-related recreational activities, such as canoeing, photography, bird watching, camping, etc., should be evaluated. Cultivated crops, pasture, hay crops, and silviculture are most likely not affected by constructed wetlands. Natural landscape (any unique areas, and open spaces), archaeological sites, and historical sites have to be preserved. The constructed wetlands should also be open for research and for public educational purposes.

Mitigation measures should be taken to avoid impacts on wetland from operation of conventional treatment facilities. The impacts include site selection, construction of interceptors and pumping stations, discharge to wetlands, included development, and increased human use of the area.

**Free water surface wetlands.** The differences between the two types of wetlands were mentioned previously. Free water surface wetlands closely resemble natural wetlands in appearance and function, with a combination of open-water areas, emergent vegetation, varying water depths, and other typical wetland features.

The main components of a FWS constructed wetland are sketched in Fig. 6.63. A typical FWS constructed wetland consists of several components that may be modified among various applications but retain essentially the same features. These components include berms to enclose the treatment cells, inlet structures that regulate and distribute influent wastewater evenly for optimum treatment, various combinations of open-water areas and fully vegetated surface areas, and outlet structures that complement the even distribution provided by inlet structures and allow the adjustment of water levels within the treatment cell. Shape, size, and complexity of design often are site characteristics rather than preconceived design criteria.

**Wetland hydrology.** The wetland water balance for a FWS constructed wetland is expressed as follows:

$$dV/dt = Q_\mathrm{i} + Q_\mathrm{c} + Q_\mathrm{sm} - Q_\mathrm{b} - Q_\mathrm{e} + (P + \mathrm{ET} + I)A \tag{6.228}$$

where
$V$ = water volume or storage in wetland
$t$ = time, hour (h) or day (d)
$Q_\mathrm{i}$ = wastewater inflow rate, $\mathrm{m^3/d}$ or $\mathrm{ft^3/d}$
$Q_\mathrm{c}$ = catchment runoff rate, $\mathrm{m^3/d}$ or $\mathrm{ft^3/d}$
$Q_\mathrm{sm}$ = snowmelt rate, $\mathrm{m^3/d}$ or $\mathrm{ft^3/d}$
$Q_\mathrm{b}$ = berm loss rate, $\mathrm{m^3/d}$ or $\mathrm{ft^3/d}$
$Q_\mathrm{e}$ = wetland outflow rate, $\mathrm{m^3/d}$ or $\mathrm{ft^3/d}$
$P$ = precipitation rate, m/d or ft/d
ET = evapotranspiration rate, m/d or ft/d
$I$ = infiltration to groundwater, m/d or ft/d
$A$ = wetland water surface area, $\mathrm{m^2}$ or $\mathrm{ft^2}$

The impact of wet weather and snowmelt on the wastewater flow is external to the water balance. Some of the terms in Eq. (6.228) may be deemed insignificant and can be neglected. $Q_b$ and $I$ can be neglected if the wetland is lined with an impermeable layer; snowmelt is important only in certain locations.

The wetland water volume $V$ can be calculated by multiplying average water depth $H$ by area $A$:

$$V = A \times H \tag{6.229}$$

Wetland porosity $n$ or void fraction is the fraction of total volume available through which water can flow. In a FWS wetland, vegetation, settled solids, litter, and peat occupy water column. Wetland porosity is difficult to measure in the field. For design purpose, it is recommended to have a porosity value of 0.65 to 0.75 for fully vegetated zones; 1.0 for wetland open-water zones; and around 0.88 for an average value.

The average wastewater flow rate is the mean of the FWS influent $Q_i$ and effluent $Q_e$ flow rates. It is expressed as:

$$Q_{ave} = (Q_i + Q_e)/2 \tag{6.230}$$

The nominal hydraulic retention time (HRT) is defined as the ratio of useable wetland water volume to the average flow rate. The theological hydraulic retention time $t$ is calculated as:

$$t = V \times n/Q_{ave} \tag{6.231}$$

The hydraulic loading rate ($q$, in m/h or $m^3/m^2/h$) is the volumetric flow rate divided by the wetland surface area and represents the depth of water distributed to the wetland surface over a specific time interval. The hydraulic loading rate can be computed as:

$$q = Q_i/A \tag{6.232}$$

Manning's equation (Eq. (4.22b)), which defines flow in open channel, is adopted to estimate head loss in FWS wetland as follows:

$$S^{1/2} = v/(1/n)(H^{2/3}) \tag{6.233}$$

where $S$ = hydraulic gradient or slope of water surface
$v$ = average flow velocity, m/s or ft/s
$n$ = Manning's resistance coefficient, $s/m^{1/3}$
$H$ = average wetland depth, m or ft

A typical slope is of 1 in 10,000, or 1 cm in 100 m. The coefficient $n$ is a function of water depth and the resistance of specific surface. The range of $n$ values in wetland are 0.3 to 1.1 s/m$^{1/3}$. Since multiple cells are recommended as good design practice to minimize short-circuiting and to maximize treatment performance, for most applications where aspect ratios (length/width) are within suggested limits of 3 : 1 to 5 : 1, or even larger.

**Design considerations.** The design models and methods have been used to predict the fate of BOD, TSS, TN, $HN_4$, $NO_3$, TP, and FCs in a FWS system. The performances of the above parameters in both FWS and VSB wetlands are discussed in detail elsewhere (US EPA, 2000). FWS constructed wetlands have usually been modeled as attached growth biological reactors, in which the plants and detrital material uniformly occupy the entire volume of the wetland.

For the purpose of providing secondary (BOD = TSS = 30 mg/L) and advanced secondary treatment of municipal wastewaters, no particular equations alone are able to accurately predict the performance of a multizone FWS constructed wetland. Even if they could be calibrated "to fit" a specific set of data, their nondeterministic basis belies their ability to fit other circumstances of operation.

The areal loading rate (ALR) method specifies a maximum loading rate per unit area for a given constituent. These methods are common in the design of oxidation ponds and land treatment systems. The ALR can be used to give both planning level and final design sizing estimates for FWS systems from projected pollutant mass loads. For example, knowing the areal BOD loading rate, the expected BOD effluent concentration can be estimated or compared to the long-term average performance data of other well-documented, full-scale operating systems. The ALR can be described as:

$$\text{ALR} = Q_\text{i}\, C_\text{i}/A \tag{6.234}$$

where $C_\text{i}$ = influent concentration of pollutant, mg/L
$Q_\text{i}$ = incoming flow, m$^3$/d or ft$^3$/d
$A$ = total area of FWS, ha or acre

The ALR method does not always correlate to a reasonable design basis, especially with regard to nutrients and pathogen removal; other mechanistic explanations are necessary. However, if typical municipal wastewaters are to be treated that have total and filtered pollutant fractionation reasonably consistent from site to site, a rational design approach can be deduced for those parameters which can be removed during the enhanced flocculation/sedimentation that occurs in the initial fully vegetated zone of a FWS constructed wetland. Therefore, based on Figs. 6.63 and 6.65, the following areal loading rates can be used for this initial zone (zone 1) of thc FWS:

| Parameter | Effluent concentration, mg/L | Zone 1 areal loading rate | |
|---|---|---|---|
| | | kg/ha · d | lb/acre · d |
| BOD | 30 | 40 | 35.7 |
| TSS | 30 | 30 | 26.8 |

The relative areal loadings imply that unless the pretreatment processes were to have a BOD concentration of greater than 1.3 times the TSS concentration, the latter would be the critical loading rate for the fully vegetated zone if secondary standards are to be met by a fully vegetated FWS system.

If the FWS system were to have significant open areas between fully vegetated zones, a better effluent quality could be attained at areal loadings, based on the entire FWS system area.

| Parameter | Effluent concentration, mg/L | Areal loading rate | |
|---|---|---|---|
| | | kg/ha · d | lb/acre · d |
| BOD | 30 | 60 | 53.6 |
| BOD | $< 20$ | 45 | 40.2 |
| TSS | 30 | 50 | 44.6 |
| TSS | $< 20$ | 30 | 26.8 |

The loading rates in the above table are based on the entire system area, not just zone 1. Therefore, with open-water zones that provide aerobic transformations and removal opportunities, a better effluent quality is achievable than with a fully vegetated FWS system. It is recommended that the minimum HRT at the maximum monthly flow in both zones 1 and 3 (fully vegetated) be 2 days. The water depth should be 0.6 to 0.9 m. There are insufficient data at this time to eliminate the need to provide effluent disinfection. The open-water zones would attract wildlife to a great degree, and would be affected by their activities.

A FWS constructed wetland is most likely to treat effluent from a stabilization or oxidation pond, or from primary-treated (settled) municipal wastewater. After the designer determines overall size of the FWS system from these BOD and TSS areal loading rates, the designer can return to evaluate the fate of other constituents. TSS removal and removal of associated BOD, organic N and P, metals, etc. occur in the initial portion of the cell (Fig. 6.65), while the subsequent zones can impact certain soluble constituents. Given sufficient dissolved oxygen in open (unvegetated) areas, soluble BOD removal and nitrification of ammonia can occur. If sufficient oxygen is present, soluble BOD can be removed very slowly by anaerobic processes.

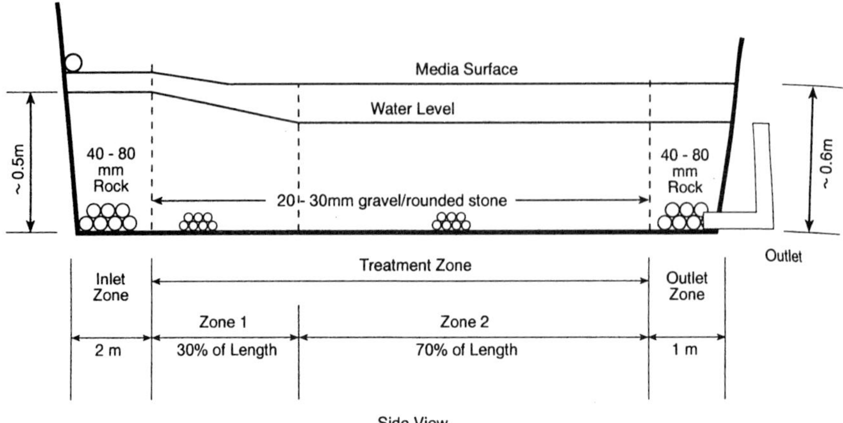


**Figure 6.65** Proposed zones in a vegetated submerged bed system (*source: US EPA 2001*).

In zone 2, which is primarily open-water, the natural reaeration processes are supplemented by submerged macrophytes during daylight periods to elevate dissolved oxygen in order to oxidize carbonaceous compounds (BOD) to sufficiently low levels to facilitate nitrification of the $NH_4$-N to $NO_3$-N. These processes require large amounts of oxygen and time in a passive system (no mechanical assistance). The maximum HRT in zone 2 is generally limited to about 2 to 3 days before unwanted algal blooms occur. The water depth should be 1.2 to 1.5 m. Then, there are three cells in each train and at least two trains (unless very small system). Therefore, more than one open zone may be required to complete these reactions. If so, the result would be a five (or more) zone design; because each open zone is followed by a fully vegetated zone. The reactions in zone 2 are essentially the same as in a facultative lagoon. Therefore, the first-order equation for FC die-off could be applied as an approximation, along with its temperature dependency as:

$$\frac{C_e}{C_i} = \frac{1}{(1 + tK_f)^N} \tag{6.235}$$

where $C_i$, $C_e$ = influent and effluent FC concentrations, respectively, cfu/100 mL

$N$ = number of open-water zones in the FWS

$t$ = HRT, days

$K_f$ = FC removal rate constant, per day
$= 2.6\,(1.19)^{T-20}$

$T$ = temperature, °C

BOD removal in the open-water zone should also follow existing equations such as:

$$\frac{C_e}{C_i} = \frac{1}{(1 + tK_b)^N} \tag{6.236}$$

where $C_i$, $C_e$ = influent and effluent BOD, respectively, mg/L
$K_b$ = specific BOD removal rate constant, per day
$= 0.15\ (1.04)^{T-20}$

The reduction in wetland volume due to settled solids, living plants, and plant detritus can be significant over the long term. The rate of accumulation of settled suspended solids is a function of the water temperature, the mass of influent TSS, the effectiveness of TSS removal, the decay rate of the volatile fraction of the TSS, and the settled TSS mass that is nonvolatile. The plant detritus buildup is a function of the standing crop and the decay rate of the plant detritus. Accumulation for emergent vegetated areas of the Arcata enhancement wetlands was measured to be approximately 12 mm/yr of detritus on the bottom due to plant breakdown and 12 to 25 mm/yr of litter forming a thatch on the surface (Kadlecik, 1996). The volume of the living plants, specifically the volume of the emergent plants, ranged from 0.005 $m^3/m^2$ (low stem density, water depth of 0.3 m) to 0.078 $m^3/m^2$ (high stem density, water depth of 0.75 m). This accumulation is more or less constant from year to year as the wetland matures. The total volume reduction under the initial vegetated zone can be estimated using a mass balance equation:

$$V_r = (V_{ss}t + V_d t)\,A \tag{6.237}$$

where $V_r$ = volume reduction over the period of analysis, $m^3$
$V_{ss}$ = volume reduction due to nonvolatile TSS and nondegradable volatile TSS accumulation, $m^3$/ha · yr
$V_d$ = volume reduction due to nonvolatile detrital accumulation as a function of annual production, $m^3$/ha · yr
$A$ = fully vegetated wetland area, ha
$t$ = period of analysis, year

**Example 1:** A reasonable default value for $V_{ss}$ when treating raw wastewater in lagoons ranges from 200 to 400 $m^3$/ha · yr (2 to 4 cm/yr). Therefore, a conservative default value of 150 $m^3$/ha · yr can be used. The loss of volume per hectare over an 8-year period for a 1-ha fully vegetated FWS wetland zone with a depth of 0.75 m can be estimated by use of this equation. One hundred percent coverage of emergent vegetation was measured to contribute 116 $m^3$/ ha · yr of bottom detritus, and 104 $m^3$/ha · yr of surface litter with a standing crop volume of 360 $m^3$/ha. Estimate the loss of volume per hectare over an 8-year period for a 1-ha fully vegetated FWS wetland zone with a depth of 0.75 m.

**solution:**

$$V_d = 116\ \text{m}^3/\text{ha}\cdot\text{yr} + 104\ \text{m}^3/\text{ha}\cdot\text{yr} = 220\ \text{m}^3/\text{ha}\cdot\text{yr}$$

Using Eq. (6.237):

$$\begin{aligned} V_r &= (V_{ss}t + V_d t)\,A \\ &= [150\ \text{m}^3/\text{ha}\cdot\text{yr} \times 8\ \text{years} + 220\ \text{m}^3/\text{ha}\cdot\text{yr} \times 8\ \text{years}] \times 1\ \text{ha} \\ &= 2960\ \text{m}^3 \end{aligned}$$

**Example 2:** Design a FWS constructed wetland to treat lagoon effluent to meet a monthly average 30 mg/L of both BOD and TSS discharge standards. Given the design conditions: population = 42,000 people; the average annual design flow $Q_{ave}$,

$$Q_{ave} = 15{,}900\ \text{m}^3/\text{d}\ (4.2\ \text{MGD}); \qquad Q_{max} = 2Q_{ave} = 31{,}800\ \text{m}^3/\text{d}$$

At $Q_{ave}$: the influent BOD = 55 mg/L, while the average TSS = 66 mg/L; at the maximum monthly flow: the BOD = 38 mg/L and TSS = 32 mg/L, respectively; the maximum ALRs are 40 kg BOD/ha · d and 30 kg TSS/ha · d for a single cell; and the maximum ALRs are 60 kg BOD/ha · d and 50 kg TSS/ha · d for with open areas.

**solution:**

Step 1a. Determine the total area required under the critical condition for a single cell

Rearranging Eq. (6.234),

$$A = Q_i\, C_i/\text{ALR}$$

For BOD yields: at $Q_{ave}$

$$\begin{aligned} A &= 15{,}900\ \text{m}^3/\text{d} \times 1000\ \text{L/m}^3 \times 55\ \text{mg/L}/(40\ \text{kg/ha}\cdot\text{d} \times 10^6\ \text{mg/kg}) \\ &= 21.9\ \text{ha} \end{aligned}$$

at $Q_{max}$

$$A = 31{,}800 \times 1000 \times 38/(40 \times 10^6) = 30.2\ \text{(ha)}$$

For TSS yields: at $Q_{ave}$

$$A = 15{,}900 \times 1000 \times 66/(30 \times 10^6) = 35.0\ \text{(ha)}$$

at $Q_{ave}$

$$A = 31{,}800 \times 1000 \times 32/(30 \times 10^6) = 33.9\ \text{(ha)}$$

Thus, the limiting factor is the TSS loading at average flow conditions where 35.0 ha are required to meet the secondary effluent standards using a fully vegetated single-zone FWS wetland system.

Step 1b. Similarly, recalculate the total area required from the higher ALRs for the FWS system with significant open areas between fully vegetated zones

For BOD yields: at $Q_{ave}$

$$A = 15{,}900 \text{ m}^3\text{/d} \times 1000 \text{ L/m}^3 \times 55 \text{ mg/L}/(60 \text{ kg/ha} \cdot \text{d} \times 10^6 \text{ mg/kg})$$
$$= 14.6 \text{ ha}$$

at $Q_{max}$

$$A = 31{,}800 \times 1000 \times 38/(60 \times 10^6) = 20.1 \text{ (ha)}$$

For TSS yields: at $Q_{ave}$

$$A = 15{,}900 \times 1000 \times 66/(50 \times 10^6) = 21.0 \text{ (ha)}$$

at $Q_{ave}$

$$A = 31{,}800 \times 1000 \times 32/(50 \times 10^6) = 20.4 \text{ (ha)}$$

The limiting condition is still the TSS areal loading at the average flow conditions. But, the total area requirement is only 21.0 ha instead of 35.0 ha.

Step 2. Compute the theoretical HRT assuming $H = 0.6$ and $n = 0.75$ in the vegetative zones (1 and 3) and $H = 1.3$ m and $n = 1.0$ in the open zone 2. The combined average depth is estimated of 0.84 m and an average $n = 0.82$. Using Eq. (6.231), for $Q_{ave}$

$$t = V \times n/Q_{ave} = A \times H \times n/Q_{ave}$$
$$= 21 \text{ ha} \times 10{,}000 \text{ m}^2\text{/ha} \times 0.84 \text{ m} \times 0.82/15{,}900 \text{ m}^3\text{/d}$$
$$= 9.1 \text{ days}$$

for $Q_{max}$ $$t = 21 \times 10000 \times 0.84 \times 0.82/31{,}800 = 4.55 \text{ days}$$

*Note*: The above implies that at $Q_{max}$, the HRT may not be adequate for the necessary treatment mechanisms to perform. For zone 1, the minimum HRT at $Q_{max}$ should be at least 2 days with 4 days at $Q_{ave}$. For zone 2, there is an upper limit that depends on climate and temperature. For most US conditions, a maximum HRT of 2 to 3 days should avoid algal blooms (a load to zone 3); although the longer the HRT, the better the removal for soluble organics, ammonia-N, and bacteria. Zone 3 should have the same considerations as zone 1. It also provides denitrification.

Therefore, the minimum HRT at $Q_{max}$ is 6 (2 + 2 + 2) days, and 12 days at $Q_{ave}$ as in this example.

Step 3. Use $t$ = 12 days to determine the area $A$ required at $Q_{ave}$

Substituting Eq. (6.229) into Eq. (6.231) and rearranging it,

$$A = t\,Q/n\,H = 12 \text{ days} \times 15{,}900 \text{ m}^3/\text{d}/(0.82 \times 0.84 \text{ m} \times 10{,}000 \text{ m}^2/\text{ha})$$
$$= 27.7 \text{ ha (say 28 ha} = 69 \text{ acres)}$$

Applying the normal additional area for the inlet, buffers, and outlet of a factor ranging from 1.25 to 1.40, the total area required for the FWS wetland is 35 ha (28 ha × 1.25) or 86.5 acres.

Step 4. Configuration: The FWS system should be designed at least with two parallel treatment trains of a minimum of three cells in each. Say for 2 trains, for this example, the area of each train would be 14 ha = 140,000 $m^2$.

The shapes of the system can be squares, rectangles, polygons, ovals, kidney shapes, and crescent shapes. Let us select rectangles. The optimum aspect ratio, or average length $L$ to average width $W$, ranges 3 : 1 to 5 : 1. If the ratio ≥ 10, we may need to calculate backwater curves. Using $L : W = 4 : 1$, then $4W^2 = 140{,}000 \text{ m}^2$. We obtain $W = 187$ m and $L = 748$ m (only for 3 cells). Say each cell has 250 m in length. By adding 25% of length for the inlet and outlet zones, the total length for the system is 937 m.

In summary, the FWS system includes 2 trains. Each train has 187-m width and contains inlet zones, 3 treatment zones (1, 2, and 3), and outlet zones. The lengths of each zone respectively are of 87, 250, 250, 250, and 90 m. The depths of each zone respectively are of 1.0, 0.6, 1.3, 0.6, and 0.6 m.

*Note*: To effectively minimize short-circuiting with baffles in the FWS system, the inlet/outlet structures should be designed for uniform distribution of inflow across the entire width of the wetland inlet and uniform collection of effluent across the entire wetland outlet width. An inlet-settling structure may be needed if there is high TSS in the influent. The outlet weir loading should be ≤ 200 $m^3$/m · d.

Multiple cells allow for redistribution of the primary cell effluent in the subsequent cell that reduces short-circuiting. Flexible intercell piping will facilitate maintenance without a major reduction in the necessary HRT to produce satisfactory effluent quality. Aspect ratios of the cells should be greater than 3:1 and adapted to the site contours and restrictions. Additional treatment will likely be required alter the FWS system to meet fecal coliform and dissolved oxygen permit requirements.

**Example 3:** Use the same design conditions as Example 2. Design a FWS wetland system to meet 20-20 (both BOD and TSS ≤ 20 mg/L) effluent standards, i.e. to meet this effluent quality, an open-water zone is required in the FWS wetland. The maximum ALRs are 45-kg BOD/ha · d and 30-kg TSS/ha · d for the system.

**solution:**

Step 1. Compute areas required at average and maximum flows as in Example 2

For BOD yields: at $Q_{ave}$

$$A = 15{,}900 \text{ m}^3/\text{d} \times 1000 \text{ L/m}^3 \times 55 \text{ mg/L}/(45 \text{ kg/ha} \cdot \text{d} \times 10^6 \text{ mg/kg})$$
$$= 19.4 \text{ ha}$$

at $Q_{max}$

$$A = 31{,}800 \times 1000 \times 38/(45 \times 10^6) = 26.9 \text{ ha}$$

For TSS yields: at $Q_{ave}$

$$A = 15{,}900 \times 1000 \times 66/(30 \times 10^6) = 35.0 \text{ ha}$$

at $Q_{max}$

$$A = 31{,}800 \times 1000 \times 32/(30 \times 10^6) = 33.9 \text{ ha}$$

The limiting factor is again the TSS loading at average flow conditions where 35.0 ha are required to meet the more stringent effluent standards.

Step 2. Compute the theoretical HRT, $t$, required for the entire 3-zone FWS system, assuming an overall average depth, $H = 0.84$ m and an average $n = 0.82$. Using Eq. (6.231)

for $Q_{ave}$

$$t = V \times n/Q_{ave} = A \times H \times n/Q_{ave}$$
$$= 35.0 \text{ ha} \times 10{,}000 \text{ m}^2/\text{ha} \times 0.84 \text{ m} \times 0.82/15{,}900 \text{ m}^3/\text{d}$$
$$= 15.2 \text{ days}$$

for $Q_{max}$

$$t = 15.2 \text{ days}/2 = 7.6 \text{ days}$$

Step 3. Determine the area required for each cell (zone)

At maximum monthly flow rate, assuming an equal minimum HRT ($t = 7.6$ days/3 = 2.53 days) in each of 3 cells

for zone 2,

$$A_2 = t\,Q/n\,H = 2.53 \text{ days} \times 31{,}800 \text{ m}^3/\text{d}/(1.0 \times 1.3 \text{ m} \times 10{,}000 \text{ m}^2/\text{ha})$$
$$= 6.2 \text{ ha}$$

Then, area for zone 1 ($A_1$) and zone 3 ($A_3$) would be

$$A_1 = A_3 = (35.0 \text{ ha} - 6.2 \text{ ha})/2 = 14.4 \text{ ha}$$

Step 4. Determine the zone dimensions

Say the total area required for the 3 treatment zones is 35 ha (86.5 acres). By adding in 25% of it, the overall FWS wetland area including inlet/outlet zones would be 43.8 ha (108 acres).

Again use two parallel trains (17.5 ha each) and aspect ratio of 4 : 1; then, $4W^2 = 175{,}000\ \mathrm{m}^2$. We obtain $W = 209$ m and $L = 836$ m (only for 3 cells). Approximately, the lengths of zones 1, 2, and 3 are 318, 200, and 318 m, respectively. By adding 25% of the length for the inlet and outlet zones, the total length for the system is 1045 m. The length of the inlet and outlet zones may be 109 and 100 m, respectively. The depths of the inlet, 3 treatment, and outlet zones respectively are of 1.0, 0.6, 1.3, 0.6, and 0.6 m.

**Vegetated submerged bed systems.** The VSB wetlands are composed of gravel beds that may be planted with wetland vegetation. Figure 6.64 illustrates a schematic sketch of a VSB wetland system. Typical VSB components include (1) inlet piping, (2) a clay or synthetic membrane lined basin, (3) loose media filling the basin, (4) wetland vegetation planted in the media, and (5) outlet piping with a water-level control system. The outlet structures are designated for regulation and distribution of wastewater flow. VSB wetlands may contain berms. In addition to shape and size, other variable factors are choice of treatment media (gravel shape and size, for example) as an economic factor and selection of vegetation as an optional feature that affects wetland aesthetics more than its performance.

The pollutant removal performance of VSB systems depends on many factors including influent wastewater quality, hydraulic and pollutant loading, climate, and physical characteristics of the systems. The main advantage of a VSB system over a free water surface wetland system is the isolation of the wastewater from vectors, animals, and humans. Concerns with mosquitoes and pathogen transmission are greatly reduced with a VSB system. Properly designed and operated VSB systems may not need to be fenced off or otherwise isolated from people and animals. Comparing conventional VSB systems to FWS systems of the same size, VSB systems typically cost more to construct, primarily because of the cost of media (Reed *et al.*, 1995).

It is not clear if it is desirable to maintain a single plant species, or a prescribed collection of plant species, for any treatment purpose. Single plant (monoculture) systems are more susceptible to catastrophic plant death due to predation or disease (George *et al.*, 2000). It is generally assumed that multiple plant and native plant systems are less susceptible to catastrophic plant death, although no studies have confirmed this assumption. Plant invasion and plant dominance further complicate the issue.

It was found that the roots do not fully penetrate to the bottom of the media and there is substantially more flow under the root zone than through it (Young *et al.*, 2000). The oxygen supply from the roots is also likely to be unreliable due to yearly plant senescence, plant die-off due to disease and pests, and variable plant coverage from year to year. Considering all of these factors, it is recommended that designers assume wetland plants provide no significant amounts of oxygen to a VSB system. The impact of wetland plants on pollutant removal performance appears to be minimal, based on current knowledge; so, the selection of plants species should be based on aesthetics, impacts on operation, and long-term plant health and viability in a given geographical area. Local wetland plants experts should be consulted when making the selection.

**Hydrology.** As in all gravity flow systems, the water level in a VSB system is controlled by the outlet elevation and the hydraulic gradient, or slope, that is the drop in the water level (head loss) over the length from the inlet to the outlet. The relationship between the hydraulic gradient and the flow through a porous media is typically described by the general form of Darcy's law (Eq. (6.238)). This model assumes laminar flow through the media finer than coarse gravel. It is recommended without modification of the general form as sufficient to estimate the water level within a VSB system. Darcy's law is a function of the flow, ALR, water depth, and hydraulic conductivity.

$$Q = KA_cS = KWD\, dh/dL \tag{6.238}$$

or, for a defined length of the VSB,

$$dh = QL/KWD \tag{6.239}$$

Substitute $L = A/W$ and rearrange the above equation to solve for minimum $W$:

$$W^2 = QA/KD\, dh \tag{6.240}$$

where $Q$ = flow rate, $m^3/d$
$K$ = hydraulic conductivity, $m^3/m^2 \cdot d$, or m/d
$A_c$ = cross-sectional area normal to wastewater flow (= $WD$), $m^2$
$W$ = width of VSB, m
$D$ = water depth, m
$L$ = length of VSB, m
$dh$ = head loss (change in water level) due to flow resistance, m
$S$ = $dh/dL$ = hydraulic gradient, m/m
$A$ = surface area of zone or of total wetland, $m^2$

The water level at the inlet of a VSB will rise to the elevation required to overcome the head loss in the entire VSB system. It should be designed to prevent surfacing. $K$ for an operating VSB varies with time and location within the media and will have a major impact on the head loss. $K$ is very difficult to determine because it is influenced by factors that cannot be easily accounted for, including flow patterns (affected by preferential flow and short-circuiting), and clogging (affected by changes in root growth/death and solids accumulation/degradation). Therefore, a value must be assumed for design purposes. Typical clean $K$ values vary for various sizes of rock and gravel. They are 6200, 21,000, 34,000, 64,000, 100,000, and 120,000 m/d for 5 mm pea gravel, 6 mm pea gravel, 5 to 10 mm gravel, 22 mm coarse gravel, 17 mm creek rock, and 19 mm rock, respectively. Based on the studies and many observed cases of surfacing in VSB systems, the following conservative values are recommend for the long-term operating $K$ values (US EPA, 2000):

Initial 30% of VSB $\quad K_i = 1\%$ of clean $K$

Final 70% of VSB $\quad K_f = 10\%$ of clean $K$

#### Design consideration

*Media.* The media of a VSB system perform several functions: They (1) are rooting material for vegetation, (2) help to evenly distribute/collect flow at the inlet/outlet, (3) provide surface area for microbial growth, and (4) filter and trap particles. For successful plant establishment, the uppermost layer of media should be conducive to root growth. A variety of media sizes and materials have been tried, but there is no clear evidence that points to a single size or type of media, except that the media should be large enough that it will not settle into the void spaces of the underlying layer.

It is recommended (US EPA, 2000) that the planting media would be 5 to 20 mm (1/5 to 4/5 in) in diameter, and the minimum depth should be 100 mm (4 in). The media in the inlet and outlet zones (Fig. 6.66) should be between 40 and 80 mm (1.6 and 3.1 in) in diameter to minimize clogging and should extend from the top to the bottom of the system. All media must be washed clean of fines and debris; more rounded media will generally have more void spaces; and media should be resistant to crushing or breakage. The hydraulic conductivity of the 20 to 30 mm diameter clean media is assumed as 100,000 m/d.

*Slopes.* The top surface of the media should be leveled or nearly leveled for easier planting and routine maintenance. Theoretically, the bottom slope should match the slope of the water level to maintain a uniform water depth throughout the VSB. However, because the hydraulic conductivity of the media varies with time and location, it is not practical to determine the bottom slope this way; the bottom slope should be designed only for draining the system, and not to supplement the hydraulic conductivity of the VSB. A practical approach is to uniformly slope the bottom along the direction of flow from inlet to outlet to allow for easy draining when maintenance is required. No research has been

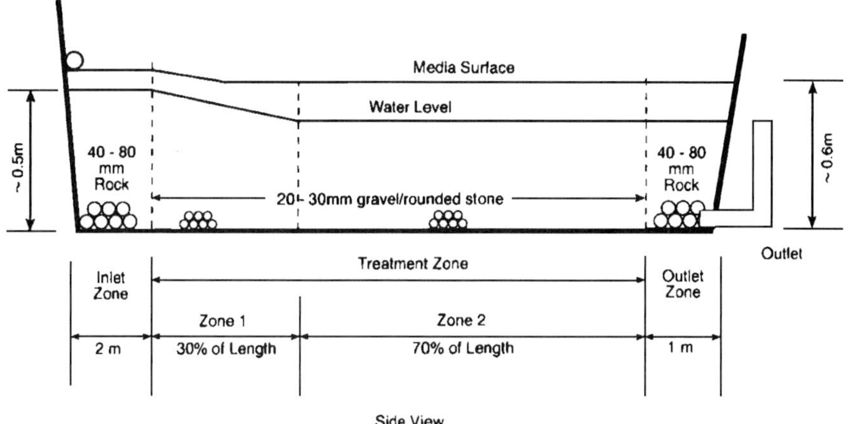


**Figure 6.66** Proposed zones in a VSB.

done to determine an optimum slope, but Chalk and Wheale (1989) recommended a slope of $\frac{1}{2}$ to 1% is for ease of construction and proper draining.

*System size.* The surface area required is calculated based upon desired effluent quality and areal loading rate. The ALRs for BOD are 6 g/m$^2$ · d (53.5 lb/acre · d) and 1.6 g/m$^2$ · d (14.3 lb/acre · d) to attain 30 and 20 mg/L effluents, respectively. The ALR for TSS is 20 g/m$^2$ · d (178 lb/acre · d) to attain 30 mg/L effluent. VSB system is not designed for nutrients (N and P) removal.

There is no optimum depth for a VSB system. But the media depth should be set equal to the maximum root depth of the wetland species to be used in the VSB. Typical average media depths in VSB systems have ranged from 0.3 to 0.7 m (12 to 28 in), and various researchers have recommended depths from 0.4 to 0.6 m (16 to 24 in). It is recommended to use a design with maximum water depth (at the inlet of the VSB) of 0.50 m (20 in). The depth of the media will be defined by the level of the wastewater at the inlet and should be about 0.1 m (4 in) higher than the water.

The width of an individual VSB is set by the ability of the inlet and outlet structures to uniformly distribute and collect the flow without inducing short-circuiting. The width can be determined using Darcy's law (Eq. (6.238)). The recommended maximum width in a TVA design manual is 61 m (200 ft). If the design produces a larger value, it should divide the VSB into several cells that do not exceed 61 m in width—use at least 2 VSBs in parallel; use an adjustable inlet device with capability to balance flows; and use an adjustable outlet control device with capability to flood and drain system.

Several researchers have noted that most BOD and TSS are removed in the first few meters of VSB, but some recommend minimum lengths ranging from 12 to 30 m (39 to 98 ft) to prevent short-circuiting. The minimum length recommended by the US EPA (2000) is 15 m (50 ft). Therefore, the aspect ratio is not a factor in the overall size design. However, the recommended values for maximum width and minimum length tend to result in individual VSB cells with a length-width ratio between 1 : 1 and 1 : 2.

**Example:** Design a VSB constructed wetland system with some assumptions or given conditions:

- The total VSB has four zones as shown in Fig. 6.65.
- The initial treatment zone 1 occupies about 30% of the total area, performs most of the treatment, and has a big decrease in hydraulic conductivity (use $K$ = 1% of clean $K$).
- The final treatment zone 2 will occupy the remaining 70% of the area and has little change in hydraulic conductivity (use $K$ = 10% of clean $K$).
- Darcy's law, while not exact, is good enough for design purposes.
- The sizing of the initial and final treatment zones follows these steps:
  - Determine the surface area, using recommended ALR
  - Determine the width, using Darcy's Law
  - Determine the length and head loss of the initial treatment zone, using Darcy's law
  - Determine the length and head loss of the final treatment zone, using Darcy's law
  - Determine bottom elevations, using bottom slope
  - Determine water elevations throughout the VSB, using head loss
  - Determine water depths, according for bottom slope and head loss
  - Determine required media depth
  - Determine the number of VSB cells

- Maximum monthly flow $Q$ = 380 m$^3$/d = 0.1 MGD
- Maximum monthly influent $C_i$ BOD = 80 mg/L = 80 g/m$^3$
- Maximum monthly influent $C_i$ TSS = 90 mg/L = 90 g/m$^3$
- Required discharge limits = 30 mg/L for both BOD and TSS
- Recommended values for VSBs are:
  - ALR for BOD = 6 g/m$^2 \cdot$ d
  - ALR for TSS = 20 g/m$^2 \cdot$ d
  - Use washed, rounded media 20 to 30 mm in diameter, clean $K$ = 100,000 m/d
  - Hydraulic conductivity of initial treatment zone 1: $K_1$ = 1% of 100,000 m/d = 1,000 m/d

- Hydraulic conductivity of final treatment zone 2: $K_2$ =10% of 100,000 m/d = 10,000 m/d
- Bottom slope: $s$ = 0.5% = 0.005
- Design water depth at inlet: $D_i$ = 0.44 m
- Design water depth at beginning of final treatment zone: $D_f$ = 0.44 m
- Design media depth: $D_m$ = 0.6 m
- Maximum allowable head loss through initial treatment zone $dh_i$ = 10% of $D_m$ = 0.06 m

**solution:**

Step1. Determine the surface area for both BOD and TSS, using Eq. (6.234)

For BOD removal:

$$A = QC_i/\text{ALR} = (380\ \text{m}^3/\text{d} \times 84\ \text{g/m}^3)/6\ \text{g/m}^2 \cdot \text{d}$$

$$= 5320\ \text{m}^2$$

For TSS removal:

$$A = (380\ \text{m}^3/\text{d} \times 96\ \text{g/m}^3)/20\ \text{g/m}^2 \cdot \text{d}$$

$$= 1824\ \text{m}^2$$

Use the large area required, i.e. 5320 m$^2$. The surface area for the initial treatment zone 1 is $A_1$ = 5320 m$^2$ × 0.3 = 1600 m$^2$. The surface area for the final treatment zone 2 is $A_2$ = 5320 m$^2$ × 0.7 = 3720 m$^2$.

Step 2. Determine the minimum width needed to keep the flow below the surface, using Eq. (6.240) and recommended values ($K_i$ = 1000 m/d, $D$ = 0.44 m, $dh$ = 0.06 m) for the initial treatment zone 1

$$W^2 = QA/KD\,dh \qquad (A = A_i = 1600\ \text{m}^2)$$

$$= 380\ \text{m}^3/\text{d} \times 1600\ \text{m}^2/(1000\ \text{m/d} \times 0.44\ \text{m} \times 0.06\ \text{m})$$

$$= 23{,}030\ \text{m}^2$$

$$W = 152\ \text{m}$$

*Note*: This width of 152 m is needed to have a head loss of 0.06 m under all given parameters.

Step 3. Compute the length $L_1$ and check the head loss of zone 1; $K_1$ =1,000 m/d

$$L_1 = A_1/W = 1600\ \text{m}^2/152\ \text{m} = 10.5\ \text{m}$$

Using Eq. (6.239)

$$dh_1 = QL_1\,/K_1WD = 380\ \text{m}^3/\text{d} \times 10.5\ \text{m}/(1000\ \text{m/d} \times 152\ \text{m} \times 0.44\ \text{m})$$

$$= 0.06\ \text{m}$$

Step 4. Compute the length $L_2$ and check the head loss of zone 2; $K_2$ = 10,000 m/d

$$L_2 = A_2/W = 3720 \text{ m}^2/152 \text{ m} = 24.5 \text{ m (is a minimum)}$$

$$dh_2 = QL_2 / K_2WD = 380 \text{ m}^3/\text{d} \times 24.5 \text{ m}/(10{,}000 \text{ m/d} \times 152 \text{ m} \times 0.44 \text{ m})$$

$$= 0.014 \text{ m}$$

Step. 5. Calculate bottom elevations

Let bottom elevation at the outlet, $E_o = 0$ as reference point for all elevations. $s = 0.005$. The elevation at the beginning of zone 2, $E_f$, is:

$$E_f = L_2 \times s = 24.5 \text{ m} \times 0.005 = 0.12 \text{ m}$$

The elevation at the inlet (the beginning of zone 1), $E_i$, is:

$$E_i = (10.5 \text{ m} + 24.5 \text{ m}) \times 0.005 = 0.18 \text{ m}$$

Step 6. Determine the water surface elevations

The water surface elevation at the beginning of zone 2, $E_{wf}$, is:

$$E_{wf} = E_f + D_f = 0.12 \text{ m} + 0.44 \text{ m} = 0.56 \text{ m}$$

The water surface elevation at the outlet, $E_{wo}$, is:

$$E_{wo} = E_{wf} - dh_2 = 0.56 \text{ m} - 0.01 \text{ m (use 0.01 instead 0.014 from Step 4)}$$

$$= 0.55 \text{ m}$$

The water surface elevation at the inlet, $E_{wi}$, is:

$$E_{wi} = E_{wf} + dh_1 = 0.56 \text{ m} + 0.06 \text{ m} = 0.62 \text{ m}$$

Step 7. Compute water depth

The depth of water at the inlet, $D_i$, is:

$$D_i = E_{wi} - E_i = 0.62 \text{ m} - 0.18 \text{ m} = 0.44 \text{ m} \qquad \text{(as designed)}$$

The depth of water at the beginning of zone 2, $D_f$, is:

$$D_f = E_{wf} - E_f = 0.56 \text{ m} - 0.12 \text{ m} = 0.44 \text{ m} \qquad \text{(as designed)}$$

The depth of water at the outlet, $D_o$, is:

$$D_o = E_{wo} - E_o = 0.55 \text{ m} - 0 \text{ m} = 0.55 \text{ m}$$

Step 8. Determine the media depth

The media depth can be designed based on two concepts: (a) level the media surface and (b) a minimum depth to water throughout the VSB system.

a. Determine the media depth based on level surface

The elevation of media top must be greater than the highest water elevation, i.e. at the inlet, $E_{wi} = 0.62$ m. Let us set the media top elevation at 0.68 m.

Then, the depth of media at the inlet,

$$D_{mi} = 0.68 \text{ m} - E_i = 0.68 \text{ m} - 0.18 \text{ m} = 0.50 \text{ m}$$

The depth of media at the beginning of zone 2,

$$D_{mf} = 0.68 \text{ m} - E_f = 0.68 \text{ m} - 0.12 \text{ m} = 0.56 \text{ m}$$

The depth of media at the outlet,

$$D_{mo} = 0.68 \text{ m} - E_o = 0.68 \text{ m} - 0 \text{ m} = 0.68 \text{ m}$$

The media depth-to-water at the inlet,

$$D_{twi} = 0.68 \text{ m} - E_{wi} = 0.68 \text{ m} - 0.62 \text{ m} = 0.06 \text{ m}$$

The media depth-to-water at the beginning of zone 2,

$$D_{twf} = 0.68 \text{ m} - E_{wf} = 0.68 \text{ m} - 0.56 \text{ m} = 0.12 \text{ m}$$

The media depth-to-water at the outlet,

$$D_{two} = 0.68 \text{ m} - E_{wo} = 0.68 \text{ m} - 0.55 \text{ m} = 0.13 \text{ m}$$

*Note*: The media depth-to-water is small at the inlet (0.06 m). However, the designer may want to add an additional layer of media in the first few meters of the initial treatment zone 1 as an added precaution against surfacing, even though the design ALR and $K$ values are very conservative. The resulting $D_{two}$ in the final treatment zone 2 would be 0.12 to 0.13 m, which should not inhibit the growth of aquatic species.

b. Determine the media depth based on a constant depth-to-water (use 0.1 m)

The elevation of media surface at the inlet,

$$E_{mi} = E_{wi} + 0.1 \text{ m} = 0.62 \text{ m} + 0.1 \text{ m} = 0.72 \text{ m}$$

The elevation of media surface at the beginning of zone 2,

$$E_{mf} = E_{wf} + 0.1 \text{ m} = 0.56 \text{ m} + 0.1 \text{ m} = 0.66 \text{ m}$$

The elevation of media surface at the outlet,

$$E_{mo} = E_{wo} + 0.1 \text{ m} = 0.55 \text{ m} + 0.1 \text{ m} = 0.65 \text{ m}$$

The depth of media at the inlet,

$$D_{mi} = E_{mi} - E_i = 0.72 \text{ m} - 0.18 \text{ m} = 0.54 \text{ m}$$

The depth of media at the beginning of zone 2,

$$D_{mf} = E_{mf} - E_f = 0.66 \text{ m} - 0.12 \text{ m} = 0.54 \text{ m}$$

The depth of media at the outlet,

$$D_{mo} = E_{mo} - E_o = 0.65\text{m} - 0 \text{ m} = 0.65 \text{ m}$$

*Note*: This approach would result in a drop in the media surface of ($E_{mi} - D_{mi}$ = 0.72 m − 0.54 m) 0.18 m over the 10.5-m length of the initial treatment zone (slope = 1.7%), which would probably not impair operation and maintenance activities.

Step 9. Determine the number of cells

It is recommended that at least two VSBs be used in parallel in all but the smallest systems, so that one of the VSBs can be taken out of service for maintenance or repairs without causing serious water quality violations. In this example, the total size of the treatment zones is 152 m wide and 35 (10.5 + 24.5) m long. Therefore, four VSB trains, each 38 m wide and 35 m long, could be used. Other combinations of length and width that have the required surface area will also work as long as the hydraulics conditions are met. Also remember that inlet (2 m in length) and outlet zones (1 m) will add to the overall length of the VSB. Thus the total length of a VSB system will also be 38 m.

In summary, to treat the 380 $m^3$/d (0.1 MGD) of wastewater flow and to meet the 30-30 standards, a VSB constructed wetland system that consists four trains is designed. Each train is 38 m wide and 38 m long (2, 10.5, 24.5, and 1 m in length, respectively, for the inlet zone, zone 1, zone 2, and outlet zone). The depths and media sizes are close to the values recommended by the US EPA (2000), as shown in Fig. 6.65.

## References


Albertson, O. E. and Davis, G. 1984. Analysis of process factors controlling performance of plastic biomedia. Presented at the 57th Annual Meeting of the Water Pollution Control Federation, October 1984, New Orleans, Louisiana.

Alessi, C. J. *et al.* 1978. Design and operation of the activated sludge process. SUNY/BUFFALO-WREE-7802. Buffalo: State University of New York at Buffalo.

Al-Layla, M. A., Ahmad, S. and Middlebrooks, E. J. 1980. *Handbook of wastewater collection and treatment: principles and practice.* New York: Garland STPM Press.

Alleman, J. E. and Irving, R. L. 1980. Nitrification in the sequencing batch biological reactor. *J. Water Pollut. Control Fed.* **52**(11): 2747–2754.

American Society of Civil Engineers (ASCE) and Water Pollution Control Federation (WPCF). 1982. *Design and construction of urban stormwater management systems.* New York: ASCE and WPCF.

American Society of Civil Engineers (ASCE) and Water Environment Federation (WEF). 1992. *Design and construction of urban stormwater management systems.* New York: ASCE & WEF.

Antonie, R. L. 1978 *Fixed biological surfaces: wastewater treatment.* West Palm Beach, Florida: CRC Press.

APHA, AWWA, and WEF. 1998. *Standard methods for the examination of water and wastewater,* 20th edn. Washington, DC: APHA.

Aqua-Aerobic Systems. 1976. *Clarifiers design.* Rockford, Illinois: Aqua-Aerobic Systems.

Autotrol Corporation. 1979. *Wastewater treatment systems: design manual.* Milwaukee, Wisconsin: Autotrol Corp.

Banerji, S. K. 1980. ASCE Water Pollution Management Task Committee report on "Rotating biological contactor for secondary treatment." In: *Proc. First National Symposium/Workshop on Rotating Biological Contactor Technology (FNSWRBCT),* Smith, E. D., Miller, R. D. and Wu, Y. C. (eds.), Vol. I, p. 31. Pittsburgh: University of Pittsburgh.

Bidstrup, S. M. and Grady, Jr., C. P. L. 1988. SSSP: simulation of single-sludge processes. *J. Water Pollut. Control Fed.* **60**(3): 351–361.

Boehnke, B., Diering, B. and Zuckut, S. W. 1997. Cost-effective wastewater treatment process for removal of organics and nutrients. *Water Eng. Mgmt.* **145**(2): 30–35.

Bruce, A. M. and Merkens, J. C. 1973. Further study of partial treatment of sewage by high-rate biological filtration. *J. Inst. Water Pollut. Control* **72**(5): 499–527.

Brenner, R. C. *et al.* 1984. Design information on rotation biological contactors. USEPA-600/2-84-106, Cincinnati, Ohio, Chaps. 2, 3, and 5.

Camp, T. R. 1946. Corrosiveness of water to metals. *J. Net Engl. Water Works Assoc.* **60**(1): 188.

Camp, T. R. 1953. Studies of sedimentation basin design. Presented at 1952 Annual Meeting, Pennsylvania Sewage and Industrial Wastes Association, Stage College Pennsylvania, August 27–29, 1952. Washington, DC: Sewage and Industrial Wastes.

Carder, C. 1997. Chicago's Deep Tunnel. *Compressed Air Magazine.*

CBI Walker, Inc. 1998. *Small egg shaped digester facilities.* Plainfield, Illinois: CBI Walker, Inc.

Cheremisinoff, P. N. 1995. *Handbook of water and wastewater treatment technology.* New York: Marcel Dekker.

Chou, C. C., Hynek, R. J. and Sullivan, R. A. 1980. Comparison of full scale RBC performance with design criteria. *Proc. FNSDWRBCT,* Vol. II, p. 1101.

Clark, J. H., Moseng, E. M. and Asano, T. 1978. Performance of a rotating biological contactor under varying wastewater flow. *J. Water Pollut. Control Fed.* **50**: 896.

Clark, J. W. and W. Viessman, Jr. 1966. *Water supply and pollution control.* New York: A Dun-Donnelley.

Clark, J. W., Viessman, Jr., W. and Hammer, M. J. 1977. *Water supply and pollution control.* New York: IEP-A Dun-Donnelley Publisher.

Clow Corporation. 1980. *Clow Envirodisc Rotating Biological Contactor System.* Florence, Kentucky: Clow Corp.

Collins, H. F. and Selleck, R. E. 1972. Process kinetics of wastewater chlorination. SERL Report 72–75. Berkeley: University of California.

Collins, H. F., Selleck, R. E. and White, G. C. 1971. Problems in obtaining adequate sewage disinfection. *J. Sanitary Eng. Div., Proc. ASCE* **87**(SA5): 549–562.

Coulson, J. M. and Richardson, J. F. 1955. *Chemical engineering* Vol. II. New York: McGraw-Hill.

Davis, M. L. and Cornwell, D. A. 1991. *Introduction to environmental engineering,* 2nd edn. New York: McGraw-Hill.

Dick, R. I. and Ewing, B. B. 1967. Evaluation of activated sludge thickening theories. *J. Sanitary Eng. Div., Proc. ASCE* **93**(SA4): 9–29.

Downing, A. L. and Hopwood, A. P. 1964. Some observations on the kinetics of nitrifying activated sludge plants. *Schweizerische Zeitschrift fur Hydrologie* **26**: 271.

Droste, R. L. 1997. *Theory and practice of water and wastewater treatment.* New York: John Wiley.

Eckenfelder, W. W. 1963. Trickling filter design and performance. *Trans. Am. Soc. Civil. Eng.* **128** (part III): 371–384.

Eckenfelder, W. W., Jr. 1966. *Industrial water pollution control.* New York: McGraw-Hill.

Eckenfelder, W. W. and Barnhart, W. 1963. Performance of a high-rate trickling filter using selected media. *J. Water Pollut. Control Fed.* **35**(12): 1535–1551.

Eckenfelder, W. W. and Ford, D. L. 1970. *Water pollution control.* Austin and New York: Jenkins.

F. W. Dodge Profile. 1994. *Perini goes underground with Chicago's Deep Tunnel project.* New York: McGraw-Hill.

Fair, G. M., Geyer, J. C. and Okun, D. A. 1966. *Water and wastewater engineering, Vol. 1. Water supply and wastewater removal.* New York: John Wiley.

Fairall, J. M. 1956. Correlation of trickling filter data. *Sewage and Industrial Wastes* **28**(9): 1069–1074.

Federal Register. 1991. Secondary treatment regulation. 40 CFR Part 133, July 1 1991, Washington, DC.

Federal Register. 1993. Standards for the use or disposal of sewage sludge; final rules. 40CFR Part 257 *et al.* Part II, EPA, *Federal Register* **58**(32): 9248–9415. Friday, February 19, 1993. Washington, DC.

Galler, W. S. and Gotaas, H. G. 1964. Analysis of biological filter variables. *J. Sanitary Eng. Div., Proc. ASCE* **90**(SA6): 59–79.

Galler, W. S. and Gotaas, H. G. 1966. Optimization analysis for biological filter design. *J. Sanitary Eng. Div., Proc. ASCE* **92**(SA1): 163–182.

Gaudy, A. F. and Kincannon, D. F. (1977). Comparing design models for activated sludge. *Water and Sewage Works* **123**(7): 66–77.

George, D. B. *et al.* 2000. Development guidelines and design equations for subsurface flow constructed wetlands treating municipal wastewater. Office of Research and Development, Cincinnati, Ohio: US EPA.

Germain, J. E. 1966. Economical treatment of domestic waste by plastic medium trickling filters. *J. Water Pollut. Control Fed.* **38**(2): 192–203.

Grady, C. P. L. and Lim, H. C. 1980. *Biological wastewater treatment: theory and application.* New York: Marcel Dekker.

Great Lakes–Upper Mississippi River Board (GLUMRB) of State Sanitary Engineers, Health Education Service. 1971. *Recommended (Ten States) standards for sewage works.* Albany, New York: Health Research, Inc.

Great Lakes-Upper Mississippi River Board of State Public Health and Environmental Managers. 1971 *Recommended standards for wastewater facilities.* Albany, New York: Health Research, Inc.

Great Lakes–Upper Mississippi River Board of State and Provincial Public Health and Environmental Managers. 1996. *Revision 5 (draft edition). Recommended (Ten States) standards for wastewater facilities: policies for the design, review, and approval of plans and specifications for wastewater collection and treatment facilities.* Albany, New York: Health Research, Inc.

Guo, P. H. M., Thirumurthi, D. and Jank, B. E. 1981. Evaluation of extended aeration activated sludge package plants. *J. Water Pollut. Control Fed.* **53**(1): 33–42.

Hammer, M. J. 1986. *Water and waste-water technology.* New York: John Wiley.

Haug, R. T. and McCarty, P. L. 1972. Nitrification with submerged niters. *J. Water Pollut. Control Fed.* **44**(11): 2086–2102.

Herzbrun, R. A., Irvine, R. L. and Malinowski, K. C. 1985. Biological treatment of hazardous waste in sequencing batch reactors. *J. Water Pollut. Control Fed.* **57**(12): 1163–1167.

Hitdlebaugh, J. A. and Miller, R. D. 1980. Full-scale rotating biological contactor for secondary treatment and nitrification. *Proc. FNSWRBCT,* Vol. I, p. 269.

Hoag, G., Widmer, W. and Hovey, W. 1980. Microfauna and RBC performance: laboratory and full-scale system. *Proc. FNSWRBCT,* Vol. I, p. 167.

Huang, C. S. and Hopson, N. 1974. Nitrification rate in biological processes. *J. Environ. Eng. Div., Proc. ASCE* **100**(EE2): 409.

Illinois Environmental Protection Agency. 1998. *Recommended standards for sewage work.* Part 370 of Chapter II, EPA, Subtitle C: Water Pollution, Title 35: Environmental Protection, Springfield, Illinois: IEPA.

Illinois Environmental Protection Agency. 1998. *Illinois recommended standards for sewage works.* State of Illinois Rules and Regulations Title 35, Subtitle C, Chapter II, parts 370. Springfield: Illinois EPA.

Institute of Water and Environmental Management (IWEM). 1988. *Unit process biological filtration: manuals of British practice in water pollution.* London: IWEM.

Jenkins, D. and Garrison, W. E. 1968. Control of activated sludge by mean cell residence time. *J. Water Pollut. Control Fed.* **40**(11): 1905–1919.

James A. Montgomery, Consulting Engineers, Inc. 1985. *Water Treatment Principles and Design.* New York: John Wiley.

Kadlecik, L. 1996. Organic content of wetland soils, Arcata Enhancement Marsh. Special report, ERE Department Wetland Workshop.

Khan, A. N. and Raman, V. 1980. Rotating biological contactor for the treatment of wastewater in India. *Proc. FNSWRBCT,* Vol. I, p. 235.

Kiff, R. J. 1972. The ecology of nitrincation/denitrification systems in activated sludge. *Water Pollut. Control* **71**: 475.

Kincannon, D. F. and Stover, E. L. 1982. Design methodology for fixed film reactors, RBCs and trickling niters. *Civil Eng. Practicing and Design Eng.* **2**: 107.

Kink, B. 1997. Touring the deep tunnel. *The Regional News,* Jan. 9.

Kinner, N. E., Balkwill, D. L. and Bishop, P. L. 1982. The microbiology of rotating biological contactor films. *Proc. FICFFBP,* Vol. I, p. 184.

Kirschmer, O. 1926. *Untersuchungen uber den Gefallsverlust an rechen.* Trans. Hydraulic Inst. 21, Munich: R. Oldenbourg.

Kraus, L. S. 1955. Dual aeration as rugged activated sludge process. *Sewage and Industrial Wastes,* **27**(12): 1347–1355.

Lawrence, A. W. and P. L. McCarty. 1969. Kinetics of methane fermentation in anaerobic treatment. *J. Water Pollut. Control Fed.* **41**(2): R1–R17.

Lawrence, A. W. and McCarty, P. L. 1970. Unified basis for biological treatment design and operation. *J. Sanitary Eng. Div., Proc. ASCE* **96**(SA3): 757–778.

Lin, S. D., Evans, R. L. and Dawson, W. 1982. RBC for BOD and ammonia nitrogen removals at Princeton wastewater treatment plant. *Proc. FICFFBP,* Vol. I, p. 590.

Livingston, E. H. 1995. Infiltration practices: The good, the bed, and the ugly. In: *National Conference on Urban Runoff Management: Enhancing urban watershed management at the local, county, and state levels,* March 30–April 2, Chicago, Illinois, EPA.625/R-95/003, pp. 352–362. Cincinnati: US EPA.

Logan, B. E., Hermanowicz, S. W. and Parker, D. S. 1987a. Engineering implication of a new trickling filter model. *J. Water Pollut. Control Fed.* **59**(12): 1017–1028.

Logan, B. E., Hermanowicz, S. W. and Parker, D. S. 1987b. A fundamental model for trickling filter process design. *J. Water Pollut. Control Fed.* **59**(12): 1029–1042.

Lyco Division of Remsco Assoc. 1982. *Lyco Wastewater Products–RBC Systems.* Marlbor, New Jersey: Lyco Corp.

Lynch, J. M. and Poole, N. J. 1979. *Microbial ecology: a conceptual approach.* New York: John Wiley.

Mancini, J. L. and Barnhart, E. L. 1968. Industrial waste treatment in aerated lagoons. In: Gloyna, E. F. and Eckenfelder, Jr., W. W. (eds.), *Advances in water quality improvement.* Austin: University of Texas Press.

Marske, D. M. and Boyle, J. D. 1973. Chlorine contact chamber design: a field evaluation, *Water and Sewage Works* **120**(1): 70–77.

Marston, A. 1930. The theory of external loads on closed conduits in the light of latest experiments. Bulletin 96, Iowa Engineering Experimental Station, Iowa City.

McCarty, P. L. 1964. Anaerobic waste treatment fundamentals. *Public Works* **95**(9): 107–112.

McCarty, P. L., Beck, L. and St. Amant, P. 1969. Biological denitrification of waste-waters by addition of organic materials. In: *Proc. of the 24th Industrial Waste Conference,* May 6–8, 1969. Lafayette, Indiana: Purdue University.

McGhee, T. J. 1991. *Water supply and sewerage,* 6th edn. New York: McGraw-Hill.

McKinney, R. E. 1962. Mathematics of complete mixing activated sludge. *J. Sanitary Eng. Div., Proc. ASCE* **88**(SA3).

Metcalf and Eddy, Inc. 1970. *Stormwater management model,* Vol. I: *Final report.* Water Pollution Control Research Series 1124 D0C07/71, US EPA.

Metcalf and Eddy, Inc. 1981. *Wastewater engineering: collection and pumping.* New York: McGraw-Hill.

Metcalf and Eddy, Inc. 1991. *Wastewater engineering treatment, disposal, and reuse.* New York: McGraw-Hill.

Miller, R. D. *et al.* 1980. Rotating biological contactor process for secondary treatment and nitrification following a trickling filter. *Proc. FNSWRBCT,* Vol. II, p. 1035.

Monod, J. 1949. The growth of bacterial cultures. *Ann. Rev. Microbiol.* **3**: 371.

Mueller, J. A., Paquin, P. L. and Famularo, J. 1980. Nitrification in rotating biological contactor. *J. Water Pollut. Control Fed.* **52**(4): 688–710.

National Research Council. 1946. Sewage treatment in military installations, Chapter V: Trickling filter. *Sewage Works J.* **18**(5): 897–982.

Nelson, M. D. and Guarino, C. F. 1977. New “Philadelphia story” being written by Pollution Control Division. *Water and Waste Eng.* **14**: 9–22.

New York State Department of Health. 1950. *Manual of instruction for sewage treatment plant operators.* Albany, New York: NY State Department of Health.

Northeastern Illinois Planning Commission. (1992). *Stormwater detention for water quality benefits.* Chicago: NIPC.

Novotny, V. and Chesters, G. 1981. *Handbook of nonpoint pollution: sources and management.* New York: Van Nostrand-Reinhold.

Novotny, V. *et al.* 1989. *Handbook of urban drainage and wastewater disposal.* New York: John Wiley.

Okun, D. A. 1949. A system of bioprecipitation of organic matter from activated sludge. *Sewage Works J.* **21**(5): 763–792.

Opatken, E. J. 1982. Rotating biological contactors: second order kinetics. *Proc. FICFFBP,* Vol. I, p. 210.

Painter, H. A. 1970. A review of literature on inorganic nitrogen metabolism. *Water Res.* **4**: 393.

Painter, H. A. 1975. Microbial transformations of inorganic nitrogen. In: *Proc. Conf. on Nitrogen as Wastewater Pollutant,* Copenhagen, Denmark.

Parcher, M. J. 1988. *Wastewater collection system maintenance.* Lancaster, Pennsylvania: Technomic.

Pasveer, A. 1960. New developments in the application of Kesener brushes (aeration rotors) in the activated sludge treatment of trade waste waters. In: *Water treatment: Proc. Second Symposium on the Treatment of Waste Waters,* ed. P. C. G. Issac. New York: Pergamon Press.

Perry, R. H. 1967. *Engineering manual: A practical reference of data and methods in architectural, chemical, civil, electrical, mechanical, and nuclear engineering.* New York: McGraw-Hill Book Co.

Pescod, M. B. and Nair, J. V. 1972. Biological disc filtration for tropical waste treatment, experimental studies. *Water Res.* **61**: 1509.

Pitt, R. and Voorhees, J. 1995. Source loading and management model (SLAMM). In: *National Conference on Urban Runoff Management,* pp. 225–243, EPA/625/R-95/003. Cincinnati: US EPA.

Poon, C. P. C, Chao, Y. L. and Mikucki, W. J. 1979. Factors controlling rotating biological contactor performance. *J. Water Pollut. Control Fed.* **51**: 601.

Pretorius, W. A. 1971. Some operating characteristics of a bacteria disc unit. *Water Res.* **5**: 1141.

Qasim, S. R. 1985. *Wastewater treatment plant – plan, design, and operation.* New York: Holt Rinehart & Winston.

Ramanathan, M. and Gaudy, A. F. 1971. Steady state model for activated sludge with constant recycle sludge concentration. *Biotechnol. Bioeng.* **13**: 125.

Recht, H. L. and Ghassemi, M. 1970. Kinetics and mechanism of precipitation and nature of the precipitate obtained in phosphate removal from wastewater using aluminum(III) and iron(III) salts. Water Pollution Control Series 17010 EKI, Contract 14-12-158. Washington, DC: US Department of the Interior.

Reed, S. C. Crites, R. W. and Middlebrooks, E. J. 1995. *Natural systems for waste management and treatment*, 2nd edn. New York: McGraw-Hill.

Rich, L. G. 1961. *Unit operations of sanitary engineering.* New York: John Wiley.

Roberts, P. V. *et al.* 1980. Chlorine dioxide for wastewater disinfection: a feasibility evaluation, Technical Report 251. San Jose: Stanford University.

Robison, R. 1986. The tunnel that cleans up Chicago. *Civil Eng.* **56**(7): 34–37.

Schroeder, E. D. 1977. *Water and wastewater treatment.* New York: McGraw-Hill.

Schultz, K. L. 1960. Load and efficiency of trickling filters. *J. Water Pollut. Control Fed.* **32**(3): 245–261.

Sepp, E. 1977. *Tracer evaluation of chlorine contact tanks.* Berkeley: California State Department of Health.

Sepp, E. 1981. Optimization of chlorine disinfection efficiency. *J. Environ. Eng. Div., Proc. ASCE* **107**(EE1): 139–153.

Sherard, J. H. 1976. Destruction of alkalinity in aerobic biological wastewater treatment. *J. Water Pollut. Control Fed.* **48**(7): 1834–1839.

Shoemaker, L. L. *et al.* 1995. Watershed screening and targeting tool (WSTT). In: *National Conference on Urban Runoff Management,* pp. 250–258. EPA/625/R-95/ 003. Cincinnati: US EPA.

Singhal, A. K. 1980. Phosphorus and nitrogen removal at Cadillac, Michigan. *J. Water Pollut. Control Fed.* **52**(11): 2761–2770.

Sommers, L., Parker C. and Meyers, G. 1981. Volatilization, plant uptake and mineralization of nitrogen in soils treated with sewage sludge. Technical Report 133, Water Resources Research Center. West Lafayette, Indiana: Purdue University.

Stanier, R. Y., Doudoroff, M. and Adelberg, E. A. 1963. *The Microbial World.* 2nd edn. Englewood Cliffs, New Jersey: Prentice-Hall.

Steel, E. W. and McGhee, T. J. 1979. *Water supply and sewerage,* 5th edn. New York: McGraw-Hill.

Stukenberg, J. R. *et al.* 1990. Egg-shaped digester: from Germany to the United States. Presented at the 63rd Annual Conference of Water Pollution Control Federation, Washington, DC, October 7–11, 1990.

Sudo, R., Okad, M. and Mori, T. 1977. Rotating biological contactor microbial control in RBC. *J. Water and Waste,* **19**: 1.

Techobanoglous, G. and Schroeder, E. D. 1985. *Water quality.* Reading, Massachusetts: Addison-Wesley.

Terstriep, M. L. and Lee, M. T. 1995. AUTO-QI: an urban runoff quality I quantity model with a GIS interface. In: *National Conference on Urban Runoff Management,* pp. 213–224. EPA/625/R-95/003. Cincinnati: US EPA.

Theroux, R. J. and Betz, J. M. 1959. Sedimentation and preparation experiments in Los Angeles. *Sewage and Industrial Wastes* **31**(11): 1259–1266.

Thirumurthi, D. 1969. Design principles of waste stabilization ponds. *J. Sanitary Eng. Div., Proc. ASCE* **95**(SA2): 311–330.

Torpey, W. N. 1971. Rotating disks with biological growths prepare wastewater for disposal as reuse. *J. Water Pollut. Control Fed.* **43**(11): 2181–2188.

Truong, H. V. and Phua, M. S. 1995. Application of the Washington, DC, sand filter for urban runoff control. In: *National Conference on Urban Runoff Management,* pp. 375–383, EPA/625/R-95/003. Cincinnati: US EPA.

US Environmental Protection Agency. 1974a. *Process design manual for upgrading existing wastewater treatment plants.* Technology Transfer, EPA 625/l-71-004a. Washington, DC: US EPA.

US Environmental Protection Agency. 1974b. *Wastewater treatment ponds.* EPA-430/9-74-001, MCD-14. Washington, DC: US EPA.

US Environmental Protection Agency. 1974c. *Water quality management planning for urban runoff.* EPA 440/9-75-004. Washington, DC: US EPA.

US Environmental Protection Agency. 1975a. *Process design manual for suspended solids removal.* EPA 625/l-75-003a. Washington, DC: US EPA.

US Environmental Protection Agency. 1975b. *Process design manual for nitrogen control.* Office of Technical Transfer. Washington, DC: US EPA.

US Environmental Protection Agency. 1975c. *Process design manual for nitrogen control.* EPA-625/1-75-007, Center for Environmental Research. Cincinnati: US EPA.

US Environmental Protection Agency. 1976. *Process design manual for phosphorus removal.* EPA 625/1-76-00/a. Washington, DC: US EPA.

US EPA. 1979. *Process design manual—Sludge treatment and disposal.* EPA 625/1-79-011, Cincinnati, Ohio: US EPA.

US Environmental Protection Agency. 1980. *Converting rock trickling filters to plastic media: design and performance.* EPA-600/2-80-120, Cincinnati: US EPA.

US Environmental Protection Agency. 1983a. *National urban runoff program,* Vol. I. NTIS PB 84-185552. Washington, DC: US EPA.

US Environmental Protection Agency. 1983b. *Municipal wastewater stabilization ponds: design manual.* EPA-625/1-83-015. Cincinnati: US EPA.

US Environmental Protection Agency. 1984. *Design information on rotating biological contactors.* EPA-600/2-84-106. Cincinnati: US EPA.

US Environmental Protection Agency. 1984a. *The ecological impacts of wastewater on wetlands—an annotated bibliography.* EPA-905/9-84-002. Chicago, Illinois: US EPA.

US Environmental Protection Agency. 1984b. *Technical report–Literature review of wetland evaluation methodologies.* Chicago, Illinois: US EPA.

US Environmental Protection Agency. 1987a. *Preliminary treatment facilities, design and operational considerations.* EPA-430/09-87-007. Washington, DC: US EPA.

US Environmental Protection Agency. 1987b. *Design manual for dewatering municipal wastewater sludges.* Washington, DC: US EPA.

US Environmental Protection Agency. 1988. *Design manual—Constructed wetlands and aquatic plant systems for municipal wastewater treatment.* EPA/625/1-88/022. Washington, DC: US EPA.

US EPA. 1989. *Design manual—Dewatering municipal wastewater sludges,* EPA 625/1-79-011, Cincinnati, Ohio: US EPA.

US Environmental Protection Agency. 1991. *Evaluating sludge treatment processes.* Washington, DC: US EPA.

US Environmental Protection Agency. 1993. *The effects of wastewater treatment facilities on wetlands in the Midwest.* EPA/905/3-83-002. Chicago, Illinois: US EPA.

US Environmental Protection Agency. 1993. *Nitrogen control.* EPA/625/R-93/010. Washington, DC: US EPA.

US Environmental Protection Agency. 1994. *A plain English guide to the EPA part 503 biosolids rule.* EPA/832/R-93/003. Washington, DC: US EPA.

US Environmental Protection Agency. 1994. *Considering wetlands at CERCLA sites.* EPA/540/R-94-019. Washington, DC: US EPA.

US Environmental Protection Agency. 1995. *Process design manual: land application of sewage sludge and domestic septage.* EPA/625/R-95/001. Washington, DC: US EPA.

University of Cincinnati, Department of Civil Engineering. 1970. *Urban runoff characteristics.* Water Pollution Control Research Series 11024DQU10/70. Cincinnati: US EPA.

Urbonas, B. and Stahre, P. 1993. *Stormwater: best management practices and detention for water quality, drainage, and CSO management.* Englewood Cliffs, New Jersey: Prentice-Hall.

Velz, C. J. 1948. A basic law for the performance of biological filters. *Sewage Works J.* **20**(4): 607–617.

Viessman, Jr., W. and Hammer, M. J. 1993. *Water supply and pollution control,* 5th edn. New York: Harper Collins.

Wanielista, M. 1990. *Hydrology and water quality control.* New York: John Wiley.

Wanielista, M. 1992. Stormwater reuse: an alternative method of infiltration. In: *National Conference on Urban Runoff Management,* pp. 363–371, EPA/625/R-95/003. Cincinnati: US EPA.

Wanielista, M. P. and Yousef, Y. A. 1993. *Stormwater management.* New York: John Wiley.

Water Environment Federation (WEF) and American Society of Civil Engineers (ASCE). 1991a. *Design of municipal wastewater treatment plants,* Vol. I. Alexandria, Virginia: WEF.

Water Environment Federation (WEF) and American Society of Civil Engineers (ASCE). 1991b. *Design of municipal wastewater treatment plants,* Vol. II. Alexandria, Virginia: WEF.

Water Environment Federation (WEF) and American Society of Civil Engineers (ASCE). 1992. *Design and construction of urban stormwater management systems.*

Water Environment Federation. 1993a. *Design of wastewater and stormwater pumping stations.* Alexandria, Virginia: WEF.

Water Environment Federation. 1993b. *Standards for the use and disposal of sewage sludge (40 CFR Part 257, 403, and 503) final rule and phase-in submission of sewage sludge permit application (Revisions to 40CFR Parts 122, 123, and 501) final rule.* Alexandria, Virginia: WEF.

Water Environment Federation. 1994. *Beneficial use programs for biosolids management.* Alexandria, Virginia: WEF.

Water Environment Federation and American Society of Civil Engineers. 1996a. *Operation of municipal wastewater treatment plants,* 5th edn., Vol. II. Alexandria, Virginia: WEF.

Water Environment Federation and American Society of Civil Engineers. 1996b. *Operation of municipal wastewater treatment plants,* 5th edn., Vol. III. Alexandria, Virginia: WEF.

Water Pollution Control Federation (WPCF). 1980. *Clean water for today: what is wastewater treatment.* Washington, DC: WPCF.

Water Pollution Control Federation (WPCF). 1983. *Sludge dewatering,* Manual of Practice No. 20. Alexandria, Virginia: WPCF.

Water Pollution Control Federation (WPCF). 1985a. *Clarifier design.* Manual of Practice FD-8. Alexandria, Virginia: WPCF.

Water Pollution Control Federation (WPCF). 1985b. *Sludge stabilization.* Manual of Practice FD-9. Alexandria, Virginia: WPCF.

Water Pollution Control Federation (WPCF). 1988a. *Sludge conditioning.* Manual of Practice FD-14. Alexandria, Virginia: WPCF.

Water Pollution Control Federation (WPCF). 1988b. *Operation and maintenance of trickling filters, RBCs and related processes.* Manual of Practice OM-10. Alexandria, Virginia: WPCF.

Water Pollution Control Federation. 1990. *Natural systems for wastewater treatment.* Manual for Practice No. FD-16. Alexandria, Virginia: WPCF.

Wehner, J. F. and Wilhelm, R. H. 1958. Boundary conditions of flow reactor. *Chem. Eng. Sci.* **6**(1): 89–93.

White, G. C. 1986. *Handbook of chlorination,* 2nd edn. New York: Van Nostrand Reinhold.

Wild, A. E., Sawyer, C. E. and McMahon, T. C. 1971. Factors affecting nitrification kinetics. *J. Water Pollut. Control Fed.* **43**(9): 1845–1854.

Wilson, T. E. and Lee. J. S. 1982. Comparison of final clarifier design techniques. *J. Water Pollut. Control Fed.* **54**(10): 1376–1381.

Appendix

# A

# Illinois Environmental Protection Agency's Macroinvertebrate Tolerance List

| Macroinvertebrate | Tolerance value |
|---|---|
| PLATYHELMINTHES | |
| TURBELLARIA | 6 |
| ANNELIDA | |
| OLIGOCHAETA | 10 |
| HIRUDINEA | 8 |
| Rhynchobdellida | |
| Glossiphoniidae | 8 |
| Piscicolidae | 7 |
| Gnathobdellida | |
| Hirudinidae | 7 |
| Pharyngobdellida | |
| Erpobdellidae | 8 |
| ARTHROPODA | |
| CRUSTACEA | |
| ISOPODA | |
| Asellidae | 6 |
| *Caecidotea* | 6 |
| *brevicauda* | 6 |
| *intermedia* | 6 |
| *Lirceus* | 4 |
| AMPHIPODA | |
| Hyalellidae | |
| *Hyalella* | |
| *azteca* | 5 |
| Gammaridae | |
| *Bactrurus* | 1 |
| *Crangonyx* | 4 |
| *Gammarus* | 3 |
| DECAPODA | |
| Cambaridae | 5 |
| Palaemonidae | |
| *Palaemonetes* | 4 |
| INSECTA | |
| EPHEMEROPTERA | |
| Siphlonuridae | |
| *Ameletus* | 0 |
| *Siphlonurus* | 2 |
| Oligoneuriidae | |
| Heptageniidae | |
| *Arthroplea* | 3 |
| *Epeorus* | 1 |
| *vitreus* | 0 |
| *Heptagenia* | 3 |
| *diabasia* | 4 |
| *flavescens* | 2 |
| *hebe* | 3 |
| *lucidipennis* | 3 |
| *maculipennis* | 3 |
| *marginalis* | 1 |
| *perfida* | 1 |
| *pulla* | 0 |
| *Rhithrogena* | 0 |

| Macroinvertebrate | Tolerance value | Macroinvertebrate | Tolerance value |
|---|---|---|---|
| *Stenacron* | 4 | Polymitarcyidae | |
| *candidum* | 1 | *Ephoron* | 2 |
| *gildersleevei* | 1 | *Tortopus* | 4 |
| *interpunctatum* | 4 | ODONATA | |
| *minnetonka* | 4 | ANISOPTERA | |
| *Stenonema* | 4 | Cordulegasteridae | |
| *annexum* | 4 | *Cordulegaster* | 2 |
| *ares* | 3 | Gomphidae | |
| *exiguum* | 5 | *Dromogomphus* | 4 |
| *femoratum* | 7 | *Gomphus* | 7 |
| *integrum* | 4 | *Hagenius* | 3 |
| *luteum* | 1 | *Lanthus* | 6 |
| *mediopunctatum* | 2 | *Ophiogomphus* | 2 |
| *modestum* | 3 | *Progomphus* | 5 |
| *nepotellum* | 5 | Aeshnidae | |
| *pudicum* | 2 | *Aeshna* | 4 |
| *pulchellum* | 3 | *Anax* | 5 |
| *quinquespinum* | 5 | *Basiaeschna* | 2 |
| *rubromaculatum* | 2 | *Boyeria* | 3 |
| *scitulum* | 1 | *Epiaeschna* | 1 |
| *terminatum* | 4 | *Nasiaeschna* | 2 |
| *vicarium* | 3 | Macromiidae | |
| Ephemerellidae | | *Didymops* | 4 |
| *Attenella* | 2 | *Macromia* | 3 |
| *Danella* | 2 | *Drunella* | 1 |
| *Isonychia* | 3 | *Ephemerella* | 2 |
| Metretopodidae | | *Eurylophella* | 4 |
| *Siphloplecton* | 2 | *Seratella* | 1 |
| Baetidae | | Tricorythidae | |
| *Baetis* | 4 | *Tricorythodes* | 5 |
| *brunneicolor* | 4 | Caenidae | |
| *flavistriga* | 4 | *Brachycercus* | 3 |
| *frondalis* | 4 | *Caenis* | 6 |
| *intercalaris* | 7 | Baetiscidae | |
| *longipalpus* | 6 | *Baetisca* | 3 |
| *macdunnoughi* | 4 | Leptophlebiidae | |
| *propinquus* | 4 | *Chloroterpes* | 2 |
| *pygmaeus* | 4 | *Habrophlebiodes* | 2 |
| *tricaudatus* | 1 | *americana* | 2 |
| *Callibaetis* | 4 | *Leptophlebia* | 3 |
| *fluctuans* | 4 | *Paraleptophlebia* | 2 |
| *Centroptilum* | 2 | Potamanthidae | |
| *Cloeon* | 3 | *Potamanthus* | 4 |
| *Pseudocloeon* | 4 | Ephemeridae | |
| *dubium* | 4 | *Ephemera* | 3 |
| *parvulum* | 4 | *simulans* | 3 |
| *punctiventris* | 4 | PLECOPTERA | |
| *Hexagenia* | 6 | Pteronarcyidae | |
| *limbata* | 5 | *Pteronarcys* | 2 |
| *munda* | 7 | Taeniopterygidae | |
| Palingeniidae | | *Taeniopteryx* | 2 |
| *Pentagenia* | 4 | Nemoundae | |
| *vittigera* | 4 | *Nemoura* | 1 |

| Macroinvertebrate | Tolerance value |
|---|---|
| Leuctridae | |
| *Leuctra* | 1 |
| Capniidae | |
| *Allocapnia* | 2 |
| *Capnia* | 1 |
| Perlidae | |
| *Acroneuria* | 1 |
| *Atoperia* | 1 |
| *Neoperia* | 1 |
| *Perlesta* | 4 |
| *placida* | 4 |
| *Perlinella* | 2 |
| Periodidae | |
| *Hydroperia* | 1 |
| *Isoperia* | 2 |
| Chloroperiidae | |
| *Chloroperia* | 3 |
| MEGALOPTERA | |
| Sialidae | |
| *Sialias* | 4 |
| Corydalidae | |
| *Chauliodes* | 4 |
| *Corydalus* | 3 |
| Corduliidae | |
| *Cordulia* | 2 |
| *Epitheca* | 4 |
| *Helocordulia* | 2 |
| *Neurocordulia* | 3 |
| *Somatochlora* | 1 |
| Libellulidae | |
| *Celithemis* | 2 |
| *Erythemis* | 5 |
| *Erythrodiplax* | 5 |
| *Libellula* | 8 |
| *Pachydiplax* | 8 |
| *Pantala* | 7 |
| *Perithemis* | 4 |
| *Plathemis* | 3 |
| *Sympetrum* | 4 |
| *Tramea* | 4 |
| ZYGOPTERA | |
| Calopterygidae | |
| *Calopteryx* | 4 |
| *Hetaerina* | 3 |
| Lestidae | |
| *Archilestes* | 1 |
| *Lestes* | 6 |
| Coenagrionidae | |
| *Amphiagrion* | 5 |
| *Argia* | 5 |
| *moesta* | 5 |
| *tibialis* | 5 |
| *Enallagma* | 6 |
| *signatum* | 6 |
| *Ischnura* | 6 |
| *Nehalennia* | 7 |
| Hydroptilidae | |
| *Agraylea* | 2 |
| *Hydroptila* | 2 |
| *Ithytrichia* | 1 |
| *Leucotrichia* | 3 |
| *Mayatrichia* | 1 |
| *Neotrichia* | 4 |
| *Ochrotrichia* | 4 |
| *Orthotrichia* | 1 |
| *Oxyethira* | 2 |
| Rhyacophilidae | |
| *Rhyacophila* | 1 |
| Brachycentridae | |
| *Brachycentrus* | 1 |
| Lepidostomatidae | |
| *Lepidostoma* | 3 |
| Limnephilidae | |
| *Hydatophylax* | 2 |
| *Limnephilus* | 3 |
| *Neophylax* | 3 |
| *Nigronia* | 2 |
| NEUROPTERA | |
| Sisyridae | 1 |
| TRICHOPTERA | |
| Hydropsychidae | |
| *Cheumatopsyche* | 6 |
| *Diplectrona* | 2 |
| *Hydropsyche* | 5 |
| *arinale* | 5 |
| *betteni* | 5 |
| *bidens* | 5 |
| *cuanis* | 5 |
| *frisoni* | 5 |
| *orris* | 4 |
| *phalerata* | 2 |
| *placoda* | 4 |
| *simulans* | 5 |
| *Macronema* | 2 |
| *Potamyia* | 4 |
| *Symphitopsyche* | 4 |
| Philopotamidae | |
| *Chimarra* | 3 |
| *Dolophilodes* | 0 |
| Polycentropodidae | |
| *Cyrnellus* | 5 |
| *Neureclipsis* | 3 |
| *Nyctiophylax* | 1 |
| *Polycentropus* | 3 |
| Psychomyiidae | |
| *Psychomyia* | 2 |

| Macroinvertebrate | Tolerance value |
|---|---|
| Glossosomatidae | |
| *Agapetus* | 2 |
| *Protoptila* | 1 |
| DIPTERA | |
| Blephariceridae | 0 |
| Tipulidae | 4 |
| *Antocha* | 5 |
| *Dicranota* | 4 |
| *Eriocera* | 7 |
| *Helius* | 2 |
| *Hesperoconopa* | 2 |
| *Hexatoma* | 4 |
| *Limnophila* | 4 |
| *Limonia* | 3 |
| *Liriope* | 7 |
| *Pedicia* | 4 |
| *Pilaria* | 4 |
| *Polymeda* | 2 |
| *Pseudolimnophila* | 2 |
| *Tipula* | 4 |
| Chaoboridae | 8 |
| Culicidae | 8 |
| *Aedes* | 8 |
| *Platycentropus* | 3 |
| *Pycnopsyche* | 3 |
| Phryganeidae | |
| *Agrypnia* | 3 |
| *Banksiola* | 2 |
| *Phryganea* | 3 |
| *Ptilostomis* | 3 |
| Helicopsychidae | |
| *Helicopsyche* | 2 |
| Leptoceridae | |
| *Ceraclea* | 3 |
| *Leptocerus* | 3 |
| *Mystacides* | 2 |
| *Nectopsyche* | 3 |
| *Oecetis* | 5 |
| *Triaenodes* | 3 |
| COLEOPTERA | |
| Gyrinidae (larvae only) | |
| *Dineutus* | 4 |
| *Gyrinus* | 4 |
| Psephenidae (larvae only) | 4 |
| *Psephenus* | 4 |
| *herricki* | 4 |
| Eubriidae | 4 |
| *Ectopria* | 4 |
| *thoracica* | 4 |
| Dryopidae | 4 |
| *Helichus* | 4 |
| *lithophilus* | 4 |
| Helodidae (larvae only) | 7 |
| Elmidae | |
| *Ancyronyx* | 2 |
| *variegatus* | 2 |
| *Dubiraphia* | 5 |
| *bivittata* | 2 |
| *quadrinotata* | 7 |
| *vittata* | 7 |
| *Macronychus* | 2 |
| *glabratus* | 2 |
| *Microcylloepus* | 2 |
| *Optioservus* | 4 |
| *ovalis* | 4 |
| *Stenelmis* | 7 |
| *crenata* | 7 |
| *vittipennis* | 6 |
| Orthocladiinae | |
| *Cardiocladius* | 6 |
| *Chaetocladius* | 6 |
| *Corynoneura* | 2 |
| *Cricotopus* | 8 |
| *bicinctus* | 10 |
| *trifasciatus* | 6 |
| *Eukiefferiella* | 4 |
| *Anopheles* | 6 |
| *Culex* | 8 |
| Psychodidae | 11 |
| Ceratopogonidae | 5 |
| *Atrichopogon* | 2 |
| *Palpomyia* | 6 |
| Simuliidae | |
| *Cnephia* | 4 |
| *Prosimulium* | 2 |
| *Simulium* | 6 |
| *clarkei* | 4 |
| *corbis* | 0 |
| *decorum* | 4 |
| *jenningsi* | 4 |
| *luggeri* | 2 |
| *meridionale* | 1 |
| *tuberosum* | 4 |
| *venustum* | 6 |
| *verecundum* | 6 |
| *vittatum* | 8 |
| Chironomidae | |
| Tanypodinae | |
| *Ablabesmyia* | 6 |
| *mallochi* | 6 |
| *parajanta* | 6 |
| *peleensis* | 6 |
| *Clinotanypus* | 6 |
| *pinguis* | 6 |
| *Coelotanypus* | 4 |
| *Labrundinia* | 4 |

| Macroinvertebrate | Tolerance value | Macroinvertebrate | Tolerance value |
|---|---|---|---|
| *Larsia* | 6 | *illinoense* | 5 |
| *Macropelopia* | 7 | *scalaenum* | 6 |
| *Natarsia* | 6 | *Pseudochironomus* | 5 |
| *Pentaneura* | 3 | *Stenochironomus* | 3 |
| *Procladius* | 8 | *Stictochironomus* | 5 |
| *Psectrotanypus* | 8 | *Tribelos* | 5 |
| *Tanypus* | 8 | *Xenochironomus* | 4 |
| *Thienemannimyia group* | 6 | Tanytarsini | |
| *Zavrelimyia* | 8 | *Cladotanytarsus* | 7 |
| Diamesinae | | *Micropsectra* | 4 |
| *Diamesa* | 1 | *Rheotanytarsus* | 6 |
| *Pseudodiamesa* | 1 | *Tanytarsus* | 7 |
| Syrphidae | 11 | Ptychopteridae | 8 |
| Ephydridae | 8 | Tabanidae | |
| Sciomyzidae | 10 | *Chrysops* | 7 |
| Muscidae | 8 | *Tabanus* | 7 |
| Athencidae | | Dolichopodidae | 5 |
| *Atherix* | 4 | Empididae | 6 |
| MOLLUSCA | | *Hemerodromia* | 6 |
| GASTROPODA | | Viviparidae | |
| *Hydrobaenus* | 2 | *Campeloma* | 7 |
| *Nanocladius* | 3 | *Lioplax* | 7 |
| *Orthocladius* | 4 | *Viviparus* | 1 |
| *Parametriocnemus* | 4 | Valvatidae | |
| *Prodiamesa* | 3 | *Valvata* | 2 |
| *Psectrocladius* | 5 | Bulimidae | |
| *Rheocricotopus* | 6 | *Amnicola* | 4 |
| *Thienemaniella* | 2 | Pleurocandae | |
| *xena* | 2 | *Goniobasis* | 5 |
| Chironominae | | *Pleurocera* | 7 |
| *Chironomus* | 11 | Physidae | |
| *attenuatus* | 10 | *Aplexa* | 7 |
| *riparius* | 11 | *Physa* | 9 |
| *Cryptochironomus* | 8 | Lymnaeidae | |
| *Cryptotendipes* | 6 | *Lymnara* | 7 |
| *Dicrotendipes* | 6 | *Stagnicola* | 7 |
| *modestus* | 6 | Planorbidae | |
| *neomodestus* | 6 | *Gyraulus* | 6 |
| *nervosus* | 6 | *Helisoma* | 7 |
| *Einfeldia* | 10 | *Planorbula* | 7 |
| *Endochironomus* | 6 | Ancylidae | |
| *Glyptotendipes* | 10 | *Ferrissia* | 7 |
| *Harnischia* | 6 | PELECYPODA | |
| *Kiefferulus* | 7 | Unionidae | |
| *Microtendipes* | 6 | *Actinonaias* | |
| *Parachironomus* | 8 | *carinata* | 1 |
| *Paracladopelma* | 4 | *Alasmidonta* | |
| *Paralauterborniella* | 6 | *marginata* | 1 |
| *Paratendipes* | 3 | *triangulata* | 0 |
| *Phaenopsectra* | 4 | *Anodonta* | 3 |
| *Polypedilum* | 6 | *Carunculina* | 7 |
| *fallax* | 6 | *Elliptio* | 2 |
| *halterale* | 4 | *Fusronaia* | 1 |

| Macroinvertebrate | Tolerance value | Macroinvertebrate | Tolerance value |
|---|---|---|---|
| *Lampsilis* | 1 | *Truncilla* | 1 |
| *Ligumia* | 1 | *Utterbackia* | 1 |
| *Margaritifera* | 1 | Sphaeridae | |
| *Micromya* | 1 | *Musculium* | 5 |
| *Obliquaria* | 1 | *Pisidium* | 5 |
| *Proplera* | 1 | *Sphaeriuni* | 5 |
| *Strophitus* | 4 | Cyrenidae | |
| *Tritagonia* | 1 | *Corbicula* | 4 |

SOURCE: Illinois Environment Protection Agency, 1987

# Appendix B

# Well Function for Confined Aquifers

| $10^{-10}$ well functions | | | | | | | |
|---|---|---|---|---|---|---|---|
| $u$ | $W(u)$ | $u$ | $W(u)$ | $u$ | $W(u)$ | $u$ | $W(u)$ |
| 1.0E − 10 | 22.45 | 3.3E − 10 | 21.25 | 5.6E − 10 | 20.73 | 7.9E − 10 | 20.38 |
| 1.1E − 10 | 22.35 | 3.4E − 10 | 21.22 | 5.7E − 10 | 20.71 | 8.0E − 10 | 20.37 |
| 1.2E − 10 | 22.27 | 3.5E − 10 | 21.20 | 5.8E − 10 | 20.69 | 8.1E − 10 | 20.36 |
| 1.3E − 10 | 22.19 | 3.6E − 10 | 21.17 | 5.9E − 10 | 20.67 | 8.2E − 10 | 20.34 |
| 1.4E − 10 | 22.11 | 3.7E − 10 | 21.14 | 6.0E − 10 | 20.66 | 8.3E − 10 | 20.33 |
| 1.5E − 10 | 22.04 | 3.8E − 10 | 21.11 | 6.1E − 10 | 20.64 | 8.4E − 10 | 20.32 |
| 1.6E − 10 | 21.98 | 3.9E − 10 | 21.09 | 6.2E − 10 | 20.62 | 8.5E − 10 | 20.31 |
| 1.7E − 10 | 21.92 | 4.0E − 10 | 21.06 | 6.3E − 10 | 20.61 | 8.6E − 10 | 20.30 |
| 1.8E − 10 | 21.86 | 4.1E − 10 | 21.04 | 6.4E − 10 | 20.59 | 8.7E − 10 | 20.29 |
| 1.9E − 10 | 21.81 | 4.2E − 10 | 21.01 | 6.5E − 10 | 20.58 | 8.8E − 10 | 20.27 |
| 2.0E − 10 | 21.76 | 4.3E − 10 | 20.99 | 6.6E − 10 | 20.56 | 8.9E − 10 | 20.26 |
| 2.1E − 10 | 21.71 | 4.4E − 10 | 20.97 | 6.7E − 10 | 20.55 | 9.0E − 10 | 20.25 |
| 2.2E − 10 | 21.66 | 4.5E − 10 | 20.94 | 6.8E − 10 | 20.53 | 9.1E − 10 | 20.24 |
| 2.3E − 10 | 21.62 | 4.6E − 10 | 20.92 | 6.9E − 10 | 20.52 | 9.2E − 10 | 20.23 |
| 2.4E − 10 | 21.57 | 4.7E − 10 | 20.90 | 7.0E − 10 | 20.50 | 9.3E − 10 | 20.22 |
| 2.5E − 10 | 21.53 | 4.8E − 10 | 20.88 | 7.1E − 10 | 20.49 | 9.4E − 10 | 20.21 |
| 2.6E − 10 | 21.49 | 4.9E − 10 | 20.86 | 7.2E − 10 | 20.47 | 9.5E − 10 | 20.20 |
| 2.7E − 10 | 21.46 | 5.0E − 10 | 20.84 | 7.3E − 10 | 20.46 | 9.6E − 10 | 20.19 |
| 2.8E − 10 | 21.42 | 5.1E − 10 | 20.82 | 7.4E − 10 | 20.45 | 9.7E − 10 | 20.18 |
| 2.9E − 10 | 21.38 | 5.2E − 10 | 20.80 | 7.5E − 10 | 20.43 | 9.8E − 10 | 20.17 |
| 3.0E − 10 | 21.35 | 5.3E − 10 | 20.78 | 7.6E − 10 | 20.42 | 9.9E − 10 | 20.16 |
| 3.1E − 10 | 21.32 | 5.4E − 10 | 20.76 | 7.7E − 10 | 20.41 | | |
| 3.2E − 10 | 21.29 | 5.5E − 10 | 20.74 | 7.8E − 10 | 20.39 | | |

| $10^{-9}$ well functions | | | | | | | |
|---|---|---|---|---|---|---|---|
| *u* | *W*(*u*) | *u* | *W*(*u*) | *u* | *W*(*u*) | *u* | *W*(*u*) |
| 1.0E − 09 | 20.15 | 3.3E − 09 | 18.95 | 5.6E − 09 | 18.42 | 7.9E − 09 | 18.08 |
| 1.1E − 09 | 20.05 | 3.4E − 09 | 18.92 | 5.7E − 09 | 18.41 | 8.0E − 09 | 18.07 |
| 1.2E − 09 | 19.96 | 3.5E − 09 | 18.89 | 5.8E − 09 | 18.39 | 8.1E − 09 | 18.05 |
| 1.3E − 09 | 19.88 | 3.6E − 09 | 18.87 | 5.9E − 09 | 18.37 | 8.2E − 09 | 18.04 |
| 1.4E − 09 | 19.81 | 3.7E − 09 | 18.84 | 6.0E − 09 | 18.35 | 8.3E − 09 | 18.03 |
| 1.5E − 09 | 19.74 | 3.8E − 09 | 18.81 | 6.1E − 09 | 18.34 | 8.4E − 09 | 18.02 |
| 1.6E − 09 | 19.68 | 3.9E − 09 | 18.79 | 6.2E − 09 | 18.32 | 8.5E − 09 | 18.01 |
| 1.7E − 09 | 19.62 | 4.0E − 09 | 18.76 | 6.3E − 09 | 18.31 | 8.6E − 09 | 17.99 |
| 1.8E − 09 | 19.56 | 4.1E − 09 | 18.74 | 6.4E − 09 | 18.29 | 8.7E − 09 | 17.98 |
| 1.9E − 09 | 19.50 | 4.2E − 09 | 18.71 | 6.5E − 09 | 18.27 | 8.8E − 09 | 17.97 |
| 2.0E − 09 | 19.45 | 4.3E − 09 | 18.69 | 6.6E − 09 | 18.26 | 8.9E − 09 | 17.96 |
| 2.1E − 09 | 19.40 | 4.4E − 09 | 18.66 | 6.7E − 09 | 18.24 | 9.0E − 09 | 17.95 |
| 2.2E − 09 | 19.36 | 4.5E − 09 | 18.64 | 6.8E − 09 | 18.23 | 9.1E − 09 | 17.94 |
| 2.3E − 09 | 19.31 | 4.6E − 09 | 18.62 | 6.9E − 09 | 18.21 | 9.2E − 09 | 17.93 |
| 2.4E − 09 | 19.27 | 4.7E − 09 | 18.60 | 7.0E − 09 | 18.20 | 9.3E − 09 | 17.92 |
| 2.5E − 09 | 19.23 | 4.8E − 09 | 18.58 | 7.1E − 09 | 18.19 | 9.4E − 09 | 17.91 |
| 2.6E − 09 | 19.19 | 4.9E − 09 | 18.56 | 7.2E − 09 | 18.17 | 9.5E − 09 | 17.89 |
| 2.7E − 09 | 19.15 | 5.0E − 09 | 18.54 | 7.3E − 09 | 18.16 | 9.6E − 09 | 17.88 |
| 2.8E − 09 | 19.12 | 5.1E − 09 | 18.52 | 7.4E − 09 | 18.14 | 9.7E − 09 | 17.87 |
| 2.9E − 09 | 19.08 | 5.2E − 09 | 18.50 | 7.5E − 09 | 18.13 | 9.8E − 09 | 17.86 |
| 3.0E − 09 | 19.05 | 5.3E − 09 | 18.48 | 7.6E − 09 | 18.12 | 9.9E − 09 | 17.85 |
| 3.1E − 09 | 19.01 | 5.4E − 09 | 18.46 | 7.7E − 09 | 18.10 | | |
| 3.2E − 09 | 18.98 | 5.5E − 09 | 18.44 | 7.8E − 09 | 18.09 | | |

| $10^{-8}$ well functions | | | | | | | |
|---|---|---|---|---|---|---|---|
| *u* | *W*(*u*) | *u* | *W*(*u*) | *u* | *W*(*u*) | *u* | *W*(*u*) |
| 1.0E − 08 | 17.84 | 3.3E − 08 | 16.65 | 5.6E − 08 | 16.12 | 7.9E − 08 | 15.78 |
| 1.1E − 08 | 17.75 | 3.4E − 08 | 16.62 | 5.7E − 08 | 16.10 | 8.0E − 08 | 15.76 |
| 1.2E − 08 | 17.66 | 3.5E − 08 | 16.59 | 5.8E − 08 | 16.09 | 8.1E − 08 | 15.75 |
| 1.3E − 08 | 17.58 | 3.6E − 08 | 16.56 | 5.9E − 08 | 16.07 | 8.2E − 08 | 15.74 |
| 1.4E − 08 | 17.51 | 3.7E − 08 | 16.54 | 6.0E − 08 | 16.05 | 8.3E − 08 | 15.73 |
| 1.5E − 08 | 17.44 | 3.8E − 08 | 16.51 | 6.1E − 08 | 16.04 | 8.4E − 08 | 15.72 |
| 1.6E − 08 | 17.37 | 3.9E − 08 | 16.48 | 6.2E − 08 | 16.02 | 8.5E − 08 | 15.70 |
| 1.7E − 08 | 17.31 | 4.0E − 08 | 16.46 | 6.3E − 08 | 16.00 | 8.6E − 08 | 15.69 |
| 1.8E − 08 | 17.26 | 4.1E − 08 | 16.43 | 6.4E − 08 | 15.99 | 8.7E − 08 | 15.68 |
| 1.9E − 08 | 17.20 | 4.2E − 08 | 16.41 | 6.5E − 08 | 15.97 | 8.8E − 08 | 15.67 |
| 2.0E − 08 | 17.15 | 4.3E − 08 | 16.38 | 6.6E − 08 | 15.96 | 8.9E − 08 | 15.66 |
| 2.1E − 08 | 17.10 | 4.4E − 08 | 16.36 | 6.7E − 08 | 15.94 | 9.0E − 08 | 15.65 |
| 2.2E − 08 | 17.06 | 4.5E − 08 | 16.34 | 6.8E − 08 | 15.93 | 9.1E − 08 | 15.64 |
| 2.3E − 08 | 17.01 | 4.6E − 08 | 16.32 | 6.9E − 08 | 15.91 | 9.2E − 08 | 15.62 |
| 2.4E − 08 | 16.97 | 4.7E − 08 | 16.30 | 7.0E − 08 | 15.90 | 9.3E − 08 | 15.61 |
| 2.5E − 08 | 16.93 | 4.8E − 08 | 16.27 | 7.1E − 08 | 15.88 | 9.4E − 08 | 15.60 |
| 2.6E − 08 | 16.89 | 4.9E − 08 | 16.25 | 7.2E − 08 | 15.87 | 9.5E − 08 | 15.59 |
| 2.7E − 08 | 16.85 | 5.0E − 08 | 16.23 | 7.3E − 08 | 15.86 | 9.6E − 08 | 15.58 |
| 2.8E − 08 | 16.81 | 5.1E − 08 | 16.21 | 7.4E − 08 | 15.84 | 9.7E − 08 | 15.57 |
| 2.9E − 08 | 16.78 | 5.2E − 08 | 16.19 | 7.5E − 08 | 15.83 | 9.8E − 08 | 15.56 |
| 3.0E − 08 | 16.74 | 5.3E − 08 | 16.18 | 7.6E − 08 | 15.82 | 9.9E − 08 | 15.55 |
| 3.1E − 08 | 16.71 | 5.4E − 08 | 16.16 | 7.7E − 08 | 15.80 | | |
| 3.2E − 08 | 16.68 | 5.5E − 08 | 16.14 | 7.8E − 08 | 15.79 | | |

| $10^{-7}$ well functions | | | | | | | |
|---|---|---|---|---|---|---|---|
| $u$ | $W(u)$ | $u$ | $W(u)$ | $u$ | $W(u)$ | $u$ | $W(u)$ |
| 1.0E − 07 | 15.54 | 3.3E − 07 | 14.35 | 5.6E − 07 | 13.82 | 7.9E − 07 | 13.47 |
| 1.1E − 07 | 15.45 | 3.4E − 07 | 14.32 | 5.7E − 07 | 13.80 | 8.0E − 07 | 13.46 |
| 1.2E − 07 | 15.36 | 3.5E − 07 | 14.29 | 5.8E − 07 | 13.78 | 8.1E − 07 | 13.45 |
| 1.3E − 07 | 15.28 | 3.6E − 07 | 14.26 | 5.9E − 07 | 13.77 | 8.2E − 07 | 13.44 |
| 1.4E − 07 | 15.20 | 3.7E − 07 | 14.23 | 6.0E − 07 | 13.75 | 8.3E − 07 | 13.42 |
| 1.5E − 07 | 15.14 | 3.8E − 07 | 14.21 | 6.1E − 07 | 13.73 | 8.4E − 07 | 13.41 |
| 1.6E − 07 | 15.07 | 3.9E − 07 | 14.18 | 6.2E − 07 | 13.72 | 8.5E − 07 | 13.40 |
| 1.7E − 07 | 15.01 | 4.0E − 07 | 14.15 | 6.3E − 07 | 13.70 | 8.6E − 07 | 13.39 |
| 1.8E − 07 | 14.95 | 4.1E − 07 | 14.13 | 6.4E − 07 | 13.68 | 8.7E − 07 | 13.38 |
| 1.9E − 07 | 14.90 | 4.2E − 07 | 14.11 | 6.5E − 07 | 13.67 | 8.8E − 07 | 13.37 |
| 2.0E − 07 | 14.85 | 4.3E − 07 | 14.08 | 6.6E − 07 | 13.65 | 8.9E − 07 | 13.35 |
| 2.1E − 07 | 14.80 | 4.4E − 07 | 14.06 | 6.7E − 07 | 13.64 | 9.0E − 07 | 13.34 |
| 2.2E − 07 | 14.75 | 4.5E − 07 | 14.04 | 6.8E − 07 | 13.62 | 9.1E − 07 | 13.33 |
| 2.3E − 07 | 14.71 | 4.6E − 07 | 14.01 | 6.9E − 07 | 13.61 | 9.2E − 07 | 13.32 |
| 2.4E − 07 | 14.67 | 4.7E − 07 | 13.99 | 7.0E − 07 | 13.59 | 9.3E − 07 | 13.31 |
| 2.5E − 07 | 14.62 | 4.8E − 07 | 13.97 | 7.1E − 07 | 13.58 | 9.4E − 07 | 13.30 |
| 2.6E − 07 | 14.59 | 4.9E − 07 | 13.95 | 7.2E − 07 | 13.57 | 9.5E − 07 | 13.29 |
| 2.7E − 07 | 14.55 | 5.0E − 07 | 13.93 | 7.3E − 07 | 13.55 | 9.6E − 07 | 13.28 |
| 2.8E − 07 | 14.51 | 5.1E − 07 | 13.91 | 7.4E − 07 | 13.54 | 9.7E − 07 | 13.27 |
| 2.9E − 07 | 14.48 | 5.2E − 07 | 13.89 | 7.5E − 07 | 13.53 | 9.8E − 07 | 13.26 |
| 3.0E − 07 | 14.44 | 5.3E − 07 | 13.87 | 7.6E − 07 | 13.51 | 9.9E − 07 | 13.25 |
| 3.1E − 07 | 14.41 | 5.4E − 07 | 13.85 | 7.7E − 07 | 13.50 | | |
| 3.2E − 07 | 14.38 | 5.5E − 07 | 13.84 | 7.8E − 07 | 13.49 | | |

| $10^{-6}$ well functions | | | | | | | |
|---|---|---|---|---|---|---|---|
| $u$ | $W(u)$ | $u$ | $W(u)$ | $u$ | $W(u)$ | $u$ | $W(u)$ |
| 1.0E − 06 | 13.24 | 3.3E − 06 | 12.04 | 5.6E − 06 | 11.52 | 7.9E − 06 | 11.17 |
| 1.1E − 06 | 13.14 | 3.4E − 06 | 12.01 | 5.7E − 06 | 11.50 | 8.0E − 06 | 11.16 |
| 1.2E − 06 | 13.06 | 3.5E − 06 | 11.99 | 5.8E − 06 | 11.48 | 8.1E − 06 | 11.15 |
| 1.3E − 06 | 12.98 | 3.6E − 06 | 11.96 | 5.9E − 06 | 11.46 | 8.2E − 06 | 11.13 |
| 1.4E − 06 | 12.90 | 3.7E − 06 | 11.93 | 6.0E − 06 | 11.45 | 8.3E − 06 | 11.12 |
| 1.5E − 06 | 12.83 | 3.8E − 06 | 11.90 | 6.1E − 06 | 11.43 | 8.4E − 06 | 11.11 |
| 1.6E − 06 | 12.77 | 3.9E − 06 | 11.88 | 6.2E − 06 | 11.41 | 8.5E − 06 | 11.10 |
| 1.7E − 06 | 12.71 | 4.0E − 06 | 11.85 | 6.3E − 06 | 11.40 | 8.6E − 06 | 11.09 |
| 1.8E − 06 | 12.65 | 4.1E − 06 | 11.83 | 6.4E − 06 | 11.38 | 8.7E − 06 | 11.07 |
| 1.9E − 06 | 12.60 | 4.2E − 06 | 11.80 | 6.5E − 06 | 11.37 | 8.8E − 06 | 11.06 |
| 2.0E − 06 | 12.55 | 4.3E − 06 | 11.78 | 6.6E − 06 | 11.35 | 8.9E − 06 | 11.05 |
| 2.1E − 06 | 12.50 | 4.4E − 06 | 11.76 | 6.7E − 06 | 11.34 | 9.0E − 06 | 11.04 |
| 2.2E − 06 | 12.45 | 4.5E − 06 | 11.73 | 6.8E − 06 | 11.32 | 9.1E − 06 | 11.03 |
| 2.3E − 06 | 12.41 | 4.6E − 06 | 11.71 | 6.9E − 06 | 11.31 | 9.2E − 06 | 11.02 |
| 2.4E − 06 | 12.36 | 4.7E − 06 | 11.69 | 7.0E − 06 | 11.29 | 9.3E − 06 | 11.01 |
| 2.5E − 06 | 12.32 | 4.8E − 06 | 11.67 | 7.1E − 06 | 11.28 | 9.4E − 06 | 11.00 |
| 2.6E − 06 | 12.28 | 4.9E − 06 | 11.65 | 7.2E − 06 | 11.26 | 9.5E − 06 | 10.99 |
| 2.7E − 06 | 12.25 | 5.0E − 06 | 11.63 | 7.3E − 06 | 11.25 | 9.6E − 06 | 10.98 |
| 2.8E − 06 | 12.21 | 5.1E − 06 | 11.61 | 7.4E − 06 | 11.24 | 9.7E − 06 | 10.97 |
| 2.9E − 06 | 12.17 | 5.2E − 06 | 11.59 | 7.5E − 06 | 11.22 | 9.8E − 06 | 10.96 |
| 3.0E − 06 | 12.14 | 5.3E − 06 | 11.57 | 7.6E − 06 | 11.21 | 9.9E − 06 | 10.95 |
| 3.1E − 06 | 12.11 | 5.4E − 06 | 11.55 | 7.7E − 06 | 11.20 | | |
| 3.2E − 06 | 12.08 | 5.5E − 06 | 11.53 | 7.8E − 06 | 11.18 | | |

| 10⁻⁵ well functions | | | | | | | |
|---|---|---|---|---|---|---|---|
| $u$ | $W(u)$ | $u$ | $W(u)$ | $u$ | $W(u)$ | $u$ | $W(u)$ |
| 1.0E − 05 | 10.94 | 3.3E − 05 | 9.74 | 5.6E − 05 | 9.21 | 7.9E − 05 | 8.87 |
| 1.1E − 05 | 10.84 | 3.4E − 05 | 9.71 | 5.7E − 05 | 9.20 | 8.0E − 05 | 8.86 |
| 1.2E − 05 | 10.75 | 3.5E − 05 | 9.68 | 5.8E − 05 | 9.18 | 8.1E − 05 | 8.84 |
| 1.3E − 05 | 10.67 | 3.6E − 05 | 9.65 | 5.9E − 05 | 9.16 | 8.2E − 05 | 8.83 |
| 1.4E − 05 | 10.60 | 3.7E − 05 | 9.63 | 6.0E − 05 | 9.14 | 8.3E − 05 | 8.82 |
| 1.5E − 05 | 10.53 | 3.8E − 05 | 9.60 | 6.1E − 05 | 9.13 | 8.4E − 05 | 8.81 |
| 1.6E − 05 | 10.47 | 3.9E − 05 | 9.57 | 6.2E − 05 | 9.11 | 8.5E − 05 | 8.80 |
| 1.7E − 05 | 10.41 | 4.0E − 05 | 9.55 | 6.3E − 05 | 9.10 | 8.6E − 05 | 8.78 |
| 1.8E − 05 | 10.35 | 4.1E − 05 | 9.52 | 6.4E − 05 | 9.08 | 8.7E − 05 | 8.77 |
| 1.9E − 05 | 10.29 | 4.2E − 05 | 9.50 | 6.5E − 05 | 9.06 | 8.8E − 05 | 8.76 |
| 2.0E − 05 | 10.24 | 4.3E − 05 | 9.48 | 6.6E − 05 | 9.05 | 8.9E − 05 | 8.75 |
| 2.1E − 05 | 10.19 | 4.4E − 05 | 9.45 | 6.7E − 05 | 9.03 | 9.0E − 05 | 8.74 |
| 2.2E − 05 | 10.15 | 4.5E − 05 | 9.43 | 6.8E − 05 | 9.02 | 9.1E − 05 | 8.73 |
| 2.3E − 05 | 10.10 | 4.6E − 05 | 9.41 | 6.9E − 05 | 9.00 | 9.2E − 05 | 8.72 |
| 2.4E − 05 | 10.06 | 4.7E − 05 | 9.39 | 7.0E − 05 | 8.99 | 9.3E − 05 | 8.71 |
| 2.5E − 05 | 10.02 | 4.8E − 05 | 9.37 | 7.1E − 05 | 8.98 | 9.4E − 05 | 8.70 |
| 2.6E − 05 | 9.98 | 4.9E − 05 | 9.35 | 7.2E − 05 | 8.96 | 9.5E − 05 | 8.68 |
| 2.7E − 05 | 9.94 | 5.0E − 05 | 9.33 | 7.3E − 05 | 8.95 | 9.6E − 05 | 8.67 |
| 2.8E − 05 | 9.91 | 5.1E − 05 | 9.31 | 7.4E − 05 | 8.93 | 9.7E − 05 | 8.66 |
| 2.9E − 05 | 9.87 | 5.2E − 05 | 9.29 | 7.5E − 05 | 8.92 | 9.8E − 05 | 8.65 |
| 3.0E − 05 | 9.84 | 5.3E − 05 | 9.27 | 7.6E − 05 | 8.91 | 9.9E − 05 | 8.64 |
| 3.1E − 05 | 9.80 | 5.4E − 05 | 9.25 | 7.7E − 05 | 8.89 | | |
| 3.2E − 05 | 9.77 | 5.5E − 05 | 9.23 | 7.8E − 05 | 8.88 | | |

| 10⁻⁴ well functions | | | | | | | |
|---|---|---|---|---|---|---|---|
| $u$ | $W(u)$ | $u$ | $W(u)$ | $u$ | $W(u)$ | $u$ | $W(u)$ |
| 1.0E − 04 | 8.63 | 3.3E − 04 | 7.44 | 5.6E − 04 | 6.91 | 7.9E − 04 | 6.57 |
| 1.1E − 04 | 8.54 | 3.4E − 04 | 7.41 | 5.7E − 04 | 6.89 | 8.0E − 04 | 6.55 |
| 1.2E − 04 | 8.45 | 3.5E − 04 | 7.38 | 5.8E − 04 | 6.88 | 8.1E − 04 | 6.54 |
| 1.3E − 04 | 8.37 | 3.6E − 04 | 7.35 | 5.9E − 04 | 6.86 | 8.2E − 04 | 6.53 |
| 1.4E − 04 | 8.30 | 3.7E − 04 | 7.33 | 6.0E − 04 | 6.84 | 8.3E − 04 | 6.52 |
| 1.5E − 04 | 8.23 | 3.8E − 04 | 7.30 | 6.1E − 04 | 6.83 | 8.4E − 04 | 6.51 |
| 1.6E − 04 | 8.16 | 3.9E − 04 | 7.27 | 6.2E − 04 | 6.81 | 8.5E − 04 | 6.49 |
| 1.7E − 04 | 8.10 | 4.0E − 04 | 7.25 | 6.3E − 04 | 6.79 | 8.6E − 04 | 6.48 |
| 1.8E − 04 | 8.05 | 4.1E − 04 | 7.22 | 6.4E − 04 | 6.78 | 8.7E − 04 | 6.47 |
| 1.9E − 04 | 7.99 | 4.2E − 04 | 7.20 | 6.5E − 04 | 6.76 | 8.8E − 04 | 6.46 |
| 2.0E − 04 | 7.94 | 4.3E − 04 | 7.17 | 6.6E − 04 | 6.75 | 8.9E − 04 | 6.45 |
| 2.1E − 04 | 7.89 | 4.4E − 04 | 7.15 | 6.7E − 04 | 6.73 | 9.0E − 04 | 6.44 |
| 2.2E − 04 | 7.84 | 4.5E − 04 | 7.13 | 6.8E − 04 | 6.72 | 9.1E − 04 | 6.43 |
| 2.3E − 04 | 7.80 | 4.6E − 04 | 7.11 | 6.9E − 04 | 6.70 | 9.2E − 04 | 6.41 |
| 2.4E − 04 | 7.76 | 4.7E − 04 | 7.09 | 7.0E − 04 | 6.69 | 9.3E − 04 | 6.40 |
| 2.5E − 04 | 7.72 | 4.8E − 04 | 7.06 | 7.1E − 04 | 6.67 | 9.4E − 04 | 6.39 |
| 2.6E − 04 | 7.68 | 4.9E − 04 | 7.04 | 7.2E − 04 | 6.66 | 9.5E − 04 | 6.38 |
| 2.7E − 04 | 7.64 | 5.0E − 04 | 7.02 | 7.3E − 04 | 6.65 | 9.6E − 04 | 6.37 |
| 2.8E − 04 | 7.60 | 5.1E − 04 | 7.00 | 7.4E − 04 | 6.63 | 9.7E − 04 | 6.36 |
| 2.9E − 04 | 7.57 | 5.2E − 04 | 6.98 | 7.5E − 04 | 6.62 | 9.8E − 04 | 6.35 |
| 3.0E − 04 | 7.53 | 5.3E − 04 | 6.97 | 7.6E − 04 | 6.61 | 9.9E − 04 | 6.34 |
| 3.1E − 04 | 7.50 | 5.4E − 04 | 6.95 | 7.7E − 04 | 6.59 | | |
| 3.2E − 04 | 7.47 | 5.5E − 04 | 6.93 | 7.8E − 04 | 6.58 | | |

| $10^{-3}$ well functions | | | | | | | |
|---|---|---|---|---|---|---|---|
| $u$ | $W(u)$ | $u$ | $W(u)$ | $u$ | $W(u)$ | $u$ | $W(u)$ |
| 1.0E − 03 | 6.33 | 3.3E − 03 | 5.14 | 5.6E − 03 | 4.61 | 7.9E − 03 | 4.27 |
| 1.1E − 03 | 6.24 | 3.4E − 03 | 5.11 | 5.7E − 03 | 4.60 | 8.0E − 03 | 4.26 |
| 1.2E − 03 | 6.15 | 3.5E − 03 | 5.08 | 5.8E − 03 | 4.58 | 8.1E − 03 | 4.25 |
| 1.3E − 03 | 6.07 | 3.6E − 03 | 5.05 | 5.9E − 03 | 4.56 | 8.2E − 03 | 4.23 |
| 1.4E − 03 | 6.00 | 3.7E − 03 | 5.03 | 6.0E − 03 | 4.54 | 8.3E − 03 | 4.22 |
| 1.5E − 03 | 5.93 | 3.8E − 03 | 5.00 | 6.1E − 03 | 4.53 | 8.4E − 03 | 4.21 |
| 1.6E − 03 | 5.86 | 3.9E − 03 | 4.97 | 6.2E − 03 | 4.51 | 8.5E − 03 | 4.20 |
| 1.7E − 03 | 5.80 | 4.0E − 03 | 4.95 | 6.3E − 03 | 4.50 | 8.6E − 03 | 4.19 |
| 1.8E − 03 | 5.74 | 4.1E − 03 | 4.92 | 6.4E − 03 | 4.48 | 8.7E − 03 | 4.18 |
| 1.9E − 03 | 5.69 | 4.2E − 03 | 4.90 | 6.5E − 03 | 4.47 | 8.8E − 03 | 4.16 |
| 2.0E − 03 | 5.64 | 4.3E − 03 | 4.88 | 6.6E − 03 | 4.45 | 8.9E − 03 | 4.15 |
| 2.1E − 03 | 5.59 | 4.4E − 03 | 4.85 | 6.7E − 03 | 4.44 | 9.0E − 03 | 4.14 |
| 2.2E − 03 | 5.54 | 4.5E − 03 | 4.83 | 6.8E − 03 | 4.42 | 9.1E − 03 | 4.13 |
| 2.3E − 03 | 5.50 | 4.6E − 03 | 4.81 | 6.9E − 03 | 4.41 | 9.2E − 03 | 4.12 |
| 2.4E − 03 | 5.46 | 4.7E − 03 | 4.79 | 7.0E − 03 | 4.39 | 9.3E − 03 | 4.11 |
| 2.5E − 03 | 5.42 | 4.8E − 03 | 4.77 | 7.1E − 03 | 4.38 | 9.4E − 03 | 4.10 |
| 2.6E − 03 | 5.38 | 4.9E − 03 | 4.75 | 7.2E − 03 | 4.36 | 9.5E − 03 | 4.09 |
| 2.7E − 03 | 5.34 | 5.0E − 03 | 4.73 | 7.3E − 03 | 4.35 | 9.6E − 03 | 4.08 |
| 2.8E − 03 | 5.30 | 5.1E − 03 | 4.71 | 7.4E − 03 | 4.34 | 9.7E − 03 | 4.07 |
| 2.9E − 03 | 5.27 | 5.2E − 03 | 4.69 | 7.5E − 03 | 4.32 | 9.8E − 03 | 4.06 |
| 3.0E − 03 | 5.23 | 5.3E − 03 | 4.67 | 7.6E − 03 | 4.31 | 9.9E − 03 | 4.05 |
| 3.1E − 03 | 5.20 | 5.4E − 03 | 4.65 | 7.7E − 03 | 4.30 | | |
| 3.2E − 03 | 5.17 | 5.5E − 03 | 4.63 | 7.8E − 03 | 4.28 | | |

| $10^{-2}$ well functions | | | | | | | |
|---|---|---|---|---|---|---|---|
| $u$ | $W(u)$ | $u$ | $W(u)$ | $u$ | $W(u)$ | $u$ | $W(u)$ |
| 1.0E − 02 | 4.04 | 3.3E − 02 | 2.87 | 5.6E − 02 | 2.36 | 7.9E − 02 | 2.04 |
| 1.1E − 02 | 3.94 | 3.4E − 02 | 2.84 | 5.7E − 02 | 2.34 | 8.0E − 02 | 2.03 |
| 1.2E − 02 | 3.86 | 3.5E − 02 | 2.81 | 5.8E − 02 | 2.33 | 8.1E − 02 | 2.02 |
| 1.3E − 02 | 3.78 | 3.6E − 02 | 2.78 | 5.9E − 02 | 2.31 | 8.2E − 02 | 2.00 |
| 1.4E − 02 | 3.71 | 3.7E − 02 | 2.76 | 6.0E − 02 | 2.30 | 8.3E − 02 | 1.993 |
| 1.5E − 02 | 3.64 | 3.8E − 02 | 2.73 | 6.1E − 02 | 2.28 | 8.4E − 02 | 1.982 |
| 1.6E − 02 | 3.57 | 3.9E − 02 | 2.71 | 6.2E − 02 | 2.26 | 8.5E − 02 | 1.971 |
| 1.7E − 02 | 3.51 | 4.0E − 02 | 2.68 | 6.3E − 02 | 2.25 | 8.6E − 02 | 1.960 |
| 1.8E − 02 | 3.46 | 4.1E − 02 | 2.66 | 6.4E − 02 | 2.23 | 8.7E − 02 | 1.950 |
| 1.9E − 02 | 3.41 | 4.2E − 02 | 2.63 | 6.5E − 02 | 2.22 | 8.8E − 02 | 1.939 |
| 2.0E − 02 | 3.35 | 4.3E − 02 | 2.61 | 6.6E − 02 | 2.21 | 8.9E − 02 | 1.929 |
| 2.1E − 02 | 3.31 | 4.4E − 02 | 2.59 | 6.7E − 02 | 2.19 | 9.0E − 02 | 1.919 |
| 2.2E − 02 | 3.26 | 4.5E − 02 | 2.57 | 6.8E − 02 | 2.18 | 9.1E − 02 | 1.909 |
| 2.3E − 02 | 3.22 | 4.6E − 02 | 2.55 | 6.9E − 02 | 2.16 | 9.2E − 02 | 1.899 |
| 2.4E − 02 | 3.18 | 4.7E − 02 | 2.53 | 7.0E − 02 | 2.15 | 9.3E − 02 | 1.889 |
| 2.5E − 02 | 3.14 | 4.8E − 02 | 2.51 | 7.1E − 02 | 2.14 | 9.4E − 02 | 1.879 |
| 2.6E − 02 | 3.10 | 4.9E − 02 | 2.49 | 7.2E − 02 | 2.12 | 9.5E − 02 | 1.869 |
| 2.7E − 02 | 3.06 | 5.0E − 02 | 2.47 | 7.3E − 02 | 2.11 | 9.6E − 02 | 1.860 |
| 2.8E − 02 | 3.03 | 5.1E − 02 | 2.45 | 7.4E − 02 | 2.10 | 9.7E − 02 | 1.851 |
| 2.9E − 02 | 2.99 | 5.2E − 02 | 2.43 | 7.5E − 02 | 2.09 | 9.8E − 02 | 1.841 |
| 3.0E − 02 | 2.96 | 5.3E − 02 | 2.41 | 7.6E − 02 | 2.07 | 9.9E − 02 | 1.832 |
| 3.1E − 02 | 2.93 | 5.4E − 02 | 2.39 | 7.7E − 02 | 2.06 | | |
| 3.2E − 02 | 2.90 | 5.5E − 02 | 2.38 | 7.8E − 02 | 2.05 | | |

| $10^{-1}$ well functions | | | | | | | |
|---|---|---|---|---|---|---|---|
| $u$ | $W(u)$ | $u$ | $W(u)$ | $u$ | $W(u)$ | $u$ | $W(u)$ |
| 1.0E − 01 | 1.823 | 3.3E − 01 | 0.836 | 5.6E − 01 | 0.493 | 7.9E − 01 | 0.316 |
| 1.1E − 01 | 1.737 | 3.4E − 01 | 0.815 | 5.7E − 01 | 0.483 | 8.0E − 01 | 0.311 |
| 1.2E − 01 | 1.660 | 3.5E − 01 | 0.794 | 5.8E − 01 | 0.473 | 8.1E − 01 | 0.305 |
| 1.3E − 01 | 1.589 | 3.6E − 01 | 0.774 | 5.9E − 01 | 0.464 | 8.2E − 01 | 0.300 |
| 1.4E − 01 | 1.524 | 3.7E − 01 | 0.755 | 6.0E − 01 | 0.454 | 8.3E − 01 | 0.294 |
| 1.5E − 01 | 1.464 | 3.8E − 01 | 0.737 | 6.1E − 01 | 0.445 | 8.4E − 01 | 0.289 |
| 1.6E − 01 | 1.409 | 3.9E − 01 | 0.719 | 6.2E − 01 | 0.437 | 8.5E − 01 | 0.284 |
| 1.7E − 01 | 1.358 | 4.0E − 01 | 0.702 | 6.3E − 01 | 0.428 | 8.6E − 01 | 0.279 |
| 1.8E − 01 | 1.310 | 4.1E − 01 | 0.686 | 6.4E − 01 | 0.420 | 8.7E − 01 | 0.274 |
| 1.9E − 01 | 1.265 | 4.2E − 01 | 0.670 | 6.5E − 01 | 0.412 | 8.8E − 01 | 0.265 |
| 2.0E − 01 | 1.223 | 4.3E − 01 | 0.655 | 6.6E − 01 | 0.404 | 8.9E − 01 | 0.265 |
| 2.1E − 01 | 1.183 | 4.4E − 01 | 0.640 | 6.7E − 01 | 0.396 | 9.0E − 01 | 0.260 |
| 2.2E − 01 | 1.145 | 4.5E − 01 | 0.625 | 6.8E − 01 | 0.388 | 9.1E − 01 | 0.256 |
| 2.3E − 01 | 1.110 | 4.6E − 01 | 0.611 | 6.9E − 01 | 0.381 | 9.2E − 01 | 0.251 |
| 2.4E − 01 | 1.076 | 4.7E − 01 | 0.598 | 7.0E − 01 | 0.374 | 9.3E − 01 | 0.247 |
| 2.5E − 01 | 1.044 | 4.8E − 01 | 0.585 | 7.1E − 01 | 0.367 | 9.4E − 01 | 0.243 |
| 2.6E − 01 | 1.014 | 4.9E − 01 | 0.572 | 7.2E − 01 | 0.360 | 9.5E − 01 | 0.239 |
| 2.7E − 01 | 0.985 | 5.0E − 01 | 0.560 | 7.3E − 01 | 0.353 | 9.6E − 01 | 0.235 |
| 2.8E − 01 | 0.957 | 5.1E − 01 | 0.548 | 7.4E − 01 | 0.347 | 9.7E − 01 | 0.231 |
| 2.9E − 01 | 0.931 | 5.2E − 01 | 0.536 | 7.5E − 01 | 0.340 | 9.8E − 01 | 0.227 |
| 3.0E − 01 | 0.906 | 5.3E − 01 | 0.525 | 7.6E − 01 | 0.334 | 9.9E − 01 | 0.223 |
| 3.1E − 01 | 0.882 | 5.4E − 01 | 0.514 | 7.7E − 01 | 0.328 | | |
| 3.2E − 01 | 0.858 | 5.5E − 01 | 0.503 | 7.8E − 01 | 0.322 | | |

| 0 well functions | | | | | | | |
|---|---|---|---|---|---|---|---|
| $u$ | $W(u)$ | $u$ | $W(u)$ | $u$ | $W(u)$ | $u$ | $W(u)$ |
| 1.0E + 00 | 0.219 | 2.1E + 00 | 0.043 | 3.2E + 00 | 0.010 | 4.3E + 00 | 0.003 |
| 1.1E + 00 | 0.186 | 2.2E + 00 | 0.037 | 3.3E + 00 | 0.009 | 4.4E + 00 | 0.002 |
| 1.2E + 00 | 0.158 | 2.3E + 00 | 0.033 | 3.4E + 00 | 0.008 | 4.5E + 00 | 0.002 |
| 1.3E + 00 | 0.135 | 2.4E + 00 | 0.028 | 3.5E + 00 | 0.007 | 4.6E + 00 | 0.002 |
| 1.4E + 00 | 0.116 | 2.5E + 00 | 0.025 | 3.6E + 00 | 0.006 | 4.7E + 00 | 0.002 |
| 1.5E + 00 | 0.100 | 2.6E + 00 | 0.022 | 3.7E + 00 | 0.005 | 4.8E + 00 | 0.001 |
| 1.6E + 00 | 0.086 | 2.7E + 00 | 0.019 | 3.8E + 00 | 0.005 | 4.9E + 00 | 0.001 |
| 1.7E + 00 | 0.075 | 2.8E + 00 | 0.017 | 3.9E + 00 | 0.004 | 5.0E + 00 | 0.001 |
| 1.8E + 00 | 0.065 | 2.9E + 00 | 0.015 | 4.0E + 00 | 0.004 | | |
| 1.9E + 00 | 0.056 | 3.0E + 00 | 0.013 | 4.1E + 00 | 0.003 | | |
| 2.0E + 00 | 0.049 | 3.1E + 00 | 0.011 | 4.2E + 00 | 0.003 | | |

SOURCE: Illinois Environmental Protection Agency (1990)

Appendix

# C

# Solubility Product Constants for Solution at or near Room Temperature

| Substance | Formula | $K_{sp}^{\dagger}$ |
|---|---|---|
| Aluminum hydroxide | $Al(OH)_3$ | $2 \times 10^{-32}$ |
| Barium arsenate | $Ba_3(AsO_4)_2$ | $7.7 \times 10^{-51}$ |
| Barium carbonate | $BaCO_3$ | $8.1 \times 10^{-9}$ |
| Barium chromate | $BaCrO_4$ | $2.4 \times 10^{-10}$ |
| Barium fluoride | $BaF_2$ | $1.7 \times 10^{-6}$ |
| Barium iodate | $Ba(IO_3)_2 \cdot 2H_2O$ | $1.5 \times 10^{-9}$ |
| Barium oxalate | $BaC_2O_4 \cdot H_2O$ | $2.3 \times 10^{-8}$ |
| Barium sulfate | $BaSO_4$ | $1.08 \times 10^{-10}$ |
| Beryllium hydroxide | $Be(OH)_2$ | $7 \times 10^{-22}$ |
| Bismuth iodide | $BiI_3$ | $8.1 \times 10^{-19}$ |
| Bismuth phosphate | $BiPO_4$ | $1.3 \times 10^{-23}$ |
| Bismuth sulfide | $Bi_2S_3$ | $1 \times 10^{-97}$ |
| Cadmium arsenate | $Cd_3(AsO_4)_2$ | $2.2 \times 10^{-33}$ |
| Cadmium hydroxide | $Cd(OH)_2$ | $5.9 \times 10^{-15}$ |
| Cadmium oxalate | $CdC_2O_4 \cdot 3H_2O$ | $1.5 \times 10^{-8}$ |
| Cadmium sulfide | $CdS$ | $7.8 \times 10^{-27}$ |
| Calcium arsenate | $Ca_3(AsO_4)_2$ | $6.8 \times 10^{-19}$ |
| Calcium carbonate | $CaCO_3$ | $8.7 \times 10^{-9}$ |
| Calcium fluoride | $CaF_2$ | $4.0 \times 10^{-11}$ |
| Calcium hydroxide | $Ca(OH)_2$ | $5.5 \times 10^{-6}$ |
| Calcium iodate | $Ca(IO_3)_2 \cdot 6H_2O$ | $6.4 \times 10^{-7}$ |
| Calcium oxalate | $CaC_2O_4H_2O$ | $2.6 \times 10^{-9}$ |
| Calcium phosphate | $Ca_3(PO_4)_2$ | $2.0 \times 10^{-29}$ |
| Calcium sulfate | $CaSO_4$ | $1.9 \times 10^{-4}$ |
| Cerium(III) hydroxide | $Ce(OH)_3$ | $2 \times 10^{-20}$ |
| Cerium(III) iodate | $Ce(IO_3)_3$ | $3.2 \times 10^{-10}$ |
| Cerium(III) oxalate | $Ce_2(C_2O_4)_3 \cdot 9H_2O$ | $3 \times 10^{-29}$ |
| Chromium(II) hydroxide | $Cr(OH)_2$ | $1.0 \times 10^{-17}$ |

| Substance | Formula | $K^{\dagger}_{sp}$ |
|---|---|---|
| Chromium(III) hydroxide | $Cr(OH)_3$ | $6 \times 10^{-31}$ |
| Cobalt(II) hydroxide | $Co(OH)_2$ | $2 \times 10^{-16}$ |
| Cobalt(III) hydroxide | $Co(OH)_3$ | $1 \times 10^{-43}$ |
| Copper(II) arsenate | $Cu_3(AsO_4)_2$ | $7.6 \times 10^{-76}$ |
| Copper(I) bromide | $CuBr$ | $5.2 \times 10^{-9}$ |
| Copper(I) chloride | $CuCl$ | $1.2 \times 10^{-6}$ |
| Copper(I) iodide | $CuI$ | $5.1 \times 10^{-12}$ |
| Copper(II) iodate | $Cu(IO_3)_2$ | $7.4 \times 10^{-8}$ |
| Copper(I) sulfide | $Cu_2S$ | $2 \times 10^{-47}$ |
| Copper(II) sulfide | $CuS$ | $9 \times 10^{-36}$ |
| Copper(I) thiocyanate | $CuSCN$ | $4.8 \times 10^{-15}$ |
| Iron(III) arsenate | $FeAsO_4$ | $5.7 \times 10^{-21}$ |
| Iron(II) carbonate | $FeCO_3$ | $3.5 \times 10^{-11}$ |
| Iron(II) hydroxide | $Fe(OH)_2$ | $8 \times 10^{-16}$ |
| Iron(III) hydroxide | $Fe(OH)_3$ | $4 \times 10^{-38}$ |
| Lead arsenate | $Pb_3(AsO_4)_2$ | $4.1 \times 10^{-36}$ |
| Lead bromide | $PbBr_2$ | $3.9 \times 10^{-5}$ |
| Lead carbonate | $PbCO_3$ | $3.3 \times 10^{-14}$ |
| Lead chloride | $PbCl_2$ | $1.6 \times 10^{-5}$ |
| Lead chromate | $PbCrO_4$ | $1.8 \times 10^{-14}$ |
| Lead fluoride | $PbF_2$ | $3.7 \times 10^{-8}$ |
| Lead iodate | $Pb(IO_3)_2$ | $2.6 \times 10^{-13}$ |
| Lead iodide | $PbI_2$ | $7.1 \times 10^{-9}$ |
| Lead oxalate | $PbC_2O_4$ | $4.8 \times 10^{-10}$ |
| Lead sulfate | $PbSO_4$ | $1.6 \times 10^{-8}$ |
| Lead sulfide | $PbS$ | $8 \times 10^{-28}$ |
| Magnesium ammonium phosphate | $MgNH_4PO_4$ | $2.5 \times 10^{-13}$ |
| Magnesium arsenate | $Mg_3(AsO_4)_2$ | $2.1 \times 10^{-20}$ |
| Magnesium carbonate | $MgCO_3 \cdot 3H_2O$ | $1 \times 10^{-5}$ |
| Magnesium fluoride | $MgF_2$ | $6.5 \times 10^{-9}$ |
| Magnesium hydroxide | $Mg(OH)_2$ | $1.2 \times 10^{-11}$ |
| Magnesium oxalate | $MgC_2O_4 \cdot 2H_2O$ | $1 \times 10^{-8}$ |
| Manganese(II) hydroxide | $Mn(OH)_2$ | $1.9 \times 10^{-13}$ |
| Mercury(I) bromide | $Hg_2Br_2$ | $5.8 \times 10^{-23}$ |
| Mercury(I) chloride | $Hg_2Cl_2$ | $1.3 \times 10^{-18}$ |
| Mercury(I) iodide | $Hg_2I_2$ | $4.5 \times 10^{-29}$ |
| Mercury(I) sulfate | $Hg_2SO_4$ | $7.4 \times 10^{-7}$ |
| Mercury(II) sulfide | $HgS$ | $4 \times 10^{-53}$ |
| Mercury(I) thiocyanate | $Hg_2(SCN)_2$ | $3.0 \times 10^{-20}$ |
| Nickel arsenate | $Ni_3(AsO_4)_2$ | $3.1 \times 10^{-26}$ |
| Nickel carbonate | $NiCO_3$ | $6.6 \times 10^{-9}$ |
| Nickel hydroxide | $Ni(OH)_2$ | $6.5 \times 10^{-18}$ |
| Nickel sulfide | $NiS$ | $3 \times 1^{-19}$ |
| Silver arsenate | $Ag_3AsO_4$ | $1 \times 10^{-22}$ |
| Silver bromate | $AgBrO_3$ | $5.77 \times 10^{-5}$ |
| Silver bromide | $AgBr$ | $5.25 \times 10^{-13}$ |
| Silver carbonate | $Ag_2CO_3$ | $8.1 \times 10^{-12}$ |
| Silver chloride | $AgCl$ | $1.78 \times 10^{-10}$ |
| Silver chromate | $Ag_2CrO_4$ | $2.45 \times 10^{-12}$ |
| Silver cyanide | $Ag[Ag(CN)_2]$ | $5.0 \times 10^{-12}$ |
| Silver iodate | $AgIO_3$ | $3.02 \times 10^{-8}$ |
| Silver iodide | $AgI$ | $8.31 \times 10^{-17}$ |
| Silver oxalate | $Ag_2C_2O_4$ | $3.5 \times 10^{-11}$ |

| Substance | Formula | $K_{sp}^{\dagger}$ |
|---|---|---|
| Silver oxide | $Ag_2O$ | $2.6 \times 10^{-8}$ |
| Silver phosphate | $Ag_3PO_4$ | $1.3 \times 10^{-20}$ |
| Silver sulfate | $Ag_2SO_4$ | $1.6 \times 10^{-5}$ |
| Silver sulfide | $Ag_2S$ | $2 \times 10^{-49}$ |
| Silver thiocyanate | $AgSCN$ | $1.00 \times 10^{-12}$ |
| Strontium carbonate | $SrCO_3$ | $1.1 \times 10^{-10}$ |
| Strontium chromate | $SrCrO_4$ | $3.6 \times 10^{-5}$ |
| Strontium fluoride | $SrF_2$ | $2.8 \times 10^{-9}$ |
| Strontium iodate | $Sr(IO_3)_2$ | $3.3 \times 10^{-7}$ |
| Strontium oxalate | $SrC_2O_4 \cdot H_2O$ | $1.6 \times 10^{-7}$ |
| Strontium sulfate | $SrSO_4$ | $3.8 \times 10^{-7}$ |
| Thallium(I) bromate | $TlBrO_3$ | $8.5 \times 10^{-5}$ |
| Thallium(I) bromide | $TlBr$ | $3.4 \times 10^{-6}$ |
| Thallium(I) chloride | $TlCl$ | $1.7 \times 10^{-4}$ |
| Thallium(I) chromate | $Tl_2CrO_4$ | $9.8 \times 10^{-13}$ |
| Thallium(I) iodate | $TlIO_3$ | $3.1 \times 10^{-6}$ |
| Thallium(I) iodide | $TlI$ | $6.5 \times 10^{-8}$ |
| Thallium(I) sulfide | $Tl_2S$ | $5 \times 10^{-21}$ |
| Tin(II) sulfide | $SnS$ | $1 \times 10^{-25}$ |
| Titanium(III) hydroxide | $Ti(OH)_3$ | $1 \times 10^{-40}$ |
| Zinc arsenate | $Zn_3(AsO_4)_2$ | $1.3 \times 10^{-28}$ |
| Zinc carbonate | $ZnCO_3$ | $1.4 \times 10^{-11}$ |
| Zinc ferrocyanide | $Zn_2Fe(CN)_6$ | $4.1 \times 10^{-16}$ |
| Zinc hydroxide | $Zn(OH)_2$ | $1.2 \times 10^{-17}$ |
| Zinc oxalate | $ZnC_2O_4 \cdot 2H_2O$ | $2.8 \times 10^{-8}$ |
| Zinc phosphate | $Zn_3(PO_4)_2$ | $9.1 \times 10^{-33}$ |
| Zinc sulfide | $ZnS$ | $1 \times 10^{-21}$ |

† The solubility of many metals is altered by carbonate complexation. Solubility predictions without consideration for complexation can be highly inaccurate.

SOURCE: Benefield, L. D. and Morgan J. S. 1990. Chemical precipitation. In: AWWA, *Water Quality and Treatment.* New York: McGraw-Hill. Reprinted with permission of the McGraw-Hill Co.

Appendix

# D

# Freundlich Adsorption Isotherm Constants for Toxic Organic Compounds

| Compound | $K$(mg/g)(L/mg)$^{1/n}$ | $1/n$ |
|---|---|---|
| PCB | 14,100 | 1.03 |
| Bis(2-ethylhexyl phthalate) | 11,300 | 1.5 |
| Heptachlor | 9,320 | 0.92 |
| Heptachlor epoxide | 2,120 | 0.75 |
| Butylbenzyl phthalate | 1,520 | 1.26 |
| Toxaphene | 950 | 0.74 |
| Endosulfan sulfate | 686 | 0.81 |
| Endrin | 666 | 0.80 |
| Fluoranthene | 664 | 0.61 |
| Aldrin | 651 | 0.92 |
| PCB-1232 | 630 | 0.73 |
| $\beta$-Endosulfan | 615 | 0.83 |
| Dieldrin | 606 | 0.51 |
| Alachlor | 479 | 0.26 |
| Hexachlorobenzene | 450 | 0.60 |
| Pentachlorophenol | 436 | 0.34 |
| Anthracene | 376 | 0.70 |
| 4-Nitrobiphenyl | 370 | 0.27 |
| Fluorene | 330 | 0.28 |
| Styrene | 327 | 0.48 |
| DDT | 322 | 0.50 |
| 2-Acetylaminofluorene | 318 | 0.12 |
| $\alpha$-BHC | 303 | 0.43 |
| Anethole | 300 | 0.42 |
| 3,3-Dichlorobenzidine | 300 | 0.20 |
| $\gamma$-BHC (lindane) | 285 | 0.43 |
| 2-Chloronaphthalene | 280 | 0.46 |
| Phenylmercuric acetate | 270 | 0.44 |

| Compound | $K(\text{mg/g})(\text{L/mg})^{1/n}$ | $1/n$ |
|---|---|---|
| Carbofuran | 266 | 0.41 |
| 1,2-Dichlorobenzene | 263 | 0.38 |
| Hexachlorobutadiene | 258 | 0.45 |
| *p* -Nonylphenol | 250 | 0.37 |
| 4-Dimethylaminoazobenzene | 249 | 0.24 |
| PCB-1221 | 242 | 0.70 |
| DDE | 232 | 0.37 |
| *m*,-Xylene | 230 | 0.75 |
| Acridine yellow | 230 | 0.12 |
| Dibromochloropropane (DBCP) | 224 | 0.51 |
| Benzidine dihydrochloride | 220 | 0.37 |
| β-BHC | 220 | 0.49 |
| *n*-Butylphthalate | 220 | 0.45 |
| *n*-Nitrosodiphenylamine | 220 | 0.37 |
| Silvex | 215 | 0.38 |
| Phenanthrene | 215 | 0.44 |
| Dimethylphenylcarbinol | 210 | 0.34 |
| 4-Aminobiphenyl | 200 | 0.26 |
| β-Naphthol | 200 | 0.26 |
| *p*-Xylene | 200 | 0.42 |
| α-Endosulfan | 194 | 0.50 |
| Chlordane | 190 | 0.33 |
| Acenaphthene | 190 | 0.36 |
| 4,4′-Methylene-bis-(2-chloroaniline) | 190 | 0.64 |
| Benzo[*k*]fluoranthene | 181 | 0.57 |
| Acridine orange | 180 | 0.29 |
| α-Naphthol | 180 | 0.32 |
| Ethylbenzene | 175 | 0.53 |
| *o*-Xylene | 174 | 0.47 |
| 4,6-Dinitro-*o*-cresol | 169 | 0.27 |
| α-Naphthylamine | 160 | 0.34 |
| 2,4-Dichlorophenol | 157 | 0.15 |
| 1,2,4-Trichlorobenzene | 157 | 0.31 |
| 2,4,6-Trichlorophenol | 155 | 0.40 |
| β-Naphthylamine | 150 | 0.30 |
| 2,4-Dinitrotoluene | 146 | 0.31 |
| 2,6-Dinitrotoluene | 145 | 0.32 |
| 4-Bromophenyl phenyl ether | 144 | 0.68 |
| *p*-Nitroaniline | 140 | 0.27 |
| 1,1-Diphenylhydrazine | 135 | 0.16 |
| Naphthalene | 132 | 0.42 |
| Aldicarb | 132 | 0.40 |
| l-Chloro-2-nitrobenzene | 130 | 0.46 |
| *p*-Chlorometacresol | 124 | 0.16 |
| 1,4-Dichlorobenzene | 121 | 0.47 |
| Benzothiazole | 120 | 0.27 |
| Diphenylamine | 120 | 0.31 |
| Guanine | 120 | 0.40 |
| 1,3-Dichlorobenzene | 118 | 0.45 |
| Acenaphthylene | 115 | 0.37 |
| Methoxychlor | 115 | 0.36 |
| 4-Chlorophenyl phenyl ether | 111 | 0.26 |
| Diethyl phthalate | 110 | 0.27 |

| Compound | $K$(mg/g)(L/mg)$^{1/n}$ | $1/n$ |
|---|---|---|
| Chlorobenzene | 100 | 0.35 |
| Toluene | 100 | 0.45 |
| 2-Nitrophenol | 99 | 0.34 |
| Dimethyl phthalate | 97 | 0.41 |
| Hexachloroethane | 97 | 0.38 |
| 2,4-Dimethylphenol | 78 | 0.44 |
| 4-Nitrophenol | 76 | 0.25 |
| Acetophenone | 74 | 0.44 |
| 1,2,3,4-Tetrahydronaphthalene | 74 | 0.81 |
| Adenine | 71 | 0.38 |
| Dibenzo[*a*,*h*]anthracene | 69 | 0.75 |
| Nitrobenzene | 68 | 0.43 |
| 2,4-D | 67 | 0.27 |
| 3,4-Benzofluoranthene | 57 | 0.37 |
| 2-Chlorophenol | 51 | 0.41 |
| Tetrachloroethylene | 51 | 0.56 |
| *o*-Anisidine | 50 | 0.34 |
| 5-Bromouracil | 44 | 0.47 |
| Benzo[*a*]pyrene | 34 | 0.44 |
| 2,4-Dinitrophenol | 33 | 0.61 |
| Isophorone | 32 | 0.39 |
| Trichloroethylene | 28 | 0.62 |
| Thymine | 27 | 0.51 |
| 5-Chlorouracil | 25 | 0.58 |
| *N*-Nitrosodi-*n*-propylamine | 24 | 0.26 |
| Bis(2-Chloroisopropyl) ether | 24 | 0.57 |
| 1,2-Dibromoethene (EBD) | 22 | 0.46 |
| Phenol | 21 | 0.54 |
| Bromoform | 20 | 0.52 |
| 1,2-Dichloropropane | 19 | 0.59 |
| 1,2-*trans*-Dichloroethylene | 14 | 0.45 |
| *cis*-l,2-Dichloroethylene | 12 | 0.59 |
| Carbon tetrachloride | 11 | 0.83 |
| Bis(2-Chloroethyoxy) methane | 11 | 0.65 |
| Uracil | 11 | 0.63 |
| Benzo[*g*,*h*,*i*]perylene | 11 | 0.37 |
| 1,1,2,2-Tetrachloroethane | 11 | 0.37 |
| 1,2-Dichloropropene | 8.2 | 0.46 |
| Dichlorobromomethane | 7.9 | 0.61 |
| Cyclohexanone | 6.2 | 0.75 |
| 1,1,2-Trichloroethane | 5.8 | 0.60 |
| Trichlorofluoromethane | 5.6 | 0.24 |
| 5-Fluorouracil | 5.5 | 1.0 |
| 1,1-Dichloroethylene | 4.9 | 0.54 |
| Dibromochloromethane | 4.8 | 0.34 |
| 2-Chloroethyl vinyl ether | 3.9 | 0.80 |
| 1,2-Dichloroethane | 3.6 | 0.83 |
| Chloroform | 2.6 | 0.73 |
| 1,1,1-Trichloroethane | 2.5 | 0.34 |
| 1,1-Dichloroethane | 1.8 | 0.53 |
| Acrylonitrile | 1.4 | 0.51 |
| Methylene chloride | 1.3 | 1.16 |
| Acrolein | 1.2 | 0.65 |

| Compound | $K$(mg/g)(L/mg)$^{1/n}$ | $1/n$ |
|---|---|---|
| Cytosine | 1.1 | 1.6 |
| Benzene | 1.0 | 1.6 |
| Ethylenediaminetetraacetic acid | 0.86 | 1.5 |
| Benzoic acid | 0.76 | 1.8 |
| Chloroethane | 0.59 | 0.95 |
| *N*-Dimethylnitrosamine | $6.8 \times 10^{-5}$ | 6.6 |

†The isotherms are for the compounds in distilled water, with different activated carbons. The values of $K$ and $1/n$ should be used only as rough estimates of the values that will be obtained using other types of water and other activated carbon.

SOURCE: Snoeyink, V. L. 1990. Adsorption of organic compounds. In: AWWA, *Water Quality and Treatment*. New York: McGraw-Hill. Reprinted with permission of the McGraw-Hill Co.

Appendix

# E

# Factors for Conversion

| U. S. Customary units | Multiply by | SI or US Customary units |
|---|---|---|
| **Length** | | |
| inches (in) | 2.540 | centimeters (cm) |
| | 0.0254 | meters (m) |
| feet (ft) | 0.3048 | m |
| | 12 | in |
| yard (yd) | 0.9144 | m |
| | 3 | ft |
| miles | 1.609 | kilometers (km) |
| | 1760 | yd |
| | 5280 | ft |
| **Area** | | |
| square inch (sq in, $in^2$) | 6.452 | square centimeters ($cm^2$) |
| square feet (sq ft, $ft^2$) | 0.0929 | $m^2$ |
| | 144 | $in^2$ |
| acre (a) | 4047 | square meters ($m^2$) |
| | 0.4047 | hectare (ha) |
| | 43,560 | $ft^2$ |
| | 0.001562 | square miles |
| square miles ($mi^2$) | 2.590 | $km^2$ |
| | 640 | acres |
| **Volume** | | |
| cubic feet ($ft^3$) | 28.32 | liters (L) |
| | 0.02832 | $m^3$ |
| | 7.48 | US gallons (gal) |
| | 6.23 | Imperial gallons |
| | 1728 | cubic inches ($in^3$) |
| cubic yard ($yd^3$) | 0.7646 | $m^3$ |
| gallon (gal) | 3.785 | L |
| | 0.003785 | $m^3$ |
| | 4 | quarts (qt) |
| | 8 | pints (pt) |
| | 128 | fluid ounces (fl oz) |
| | 0.1337 | $ft^3$ |

| U. S. Customary units | Multiply by | SI or US Customary units |
|---|---|---|
| million gallon (Mgal) | 3785 | $m^3$ |
| quart (qt) | 32 | fl oz |
| | 946 | milliliters (mL) |
| | 0.946 | L |
| acre-feet (ac ft) | $1.233 \times 10^{-3}$ | cubic hectometers ($hm^3$) |
| | 1233 | $m^3$ |
| | 1613.3 | cubic yard |
| **Weight** | | |
| pound (lb, #) | 453.6 | grams (gm or g) |
| | 0.4536 | kilograms (kg) |
| | 7000 | grains (gr) |
| | 16 | ounces (oz) |
| grain | 0.0648 | g |
| ton (short) | 2000 | lb |
| | 0.9072 | tonnes (metric tons) |
| ton (long) | 2240 | lb |
| gallons of water (US) | 8.34 | lb |
| Imperial gallon | 10 | lb |
| **Unit weight** | | |
| $ft^3$ of water | 62.4 | lb |
| | 7.48 | gallon |
| pound per cubic foot ($lb/ft^3$) | 157.09 | newton per cubic meter ($N/m^3$) |
| | 16.02 | kg force per square meter ($kgf/m^2$) |
| | 0.016 | grams per cubic centimeter ($g/cm^3$) |
| pound per ton | 0.5 | kg/metric ton |
| | 0.5 | mg/kg |
| **Concentration** | | |
| parts per million (ppm) | 1 | mg/L |
| | 8.34 | lb/million gal (lb/Mgal) |
| grain per gallon (gr/gal) | 17.4 | mg/L |
| | 142.9 | lb/mil gal |
| **Time** | | |
| day | 24 | hours (h) |
| | 1440 | minutes (min) |
| | 86,400 | seconds (s) |
| hour | 60 | min |
| minute | 60 | s |
| **Slope** | | |
| feet per mile | 0.1894 | meter per kilometer |
| **Velocity** | | |
| feet per second (ft/sec) | 720 | inches per minute |
| | 0.3048 | meter per second (m/s) |
| | 30.48 | cm/s |
| | 0.6818 | miles per hour (mph) |
| inches per minute | 0.043 | cm/s |
| miles per hour (mi/h) | 0.4470 | m/s |
| | 26.82 | m/min |
| | 1.609 | km/h |
| knot | 0.5144 | m/s |
| | 1.852 | km/h |

| U. S. Customary units | Multiply by | SI or US Customary units |
|---|---|---|
| **Flowrate** | | |
| cubic feet per second ($ft^3/s$, cfs) | 0.646 | million gallons daily (MGD) |
| | 448.8 | gallons per minutes (gpm) |
| | 28.32 | liter per second (L/s) |
| | 0.02832 | $m^3/s$ |
| million gallons daily (MGD) | 3.785 | $m^3/d$, (CMD) |
| | 0.04381 | $m^3/s$ |
| | 157.7 | $m^3/h$ |
| | 694 | gallons per minute |
| | 1.547 | cubic feet per second ($ft^3/s$) |
| gallons per minute (gpm) | 3.785 | liters per minute (L/min) |
| | 0.06308 | liters per second (L/s) |
| | 0.0000631 | $m^3/s$ |
| | 0.227 | $m^3/h$ |
| | 8.021 | cubic feet per hour ($ft^3$ /h) |
| | 0.002228 | cubic feet per second (cfs, $ft^3/s$) |
| gallons per day | 3.785 | liters (or kilograms) per day |
| MGD per acre-ft | 0.4302 | gpm per cubic yard |
| | 0.9354 | $m^3/m^2 \cdot d$ |
| acre-feet per day | 0.01427 | $m^3/s$ |
| **Application or loading rate** | | |
| cubic feet ($ft^3$) per gallon | 7.4805 | $m^3/m^3$ |
| $ft^3$ per million gallons | 0.00748 | $m^3/1000\ m^3$ |
| $ft^3$ per 1000 $ft^3$ per minute | 0.001 | $m^3/m^3 \cdot min$ |
| $ft^3$ per $ft^2$ per hour | 180 | $gal/ft^2 \cdot d$ |
| $ft^3$ per minute per foot | 0.00748 | $m^3/min \cdot m$ |
| gallons per foot per day | 0.0124 | $m^3/m \cdot d$ |
| gallons per square feet per minute | 40.7458 | $L/m^2 \cdot min$ |
| | 0.04075 | $m^3/m^2 \cdot min$ |
| | 2.445 | $m^3/m^2 \cdot h$ |
| | 58.6740 | $m^3/m^2 \cdot d$ |
| gallons per acre | 0.00935 | $m^3/ha$ |
| MGD per acre-feet | 0.430 | $gpm/yd^3$ |
| square root of qpm per square foot (gal/min) $0.5/ft^2$ | 2.70 | (L/s) $0.5/m^2$ |
| pound per acre (lb/a) | 1.121 | kilograms per hectare (kg/ha) |
| pound per pound per day | 1 | $kg/kg \cdot d$ |
| pound per day | 0.4536 | kg/d |
| $lb/ft^2 \cdot h$ | 4.8827 | $kg/m^2 \cdot h$ |
| $lb/1000\ ft^2 \cdot d$ | 0.0049 | $kg/m^2 \cdot d$ |
| pound per acre per day | 1.1209 | $kg/ha \cdot d$ |
| $lb/ft^3 \cdot h$ | 16.0185 | $kg/m^3 \cdot h$ |
| $lb/1000\ ft^3 \cdot d$ | 0.0160 | $kg/m^3 \cdot d$ |
| pound per 1000 gallons | 120.48 | $kg/1000\ m^3$ |
| pound per million gallon | 0.12 | mg/L |
| pounds per foot per hour($lb/ft \cdot h$) | 1.4882 | $kg/m \cdot h$ |
| pounds per horse power per hour ($lb/hp \cdot h$) | 0.608 | $kg/kw \cdot h$ |

| U. S. Customary units | Multiply by | SI or US Customary units |
|---|---|---|
| **Force** | | |
| pounds | 0.4536 | kilograms force (kgf) |
| | 453.6 | grams |
| | 4.448 | newtons (N) |
| **Pressure** | | |
| pounds per square inch ($lb/in^2$, psi) | 2.309 | feet head of water |
| | 2.036 | inches head of mercury |
| | 51.71 | mmHg |
| | 6895 | newtons per square meter ($N/m^2$) = pascal (Pa) |
| | 703.1 | $kgf/m^2$ |
| | 0.0703 | $kgf/cm^2$ |
| | 0.0690 | bars |
| pounds per square foot ($lb/ft^2$) | 4.882 | $kgf/m^2$ |
| | 47.88 | $N/m^2$ (Pa) |
| pounds per cubic inch | 0.01602 | $gmf/cm^3$ |
| | 16.017 | gmf/L |
| tons per square inch | 1.5479 | $kg/mm^2$ |
| millibars (mb) | 100 | $N/m^2$ |
| inches of mercury | 345.34 | $kg/m^2$ |
| | 0.0345 | $kg/cm^2$ |
| | 0.0334 | bar |
| | 0.491 | psi ($lb/in^2$) |
| inches of water | 248.84 | pascals (Pa) |
| atmosphere | 101,325 | Pa |
| | 1013 | millibars (1 mb = 100 Pa) |
| | 14.696 | psi ($lb/in^2$) |
| | 29.92 | inches of mercury |
| | 33.90 | feet of water |
| pascal (SI) | 1.0 | $N/m^2$ |
| | $1.0 \times 10^{-5}$ | bar |
| | $1.0200 \times 10^{-5}$ | $kg/m^2$ |
| | $9.8692 \times 10^{-6}$ | atmospheres (atm) |
| | $1.40504 \times 10^{-4}$ | psi ($lb/in^2$) |
| | $4.0148 \times 10^{-3}$ | in. head of water |
| | $7.5001 \times 10^{-4}$ | cm head of mercury |
| **Mass and density** | | |
| slug | 14.594 | kg |
| | 32.174 | lb (mass) |
| pound | 0.4536 | kg |
| slug per $foot^3$ | 515.4 | $kg/m^3$ |
| density ($\gamma$) of water | 62.4 | $lb/ft^3$ at 50°F |
| | 980.2 | $N/m^3$ at 10°C |
| specific wt ($\rho$) of water | 1.94 | $slugs/ft^3$ |
| | 1000 | $kg/m^3$ |
| | 1 | kg/L |
| | 1 | gram per milliliter (g/mL) |
| **Viscosity** | | |
| pound-second per $foot^3$ or slug per foot second | 47.88 | newton second per square meter ($N\ s/m^2$) |
| square feet per second ($ft^2/s$) | 0.0929 | $m^2/s$ |

| U. S. Customary units | Multiply by | SI or US Customary units |
|---|---|---|
| **Work** | | |
| British thermal units (Btu) | 1.0551 | kilo joules (kj) |
| | 778 | ft lb |
| | 0.293 | watt-h |
| | 1 | heat required to change 1 lb of water by 1°F |
| **Work** | | |
| Btu per pound | 2.3241 | kJ per kg |
| Btu/ft$^2$ · °F · h | 5.6735 | W/m$^2$ · °C · h |
| hp-h | 2545 | Btu |
| | 0.746 | kW-h |
| kw-h | 3413 | Btu |
| | 1.34 | hp-h |
| hp per 1000 gallons | 0.1970 | kW per m$^3$ |
| **Power** | | |
| horsepower (hp) | 550 | ft lb per sec |
| | 746 | watt |
| | 2545 | Btu per h |
| kilowatts (kW) | 3413 | Btu per h |
| Btu per hour | 0.293 | watt |
| | 12.96 | ft lb per min |
| | 0.00039 | hp |
| **Temperature** | | |
| degree Fahrenheit (°F) | (°F − 32) × (5/9) | degree Celsius (°C) |
| (°C) | (°C × (9/5) + 32 | (°F) |
| | °C + 273.15 | Kelvin (K) |

# Index